Deep-Sea Fishes

Biology, Diversity, Ecology and Fisheries

The technological advances of the last twenty years have brought huge advances in our understanding of the deep sea and of the species inhabiting this elusive and fascinating environment. Synthesising the very latest research and discoveries, this is a comprehensive and much-needed account of deep-sea fishes. Priede examines all aspects of this incredibly diverse group of animals, reviewing almost 3,500 species and covering deep-sea fish evolution, physiology and ecology as well as charting the history of their discovery from the eighteenth century to the present day. Providing a global account of both pelagic and demersal species, the book ultimately considers the effect of the growing deep-sea fishing industry on sustainability.

Copiously illustrated with explanations of the deep-sea environment, drawings of fishes and information on how they adapt to the deep, this is an essential resource for biologists, conservationists, fishery managers and anyone interested in marine evolution and natural history.

Imants G. Priede is Professor Emeritus at Oceanlab, a field research station of the Institute of Biological and Environmental Sciences, University of Aberdeen. He has over 40 years of experience studying deep-sea fish, from longlining and trawling on the Royal Research Ship *Challenger* in the 1970s to participating in and leading expeditions to the Pacific Ocean, Atlantic Ocean, Mediterranean Sea and the Mid-Atlantic Ridge. The founder of Oceanlab, he also pioneered satellite tracking of sharks in the 1980s and is the recipient of the 2011 Beverton Medal of the Fish Society of the British Isles.

Deep-Sea Fishes

Biology, Diversity, Ecology and Fisheries

IMANTS G. PRIEDE

Professor Emeritus, Oceanlab, University of Aberdeen,
Scotland, UK

CAMBRIDGE
UNIVERSITY PRESS

University Printing House, Cambridge CB2 8BS, United Kingdom

One Liberty Plaza, 20th Floor, New York, NY 10006, USA

477 Williamstown Road, Port Melbourne, VIC 3207, Australia

314-321, 3rd Floor, Plot 3, Splendor Forum, Jasola District Centre, New Delhi - 110025, India

79 Anson Road, #06-04/06, Singapore 079906

Cambridge University Press is part of the University of Cambridge.

It furthers the University's mission by disseminating knowledge in the pursuit of education, learning and research at the highest international levels of excellence.

www.cambridge.org
Information on this title: www.cambridge.org/9781107083820
DOI: 10.1017/9781316018330

First published 2017

A catalogue record for this publication is available from the British Library

Library of Congress Cataloging in Publication data
Names: Priede, I. G., author.
Title: Deep-sea fishes : biology, diversity, ecology and fisheries / Imants G. Priede, professor emeritus, Oceanlab, University of Aberdeen, Scotland, UK.
Description: Cambridge, United Kingdom ; New York, NY : Cambridge University Press, 2017. | Includes bibliographical references and index.
Identifiers: LCCN 2017006037 | ISBN 9781107083820 (Hardback : alk. paper)
Subjects: LCSH: Deep-sea fishes. | Deep-sea biology. | Deep-sea ecology.
Classification: LCC QL620 .P75 2017 | DDC 597/.63–dc23 LC record available at https://lccn.loc.gov/2017006037

ISBN 978-1-107-08382-0 Hardback
ISBN 978-1-107-44452-2 Paperback

To my wife, Maria.

CONTENTS

PREFACE

My first direct experience of deep-sea fishes was during November 1974 when I joined the Royal Research Ship *Challenger* for an expedition in the NE Atlantic under the leadership of Malcolm Clarke (Marine Biological Association Laboratory Plymouth, MBA) and John Blaxter (Scottish Marine Biological Association, Oban, SMBA). Also on board were Bob Foster (MBA) and John Gordon (SMBA), and the aim was to test the fishing capabilities of the new ship using long lines, midwater trawls and bottom trawls. It was a great privilege to sail with this eminent team, and I was immediately enthralled by the extraordinary diversity of deep-sea fishes as I encountered my first morids, grenadiers, deep-sea sharks, lantern fishes, hatchet fishes, smooth-heads, dragonfishes and many other species. Having previously specialised in research on salmon and trout, this was all new to me. This volume is to some extent addressed to my younger self, providing information I wish I had known then. However much has changed since that expedition, almost 1,000 additional deep-sea species have been described since then. We were fishing in a pristine environment not yet touched by commercial fishing activity, and there was no GPS for navigation or 3-D sonar, so bottom fishing was relatively blind with nets often damaged or lost. Years later when I returned on the same ship as the cruise leader, it became clear that the abundance of many dominant demersal species had been significantly reduced by commercial fishing, and by the 1990s we were working in an environment that had become modified by human activity, possibly motivated by reports of the earlier exploratory fishing activity.

Science has advanced greatly in the intervening years, and in addition to those named earlier, I am grateful to key persons who have increased my knowledge of the deep sea and enabled me to develop my own research. George Somero first invited me to the Scripps Institution of Oceanography, where I learned of his work on biochemical adaptations of fishes to high pressure and others doing interesting work at the time. Amongst these was Ken Smith Jr., with whom I started a most fruitful collaboration studying the behaviour of living fish in their natural environment at abyssal depths in the Pacific Ocean. In the Atlantic Ocean Tony Rice (National Oceanography Centre, NOC) and Peter Herring (NOC) invited me on their cruises aboard the RRS *Discovery* off Madeira and West Africa and working with Julian Partridge (University of Bristol), Ron Douglas (City University, London) and Hans-Joachim Wagner (University of Tübingen) taught me much about adaptations of visual and neural systems of fishes. With Anastasios Tselepidis (Hellenic Centre for Marine Research, HCMR) I worked on RV *Philia* in the Eastern Mediterranean and found out about fishes in the warm deep sea. Meetings with Katsumi Tsukamoto (University of Tokyo) gave me insights into the biology of eels and links with Japan that enabled my team to begin work in hadal trenches of the Pacific Ocean. Nigel Merrett (Natural History Museum, London) has been most influential, acting as a fount of knowledge on deep-sea fishes and as the scientific trawl skipper on several of my cruises, training his successors in arts of fishing at great depths in the years before he retired from active field work. The Census of Marine Life programme (2000–2010) was most important for many marine biologists, and I am particularly grateful to Odd Aksel Bergstad (Institute of Marine Research, Norway) for our collaboration on the Mid-Atlantic Ridge and colleagues I worked with during this time, including

Tracey Sutton for his knowledge of pelagic species (Nova Southeastern University, Fort Lauderdale, Florida) and David Shale whose photographs feature in this volume.

The aim of this book is an account of the deep sea and the fish that live there. Chapter 1 defines deep-sea fish, outlines the structure of the ocean basins and their history over geological time, the oceanography necessary to understand the environment and food supply to the deep. The history of the discovery of deep-sea fishes is described from before Linneaus, through great expeditions of the nineteenth century to the present day. The methods for capture and investigation of deep-sea fishes are explained. Chapter 2 describes the colonisation of the deep sea in relation to the history of evolution of fishes, survival through major extinction events and recent evidence from molecular genetics. Chapter 3 considers the physiological and other adaptations necessary to survive in the deep and concludes with a discussion of the maximum depth limits of fishes. Chapter 4 gives a description of all deep-sea fishes arranged systematically. Two hundred and twenty two families are identified as contributing to the global deep-sea fish fauna. The aim has been to create a reference work in which a fish caught in the deep sea anywhere in the world can be looked up. The reader should find some information at least down to genus level with an account of the most important or interesting species. To achieve this aim, a rather broad definition of deep-sea fishes has been adopted, but nevertheless species such as some lamnid sharks and tunas that are capable of diving deep here are not considered as true deep-sea species. Line drawings are provided of at least one representative from every significant family. The drawings are diagrammatic, inspired by the thumbnail diagrams in the excellent *Fishes of the World* by Nelson (2006) but with slightly different conventions. As in Nelson, details of the texture of the skin or scales are not shown, but here the fin rays are drawn. The details of branches in soft fin rays are not shown; caudal and other soft fin rays are depicted by a single bold line. Spiny fin rays are represented accurately within the limitations of image size. All the drawings are the same length, but a scale bar indicates the true size of a typical large adult specimen. Neither the illustrations nor the text are intended as a means of species identification; there are excellent guides published by FAO and regional keys that fulfil that role. Chapter 5 cuts across taxonomy to describe the assemblages or communities of fishes, often comprising species distantly related to one another that live together in the main deep-sea environments from just beneath the surface layers to deep-ocean trenches and their regional differences. Chapter 6 is concerned with exploitation of deep-sea fishes from global trends in commercial fishing through to consideration of over 250 species recorded as exploited and the environmental impacts of such fisheries. The analysis in this chapter is derived from official statistics compiled by the Food and Agriculture Organisation of the United Nations (FAO) and does not take account of illegal, unreported and unregulated landings (IUU), except where specifically mentioned. Chapter 7 makes some concluding remarks.

This book has depended greatly on data from FishBase. I am grateful to Rainer Froese, Nicolas Bailly and the staff at FishBase for their help in accessing the necessary information. I have also drawn on the work and help of my own research team, foremost in which were engineers, Phil Bagley, with whom I set up the Oceanlab facility in the University of Aberdeen, and Alan Jamieson, who is now leader of research at hadal depths. The following former postdoctoral fellows and research students have all made their distinctive contributions: John Armstrong, David Bailey, Luke Bullough, Martin Collins, Nicola Cousins (Neé, King), Jessica Craig, Simon Creasey, Toyo Fujii, Jasmin Godbold, Amy Heger, Camila Henriques,

Emma Jones, Kirsty Kemp, Thom Linley, Kirsty Morris, Tomasz Niedzielski, Mark Shields, Martin Solan, Alastair Smith, Ben Wigham and Cynthia Yau.

I also thank Rupert Baker (Royal Society of London) for granting access to the Francis Willughby's *Historia Piscium* (1686), Gilbert T. Rowe (Texas A&t M University) and Chih-Lin Wei (University of Taiwan) for access to their biomass data, Bruce Robison (Monterey Bay Research Institute) for advice on underwater vehicles and other issues, Poly Hadziabdic (British Oceanographic Data Centre) for access to CTD data, Julien Claes (University of Louvain) and Dave Ebert (Moss Landing Marine Laboratory) for advice on bioluminescent sharks and Jeff Drazen (University of Hawaii) for helpful discussions by email.

Georges Cuvier wrote in 1828 that positive natural history requires work and expense that cannot be afforded without patronage. He points out that in the fourth century BC, Alexander the Great gave Aristotle eight hundred talents, sufficient to employ several thousand assistants to work on his *Historia Animalium*; possibly the largest research grant in the history of zoology expressed as a percentage of the total economy. Cuvier argued that progress in ichthyology requires more Alexanders. The Alexanders that provided most of the funding for my research have been the UK Natural Environment Research Council (NERC) and the Framework Programme of the European Union. It was a privilege to be appointed by NERC as the scientist on the project board for construction of the RRS *James Cook*, which in 2007 replaced the RRS *Challenger* on which I had started my deep-sea career. The RRS *James Cook* can deploy sophisticated equipment with an accuracy of a few metres in mid-ocean, something we could only dream of 30 years previously.

I have written this book during tenure of an Emeritus Professorship at the University of Aberdeen, Scotland. I thank the Heads of the School of Biological Sciences, Professor Elizabeth Baggs and Professor Graeme Paton for granting me access to necessary facilities. Much of the work has been done at the Hellenic Centre for Marine Research (HCMR) in Crete, Greece, and I thank the Director of the Institute of Oceanography of HCMR, Vassilis Lykousis, for supporting this work and George Petihakis for accommodating me in his team. The entire work has been read by Margaret Eleftheriou, whom I thank for her hard work and support throughout. I thank Dominic Lewis and Jenny van der Meijden of Cambridge University Press for nursing this text through commission and production.

Finally, I thank my family for their support through my long absences at sea and preoccupation with work and writing while at home: my late wife Lindsay, wife Maria, daughters Susannah, Camilla and Nanouyia and sons-in-law Nikolas and Tim. Although in some respects this book reflects my life's work, it has been mostly produced during 30 months of intense reading and writing aided by Maria's motivation and sustenance, without which it would not have been completed. Granddaughters Zoelia and Rosalind have provided welcome distractions from the task, greatly easing the burden of work.

1 Introduction

1.1 Definition of Deep-Sea Fishes

Life on earth is dependent on energy from the Sun, which warms the biosphere and provides direct input through photosynthesis creating the chemical energy that supports living processes. Deep-sea fishes are species that live at depths greater than 200 m beyond the effective range of solar radiation. The heat of the Sun is absorbed in the uppermost one or two metres of the ocean; sufficient light for photosynthesis reaches a maximum depth of 100–150 m in clear open ocean waters (Ryther, 1956). Deep-sea fishes hence live remote from their main source of energy and must somehow indirectly access surface-derived food (Herring, 2002), with many species developing extraordinary adaptations to cope with the reduced food supply at depth.

Defining a strict upper depth boundary for deep-sea fishes is problematic. Below 1000 m, the oceans are totally devoid of solar light, the temperature is generally less than 4°C and there is no doubt that fishes living in these cold dark conditions can be considered to be deep-sea species. However between 200 and 1000 m there is a transition zone where, although there may be sufficient light for vision using highly sensitive specially adapted eyes, life is more or less untenable for normal surface-dwelling fishes. This zone contains some of the most interesting deep-sea fishes that are obviously different from shallow species. Despite living deep, many species have buoyant eggs that float to the surface so that larvae can develop in the plentiful food supply in the surface layers. In the open ocean many deep-sea fishes migrate towards the surface at night and descend again at dawn to pass the day in darkness and the cold waters below. Thus deep-sea fishes do not necessarily pass their entire life cycle in the deep sea, and adults may not be restricted entirely to depths greater than 200 m.

Conversely many shallow-water species are capable of moving deeper than 200 m. For example, although the Atlantic Cod (*Gadus morhua*) has been recorded as far down as 600 m_{fbd} depth and the Atlantic Herring (*Clupea harengus*) down to 364 m_{fbd}, these species would never be considered to be deep-sea fishes. These are the maximum depths recorded for the species in the online database FishBase (Froese and Pauly, 2016), which in this volume are denoted by the suffix 'fbd', denoting '*fish base depth*'.

Early researchers recognised the difficulties of defining a fixed depth horizon for deep-sea fishes. Albert Günther (1887) of the London Natural History Museum, for the purposes of his report on deep-sea fishes captured during the voyage of the HMS *Challenger* around the world 1873–1876, had initially agreed with Professor Charles Wyville Thomson of Edinburgh University that a depth of 300–350 fathoms (572–640 m) should be considered as the boundary between surface and deep-sea fishes. However, as species that were unmistakably deep-sea species were being discovered at shallower depths in the North Atlantic, he finally adopted 100 fathoms (183 m) as the upper boundary delineating the deep sea, not very different from the present defined limit of 200 m. Goode and Bean (1895) wrote that the limit of 100 fathoms is ordinarily used to define deep-sea fishes and the 200 m limit is now recognised by the United Nations, Food and Agriculture Organisation, Fisheries and Aquaculture Department who define deep-sea fisheries as taking place at great depths (between 200 and 2000 m; FAO, 2014).

For bottom-living or demersal fishes the 200 m depth horizon coincides with the depth of the shelf break, the edge of the continental shelf where the seafloor begins to slope off down to the abyss. Thus 200 m defines the boundary between coastal shelf-dwelling neritic species and deep-sea species of the continental slope. This distinction is clearest in the North Atlantic in areas such as the Celtic Sea area, west of the British Isles, where the depth reaches 200 m over 300 km offshore (Fig. 1.1a), defining a clear boundary between productive shelf seas and the deep sea. However elsewhere, the 200 m depth boundary is less clear. The most important fishing area in New Zealand waters is the Chatham Rise off the east coast of South Island (Fig. 1.1b). Much of the Rise is about 400 m deep and is populated by shelf-seas species with truly deep-sea species *Centrophorus squamosus* (Leafscale gulper shark) and *Mora moro* (Common mora) occurring at the outer edges of the Chatham Rise at depths greater than 500 m (McMillan et al., 2011). Off the east of Japan there is a shelf break at around 200 m before the seafloor slopes off down to great depths in the Japan Trench (Fig. 1.1c). However at oceanic islands such as Oahu in the Hawaiian Islands there is no distinctive 200 m depth horizon; the seafloor slopes down from the beach at Honolulu and reaches 500 m depth approximately 5 km offshore (Fig. 1.1d).

Confusingly, in the English language the term 'deep-sea fishing' is also often used to refer to fishing distant from the shore, despite the fact that the fish being captured are usually shallow or surface-dwelling species. Holdsworth (1874) stated that British deep-sea trawlers rarely fish so much as 50 fathoms (92 m) deep. From the 1890s onward steam-powered trawlers from England began to fish for cod off Iceland, Greenland, Bear Island, the White Sea off Russia, and across the Atlantic off Newfoundland and were known as deep-sea trawlers; the leading British fishermen's welfare organisation (founded 1881) is still known as the Royal National Mission to Deep Sea Fishermen. Few of the fishermen supported by the organisation were ever engaged in catching truly deep-sea fish. Similarly recreational sports fishermen often are said to engage in deep-sea fishing which Coghlan (2008) defines as a form of marine tourism involving overnight travel offshore. The species sought by such recreational anglers are the big-game fish: marlin, billfish, mackerel, tuna, dolphin-fish and sailfish, all open ocean surface-dwelling fishes (Pepperell, 2010).

For the purposes of this text, deep-sea fishes are considered to be species that live a large proportion of their lives at depths greater than 200 m. Nevertheless depth overlaps between shallow and deep-sea species do occur, and some deep-sea fisheries may exploit a mixture of shallow and deep-sea species. Only fishes of the globally interconnected oceans and seas including the Arctic Ocean, Red Sea and Mediterranean Sea are considered. Inland seas and lakes, although deep (e.g. Caspian Sea, maximum depth 1025 m, and Lake Baikal, maximum depth 1,642 m) are not considered.

1.2 The Structure of the World's Ocean Basins

1.2.1 Ocean Depth Zones

Zonation of species with increasing altitude is a familiar phenomenon on land, from low-lying grasslands to upland forests and increasingly barren slopes above the treeline up to the snow-capped peaks of mountains. When deep-sea fishes were first discovered by Risso (1810), it was immediately apparent that a similar zonation occurs in the deep sea with different species of fishes living at different depths. Indeed, Risso (1826) went on to describe the zonation of flora and fauna from the summits of the Alps to the depths of the Mediterranean Sea. Deep-sea fishes show remarkable fidelity to their preferred depth zones (Merrett and Haedrich, 1997), so in order to understand their distribution it is important to know about the shapes of the ocean basins. In particular, (a) How much area is there of different depths? (b) Is the area fragmented or continuous? (c) What is the distance between patches of the same

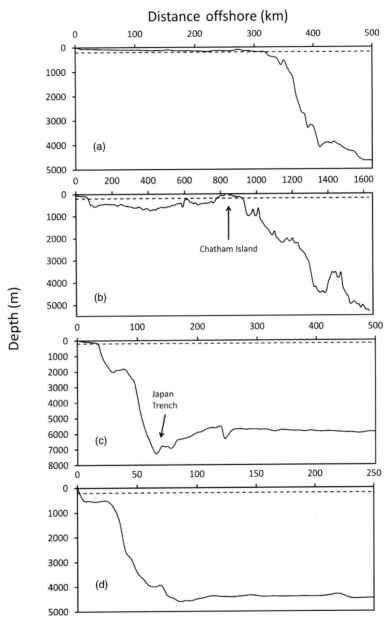

Figure 1.1 Seafloor depth profiles at different locations around the world. The horizontal dashed line is the generally recognised 200 m upper depth limit for deep-sea fishes. The solid line is the seafloor.

(a) Celtic Sea, NE Atlantic Ocean, Transect from Land's End (50.065°N 5.716°W), Cornwall, England, heading 234°.

(b) Chatham Rise, SW Pacific Ocean, Transect from Banks Peninsula (43.820°S 173.103°E) Christchurch, South Island, New Zealand, heading 94° passing through the Chatham Islands.

(c) Japan Trench, NW Pacific Ocean, Transect from Coshi (37.713°N 140.878°E) Honshu Island, Japan, heading 102°.

(d) Hawaii, N Pacific Ocean, Transect from Honolulu (21.278°N 157.835°E), Oahu Island, Hawaiian Islands, heading 215°.

depth? This section of the book seeks to answer those questions and to describe the characteristics of the seafloor.

The oceans and seas occupy 360.663 million km^2 of the earth's surface, equivalent to 70.55 per cent of the area of the planet (Costello et al., 2010). Most of this is deep sea, with the total area deeper than 200 m amounting to 65.26 per cent of the earth surface area. The average depth of the ocean is 3729 m, and its total volume is 1.335819×10^9 km^3. It is convenient to divide the seafloor into four depth categories: 0–200 m – shallows around land masses, mostly continental shelves; 200–3,000 m – the bathyal zone comprising slopes around continents, islands, seamounts and ridges; 3,000–6,000 m – the abyssal zone, mostly abyssal plains; depths greater than 6,000 m – the hadal zone, trenches and deep basins. The hypsographic curve (Fig. 1.2) shows that the abyssal

zone is the largest and accounts for 53.2 per cent of the surface of the earth. The hadal zone, between 6,000 m and its maximum ocean depth 10,990 m at the Challenger Deep in the Pacific Ocean, accounts for 45 per cent of the ocean depth range but less than 0.5 per cent of total ocean area.

The boundaries between these zones cannot be sharply defined, and terminology varies between different authorities and according to whether physiographic or ecological principles are applied (Fig. 1.3). Gage and Tyler (1991) used the ecological depth zones; bathyal (200–2000 m), abyssal (2000–6000 m) and hadal (>6000 m). For describing the depth distribution of demersal fishes around the North Atlantic basin Haedrich and Merrett (1988) developed a scale with 750 m depth increments, upper middle and lower slope extending down to 2250 m and the rise (upper, middle and lower) extending

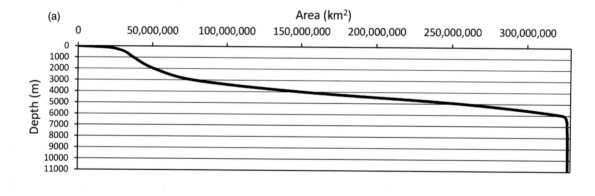

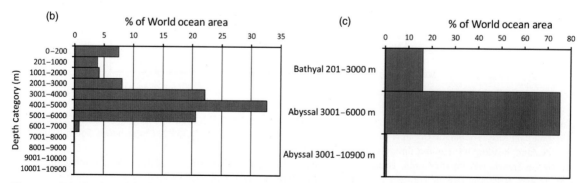

Figure 1.2 Distribution of depths in the oceans. (a) The hypsographic curve based on the GEBCO global bathymetry showing cumulative area with increasing depth. (b) Percentage of world ocean floor area at 1000 m depth increments (except the 0–200 m continental shelf, nominally continental shelf). (c) Percentage of world ocean floor in the three main deep-sea depth categories.

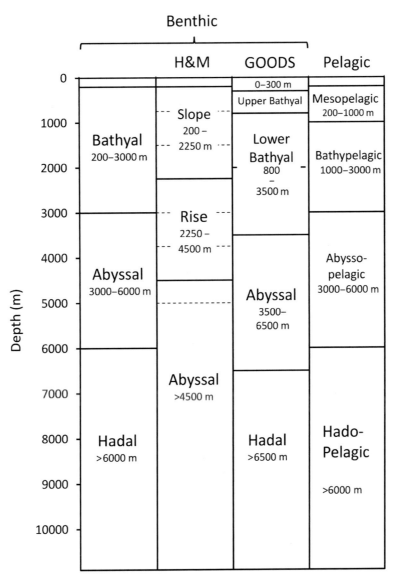

Figure 1.3 Deep-sea depth zones; nomenclature according to different authorities. The first column shows the benthic scheme used in this volume. H&M is according to Haedrich and Merrett (1988) with upper middle and lower subdivisions within each depth band. GOODS is according to UNESCO (2009). Pelagic nomenclature is in the final column.

down to 4500 m. All depths greater than 4500 m were considered abyssal, and there was no hadal zone to consider in their study. This scheme accords well with the observed boundaries between deep-sea fish species assemblages and is appropriate for passive ocean margin typical of the North Atlantic (Fig. 1.3). The Global Open Ocean and Deep Seabed (GOODS) – Biogeographic Classification, (UNESCO, 2009) offers

an alternative scheme in which the boundaries are defined by differences in mainly invertebrate faunal assemblages. The deep sea is defined as the seafloor below 200–300 m, and owing to the detection of bathyal and abyssal fauna at greater depths than previously recognised, the boundaries of zones have been moved 500 m deeper to 3500 m and 6500 m, respectively (Fig. 1.3).

For the pelagic realm, the ocean remote from the shore and not in contact with the seafloor, it is convenient to use similar depth limits as for the seafloor. Thus the epipelagic zone is from the surface to 200 m depth within which during daylight photosynthesis occurs, fish can hunt their prey by conventional visual means and many species have coloured patterns. The mesopelagic zone, from 200 to 1000 m, is a twilight zone reaching down to the maximum depth of penetration of solar light in clear oceanic water. Low-intensity diffuse monochromatic down-welling light, insufficient for photosynthesis, is nevertheless sufficient for visual function by many organisms, including mesopelagic fishes. Below 1000 m there is total darkness other than occasional bioluminescent flashes with the Bathypelagic zone from 1000 to 3000 m and the Abyssopelagic zone from 3000 to 6000 m (Merrett and Haedrich, 1997). However, in practice there is no distinct boundary in midwater at 3000 m, and unless otherwise qualified the term 'bathypelagic' commonly refers to any depth greater than 1000 m. Hado-pelagic is used to refer to depths greater than 6000 m. (Fig. 1.3)

1.2.2 Ocean Basin Formation, Slopes, Plains, Ridges, Islands and Seamounts

The shape of the ocean basins derives from the basic structure of the earth's crust, with the relatively low-density continental crust floating on denser mantle rocks and surrounded by thinner oceanic crust (Fig. 1.4). The continental crust, made up of granitic rocks, is of ancient origin and occupies about 40 per cent of the earth's surface. The oceanic crust, made up of basaltic rock, is continuously renewed at the mid-ocean ridges so that most of the ocean floor is less than 150 Ma old. Upward extrusion of magma raises the seafloor in mid-ocean creating shallow areas and volcanic islands such as the Azores and Iceland in the Atlantic Ocean. Summits that do not reach the sea surface are termed seamounts. Newly created seafloor moves away symmetrically either side of the mid-axial rift valley, and, as it cools the crust thins, deepening the ocean with increasing distance from the mid-ocean spreading centre. Volcanic islands and seamounts created by eruptions at the spreading centre are transported away from the mid-ocean ridge as if on a conveyor belt, gradually decreasing in altitude (or increasing in depth) as the crust sinks and the summit is eroded over time (Fig. 1.5). Chains of large seamounts can be formed where the moving oceanic crust passes over a hotspot or mantle plume where successive eruptions raise one volcanic summit after another. In the North Atlantic, the New England seamount chain with more than 20 peaks rising from the abyssal seafloor stretches over 1000 km from the Mid-Atlantic Ridge to Georges Bank. In the North Pacific Ocean the Hawaiian and Emperor seamount

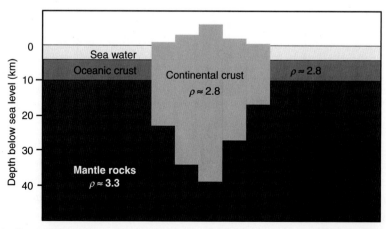

Figure 1.4 Schematic diagram of a section through a continent showing continental crust floating on dense mantle rocks surrounded by thinner oceanic crust creating deep ocean of average depth ≈ 4000 m. Continental slopes and sediment deposits are omitted for simplicity.

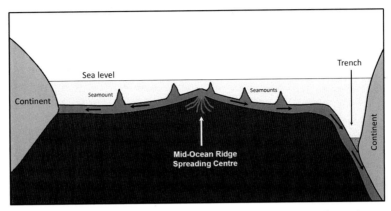

Figure 1.5 Continental Drift. Schematic cross-section of an ocean basin showing motion of oceanic crust away from the mid-ocean ridge spreading centre. Where oceanic crust moves beneath the continent a deep-sea trench is formed (right).

chain, including the Hawaiian Islands, has more than 80 summits extending over 5,800 km. The rate of seafloor spreading varies between 1 and 10 cm per year. At a passive ocean margin there is no relative motion between the oceanic crust and the continental crust; the continent moves away from the mid ocean at the same rate as the seafloor spreads. At the junction between the oceanic and the continental crust the slope of the passive margin tends to become filled with sediments eroded from the continental land mass, creating a zone of sedimentary rocks, which often harbours hydrocarbon deposits (Fig. 1.6a). This type of margin is characteristic of the Atlantic Ocean, where North and South America are moving away from Europe and Africa, respectively. The deepest parts of the Atlantic Ocean are generally in the abyssal plains on either side of the mid-ocean ridge. The mid-ocean ridge axis does not form a continuous straight line; the rate of spreading varies in different parts and because the earth is spherical, a strict linear geometry is not possible. Discontinuities in the ridge appear as transverse fracture zones where the axis of the ridge is often displaced laterally. The fracture zone may also include one or more transverse canyons crossing from one side of the ridge to the other. These are important as areas of exchanges of deep water between the two sides of the mid-ocean ridge and also provide possibilities for deep-sea species to move across what might otherwise be a barrier to dispersal.

Continuous production of new oceanic crust at mid-ocean spreading centres requires equivalent consumption of an equal area of seafloor elsewhere in the ocean. This occurs by subduction at active ocean margins where the oceanic crust moves either directly beneath the continental crust (Fig. 1.6b) or under an offshore volcanic island arc (Fig. 1.6c). The area of subduction is marked by a depression in the seafloor, a trench parallel to the coast. Between the island arc and the continent, a back arc basin may be formed by a distinctive kind of seafloor spreading. Tension is created in the seafloor, either by movement of the continent entrained by the subducting oceanic crust (Van Dover, 2000) or by movement of the island arc towards the ocean by a process known as trench roll-back (Becker and Faccenna, 2009). The resulting tension in the intervening seafloor results in faulting that enables the ascent of magma from the mantle, which then creates a subsidiary seafloor spreading centre and expansion of the back arc basin. This process is most prevalent in the western Pacific Ocean, a notable example being the Sea of Japan (mean depth 1752 m, maximum depth 3742 m). Owing to differences in the process of formation and density of rocks, back arc basins are shallower than the ~4000 m equilibrium depth of oceanic basins (Schopf, 1980) but nonetheless provide significant areas of deep-sea floor. Between the trench and continent or island arc there may be a fore arc basin formed as part of the accretionary wedge of material

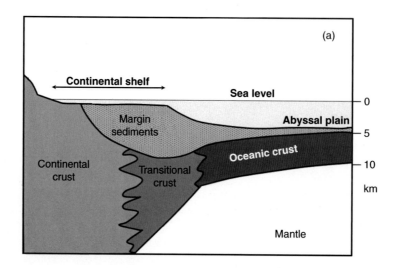

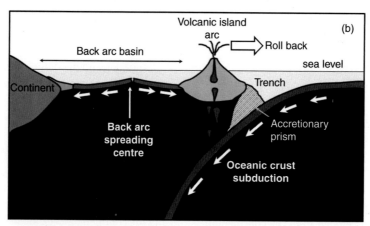

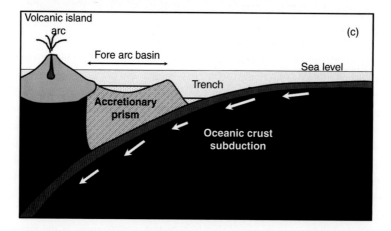

Figure 1.6 Types of Ocean Margin.

(a) Passive Margin, where there is no relative motion between continental and oceanic crust sedimentary deposits accumulate forming layered structures extending out to the abyssal plains.

(b) Active margin with a back arc deep-sea basin. Subducting oceanic crust melts and molten magma ascends to form a volcanic island arc parallel to the trench axis. Movement of the islands towards the trench axis creates a back arc spreading centre with a new sea area bounded by the continent and offshore string of islands.

(c) Active margin with a fore arc deep-sea basin. Folding of the accretionary prism of material trapped as oceanic crust subducts creates a fore arc basin inshore of the trench axis.

accumulating adjacent to the subducting ocean crust. On the Pacific coast of Japan such basins lie between the coast and the trenches further offshore (Fig. 1.6c).

At active continental margins, relative motion of the crust at the edge of the plates creates frequent earthquakes. Volcanic activity is common, bringing material to the surface that creates the offshore island arc or erupts on the edge of the continent. Active margins surround the Pacific Ocean making up the so-called Ring of Fire with active volcanoes, earthquakes and the presence of most of the world's deep trenches with depths in excess of 6000 m (Fig. 1.7). Passive margins are found predominantly around the Atlantic Ocean, the Indian Ocean and the Arctic Ocean (Fig. 1.7). On a passive ocean margin the seafloor generally descends quite steeply beyond the shelf break, reflecting the step down from continental to oceanic crust. This segment is known as the continental slope and extends down to about 2000 m, although subdivisions are often recognised. The continental slope may be bisected by canyons, which add further complexity to the seafloor (Plate 1c). Apart from bare rock in areas of strong currents or where the topography is very steep, the slope is draped

with fine sediment of largely terrigenous origin. The gradient then decreases on the continental rise out to the almost flat abyssal plains. In graphic displays the vertical scale is usually exaggerated giving a misleading impression of the true gradient. The average continental slope gradient between 200 and 2000 m depth in the NE Atlantic (Fig. 1.1a) is 3.4 per cent and on the continental rise the gradient decreases to 2.8 per cent. The surface of the continental rise is usually made up of successive layers of sediment that have slumped down from the slopes, been transported through the canyons from the continental shelf or slowly deposited over time. Occasional turbidite flows of suspended sediments triggered by earthquakes or slope failure can be violent events both eroding the sediment surface and depositing new layers on old. The largest such slides deposit material on areas in excess of 100,000 km extending over the continental slope, rise and abyssal plains (Mienert et al., 2003). With increasing distance offshore, the sediment of terrestrial origin gradually disappears and is replaced by oceanic sediment derived from shells of planktonic organisms. Wind-blown dust from human sources, deserts and volcanoes precipitated out of the

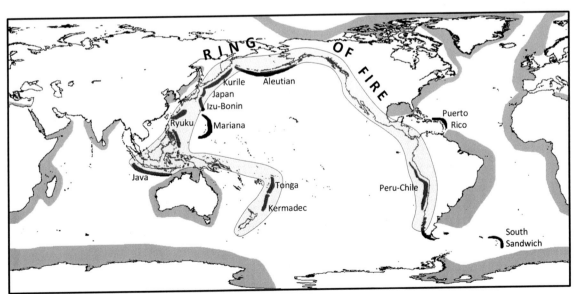

Figure 1.7 Distribution of Active and Passive Margins around the World's Ocean. The Pacific Ocean is surrounded by active ocean margins of the 'Ring of Fire' characterised by seismic activity, major volcanoes and deep trenches (black and labelled with names) created by subduction of oceanic crust. Passive ocean margins (grey shading) occur around most of the rest of the oceans. Based on information from Kious and Tilling (1996) for the active margins and Pinto (2007) for the passive margins.

atmosphere and deposited on the seafloor is an additional source of sediment.

Most of the ocean floor (>90 per cent) has never been directly mapped by soundings from ships. Continental slopes and adjacent areas have been surveyed relatively well in recent years to assess resources and to support offshore sovereignty claims by coastal states under the United Nations Convention on the Law of the Sea article 76 (UN, 1982). However, in the open ocean, outside specially surveyed areas, bathymetric maps are based on satellite gravimetry interpolated with available ship-borne data (Smith and Sandwell, 1997). The highest resolution achieved in this way for global mapping is 30 arc second (Becker, 2009; GEBCO, 2014) or approximately 1 km (Wessel et al., 2010). Using such data, Yesson et al. (2011) estimated the global number of seamounts to be 33,452 plus 138,412 knolls. A computer algorithm was used to search for conical features defined as seamounts in which the average height of the summit is >1000 m above the surrounding seafloor (Morato et al., 2008). Small features in which the average height of the summit is between 200 and 1000 m above the surrounding seafloor are termed knolls (Yesson et al., 2011) or small seamounds (Morato et al., 2008). The existence of most of these features has not been directly verified by ship's soundings, and the estimated number of seamounts differs between different studies. There seems to be convergence towards an estimate of *ca.* 15,000 large seamounts with heights >1500 m (Wessel, 2001; Yesson et al., 2011). However, Wessel et al. (2010) speculate that there may be 100,000 uncharted seamounts >1000 m in height and 25 million knolls or small seamounts >100 m in height.

Yesson et al. (2011) define summits with a depth less than 1500 m below the sea surface as 'productive' and identify 9,239 productive seamounts and 10,185 productive knolls worldwide. Whilst 1500 m is beyond the depth at which photosynthesis is normally possible, these are depths at which cold-water corals, sponges and other reef- building organisms are likely to thrive, and these depths are accessible to deep water fishing activity. There are therefore probably about 20,000 submarine summits worldwide that can be considered to be of economic significance (Fig. 1.8).

Seamounts and mid-ocean ridges are made up of basaltic rocks reflecting their volcanic origin, and this hard substrate is usually exposed on summits, cliffs, steep slopes (gradient >45 per cent) or in areas of strong currents. However, intervening flat areas and gentle slopes are generally covered with soft sediment of pelagic origin (Hughes, 1981). Animal tracks or lebenspüren are conspicuous in the soft foraminiferal sand covering of deep seamounts in the Central North Pacific (Kaufman et al., 1989). At lower bathyal

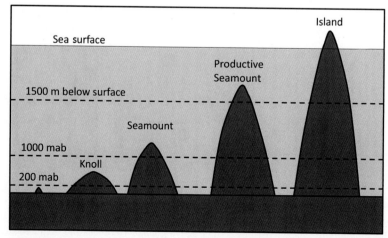

Figure 1.8 Classification of seamount summits according to height above the surrounding seafloor and depth below the sea surface. mab = metres above bottom.

depths more than 95 per cent of the North Mid-Atlantic Ridge is sediment-covered (Priede, Bergstad et al., 2013). Thus, like the continental slopes, offshore bathyal seafloor is mainly composed of fine soft sediment punctuated by patchily distributed hard substrate, either the exposed bedrock, boulders, biogenic structures (e.g. corals) or anthropogenic (e.g. shipwrecks and litter).

1.2.3 The Bathyal Slopes (200–3000 m Depth)

From the point of view of bathyal deep-sea fishes, their habitat is distributed between the continental slopes and offshore locations (Figs. 1.9 and 1.10). Priede, Bergstad et al. (2013) found that the offshore component comprising mid-ocean ridge, seamounts and island slopes accounts for 49.4 per cent of the lower bathyal habitat in the North Atlantic Ocean. Etnoyer et al. (2010) estimated the global area of seamounts to be 28.8 million km². It can be concluded that generally about half of all bathyal seafloor lies offshore, remote from the continents. Conversely, bathyal basins occur close to the coast in fjords, which have been eroded by glaciers, e.g. Sogne Fjord in Norway 1,308 m deep, Canal Messier in Chile 1,270 m deep and Skelton Inlet in Antarctica 1933 m deep; however, the area is small. The Skaggerak is a channel with a maximum depth of 700 m off the south coast of Norway, which supports a significant deep-water fishery for *Coryphaenoides rupestris* (Bergstad et al., 2013).

In the Atlantic Ocean there is a continuous mid-ocean ridge from the Arctic to around 55°S where the Mid-Atlantic Ridge meets the Antarctic Ridges. At 40°S the Walvis Ridge in the east and Rio Grande Rise provide bathyal connections to the slopes of Africa and South America, respectively. The Antarctic is circumscribed by a ridge system comprising the North and South Scotia ridges, The Chilean Rise, the Pacific Antarctic Ridge, the South East Indian Ridge and the South West Indian Ridge, which extends south of Africa to meet the Atlantic-Antarctic Ridge. The Pacific Ocean has numerous scattered seamounts and island chains, but the main oceanic spreading centre is the East Pacific Rise, a northward extension of the Pacific Antarctic Ridge that meets the coast of North America off California. In the Indian Ocean the Central Indian Ridge forms the stem of an inverted 'Y' shape with the SE and SW Indian ridges. This becomes the Carlsberg Ridge in a North-Westward extension to Arabia. To the east lie two north–south ridges, the Chagos-Lacadive Ridge and the Ninety East Ridge, which are not part of the mid-ocean ridge system. The global mid-ocean ridge system, despite

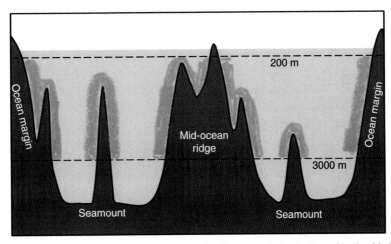

Figure 1.9 Schematic cross section of an ocean basin showing the fragmented distribution of bathyal habitat between ocean margin and offshore seamount and mid-ocean ridges.

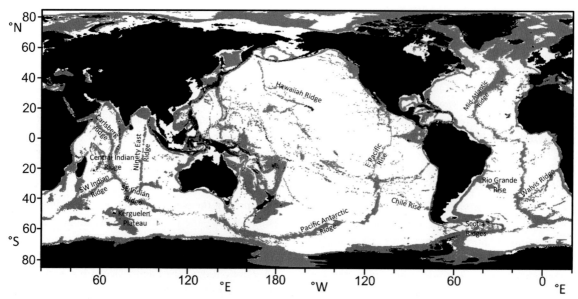

Figure 1.10 Distribution of bathyal (depth 200–3000 m) seafloor in the world's oceans indicated by grey shading. Major ridge systems are labelled. Numerous seamounts reaching bathyal depth are too small to be visible at this global resolution.

interruptions of fracture zones, can be considered to be a single structure about 65,000 km long girdling the planet (Searle, 2013).

1.2.4 Hydrothermal Vents and Cold Seeps

Wherever volcanic activity and geothermal heating of the seafloor occur, on mid-ocean ridges, seamounts or subduction zones, there is the possibility of the existence of hydrothermal vents. These are the marine equivalent of terrestrial geysers or hot springs and are characterised by buoyant plumes of heated water emitted from the seafloor. The temperature can vary between slightly above ambient for diffuse flows to more than 300°C in intense flows of sulphide-rich black smokers (Van Dover, 2000). The often opaque colouration of hydrothermal vent water is caused by precipitation of dissolved minerals as the plume water cools on contact with sea water. The precipitate can also form rigid tubular chimney-like structures around the vent with the plume emanating from the top at a height of tens of metres above the surrounding seafloor. The vent fluid is sea water that has diffused into the oceanic crust, has become chemically modified, heated and is then expelled

through the vent with some magmatic water originating from rocks deep below the seafloor. The properties vary according to the site of the vent, temperature and flow rate. There is great interest in hydrothermal vents owing to the presence of chemosynthetically supported life creating oases of high biomass that thrive in the deep, independent from food supply from the surface. Whilst hydrothermal vents occur worldwide in a variety of settings and can harbour locally high biomass (Gebruk et al., 2000), their total area is small and can only contribute a very small fraction of total deep-sea productivity. Nevertheless some deep-sea fish species are known to exploit these areas.

Oases of life chemosynthetic life similar to hydrothermal vents also occur at cold seeps (Bernardino et al., 2012). These are sites of oil or methane gas emission from the seafloor, often associated with pock marks, mud volcanoes or brine pools in a sedimentary setting in contrast to hydrothermal vents, which are sediment-free. The methane is largely derived from deeply buried organic carbon, which undergoes microbially mediated methanogenesis (Boetius and Wenzhöfer, 2013). Cold seeps are not directly dependent on volcanic activity, although seismicity can be important in creating fissures through which fluid rises towards the surface. Seep communities have been found in the Japan Trench down to hadal depths (Watanabe et al., 2010).

1.2.5 The Abyss (3000–6000 m Depth)

The abyssal zone 3000–6000 m covers a large area of the planet (Fig. 1.11). It is possible to distinguish different basins such as the eastern and western basins of the Atlantic Ocean delineated by intervening mid-ocean ridges. However such topographical features cannot be considered barriers and may not correspond to zoogeographic province boundaries. Basins in the West Pacific Ocean including the Japan Sea (maximum depth 3,742 m), the South China Sea (maximum depth 5,016 m), the Sulu Sea (maximum depth 5,600 m), the Celebes Sea (maximum depth 6,200 m) and the Banda Sea (maximum depth 7300 m) are however truly isolated basins with no direct abyssal connections between them or to the wider ocean. Similarly the Gulf of Mexico and the Caribbean Sea are abyssal seas with connections to the Atlantic Ocean through the Strait of Florida (sill depth 1800 m) and other channels (Sturges, 2005).

The Arctic Ocean has three Abyssal basins: the Canada Basin (maximum depth 3800 m), the Makarov Basin (maximum depth 4025 m) and the Eurasian Basin (maximum depth 5450 m) connected to the Atlantic Ocean via the Fram Strait (sill depth 2600 m), the Greenland Basin (maximum depth 4,846 m) and the Norwegian Basin (maximum depth 3970 m) (Fig. 1.12). The boundary between these northern seas and the North Atlantic Ocean Basin is marked by a chain of ridges between Greenland and Europe. Flow across this ridge system is the main route of exchange between the Arctic Ocean and the rest of the world's oceans. The Bering Strait between the Arctic Ocean and the Pacific Ocean, just 84 km wide and maximum depth 55 m, allows only minimal surface water exchange.

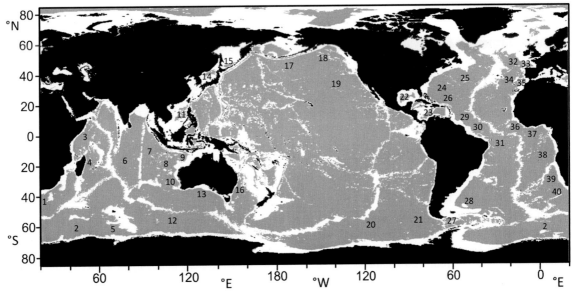

Figure 1.11 Distribution of Abyssal (depth 3000–6000 m) seafloor in the world's oceans indicated by grey shading. Major Abyssal Plains are marked by numbers: 1 – Agulhas, 2 – Enderby, 3 – Somali, 4 – Mascarene, 5 – Valdivia, 6 – Mid Indian, 7 – Cocos, 8 – Gascoyne, 9 – North Australian, 10 – Perth, 11 – South China Sea, 12 – South Indian, 13 – South Australia, 14 – Japan, 15 – Okhotsk, 16 – Tasman, 17 – Aleutian, 18 – Alaska, 19 – Cascadia, 20 – Amundsen, 21 – Bellinghausen, 22 – Sigsbee, 23 – Colombian, 24 – Hatteras, 25 – Sohm, 26 – Nares, 27 – Burdwood, 28 – Argentine, 29 – Demerara, 30 – Ceara, 31 – Pernambuco, 32 – Porcupine, 33 – Biscay, 34 – Madeira, 35 – Seine, 36 – Sierra Leone, 37 – Guinea, 38 – Angola, 39 – Namibia, 40 – Cape. Bathymetry from GEBCO, names of Abyssal Plains after Weaver et al. (1987)

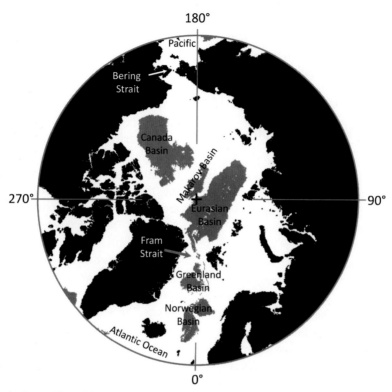

Figure 1.12 The Arctic Ocean Abyssal Basins. Areas over 3000 m deep are shaded grey.

The Mediterranean Sea is connected to the Atlantic Ocean by the Strait of Gibraltar, 14.3 km wide at the narrowest point and with a sill depth of 280 m and is divided into Western and Eastern Basins by the strait between Sicily and Tunisia, where the sill depth is *ca.* 500 m (Fig. 1.13). The Western basin is predominantly at bathyal depths <3000 m, but in the Tyrrhenian Sea it reaches a maximum depth of 3785 m. Most of the abyssal area of Mediterranean Sea is in the Eastern basin with a maximum depth of 5272 m in the Ionian Sea and further deep areas in the Rhodes and Herodotus Basins (Ben-Avraham et al., 2006). Other peripheral seas make no contribution to the abyssal habitat. The Black Sea has a maximum depth of 2,212 m and the Red Sea a maximum depth of 2211 m within a deep central axial trough comprising less than 15 per cent of the total area.

Generally the abyssal seafloor is covered with deep muddy sediments with little hard substrate. Hard polymetallic or manganese nodules up to 10 cm diameter occur in all the world's oceans and are typically found at abyssal depths 4000–6000 m depth. These together with sharks' teeth and skeletal remains of large fishes and marine mammals (Murray and Hjort, 1912) form conspicuous features in an otherwise monotonous terrain. Rocks originating from melting icebergs can also be deposited on the seafloor at high latitudes. Increasing knowledge from deep-sea bathymetry has shown that the previously assumed featureless flat abyssal plains are punctuated by abyssal hills, knolls or mounds, which are becoming the subject of biological research. If the prediction of 25 million such features worldwide is confirmed (Wessel et al., 2010), then considerable revision of the classic view of the monotony of abyssal plains will be necessary.

In contrast to the extensive abyssal plains, small pockets of abyssal habitat occur in the axial valleys of mid-ocean ridges. At 50°N in the Atlantic Ocean an abyssal central rift valley with depth over 3500 m and width less than 10 km has been recorded (Priede,

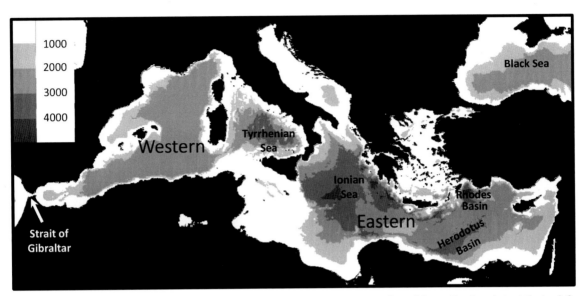

Figure 1.13 The Mediterranean Sea. The Western and Eastern Basins and the main abyssal deeps. Depth scale is at the top left.

Billett, et al., 2013), harbouring typical abyssal demersal fish species (King et al., 2007).

1.2.6 The Hadal Zone (>6000 m Depth)

The hadal zone (depths greater than 6000 m) comprises a very small proportion of the earth's surface. Analysis using GEBCO 30 arc second bathymetry shows that there are 95 distinct basins or trenches deeper than 6000 m, within which there is a depth of 7000 m or more.[1] The total area of these deeps is 1,455,463 km^2, equivalent to 0.404 per cent of the global ocean floor. Of this, 78 per cent is contained in 16 generally recognised trenches in subduction zones accounting for 0.32 per cent of the total ocean floor. The rest of the hadal seafloor is in depressions, small basins and fracture zones (Table 1.1). The largest of these is the NW Atlantic Deep (26°N, 58°W) on the western flanks of the Mid-Atlantic Ridge where deep pockets are found in fractures traversing the ridge. Similar deeps occur in the Mendocino Fracture Zone of the Pacific Ocean (Fig. 1.14). It is questionable whether such features are

of ecological significance, but theoretically these could act as stepping stones for the dispersal of obligatory hadal species. The Weber Deep is a small trench in a subduction zone in the Banda Sea. The hadal habitat is dominated by five major trenches that reach depths greater than 10 km; the Japan-Kurile-Kamchatka, Kermadec, Tonga, Mariana and Philippines Trenches. All these lie along the western boundary of the Pacific Ocean. Other important trenches are the Aleutian and Peru-Chile Trench in the Pacific Ocean, the Java Trenches in the Indian Ocean and the Puerto Rico and South Sandwich Trenches in the Atlantic Ocean. Trenches are characterised by very steep slopes descending to a sediment-filled flat plain in the central valley floor. The deepest point in the world's oceans, the Challenger Deep (10,990 m), is in the southern part of the Mariana Trench.

It will take many decades before the deep sea is fully mapped. Whilst Smith and Sandwell (1997) were responsible for an extraordinary leap forward in our understanding of the structure of the oceans through the use of satellite gravimeter data, further progress can only be made by piecemeal surveys using surface ships and underwater vehicles to reveal high-resolution detail. Outwith well-surveyed areas, deep-sea demersal ichthyologists can expect to do their

---

[1] Unpublished analysis by Tomasz Niedzielski, University of Wrozlaw.

Table 1.1 **List of the Major Hadal Trenches and Basins with the Area Exceeding 6 km Depth. Ranked by Area. Depth Is the GEBCO Maximum Depth. Basins Are Grouped or Separated According to the Continuity of the 6 km Depth Contour.**
Analysis by Tomasz Niedzielski, University of Wrozlaw from GEBCO 30 arc second bathymetry

	Name	Global Rank	Area (km²)	Depth (m)	Ocean
Trenches	Japan-Kuril-Kamachatka	1	271228	10272	Pacific
	Aleutian	2	104409	7877	Pacific
	Kermadec	3	102780	10177	Pacific
	Tonga	4	97996	10799	Pacific
	Mariana South	7	84215	10977	Pacific
	Mariana North	17	20043	8744	Pacific
	Philippine	8	82090	10041	Pacific
	Peru-Chile	11	43820	8123	Pacific
	Ryukyu	12	38370	8109	Pacific
	Idzu-Bonin	13	37374	8172	Pacific
	New Britain	15	29713	9023	Pacific
	Yap	23	9495	8524	Pacific
	Puerto Rico	5	85584	8648	Atlantic
	South Sandwich	10	45393	8185	Atlantic
	Java West	9	51272	7269	Indian
	Java East	19	14968	7291	Indian
Basins	Emperor Trough	16	25738	8167	Pacific
	Mendocino FZ West	20	13920	7084	Pacific
	Chinook Trough West	21	10992	8053	Pacific
	North West Atlantic Deep	6	84554	7196	Atlantic
	North East Atlantic Deep	22	10812	7315	Atlantic
	Wharton Basin	14	34253	7236	Indian
	Weber Basin	18	15711	7351	Indian

own bathymetric surveys before sampling, especially in complex terrain (Priede, Billett, et al., 2013).

1.3 History of the Ocean Basins

Over the past 500 million years the area of the continents and the deep sea has remained more or less constant; however, the major continental tectonic plates have moved considerably with corresponding changes to the location and shape of the major ocean basins. From the first appearance of fish remains in the fossil record during the Cambrian there has been a cycle of aggregation of land masses into a single continent, Pangaea, which then disintegrated to produce the present-day distribution of land masses. Figure 1.15 shows the sequence diagrammatically using base maps from Scotese et al. (1979) for the Palaeozoic (Cambrian to the Late Permian) and Scotese (2002) for the more recent Mesozoic and Cenozoic maps. Scotese (2002) shows details of mountain ranges, shorelines, active plate boundaries, and the extent of paleoclimatic belts. Whilst there is close agreement on recent earth history over the past 150 million years, reconstructions of Palaeozoic geography are to

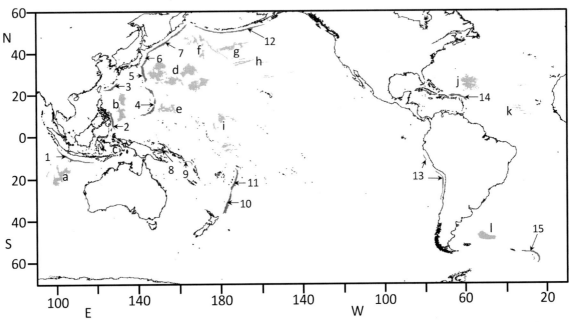

Figure 1.14 Global distribution of Hadal Depths (>6000 m). Trenches, dark grey: : 1 – Java, 2 – Philippines, 3 – Ryukyu, 4 – Marianas, 5 – Izu-Bonin, 6 – Japan, 7 – Kuril-Kamchatka, 8 – New Britain, 9 – South Solomon, 10 – Kermadec, 11 – Tonga, 12 – Aleutian, 13 – Peru-Chile, 14 – Puerto-Rico, 15 – South Sandwich. Hadal Basins, pale grey: a – Wharton Basin, b – Philippine Basin, c – Weber Basin, d – North West Pacific Basin, e – East Mariana Basin, f – Emperor Trough, g – Chinook Trough, h – Mendocino Fracture Zone, i – Central Pacific Basin, j – Nares Abyssal Plain, k – Gambia Abyssal Plain, l – Argentine Basin. Bathymetry from GEBCO; cylindrical equal area projection but some trenches have been increased in width to make them visible at this scale.

varying degrees controversial, and alternative reconstructions can be seen in Blakey (2011).

In the late Cambrian (Fig 1.15a) a supercontinent, Gondwana, had just formed, and parts of the modern northern continents, North America, Europe and Asia were distributed around the equator. The open ocean surrounding the continents has been named Panthallasa (from Greek πᾶν 'all' and θάλασσα 'ocean'). It is presumed that there were subduction zones and consequently trenches around some of the continental margins. By the late Ordovician (Fig 1.15b) enclosed oceans, Lapetus and Palaeothethys, separated Gondwana from Laurentia, Baltica and Gondwana. In the middle Silurian, Baltica had moved northward to join Laurentia and Siberia, opening up the Rheic Ocean with Gondwana and a subduction zone on its southern shores (Fig. 1.15c). During the Devonian, the so-called Age of Fishes, continents moved closer together (Fig. 1.15d), so that by the Early

Carboniferous (Fig. 1.15e) the ancient oceans separating the continents had largely disappeared, and in the late Carboniferous the supercontinent Pangaea had formed with the semi-enclosed Palaeo-Tethys Ocean to the East (Fig. 1.15f). During the Permian, a microcontinent named Cimmeria, which includes modern Turkey, Iran and Tibet, started to break away from Gondwana with the opening up of a rift valley and spreading centre (Fig. 15g). The end of the Permian marks the end of the Paleozoic era with major extinctions of flora and fauna occurring all over the planet.

During the Triassic, Cimmeria continued to move northward (Fig. 1.15h) with a new seafloor of the Tethys Ocean opening to the south replacing the Palaeo-Tethys seafloor, much of which subducted under China. By the early Jurassic (Fig. 1.15i) Cimmeria had collided with Asia, and the formation of the Tethys Sea in the previous location of the Palaeo-

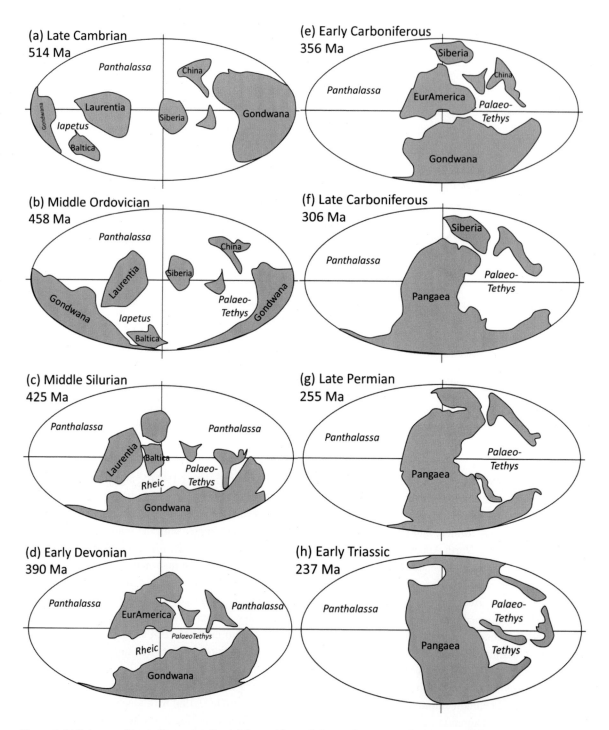

Figure 1.15 Palaeocontinents. Reconstruction of the positions of the continents over the last 514 Million Years since fishes first appeared in the oceans. The outlines show the continental margins, and white area is therefore deep sea. Shallow seas that flooded varying areas of the continents are not shown. The area and mean depth of the deep sea has remained more or less constant throughout this time.

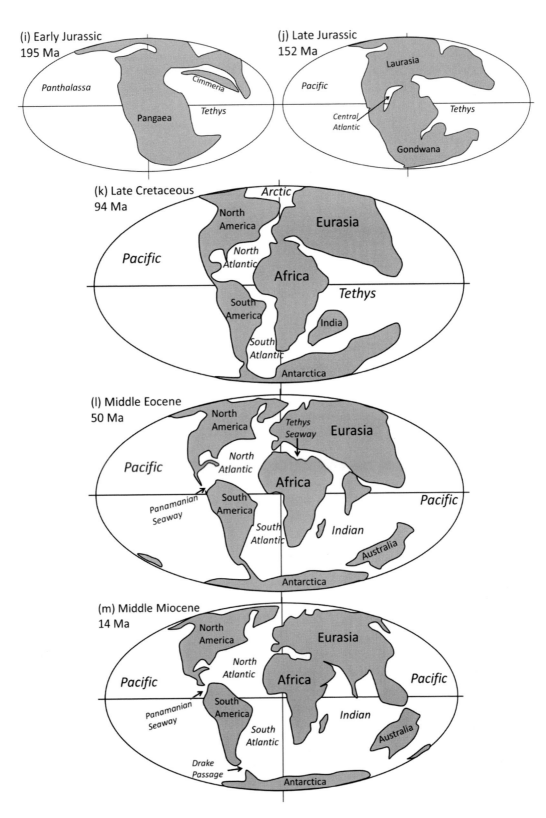

Figure 1.15 (*cont.*)

Tethys was complete. Pangaea began to break up around this time, and by the late Jurassic (Fig. 1.15j) the Central Atlantic Ocean Rift opened up, separating North America from Africa, and Gondwana began to split apart, separating Africa from eastern Gondwana (Antarctica, India and Australia). The new Atlantic Ocean Basin was connected by shallow seas to the Tethys Ocean. The Pacific Ocean is recognised for the first time, bordered in the east by a subduction zone along the coasts of North and South America and with the vast bay of the Tethys Ocean to the West. In the Cretaceous (Fig. 1.15k), the South Atlantic opened up, the Gulf of Mexico and Proto Caribbean Basins formed, and there was a deep water connection between the Atlantic and Tethys Oceans around the south of Africa. Australia remained attached to Antarctica, but India began to separate from Madagascar and move northward across the Tethys Ocean.

By the end of the Cretaceous, 66 Ma, the Mesozoic era is marked by the K-T event caused by a comet impacting the earth in the area of the Gulf of Mexico, which resulted in global mass extinctions. By the Middle Eocene (Fig. 1.15l) India had begun to collide with Asia, and most of the Tethys Ocean floor had been replaced by a new Indian Oceanic crust. Australia began to separate from Antarctica. The present-day spreading centres such as the East Pacific Rise, Mid-Atlantic Ridge and Indian Ocean Ridges can be recognised. At 50 Ma there was a deep-water connection, *ca.* 4000 m depth (Hamon et al., 2013), between the Northern Indian Ocean and the Atlantic through the Tethys Seaway. Part of this corresponds to the present-day Mediterranean Basin with vestiges of the Tethys ocean floor. The Tethys Seaway is assumed to have closed around 14 Ma creating a major reorganisation of oceanic circulation (Hamon et al., 2013). Conversely at 50 Ma the Drake Passage between South America and Antarctica was closed; however, tectonic movements were beginning to separate the two continents and shallow basins may have begun appearing in the Middle Miocene with a deep-water connection being established between 34 and 30 Ma (Livermore et al., 2005). The separation between

Australia and Antarctica created the deep waters (>2000 m depth) of the Tasman Sea around 32 Ma allowing circumglobal circulation of deep water around the Antarctic (Lawver & Gahagan, 2003). Deep-water connection between the Arctic basins, the Norwegian Sea and the North Atlantic became established in the early Oligocene 35 Ma (Davies et al., 2001). The Panamanian Seaway began to close between 10.5 and 9 Ma, major deep-water connection was lost around 7 Ma and the seaway finally closed 3 Ma (Schmittner et al., 2004).

The Mediterranean Sea can be regarded as a senescent ocean containing the vestiges of the Tethys Ocean floor in the Ionian Sea and the east Mediterranean basins, ranging in age from about 270 Ma (Late Permian) to 230 Ma (Middle Triassic), the oldest oceanic crust remaining on the planet (Müller et al., 2008). The sea is gradually narrowing as the African plate moves towards Europe. The Eastern Mediterranean will cease to exist about 6–8 Ma from this period, followed by the Western Mediterranean some millennia later (Ben-Avraham et al., 2006). During the Messinian Salinity Crisis 6 Ma, in the late Miocene the Mediterranean was entirely closed off from adjacent oceans (Hsü et al., 1973). Salinity increased as water evaporated, salt was deposited on the floors of the great basins and eventually the sea dried out entirely. Tectonic movements opened the Strait of Gibraltar at the end of the Miocene 5.33 Ma and the Mediterranean refilled with water from the Atlantic Ocean over several thousand years, although it is likely that 90 per cent of the water was transferred within a two-year period during a catastrophic event known as the Zanclean flood (Garcia-Castellanos et al., 2009). Dissolution of salt from the Messinian evaporites beneath the seafloor (Westbrook and Reston, 2002) has created very high salinity brine pools in deep basins of the Eastern Mediterranean. All marine life in the Mediterranean was extinguished during the Messinian salinity crisis, and present-day fauna, including deep-sea fishes, is the result of recolonisation during the last 5.33 Ma.

Throughout the history of the planet since the first appearance of fishes over 500 Ma the deep sea has existed with a typical depth as at present of around 4000 m. Continental slopes, mid-ocean ridges and subduction zones have also existed; however, their configuration has changed, and whether the deep sea has been continuously habitable by fishes will be discussed in subsequent chapters.

1.4 The Deep-Sea Environment

In order to survive in the deep sea, fish require oxygen and food, both of which must be derived from the upper ocean, oxygen from the atmosphere and food from photosynthesis in the solar-lit surface layers of the ocean. If we imagine the ocean as a vast basin filled with water, the surface tends to be heated by radiation from the Sun creating a layer of warm low-density water floating on top of denser cold water below (Fig. 1.16a). Any rainfall or river runoff from land lowers surface salinity and further reduces density. This stratification is stable and leads to stagnation (Schopf, 1980); any oxygen present in deep water is gradually consumed, mainly by bacterial oxidation of organic particles (Keeling et al., 2010), which results in anoxia in the absence of replenishment from the surface. This is the condition of the Black Sea, which is anoxic, rich in hydrogen sulphide and devoid of metazoan life below 50 m (Murray et al., 1989). Similar stagnation has occurred on a large scale during seven global anoxic events in the last 500 Ma (Takashima et al., 2006) during which large volumes of the deep sea would have been uninhabitable to fishes. In the modern oceans, anoxic conditions are rare (Keeling et al., 2010), but life in the deep sea depends on a continuous supply of oxygenated water from the surface.

1.4.1 Deep Water Formation, Circulation and Oxygenation of the Deep

The main mechanism transporting oxygenated water to the deep sea is by cooling of surface water coupled with ice formation that removes pure water so that salinity increases. In this way, cold saline water is produced, the density of which exceeds that of the

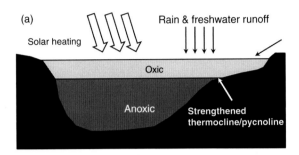

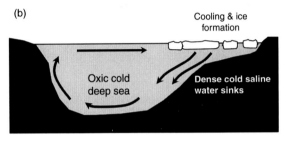

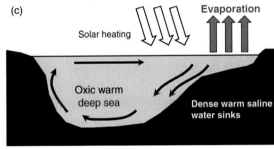

Figure 1.16 Oxygenation of the deep sea, contrasting patterns of deep-water formation.
(a) Stagnation. A strong thermocline with warm low-density water above, possibly further reduced in density by freshwater input is stable with no tendency for ventilation of the deep basin, which becomes anoxic. This appears to have occurred globally several times during earth history (Takashima et al., 2006) and is the present status of the Black Sea.
(b) Thermohaline circulation by cooling. In cold regions high-salinity cold water is formed that sinks. This is the main mechanism currently ventilating the global ocean resulting in a cold deep water.
(c) Thermohaline circulation by evaporation. In warm regions high-salinity warm water is formed that sinks ventilating the deep-sea basin with warm oxygenated water. This occurs in the Mediterranean and Red Seas.

underlying water and therefore sinks (Fig. 1.16b). There are two main sites of such deep water formation, the northern margin of the North Atlantic

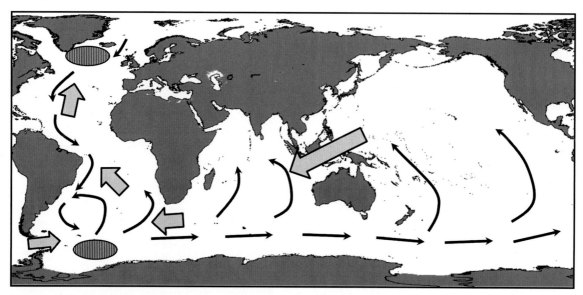

Figure 1.17 The global great ocean conveyor transporting oxygenated water into the deep sea. Deep water is formed by cooling in two main locations indicated by vertically shaded ellipses. Solid black arrows indicate deep water flow distributing oxygenated water around the world. There is mixing in the South Atlantic between water of Arctic and Antarctic origin. The box arrows indicate the likely path of return flow at the surface.

Ocean, including some Arctic Ocean outflow, forming the North Atlantic Deep Water (NADW), and the Weddell Sea forming the Antarctic Bottom Water (AABW) (Stommel and Arons, 1960). The NADW sinks to depths of around 4000 m and flows southward along the western boundary of the Atlantic Ocean. The AABW flows northward on the bottom of the South Atlantic into the Argentine Basin and in the vicinity of the Rio Grande Rise mixes with NADW, creating a blend of North Atlantic and Antarctic water that flows eastward around the Antarctic Ocean with branches of northward flow into the Indian Ocean and Pacific Ocean along their western boundaries (Fig. 1.17). Thus the deep sea is ventilated by a great ocean conveyor (Broecker, 1991) fed by deep water formation in the Atlantic sector of the world ocean. In the Indian and Pacific Oceans salinities are too low for cooling to create water that is dense enough to sink beneath the surface. The return flow to the Atlantic Ocean from the Pacific Ocean is twofold: warmed water flows at the surface through the Indonesian archipelago and across the Indian Ocean, and cold deep water flows through the Drake Passage to form

the Atlantic Intermediate Water (AIW). The Indian Ocean return flow is made up of cold deep water that passes around the Cape of Good Hope and contributes to the AIW (Oppo and Curry, 2012). The oxygen in the deep water is gradually consumed, and this is reflected in the decreasing abyssal oxygen concentration with increasing distances from the source, 7 ml.l^{-1} in the North Atlantic, 4–5 ml.l^{-1} over much of the Southern Hemisphere, to <3 ml.l^{-1} in the North Pacific (Mantyla and Reid, 1983). The renewal of oceanic deep water in this way requires a duration of several centuries to a millennium. The time elapsed since the water was last at the surface (and hence oxygenated) is 300–500 years in the Atlantic and Antarctic Oceans and over 1000 years in the Northern Indian Ocean and the North Pacific Ocean, according to inferences from radio-carbon dating (Gebbie and Huybers, 2012). Deep-sea fishes in the North Pacific Ocean are therefore consuming oxygen that was dissolved in the seawater from the atmosphere over 1000 years ago.

This cooling-driven thermohaline circulation not only supplies oxygen to the deep but also results in

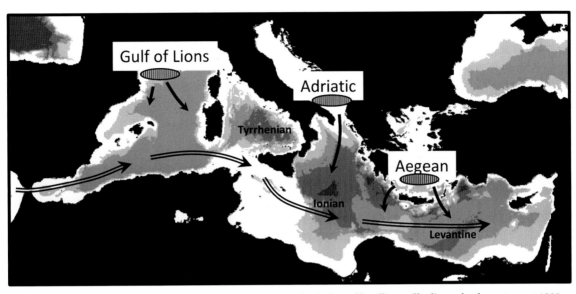

Figure 1.18 Areas of deep water formation in the Mediterranean Sea indicated by ellipses. Shading – depth contours at 1000 m intervals. Solid arrows – deep water flow. Open Arrows – surface water flow from the Atlantic Ocean.

low temperatures in the deep sea. The temperature of abyssal water is typically below 0°C in the Arctic and Antarctic Oceans 2–2.5°C over much of the Atlantic Ocean but colder, 0.25–1.5°C in South Atlantic areas influenced by Antarctic bottom water such as the Argentine Basin. In the Abyssal Indian and Pacific Oceans there is an increase in temperature from south to north from 0.5°C around the Antarctic, ~1°C at the equator to 1.5°C in the north (Mantyla and Reid, 1983). It is important to note that decrease in temperature is not an inevitable consequence of increase in depth; on the contrary, increase in pressure at depth causes adiabatic heating by compression of sea water, which results in an increase of temperature of ~1°C between 5000 m and 10,000 m depth in hadal trenches of the Pacific Ocean (Jamieson et al., 2010).

In contrast to the major oceans, deep water temperatures in the Mediterranean Sea are high, about 12.8°C–13.5°C in the western basin and 13.5°C to 15.5°C in the eastern basin (Danovaro et al., 2010). This arises from a rather different mechanism responsible for the formation of deep water and ventilation of the deep basins (Fig. 1.16c). Relatively low-salinity water from the Atlantic Ocean enters through the Straits of Gibraltar and flows eastward on the surface. Bottom water is produced in three main locations, the Gulf of Lions in the Western Mediterranean, the Southern Adriatic and the Aegean in the East (Fig. 1.18). In these areas evaporation creates high-salinity water, which, aided by cooling in the winter months, increases in density and sinks, forming deep water that generally flows westward, exiting the Mediterranean through the Straits of Gibraltar beneath the incoming surface Atlantic water (Robinson et al., 2001). The water flowing into the Atlantic Ocean becomes a distinct warm saline Mediterranean Intermediate Water (MIW) layer that is widely traceable over the North East Atlantic Ocean. The deep water oxygen concentration in the Mediterranean Sea is ~4.5 ml.l^{-1} in the western and 4.2 ml.l^{-1} in the eastern basin. A similar mechanism for ventilation pertains in the Red Sea with creation of high-salinity warm (21°C) deep water by evaporation in the north which flows southward towards the Indian Ocean (Cember, 1988).

In addition to deep water formation in the open ocean, episodic cascades of water descend into the deep sea from continental shelf areas. By their very nature such dense shelf water-cascading events are difficult to observe but Ivanov et al. (2004) catalogued 61 such cases worldwide, from the Arctic, midlatitudes, subtropics, tropical regions and the Antarctic. During a preconditioning phase dense water produced by cooling, evaporation or removal of water by ice formation accumulates in a shallow shelf sea area. This is followed by an active phase when the water has reached a critical density, enabling it to accelerate down the slope often through one or more canyons. The water may pick up a sediment load further increasing its density before it spreads out over the seafloor or at an appropriate mid-water density stratum. In the Japan Sea, in particularly cold winters, coastal surface water freezes, leaving a high-density cold brine that cascades downward reoxygenating the deep-sea basin down to over 3000 m depth (Talley et al., 2003). The best-documented cascades are in the Gulf of Lions in the NW Mediterranean Sea, where a combination of oceanographic instrumentation and deep-sea neutrino telescope infrastructure has made possible continuous measurements over several years. In this region, dry northerly winds cause cooling, heat loss, evaporation and the mixing of waters in winter months. In certain years, such as the winter of 2004–2005, low temperatures and reduced freshwater flow from rivers further increased shelf water density, triggering a dense shelf water cascade event that started in late February and, lasting 40 days, transported water down to over 1000 m depth at velocities of up to 85 cm.s^{-1} (Canals et al., 2006). Such events have far-reaching effects, influencing deep-sea ecology. The neutrino telescope photo-detectors at over 2000 m depth in the Gulf of Lions detected over 100 times the background levels of photon accounts in the months following cascading in 2009 and 2010 (Tamburini et al., 2013). This was attributed to deep 'bioluminescent blooms' stimulated by the organic matter transported into the deep. Proliferation of bioluminescent bacteria on organic detritus and increased abundance of bioluminescent fauna produced the increased light levels at depth. Company et al. (2008) found that commercial catches of deep-sea shrimps (*Aristeus antennatus*) in the Gulf of Lions region decreased following strong cascade winters but then recovered in intervening years with

recruitment apparently enhanced by the organic enrichment and oxygenation of deep waters following the cascade.

Currents near the deep-sea floor are weak, typically less than 10 cm.s^{-1} usually with tidal changes in speed and direction. Kemp et al. (2004) recorded a mean speed of 4. 9 cm.s^{-1} with variation between 0.2 and 14.8 cm.s^{-1} with a semidiurnal (12.4 h) rhythm at 2555 m depth in the Porcupine Seabight in the NE Atlantic Ocean. The tidal component becomes weaker with increasing depth but remains evident at great depths in the Madeira Abyssal Plain, 5300 m depth (Van Haren, 2004) and the Central North Pacific (5900 m; Priede et al., 1990). Over abyssal plains the currents are rather incoherent with little correlation between quite closely spaced sampling locations (Van Haren, 2004). In areas of strong topography, such as the flanks of Mid-Atlantic Ridge at 2500 m depth the tidal component runs parallel to the prevailing pattern of terraces and valleys (Priede, Bergstad, et al., 2013). At hadal depths, Jamieson et al. (2009) reported near sea bed currents of (8–14 cm s^{-1}) in the Kermadec Trench, and (3–7 cm s^{-1}) in the Japan Trench.

At high latitudes close to the sites of deep water formation, the temperature depth profile is rather uniform with more or less constant temperatures from the surface to the deep-sea floor and oxygen concentrations close to saturation throughout the water column (Fig. 1.19a). At temperate latitudes increased solar radiation in summer months creates a warm surface layer above a thermocline or temperature gradient that extends down to about 1000 m depth (Fig. 1.19b). In the winter increased storm activity results in mixing that obliterates some of the vertical temperature gradient. In the tropics (Fig. 1.19c) the warm surface layer is a constant feature with cold oxygenated water below. In the Mediterranean Sea the temperature profile is noticeably different; there is a warm surface layer and a temperature minimum at about 1000 m, but below this depth the temperature tends to increase towards the seafloor reflecting the high temperature origin of the deep water (Fig. 1.19d). In all regions surface water is well oxygenated, close to air saturation value in equilibrium with the atmosphere. However with

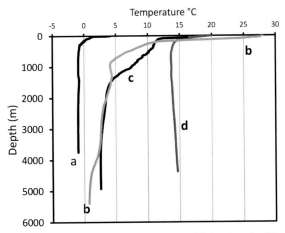

Figure 1.19 Temperature-depth profiles. (a) Arctic 76°15′N 2°30′W, (b) Tropical Atlantic Ocean 6°39′S 25°00′W, (c) Temperate North Atlantic Ocean 48°50′N 16°30′W, and (d) Eastern Mediterranean, Rodos basin. 35°59′N 28°24′E. Data from British Oceanographic data Centre and Hellenic Centre for Marine Research.

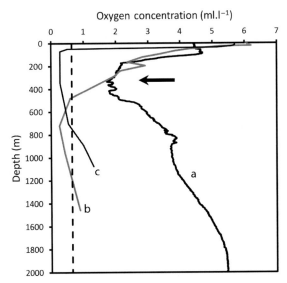

Figure 1.20 Oxygen-Depth Profiles. The vertical dashed line is at oxygen concentration 0.64 ml.l^{-1}, the threshold for definition of an Oxygen Minimum Zone or OMZ. (a) Tropical Western Atlantic Ocean with depth of oxygen minimum at 335 m (arrow) but no OMZ. (b) East Pacific Ocean off Peru, OMZ from 500–1200 m depth. (c) East Pacific Ocean off Northern California OMZ from 50 to 750 m depth. Pacific Data from Levin (2000), Atlantic Data from BODC.

increasing depth below the surface, oxygen concentration declines owing to consumption by microbial activity and other biota. Eventually an oxygen minimum depth is reached between 200 and 1000 m beneath which oxygen concentration increases again as deep water is encountered (Fig. 1.20). In most of the Atlantic Ocean the concentration of oxygen at the oxygen minimum depth is above 2.5 ml.l^{-1} (Keeling et al., 2010; Fig. 1.21b), but in several parts of the world the oxygen concentration falls below 0.64 ml.l^{-1} (20 μmol.l^{-1}) forming an Oxygen Minimum Zone (OMZ) within which there is insufficient oxygen to support normal aerobic respiration (Ramsing and Gundersen, 2013; fig. 1.22a). Such OMZs tend to occur on the eastern boundaries of the major oceans, in the Pacific Ocean off North America, Mexico, Peru and Chile, in the Atlantic Ocean off Northwest-African and Namibia and the Arabian Sea area of the Indian Ocean. The total volume of water affected by such anoxia is estimated as 1–7 per cent of global ocean volume (Wright et al., 2012) corresponding to over 10^6 km^2 of seafloor (Levin 2000). The OMZs occur in areas of coastal upwelling of deep water that feeds nutrients to the surface enhancing biological

productivity in the surface photosynthetic layer. Excess organic matter produced results in elevated rates of utilisation of dissolved oxygen in the water column below and hence the anoxia. The upwelling is driven by prevailing winds that drive surface water offshore drawing deep water upward near the coast. Areas where this occurs are known as Eastern Boundary Upwelling Ecosystems (Chavez and Messié, 2009) because due to the rotation of the earth they tend to occur on the eastern side of major ocean basins. In the upwelling area, circulation is associated with a surface current flowing towards the equator and a deep current on the continental slope below the shelf edge flowing in the opposite direction towards the poles (Stramma et al., 2010). Eastern Boundary Upwelling Ecosystems, notably off Peru and Namibia, support some of the world's most productive epipelagic fisheries.

Periodically the upwelling off Peru fails owing to an influx of warm surface water from the central tropical Pacific Ocean. This climatic event, known as

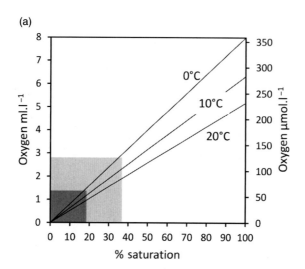

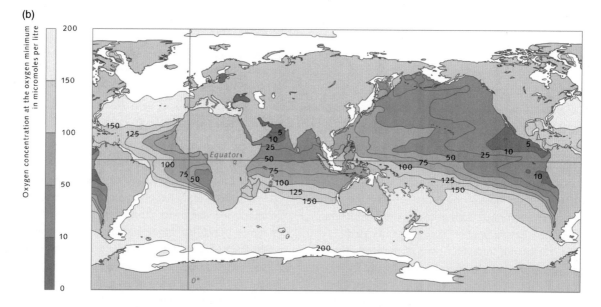

Figure 1.21 Dissolved Oxygen Concentration.

(a) Dissolved oxygen concentration in seawater (salinity 35 $^o/_{oo}$) at 0, 10 and 20°C. Calculated from Ramsing and Gundersen (2013). Dark shading – anoxic conditions probably lethal to fishes, Pale shading – hypoxic conditions limiting fish activity.

(b) Global oxygen concentrations at the depth of minimum oxygen. Reproduced from Keeling et al. (2010) with permission.

El Niño, occurs around December every two to seven years and results in a crash in regional fish production off the Pacific coast of South America. It is part of large-scale periodic event known as the El Niño southern oscillation (ENSO), which together with a North Atlantic Oscillation (NAO) causes large-scale climatic and ecological changes around the planet (Stenseth et al., 2002).

The Arabian Sea in the Northern Indian Ocean has a rather persistent OMZ between 150 and 1250 m depth, with oxygen concentrations reaching values of <0.05 ml.l^{-1} over large open ocean areas (Reichart

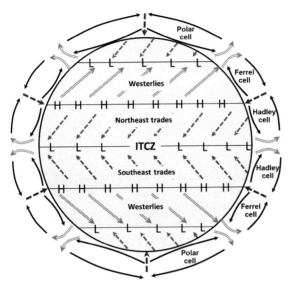

Figure 1.22 Global Wind patterns after Maury (1855) and others. The arrows around the periphery indicate the vertical circulation pattern with, e.g., warm air rising at the equator and sinking at intermediate latitudes. Prevailing surface winds are indicated on the outline of the globe. H – High pressure, ITCZ – Intertropical convergence zone, L – Low pressure.

et al., 1998). This is caused by high surface productivity and poor ventilation of the depths around the thermocline. Where the OMZ impinges on the slopes of the Pakistan margin, extensive areas of seafloor sediments are rendered anoxic with an exceptional accumulation of organic matter (Cowie and Levin, 2009). Some species of myctophid fishes, *Diaphus arabicus* and *Benthosema pterotum*, appear to be adapted to survive the low oxygen conditions prevailing in the Arabia Sea mesopelagic (Catul et al., 2011).

1.4.2 Surface Ocean Circulation and Gyres

Upwelling, ENSO and NAO are all influenced by the action of winds on the surface layers of the oceans. The pattern of prevailing winds on the planet was first described by Maury (1855) based on meticulous compilation of information from US Naval ship log books. Solar heating around the equator in the intertropical convergence zone causes warm air to rise, creating a belt of low atmospheric pressure,

known as the doldrums. The rising warm air circulates by convection, is cooled and sinks at a latitude of 30°N and 30°S to create a belt of high pressure around the planet known as the subtropical high or horse latitudes. Surface air flows from the high pressure towards the equatorial low, deflected in direction by Coriolis forces, to create the well-known North East and South East Trade Winds (Fig. 1.22). North of the horse latitudes prevailing westerlies blow across the sea surface towards the subpolar low pressure zones at around 60°N and 60°S. At the poles cold air sinks creating a high pressure zone, which drives polar easterly winds that meet the westerlies at the subpolar low. This pattern of three wind zones and three atmospheric convection cells in the Northern and Southern Hemispheres is fundamental property of the size and rate of rotation of planet earth and has probably been a constant feature throughout much of earth's history, although variations are important in producing fluctuations in climate (Diaz and Bradley, 2004). The surface winds from these drive the surface circulation of the oceans (Fig. 1.23) with a major gyre in each hemisphere driven by the trade winds at low latitudes and westerlies at high latitudes. This process results in a clockwise gyre in the Northern Hemisphere and an anticlockwise gyre in the Southern Hemisphere, e.g. in the North Atlantic there is a gyre comprising the North Equatorial Current, the Gulf Stream, the North Atlantic Drift and the Canary Current. Owing to the rotation of the earth the most intense currents are in the western boundary of the ocean basins, carrying warm water from low latitudes to high latitudes, e.g. the Gulf Stream, Brazil Current, Agulhas Current and Kuroshio Current. However, the current flow in the eastern parts of the basin is more diffuse. In the southern ocean there is a continuous Antarctic Circumpolar Current driven by westerly winds; with the flow unimpeded through the Drake Passage and the Tasman Sea (Broecker, 1991).

1.4.3 Ocean Biomes

With the advent of satellite technology, especially the detection of the sea surface chlorophyll concentration

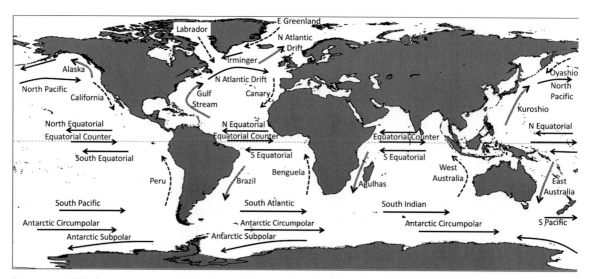

Figure 1.23 Global ocean surface currents. Dashed lines – cold eastern boundary currents, Double lines – warm western boundary currents.

by remote sensing (Hovis et al., 1980), it has become possible to build up a global image representing ocean biological productivity (Feldman et al., 1989). An inspection of Plate 1a reveals that in the centre of each of the oceanic gyres productivity is low and that highest chlorophyll concentrations are in coastal areas, the westerly wind latitudes and the eastern boundary upwelling areas. Based on such remote sensing images and extensive analysis of ship-borne data, Longhurst (1989) proposed the division of the oceans into twelve major biomes, characterised by the pattern of productivity (Table 1.2). These are based on four primary biomes: the Westerlies Biome, Trades Biome, Polar Biome and Coastal Biome, each with different characteristics of the surface mixed layer and consequent patterns of primary production (Fig. 1.24). The Trades Biomes is essentially the world's tropical ocean in which there is a strong permanent thermocline and diminishing Coriolis forces towards the equator. Occupying 45 per cent of the total ocean surface this is by far the largest oceanic Biome. Diel variation is greater than annual variation but shifting of the trade wind zone north and south of the equator does give rise to a perceptible annual cycle. In the Indian Ocean there is an exceptionally strong seasonal signal driven by the monsoon winds. During June to

Table 1.2 **List of Ocean Biogeographic Biomes According to Longhurst (1989)**

1. Antarctic Polar
2. Antarctic Westerly Winds
3. Atlantic Coastal
4. Atlantic Polar
5. Atlantic Trade Wind
6. Atlantic Westerly Winds
7. Indian Ocean Coastal
8. Indian Ocean Trade Wind
9. Pacific Coastal
10. Pacific Polar
11. Pacific Trade Wind
12. Pacific Westerly Winds

September the Southwesterly monsoon blows across the Arabian Sea bringing rain onto the Indian subcontinent. In September this pattern reverses during the North Easterly monsoon producing seasonal upwelling along the northern coasts and a changes in currents in the Northern Indian Ocean (Schott and McCreary, 2001).

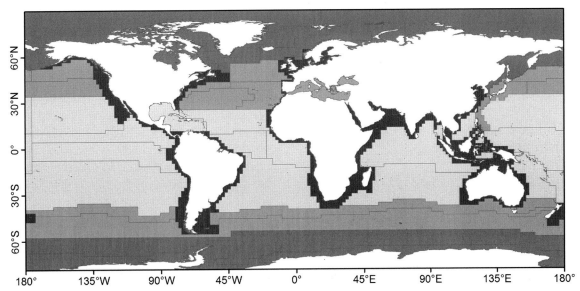

Figure 1.24 Oceanic ecological provinces defined by Longhurst (1989). Colours indicate the four main biomes, Very Dark grey – Coastal, Dark Grey – Polar, Grey – Westerly, Pale Grey – Trade Wind.

The Westerlies Biome is marked by strong seasonality particularly notable in the North Atlantic Spring phytoplankton bloom triggered by increased solar radiation following mixing and surface enrichment with nutrients by winter storms. The seasonal annual pulse of productivity in the North Atlantic is one of the major biological phenomena on the planet clearly visible from space and revealed by the CZCS (coastal zone color scanner) images. Western boundary currents such as the Kuroshio in the Pacific Ocean and Gulf Stream in the Atlantic Ocean are strong and stable sustained by Coriolis forces. The Polar Biome is characterised by the influence of ice cover in winter, surface freshwater as the ice melts in spring and a seasonal bloom of phytoplankton. However at cold temperatures herbivorous phytoplankton cannot complete a full life cycle in a single season, which has important consequences for the behaviour of the ecosystem. The Polar Biome at 6 per cent of total ocean surface is the smallest of the oceanic biomes, albeit appearing large in map projections such as Mercator. The Coastal Biomes are very diverse in their characteristics with important differences compared with open ocean settings.

Longhurst (1989) further subdivided the biomes into 51 distinct provinces, including 22 coastal provinces. Boundaries between the provinces are specified by straight longitudinal and latitudinal lines at a resolution of 1° the coordinates of which can now be downloaded from a number of sources (VLIZ, 2009) for use in GIS and other mapping applications. The advantage of this scheme is that the areas are clearly identified and reproducible. There is a continuing debate on the location and number of biogeographic zones in the oceans. UNESCO (2009) offers an alternative map with 30 oceanic pelagic provinces; however, any scheme must be flexible and acknowledge the shifting nature of oceanic boundaries compared with boundaries on land. A further difficulty is that highest biological productivity often occurs in the frontal zone at the boundaries between waters of different temperatures (Taylor and Ferrari, 2011) or in dynamic eddies where different waters are mixed (Priede, 2014).

UNESCO (2009) proposes schemes of benthic provinces in the deep sea, lower bathyal (14 provinces worldwide), abyssal (14 provinces) and hadal (10 provinces). However, the divisions (e.g. South,

Central, North Central and North Pacific Ocean) largely correspond to surface biomes proposed by Longhurst (1998). In Chapter 5 a global scheme of 33 mesopelagic ecoregions developed by Sutton et al. (2017) is described (Fig. 5.2.)

1.5 Food Supply to the Deep and the Distribution of Biomass

Deep-sea fishes derive their food supply from three sources: particulate organic matter that can be considered to fall like rain from the surface, which most fish cannot utilise directly; large food falls reaching the seafloor, which some scavenging species can exploit; and most commonly foraging on prey organisms living on the seafloor, within sediments or in the water column. In this section we consider the processes that drive trends in food availability.

1.5.1 Export of Particulate Organic Matter from the Surface

The defining characteristic of deep-sea fishes is that they live beneath the sunlit layers of the ocean, remote from the source of food on which they depend. The process by which organic carbon from biological production in the surface layers is transported downward into the ocean interior is termed the 'biological pump' in contrast to the smaller 'solubility pump' that transports dissolved carbon dioxide from the atmosphere into the ocean (Passow and Carlson, 2012). The biological pump starts with absorption of dissolved carbon dioxide by phytoplankton and its fixation into organic matter by photosynthesis. When phytoplankton die, together with bacteria and exuded exopolymers, they form gelatinous aggregations known as marine snow, which sinks by gravity through the water column at 80–330 m. day^{-1} accelerating with depth as buoyancy from organic carbon is lost (Berelson, 2002). The material arrives on the abyssal seafloor as phytodetritus, fluff or snow, in some regions of the world providing a strong seasonal pulse of organic enrichment to the deep-sea floor about one month after the spring phytoplankton

bloom at the surface (Billett et al., 1983; Beaulieu, 2002).

In the surface layer some of the phytoplankton is grazed and consumed by herbivorous zooplankton, which through their vertical movements contribute an active component to the downward transport of organic matter. The herbivores in turn may be consumed by carnivores including fish (Radchenko, 2007), which further contribute to the biological pump. The grazers and consumers package organic matter into faeces, which add to the rain of particulate matter falling towards the seafloor. These faeces together with exoskeletons and carcasses of zooplankton are often found in the mass of marine snow arriving on the seafloor (Beaulieu, 2002). In the Bermuda time series experiment in the North Atlantic, around the time of the spring bloom, transport by diel vertical migration of zooplankton can be equal to 70 per cent of mean particulate organic carbon (POC) flux compared with a year-round average of 8 per cent (Ducklow et al., 2001). Gelatinous zooplankton such as salps are abundant at certain times and are very effective in producing large faecal pellets that fall more rapidly through the water column (up to 2700 m d^{-1}) than the much smaller faeces of copepods (Bruland and Silver, 1981).

Honjo et al. (2008) show that the magnitude of the particulate organic matter flux varies according to surface productivity in the different biogeographic provinces identified by Longhurst (1989) from a minimum of 25 mmol C m^{-2} yr^{-1} in the Pacific Warm Pool to a maximum of 605 mmolC m^{-2} yr^{-1} in the Arabian Sea. Such vast differences in organic matter input are reflected in regional differences in the biomass and biodiversity of the deep-sea benthic fauna (Rex and Etter, 2010). Furthermore the particulate organic carbon flux at abyssal depths is temporally correlated with satellite-based estimates of surface productivity with a time lag of 1–2 months from the surface to the deep-sea floor (Lampitt et al., 2001; Smith et al., 2001). This results in pulsed input to the deep-sea floor reflecting the seasonal pattern of surface productivity. Thus there is clear regional and temporal variation in food supply to the seafloor with quite close coupling between the surface and the deep sea.

A large proportion of organic matter from primary production is reutilised in the upper 1000 m layers of the ocean; otherwise upper ocean pelagic ecosystems would tend to collapse. Export downward towards the deep sea must be equal to new production, which amounts to approximately 20–50 per cent of total production (Ducklow et al., 2001; Buesseler et al., 2007). However organisms constituting the biological pump themselves consume and respire much of the organic matter so that the amount arriving at the abyssal seafloor is equal to less than 5 per cent of surface production (Smith et al., 2009; Rex and Etter, 2010).

1.5.2 Food Falls

In the NE Pacific Ocean, Smith (1987) noted a discrepancy between supply of organic matter to the deep-sea floor and the observed consumption on the abyssal seafloor measured as benthic respiration. There was insufficient particulate organic matter being exported from the surface. The funnel-shaped traps used to measure particulate organic matter flux to the deep-sea floor (Honjo et al., 2008) do not intercept large carcasses or carrion falls, which are infrequent episodic events that bypass the small sampling aperture of the sediment trap and could account for at least part of the discrepancy in the carbon budget. Such large food falls may be of particular significance to scavenging deep-sea fishes, which would generally not be expected to utilise particulate organic matter directly. Stockton and De Laca (1982) point out that food falls represent extremely large local energy enrichment that is not measured by global flux measurements. Britton and Morton (1994) classified carrion falls by size, from the largest; Megacarrion ~100,000 kg (e.g. blue whale), Macrocarrion 100 kg (e.g. seal or dolphin), Mesocarrion 1 kg (bird or fish) to nanocarrion and ultracarrion representing the constituents of marine snow. Megacarrion falls are sufficiently large to form distinctive ecosystems that persist over years or decades before all the organic matter has been utilised. Whale falls show an ecological succession from an initial mobile

scavenger stage when soft tissue is removed by fishes and invertebrates, to final anaerobic breakdown of lipids in skeletal remains by bacteria and development of a complex specialist community living on the bones (Smith and Baco, 2003). Macrocarrion falls are consumed by fishes and other scavengers on a timescale of weeks to months as demonstrated by experiments with dolphins at abyssal (Jones et al., 1998) and bathyal (Kemp et al., 2006) depths. Mesocarrion falls, such as individual fishes, are intercepted and consumed by scavengers including fishes, often within hours. The response time of scavengers varies between a few minutes and a few hours according to depth and locality but in all areas of the deep sea investigated, food falls are reliably intercepted and consumed (King et al., 2007). In contrast to individual food falls, carcasses of large gelatinous zooplankton such as medusae following a bloom can be very abundant, covering a large proportion of the seafloor and providing a major input of organic matter to the seafloor (Billett et al., 2006; Yamamoto et al., 2008). Smith et al. (2014) found salp tunics covering up to 98 per cent of the seafloor at 4000 m depth following a surface spring bloom in the NE Pacific Ocean.

1.5.3 Distribution of Biomass in the Deep Sea

It has been evident since scientific sampling first began in the deep sea that life is sparser with increasing depth (Wyville Thompson, 1873) but nevertheless animal life comprising holothurians, amphipods and xenophyophores has been observed even at the deepest point in the world's ocean in the Mariana Trench at 10918 m depth (Gallo et al., 2015). For benthic biomass, thousands of samples have been taken by dredge, trawl and corers from which data have been compiled to discern global trends (Rowe, 1983; Rex et al., 2006; Rex and Etter, 2010; Wei et al., 2010). In these studies macrofauna are defined as small invertebrates living in or on the ocean floor sediments, captured in sediment cores and retained by sieves with 250–520 μm mesh size. Megafauna are large invertebrates and demersal fishes, usually larger than one centimetre, that can

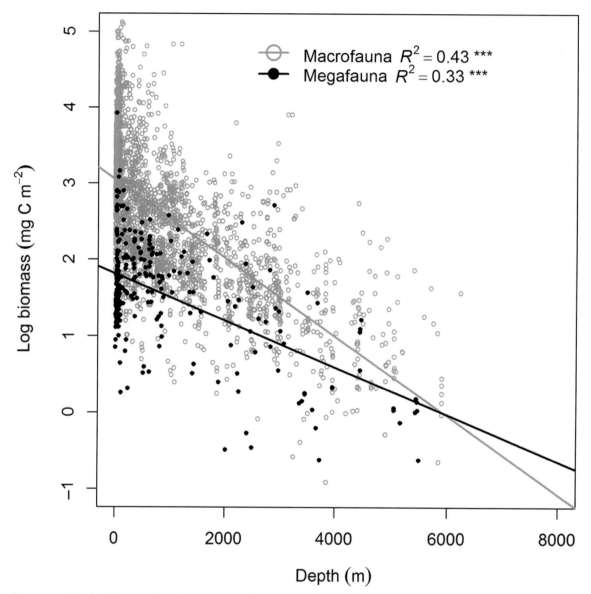

Figure 1.25 Trends of deep-sea benthic biomass with depth for macrofauna and megafauna (Wei et al., 2010).

be seen in seafloor images and are captured by bottom trawls or dredges. These two categories of biomass probably reflect the size range of organisms of significance as potential diet of fishes and biomass is expressed in terms of organic carbon per unit of seafloor. They both show a highly significant ($P < 0.0001$) logarithmic decline in biomass with depth (Wei et al., 2010):

(1) Macrofauna $Y_b = 3.052 - 0.000515\,X$ ($n = 2552$)
(2) Megafauna $Y_b = 1.812 - 0.000307\,X$ ($n = 282$)

Where Y_b = benthic biomass, Log_{10} (mg. C. m^{-2}), X = depth m. n = number of samples (Fig. 1.25)

It is instructive to convert these trends into ratios compared with the 200 m depth horizon that defines the upper limit of the deep sea. At 2000 m depth the macrofauna biomass is 12 per cent of the biomass at 200 m, 1.1 per cent at 4000 m, 0.1 per cent at 6000 m and extrapolating to 10000 m in hadal trenches the predicted biomass is 0.001 per cent of that at 200 m depth. For megafauna the decline is less steep, and corresponding predictions are 28 per cent at 2000 m, 7 per cent at 4000 m, 1.6 per cent at 6000 m and 0.1 per cent at 10000 m.

In addition to benthic biomass on the seafloor there is the pelagic biomass throughout the water column comprising zooplankton, fishes and other nekton. From net tows at different depths Roe (1988) found that the pelagic biomass at depths between 1000 m and over 5000 m follows a logarithmic trend with slopes between –0.0036 and –0.00076 (Fig. 1.26).

Priede et al. (2013) showed that taking average values, this can be approximated to:

(3) Pelagic biomass $Y_p = 0.1614 - 0.00048\ X$

Where Y_b = benthic biomass, Log10 (mg. C. m^{-3}), X = depth m.

Thus pelagic biomass decreases with depth in a very similar way to the benthic biomass except that it is expressed per unit volume (m^3) rather than per unit area (m^2). At 2000 m depth the pelagic biomass is 19 per cent of the value at 200 m, 2 per cent at 4000 m, 0.23 per cent at 6000 m and 0.003 per cent at 10000 m. A given amount of export production from the surface layers appears to be able to support an approximately fixed amount of biomass. In deep water there is a longer water column filled with pelagic biomass with a correspondingly small benthic biomass, whereas in shallow water the pelagic biomass is small and the benthic biomass is larger. Priede et al. (2013) showed this algebraically. If the pelagic biomass equation (3) is expressed in conventional mathematical form:

(4) $P(z) = P_0 \cdot 10^{-kz}$

Where $P(z)$ = pelagic biomass at depth z and P_0 and k are constants. P_0 is the intercept, i.e. theoretical surface biomass density (1.45 mg. C. m^{-3}, antilog of 0.1614), and k is the rate of decrease in biomass

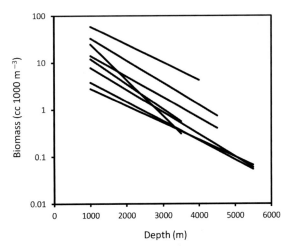

Figure 1.26 Trends of deep-sea pelagic biomass with depth in the NE Atlantic Ocean. Data from Roe (1988).

with depth (–0.00048). Then taking a depth of 800 m as a baseline, the total biomass below that depth is made up of pelagic biomass down to the seafloor at depth H:

(5) $\dfrac{\text{Total pelagic biomass}}{\text{per unit surface area}} = \displaystyle\int_{800}^{H} P_0 \cdot 10^{-kz}\,dz$

plus the benthic biomass at depth H:

(6) $\dfrac{\text{Total benthic biomass}}{\text{per unit surface area}} = \displaystyle\int_{H}^{\infty} P_0 \cdot 10^{-kz}\,dz$ (Fig. 1.27).

As depth H varies, the total biomass per unit area is constant with benthic biomass replacing pelagic biomass in shallow water and vice versa in deep water. P_0 the theoretical biomass at the sea surface will vary according to surface productivity. The value 1.45 mg. C. m^{-3} is an average value that provides a best fit to global deep-sea biomass data sets. Integrating only depths >800 m avoids effects of patchiness and discontinuities in the upper ocean (Roe 1988; Fig. 1.27)

These biomass trends can be taken as indicative of food availability and hence likely abundance of deep-sea fishes. Thus at 4000 m depth in the bathypelagic (equations 3, 4) the biomass is 0.017 mg. C.m^{-3}, however on an area of seafloor at the same depth the biomass would be 13.6 mg. C.m^{-2} (equation 6). The seafloor has a concentrating effect. Biomass that

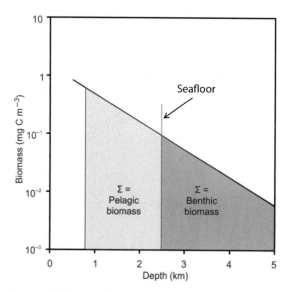

Figure 1.27 The relationship between deep-sea pelagic and benthic biomass. The line shows the general trend of decrease in pelagic biomass with depth. The pale shaded area shows the total integrated pelagic biomass above the local seafloor depth. The benthic biomass is the darker shaded area integrated under the line extrapolated beneath the seafloor, i.e. for a given surface export pelagic + benthic biomass ≈ constant (Priede et al., 2013).

would have been spread out over thousands of metres of water column, if the water were deeper, is concentrated into what Rex and Etter (2010) term the 'thin veneer of benthic life' generally reaching no more than 20–30 cm above the surface and concentrated in the upper few centimetres of sediment. Furthermore the seafloor acts as the ultimate sediment trap (Rex and Etter, 2010) intercepting material falling from the surface and serving as the final resting place for carrion falling from the surface, further increasing the potential food supply. In some canyons and trenches this effect may be further enhanced as organic matter and detritus are captured by the topography, accumulating in high concentrations in the axial valley enabling the habitat to support a higher abundance of fishes than would otherwise be expected (De Leo et al., 2010; Jamieson et al., 2010).

Currents passing over the seafloor can resuspend organic and inorganic particles into the water column (Thomsen and McCave, 2000). This often creates a layer of enriched pelagic or benthopelagic biomass in the benthic boundary layer (BBL) extending possibly tens or hundreds of metres above the seafloor (Wishner, 1980; Smith, 1982). This effect, however, is not universal, and it cannot be assumed that there is inevitably an increase in biomass in the benthic boundary layer (Craig et al., 2015). At bathyal depths on continental slopes, concentrations of pelagic prey can occur in the BBL through truncation of their vertical distribution by impingement on the seafloor, a crowding effect that is exploited by benthopelagic predatory deep-sea fishes (Mauchline and Gordon, 1991).

Wei et al. (2010) used a machine-learning algorithm, Random Forests, to model and predict seafloor biomass from surface primary production, water-column integrated and export particulate organic matter, seafloor relief, and bottom water properties. This produced a global map (Plate 1b). The pattern is heavily influenced by surface production, and there are obvious similarities with the global map of surface chlorophyll (Plate 1a). The lowest biomass prediction was between 30 and 80 mg C per square metre on the abyssal plains of the Pacific, Atlantic, and Indian Oceans with a bias towards higher biomass in the eastern Pacific and Atlantic influenced by coastal upwelling. The highest deep-sea biomass predictions were at 1200–3000 m water depths in the northern Arabian Sea with 1.3 and 2.5 g C per square metre. This corresponds to the areas with extremes in particulate organic matter flux from the surface observed by Honjo et al. (2008). Whilst the map successfully identifies some global patterns, it does not include food supply derived from carrion falls and deviates from the observations in some areas, such as the Gulf of Mexico, in which observed biomass is four times higher, and in canyon areas off New Zealand, in which observed biomass is 100 times higher. Conversely the map also includes smaller size fractions, bacterial, archaea (but not viruses) and meiofauna (>20–74 μm). The static map also does not take account of temporal change. Drazen et al. (2012) show that fluctuations in epipelagic fish abundance, and consequent changes in carrion falls to the deep sea have resulted in a sixfold increase in biomass of abyssal grenadiers living at 4100 m depth over a period of 17 years.

1.6 History of Discovery of Deep-Sea Fishes

1.6.1 From the Beginnings of Ichthyology to the First Deep-Sea Discoveries, 1554–1860

Deep-sea fishes appear to have been unknown to science before 1800. In his introduction to *Natural History of Fishes* (Cuvier and Valenciennes, 1828) the great French naturalist Georges Cuvier gave a chronological account of the development of ichthyology from its earliest beginnings by ancient Egyptians, through the writings of Aristotle, to the development of modern taxonomy and his own collections of material for the *Natural History*. Nowhere in this account are deep-sea fishes mentioned, and editorial notes in the English translation (Cuvier and Pietsch, 1995) state that deep-sea fishes were unknown. However, it seems very unlikely that humans could have been totally unaware of the existence of deep-sea fishes. For example, in the Straits of Messina, between Sicily and the mainland of Italy, during certain weather and sea states, upwelling conditions are created that bring deep-sea fishes to the surface, stranding them on adjacent beaches and attracting seabirds and other predators (Zahl, 1953). The fishes found in such events are typical mesopelagic and bathypelagic species, *Chauliodus sloani* (Bloch and Schneider, 1801), *Argyropelecus, Diaphus, Chlorophthamus* and *Cyclothone* that were not scientifically described until the early nineteenth century, though mariners and coastal-dwelling peoples must have known of them for millennia. Günther (1880) points out that deep-sea fish, particularly species with swim bladders, if accidentally displaced upward tend to float uncontrollably upward as internal gases expand and arrive at the surface dead or dying. He suggests that this must happen very often. Lowe (1843–1860) and Vaillant (1888) describe the capture of deep-sea fishes by traditional artisanal fishermen off the coasts of Madeira and Portugal, respectively, an activity that must have long predated scientific 'discovery' of deep-sea fishes. It is inconceivable that, around the Pacific Ocean where the seafloor tends to slope off to very great depths close to the shore, coastal people did not encounter or capture deep-sea fishes. Discovery of early Holocene fish hooks and fish bones (O'Conner and Veth, 2005) at human settlements on East Timor indicates the existence of line fishing 10,000 years ago, and off the coast of British Columbia early human settlers 10,600 years ago were exploiting a full range of marine species including Pacific halibut, *Hippoglossus stenolepis* (Schmidt, 1904; 1200 m_{fbd}), Lingcod *Ophiodon elongatus* (Girard, 1854; 475 m_{fbd}), and Cabezon, *Scorpaenichthys marmoratus* (Ayres, 1854; 200 m_{fbd}; Erlandson, 2008).

Guillaume Rondelet (1554) is acknowledged as the leading figure who established the study of ichthyology during the Renaissance with accurate illustrations from new material.

Scientific description of fishes and their classification based on comparative morphology rather than habitat began in England with the work of Francis Willughby (1686), *De Historia Piscium*. The book was completed, after Willughby's death in 1672, by John Ray (Kusukawa, 2000). Fishes were defined as animals with '*hairless skin and fins, lacking feet and being unable to live freely or long without water*'; this excluded the hippopotamus and the crocodile but included cetaceans, whales and dolphins, which were placed into Class I, *Cetacei* alongside Class II *Cartilaginei* (Chondrichthyes) and Class III *Spinosi* (Osteichthyes). The book, a publication published by the Royal Society of London with lavish illustrations, apparently depleted funds that might otherwise have supported Isaac Newton. Hagfish (Myxine), which were not regarded as a fish, were absent from the text, and Holocephali were unknown. Several species whose range is now known to extend into deep water were described with plates drawn from fresh specimens: angular roughshark (*Centrine* in

Willughby) *Oxynotus centrina* (Linnaeus, 1758)[2]
777 m_{fbd}, long-nosed skate (*Raja oxyrhynchus major*
in Willughby) *Dipturus oxyrinchus* (Linnaeus, 1758)
900 m_{fbd}, halibut (*Hippoglossus* in Willughby)
Hippoglossus hippoglossus (Linnaeus, 1758) 2000
m_{fbd}, Ling (*Asellus longus* in Willughby) *Molva molva*
(Linnaeus, 1758) 1000 m_{fbd} and Angler (*Rana
Piscatrix* in Willughby) *Lophius piscatorius* Linnaeus,
1758 1000 m_{fbd}. The species descriptions in Willughby
(1686) list authorities including Gessner (1555),
Rondelet (1554), and Aldrovandi (1613) who are
reviewed in Cuvier and Pietsch (1995). Recent
discoveries from North and South America were also
described.

Subsequently in Sweden, Peter Artedi (1738)
further developed the classification of fishes, listing
242 species organised into 52 genera. Carl Linnaeus
oversaw the posthumous publication of this work,
which he also incorporated into his *Systema Naturae*
(Linnaeus, 1735). *Systema Naturae* established the
modern binomial nomenclature for species and their
organisation into Classes, Orders and Genera. Over
subsequent editions the work was expanded until in
the tenth edition (Linnaeus, 1758) the cetaceans were

removed from the fishes and included in the class
Mammalia (Cuvier and Pietsch, 1995). This tenth
edition of 1758 is conventionally used as the starting
date for modern zoological nomenclature (ICZN,
1999). The hagfish *Myxine glutinosa* Linnaeus, 1758,
1200 m_{fbd} was listed in the Vermes (worms), not being
considered a fish until grouped with the lampreys in
the taxon 'Cyclostomi' by Duméril (1806). Bizarrely,
the lower fishes including the lampreys (*Petromyzon*),
Chondrichthyes (sharks and rays) and the angler
(*Lophius piscatorius*) were classified by Linnaeus as
'*Amphibia Nantes*' or swimming amphibians. Gray
(1789) commented that though far superior to what
any other person has done, the work of Linnaeus was
evidently done in a hurry containing false references
and careless descriptions. He states '*every anatomist
now agrees that the Amphibia Nantes are not
furnished with lungs and every naturalist is convinced
of the propriety of removing them from the class of
amphibian to that of fishes*'. Within the *Amphibia
Nantes*, Linnaeus also describes *Chimaera monstrosa*
Linnaeus, 1758, 1000 m_{fbd}, which had been absent
from the work of Artedi, but Linnaeus erroneously
synonymises with *Galeus acanthias* of Willughby,
who clearly referred to the picked dogfish *Squalus
acanthias* Linnaeus, 1758. According to Cuvier and
Pietsch (1995), the Chimaera had been previously
recorded by Carolus Clusius of Leiden in his *Exoticum
libri X*, published in 1605, but this appears to have
been forgotten. Linnaeus also described the deep
water velvet belly shark *Etmopterus spinax* (Linnaeus,
1758) 2490 m_{fbd}, which he placed in the genus
Squalus. Linnaeus, in common with other
taxonomists of the time gave only very broad habitat
information, e.g. '*in Europa, in Mari Atlantico* and *in
Mari Americano*' with no indication of depths of
occurrence.

By the end of the eighteenth century more than
40 species had been described with FBD greater than
1000 m, including deep demersal elasmobranchs, the
spiny eel *Notacanthus chemnitzii* Bloch, 1788, the
macrourid, *Coryphaenoides rupestris* Gunnerus, 1765,
and some pelagic species such as *Sternoptyx diaphana*
Hermann, 1781, and *Regalecus glesne* Ascanius, 1772.
However these were isolated taxonomic descriptions

---

[2] The International Code of Zoological Nomenclature (ICZN,
1999) defines how scientific names are given to animals.
Founded on the principles of Linneaus (1758) each species
is given a binomen comprising the genus and species
names such as, '*Macrourus berglax* Lacepède, 1801' the
roughhead grenadier. The full name includes the authority
and date. Here it is denoted that Bernard Germain de
Lacépède described and named this species in 1801.
'*Coryphaenoides armatus* (Hector, 1875)', the abyssal
grenadier was described by James Hector in 1875 but the
brackets indicate there has been a revision; the original
name given by Hector was *Macrurus armatus*. All species
listed by Linneaus are given his name and date such as the
thornback ray *Raja clavata* Linnaeus, 1758, although this
species was known previously and listed by Willughby
(1686) as *Raja clavata*. Linneaus tightened up the rules of
a naming system which was already evolving to that we
use today.
In this book the authorities and dates are generally
omitted to avoid making the text too cumbersome to read.
In this section however they are included to make clear
when each species was discovered.

with no depth information and hence no apparent recognition that they formed part of a deep-sea ichthyofauna.

The first person to describe deep-sea fishes scientifically as a distinctive fauna was Risso (1810, 1826). Giuseppe Antonio Risso or Antoine Risso (Fig. 1.28), as he later became known, was a native of the city of Nice in the Duchy of Savoy on the border between France and Italy. In his *Ichthyologie de Nice*, (Risso,1810) he described the seafloor as extending to 2000 metres in depth in the Bay of Nice and to immeasurable depths further offshore. Amongst the new species he found there, *Mora moro* (Risso, 1810), he wrote, is '*very common at great depths with tender white flesh and good flavour*'. Later he wrote, '*The deep basins of the sea are frequented by alepocephalids, epigonids, chimaeras and macrourids. Lesser depths are the habitat of whitings, lings, hakes, soles, citulas, seriolas*' (Risso, 1826). Risso's work was important from four points of view in that: he established that fish live at depths possibly down to 2000 m and beyond; he described a number of deep-sea fish species; he identified depth zonation with different species living in discrete depth strata; and he noted the commercial potential for the exploitation of deep-sea fishes including descriptions of capture methods. In addition to these demersal fishes, Risso (1810) also described the bathypelagic Stomiiform fish, *Stomias boa boa* (Boa dragonfish, 1500 m$_{fbd}$), which he named *Esox boa* stating that the abdomen is marked with four rows of golden dots, features now known to be bioluminescent photophores. Cuvier praised the work of Risso for the new and precise details of habits of fishes in the Mediterranean but did not mention the depths at which they were discovered, commenting only on their classification.

Deep-sea fishes in the Atlantic Ocean came to the attention of the scientific community when Robert T. Lowe, began publishing a series of papers (Lowe, 1833) on fishes found around the island of Madeira, including the Black Scabbardfish, *Aphanopus carbo* (Lowe, 1839). He gave an account of how fishermen retrieve Wreck-fish, *Polyprion americanus* (Bloch and Schneider, 1801) from 550 to 730 m depth using baited lines (Lowe, 1843–1860). From 1851 James

Figure 1.28 Antoinne Risso (1777–1845) of Nice, who first described a deep-sea fish fauna in his book *Ichthyologie de Nice*, published in 1810.

Yate Johnson continued the work off Madeira (Johnson 1862, 1863) naming important species such as *Synaphobranchus kaupii* (Johnson, 1862). The material collected from Madeira by Lowe and Johnson was sent to the British Museum in London, where Albert Günther was working on the catalogue of fishes. For Günther these specimens served as his first introduction to the special adaptations of fishes to life in the deep sea (Günther, 1887).

In 1843 Edward Forbes at the meeting of the British Association for Advancement of Science in the city of Cork, Ireland, reported results from bottom dredging in the Aegean Sea (Eastern Mediterranean), which indicated that the abundance of invertebrate life decreases with depth, reaching a '*Zero of Animal Life probably about 300 fathoms*' (see page 170 in Forbes, 1844). This azoic hypothesis is often cited as evidence that it was widely thought that no life could exist at depths greater than 550 m (Anderson and Rice, 2006, Kunzig, 2003). However an account of European seas (Forbes and Goodwin-Austen, 1859), completed by his editor after Forbes's untimely death in 1854, on pages 194–196 cites the works of Risso (1810, 1830), stating that Alepocephalids and other fishes occur at '*depths of 2000 feet and upwards*' (>610 m) directly contradicting the concept of zero animal life. The azoic hypothesis applied to the especially sparse populations of invertebrates in the Aegean, and there is no evidence that Forbes or his contemporaries seriously doubted the existence of fishes in the deep sea. The story of Forbes and his azoic hypothesis is a didactic device, much used by teachers of deep-sea biology but probably misrepresents scientific thinking in the mid-nineteenth century.

1.6.2 Great Voyages of Discovery, 1860–1914

Following the publication of *The Origin of Species* (Darwin, 1859), ichthyology entered its early modern phase as described by Jordan (1919), embracing the concepts of evolution and genetics in addition to morphology as the fundamental basis for classification. Biological sampling by British ships, the HMS *Porcupine* and HMS *Lighting*, during 1868 to 1870 in the NE Atlantic provided convincing evidence of life in the deep sea but gave no information on fishes (Wyville Thompson, 1873). There was tremendous excitement about the new findings; Edward Perceval Wright (1870) of Dublin University wrote, '*So much is now known of a deep sea fauna that one is very apt to forget how little was known just 18 months ago*'. He was studying deep-sea sponges at depths down to 180 m but was surprised to discover that Portuguese fishermen routinely captured sharks off Setubal at depths of 550–730 m.

A major landmark in knowledge of life in the deep sea is the voyage of the HMS *Challenger*, which circumnavigated the globe between 1872 and 1876 under the scientific leadership of Charles Wyville Thomson assisted by John Murray. This expedition successfully sampled the deep sea across the Atlantic, Pacific, Indian and Antarctic Oceans, laying the foundations for modern oceanography. A maximum depth of 8,184 metres (with no biological samples) was recorded in the Pacific Ocean between Guam and Palau in a location now known to be the southern end of the Marianas Trench. Biological samples taken in 151 open water trawls and 133 bottom dredges

Figure 1.29 Albert K. L. G. Günther (1830–1914) of the Natural History Museum, London, author of 'Report on the Deep-Sea Fishes collected by the H.M.S. *Challenger*, 1873–1876'.

captured fish for the first time at abyssal depths. Reports of the scientific results of the voyage were published in 50 volumes between 1880 and 1895 but also in short papers by appropriate invited experts (Rice, 1986). For example, a collection of five new species of fishes from 730 m depth in New Zealand waters was described by James Hector, Director of the Geological Survey of New Zealand, and published a year before the end of the *Challenger* voyage (Hector, 1875). A notable species in this collection was *Macrurus armatus*, now known as *Coryphaenoides armatus* (Hector, 1875), and found at abyssal depths throughout most of the world.

Wyville Thomson died in 1882, so it fell to John Murray to complete editing the reports. The deep-sea fish samples had been entrusted to Albert Günther, who was by then Keeper of the Department of Zoology in the British Museum (Fig. 1.29). In Chapter 21 of his book, *Introduction to the Study of Fishes*, Günther (1880) gave a preliminary account of his findings, indicating that fish had been found down to at least 2900 fathoms (5300 m) depth. Notwithstanding the importance of the new data from the *Challenger*, on page 297 he added, '*the existence of fishes peculiarly adapted for the deep sea has been a fact maintained and admitted for some time in Ichthyology*'. His duties did not permit him to start fully analysing the *Challenger* samples until 1884, and the full account was not published until 1887 with 66 illustrative plates plus appendixes on anatomy by other authors, including an account of light organs (Günther, 1887). The species descriptions were substantially revised so that, for example, *Marcurus armatus* was redescribed on the basis of 26 specimens including the single one available to Hector (1875).

Günther (1887) based his report on 610 specimens obtained during the *Challenger* voyage plus 184 from other expeditions that immediately followed, resulting in the description of 154 new species, 144 of which were from the *Challenger* samples. Prior to the *Challenger* voyage only 37 deep-sea fish species were known. Günther showed that the number of species decreased with depth, recording 232 at 183–550 m depth and 23 at 3660–5300 m. He enumerated 385 deep-sea species of which he considered 230 that occur deeper than 550 m as truly abyssal forms, noting that macrouridae were the most abundant.

After the *Challenger* expedition, voyages of exploration were launched by other nations. The Norwegian North Sea Expedition (1876–1878) reported its findings before the *Challenger* results were published (Collett, 1880). In the North East Atlantic the French vessels *Travailleur* (1880, 1881 and 1882) and *Talisman* (1883) confirmed the existence of a distinctive deep-sea ichthyofauna (Vaillant, 1888). In his introduction Vaillant comments on the long-standing traditional fishery for deep-sea sharks off Setubal of which the scientific world had been ignorant.

In the North Western Atlantic Alexander Aggasiz of the Harvard Museum of Comparative Zoology sponsored a series of voyages by US Fish Commission Steamers *Blake*, *Albatross* and *Fish Hawk* (1877–1884), which resulted in a treatise on Oceanic Ichthyology with oceanic fishes defined as species living deeper than 500 feet (152 m; Goode and Bean, 1895). George B. Goode and Tarleton H. Bean of the Smithsonian Institution in Washington, DC, went on to describe over 80 new species of deep-sea fishes in a series of papers. The USFCS *Albatross* had been purpose-built in 1882 by the US Fish Commission as a scientific research vessel to investigate fish populations and make hydrographic surveys. In 1891 Agassiz led a voyage of the *Albatross* in the Pacific Ocean from Panama to the Galapagos Islands, Mexico and the Sea of Cortez trawling at great depths (Summers et al., 1999). The collections were consigned to Samuel Walton Garman of the Harvard Museum of Comparative Zoology, resulting in a monograph with colour plates describing over 100 new species of deep-sea fishes (Garman, 1899). Further expeditions on the *Albatross* by Charles Henry Gilbert of Stanford University, to Alaska, Californian waters, the Hawaiian Islands and the Japanese Archipelago resulted in descriptions of over 200 deep-sea fish species, including a report on deep-sea fishes of Hawaii (Gilbert, 1905). The series of voyages by the *Albatross* in previously unexplored regions discovered more deep-sea fish species than ever achieved by any other vessel (Fig. 1.30). The Russian Imperial vessel *Vitiaz* circumnavigated the globe under the command of Stepan Makarov during 1886–1889, making valuable oceanographic observations in the North Pacific Ocean; however, no biological sampling was undertaken.

Germany immediately followed the *Challenger* expedition with the voyage of the *SMS Gazelle,* which sailed from Kiel in 1874 and circumnavigated the globe, meeting with *Challenger* in South America before returning in 1876. Although this expedition helped to define the nature of deep-water circulation, more significant progress in the study of fishes came with the German Deep Sea Expedition 1898–1899 on the steamship *Valdivia*, which circumnavigated under the leadership of zoologist Carl Chun. The deep-sea fishes were described by August Brauer in Volume 15 of the expedition reports; Part 1 on systematics (Brauer, 1906) and Part 2, on anatomy (Brauer, 1908), illustrated with colour plates by Fritz Winter (Fig. 1.31). Although only 22 years had elapsed between the end of the *Challenger* voyage and the sailing of the *Valdivia*, extraordinary technical progress had been made. The *Challenger* was a wooden sailing ship, with an auxiliary steam engine,

Figure 1.30 Charles Henry Gilbert (1859–1938) of Stanford University, who between 1888 and 1906 participated in voyages of the U.S. Fish Commission Steamer *Albatross* in the Pacific Ocean describing more species of deep-sea fishes than any other taxonomist. (Photo courtesy, Theodore W. Pietsch)

Figure 1.31 *Melanocetus johnsonii* Günther, 1864. A Ceratioid Angler Fish of the Family Melanocitidae (Black Seadevils). Illustration by Fritz Winter in 'The Deep-Sea Fishes' (Brauer, 1906) from the voyage of the *Validivia*, 1898–99, edited by Carl Chun. Part of Plate XV, labelled, *Melanocetus krechi* Brauer, 1902 in the original illustration.

that travelled mainly under sail and handled nets and sampling instruments using hemp ropes. Dredging on the seafloor was accomplished by allowing the ship to drift sideways to the wind and waves (Rice, 1986). The *Valdivia* by contrast was a steel steamship with no sails, able to travel under its own power and tow nets in mid-water and on the bottom using steel wires and steam-powered winches, achieving a much higher rate of work than had been possible on the *Challenger*. The major contribution of the *Valdivia* expedition to deep-sea ichthyology was the recognition of the importance of the deep pelagic fauna and that ceratioid angler fishes were mesopelagic and bathypelagic life-forms rather than bottom-living fishes as previously thought.

While in Great Britain there was a decrease in activity following the *Challenger* expedition, the British colonial government in India commissioned a new paddle steamer, the Royal Indian Marine

Steamship RIMS *Investigator*, launched in Bombay in 1881 (Rice 1986). This ship undertook a regular programme of deep-sea dredging in the Northern Indian Ocean at depths from 182 m to 3652 m during 1885–1899. From this collection, 169 deep-sea fish species were identified of which 126 were new species (Alcock, 1899). In his report Alcock notes that of the 43 species found elsewhere, 23 are common to the Atlantic and the remaining 20 are Indo-Pacific species, some of the earliest observations of global zoogeography of deep-sea fishes.

From 1885, Prince Albert of Monaco conducted work in the Mediterranean Sea and the Atlantic Ocean from his yachts, *Hirondelle* (1885–1890), *Princess Alice* (1891–1897), *Princess Alice II* (1898–1910) and *Hirondelle* II (1911–1915). An important species discovered during these voyages was the Orange roughy *Hoplostethus atlanticus* Collett, 1889. With each successive ship, the technology for deep-water bottom and pelagic trawling improved using steel wire and steam-driven winches. Sampling extended from the Northern coast of Norway to the tropical Atlantic and the coast of Brazil. In 1894 Prince Albert compared results from bottom trawling (using an Agassiz beam trawl) and a baited cage trap at 1674 m and found that whereas the trawl caught predominantly slow-moving invertebrates such as crustacea and echinoderms, the trap caught mainly fishes. The *Princess Alice II* regularly sampled at depths greater than 4000 m and successfully trawled to 6037 m depth SW of the Cape Verde Islands on 1 August 1901, a record that stood until 1947 (Mills, 1983).

By this time certain forms of deep-sea fishes had become clearly recognisable, and Arthur Smith Woodward, Keeper of the Geological Department at the British Museum, was able to identify several Cretaceous fossils as deep-sea fish species providing first information on the antiquity of the deep-sea ichthyofauna (Woodward, 1898).

During the late nineteenth and early twentieth centuries permanent marine biological or oceanographic laboratories began to be established in different parts of the world, e.g., in the Mediterranean; Naples, Italy (1872), Villefranche, France (1882) and Monaco (1910), the Atlantic; Plymouth England (1888) and Woods Hole USA (1888) and the Pacific Ocean, Scripps, La Jolla USA (1903) and Nanaimo, Canada (1905). These provided bases for systematic sampling of adjacent sea areas. With access to near-shore deep water, Shigeho Tanaka of the Imperial University of Tokyo described a range of species of deep-sea fishes, both pelagic, e.g. the Myctophid *Notoscopelus japonicus* Tanaka, 1908 and demersal, e.g. the Ophidiiform *Encheliophis sagamianus* Tanaka, 1908.

In 1909 Sir John Murray wrote to Johan Hjort, Director of the Norwegian Institute of Marine Research in Bergen, offering to defray the costs of an expedition in the North Atlantic using the Norwegian government Research Steamer *Michael Sars*. Murray had become wealthy from interests in mining phosphate deposits he had discovered on Christmas Island during the voyage of the *Challenger*. Murray joined the *Michael Sars* in Plymouth in April 1910 for a four-month voyage, which completed two transects across the Atlantic Ocean from the Canary islands via the Azores to Newfoundland and then across to

Ireland on the return. The results were presented in what became a best-selling book, *The Depths of the Ocean* (Murray and Hjort, 1912), in which the Mid-Atlantic Ridge was identified as dividing the Atlantic Ocean into eastern and western basins from Iceland to at least 53°S, and extensive observations were made of deep-sea fishes, their eyes, light organs, reproduction and larval stages, e.g. pelagic young of bathydemersal macrouridae. The cruise was marked by very effective towing of fine-meshed pelagic nets that successfully collected very high quality samples of mesopelagic and bathypelagic fishes.

The *Michael Sars* cruise formed part of a trend towards international collaboration in research on the high seas, but like the work of Alexander Agassiz and Prince Albert, depended on sponsorship by a wealthy individual. In 1902 an agreement between eight nations, Denmark, Finland, Germany, the Netherlands, Norway, Sweden, Russia, and the United Kingdom established the International Council for the Exploration of the Sea (ICES) providing the basis for international cooperative research sponsored by the governments of Western Europe. Regular trawl surveys were established to determine the distribution and abundance of fishes. At the tenth anniversary of the founding of ICES (Pettersson and Dresschel, 1913), a proposal was addressed to governments for resources to undertake regular sampling across the Atlantic Ocean to a depth of 1000 m through a series of quarterly cruises. Plans for this, the work of Prince Albert and much else were curtailed by the onset of the First World War (1914–1918). ICES continued throughout the political upheavals of the twentieth century and is today the world's oldest intergovernmental science organization playing an important role in fisheries management of the NE Atlantic Ocean and Baltic Sea.

1.6.3 To the First Manned Descents into the Deep, 1914–1945

Mills (1983) states that after 1914 deep-sea biology became a side issue with little activity, a relic of nineteenth-century thought that did not develop again until after the Second World War (1939–1945).

Certainly the disruption of the war years and economic recession inhibited major research in the deep sea. In Great Britain, funding was dominated by research related to commercial whaling in the Southern Ocean under the auspices of the *Discovery* committee that sponsored fundamental studies around Antarctica including sampling of fishes (Rice, 1986). Individual fish taxonomists nevertheless were active in museums and institutes working on old collections and receiving new material from around the world. David Starr Jordan (1919, 1917–1920) of Stanford University consolidated the advances since Linnaeus by cataloguing all the genera of fishes. Charles Tate Regan, who was at the Natural History Museum in London from 1901 to 1943, is best known for the discovery of dwarfed males and sexual parasitism in ceratioid angler fishes derived from examination of material from the *Dana* expedition (Regan, 1925) but was also prolific in descriptions of deep-sea species (Regan and Trewavas, 1932). In South Africa, a new research vessel, the SS *Pickle*, commissioned in 1920 collected numerous new species of deep-sea fishes Gilchrist (1922). During 1933–1934, the Egyptian research vessel *Mabahiss* was used for an expedition in the NW Indian Ocean funded by the John Murray bequest. Germany commissioned a new research ship, the *Meteor*, equipped with novel acoustic sounding equipment and during the German Atlantic Expedition 1925–1927 discovered the Meteor Bank in the South Atlantic (Schlee, 1973) and on subsequent voyages, the Great Meteor Seamount in the North Atlantic, which many years later became the subject of fish diversity studies (Fock et al., 2002).

Exceptions to the decrease in biological exploratory activity are the expeditions of the *Dana I* mainly in the North Atlantic (1920–1922) and the *Dana II*, which circumnavigated the globe (1928–1930) under the leadership of Johannes Schmidt. These voyages took advantage of passage through the Panama Canal, which had been opened in 1914. Samples from *Dana I* sent by Schmidt to London provided two of the species of Ceratioid with dwarf males described by Regan (1925), who named one of them *Edriolychnus schmidti* in his honour. Later this species was found to

have been previously described by Brauer from the *Vildivia* collections and is now known as *Haplophryne mollis* (Brauer, 1902). Brauer had been unaware of dwarf males (Pietsch, 2009). The *Dana II* expedition was sponsored by the Carlsberg Foundation (of the famous brewery) that continued to fund the publication of the *Dana Reports* for 40 years after the voyages had ended (Wolff, 2002). The *Dana* voyages produced a prodigious quantity of mainly pelagic samples. Regarding deep-sea fishes, the major advances were in four bathypelagic families, the Notosudidae, Paralepididae, Ceratioidae (superfamily) and Giganturidae.

In 1932, William Beebe and Otis Barton descended to 922 m depth in the NW Atlantic Ocean off Bermuda inside a steel diving capsule with quartz windows known as the *Bathysphere*, the first successful attempt by humans to directly view life in the deep sea. During this and previous shallower dives, Beebe (1934) observed copious pelagic bioluminescence and was able to identify fishes such as *Cyclothone* and *Argyropelecus* through the viewing ports. He also described four putative new fish species based on his observations of characteristic light patterns. This attempt to describe new genera and species in the absence of type specimens or any permanent record such as a photograph was viewed with scepticism by contemporary ichthyologists (Hubbs, 1935). These species have not been subsequently confirmed, and the names have disappeared from authoritative listings. In this sense the *Bathysphere* made no scientific progress, but there is no doubt about its pioneering significance, particularly in observations of bioluminescence in situ (Pietsch, 2009). The scientific reputation of Beebe in deep-sea ichthyology rests on conventional descriptions of deep-sea species he captured around his research base in Bermuda (e.g. Beebe, 1932) and his major role in popularising ocean science.

1.6.4 The Great Twentieth-Century Scientific Expansion from 1945

Following the Second World War the first major deep-sea expedition was the Swedish Deep Sea Expedition of the *Albatross*, which circumnavigated the globe during 1947/48. The *Albatross* featured a specially built deep-sea electrically driven winch and a deep-sea echo sounder. Most of the voyage was devoted to bottom coring, but in the last three months of the voyage Orvar Nybelin conducted 14 bottom trawls at depths greater than 4000 m, reaching 7900 m in the Puerto Rico Trench in August 1948, thus breaking the record set by Prince Albert of Monaco in 1901 (Mills, 1983). There were no fishes in the deepest trawl, but significant deep demersal fish samples were obtained in other trawls (Nybelin, 1957).

After the *Albatross* returned, the winch and other expedition equipment were sold to a Danish team who were reequipping an ex-British navy ship renamed the *Galathea* for a round-the-world voyage. The *Galathea* expedition was led by Anton Bruun, who had served on the *Dana* voyages, and its primary aim was not only to collect fauna from the deepest trenches but also to capture active animals of the abyss. During 1950–1952 the *Galathea* trawled in five trenches, and one major success was finding life at over 10,190 m depth in the Philippine Trench, which led Bruun (1951) to postulate the presence of a distinctive hadal fauna endemic to trenches deeper than 6000 m.

From 1949 to 1959 the Soviet Research Vessel *Vitiaz* conducted sampling at abyssal and hadal depths in the North West Pacific Ocean. In contrast to the ad hoc adaptation of *Albatross* and *Galathea*, the *Vitiaz* was a purpose-built vessel 109 m long with 14 laboratories capable of supporting 53 researchers; the flagship of a Soviet fleet of research vessels representing a major commitment to ocean research built up by the USSR since 1922 (Suziumov, 1970–1979). Anatole Petrovich Andriashev made extensive studies of deep-sea fishes and in 1935 proposed the concept of categorising the deep-sea fish fauna into ancient or true deep-water forms and secondary deep-sea forms. The former have a long history of adaptation to the deep sea whereas the latter are recent immigrants with fewer specialisations. The theory gained wide attention after the Second World War when an account (Andriashev, 1953) was translated into English (Fig. 1.32). Over an extraordinary career spanning 74 years between his first and last paper, Andriashev's contribution to

Figure 1.32 Anatole Petrovich Andriashev (1910–2009) of Zoological Institute St Petersburg was an influential researcher with a prolific output from 1934 to 2008. He developed concepts of deep-sea zoogeography. (Photo courtesy, Natalia Chernova, ZIN)

deep-sea ichthyology is unparalleled (Pavlov et al., 2009; Stein and Chernova, 2009).

Andriashev (1955) described a new species of hadal liparid *Careproctus (Pseudoliparis) amblystomopsis* captured at 7230 m in the Kurile-Kamchatka Trench. Further specimens were captured and an individual captured by the *Vitiaz* from 7529 m in the Northern part of the Japan Trench (Zenkevitch and Birstein, 1960) was for some years regarded as the world's deepest fish (Nielsen, 1964). On 21 January 1970, during one of a series of deep-sea expeditions by the RV *John Elliot Pillsbury* of the University of Miami, a fish was captured in a 41-foot (12.5 m) otter trawl from 8370 m depth in the Puerto Rico Trench. Identified as *Bassogigas profundissimus* (Roule, 1913), it was embedded in fine mud, which was taken to indicate that it had been retrieved from the seafloor and was recorded as the deepest living vertebrate (Staiger, 1972). The specimen was subsequently examined by Nielsen (1977), who reassigned it to a new genus and species, *Abyssobrotula galatheae*, based on comparative samples taken by the *Galathea, Vitiaz* and other ships, and has since been widely recognised as the world's deepest fish.

In the years after the Second World War, major nations followed the USSR in acquiring research vessels capable of worldwide operations and sampling the deep sea. The USA commissioned the *Atlantis II* (1962, 2,300 tons) and *Melville* (1969, 2,075 tons), France the *Jean Charcot* (1965, 2,200 tons), UK the *Discovery* (1962, 2,800 tons) and Japan the *Haku Maru* (1967, 3,225 tons). Germany followed later with the RV *Meteor* (1986, 3990 tons) but had smaller fishery research vessels. Most maritime nations also had regionally-operating fishery survey vessels, which all added up to a vast expansion in sampling capability around the world. This is reflected in the acceleration in the rate of discovery of new species of deep-sea fishes from 1950 onward, exceeding that achieved in the period 1890–1910 (Fig. 1.33). A new generation of taxonomists began their careers (Table 1.3), but their pattern of work differed from that of their predecessors. Whereas Garman, Alcock and Gilbert embraced a wide range of taxa from a geographical area, Nielsen, Iwamoto and Stein each tend to specialise in a given group (Ophidiiformes, Macrouridae and Liparidae, respectively) but through air travel and improved communications are able to work on samples derived worldwide. However even the most prolific taxonomists have been responsible for no more than 5 per cent of the deep-sea ichthyofauna; cataloguing the world's deep-sea fishes has been a collaborative effort involving hundreds of individuals, too numerous to name here.

Fish sampling shifted from voyages of exploration aimed at discovery of new species to regular and quantitative sampling to measure the biomass and distribution of known species.

From 1963 and 1976 Pearcy et al. (1982) trawled for benthic fishes at 400–5180 m depth in the NE Pacific Ocean off the west coast of the USA from the continental slope to the Cascadia and Tufts Abyssal plains. In the NE Atlantic a similar series of studies was done off Northwest Africa (Merrett and Marshall, 1981) and the west of Ireland (Merrett

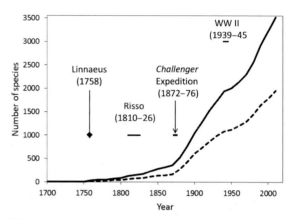

Figure 1.33 Cumulative number of deep-sea fish described from 1700 onward. Solid line – species ≥500 m depth, Dashed line – species ≥1000 m depth (FishBase depth).

et al., 1991a, 1991b). Accumulating data from different areas, Haedrich and Merrett (1988) were able to assemble an atlas showing the depth distribution of deep demersal species around the entire North Atlantic Basin. Koslow (1994) used a similar approach off SE Australia. Regular sampling in this way, providing statistically appropriate samples sizes, also resulted in studies on diet (Pearcy and Ambler, 1974), growth (Bergstad, 1995) and life history investigations (Stein and Pearcy, 1982). In addition to academically driven research, ground-fish surveys in many parts of the world extended into greater depths in support of development or management of deep-sea fishing. In 1962 the USSR began surveys to potentially target deep-sea species in the North Atlantic (Shibanov and Vinnichenko, 2008) and North Pacific Ocean (Tuponogov, 2008). The USA and Canada established a joint East Coast of North America groundfish survey from 1975 (Brown et al., 1996).

Regular sampling of deep-sea fishes has provided opportunities for biological studies on structure, function, physiology and biochemistry. Freshly caught specimens could be studied in on-board laboratories at sea or preserved for analysis on shore. From 1966 to 1979 the Scripps Institution of Oceanography operated the RV *Alpha Helix*, a specially equipped 500 tonne vessel for experimental biology at sea built under the guidance of the

Table 1.3 **List of Taxonomists Who Have Named the Greatest Number of Deep-Sea Species with FishBase Depths ≥ 500 m. Risso Is Included as the Founding Father of Deep-Sea Ichthyology. (Authors Are Ranked in Chronological Order of Their First Significant Publication.)**

Name	Forename(s)	Country	Dates	Institution	Species	Major Contribution
Risso	Antoine (Giuseppe Antonio)	Nice (France/ Italy)	1777–1845	Lycée Impérial de Nice	21	First described deep-sea fish species in the Mediterranean Sea and their depth zonation.
Günther	Albert Karl Ludwig Gotthilf	Britain (German)	1830–1914	Natural History Museum London	140	Deep-Sea Fishes from the voyage of the HMS *Challenger*
Garman	Samuel Walton	USA	1843–1927	Harvard Museum of Comparative Zoology.	118	Fishes from 1891 cruise of the *Albatross* in the East Pacific Ocean
Goode	George Brown	USA	1851–1896	US Fish Commission	84	Goode and Bean published together on fishes from the voyages of the *Blake*, *Albatross* and *Fish Hawk* in the NW Atlantic Ocean
Bean	Tarleton Hoffman	USA	1846–1916	Smithsonian Institution, Washington, DC	97	
Alcock	Alfred William	British India	1859–1933	Indian Museum Calcutta	82	Deep-sea fishes collected by the RIMS *Investigator*, in the Indian Ocean.
Gilbert	Charles Henry	USA	1859–1928	Stanford University, California	227	Voyages of the *Albatross* in the Pacific Ocean, deep-sea fishes of Hawaii.
Regan	Charles Tate	Britain	1878–1943	Natural History Museum London	115	Discovered dwarf parasitic males of oceanic angler-fishes and described many deep-sea species
Andriashev	Anatoly Petrovich	USSR (Russia)	1910–2009	Zoological Institute, St. Petersburg	124	Fishes of northern seas, Antarctica, *Vitiaz* voyages.
Nielsen	Jørgen G.	Denmark	1932–	Zoological Museum, University of Copenhagen	75	Fishes of the *Galathea* expedition, specialist in Ophidiiformes

Table 1.3 (*cont.*)

Name	Forename(s)	Country	Dates	Institution	Species	Major Contribution
Iwamoto	Tomio	USA	1939–	California Academy of Sciences, San Francisco	88	Specialist in Macrouridae
Stein	David L.	USA	1944–	Dept. of Fisheries and Wildlife, Oregon State University	114	Specialist in Liparidae

comparative physiologist Per Scholander (Schmidt Nielsen, 1987). Important early insights from this era concerned changes in enzyme activities with depth (Childress and Somero, 1979), buoyancy mechanisms of bathypelagic species (Denton and Marshall, 1958), and the function of bioluminescence (Denton et al., 1972).

Following the Second World War, a revolution took place in the scientific understanding of the deep-sea floor. Since the publication of the first edition of the General Bathymetric Chart of the Oceans (GEBCO) in 1905 under the direction of Prince Albert of Monaco, there had been gradual progress in mapping of the deep-sea floor. However, by the 1950s continuous echo-sounding lines began to reveal unprecedented detail of seabed features. With access to US Defense Department and cable industry data, from 1957 Bruce Heezen and Marie Tharp began producing a series of 'physiographic diagrams' of the ocean floor (Doel et al., 2006). These maps, with a three-dimensional presentation of the seafloor by the Austrian artist Heinrich Berann, were published by the National Geographic (1967–1969) and became well- known iconic images of the seafloor, showing the extent of the mid-ocean ridge system that spanned the globe. Recognition that the central rift valleys of the ridges were the sites of seafloor spreading, together with much other evidence from numerous workers, resulted in a general acceptance of the theory of Continental

Drift (Le Pichon, 1968; Heezen, 1969). For ichthyology, these developments revealed the complexity of deep-sea habitats but also the theory of Continental Drift, and new knowledge of the recent origin of the Atlantic Basin profoundly influenced oceanic zoogeography.

The Heezen and Tharp maps were criticised for their lack of precision, as many details were inserted by interpolation with no direct evidence for their existence or location. GEBCO responded with the publication of a fifth edition in 1982 at a scale of 1:10,000,000, which for a decade provided the basis for planning of deep-sea sampling voyages. A further major step forward was achieved by Smith and Sandwell (1997), who used satellite altimetry data combined with soundings to produce a world map at a two-minute grid resolution. This provided unbiased sampling across the globe, revealing seamounts and other mid-ocean features with a new degree of certainty. Widespread adoption of the satellite-based Global Positioning System (GPS), which became fully operational in 1995, enabled research vessels to locate and sample such discrete features in mid-ocean. In 2009 GEBCO released a new digital global chart at 30 arc-second resolution, providing a view superficially similar to those shown by Heezen and Tharp a half century earlier, but with features accurately located enabling discrimination of the true distribution of seamounts across the oceans (Yesson et al., 2011). Seamount ecology (Rogers, 1994) and the fishes and fisheries associated with them (Froese and Sampang,

2004) began to emerge as a dominant theme in deep-sea research.

Surveying of the seafloor was still hampered by the single line of depth data obtained by conventional echo sounders. For detecting summits of seamounts multiple passes might be necessary to ensure that the true summit had been found. In 1977 the first commercially available multibeam echo sounder was installed in a research vessel producing a display of a broad strip of seafloor beneath the ship, so-called swath bathymetry (Farr, 1980). Such systems have since been installed on many research vessels enabling rapid mapping of areas of interest with identification of substrate characteristics and habitat types (Connelly et al., 2012).

In parallel with improved knowledge of the seafloor, the launch of the Coastal Zone Color Scanner (CZCS) on board a polar-orbiting Nimbus satellite in 1978 (Hovis et al., 1980) marked the beginning of global mapping of ocean chlorophyll concentration and sea surface temperature by remote sensing. Using these data and ship-based sampling, Longhurst (1989) defined biogeographic zones of the oceans characterised by different patterns of surface productivity. The role of the ocean in the global carbon cycle and ocean biochemistry emerged as important areas of research with the founding of the Joint Global Ocean Flux Study (JGOFS) in 1987 (Fasham, 2003). Measurement of export flux of organic carbon from the ocean surface to the deep sea below (Smith et al., 2009) has further influenced our understanding of the processes determining the nature of life in the water column and the seafloor below (Priede, 2013).

A fundamental problem remained in deep-sea ichthyology that with the exception of Beebe (1934), fishes had never been seen alive in their natural environment. The first images of deep-sea fishes began to be recorded by geologists inspecting the seafloor using bottom-contact-triggered cameras lowered on wires. Hartman and Emery (1956) reported a few fishes at depths greater than 600 m off California using a device known as a Benthograph with a camera mounted inside a steel sphere derived from the bathyscaph (Robison, 2000). Heezen and Holister (1971) compiled a remarkable set of deep-sea images from around the world, which included a view of *Bathypterois bigelowi* in a typical tripod-fish pose on the floor of the Gulf of Mexico. However, such images could only be random incidental events. The first trials of manned submersibles capable of descending to great depths had begun in 1948 under the direction of August Piccard with the launch of the first bathyscaphe. The human occupants were secured at one atmosphere pressure inside a sphere supported by a buoyancy tank filled with gasoline. In 1954 a French Navy bathyscaphe reached a depth of 4,050 m in the NE Atlantic Ocean. The US Navy bathyscaphe *Trieste* reached the deepest point in the world's oceans on January 1960 when Jacques Piccard and Don Walsh descended to the Challenger Deep at 10912 m in the Mariana Trench. They described seeing a flatfish on the seafloor, disturbed by the arrival of the submarine (Piccard and Dietz, 1961), an observation that is now considered erroneous (Wolff, 1961; Jamieson and Yancey, 2012). In 1964 the French bathyscaphe *Archimède* encountered 200 liparids, probably of the genus *Careproctus*, at 7300 m depth in the Puerto Rico Trench: small individuals 10–12 cm long and a larger size class up to 25 cm long (Pérès, 1965). In addition a zoarcid and a macrourid were seen, the first observations of living fish at hadal depths. On these dives the camera system was not functioning, so no permanent record was made of these observations. Also from *Archimède*, Wolff (1971) describes encountering macrourids, *Coryphaenoides armatus* at 4160 m depth on the Madeira abyssal plain and their capture on baited hooks. Barham et al. (1967) estimated population densities of two fish species *Anoplopoma fimbria* and *Sebastolobus altivelis* at 1243 m depth in the San Diego Trough surveyed by bathyscaphe *Trieste* in 1962, taking cine and stills film camera images of transects.

Following these pioneering studies, the Woods Hole Oceanographic Institute commissioned the Deep Submergence vehicle *Alvin* in 1964 to carry a crew of three, a pilot and two scientists. Observations on abundance, behaviour (Grassle et al., 1975), activity (Cohen, 1977) and metabolism (Smith and Hessler, 1974; Smith, 1978) of deep-sea fishes were important early achievements. *Alvin* was followed by *Nautile* in France (1984), two *Mir* submersibles in Russia (1987) and *Shinkai* in Japan (1990). Such submersibles were essential in the discovery and exploration of chemosynthetically supported life around hydrothermal vent sites, first on the East Pacific Rise using *Alvin* (Corliss and Ballard, 1977) and later at numerous locations around the global ocean ridge system (Van Dover, 2000). New species of fishes were observed associated with these sites (Rosenblatt and Cohen, 1986).

Much progress has also been made with unmanned systems. Isaacs and Schwartzlose (1975) recorded images of active fishes at abyssal depths attracted to bait placed within view of their 'monster camera', a free-fall system activated by timers. Such free-fall cameras, or landers, have since been used in a wide range of studies on fishes around the world (King and Priede, 2008), culminating in recordings of images of hadal fishes at depths greater than 7000 m depth in the Japan and Kermadec Trenches in the Western Pacific Ocean (Jamieson et al., 2009). These studies showed that in most deep-sea environments there exists an active scavenging demersal fish fauna capable of intercepting carrion falls from the surface. From the 1970s onward remotely operated vehicles (ROVs) have gradually become increasingly available to the scientific community. Powered and controlled by a cable tether, they can manoeuvre through the water and across the seafloor providing an efficient means of observing, collecting and manipulating samples in the deep. Widely used in industry, they have become effective alternatives to manned submersibles and with real-time high-resolution imagery are replacing trawl surveys in deep-sea fish abundance studies, especially in complex habitats (Baker et al., 2012). From 1995, full ocean depth ROVs, the Japanese ROV *Kaiko* and the US Hybrid ROV *Nereus* from 2009 have revisited the Challenger Deep at 10,900 m (Nakanishi and Hashimoto, 2011) in the Mariana Trench. There have been no sightings of any fishes at these extreme depths (Jamieson and Yancey 2012). By the end of the twentieth century a wide diversity of submersibles including autonomous underwater vehicles (AUVs) were greatly influencing the course of deep-sea research through a process of coevolution of research questions and technical capabilities (Robison, 2000).

During the second World War reports emerged of a 'false bottom' in echo-sounding traces caused by the scattering of sound in a so-called Deep Scattering Layer (Eyring et al., 1948). The fact that this was a false bottom echo became obvious as the layer moved closer to the surface during the night and descended during the night, leading Johnson (1948) to postulate that the layer was made up of living organisms displaying a circadian vertical migration behaviour. This hypothesis was confirmed by direct observation from a manned submersible that dove through the Deep Scattering Layer in the Pacific Ocean off Baja California, demonstrating that myctophid fishes formed an important component of the migrating fauna (Barham, 1966). In the NE Atlantic Ocean using DSRV *Alvin*, Backus et al. (1968) observed the structure and movement of shoals of the myctophid *Ceratoscopelus maderensis* at around 600 m depth. They were able to detect the shoals using a scanning sonar system on *Alvin*, obtain close-up photographs and capture specimens for confirmation of identification, a major advance over the controversial observations by Beebe (1934). From early observations of fish shoals in echo-sounder traces during the 1920s to the discovery of Deep Scattering Layers in the 1940s and subsequent developments of multifrequency techniques, active acoustics or SONAR has become an important technique in marine biology research enabling biomass distribution in the vertical and horizontal planes to be recorded from a ship whilst underway (Fornshell and Tessai, 2013). The demonstration by Foote (1983) that integrated acoustic backscatter is linearly related to biomass provided the basis for such calculations. For deep-sea studies, ship-borne systems are limited by the maximum range of the acoustic signal to 500–1000 m depth, depending on the frequency used. Koslow et al. (1997) showed how by combining net tows with acoustic surveys large-scale studies of deep pelagic fishes could be realised at mid-continental slope depths off Australia. Irigoen et al. (2014) estimated the global biomass of mesopelagic fishes between 200 and 1000 m depth based on data from the Spanish Circumnavigation Expedition (December 2010–July 2011), using a 38kHz echo sounder.

The second half of the twentieth century saw an extraordinary explosion in the development of scientific methods and instrumentation. The invention of the Polymerase Chain Reaction (PCR) in 1983 (Barlett and Stirling, 2003) to amplify fragments of DNA opened up the field of molecular phylogeny, population genetics and the use of genetic barcodes for species identification (Hebert et al., 2003). In addition to new chemical and biological analytical methods, the electronics and communications revolution has greatly altered methods of working. Computers make possible statistical analyses and data processing that were previously extraordinarily laborious. FishBase (Froese and Pauly, 2016), the now widely used global database on fin fishes, was proposed by Daniel Pauly in 1988, first published as a CD-ROM in 1995, appeared on the internet in 1998 and by the year 2000 was considered to have entries for all known fish species (Froese, 2014). FishBase is now part of a suite of complementary online databases that have become essential tools for research and dissemination of information. Catalog of Fishes (Eschmeyer et al., 2014) first published in 1990 gives an authoritative list of fish names, OBIS (Grassle, 2000) Ocean Biogeographic Information System incorporates mapping tools to provide geospatial information for all marine species and World Register of Marine Species (WoRMS; Glover et al., 2014) covers the taxonomy of all marine species.

Analysing entries in OBIS, Mora et al. (2008) estimated that 2081 species of bathydemersal and 1275 bathypelagic marine fishes had been described; a total of 3356 deep–sea species compared with the small number known at the time of Risso (1810) about 200 years earlier. Species accumulation curves predict the ultimate number of expected species to be 3719 (3594–3843, 95 per cent CL) bathydemersal species

and 1670 (1591–1749) bathypelagic species. The species inventory is less complete for demersal species (56 per cent) than for bathypelagic species (76 per cent), indicating there are about 1638 and 395 species, respectively, remaining to be described. Ever since the work of Linnaeus (1758) and his predecessors, the description of new species appears to have been a never-ending task. However, Costello et al. (2013) suggest that the total number of species on earth has been overestimated in the past and that by concerted action all remaining species on the planet could be described within 50 years. In common with other vertebrates, the deep-sea fishes are relatively well known, and given appropriate sampling effort, it seems that the inventory of species could be essentially complete within the lifetime of many readers of this work.

Over the time span since scientific description of fishes began, the language used by authors has changed. Willughby (1686), Artedi (1738) and Linnaeus (1735, 1758) all wrote in Latin with the aim that their work could be universally understood. From 1800 onward there was a tendency to write in a leading national language, e.g. Risso (1810, 1826), Cuvier and Vallenciennes (1828) and Vaillant (1888) wrote in French; Lowe (1833), Günther (1887), Garman (1899) and Gilbert (1905) in English; Brauer (1906) in German and Andriashev (1953) in Russian. In the last 50 years English has achieved ascendancy as a universal language of scientific discourse regardless of the national origins of the author.

1.7 Collection and Methods

1.7.1 Accessing the Deep Sea

Most methods for capture of deep-sea fishes are a simple extension of those used in shallow water and include traditional techniques such as hooks, traps and nets. However one of the major problems of working in the deep sea is the weight and strength of cables that are needed to lower and recover equipment at great depths. A 10 km length of steel-armoured oceanographic electrical cable, 11mm diameter has a breaking strength of 6.5 tonnes and weighs 3.3 tonnes in water (Tyco, 2012). This means that at full ocean depth (ca. 11 km), the weight of the cable itself exceeds 50 per cent of its breaking strength, whereas normally on land it would be considered unsafe to lift loads exceeding 10–20 per cent of the breaking strength. Adding the weight of underwater equipment, hydrodynamic drag and variable loading owing to the ship's motion in rough weather, it is easy to understand that loss of equipment in the deep sea through cables parting is not unusual. These problems can be mitigated by the use of tapered cables and novel materials such as titanium or Kevlar, but the fundamental difficulty remains. A feature of deep-sea operations is the need for constant monitoring of the cable tension while fishing to ensure that the small margin for safety is not exceeded. For trawling at abyssal depths, 10–15 km of steel wire are required, weighing over 10 tonnes. The wire is wound onto a storage drum under tension, which creates a lateral force of ca. 300 tonnes on the end plates of the winch drum (Currie, 1983), a source of multiple failures in early attempts to construct deep-sea winches (Murray and Hjort, 1912). Stein (1985) reported damage to fishing winches when loading the long lengths of wire necessary for trawling in the abyssal NE Pacific Ocean. Deep-sea winches are therefore massive pieces of machinery that, together with generators necessary to power them, are an essential feature of most deep-sea research vessels.

The research vessel must provide a safe stable platform for work at sea (Jamieson et al., 2013). The RRS *Challenger*, operated by the UK Natural Environment Research Council (NERC) from 1973 to 2002, was 54.3 m long, displaced 988 tonnes, could carry 14 scientists (Gage and Tyler, 1991) and is typical of a small 'regional class 'research vessel suitable for deep-sea fishing by bottom trawl. Capable of towing nets to a depth of 5000 m, it was extensively used for deep-sea demersal fish studies in the slope and abyssal plain environments of the NE Atlantic (Merrett et al., 1991a, 1991b). For mid-ocean studies, the transit time to the working area may take one week or more, so larger vessels of greater endurance and ability to ride out the worst storms are necessary. Vessels in the 90 to >100 m size class have space for multiple sets of equipment including remotely operated vehicles as well as space for > 30 scientists. Satellite navigation, real-time sonar mapping of the seafloor and ability to manoeuvre using computer-controlled multithruster systems allow such vessels to manoeuver in mid-ocean to an accuracy of a few metres in relation to submarine features. Modern vessels such as the RRS *James Cook* (length 89.2 m, 5401 tonnes) delivered to NERC in 2007 are also built to minimise underwater noise emissions from the machinery and propellers so that there is no disturbance to fish behaviour more than 20 m from the ship (Mitson and Knudsen, 2003).

Deep-sea research can be successfully conducted from much smaller vessels where research is in areas that are close to shore and where long transit voyages are therefore unnecessary. From 1988 to 2011 the Monterey Bay Aquarium Research Institute operated the ROV *Ventana* down to 1500 m depth from the RV *Point Lobus* (length 33.5 m, displacement 400 tonnes) on the slopes and canyons off California. Jamieson et al. (2013) retrieved samples and imagery from hadal depths >6000 m in the Kermadec Trench off New Zealand from an even smaller vessel, the RV *Kaharoa* (length 28 m, displacement 300 tonnes), using autonomous systems with no cable connection to the ship.

To avoid the use of long cables it is possible to allow equipment to sink to the seafloor with ballast attached to a release mechanism, which may be a soluble link, an electro-mechanical timer (Isaacs and Schick, 1960) or an acoustic device that responds to a coded sonar command transmitted from the ship (Jamieson et al., 2013). When the ballast is released the equipment then returns to the surface by virtue of its buoyancy (Fig. 1.34). Conspicuous flags, colours, flashing lights and radio beacons aid recovery on the surface. Such Deep Free Vehicle, free-fall or pop-up systems have the added advantage that while the equipment is on the seafloor it is secure and cannot be affected by shipping or bad weather. Furthermore the ship can deploy multiple systems achieving seafloor sampling time vastly exceeding the available ship

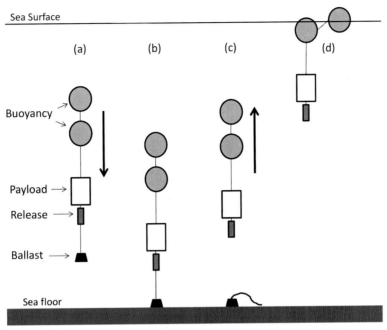

Figure 1.34 The Free-Fall, Lander or Pop-up method for deployment and recovery of equipment in the deep sea.
(a) Descent by free-fall after deployment from the ship.
(b) System operating autonomously on the seafloor for hours, days or months.
(c) Ascent, ballast release triggered by time delay or acoustic command from the surface.
(d) System on the surface ready for recovery by a surface ship.

time. The key elements are the buoyancy and the release mechanism. Evacuated glass spheres are now most commonly used with 6000 m depth-rated floats available off the shelf from several suppliers. Acoustic releases have largely replaced the time releases and soluble links that were used in early studies, e.g. Isaacs and Schick (1960). These developments together with other novel autonomous systems have the potential to greatly increase the scope for deep-sea research and thus avoid the need for long cables. More elaborate deep-free vehicle platforms that descend to the seafloor with a suite of instruments are now often known as deep-sea landers (Jamieson et al. 2013).

1.7.2 Methods of Fish Capture

1.7.2.1 Static Gears

Static or passive fishing gears are placed in a fixed position on or above the seafloor and depend on the movements of fish, whether attracted to bait or their natural activity, enabling them to be captured.

Baited Hooks Fishing with baited hooks has a long history and was one of the methods used by Prince Albert I for collection of deep-sea fishes during the voyages of his yachts between 1885 and 1915. He adapted the technique used by the Portuguese deep-sea shark fishermen of Setubal, setting 150 hooks on a 300 m length of line stretched out on the seafloor and in 1908 was successful in fishing at a depth of 5940 m (Mills, 1983). Forster (1964) used a buoyant vertical polyethylene line between a surface buoy and a weight on the seafloor to which was attached a ground line with 100 hooks spaced at 14 m intervals. The hooks were baited with squid or mackerel, attached to 2 m long snoods clipped onto the ground line (Forster, 1968). The gear was handled with a powered capstan for hauling the line and hand-

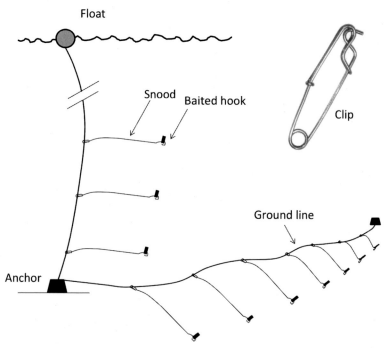

Figure 1.35 Long line fishing. Baited hooks can be attached to the vertical line, the ground line or a horizontal line at any depth below the surface. Hundreds or thousands of hooks may be used. The snoods or traces are attached to the main line at intervals using clips (example of one type illustrated).

wound onto a storage drum. This method captured numerous elasmobranchs including the deepest known record of a shark, *Centroscymnus coelolepis*, from 3658 m depth in the SE corner of the Bay of Biscay (Forster, 1973). Numerous versions of the technique depending on the facilities of different vessels have been used. Murray and Hjort (1912) describe how on the *Michael Sars* long lines were deployed and allowed to sink from the stern of the vessel and hauled in amidships using a small winch.

Fossen (2008) describe results using a technologically-advanced long-lining vessel, the M/S *Loran* (length 51 m) for a survey of demersal fishes on the Mid-Atlantic Ridge. This vessel had automatic equipment that could deploy hooks at a rate of 250 min^{-1}. A horizontal line with hooks 1.2 m apart was deployed on the seafloor followed by a vertical line with hooks 0.6 m apart to sample at different heights above the sea bed. Up to 3470 hooks per set were baited with squid and mackerel at depths from 400 to 4300 m. Consistent with earlier sampling work,

Chondrichthyes dominated the catches. Foster spaced the hooks out so that fish could be handled individually, whereas the automated equipment used by Fossen (2008) brought the catch rapidly into a processing facility on board the vessel. Long-lining causes less damage to the environment than trawling (Pham et al., 2014) but is selective; Fossen et al. (2008) retrieved 40 different species, whereas Bergstad et al. (2008) found 80 species in trawls working over the same area (Fig. 1.35).

Baited Traps Prince Albert I advocated traps as a means of capturing large mobile predatory deep-sea species. He used traps over 2 m long, containing smaller traps inside and deployed them on 500 m lengths of wire rope linked together by swivels to make up the necessary length to be connected to a surface buoy and a light. This system was operated successfully down to over 5000 m depth (Mills, 1983). Traps have since been widely used in deep-sea sampling including capture of the deep-sea eel

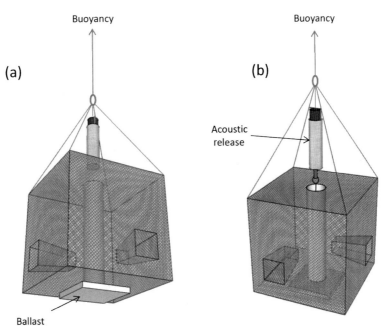

Figure 1.36 Baited deep-sea fish trap rigged for free-fall deployment using an acoustic release. One-metre cube cage trap with two entrances and 25 mm mesh walls (Priede et al., 1994). (a) View from below showing the ballast plate. (b) View from above showing the ballast strop passing through a tube and attached to the acoustic release. This rig allows the trap to be close to the seafloor to maximise catches of bottom-living fishes.

Ilyophis saldanhai associated with hydrothermal vents on the East Pacific Rise (Causse et al., 2005). All traps have in common a main chamber, which may be subdivided in which the bait is contained, usually in a mesh bag. One or more tapered entrance funnels are provided with an opening large enough for the size of animals expected. Isaacs and Schick (1960) applied the free-fall vehicle principle to a lightweight pyramid-shaped trap with 1.83 m sides. Buoyancy consisted of plastic containers filled with gasoline and a corrodible magnesium link provided the delayed ballast release. This low-cost system successfully captured fish from 2025 m depth. Priede et al. (1994) used a 1 m cube trap made of galvanised steel with a 25 mm mesh at depths from 314 to 3976 m in the NE Atlantic Ocean with glass floats and an acoustic release (Fig. 1.36). Only two species of fish were captured compared with 18 species seen at baited cameras and 71 species captured in trawls in the same area. Small traps (e.g. 10 cm diameter and 30 cm long) designed for capture of amphipods are capable of

catching small fishes and have been effective for capturing liparids at hadal depths (Jamieson et al., 2010).

To capture the suite of scavengers attending a dolphin carcase on the seafloor at 4000 m depth in the NE Atlantic Ocean, Jones et al. (1998) used a drop trap 2 m long, 1 m wide and 0.7 m high that fell into place and enclosed the carcase. The dolphin was tied to a mesh platform on the base of a lander frame, and the descent of the box trap was triggered by the ballast release acoustic command. The same principle was applied by Bailey et al. (2002) to capture fish in a respirometer chamber to measure their oxygen consumption. The drop trap has the advantage that the fish do not have to enter a funnel or maze; all fish attracted to the bait are captured.

Drazen et al. (2001) used an alternative means to overcome the trap shyness problem with tube traps mounted on a lander to catch slope-dwelling fish in the NE Pacific Ocean. The fish were initially attracted

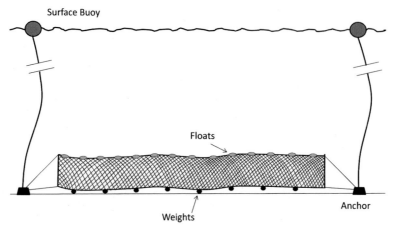

Figure 1.37 Gill net set on the deep-sea floor (not to scale).

to bait on a hook outside the open end of each tube. Movement from a hooked fish triggered an elastic spring that pulled it into the trap chamber and closed the entrance. This method was very effective at catching large-size classes of marcourids, *Coryphaenoides acrolepis* and *Albatrossia pectoralis* that were not well represented in trawl samples. Drazen et al. (2005) subsequently applied the same principle to the capture of fish at ambient temperature and pressure and retrieving them alive to the surface in a hyperbaric trap respirometer mounted on a large lander frame.

Fixed Nets Fixed nets are deployed like a curtain, with the bottom of the net on or at a fixed height above the seafloor and the top of the net held up with floats. Fish are captured as they swim through the mesh and are retained according to their girth. The net can be deployed in a similar way to long-lines or strings of traps with a vertical line to a surface buoy to aid recovery (Fig. 1.37). Halliday et al. (2012) surveyed the deep-sea demersal fishes on the continental slope in the NW Atlantic at depths from 500 to 2800 m using bottom gill nets. Because gill nets are size selective, a range of mesh sizes was used from 6 to 27 cm (stretched mesh) in approximately 100 m lengths. Twenty-one different species of fish were caught, but *Coryphaenoides rupestris* and *Alepocephalus agassizi*, which were dominant in

trawl catches from the area, were completely absent from the gill net catches. The large mesh size used was effective at catching large chondrichthyans that were not captured by small slow-moving scientific trawls.

1.7.2.2 Towed Gears

Towed or active fishing gears share a common characteristic in that they sweep over a defined geometric space, area of seafloor or volume of sea water, and results are hence amenable to quantitative analysis (Hayes et al., 2013)

Dredge Dredges occupy a special position in the history of deep-sea biology because this was the principal instrument that first enabled the scientific sampling of animals from the deep sea. Wyville Thompson (1873) describes early designs of scientific dredges originating from Scandinavia in the 1750s. The scientific dredge usually consists of a rectangular metal frame towed along the seafloor by a rope attached to the apex of a triangular tow bar (Fig. 1.38) The dredge frame is symmetrical so that it does not matter whichever way up it lands on the seafloor. The edge that lands on the seafloor acts as a scraper blade removing the surface layer of sediment, and the catch then falls back into a bag attached to the frame. The bag can be chain mail or netting with an appropriate

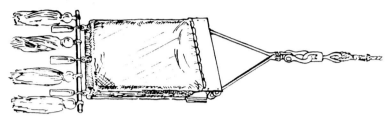

Figure 1.38 Scientific dredge. This illustration from Sigsbee (1880) shows tassels and weights attached to the bar behind the net bag. These are not used in modern dredges but otherwise the design has not changed.

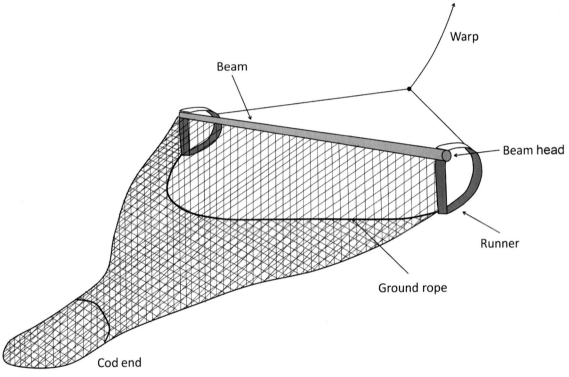

Figure 1.39 Beam Trawl.

mesh for retaining the catch. A layer of armouring of stronger netting or sheet material such as canvas, leather or plastic to withstand abrasion may be attached to the outside. The triangular tow bar normally has a weak link on one arm that breaks if the frame catches on an obstruction, thus allowing the dredge to be brought to the surface safely by pulling on the other arm.

Beam Trawl Various designs of beam trawl operate in a similar way to the dredge though without the heavy frame with bottom scrapers. Rather, the net is supported by runners on either side that slide on the seafloor like skis. One or more cross beams hold the mouth of the net open but do not make contact with the seafloor (Fig. 1.39). The net has a rectangular opening reinforced with rope, which glides on or just above the seafloor. The net is designed to capture everything within a volume defined by the width and height of the opening and the length of tow on the seafloor. However, more active large fishes such as sharks may readily accelerate away and avoid

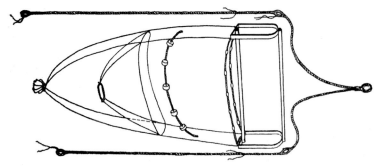

Figure 1.40 Agassiz, Sigsbee or Blake trawl. Illustration from Sigsbee (1880).

capture. On the voyage of *Challenger* a beam trawl with a 3 m (10 foot) beam was preferred for deep-sea use. The same net was subsequently used on board the *Blake* by Agassiz and Sigsbee, who introduced a modification to the runners so that they worked equally well whichever way up the net landed on the seafloor (Sigsbee, 1880; Fig. 1.40). Standardised at a width of 3 m, and variously known as the Agassiz, Sigsbee or Blake trawl, it continues to be widely used. Normally it is equipped with a 20 mm main mesh and 10 mm mesh in the cod end to retain small organisms (Gage and Tyler, 1991). A larger trawl with a 6 m long wooden beam known as the Chalut à Perche is operated at abyssal depths by French researchers (Jamieson et al., 2013).

Normally the frame of the beam trawl is ballasted (to a weight of 100 kg for an Agassiz trawl in the deep sea) to ensure it sinks through the water column and stays on the seafloor. Sigsbee (1880) recommended paying out the tow wire to a length of twice the depth at <650 m and depth plus one-third at greater depths. Such trawls have been used at great depths in hadal trenches (Jamieson, 2013).

Otter Trawl Because life becomes sparser with increasing depth in the ocean, it is desirable to use the largest nets possible at great depths. This is best achieved with otter trawls in which the mouth of the net is held open by two otter boards (doors or hydroplanes) that are rigged to produce lift sideways to the direction of tow. Vertical opening is maintained by the shape of the net, floats on the head rope and

weights and on the foot rope (Fig. 1.41). The net is normally towed on two warps, which are stored on the twin drums of a trawl winch. Ever since the voyage of Murray and Hjort (1912) on board the *Michael Sars,* deep-sea scientists have used smaller versions of the otter trawls used by commercial fishermen. Merrett et al. (1991a) compare three different otter trawls, the largest (BT200) of which had a head line length of 39.1 m, wing spread of 24 m and headline height 6.0 m and the smallest (Granton trawl), 20.6, 12.8 and 1.8 m, respectively. The mouth area, even on the smallest trawl, is much larger than for any scientific beam trawl. A towing speed of 4 knots, about twice the speed that is normal for Agassiz trawls, enabled the capture of more active fish species and mesh size graded from 80–140 mm on the wings to 12 mm in the cod end liner. Wenneck et al. (2008) describe fishing down to over 3000 m depth on the Mid-Atlantic Ridge from the *G.O. Sars* with twin warps using a Campelen 1800 shrimp trawl with 17 m wing spread and 2.5 m head line height. The otter boards weighed 2.25 tonnes each, and ground gear was equipped with 35 cm diameter 'rock hopper' rollers for towing over rough ground. Tow speed was ca. 2 knots.

Abundance in the catches is normally calculated from the wing spread and distance towed on the seafloor, but Merrett et al. (1991a) showed that there are significant differences between different trawls. In shallow water, fish are herded in front of a trawl by a visual reflex so that the entire fishing gear, including wings, bridles, sweeps, otter boards and warps, all cause fish to converge into the centre of

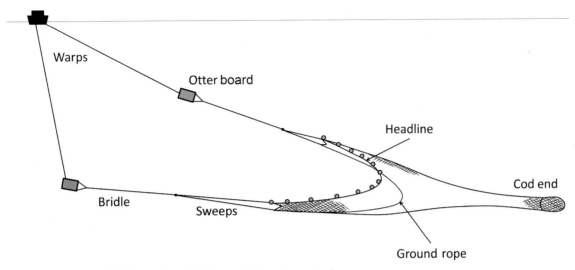

Figure 1.41 Otter trawl. Redrawn from Løkkeborg (2005) with permission.

the trawl mouth until they are exhausted and fall back into the cod end (Wardle, 1983). Tow speed therefore must be faster than the sustainable swimming speed of the fish. In the deep sea, in the absence of sun or moonlight such visual reflexes cannot occur although bioluminescence can make a trawl gear very conspicuous in darkness (Jamieson et al., 2006). It is doubtful if abyssal fishes have the necessary reflexes to respond to nets in this way; it is likely that they may be filtered out of the water by the meshes rather than scared into a stampede in front of the net.

It is usually not possible to fish at abyssal depth much beyond 3000 m using twin trawl warps. No ships are equipped with two drums of exactly matching wire of the necessary length. Even if this were achieved, the difficulties of paying out two cables of up to 15 km length to be exactly equal in length and perfectly parallel are formidable. This problem can be overcome by using a single trawl warp. The otter trawl is first of all shot in the conventional way, using two winches and towed behind the ship, which is kept continually moving to keep the mouth of the net open. Bridles from the otter boards are then transferred to a swivel on the end of the single tow wire. The net can then be paid out on the single warp until the seafloor is reached

(Fig. 1.42). This method was first used extensively by Murray and Hjort (1912). On 31 May 1910, a trawl on the *Michael Sars* to 5160 m depth took 14 h 15 min, using up most of the working day, a characteristic of deep sampling that has not changed much over the years. A long wire is necessary, 2.5–3.1 times the depth, and the net is towed on the bottom for up to 3 h at 2–2.5 knots. On a modern vessel the total time necessary for a trawl at 5000 m depth is 12 h (Jamieson et al., 2013). Murray and Hjort (1912) used a small otter trawl with 15.24 m head rope, slightly larger than the current standard OTSB (otter trawl semiballoon), which is widely used for deep-sea sampling. The OTSB, derived from a Gulf of Mexico shrimp trawl, has a headrope length of 13.7 m, wing spread of 8.6 m and headline height of 1.5 m (Merrett et al., 1991b; Jamieson, 2013). Stein (1985) fished a much larger commercial-size otter trawl (29.6 m headrope length) to 4300 m depth on the abyssal plains of the NE Pacific Ocean on a single warp. The collection of unique specimens of female *Coryphaenoides armatus* with ripening ovaries suggests such very large nets may enable the capture of parts of fish life cycles otherwise not amenable to investigation. However the equipment on the vessel was stressed up to and beyond safety limits, and this remains an experiment that has not been repeated.

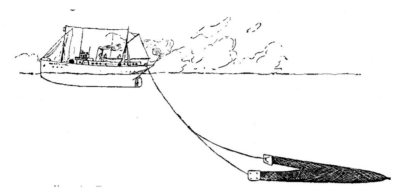

Figure 1.42 Operation of an Otter Trawl with a single warp. Illustration of the RV *Michael Sars* towing a trawl in this way from Murray and Hjort (1912).

The engineering effort and ship time required make large trawl deployment at abyssal depths prohibitively expensive.

Though bottom trawl equipment has changed little over the past century apart from the introduction of synthetic materials for ropes and nets, major advances have been made with underwater telemetry and instrumentation of the trawl. On their otter trawls Wenneck et al. (2008) used a full suite of wireless sensors that provide real-time information on otter board spread (width of trawl), depth, otter board angle, water temperature and an echo sounder on the trawl monitored ground-gear clearance, headrope distance from the bottom as well as numbers of fish entering the net. This was presented as a display on the control centre on the bridge so that trawl geometry could be continuously monitored and time and distance run on the seafloor accurately ascertained. Such full telemetry can be difficult to achieve with a net as much as 10 km behind the ship. However, a simple high-power acoustic pinger on the net can allow determination of height above the seafloor and together with the wire tension gauge to detection of time of contact and lift off from the seafloor (Backus, 1966; Rowe and Menzies, 1967).

There is no doubt that of all the fishing gears used for demersal fish sampling, bottom trawls catch the most comprehensive samples with the largest number of species represented. It is because of the indiscriminate nature of the catch and the acknowledged damage to vulnerable marine ecosystems that their use for commercial fisheries is being phased out or severely restricted in several jurisdictions. It is likely that scientific bottom trawling will also be restricted in the future.

1.7.2.3 Pelagic Sampling

The tow net used for obtaining pelagic samples on the voyage of the *Challenger* was the classic conical plankton net with the mouth held open by a 25–40 cm diameter ring to which was attached a triple towing bridle. These nets were used both for horizontal tows and vertical tows. The basic conical design has been scaled up to filter much larger volumes of water, and as early as 1898–1899 on voyage of the *Valdivia*, Carl Chun was using a 3 m diameter net with a mechanism that would enable the net to be opened and then closed at predetermined depths (Fig. 1.43). These early designs are described by Murray and Hjort (1912). Ring nets are still widely used, usually equipped with a cod end bucket for retaining the samples. In 1991, the International Council for Exploration of the Sea reverted to the use of 2 m diameter mid-water ring net as the standard gear for the sampling of fish larvae in the North Sea. The gear is robust and can be easily handled in rough weather often encountered in NE Atlantic winters (ICES, 2012). One concern with the basic ring net is that the bridle in front of the mouth

Figure 1.43 Classic conical plankton net. Cone of netting material and a collecting bottle at the codend.

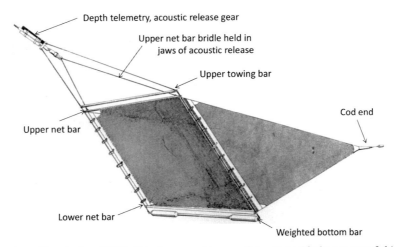

Figure 1.44 Rectangular mid-water trawl (RMT-8) in the normal tow configuration with the net open fishing with a mouth at an angle of 45° giving a sampled area of 8m². (From Clarke, 1969).

of the net may stimulate some of the active swimmers to avoid the approaching net.

To overcome the problem of the towing bridle deterring potential catch, Tucker (1951) developed a net with a square mouth (1.83 × 1.83 m), which was towed by two bridles attached to a cross bar on the upper edge of the net; the bottom of the net was held open with a second bar that was heavily weighted. The sides of the net were supported by wires. There were therefore no wires or ropes directly in front of the mouth of the net. Davies and Barham (1969) in San Diego, California, modified this basic net so that it could be opened and closed at the mouth by time-release mechanisms. The net was launched closed, with the upper and lower net bars held together by bridles linked to the time-release hook. When the first timer was released, the bottom bar dropped down, opening the net. When the second timer was released the upper bar dropped down to close the net. Various workers, including Tucker, had also experimented

with opening and closing cod end buckets, but there was always the danger of the contamination of samples with material that was in the net funnel before the cod end opened. The opening and closing mouth gives an uncontaminated sample from a discrete depth stratum that could be verified by a depth recorder on the net.

Clarke (1969) describes a similar modification of the Tucker design net known as the rectangular mid-water trawl (RMT-8). The mouth opening was 2.83 m wide and 4 m high, which, when fished at an angle of 45° to the direction of tow, gave a sampled area of 8 m² (Fig. 1.44) Opening and closing was controlled by acoustic command from the ship. This system successfully obtained discrete samples down to over 2000 m depth at a tow speed of 2.0–2.2 knots. The mesh size was 5 mm, and the cod end bucket was 80 cm long and 21 cm in diameter. Roe and Shale (1979) produced a multinet version of the system (RMT-8M) that had three opening and closing nets

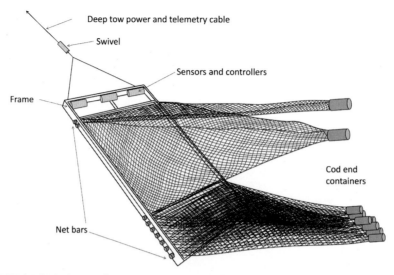

Figure 1.45 MOCNESS (Multiple Opening/ Closing Net and Environmental Sampling System). Up to 20 nets can be mounted on a rigid frame. The net mesh is not to scale, typical mesh size is 3 mm. This shows the configuration with the penultimate net open near the end of a tow. Previous discrete samples are in the closed nets at the bottom of the frame. (Redrawn from Gulf of Maine Research Institute (www.gmri.org) with permission).

stacked on top of one another. They describe the use of this system down to 4500 m depth. It was normally used in tandem with a 1m^2 RMT-1 with 320 μm mesh for taking simultaneous micronekton samples, a combination known as the RMT 8+1M. Larger versions of the system were also produced: the RMT 25 and the RMT 90. The RMT 90 had to be towed from twin warps and was not a multinet.

A parallel development to these modifications of the Tucker Net is the MOCNESS (Multiple Opening/ Closing Net and Environmental Sensing System (Wiebe et al., 1985) from workers at the Woods Hole Oceanographic Institution. The mouth of the MOCNESS is supported by a rigid rectangular aluminium H section frame that is ballasted at the bottom (Fig. 1.45). This has been produced in a series of sizes denoted by the effective mouth area in m^2: ¼, 1, 2, 4, 10 and 20 (Sameoto et al., 2013). The MOCNESS is towed on a conducting cable enabling real-time data on flow, depth, temperature, etc. to be transmitted to the ship and also commands to open or close nets are sent from the shipboard controller.

The Isaacs-Kidd Mid-Water Trawl (IKMT), first developed in 1950 in San Diego, California, is now produced in numerous versions worldwide. The IKMT features a depressor hydroplane that holds the vertical rectangular mouth open (Fig. 1.46). Multiple cod end mechanisms have been added to the basic design (Pearcy and Hubbard, 1964). Clarke (1969) found that the quality of samples retrieved by the RMT depended to a very large extent on the detailed design of the cod end. Childress et al. (1978) showed that a major cause of mortality and deterioration of catches is the temperature change occurring on transfer from cold deep water to warm surface waters (at low latitudes). To ameliorate this effect, thermally insulated cod end tanks have proved very successful on various types of Tucker, RMT, MOCNESS and IKMT.

All the ring nets and various multinets have the disadvantage that when they are scaled up to the size necessary to sample sparse bathypelagic fauna they become impossible to handle owing to the size and weight of the rigid components. In order to sweep very large water volumes, pelagic otter trawls have been used ever since reliable steam propulsion has been available to tow them (Murray and Hjort, 1912). In 1963 a pelagic Engels trawl, mouth area 40 × 20 m = 800 m^2, was deployed from the RRS

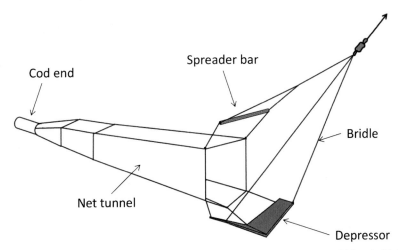

Figure 1.46 Isaacs-Kidd Mid-water Trawl (IKMT). The depressor hydroplane creates downward lift to hold the mouth of the net open and perpendicular to the direction of tow.

Discovery (Clarke, 2003), and Stein (1985) fished a 107 m² mouth area pelagic trawl down to over 2000 m depth. However there are great difficulties in fishing large commercial-size type nets from oceanographic research vessels with crews having limited fishing skills. A good example of current best practice is given by Wenneck et al. (2008) on board the Norwegian RV *G.O. Sars,* who used a range of pelagic fishing gears operated by a crew with the necessary specialist skills. Three different pelagic trawls were used. A macrozooplankton trawl (krill trawl) with a 6 m × 6 m rectangular mouth fished down to over 3000 m depth at 2 knots equipped with five remotely controlled cod ends. A medium-sized pelagic fish trawl (Aakratrawl) vertical net opening 20–35 m, door spread 110 m fished down to a maximum depth of 2600 m at a mean speed of 2.8 knots equipped with three remotely controlled cod ends. A large mid–water fish trawl (Egersundtrawl) was used on a few occasions with 90–180 m vertical opening, door spread 150 m, fished to a maximum depth of 2015 m at 2.9 knots with a conventional single cod end. All the gear used on the *G.O. Sars* was fully instrumented with wireless sensors so that the location and disposition of the net in 3-D space could be continuously monitored.

Webb et al. (2010) in a provocative paper pointed out the paucity of deep pelagic samples and the chronic underexploration of the bathypelagic zone. This reflects the real practical difficulties of sampling at great depths. None of the large pelagic trawls has been routinely used at depths greater than 3000 m.

1.7.3 Imaging Systems

As an alternative to capture of fish specimens, there is increasing use of manned and unmanned imaging systems.

The availability of digital cameras and data storage have revolutionised the capabilities of imaging systems in the deep sea. Free-fall landers equipped with cameras that descend to the seafloor, take images at preprogrammed intervals and then ascend to the surface by acoustic command from a surface ship have been used throughout the world's oceans. There are two basic forms (Fig. 1.47): fixed frames such as tripods that rest directly on the seafloor (Lampitt and Burnham, 1983; Smith et al., 1993) and tethered systems, which are suspended above the seafloor by buoyancy (Priede and Smith, 1986). In the former, the camera gives a close-up horizontal or oblique view of the seafloor, which is ideal for recognising diagnostic characters of species. The tethered vehicle with a single point of

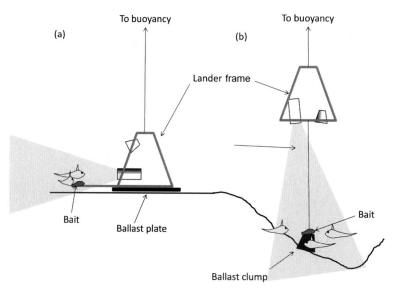

To buoyancy

(a)

To buoyancy

(b)

Lander frame

Bait

Ballast plate

Bait

Ballast clump

Figure 1.47 Comparison of baited camera lander configurations. Details of releases and other hardware omitted for clarity. (a) Lander frame resting close to the seafloor on a flat ballast plate. (b) Lander frame held above the seafloor by a tension mooring.

contact on the seafloor is capable of landing on sloping or rough terrain and can view a relatively large area of the seafloor without the need for a large supporting frame. The images, however, are mainly dorsal views of fishes, which can be difficult to interpret

A 15-year study that viewed a 20 m^2 area of the seafloor at one image per hour from a camera tripod at 4100 m depth in the Pacific Ocean Vardaro et al. (2007) recorded approximately one abyssal grenadier fish (*Coryphaenoides* spp.) passing through the field of view every two days. For fish faunal studies it is usual to use bait to attract fishes from a much wider area into the field of view. Early studies tended to use protected bait within a can or mesh so that odour was released but the fish were unable to consume the bait (Isaacs and Schwartzlose, 1975; Lampitt et al., 1983) However. Priede and Smith (1986) adopted a standard unprotected bait that fish can access, in order to mimic more accurately natural food falls. The time delay before the first fish arrives at a bait is related to local abundance, in principle in accordance with the inverse square law:

$$A = k/t_{arr}^2$$

where A is the species abundance in number of individuals per unit area, k is a constant and t_{arr} is the arrival time of the first fish (Priede and Merrett, 1996). Departure of fish from the bait is dependent on the rate of consumption and the average staying time of individuals (Priede et al., 1990). The baited underwater camera technique provides information on species presence or absence, size distribution, behaviour and abundance (Bailey et al., 2007).

For all observations of life in the deep sea, there is a concern that the presence of lights may deter some fishes. The white light typically used is many times brighter than any sources that animals at bathyal and abyssal depths might experience naturally, and indeed Herring et al. (1999) suggested that deep-sea animals may be permanently blinded by the lights of scientific submersibles. To overcome this problem Widder et al. (2005) used red light (660 and 680 nm wavelength) to which deep-sea animals are insensitive and found that five times more sable fish (*Anoplopoma fimbria)* were seen around a bait site at 520 m depth compared with sampling using normal white light illumination. Other studies have shown that the effect of presence of lights is

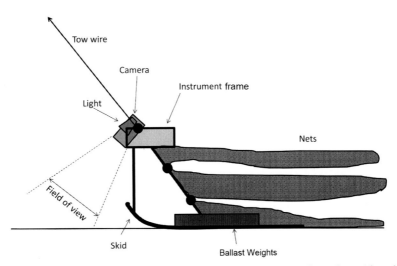

Figure 1.48 Epibenthic sledge, schematic side view. The sledge is towed across the seafloor often with multiple nets sampling at different heights above the seafloor. The camera records images in an undisturbed strip of seafloor in front of the sledge imaging fauna that are subsequently captured.

minimal at abyssal depths and the effect may be most pronounced for species living at bathyal depths for whom light has some ecological significance (Bailey et al., 2007). The use of high-sensitivity cameras in conjunction with red LED (light-emitting diode) illumination results in rather low-resolution monochrome images, making species discrimination difficult. The amount of information in an image is proportional to the power of the number of colour channels. Thus assuming 8 bit coding with 2^8 or 256 intensity levels, three colour channels increase the states that can be distinguished from 256 to 256^3 or 16777216, i.e. theoretically 65536 times more information than in an equivalent monochrome image (Priede, 1983). It is because of this overwhelming advantage of colour images that few studies have adopted red-only illumination. Ideally no lights should be used at all; Gillibrand et al. (2007) and Heger et al. (2007) observed natural bioluminescence stimulated by deep-sea fishes in the NE Atlantic Ocean. However, even in these studies, short periods of white-light illumination were necessary to interpret the results.

To view a large area of seafloor it is necessary to mount cameras on moving platforms. Possibly the simplest systems are epibenthic sledges, which are essentially beam trawls with cameras and lights added to the frame (Fig. 1.48). Rice et al. (1982) used a system with acoustic telemetry to provide real-time information on distance towed using an odometer wheel in contact with the seafloor. The frame was 2.3 m wide, and the camera was mounted looking forward at the seafloor in front of the sledge. There was a wide variation in abundance estimates of fishes from the camera and the net catches; for example, the eel *Synaphobranchus kaupii* was captured in significant numbers but not seen in any photographs, whereas *Lepidion eques* were more abundant in photographs than in net captures. The system was used down to abyssal plain depths (4800 m) in the NE Atlantic Ocean. The area photographed per tow (ca. 50–250 m^2) was considerably smaller than the area fished by the nets. Bailey et al. (2006) analysed from the 4100 m depth in the NE Pacific Ocean from a camera sledge producing a continuous mosaic of photographs that represented 2500 m^2 of seafloor long per tow (Lauerman and Kaufmann, 1998). With this larger survey area reliable fish abundance estimates were obtained.

Camera sledges can only be used where the sea bed topography is well known and relatively smooth. For rough or unknown terrain, towed camera frames of various kinds have been used, suspended above the seafloor. These essentially produce a sequence of photographs of the seafloor similar to the camera sledge (Jamieson et al., 2013) enabling the distribution of habitat type and fishes to be surveyed. Parallel lasers mounted on the frame provide illuminated spots on the seafloor at a known distance apart for estimating the size of objects, and nonparallel laser beams or sonar can be used to measure altitude above the seafloor. Often a weight hanging on the frame within the field of view of the camera provides a simple scale and physical measure of altitude and can be used to trigger cameras by bottom contact. Oceanographic electrical conducting cables can transmit data from the platform to the surface but as the bandwidth is usually insufficient for high-definition (HD) real-time video, images are often recorded on board the vehicle and retrieved at the end of the mission (Jones et al., 2009). Electro-optical cables have sufficient bandwidth to transmit real-time images from multiple HD cameras, which can be monitored at the surface. McIntyre et al. (2013) describe a system for surveys of fishes. Towed at a speed of ca. 3 knots, the area viewed per tow could exceed 100,000 m^2, similar to swept areas in traditional bottom trawl surveys. It was operated in marine protected areas, which are closed to bottom trawling.

Various manned and unmanned scientific submersibles provide subsea platforms that can manoeuver in three dimensions throughout the water column and down to the sea bed. Robison (2000) provides an excellent review of the capabilities of such vehicles. Manned systems or human-occupied vehicles (HOVs) operate free of the surface ship and are powered by on-board batteries, whereas unmanned remotely operated vehicles (ROVs) are tied to the ship by a tether that provides electrical power and data transmission. For safety reasons HOVs nevertheless require a ship in continuous attendance, good weather conditions and dive duration of typically less than 12 h. ROVs are capable of remaining submerged for days continuously operated by shifts of personnel on board ship.

With modern electro-optical cables power and bandwidth for multiple HD cameras are essentially unlimited (Jamieson et al., 2013). A typical scientific

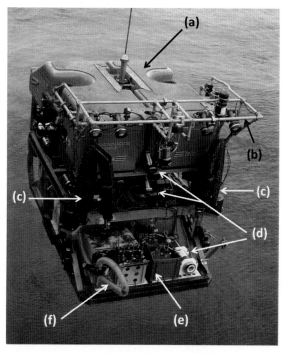

Figure 1.49 Remotely Operated Vehicle (ROV) *ISIS*. Length, 2.7 m Width: 1.5 m, Height: 2.0 m Weight in air 3.4 T. Maximum working depth 6500 m. (a) Buoyancy module, (b) bar with forward-looking lights mounted, (c) hydraulic manipulator arms (two), (d) cameras, (e) tool tray with sample storage boxes and other items and (f) suction sampler tube.

ROV has fixed cameras as well as pan, tilt and zoom cameras allowing close-up images of objects of interest to be obtained (Fig. 1.49).

Submersibles can be used in three kinds of dives: exploratory, manipulation and transect. Exploratory dives are generally used for initial studies in an unknown area. The dive principal is free to manoeuver the vehicle and change speed, altitude and direction while altering camera angle and focus to examine objects of interest. Initially a predefined search pattern may be adopted to find a feature such as a vent site, the approximate location of which may have been inferred from thermistor or sonar data. Apart from locating and identifying objects, data from exploratory dives are often not amenable to statistical analysis. Most manned submersibles and ROVs have manipulator arms, tools and suction pumps that can be used for sample collection. In manipulation dives the priority concerns the job to be done, e.g. placing an experiment on the seafloor or collecting species. The dive principal manoeuvers the submersible, adjusts the cameras and uses the tools as required to achieve the objective. In a biological collecting dive the duration of work is usually curtailed when the sample containers are full or if there is a need to retrieve fresh samples to the surface. Samples or equipment can also be sent to the surface in an elevator, a special lander with a sample tray to enable the ROV to remain on the seafloor.

ROVs are increasingly used for video surveys to estimate abundance of fishes as an alternative to trawl surveys and to map habitat distribution. It is important that the statistical design of such surveys is well thought out in accordance with line transect theory (Buckland et al., 2001). The ROV should be operated at a constant speed, at a constant altitude with a fixed camera position to generate swathe imagery of the seafloor of sufficient area and replication to generate statistically significant data (Jamieson et al., 2013). Postanalysis of segments of wandering tracks characteristic of exploratory or manipulatory dives is not recommended. Even in well-designed surveys the reaction of fishes to the bright lights and noise of ROVS moving at slow speed close to the seafloor (Stoner et al., 2008) can introduce significant survey bias. There is no doubt that ROVs and HOVs are extremely powerful tools enabling the fish biologist to study living deep-sea fish in their natural environment, but their use for surveys is very expensive in ship time, ROV time and labour.

As an alternative to towed systems, HOVs or ROVs, Autonomous Underwater Vehicles (AUVs), which can operate independent of a surface ship for up to several days, are increasingly accepted as a means of large-scale survey of the seafloor using high-resolution sonars (Jamieson et al., 2013). There is currently rapid development of camera systems for such AUVS, and Morris et al. (2014) demonstrated the use of the Autosub6000 ROV fitted with digital still cameras to produce photomosaic imagery of large areas of the Porcupine Abyssal Plain at 4800 m depth in the NE

Atlantic Ocean. With a streamlined torpedo shape the Autosub6000 could move at 1.7 m.s^{-1}, 7–10 times faster than the standard survey speed of an ROV/HOV whilst maintaining a constant altitude of 3.2 m above the seafloor and taking an image every 0.87 seconds (Fig. 1.50). The abundance estimates for macro-invertebrates were 10–50 times more accurate than those from deep-sea otter trawling in the same area. Milligan et al. (2016) used the same system to survey abyssal demersal fishes in relation to sea bed topography at the same location and obtained improved estimates of fish abundance. Whereas trawls accumulate the collected animals into a single sample that is collected at the end of the tow, the AUV data reveals the patchiness of distribution of species within the area sampled. It is likely that the high speed, low noise and intermittent strobe lighting used by the AUV may be less disturbing to fishes than typical ROVs. Using a single AUV, the seafloor survey time was more than double the ship time required for operations.

The high speed of the Autosub6000 AUV and similar torpedo-shaped AUVs makes them unsuitable for surveys in rough terrain, where towed and ROV systems continue to be most efficient for imagery surveys. It is likely that slow speed AUV systems will be developed specifically for such environments.

1.7.4 Tagging and Tracking of Fishes

Tagging of fishes for estimation of population size, growth and movement patterns has a long history in fisheries science. However such methods have limited application for deep-sea fishes because they can rarely be retrieved to the surface alive and in sufficiently good condition for tagging. The most successful approach has been in situ tagging. Stalking individual fish using the manned submersible *Jago*, Schauer et al. (1997) tagged Coelacanths (*Latimeria chalumnae*) during daylight when they were inside their caves at ca. 200 m depth using a compressed air dart gun to attach acoustic tags externally. For pelagic fish Sigurðsson, Thorsteinsson et al. (2006) used a device known as the UTE (underwater-tagging equipment), which was attached to a trawl and

Figure 1.50 Autosub6000. A torpedo-like AUV (Autonomous Underwater Vehicle), 5.5 m long and 0.9 m diameter equipped with cameras in the nose cone for sea bed imagery. The arrows indicate the direction of view of downward-looking vertical and oblique (35° below horizontal) forward-looking cameras. The AUV is being held by its launch and recovery gantry.

diverted deep-sea redfish, *Sebastes mentella*, from the cod end through an automatic tagging device. This gripped the individual fish, made a ventral incision into the peritoneal cavity, squeezed an electronic tag into the body cavity and left a protruding conventional spaghetti tag for external identification by fishermen. The UTE is 3 m long, 1.4 m wide and weighing 650 kg is designed to be operated with commercial-size demersal and pelagic trawls. Video of operation of the UTE is transmitted to the trawler and precise digital images are recorded of each fish for measurement and identification. The tag return rate from the commercial fishery was 3.9 per cent.

Priede and Smith (1986) found that abyssal demersal grenadier fish could be induced to ingest acoustic transmitters embedded in small packets of bait deployed by landers on the seafloor (Fig. 1.51). The method was first used for the macrourid *Coryphaenoides yaquinae* at 5800 m depth in the Pacific Ocean and has since been used to tag *Coryphaenoides armatus,* the morid *Antimora rostrata*, the shark *Centroscymnus coelolepis* and

other species in the NE Atlantic (Priede and Bagley, 2000). Because the range of the transmitters (ca. 1000 m) is much less than the depth of the water, the fish were tracked using sonar equipment mounted on the lander used for deployment. Images from cameras on the lander verified the species ingesting the bait. Three versions of acoustic transmitters were used: MKI, simple pingers that emit a pulse at regular intervals; MkII, transponders interrogated by a sonar on the lander to measure precise range; and MkIII, Code Activated transponders that enable several individually-identifiable fish to be tracked simultaneously (Fig. 1.52). All the transmitters used a frequency of 75 kHz, which is inaudible to fishes. The experiment was limited by the time before the electronic tag was regurgitated, likely to be a few weeks (Armstrong et al., 1992), and the range is limited by the extent of the receiving equipment deployed on the seafloor. These methods provided information on swimming speeds and behaviour but can only be used for necrophagous species that voluntarily ingest large food parcels.

Figure 1.51 An abyssal grenadier *Coryphaenoides armatus* ingesting an acoustic transmitter embedded in a bait parcel. This particular fish also has an ectoparasitic copepod just behind its dorsal fin.

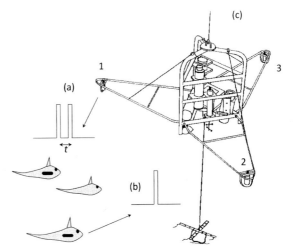

Figure 1.52 Tracking of deep-sea fishes using ingestible code-activated transponders (CAT).

(a) A double acoustic pulse is emitted by the interrogator hydrophone 1, on the apex of one of the lander arms.

(b) The signal is received by CATs inside the fish and if the time interval (*t*) between the pulses corresponds to the individual code it responds with a single pulse.

(c) The pulse is received and timed accurately at each of the three arms of the lander, 1, 2 and 3 enabling location of the fish relative to the lander to be calculated.

The CATs are deployed in bait packets attached to the ballast on the seafloor. The lander is equipped with cameras and lights to observe the fishes.

Because radio signals cannot travel through sea water and the range of acoustic tags is limited (Priede, 1992), long-distance tracking of deep-sea fishes remains elusive. However, some progress is being made with use of data-logging tags that record temperature and depth at regular intervals and pop up to the surface to transmit data to a satellite after a predetermined time interval. The pop-up location provides a final tag location, and temperature-depth data allow some inferences regarding the intervening route taken. The technique has been applied to silver freshwater eels (*Anguilla anguilla* and *Anguilla rostrata*) during their mesopelagic return migration across the Atlantic Ocean to spawn in the Sargasso Sea (Aarestrup et al., 2009; Béguer-Pon et al., 2015) and to deep-sea sharks on the continental slopes of the Bay of Biscay (Rodríguez-Cabello and Sánchez, 2014).

2 Colonisation of the Deep Sea by Fishes

2.1 Introduction

Fishes first appear in the fossil record in the late Cambrian, about 500 Ma, but the species prevalent in the deep sea today are dominated by a diverse assemblage of types that originated in freshwater and shallow seas during the last 100 Ma since the mid-Cretaceous. Colonisation of the deep sea by these species has been contemporary with the evolution of birds and mammals and forms part of a wider explosion in vertebrate diversity, including fishes, that has occurred since the extinction of dinosaurs 66 Ma at the end of the Cretaceous. In the last 100 Ma the total number of families of fishes has increased from a long-term average in the fossil record of less than 50 families during the previous 300 Ma to over 300 in the present day (Benton, 2005). This recent invasion of the deep sea by fishes coincides with the formation of the modern oceans as we know them today, following the breakup of Pangaea and the opening up of the Atlantic Ocean (Fig. 1.16). There are broadly three hypotheses regarding the history of colonisation of the deep sea.

1. *Azoic Pre-Cretaceous Deep sea*: This is most clearly stated by Wells et al. (1931), who wrote, 'Most authorities believe that the invasion of the great deeps began no earlier than Cretaceous times' and 'the marine abyss was such a difficult region to colonise that it stood untenanted for most of geological time.'
2. *Successive Colonisations and Extinctions.* Priede and Froese (2013) suggested that the deep sea has been colonised by fishes whenever conditions were favourable from the Cambrian onwards but that these ancient deep-sea faunas were extinguished

by successive extinction events culminating in several global Oceanic Anoxic Events (OAEs) during the Cretaceous. Thus the modern deep-sea fish fauna is entirely the result of recolonisation from shallow water since the Cretaceous.
3. *Persistence of a Deep-Sea Fish Fauna through Extinction Events.* There is new evidence that some groups of fish have used deep-sea refuges to survive through major extinction events (Guinot et al., 2013). This raises the possibility of continuous lineages in the deep sea through geologic time.

To understand the extent and time-course of the colonisation of the deep sea, we first examine what is known of the present-day depth distribution of fishes before moving on to speculate how this pattern may have arisen historically.

2.2 Present-Day Depth Distribution of Fishes

The deep sea is an extreme environment in which most fishes are unable to survive. The greatest diversity of fishes occurs in shallow seas; over 70 per cent of marine species have a maximum depth of occurrence of less than 500 m, and only 3 per cent of species venture beyond 3000 m depth (Fig. 2.1.). Woodward (1898) commented that the most favourable conditions for the process of evolution occur in shallow seas, and forms that cannot compete in the intense struggle for life along the shore retreat either into freshwater or the deep sea. In fact freshwaters have been the site of great diversification of fish species, but the concept of the deep sea as a

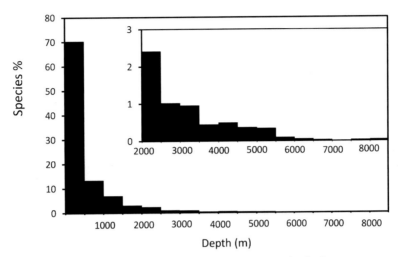

Figure 2.1 The depth distribution of marine fish species according to maximum depth of occurrence recorded in FishBase (Froese and Pauly, 2016). Inset: Data for depths greater than 2000 m on an expanded vertical scale.

refuge into which species originating from shallow-water retreat does retain some validity. Present-day marine fishes belong to four classes, Myxini (hagfishes), Chondrichthyes (cartilaginous fishes including chimaeras, sharks and rays), Actinopterygii (ray-finned fishes) and Sarcopterygii (lobe-finned fishes, Coelocanths). Fig. 2.2 shows the distribution of maximum lengths and maximum depth for all species of the four classes recorded in FishBase.

Myxini are found down to a maximum depth of 3003 m, the deepest species being *Eptatretus carlhubbsi* (Yeh and Drazen, 2009). Of the 78 known species, 26 occur at depths less than 500 m, 27 at 500–999 m with a logarithmic decrease in species number down to the maximum depth of occurrence (Fig. 2.3). There is a slight increase in size with depth (see Fig. 3.3); mean maximum total length is 51·2 ±18·5 cm (Fig. 2.3). A single reported capture of *Petromyzon marinus* (sea lamprey) at 4099 m depth is excluded from this analysis (Haedrich, 1977).

For the Chondrichthyes, 913 maximum depth records are available for 1140 species. The deepest species is Bigelow's ray *Rajella bigelowi* recorded at 4156 m. More than 50 per cent of species (475 records) occur at depths shallower than 500 m, and there is a clear logarithmic decrease in species number with depth with only three species recorded at abyssal

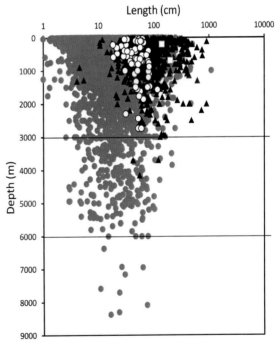

Figure 2.2 Scatterplot of maximum lengths (cm) and maximum reported depth (m) of occurrence of marine fish species. Open circles – Myxini, Black Triangles – Chondrichthyes, Grey circles – Actinopterygii, Open squares – Sarcopterygii. The horizontal lines indicate at 3000 m the upper limit of the abyssal zone and at 6000 m the upper limit of the hadal zone.

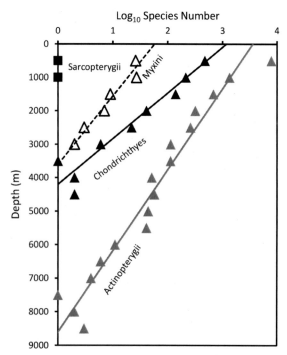

Figure 2.3 Number of species maximum depths per 500 m depth increment for each of the classes of marine fishes with fitted regression lines. Open triangles – Myxini (hagfishes) $\log_{10}N = 0.000488x + 1.760000$ ($r^2 = 0.965$), Black triangles – Chondrichthyes (Cartilaginous fishes) $\log_{10}N = -0.000731x + 3.070000$ ($r^2 = 0.954$), Grey triangles - Actinopterygii (ray-finned bony fishes) $\log_{10}N = -0.000413x + 3.550000$ ($r^2 = 0.944$) where, N = number of species and x = depth (m).

depths (>3000 m). Apart from these few records, chondrichthyes are essentially absent from the abyss (Priede et al., 2006). The mean maximum length is 108·5±119·3 cm with a slight increase in size with depth (Fig. 3.3).

The Actinopterygii is by the far the most abundant group of fishes with 10,954 marine species for which maximum depth and length are recorded in FishBase (Fig. 2.2). The greatest size range is at shallow depths reflecting the wide variety of niches from tiny coral-reef-dwelling species barely 1 cm long to the mesopelagic giant oarfish (*Regalecus glesne*) recorded at 11 m total length. With increasing depth, size diversity decreases, converging on an apparent optimum size in the deep sea close to the 28 cm average size for Actinopterygii. There is a small but

significant increase in size with depth, which is discussed in Chapter 3. The deepest Actinopterygian recorded is *Abyssobrotula galatheae* from 8370 m. The single depth observation for this species (Nielsen, 1977) remains unconfirmed, but several other species have been recorded and verified close to 8000 m, which can be considered to be the maximum depth limit for fishes (Linley, Stewart, et al., 2016; Jamieson and Yancey, 2012; see Table 3.5). Apart from a few stray individuals of other classes venturing beyond the bathyal, depths greater than 3000 m in the abyssal and hadal zones of the ocean are the exclusive habitat of the Actinopterygii.

Sarcopterygii are represented by two species, the African or West Indian Ocean coelacanth, *Latimeria chalumnae*, famously discovered off South East Africa in 1938, and the Sulawesi Coelacanth, *Latimeria menadoensis*, discovered in 1997 (Erdman, 1999). Using ultrasonic tags *Latimeria chalumnae* have been tracked down to a maximum depth of 698 m and were found to spend most of their time between 200 and 300 m depth, within the subphotic zone (Hissman et al., 2000). There is less information available for *Latimeria menadoensis*, but they have been captured over deep water off the edge of reefs.

Comparing these four classes, there is a clear trend that those groups with the greatest number of species have attained the greatest depths in the ocean (Fig. 2.4). This superficially supports Woodward's hypothesis implying that greater intensity of competition within a group results in higher probability of retreat into the deep sea. The slopes of the lines for decrease in species number with depth (Fig. 2.3) in the Actinopterygii (−0.000413) and Myxini (−0.000488) are similar to the slope for rate of decrease in biomass with depth in the oceans (−0.00048; see Chapter 1). This led Priede and Froese (2013) to propose that the number of species is correlated with, and limited by, food supply in the deep. Sparse resources in the deep can support only a few species. However, for the Chondrichthyes, the trend of decline in species number is much steeper (−0.000731) than for the other classes, suggesting that some other mechanism is preventing these fishes from colonising the deep sea. Possible physiological limitations of the Chondrichthyes are discussed in Chapter 3.

The Chondrichthyes are divided into three main groups: the subclass Holocephali (chimaeras) and subdivisions Selachii (sharks) and Batoidea (rays). These all span the entire bathyal depth range, 200–3000 m and reach maximum depths of 3000 m, 3700 m and 4167 m, respectively. There are important differences in their patterns of depth distribution;

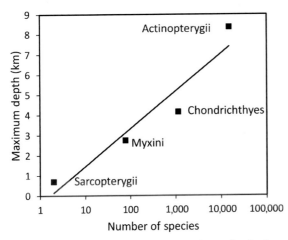

Figure 2.4 The relationship between maximum depth of occurrence and number of species within the four classes of marine fishes.

the Holocephali show a peak of species occurrence deeper than 500 m, whereas in the Selachii and Batoidea 50 per cent or more of species are limited to depths less than 500 m (Fig. 2.5). With a mean maximum depth of occurrence of 1168 m (SD = 657 m), the Holocephali may be regarded as a deep-sea endemic group, whereas the Selachii and Batoidea appear to be deep-sea invasive.

Within the Actinopterygii there is considerable variation in depth distribution within different families and, as for the Chondrichthyes, can be divided into two kinds, deep-sea endemic and deep-sea invasive. The family Alepocephalidae (slickheads) can be regarded as a deep-sea endemic family (Priede and Froese, 2013). The peak of distribution is clearly in the deep sea with mean maximum depth of 2048 m, and the deepest species has been recorded at 5850 m depth (Fig. 2.6a). In contrast the family Ophidiidae (cusk-eels) is a deep-sea invasive family. The peak of distribution is in shallow seas with over 30 per cent of species occurring at less than 500 m depth. Nevertheless, the tail of the depth distribution extends throughout the bathyal and abyssal zones into the hadal zone >6000 m depth (Fig. 2.6b). Defined by their depth distribution, the deep-sea endemics and deep-sea invasives roughly correspond to what Andriashev (1953) termed ancient and

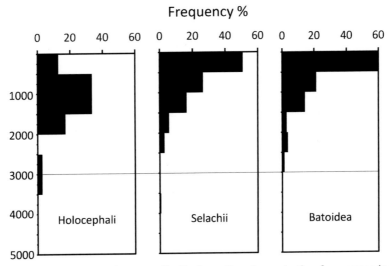

Figure 2.5 Chondrichthyes, comparison of frequency distributions of maximum depths of occurrence in the Holocephali (Chimaeras), Selachii (Sharks) and Batoidea (Skates and Rays).

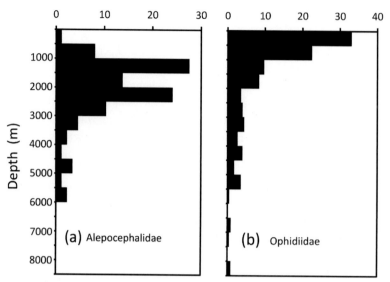

Figure 2.6 Comparison of frequency distributions of maximum depths of occurrence in two families of Actinopterygii. (a) Alepocephalidae – A deep-sea endemic or ancient family with the peak of distribution in the deep sea. (b) Ophidiidae – A deep-sea invasive or secondary family with the peak of distribution in shallow waters but some species that occur at great depth.

secondary deep-sea forms, respectively. The ancient deep-water forms colonised the deep sea early in evolutionary history. They tend to be morphologically specially adapted to deep-sea life, and many families no longer have any shallow-water representatives. In contrast, the secondary deep-sea families show few adaptations to the deep, have colonised recently and show evidence of continuing diversification from shallow to deep waters. It is interesting to note that despite their lack of overt specialist morphological adaptations to the deep sea, several invasive families have species that are very successful at great depths, e.g., Ophidiidae, Liparidae (snail fishes) and Zoarcidae (eel pouts). These successful deep-sea invasive families are all benthic fishes, whereas the ancient or endemic families predominate in the bathypelagic realm.

2.3 The History of Deep-Sea Fishes

At the time of the first appearance of fishes in the fossil record about 500 Ma, deep ocean basins with an average depth of around 4000 m were in existence,

occupying about the same fraction of the earth as at present but with a very different disposition of the continental land masses compared with the present day (Fig.1.16). The earliest fishes were the jawless fishes, Agnatha, represented by abundant tooth-like fragments in the late Cambrian deposits known as Conodonts. These were only recently identified as fishes from complete fossils found in the Carboniferous (Benton, 2005), and they persist in the fossil record to the end of the Triassic. Conodonts are regarded as more derived than the living representatives of the Agnatha, lampreys and hagfishes. During the Ordovician and Silurian the Agnatha diversified into a wide variety of forms loosely grouped together as the 'Ostracoderms', characterised by varying degrees of body armouring and often a head shield. By the Devonian these had been joined by the fishes with jaws (Gnathostomes) belonging to the extinct Acanthodii and Placodermi as well as early representatives of the Chondrichthyes and Osteichthyes. The Devonian (419.2–358.9 Ma) is therefore often termed the 'Age of Fishes', reflecting the presence of so many major groups of fishes (Fig. 2.7).

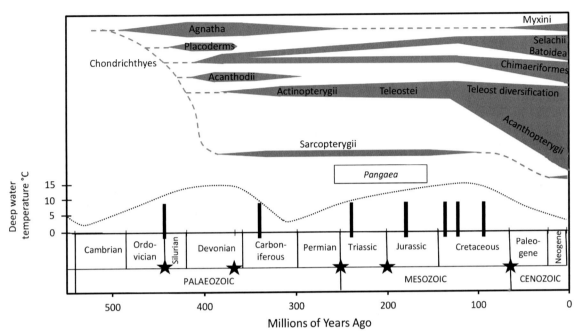

Figure 2.7 The evolution of fishes in relation to the geological time scale. Stars denote the five global mass extinction events, Ordovician–Silurian, 443.4 Ma; Late Devonian, 367 Ma; Permian–Triassic, 252 Ma; Triassic–Jurassic, 201.3 Ma; and Cretaceous–Paleogene, 66 Ma (also often known as the K/T event). The vertical black bars indicate Oceanic Anoxic Events (OAEs) (Takashima et al., 2006). Deep water temperature is according to (Schopf, 1980). The box indicates the period when all the land masses coalesced into one supercontinent, Pangaea.

It seems likely, given the high diversity of fishes in the Devonian, that at least some of these species did occupy the deep sea. Living representatives of the Agnatha and Chondrichthyes are well-known scavengers at continental slope depths of 500–1500 m, and there seems no reason why their Devonian predecessors should not have ventured to similar depths, feeding on benthic fauna and carrion falling from above. Modern hagfishes are able to tolerate severe anoxia (Hansen and Sidell, 1983; Cox et al., 2011), suggesting that Agnatha may have occupied the deep ocean margins throughout their history in the face of the advance and regression of deep-water oxygenation.

There may have been impediments to the colonisation of the abyssal open ocean at depths beyond 3000 m. Firstly, amongst the modern fishes it is only the advanced Actinopterygii belonging to the Teleostei that have evolved the physiological mechanisms necessary to survive at great depths. As

Teleosts do not appear in the fossil record until about 180 Ma in late Jurassic freshwater deposits, it is possible that throughout the Palaeozoic and much of the Mesozoic there were no marine fish capable of surviving at abyssal depths. Secondly, life in the open ocean abyss is dependent to a large extent on the rain of particulate organic matter derived from primary production in the surface layers. In the modern ocean, diatoms are the dominant primary producers and copepods the main consumers in the plankton that, respectively, produce and package organic matter as faecal pellets falling to the seafloor. Diatoms first appear in the fossil record just before the Teleosts in the mid-Jurassic 185 Ma, although molecular clock data indicate a possible earlier origin close to the Permian–Triassic boundary (Kooistra and Medlin, 1996). Copepods do not appear until the early to mid-Cretaceous (Rigby and Milsom, 2000). In the absence of diatoms and copepods, it cannot necessarily be assumed that in the Palaeozoic Ocean there would have been a reliable food supply to the abyss as in the modern ocean. However, there is evidence of faecal pellets very early in the fossil record from 1900 Ma, including pellet microfossils on the deep-sea floor from 800 Ma (Robbins et al., 1985). It is suggested that from around 550 Ma faster sinking of the faecal pellets established an effective biological pump removing organic matter from the water column. During the early evolution of fishes in the Ordovician and Silurian, planktonic diversity was well established (Rigby and Milsom, 2000) together with the efficient downward transport of particulate organic matter, which was presumably capable of supporting life in the abyss. Early in the Devonian there was a marked change in the composition of plankton, and extinctions in the Late Devonian resulted in a crisis of low diversity in the plankton that lasted until the Jurassic. This is often linked with the development of plant life on land that may have retained nutrients that had previously been washed into the sea. Nevertheless, many groups of plankton including jellyfish persisted, and it is possible that food supply to the deep sea was not curtailed.

It is therefore reasonable to make a provisional assumption that some Palaeozoic fishes may have been capable of colonising the deep sea particularly during the Devonian, the so-called Age of Fishes. However, for continuous occupation of the deep sea by fishes through to the present day they would have had to survive through five major extinction events that have punctuated earth history: (1) Ordovician-Silurian, 443.4 Ma, (2) Late Devonian, 367 Ma, (3) Permian-Triassic, 252 Ma, (4) Triassic-Jurassic, 201.3 Ma and (5) Cretaceous-Paleogene, 66 Ma, all of which resulted in a massive loss of marine life (Bambach et al., 2004).

The Ordovician-Silurian resulted in loss of 42 per cent of marine genera and is associated with the movement of Gondwana to the South Pole and global cooling, although deep-water temperatures appear to have been high at that time. There was a major decline in conodont early fish fossils (Friedman and Salan, 2012). During the Devonian, there was a great proliferation of families of fishes belonging to the Agnatha, Placoderms, Chondrichthyes and

Acanthodia, as well the appearance of early bony fishes, the Actinopterygii and Sarcopterygii. This was followed by the Late Devonian extinction event, which resulted in the loss of 22 per cent of marine invertebrate families and a much higher proportion of fishes. From 377 Ma to 362 Ma, 51 of the 70 families of fishes present in the Late Devonian died out, representing an extinction rate of 73 per cent (Benton, 2005). The Agnatha and Placoderms were severely reduced and never recovered. The late Devonian extinction can be regarded as an evolutionary bottleneck for fishes, the aftermath of which was great diversification of the chondrichthyans and actinopterygians that have dominated all subsequent marine fish faunas (Sallan and Coates, 2010).

The Permian–Triassic event was the most severe of the five major extinctions, resulting in the loss of 58 per cent of marine genera and marking the transition from the Palaeozoic to the Mesozoic era. The effect on fishes, however, is unclear, with little evidence of major taxonomic shifts. It is even suggested that some groups of fishes may have benefitted from feeding on the invertebrate victims of extinction. The Chondrichthyes appear to have been the most affected, with Palaeozoic forms being replaced by modern Elasmobranchs around this time. Diverse durophagous (hard shell-eating) holocephalans, characteristic of the Permian, disappeared. Holocephalans are subsequently absent from the fossil record until the late Triassic, when Chimaeriformes, very similar to the modern living forms, reappeared. Guinot et al. (2013), however, show that at least one group of Chondrichthyes, the Palaeozoic Cladodontomorph sharks, previously thought to have become extinct, appear to have survived the Permian-Triassic extinction by shifting to a deep-sea refuge. They were found in deep-water deposits of the early Cretaceous, 120 Ma after the Permian-Triassic transition.

The Triassic-Jurassic extinction resulted in the loss of over 30 per cent of marine genera, but the effect on fishes appears to have been minimal except for the final disappearance of the conodonts from the fossil record. Subsequently in the early Jurassic there was an apparent opportunistic expansion of the neoselachians, Selachii and Batoidea, in response to ecological restructuring after the mass extinction event. This is associated with small body size, short life spans and the development of oviparity (Kriwet et al., 2009).

The Cretaceous–Paleogene extinction (K–Pg), also known as the Cretaceous–Tertiary (K–T) event, marks the end of the Mesozoic era. It is widely accepted this was triggered by a massive asteroid impact at Chicxulub in Mexico. The debris thrown up by this impact altered the global climate, leading to an abrupt decrease in plankton productivity and disruption of the biological pump delivering organic matter to the deep sea (Schulte et al., 2010). Large reptiles, Pterosaurs, Dinosaurs and Plesiosaurs famously suffered 100 per cent extinction. However, overall fishes were only modestly affected with the loss of merely 15 per cent of families (Benton, 2005), although within the Chondrichthyes some high extinction rates have been detected: 17 per cent of families were lost, 34 per cent of genera and 45 per cent of species. Extinctions were most severe among the Batoidea. Deep-sea chondrichthyes, notably the Squalids such as *Centrophorus*, *Dalatias*, *Isistius*, and *Squalus* spp. were little affected and appear to have avoided the influence of the K–Pg event (Kriwet and Benton, 2004). Amongst the Actinopterygii, the pelagic marine families were most affected by extinction, whereas deep-sea and freshwater families escaped relatively unscathed (Cavin and Martin, 1995). The potential importance of the deep sea as a refuge from major extinction events is further reinforced by the observation that the Coelocanths, which in the Cretaceous were widespread in shallow shelf seas, appear to have survived the K–Pg event by a contraction of their habitat to deeper slope environments (Cavin, 2001; Guinot et al., 2013).

In addition to the five major global extinction events, there have been seven OAEs associated with the breakdown of thermohaline circulation (Takashima et al., 2006). Whilst the sea surface and shelf areas are likely to remain oxygenated (as in the present Black Sea) the deep sea below the thermocline or halocline would have become uninhabitable. In addition to the global OAEs there were regional OAEs.

During the Cretaceous there were three major global OAEs and seven regional OAEs. Despite the high frequency of OAEs in the Jurassic to Cretaceous time period, Friedman and Sallan (2012) found no evidence of associated major extinction or faunal turnover amongst fishes that could be attributed to these events.

Carrete Vega and Wiens (2012) show that within the Actinopterygii there are far fewer species in the sea than would be expected, given the vast area and volume of the oceans compared with freshwater. They argue that this apparent paucity of marine species is the result of depletion by successive ancient extinctions and argue that the seas have been recolonised relatively recently from lineages originating in freshwater. Thus colonisation of the deep sea by relatively recent forms merely reflects the general pattern of colonisation of the oceans by fishes. However, domination of the present-day deep-sea fish fauna by groups of post-Cretaceous origin does not preclude the possibility of a significant contribution from deep-sea lineages of more ancient origin.

Studies on the origins of deep-sea invertebrates parallel the debate concerning the origin of deep-sea fishes. Jacobs and Lindberg (1998) observe that most higher marine taxa originate in shallow waters and retreat into deep-water environments. They place great emphasis on mass extinctions during mid-Cretaceous anoxic events and conclude that probably modern deep-sea faunas did not evolve in situ for long periods of time but colonised the deep sea since the late Cretaceous and early Palaeogene. McClain and Hardy (2010) state that most current hypotheses for the origins of deep-sea fauna centre on such 'extinction and replacement', the present-day deep-sea fauna being mostly the result of recolonization since the last widespread anoxic event in the Palaeogene ca. 34 Ma. However, they point out that many clades have survived through anoxic periods, and Rogers (2000) argues that the expansion of oxygen minimum zones creates environmental gradients and barriers to dispersal that can enhance rather than inhibit speciation. Wilson (1999) states that even in the most severe global OAEs, complete anoxia is unlikely because localised pockets of bottom water oxygenation can occur wherever high-density water is created at the surface, even in equatorial regions, and cites evidence of deep water as an effective refuge for the survival of taxa through mass extinction events. Using molecular clock data, Lins et al. (2012) show that asellotan isopods, a very diverse group of exclusively deep-sea crustacea, first colonised the deep sea around 272 Ma during the early Permian and have survived throughout the Mesozoic and Cenozoic. Other isopod families colonised the deep sea at 164 Ma and 173 Ma, further contradicting the extinction and replacement hypothesis. McClain and Hardy (2010) point out that there is no common time of origin of the present-day deep-sea fauna: 5 per cent of deep-sea foraminifera are of early Palaeozoic origin, bivalve molluscs first appeared in the Ordovician, gastropod molluscs appeared at various stages throughout the Palaeozoic, hexactinellid sponges (40–80 per cent), bryozoans (25–67 per cent) and brachiopods (10–20 per cent) all appeared during the Mesozoic. The general conclusion is that throughout the time since metazoan life first appeared in the seas, the deep sea has been colonised and recolonised with significant persistence of clades in the deep sea. It is therefore reasonable to suppose that the same applies to fishes; there has probably been opportunistic colonisation of the deep sea throughout history, only the late Devonian extinction event appears to have resulted in catastrophic loss of fish taxa.

2.4 Origins of the Modern Deep-Sea Fish Fauna

2.4.1 Agnatha

In the modern deep oceans, the jawless fishes are represented by just one class, Myxini, the hagfishes. The sister class, Petromyzontida, the lampreys, are largely confined to freshwater. The great diversity of Agnatha present in the Devonian belonging to the classes Pteraspidomorphi, Anapsida, Thelodonti and Cephalaspidomophi (Nelson, 2006) all disappeared in the late Devonian extinction event, 367 Ma, with only

the Conodonts surviving until the late Triassic. An early fossil myxinid, *Myxinikela siroka*, has been found from about 300 Ma in the late Carboniferous (Bardack, 1991). The general body form is very similar to modern species of hagfish, indicating little change over geological time. The specimen has relatively well-developed eyes and was obtained from a coastal estuarine setting, suggesting a shallow-water habitat. However, because it was a single specimen amongst a diverse fossil assemblage, it is suggested that it was a rare visitor to this environment. The origins of the hagfishes have been much debated (Janvier 2010), with morphological evidence indicating that they are basal vertebrates. However, microRNA analysis shows they have undergone a process of degeneration from an anatomically more complex common vertebrate ancestor (Heimberg et al., 2010). Molecular clock data indicates that hagfishes diverged from the lampreys 470–390 Ma in the Ordovician, Silurian or Devonian periods. Within the hagfishes the split between the two subfamilies, Myxininae and Eptratretinae, occurred 90–60 Ma in the Cretaceous–Paleogene Periods (Fig. 2.8), possibly associated with the time of the K–Pg extinction event. The fossil *Myxinikela siroka* is a putative outgroup from the extant hagfishes (Kuraku and Kuratani, 2006). The time tree in Fig. 2.8 defines the relationship and time of divergence between extant forms but gives no information on extinct groups. It is likely that the modern Myxinii are survivors of a process of pruning of lineages that has taken place through global and oceanic extinction events.

2.4.2 Chondrichthyes

2.4.2.1 Holocephali (Chimaeras)
The Chimaeriformes, with about 39 species, is the only surviving order of a formerly very diverse and successful subclass of the Chondrichthyes with 12 extinct orders (Nelson, 2006). The Holocephali diverged from the subclass Elasmobranchii during the Devonian, possibly as early 410 Ma in the lower Devonian if the fossil *Stensioella heintzi* Broili, 1933, initially identified as a placoderm, is accepted as the earliest Holocephalan (Licht et al., 2012). Certainly by

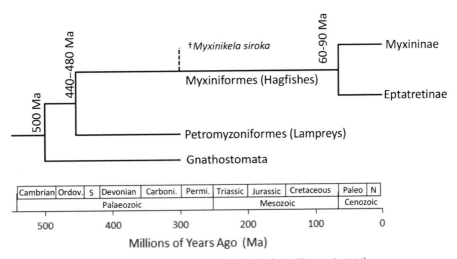

Figure 2.8 Time scale of evolution of Myxini, the hagfishes (after Kuraku & Kuratani, 2006).

the late Devonian the holocephali had emerged as a distinctive group and achieved their greatest diversity during the Carboniferous with species ranging in size from 10 cm long debeerids to 4 m long orodontids and a variety of body forms including the eel-like *Chondrenchelys* and skate-like *Janassa* (Helfman et al., 2009). The earliest example of a possible modern form of Holocephalan with typical fins and jaws of a chimaeroid is *Helodus*, found in freshwater deposits of the Carboniferous (Benton, 2005). It is suggested that holocephali survived the subsequent Permian–Triassic extinction by retreating into deep-sea refuges (Grogan and Lund, 2004). Typical modern forms of chimaeras appear in the Jurassic. Three families of Chimaeriformes from the Palaeozoic and Mesozoic are extinct, leaving the three extant families, Callorhinchidae (ploughnose chimaeras) of the Southern Hemisphere, Rhinochimaeridae (longnose chimaeras) and Chimaeridae (shortnose chimaeras) as present-day representatives of the group. Molecular clock studies indicate that the Callorhinchidae diverged in the early to mid-Jurassic (190–167 Ma), and the divergence time between Rhinochimaeridae and Chimaeridae was during the Late Jurassic to Mid-Cretaceous (159–122 Ma; Inoue, Miya, Lam, et al., 2010). The Callorhincidae have undergone a recent diversification during 18–8 Ma when the present-day genera and species diverged.

The Rhinochimaeridae and Chimaeridae show a series of divergences from 109 Ma in the Cretaceous to 15 Ma in the Neogene (Licht et al., 2012; Fig. 2.9). It is likely that Chimaeriformes have been present at bathyal depths around continents, islands, sea mounts and ridge systems since the Jurassic and have survived through anoxic events and the K–P/K–T mass extinction more or less in situ.

2.4.2.2 Selachii (Sharks)

Sharks are found in the fossil record from the Devonian Period onwards, notably the well-known *Cladoselache*, which grew up to 2 m long in the Upper Devonian ca. 380 Ma (Benton, 2005). Following the late Devonian extinction event when many early groups of fishes disappeared, there was a major diversification of sharks from the Carboniferous onwards. Some of the Carboniferous sharks, especially the hybodonts, survived into the Mesozoic but became extinct by the end of the Cretaceous. These early sharks were all replaced by the emergence of a sister group, the Neoselachii, which includes all modern sharks and rays (Nelson, 2006). The Neoselachii split into two subdivisions, Selachii (sharks) and Batoidea (rays), in the late Triassic (200–230 Ma) according to Aschliman et al. (2012) or as early as the end of the Devonian (364 Ma) according to Sorensen et al. (2014).

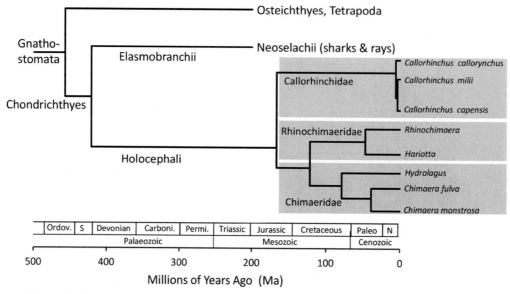

Figure 2.9 Time scale of evolution of Holocephali (chimaeras) derived using molecular clock methods (Inoue et al., 2010). The three grey boxes indicate the three extant families, Callorhinchidae, Rhinochimaeridae and Chimaeridae. Fossil diversity is not shown by this method. The list of species and genera to the right is not complete but is shows typical divergence times within each family.

By the time of the end of Permian extinction events, the Selachii had divided into two superorders, the Squalomorphi and the Galeomorphi, which both began diversification reflecting widespread recovery of marine fauna during the Triassic (Fig. 2.10).

The Squalomorphi are predominantly deep-sea species, and all the modern orders had emerged by the early Triassic: Hexacanthiformes (six-gilled sharks), Echinorhiformes (bramble sharks), Squaliformes (dogfishes), Squatiniformes (angel sharks) and Pristiophoriformes (saw sharks). Within these orders the modern families originated in the late Jurassic and early Cretaceous. Bioluminescence, a characteristic feature of adaptation to the deep sea, is found in the Dalatiidae and Etmopteridae and thus appears to have evolved twice in this group. However in alternative phylogenetic trees bioluminescence is hypothesised to have arisen just once in the Squalomorphii about 110 Ma in the mid-Cretaceous (Klug and Kriwet, 2010). The Squalomorphi includes most well-known deep-sea sharks and is a lineage that has survived from the Triassic and through the end Cretaceous extinction

by virtue of deep sea refuges (Kriwet and Benton, 2004). Diversification has continued within families, for example, the genus *Centroscymnus* split from *Oxynotus* ca. 40 Ma, and within the genus *Squalus* new speciation is evident during the last 5 Ma (Sorensen et al., 2014).

The Galeomorphi comprises about 250 species of relatively large tropical and warm temperate sharks living mainly in shallow and surface waters, including the Heterodontiformes (bullhead sharks), Orectolobiformes (carpet sharks), Lamniformes (mackerel sharks) and Carchariformes (ground sharks). The Carchariformes includes important large epipelagic species such as the requiem and hammerhead sharks but also the family of small demersal cat sharks, Scyliorhinidae. The Scyliorhinidae show a radiation into deep water since the late Cretaceous, the genus *Apristurus* about 100 Ma, *Galeus* 80 Ma, with *Scylliorhinus* and *Cephaloscyllium* splitting from one another after the end of the Cretaceous. The maximum depth is 2200 m for *Apristurus microps*, and mean maximum depth for the family is 718 m.

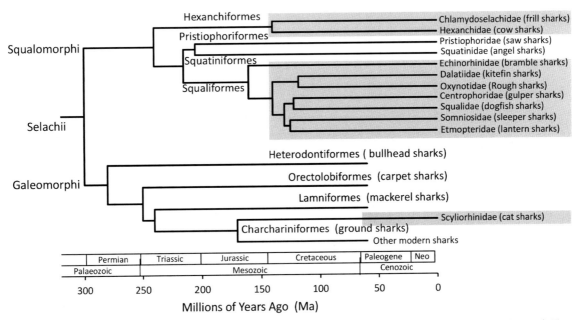

Figure 2.10 Time scale of evolution of Selachii (sharks) derived using molecular clock methods (Sorensen et al., 2014). The grey boxes indicate deep-sea clades. The Superorder Squalimorphi, originated in the deep sea and began diversification around the time of the Permian–Triassic extinction events. The family Scyliorhinidae (cat sharks) within the shallow-water Superorder Galeomorphi, have diversified into the deep sea after the Cretaceous–Paleogene extinction.

2.4.2.3 Batoidea (Rays)

Following their divergence from the Selachii, the major lineages of the Batoidea were derived by about 140 Ma, in the early Cretaceous. This is thought to have occurred in shallow continental seas surrounding the Tethys Ocean, and most Batoidea continue to occupy shallow waters. Possibly owing to their shallow-water distribution, the Batoidea seem to have been greatly affected by the end Cretaceous extinction event with the loss of 65 per cent of genera (Kriwet and Benton, 2004). The main group to invade the deep sea is the family Rajidae, which are likely to have been displaced from shallow waters by diversification of the Mylobatoidei (stingrays). Occupation of the deep sea by the Rajidae has occurred since the end of the Cretaceous across two subfamilies, the Rajinae (hardnose skates) and Arhynochobatinae (softnose skates; Aschliman et al., 2012). Within the Arhynochobatinae, the deep-sea ray genus *Bathyraja* diverged from a shallow-water clade around the Cretaceous/Paleogene boundary ca. 65 Ma

and within the Rajinae the genus *Rajella* (which includes the deepest-living Chondrichthyan, *R. bigelowi*) split from the sister genus of deep-sea skates *Amblyraja* ca. 10 Ma (Fig. 2.11). Compared with the other Chondrichthyes, occupation of the deep sea by rays has been a recent phenomenon.

2.4.3 Actinopterygii

The ray-finned bony fishes, or Actinopterygii, first appear in the late Silurian (430 Ma). They are well represented in the fossil record during the Devonian and greatly increased in diversity from the Carboniferous onwards until they became the dominant fishes in today's oceans (Benton, 2005). The small number of extant species belonging to archaic lineages such as Chondrostei (sturgeons, paddlefishes) and Holostei (bowfins) are now confined to freshwater or coastal waters. The deep sea is now entirely populated by members of the Teleostei, the main modern group of Actinopterygii that originated about

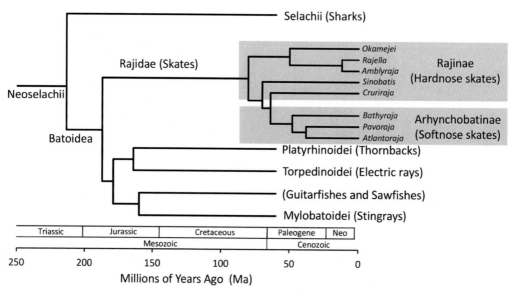

Figure 2.11 Time scale of evolution of Batoidea (skates and rays) derived using molecular clock methods (Aschliman et al., 2012). The grey boxes indicate two deep-sea clades within the Rajidae. Fossil diversity is not shown by this method. The list of genera to the right is not complete but is intended to show typical diversification times

300 Ma in the Carboniferous. It is reasonable to assume that some of the great diversity of basal Teleosts that thrived during the Jurassic and early Cretaceous may have entered the deep sea. Amongst the extant lineages that survive through to the present day, it is possible to recognise at least 25 separate diversifications into the deep sea (Fig. 2.12). The application of molecular phylogenetic studies to large datasets has resulted in many recent changes to our understanding of the classification and evolutionary history of the Actinopetrygii. Here we follow the new classification and phylogeny proposed by Betancur-R et al. (2013) with additional information from Near et al. (2012) and Near et al. (2013).

The Elopomorpha diverged from the rest of the Teleosts about 260 Ma in the Permian and have a distinctive ribbon-like larval form known as a Leptocephalus, which is totally unlike the larvae of any other fishes. Within the Elopomorpha there are three main groups of deep-sea species. The Halosaurs (16 species) occur worldwide, have a characteristic elongated body, all live deep and adults grow to over 60 cm in length. Woodward (1898) recognised the fossil *Echidnocephalus* from Cretaceous deposits in

Westphalia, Germany, as a deep-sea Halosaur. The Notacanthidae (spiny eels, 11 species) is a sister group that are also all deep-sea with a worldwide occurrence. The deep-sea Halosauridae and Notacanthidae appear to have diverged into deep water from the shallow-water Albuliformes (bone fishes) during the mid-Cretaceous. The Anguilliformes (eels, ca. 900 species) occupy a great diversity of habitats from freshwater to the deep sea. Within the group there are two main radiations into the deep sea, demersal and pelagic. The Synaphobranchidae (cutthroat eels, 48 species) are relatively robust bottom-living species that occur from the upper slope down to the abyss where *Histiobranchus bathybius* has been recorded at 5440 m$_{fbd}$. The second radiation is an assemblage of deep pelagic eels (ca. 50 species) with very slender bodies and large mouths, Cyematidae (bobtail snipe eels), Saccopharyngidae (swallowers), Eupharingidae (gulpers, pelican eels), Monognathidae (one jaw gulpers), Nemichthidae (snipe eels) and Serrivomeridae (sawtooth eels). Surprisingly the freshwater eels, Anguillidae, are most closely related to this deep-sea pelagic group and have, despite migration into freshwater, retained the

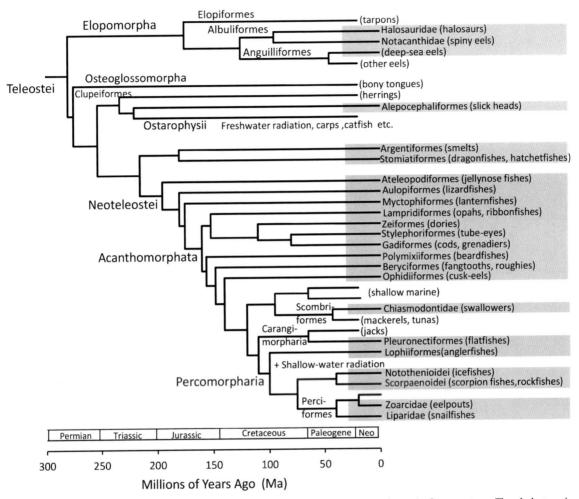

Figure 2.12 Time scale of evolution of Teleostei (Actinopterygii) with emphasis on the main deep-sea taxa. The phylogeny is based on Betancur-R et al. (2013) and Near et al. (2012). Shallow and freshwater lineages are omitted or truncated for simplicity. The grey boxes indicate groups that are predominantly deep-sea or contain deep-sea species.

deep-sea pelagic spawning habit (Inoue, Miya, Miller, et al., 2010). Miller et al. (2013) show that the leptocephalus larvae of the Japanese freshwater eel (*Anguilla japonica*) feed in the mesopelagic on particles of marine snow. It seems that within the Angulliformes, adaptation of all life history stages (eggs, larvae and adults) to the deep sea became established during the Cretaceous.

Amongst the non-Elopomorph teleosts, the Osteoglossomorpha (bony tongues) are an entirely freshwater species, and the Clupeiformes (herrings, anchovies and sardines) are all surface-dwelling or shallow-water species. The Ostarophysii comprises about two-thirds of all known freshwater fishes or about 8,000 species. Recent molecular phylogenies (Near et al., 2012; Betancur-R et al., 2013) are in agreement that the deep-sea Alepocephaliformes is a sister group to the Ostarophysii. The Alepocephaliformes group includes the Platytroctidae (tubeshoulders), Bathylaconidae and Alepocephalidae (slickheads). There are over 100 species that live exclusively in the deep sea, and many possess light organs. The Alepocephaliformes diverged from the Ostarophysii ca. 220 Ma in the Triassic and can be regarded as a long-established deep-sea endemic group.

The Argentiformes and Stomiatiformes fall into a group together with the freshwater or anadromous Esociformes (pikes) and Salmoniformes (salmons and trouts) but diverged ca. 210 Ma in the Jurassic. They include many interesting highly specialised deep-sea families. Amongst the Argentiformes (marine smelts) there are Opisthoproctidae (barrel eye or spookfishes, 11 species) with upward-looking tube eyes, Microstomidae (pencil smelts, 38 species) and Bathylaginae (deep-sea smelts, 20 species), which are slender bathypelagic species often found at great depths. The Stomiatiformes includes many iconic deep-sea species with a variety of patterns of light organs: Gonostomatidae (bristlemouths), Sternoptychidae (hatchetfishes, over 180 species), Phosichthyidae (lightfishes, 20 species) and Stomiidae (barbeled dragonfishes, over 250 species). The bristlemouths, *Cyclothone* and *Vinciguerria*, are small fishes that occur in large numbers throughout vast volumes of the mesopelagic and bathypelagic zone and are considered to be the most abundant vertebrates on the planet.

The Neoteleostei includes all the higher groups of fishes that have appeared from the Jurassic onwards. The Ateleopodiformes comprises just one family, the Ateleopodidae (jellynose fishes, 13 species), which are characterised by a large head and long thin tapering body and grow up to 2 m in length. They are highly adapted deep-sea fishes with a widespread distribution. The Aulopiformes (lizardfishes) is a diverse group with a number of highly specialised deep-sea families and worldwide distribution: Bathysauridae (one species), Chlorapthalmidae (greeneyes, 19 species), Bathysauropsidae (3 species), Notosudidae (waryfishes, 19 species), Ipnopidae (deep-sea tripod fishes, 28 species), Scopelarchidae (pearleyes, 17 species), Evermannelidae (sabretooth fishes, 7 species), Alepisauridae (lancet fishes, 2 species), Paralepididae (barracudinas, 56 species), Bathysauridae (deep-sea lizardfishes, 2 species) and Giganturidae (telescope fishes, 2 species).

The Myctophiformes is dominated by one family, the Myctophidae (lantern fishes, 240 species) that are very abundant in the mesopelagic zone of the ocean and form an important part of the biomass in the Deep Scattering Layer. They are small silvery fishes typically up to 10 cm long with characteristic light organs mostly on the ventral surface.

The Acanthomorphata (spiny rayed fishes) arose in the late Jurassic and are recognised by the presence of true spines in the dorsal, anal and pelvic fins in most species, as opposed to the jointed fin rays found in the lower teleosts. The Lampriformes includes several oceanic pelagic families, which are found below 200 m depth: Lophotidae (crestfishes, 3 species), Radiicephalidae (taper tails, 1 species), Trachipteridae (ribbonfishes, 12 species) and Regaliciadae (oarfishes, 2 species). *Regalecus glesne* (King of Herrings) that grows up to 8 m in length is the longest of the bony fishes. The Opahs are laterally flattened disc-shaped oceanic fishes up to 2 m long that are found down to 500 m depth.

The Zeiformes (dories) are laterally flattened oceanic fishes with large mouths and protrusible jaws found in the upper 1000 m of the ocean. There are six families, Cyttidae (lookdown dories, 3 species), Oreosomatidae (Oreos, 10 species), Parazenidae (smooth dories, 3 species), Zeniontidae (armoreye dories, 7 species), Grammicolepididae (tinselfishes, 2 species) and Zeidae (dories, 5 species). Fossil Zeiformes are known from the upper Cretaceous. The Stylephoriformes is represented by just one species, *Stylephorus chordatus* (tube eye) that grows up to 30 cm long and lives circumglobally in deep pelagic tropical waters. Formerly classified with the Lampridiformes, molecular methods place this species alongside the Gadiformes in this phylogeny. The Gadiformes comprise over 600 species ranging from the freshwater burbot (*Lota lota*) to the deep-sea grenadier (*Coryphaenoides yaquinae*) that has been found at depths over 6000 m in hadal trenches of the Western Pacific Ocean. They are relatively robust fleshy fish with desirable white muscle and oily livers leading to their importance commercially, including exploitation in deep-sea fisheries. Molecular phylogeny shows three main lineages. The family Muraenolepididae (eel cods, 9 species) found in the Southern Hemisphere at depths down to 3000 m is considered to be the most ancient basal surviving lineage (Roa-Varón and Ortí, 2009). The remaining families are divided between the Suborder Macrouroidei and the Suborder Gadoidei. The Macrouroidei is dominated by the family Macrouridae (grenadiers or rattails, comprising 394 species), which is considered to have originated during the Cretaceous and acquired the deep-water habitat at the beginning of the Eocene, ca. 55 Ma. They are all deep-sea demersal fishes with a few bathypelagic representatives (Iwamoto, 2008a). The Gadoidei includes six families, Bregamacerotidae (codlets, 15 species), Moridae (deep-sea cods, 105 species), Melanidae (pelagic cods, 2 species), Merlucidae (hakes, 13 species), Phycidae (hakes and rocklings, 25 species) and Gadidae (cods, 22 species). They are mostly demersal species, but there are important deep pelagic species such as *Micromesistius poutassou* (Blue whiting) that form the basis for a major fishery in the northeastern Atlantic. The Gadoidei are generally bathyal species associated with continental slopes, ridges and seamounts. This influences their biogeography with strong patterns of regional distribution. Howes (1991) points out that the phylogenetic trend within the Gadoidei is from deep to shallow with the more ancestral taxa occupying deep oceanic habitats and the derived supragadoid taxa mainly confined to shelf waters rarely deeper than 500 m. This fits with the association with *Stylephorus,* indicating potential deep water ancestry in the late Cretaceous and subsequent distribution influenced by continental drift during the Paleogene.

The Polymixiiformes comprise just one family, Polymixiidae (beardfishes, 10 species), with one genus, *Polymixia*, of tropical and subtropical demersal fishes that live on the upper slopes generally deeper than 200 m, where they are locally exploited. They are well represented by four genera of fossil species in the upper Cretaceous. The Beryciformes here includes the Stephanoberyciformes and the Cetomimiformes, giving a total of 263 species with an average maximum depth of over 1000 m. The clade originated in the early Cretaceous but shows considerable diversification since the K–Pg mass extinction. There are 12 deep-sea families of mainly pelagic, often rare but some abundant species: Melamphidae (bigscale fishes, 36 species), Stephanoberycidae (prickle fishes, 4 species), Hispidoberycidae (hispidoberycids, 1 species), Gibberichthyidae (gibberfishes, 2 species), Rondeletiidae (redmouth whalefishes, 2 species), Barbourisiidae (red whale fishes, 1 species), Cetomimidae (flabby whalefishes, 29 species), Mirapinnidae (tapertails, 5 species), Megalomycteridae (largenose fishes, 5 species), Anoplogasteridae (fangtooths, 2 species), Diretmidae (spinyfins, 4 species) and Trachichthyidae (roughies, 50 species). The Trachichthyidae include the well-known orange roughy (*Hoplosthethus atlanticus*), which is the subject of major deep-water fisheries in several parts of the world.

The Ophidiiformes (cusk-eels) are mostly demersal fishes ranging in depth from inshore shallow reefs to hadal depths. There are over 500 species divided between five families: Carapidae (pearlfishes, 35 species), many species of which hide inside other animals, a few occurring in the deep sea; Ophidiidae (cusk-eels, 250 species) that occur at all depths; Bythidae (viviparous brotulas, 200 species) distributed from inshore locations down to 2500 m depth; Aphyonidae (23 species) that are all deep-sea species; and Parabrotulidae (false brotulas, 3 species) that are all bathypelagic.

The Chiasmodontidae (swallowers, 32 species) is a family of deep-sea fishes grouped phylogenetically amongst families of predominantly shallow-dwelling spiny-rayed fishes. They are bathypelagic, body length typically less than 20 cm long and some species reach depths >4000 m. One genus, *Pseudoscopelus,* possesses light organs. The Chiasmodontidae diverged from the mainly epipelagic Scombriformes (mackerels), as defined by Bentancur-R et al. (2013), in the Paleogene about 45 Ma and have a basal position in this clade. The Pleuronectiformes (flatfishes, 719 species) is a monophyletic group, descended from a common ancestor that lived about 77 Ma in the late Cretaceous and related to the Carangiform fishes such as jacks, horse mackerel and sword fish (Campbell et al., 2013). Most species live in shallow shelf seas; however, nine families have significantly extended their distribution into the deep sea: Cithardidae (large-scale flounders, 7 species) found in the Mediterranean and Indo-West Pacific with 4 species considered bathydemersal; Scophthalmidae (turbots, 9 species) from the North Atlantic, Mediterranean, Baltic and Black Seas includes *Lepidorhombus* (megrims) that are captured in mixed commercial deep-sea fisheries down to 800 m; Paralichthyidae (sand flounders, 109 species) of which a few are deep-sea, including *Citharichthys dinoceros* (spined whiff) recorded down to 2000 m in the western Atlantic; Bothidae (lefteye flounders, 164 species) with a bathydemersal genus *Chascanopsetta* (pelican flounders) found down to over 3000 m; Achiropsettidae (southern flounder, 4 species) found in the Antarctic and sub-Antarctic to over 1000 m depth; Pleuronectidae (right eye flounders, 98 species) a family that includes halibuts that occur down to 3000 m depth and forms the basis for important deep-water fisheries; Samaridae (crested flounders, 27 species) from the tropical and subtropical Indo-West Pacific; Soleidae (soles, 150 species) with the genus *Bathysolea* occurring down to over 1000 m depth and Cynoglossidae (tongue fishes, 130 species) with bathydemersal species found down to 1000 m depth including some endemic to hydrothermal vents.

The Percomorpharia represent an important recent diversification of the bony fishes that began about 100 Ma but has continued through the Paleogene and Neogene with thousands of species contributing to the 'bush at the top' of the teleost evolutionary tree (Near et al., 2013). Amongst this diversity there are numerous families with species distributed deeper than 200 m and that are considered to be deep-sea fishes. Here we identify just four groups within which there are important deep-sea fishes.

The Notothenioidei have a primarily Antarctic distribution; there are eight families of which four can be considered deep-sea. The Antarctic ice fishes diverged from the shallow and freshwater Notothenioidei of Australia, New Zealand and South America, Bovichtidae, Pseudaphritidae and Eleginopidae at 75 Ma, 60 Ma and 30 Ma, respectively (Betancur-R et al., 2013). The family Nototheniidae (cod icefishes, 57 species) occurs from brackish inshore waters down to 3850 m, the maximum depth

recorded for *Dissostichus eleginoides*; the Patagonian toothfish that forms the basis of an important fishery in the Southern Ocean. The Artedidraconidae (barbelled plunderfishes, 28 species) are all Antarctic deep-water fishes, the Bathydraconidae (Antarctic dragonfishes, 17 species) occur down to over 2500 m and the Channichthyidae (crocodile icefishes, 25 species) reach a maximum depth of 2000 m and can survive without haemoglobin in the cold oxygenated waters of the Antarctic.

The Scorpaenoidei originated ca. 90 Ma and have diversified into four separate clades within the family Scorpaenidae plus the family Sebastidae, which split off ca. 55 Ma. The Scorpaenidae (scorpion fishes and rockfishes, 209 species) are demersal fishes living from inshore to a maximum reported depth of over 900 m with 15 species considered bathydemersal. The Sebastidae (rockfishes, ocean perches, rockcods, 136 species) have a predominantly greater depth of occurrence down to 2500 m, and 28 species are considered bathydemersal. Several species are important deep-water fishery resources: *Sebastes norvegicus* (Golden redfish) in the North Atlantic, *Helicolenus dactylopterus* (Blackbelly rosefish) in the Mediterranean Sea and *Sebastes alutus* (Pacific ocean perch) in the Pacific Ocean. Within the Scorpaeniformes the closely related family Anoplopomatidae (sablefish, 2 species) is endemic to the North Pacific Ocean, and *Anoplopoma fimbria* is an important fishery in the Gulf of Alaska and off Japan ranging down to over 2700 m.

The family Zoarcidae (eelpouts, 288 species) diverged from its nearest relations about 20 Ma and have colonised from the Arctic to the Antarctic and from inshore to abyssal depths over 5000 m. These mainly demersal species have shown a remarkable dispersion into the deep sea over a relatively short period of time with over 50 per cent of species considered to be deep-sea. The Liparidae (snailfishes, 382 species) diverged from the Cyclopteridae (lumpsuckers) about 45 Ma and comprising 29 genera have diversified into shallow habitats and all depthsdown to over 7000 m from the Arctic to the Antarctic. 250 species are considered to be deep-sea, and there are hadal endemic species confined to individual trench systems in the Pacific Ocean.

The Lophiiformes (anglerfishes, 321 species) is a diverse array of marine fishes, many of which have extraordinary adaptations to life in the deep sea and are characterised by the prehensile first dorsal fin ray used as a lure. Their position in the overall classification of fishes has been greatly revised in the light of new molecular evidence. They are considered to have diverged from related percomorphs early in the Cretaceous and diversified over a relatively short time period in the mid to late Cretaceous (100–130 Ma), giving rise to five suborders and all the major families (Miya et al., 2010). The Suborder Lophoidei contains one family, the Lophiidae (goosefishes, monkfish or anglerfish, 28 species) distributed from coastal waters to over 1500 m depth; these are robust fish considered valuable in deep-water fisheries. The Antennarioidei are shallow or surface-dwelling species. The Chaunacoidei contains one family, the Chaunacidae (coffin fishes or sea toads, 16 species) that occur down to over 2000 m. The Suborder Ogcocephaloidei has one family, Ogcocephalidae (batfishes, 73 species), which have a distinctive shape, walk on the bottom with their large pectoral fins and are found down to 4000 m depth. The largest suborder is the Ceratioidei with 11 families and 160 species of deep-sea angler fishes characterised by extreme sexual dimorphism with dwarfed males that attach themselves to the giant females. The Ceratioidei form a distinctive and highly diverse component of the bathypelagic ichthyofauna.

2.4.4 Sarcopterygii

The Coelocanths diverged from freshwater lungfishes around 400 ma in the early Devonian and have a rich evolution history with just two relict species alive today, *Latimeria chalumnae* and *L. menadoensis*, living on opposite sides of the Indian Ocean. These two species diverged from one another about 20 Ma (Sudarto et al., 2010), and there is evidence of continuing recent differentiation between populations of *L. chalumnae* in the Western Indian Ocean (Nikaido et al., 2011).

2.5 Ancient and Secondary Deep-Sea Fishes

Andriashev (1953) proposed that deep-sea fishes can be divided into two kinds: ancient (or true deep-water) and secondary deep-water forms. The ancient deep-water families, morphologically specialised for life in the deep sea (body form, light organs, etc.), belong to the lower phylogenetic groups of Teleosts and have an extremely wide, often worldwide distribution in the oceans. Conversely the secondary deep-water forms are not generally specialised for life in the deep sea, belong to the younger groups of Teleosts (mainly Perciformes) and are restricted in their distribution. These trends are apparent at the opposite ends of the phylogenetic spectrum; the Alepocephalidae are peculiarly deep-sea fishes with light organs, very ancient and widespread, e.g., *Xenodermichthys copei*, which occurs throughout most of the world's oceans. In contrast, the Nototheniidae show no major deep-sea specialisations, are of recent origin and are localised in the Southern Ocean. Priede and Froese (2013) studied depth distribution of different families and termed those with no shallow-water representatives as deep-sea endemic and those with many shallow-water species as deep-sea invasive. They suggested that the endemic and invasive families may be regarded as synonymous with Andriashev's ancient and secondary deep-sea forms, respectively. However, it did not prove possible to divide deep-sea families clearly between endemics and invasives, there being a continuum between the two kinds. The boundary between ancient and secondary in Andriashev (1953) has also become somewhat confused as fish classification has changed. It seems convenient to define the boundary between ancient and secondary at the Cretaceous-Paleogene boundary (K–Pg); families that appeared before that time (65 Ma) are ancient and more recent ones secondary. Much attention has been directed to determine whether the super-radiation of recent spiny-rayed teleost fishes was influenced by the K–Pg mass extinction with suggestions that following the event new niches opened up and the diversification rate increased. Near et al. (2013) analysed the phylogenetic diversification rate within the acanthomorphs and could find no change at the K–Pg boundary; there seems to have been a smooth transition through the K–Pg. Neither does habitat influence diversification rate; deep-sea Liparids and freshwater Cichlids have equally high rates. It does not seem to be possible to define a boundary between ancient and secondary or endemic and invasive, although trends between the extremes are discernible.

The extreme view of an azoic pre-Cretaceous deep-sea as expressed by Wells et al. (1931) is no longer tenable; there is copious evidence of deep-sea life in the Palaeozoic and Mesozoic. The concept of extinction and replacement (McClain and Hardy, 2010; Priede and Froese, 2013) is now tempered by the realisation that lineages do persist through extinction events, possibly benefitting from the deep sea as a refuge. Examining Fig. 2.12, it is evident that most modern lineages have their origins in Jurassic and Cretaceous, and the OAEs and the K–Pg extinction have not had a catastrophic effect on deep-sea lineages. The modern deep-sea fish fauna (like all fish faunas) is dominated by recent diversification during Cretaceous to Paleogene periods termed by Near et al. (2012) as the 'Second Age of Fishes'.

3 Adaptations to the Deep Sea

3.1 Introduction

The deep sea is an extreme environment characterised by high pressure, absence of solar light, low temperature (except in the Mediterranean and Red Seas), poor food supply and zones of low dissolved oxygen concentration (see Chapter 1). Pressures of 100 to 200 atmospheres, equivalent to 1000–2000 m depth, are lethal to shallow-water fishes (Sébert, 1997). The other factors are within ranges experienced by fishes elsewhere in aquatic environments but, taken together with the three-dimensional vastness of the oceans, present unique challenges for survival. The purpose of this chapter is to describe the special adaptations used by different species to survive in the deep.

There are divergent responses to extreme environments, either as compensation to maintain ability to function in the face of increasingly adverse conditions, or conforming by abandoning or curtailing nonessential functions. For example, some groups of fishes adapt to low light levels by increasing the size and complexity of the eyes to sustain visual function, whereas others, living at the same depths, have reduced or degenerate eyes. Within the eel species, the two extremes are exemplified by the contrast between *Histiobranchus bathybius* (Deep-water arrowtooth eel), which grows to over one metre long and is an active wide-ranging predator or scavenger, whereas the Monognathidae (one jaw gulpers) are small and functionally degenerate with rudimentary sense organs and reduced skeleton. Both *Histiobranchus* and *Monognathus* species survive in the abyss, though with very different evolutionary adaptations.

Because fish bodies are filled with more or less incompressible fluids and solids, pressure has no gross effect on the anatomy such as occurs in the cephalopod molluscs (cuttlefish and nautiluses) in which the gas-filled shell implodes when a critical depth is reached (Sherrard, 2000). In fishes with a swim bladder, the gas-filled chamber is flexible and simply decreases in size with increase in pressure. However, when a fish is raised from depth it generally cannot remove gas fast enough to avoid expansion and possible rupture of the bladder before reaching the surface. Such fishes commonly arrive on the deck of the ship with bulging eyes and everted stomachs pushed outward by expansion of the swim bladder. Even in species with no gas-filled bladder, retrieval from depth is generally lethal but can be greatly improved by thermal insulation of the collecting apparatus and maintenance at high pressure (Drazen et al., 2007). The effects of pressure in fishes generally operate at the molecular and ultra-structural level expressed by symptoms such as hyperactivity, loss of coordination and immobilisation.

3.2 Energy Metabolism

Deep-sea fishes are generally characterised as having a low metabolic rate. Smith and Hessler (1974) first measured oxygen consumptions of deep-sea fish at 1230 m depth and found that the respiration rate of *Coryphaenoides acrolepis* (Pacific grenadier, 3700 m_{fbd}) was 4.5 per cent of that of the shallow-water *Gadus morhua* (Atlantic Cod) living at similar temperatures. The respiration rate of *Eptatretus deani* (Black hagfish, 2743 m_{fbd}) was 22 per cent of that of shallow-water congeners. Torres et al. (1979) measured rates for 23 pelagic species at depths down to 1000 m and found a significant trend of decrease in metabolic rate with depth (Fig. 3.1) with average metabolism at 1000 m depth 10 per cent of that for surface-dwelling species but with wide variation; the Alepocephalid, *Bajacalifornia burragei* (Sharpchin slickhead, 1080 m_fbd) is just 1.5 per cent of the shallow-water mean. A similar trend is observed in benthopelagic species such as macrourids in which the metabolic rates are lower than for pelagic species, and predicted rate at 1000 m depth is 18 per cent of the value for shallow-water species (Drazen and Seibel, 2007; Drazen and Yeh, 2012). Benthic species generally have the lowest metabolic rates, and the trend for decrease with depth in bottom-living fishes is not very strong with the predicted rate at 1000 m depth just 39 per cent of the shallow-water mean. Data are sparse, but the measured metabolic rate of the zoarcid *Pachycara gymninium* (Nakednape eelpout, 3225 m_fbd) is equal to or exceeds that of many shallow-water species (Drazen and Yeh, 2012). Seibel and Drazen (2007) argued that for benthic species there may be no significant difference between metabolic rates of shallow- and deep-water fishes.

As an alternative to measuring the metabolic rate of intact living fishes insights can be gained into their metabolic capacities by measuring the activities of metabolic enzymes extracted from muscle samples. Figure 3.2 shows a compilation of such results for lactate-dehydrogenase (LDH) of the anaerobic metabolic pathway, and citrate synthase (CS) of the Krebs cycle for aerobic metabolism (Childress and Somero, 1979; Childress and Thuesen, 1995; Drazen and Seibel, 2007; Drazen et al., 2015). Pelagic fishes show a very steep decline in LDH activity with depth (Fig 3.2a) so that species living at 1500 m depth in the bathypelagic have a predicted anaerobic capacity just 3 per cent of that of epipelagic species near the surface. The equivalent values are 13 per cent for benthopelagic species and 46 per cent for benthic species. CS activity also decreases most steeply in the pelagic species (Fig 3.2b) so that predicted aerobic capacity at 1500 m depth is 9 per cent of the surface values, 28 per cent for benthopelagic species and 37 per cent for benthic

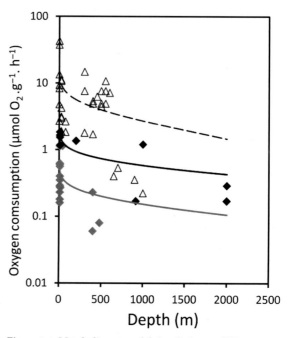

Figure 3.1 Metabolic rates of fishes living at different depths. Each point generally represents a different species. Open triangles: Pelagic fishes, from measurements of oxygen consumption at 5°C (deep-living) and 10°C (surface dwelling) solid black line – fitted curve $Y = 12.81X^{-0.494}$ $R^2 = 0.923$ (Torres et al., 1979). Grey diamonds: Benthopelagic fishes corrected to 5°C, solid grey line – fitted curve $Y = 12.81X^{-0.494}$, $R^2 = 0.6499$ (Drazen and Seibel, 2007). Black diamonds: Benthic fishes corrected to 5°C dashed line – $Y = 0.6167X^{-0.206}$ $R^2 = 0.195$ (Drazen and Seibel, 2007; Drazen and Yeh, 2012). $Y =$ oxygen consumption (μmol O_2. g^{-1}. h^{-1}) $X =$ depth (m).

species confirming the trends observed in oxygen consumption measurements. For purposes of comparison the data in Figure 3.2 are plotted with respect to minimum depth of occurrence but Drazen et al. (2015) show that when analysed with respect to median depth of occurrence the trends of decline in metabolic enzyme activity of demersal fishes, both benthic and benthopelagic, are more significant.

Torres et al. (1979) estimated that 30 per cent of the decrease in metabolism at depth in pelagic fishes could be accounted for by an increase in body water

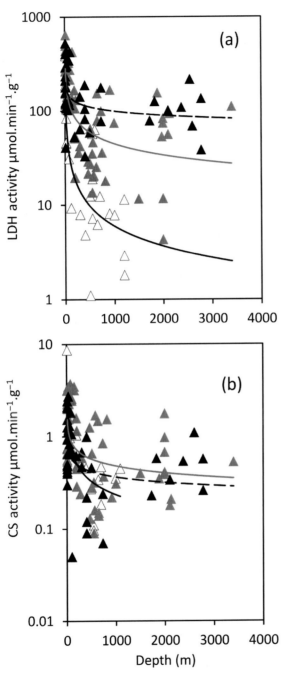

Figure 3.2 Muscle metabolic enzyme activities of species of fishes living at different depths (Minimum depth of occurrence). Benthic and Benthopelagic data are from Drazen and Seibel (2007) and Drazen et al. (2015) redrawn with permission.

(a) Lactate Dehydrogenase (LDH) activity. Open triangles-Pelagic fishes solid black line – fitted curve

content. Many deep-sea fishes have soft watery muscles, hence with a smaller quantity of metabolically active tissue. Blaxter et al. (1971) showed that mesopelagic fishes without gas-filled swim bladders living at 100–1000 m depth have very high water content in the range 88–95 per cent, *Xenodermichthys*, *Bathylagus*, *Chauliodus*, *Eurypharynx*, *Gonostoma elongatum*, *Photostomias*, *Melanocoetus* and *Leptostomias*. Conversely, for deep-sea demersal species Crabtree (1995) only found a trend of increase in water content with depth in benthopelagic species with swim bladders, such as Gadidae, Ophidiidae and the Macrouridae. Within the Order Aulopiformes (Lizardfishes), the benthic suborder Chlorophthalmoidei shows no change in water content from the shallow-water greeneye, *Chlorophthalmus agassizi* to the abyssal tripod fish, *Bathypterois longipes* at over 5000 m depth. For a selection of demersal species from the NE Pacific, Drazen (2007) found a significant increase in water content of white muscle with depth and that this was correlated with a decrease in protein content.

There is a general underlying trend of decrease in metabolic rate with depth in fishes. The reasons for this can be divided into three categories:

I) Effects of pressure and other factors on biochemical processes limiting the metabolic rates attainable in the deep sea

ii) Adaptation to a low food supply

iii) Adaptation to lower energy demand for predator–prey behavioural interactions in the absence of solar light.

$Y = 773.63X^{-0.704}$, $R^2 = 0.63$ (data from Childress and Somero, 1979). Grey triangles: Benthopelagic fishes, solid grey line – fitted curve $Y = 713.31X^{-0.401}$, $R^2 = 0.43$, Black triangles: Benthic fishes dashed line – $Y = 285.66X^{-0.0.153}$, $R^2 = 0.199$.

(b) Citrate synthase (CS) activity. Open triangles – Pelagic fishes solid black line – $Y = 6.37X^{-0.478}$, $R^2 = 0.59$ (data from Childress and Thuesen, 1995). Grey triangles: Benthopelagic fishes, solid grey line – $Y = 2.75X^{-0.252}$, $R^2 = 0.20$, Black triangles: Benthic fishes dashed line – $Y = 1.41X^{-0.196}$, $R^2 = 0.18$.

3.3 Effects of Pressure

Biochemical reactions involve a volume change between the start of the reaction and its conclusion. If there is an increase in volume of the reactants, then elevated hydrostatic pressure tends to inhibit the reaction, and conversely if the volume decreases the reaction is accelerated (Somero, 1992). Proteins that catalyse biochemical reactions have complex three-dimensional structures held together by numerous weak bonds, including ionic and hydrophobic bonds, which usually involve an increase in volume during their formation. Pressure therefore disrupts these bonds, affecting protein polymerisation and enzyme reactions. Very high pressures (> 100 MPa) denature proteins completely (Gibbs, 1997). In order to survive at depth, fishes must adapt at the molecular level to sustain normal function.

Cell membranes are formed of a bilayer of phospholipid molecules with the hydrophilic head on the outside and hydrophobic fatty acid chains on the inside. To act as an electrical insulator and separator of cellular compartments, the bilayer needs to be in a state of transition between the fluid and solid phases. Increase in pressure and decrease in temperature reduce membrane fluidity. Membrane-based processes, notably trans-membrane ion fluxes, are extremely pressure-sensitive manifested, for example, as a slowing in the action potential in nerve fibres when pressure is applied. Harper et al. (1987) showed that action potentials in the nerves of *Gadus morhua* and *Mora moro* from < 900 m depth were inhibited by pressures >10 MPa, whereas those of deep-water species from 4000 m depth *Bathysaurus molis* and *Coryphaenoides armatus* showed optimal function at 41 MPa. The nerves of the deep-sea species are piezophilic, with loss of function when brought to the surface. In deeper-living species there appears to be natural selection for phospholipids with high inherent fluidity. Liver mitochondrial membranes of deep-sea fishes captured at 4000 m depth contain a greater proportion of unsaturated fatty acids than shallow-living species (Cossins and MacDonald, 1986).

Cell membranes also host enzymes such as sodium-potassium ATPase (Na$^+$/K$^+$-ATPase), a molecule that traverses the thickness of the lipid bilayer and is responsible for pumping three Na$^+$ ions out of a cell in exchange for two K$^+$ ions entering thus generating the normal trans-membrane electrical potential of the cell. In the gills, this so-called sodium pump plays an important role in osmoregulation. The Na$^+$/K$^+$-ATPase activity of the deep-sea species *Coryphaenoides armatus* is much less pressure-sensitive than that of shallow-water species, and to a large extent this is determined by the gill membrane fluidity protecting the enzyme function. Nevertheless, despite this pressure tolerance, the estimated Na$^+$/K$^+$-ATPase activity in gills of deep-water species is at least one order of magnitude lower than in shallow-water species, reflecting their relatively sluggish lifestyle (Gibbs and Somero, 1990). However, membrane lipid changes are not solely responsible for the pressure adaptation; structural adaptations of the enzyme molecule itself also play a role.

Important evidence that proteins of deep-sea fishes are structurally modified arose from the observation by Sweezey and Somero (1982) that actin extracted from muscle fibres of deep-sea fishes is extremely heat resistant with properties equivalent to actins from heat-loving desert lizards. The proteins are structurally strengthened to resist pressure by modification of bonds in the tertiary structure shaping the molecule. This has the side effect of conferring high thermal stability. Wakai et al. (2104) show that two amino acid differences in muscle actins of *Coryphaenoides armatus* and *C. yaquinae* compared with shallow-water fishes are sufficient to stabilise ATP binding and subunit arrangement. Evidence of pressure tolerance has also been found in key enzymes involved in muscle anaerobic metabolic pathways, Lactate dehydrogenase (LDH), malate dehydrogenase (MDH) and Glyceraldehyde-3-phosphate dehydrogenase (GAPDH), with those from species living deeper than 500–1000 m being insensitive to pressure (Somero, 1992). Citrate synthase (CS), which is a key enzyme of the Krebs cycle for aerobic metabolism, has been found to have a pressure optimum of 200–250 atmospheres in gill tissue taken from *Antimora rostrata* living at 2010 m depth (Hochachka et al., 1975). This pressure tolerance, however, is realised at the expense of reaction rate, which is half to a fifth of what could be maintained at the surface. Somero and Siebenaller (1979) point out that because of the reduced structural flexibility of the metabolic enzyme protein molecules of deep-sea fishes, they have low catalytic efficiency, which may partly explain their low metabolic rates. However, the gills of *Antimora* have three to four times higher concentrations of citrate synthase than shallow-water species, which compensates for the underlying inefficiency.

The function of proteins is also influenced by the composition of the intracellular fluid environment within which they are contained. The need for an appropriate concentration of inorganic ions such as K^+ is well known, but it is increasingly recognised that organic osmolytes such as carbohydrates, amino acids, methylamines and urea also modulate protein function. Urea, for example, commonly found in high concentrations in Chondrichthyes, generally destabilises proteins. Tri-methyl amine oxide (TMAO) counteracts the urea effect enabling Chondrichthyes to maintain osmotic equilibrium with the environment using urea. A molecular ratio of 2 (urea): 1(TMAO) is optimal for maintaining protein function. TMAO can be regarded as a 'chemical chaperone', which appears to operate by structuring the water molecules around a protein molecule so that the peptide backbone is inhibited from dissolving in water and is driven to fold in on itself forming an α-helix. TMAO is the strongest stabilizer amongst common osmolytes and in teleost fishes is found at increasing concentrations with increasing depth of occurrence (Yancey and Siebenaller, 2015).

In summary, pressure has multiple effects on metabolism but is compensated for by changes in membrane phospholipid composition, intrinsic modifications of proteins and extrinsic conservation of protein function by accumulation of protective osmolytes such as TMAO. These pressure-counteracting osmolytes that confer protection for proteins that are not fully pressure resistant are collectively known as piezolytes. There is some evidence that pressure resistance may be at the expense of maximum rates of reactions.

3.4 Size and Shape

A general phenomenon in deep-sea biology is decrease in body size with depth, so-called dwarfism. This is most clearly observed in benthic invertebrate macrofauna and was interpreted by Thiel (1975) as an adaptation to low food availability enabling population size to be maintained through reduction of food requirement per individual (Rex and Etter, 2010; Van der Grient and Rogers, 2015). Converse trends (gigantism) are observed in some taxa, e.g. large gastropod molluscs (McClain et al., 2005) with regional variation related to food supply; nevertheless the general global trend towards reduction in body size with depth is highly significant (Van der Grient and Rogers, 2015). It has already been pointed out in Chapter 2 that there is no such trend in the three main classes of fishes. Inspecting Figure 2.3 the main feature is decrease in diversity of fish size with depth so that deep-sea fishes converge towards a relatively narrow, presumably optimum, body size range at depth. Within demersal species an increase in body size with depth is often observed. This phenomenon was first described in Heinke (1913), who found in trawl surveys of plaice (*Pleuronectes platessa*) that the average size of individuals increases with depth. This has come to be known as Heinke's law and has been observed in a number of deep-sea fish species (Merrett and Haedrich, 1997). In the NE Atlantic, Priede and Bagley (2000) show that within species occurring at a succession of increasing depths *Synaphobranchus kaupii* (800–1800 m) *Antimora rostrata* (1400 m–2500 m) and *Coryphaenoides armatus* (2500–4100 m) the smaller individuals occur at the shallowest depths. This is possibly related to the lifecycle patterns whereby juveniles first appear on the seafloor at the shallow end of the species' depth range, and individuals migrate deeper as they grow. Merrett and Haedrich (1997) argue that evidence for a more general bigger-deeper Heinke's law trend is weak and suggest that it may be the result of a sampling artefact caused by faster-swimming large individuals that escape from trawls at upper slope depths.

However within the global data set for all three major classes Myxine, Chondrichthyes and

Actinopterygii, there is a slight trend towards increase in maximum body size with depth (Figure 3.3a). The R^2 values of the fitted regression lines are low, indicating that depth accounts for a small part of the variation in fish size. In the Kermadec and Mariana Trenches at depths of 6500 to 8000 m a smaller-deeper trend is observed in the endemic snailfishes (Liparidae; Linley et al., 2016) contrary to the predicted trend for Actinopterygii in Figure 3.3a.

In a survey of demersal fishes living in the NE Atlantic, Collins et al. (2005) detected opposing trends in different functional groups of deep-sea fishes. Scavenging species, identified as those that regularly appear at baited cameras, increase in size with depth whereas nonscavenging species decrease in size over the same depth range (Figure 3.3b). These trends were interpreted as responses to differences in characteristics of food supply by the two groups. The scavenging species depend on large, randomly distributed packages of carrion falling from the overlying layers of the ocean. The larger body size permits higher swimming speeds and greater endurance allowing the fish to move efficiently between relatively rare feeding events in the abyss. For nonscavengers, depending on more evenly distributed continuous benthic food supply, there is an advantage in small body size as predicted in the classic arguments for dwarfism in the deep.

Neat and Campbell (2013) observed differences in body shape of fishes captured at different depth on bathyal slopes in the NE Atlantic (Figure 3.3c). With increasing depth the average form factor ($\alpha_{3.0}$) of the assemblage of fishes caught per trawl haul decreased, indicating a higher proportion of fishes with slimmer elongated bodies such as Macrouridae, halosaurs, chimaeras and Anguilliformes at greater depths. The trend for elongation of body form with depth is most pronounced in the Gadiformes and Argentiniformes. It is suggested that an elongated body is more efficient for the predominantly low-speed anguilliform swimming modes used by deep-sea fishes. There may also be advantages in having a longer lateral line sense organ for sensing vibrations in the darkness. However, there are numerous exceptions to the trend towards slender bodies in the deep sea, most

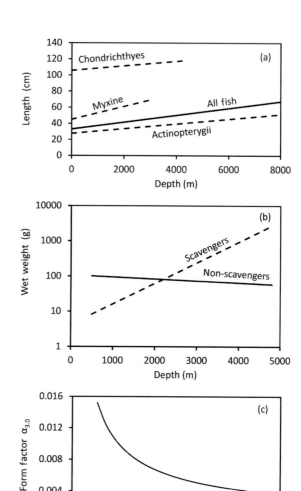

Figure 3.3 Fish size and shape.
(a) Underlying trends in maximum total length as a function of maximum depth of species occurrence from data shown in Figure 2.3. Class Myxine. $y = 0.00806x + 44.95$, $R^2 = 0.076$, $P \ll 0.001$, Class Chondrichthyes $y = 0.00287x + 105.78$, $R^2 = 0.00028$, $P = 0$. Class Actinopterygii $y = 0.00297x + 27.58398$, $R^2 = 0.00439$, $P = 0$.
(b) Comparison of trends in body weight of scavengers and nonscavenging fishes. Scavengers $y = 10^{(0.000586x + 0.622)}$ ($p < 0.001$, $r^2 = 0.75$), nonscavengers $y = 10^{(-0.0000564x + 2.033)}$. (After Collins et al., 2005).
(c) Change in average shape of fishes captured in demersal trawls in the NE Atlantic. High values of form factor ($\alpha_{3.0}$) indicate a short fat shape. $y = e^{-0.683\ln x - 0.29}$. (After Neat and Campbell, 2013). Y is fish length, weight or shape. X is depth (m).

obviously short-bodied species such as orange roughy (*Hoplostethus atlanticus*), Alfonsinos (*Beryx* sp.) and the Ceratioid anglerfishes.

3.5 Buoyancy

The amount of energy required for swimming can be greatly reduced if a fish is neutrally buoyant. Some demersal fishes such as the flatfishes (Pleuronectiformes) are usually negatively buoyant so as to be able to anchor to the seafloor and avoid being swept away by bottom currents. Priede and Holliday (1980) showed that for the plaice (*Pleuronectes plattessa*) creating the lift necessary to overcome such negative buoyancy accounts for 16 per cent of the energy cost of swimming at maximum speed. Some deep-sea species such as the tripod fish *Bathypterois* (Family Ipnopidae) are likely to be slightly negatively buoyant to be able to hold station in their characteristic feeding posture. Conversely, for fish that swim at high speed such as Scombroids (tunas and mackerels), the penalty of increased drag forces resulting from the added volume occupied by the buoyancy organ makes neutral buoyancy energetically disadvantageous (Alexander, 1990). Such fishes can readily generate lift hydrodynamically from the action of the fins and body shape during swimming. For fish with a body mass of 1 kg, Alexander (1990) calculated that hydrodynamic lift is more economical than buoyancy at speeds greater than 0.75 m.s^{-1} for fish with a swim bladder and 0.45 m.s^{-1} if lipids are used for buoyancy. Because deep-sea benthopelagic fishes generally swim at speeds of no more than 0.15 m.s^{-1} (Bagley et al. 1994) neutral buoyancy is clearly advantageous.

Neutral buoyancy can be achieved either through a general reduction in the density of body components or the presence of specialised buoyancy organs.

3.5.1 The Swim Bladder and Gas Gland

The swim bladder is a most conspicuous organ located in the body cavity beneath the kidney near the centre of gravity of the body of teleost fishes (Fig. 3.4a). It originates during development as an outgrowth from the foregut. In physostome species, belonging to the lower orders of fishes, the swim bladder, if present, retains its connection to the gut through a pneumatic duct. In more advanced teleosts known as physoclists, there is no pneumatic duct. Shallow-water physostome species can inflate the swim bladder by gulping air at the surface. This is clearly not possible in deep-sea species, and whilst air can be expelled through the pneumatic duct in physostomes, inflation of the swim bladder in deep-sea fishes must occur through the bloodstream.

Gas is transferred from the arterial circulation into the swim bladder through the action of the gas gland. During gas secretion, blood flows from the arterial system via the *rete mirabile* to the gas gland, which lies in the wall of the swim bladder. Blood from the gas gland returns through the *rete mirabile* into the retial vein (Fig.3.4b). Within the *rete mirabile*, the artery branches into numerous capillaries, which run parallel to corresponding venous capillaries in which blood flows back counter to the arterial flow. In the *rete*, there are up to 100,000 capillaries, closely juxtaposed to one another with no more than 1–2 μm diffusion distance between the arterial and venous capillaries. In many deep-sea species the *rete* is a distinctive sausage-shaped organ in which the capillaries join together at the swim bladder end to form one or more large arteries and veins connecting to the gas gland as in *Astronesthes niger* shown in Figure 3.4 known as a bipolar *rete mirabile*. In other species the *rete mirabile* is unipolar, and capillaries pass directly to the gas gland tissue. Often there may be several parallel *rete* organs. The gas gland comprises a mass of glandular cells that vary in size from small cells, 10–25 μm diameter that surround the blood capillaries (e.g. in *Myctophum, Diaphus, Opisthoproctus*) to giant cells, 50–100 μm diameter in which the capillaries pass through the cells (e.g. *Vinciguerra, Sternopteryx*). During gas secretion the gas gland swells to a large size, and both the venous and arterial capillaries of the *rete mirabile* are expanded to maximum diameter as blood flows to and from the gas gland (Figs. 3.5a, b).

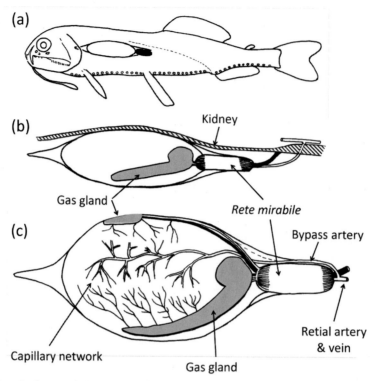

Figure 3.4 Swim bladder of a deep-sea fish dragonfish (*Astronestes niger*). (After Marshall, 1960). Veins are shown black and arteries white.

(a) Location of the swim bladder in the body cavity with the *rete mirabile* to the posterior.

(b) Lateral view from the left.

(c) Ventral view. The gas gland is split into two parts in this species, a major lobe to the right and minor lobe with its own artery and vein to the left. The capillary network is the area where gas is resorbed.

During gas reabsorption, arterial blood flow bypasses the *rete mirabile* and flows directly to a resorption capillary network in the wall of the swim bladder. Return blood flows back through the venous capillaries of the *rete mirabile* or through other venous routes. Devoid of arterial blood supply, the arterial capillaries within the rete collapse, and the gas gland shrinks in volume (Figs. 3.5.c, d). The resorptive tissue correspondingly expands in volume. Gas is removed from the swim bladder by diffusion into the resorption capillaries, and ultimately excess gas in the bloodstream escapes via the gills into the surrounding sea water. Gas removal from the swim bladder can also be regulated by the presence of a gas-impermeable membrane extension of the swim bladder wall known as the oval. The membrane covers the resorptive tissue preventing gas escape and has an elliptical aperture that is opened or closed by the action of smooth muscles. This is found in many gadiformes.

Gas deposition into the swim bladder occurs by passive diffusion down a concentration gradient from a very high partial pressure generated in the gas gland. The process is driven by anaerobic metabolism of glucose in the gas gland cells that releases lactic acid and CO_2 into the gas gland capillaries (Pelster, 1997). This acidifies the blood and has two distinct effects on the haemoglobin in the red blood cells that cause release of oxygen into solution. The lowering of pH shifts the oxygen dissociation curve to the right

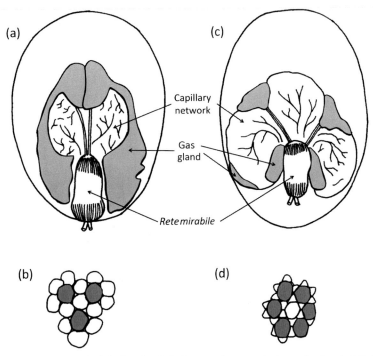

Figure 3.5 Appearance of the gas gland of the hatchet fish *Argyropelecus aculeatus* and *rete mirabile* (a) and (b) during gas secretion, (c) and (d) during gas resorption.
(a) The gas gland expanded and the capillary network relaxed.
(b) Cross section of rete capillaries, both veins and arteries expanded.
(c) The resorbent capillary network expanded and gas glands contracted.
(d) Cross section of rete capillaries. Arteries collapsed but veins full. (After Marshall, 1960).

(Fig. 3.6), the Bohr effect causing oxygen to be released into solution. However in teleost fishes the Root effect is much more important, releasing typically 40 per cent and up to 80 per cent (in *Gadus*) of the oxygen from the haemoglobin. Berenbrink et al. (2005) argue that evolution of the Root effect in teleost lineages about 130–140 mya was a prerequisite for invasion of deep-sea habitats by fishes with swim bladders. The presence of a Root effect is correlated with presence of the *rete mirabile*.

The *rete mirabile* amplifies the high oxygen partial pressure generated in the gas gland by virtue of the countercurrent exchanger principle. Venous blood flowing out from the gas gland is enriched with oxygen and lactate. Within the *rete mirabile* the oxygen and lactate diffuse across the capillary walls into the incoming arterial blood flow (Fig. 3.7). This

blood then receives more oxygen in the gas gland, which is in turn added to incoming blood, gradually building up the concentration in the gas gland until the partial pressure exceeds the hydrostatic pressure at the prevailing depth and oxygen bubbles out of solution into the swim bladder. The gas gland may secrete surfactants to aid bubble formation. The pressure generated is dependent on the length of the parallel juxtaposed capillaries within the *rete mirabile*. The length of these capillaries increases with mean depth of occurrence of the species (Fig. 3.8a) Marshall (1972).

CO_2 is also deposited into the swim bladder alongside oxygen; however, because it is highly soluble in blood plasma, it preferentially diffuses out resulting in a concentration of 1–2 per cent under steady state conditions. Lactate in the gas gland

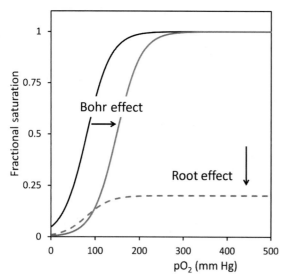

Figure 3.6 The effect of blood pH on oxygen saturation of fish haemoglobin. Theoretical oxygen dissociation curves calculated using the logistic equation; $S = 1/(1+ e^{-P})$. pO_2 mm Hg is the partial pressure of oxygen in the blood. 160 mm Hg is approximately the normal value for air. Fractional saturation value of 1 indicates the haemoglobin is fully saturated with oxygen. Decrease in pH shifts the dissociation curve to the right (Bohr effect) and decreases the maximum saturation value (Root effect). The two effects summed together enable secretion of oxygen into the swim bladder.

releases small amounts of nitrogen by a salting-out effect and inert gases such as argon, neon and helium also diffuse into the swim bladder. Because these are relatively insoluble, they tend to accumulate in the swim bladder.

The wall of the swim bladder is composed of outer fibrous layers, a layer of smooth muscle and an inner epithelial layer. Within the fibrous layers there are numerous needle-like guanine crystals that act as a barrier to gas diffusion. The deeper the fish lives, the higher the concentration of guanine in the swim-bladder wall (Fig. 3.8b).

Thus by virtue of increase in length of capillaries in the *rete mirabile*, the Root effect and high guanine content in the swim bladder wall, deep-sea fishes are able to maintain near-neutral buoyancy even at abyssal depths. However, at these great depths the density of compressed gas, commonly

safely ignored at shallow depths, becomes significant. Marshall (1960) calculated the density of gas in the swim bladder using the ideal gas equation

$$PV = nRT$$

where n = the number of moles of gas, R = the universal gas constant (8.314 J.mol^{-1}.K$°$), T = the temperature (K$°$) and P = ambient hydrostatic pressure at depth. For oxygen the theoretical density of compressed gas becomes equal to that of squalene (0.860), the widely-occurring buoyancy lipid, at a depth 5870 m and equal to sea water at 6990 m providing no buoyancy at greater depths (Fig. 3.9a). However, Alexander (1966) pointed out that such estimates are wrong because at high pressures the behaviour of gases deviates considerably from the ideal gas equation. Van der Waals equation takes into account repulsive forces between gas molecules that resist compression:

$$\left(P + \frac{n^2 a}{V^2}\right)(V - nb) = nRT$$

where a and b are Van der Waals constants. Solving this equation with the aid of Calistry (2017), it can be seen (Fig. 3.9a) that oxygen provides significant buoyancy even at full ocean depth where it reaches a density of around 0.7. The deepest-occurring fish with gas-filled swim bladders are probably species of the genus *Bassozetus* that have been observed down to 6898 m depth (Linley et al., 2016) and grenadiers *Coryphaenoides yaquinae* at 6945 m (Jamieson et al., 2012; Fig. 3.9a).

The partial pressure of oxygen dissolved in sea water in equilibrium with the atmosphere (i.e. an air-saturated solution) is equal to the proportion of oxygen in the atmosphere (20 per cent) times the atmospheric pressure (i.e. 0.2 bar, 20 kPa or 152 mmHg). This does not change with depth so despite very high hydrostatic pressures (61.6 MPa at 6000 m depth) the partial pressure of oxygen remains at 20 kPa. In fact, as we have seen in Chapter 1 deep-sea water is often depleted in oxygen so that the partial pressure will generally be below 20 kPa. To

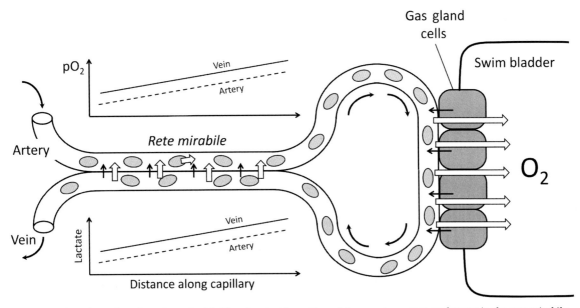

Figure 3.7 Secretion of gas into the swim bladder showing the action of the countercurrent exchanger in the *rete mirabile*. Only one arterial-venous capillary pair is shown for clarity. Curved arrows – direction of blood flow. Straight arrows – lactate diffusion. Open arrows – oxygen diffusion. Ellipses – red blood cells. The graphs show the relative concentrations of oxygen and lactate along the lengths of the capillaries.

fill the swim bladder, oxygen must be pumped up the pressure gradient from the maximum of 20 kPa available at the gills to a gas pressure inside the swim bladder equal to that of the surrounding water (61.6 MPa in this case). Ludwig and MacDonald (2005), however, point out that there is a widely forgotten effect of hydrostatic pressure on dissolved oxygen partial pressure. High hydrostatic pressures tend to squeeze dissolved oxygen out of the water, effectively increasing the partial pressure by 14 per cent per 10 MPa. At the maximum abyssal depth of 6000 m (61.6 MPa) this implies an increase in partial pressure of oxygen from 20 kPa to 37 kPa. For oxygen diffusion in and out of body fluids (water solutions) this is not important because the effect is equal in sea water and blood plasma. However, when moving oxygen from the dissolved state into the gas phase within the swim bladder this does significantly facilitate gas secretion (Enns et al., 1965) reducing the energy required to fill the swim bladder by 5 per cent at 3000 m depth and 8.5 per cent at 6000 m.

The energy required to inflate the swim bladder can be calculated from the gas equation:

$$E = n.R.T.\ln(P_2/P_1)$$

where n = the number of moles of gas, R = the universal gas constant (8.314 J.mol^{-1}.K°), T the temperature (K°), ln = natural logarithm, P_2 = ambient hydrostatic pressure at depth, P_1 = partial pressure of oxygen in the water (here assumed to be 0.2 bar, but adjusted for hydrostatic effect at depth). The amount of energy required to inflate the swim bladder increases with depth (Fig. 3.9b) so that at the typical abyssal depth of 4000 m the energy required to generate 1 kg of buoyancy is 695 kJ. The energy content of sufficient squalene oil to produce the same amount of hydrostatic lift is 264000 kJ. The swim bladder is therefore remarkably energy efficient requiring just 0.26 per cent of energy compared with lipid buoyancy. Kanwisher and Ebeling (1957) suggest that the gas secretion process may be approximately 25 per cent efficient and continuous pumping may be necessary to compensate for loss of gas so the

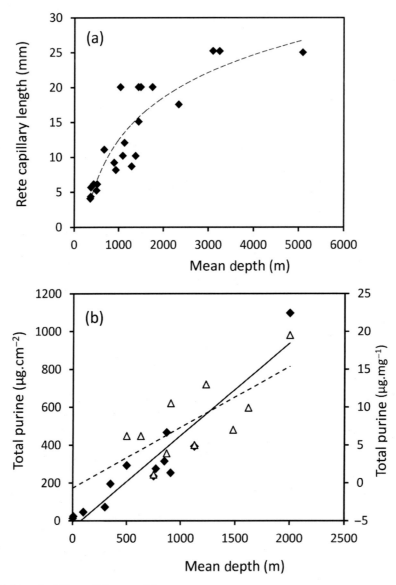

Figure 3.8 Swim bladder adaptations of deep-sea fishes.
(a) The lengths of the rete capillaries in relation to mean depth of species occurrence. Data from Marshall (1972).
(b) The quantity of guanine crystals in the swim bladder wall in relation to mean depth of species occurrence. Guanine is assayed as purine content cm^{-2} (diamond symbols and solid line) and mg^{-1} (triangles and dashed line) of swim bladder wall. Data from Ross and Gordon (1978).

difference in energy cost may not be as great as suggested by these theoretical calculations.

If a fish finds itself in hypoxic conditions the energy required to fill the swim bladder is increased, e.g. at 1000 m depth, typical of oxygen minimum zones, oxygen at 25 per cent saturation increases the energy required by 23 per cent and at 10 per cent saturation the energy required increases by 38 per cent.

The deepest-occurring fish with gas-filled swim bladders are probably species of the genus

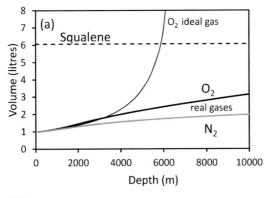

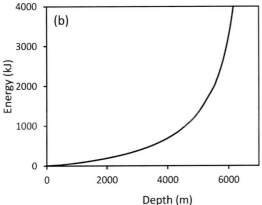

Figure 3.9 Volume of gas and energy required to generate 1 kg of buoyancy in relation to depth.

(a) The volumes of oxygen and nitrogen required to produce 1 kg of buoyancy. O_2 and N_2 limit – the maximum depths at which the gas provides buoyancy. The horizontal line indicates the volume of squalene oil required to produce 1 kg buoyancy assuming constant density of 0.860. Note the discrepancy in the volume of O_2 required calculated using the ideal gas equation compared with Van der Waals' equation for real gases.

(b) The energy required to pump enough oxygen into a swim bladder to generate 1 kg of buoyancy.

Bassozetus that have been observed down to 6898 m depth (Linley et al., 2016) and grenadiers *Coryphaenoides yaquinae* at 6945 m (Jamieson et al., 2012). This is very close to the depth at which oxygen theoretically provides no buoyancy. Even at these high pressures there is an energetic advantage compared with lipid buoyancy. It would be interesting to find out if these extremely deep-living

fishes with swim bladders accumulate nitrogen preferentially as a buoyancy gas because deep-sea fishes generally appear to have relatively high oxygen concentrations (Kanwisher and Ebeling, 1957).

In his survey of swim bladders in deep-sea fishes Marshall (1960) shows that in the mesopelagic zone most species of mytophids (lanternfishes) and Sternoptychidae (hatchetfishes) have swim bladders, but in some species it is reduced or filled with lipid. Amongst demersal species Macrourids, Brotulids and Halosaurs, although fully functional swim bladders are found even at the greatest abyssal depths, the swim bladder is nevertheless absent in many species. In bathypelagic species such as Stomatoids, Ceratioid anglerfishes and Gonostomatidae (Bristlemouths) the gas-filled swim bladder is invariably absent. For the deeper-living vertically migrating mesopelagic fishes there is insufficient time within the 24 h cycle to secrete gas to maintain buoyancy during descent and then to reabsorb during ascent the following evening. For example (Kanwisher and Ebeling, 1957) calculated that it would take 33 h for a 10 g fish to fill its swim bladder after descending to 400 m from the surface, and this would require one-third of the fish's metabolic capacity. In many Stomatiform (e.g. *Cyclothone* spp., *Gonostoma elongatum*, *Stomias* spp.) and myctophid species (e.g. *Stenobrachius leucopsarus* and *Diaphus theta*), the swim bladder shrinks during development and becomes progressively filled with fat.

Bone (1973) groups myctophids into four functional groups according to their buoyancy characteristics:

i) species with low lipid content (< 5 per cent), gas-filled swim bladders and large pectoral fins, which are neutrally buoyant at the top of their vertical range but need to generate lift to overcome negative buoyancy at deeper levels (*Myctophum punctatum, Hygophum reinhardtii, Protomyctophum thompsoni, P. crokeri, Symbolophorus evermanni, S. californiensis*)

ii) species with higher lipid levels (*ca.* 8 per cent) and gas-filled swim bladders, which are probably neutrally buoyant over part of their vertical range (*Diaphus rafinesquii, Ceratoscopelus maderensis*)

iii) species with high lipid levels (14–22 per cent), small pectoral fins and no gas in the swim bladder that are probably neutrally buoyant over the entire depth range (*Diaphus theta, Stenobrachius leucopsarus, S. nannochir, Triphoturus mexicanus, Nannobrachium ritteri*)

iv) species with low lipid levels with or without a swim bladder that are always negatively buoyant (*Nannobrachium regale, Notoscopelus kroyeri, Tarletonbeania crenularis*).

It seems that those species that make the greatest vertical migrations (groups iii) and iv) have abandoned the use of a swim bladder for buoyancy.

3.5.2 Lipid Buoyancy

Apart from gas-filled chambers, lipids represent the lowest density organic compounds that can be used by fishes for buoyancy. Amongst the lipids accumulated in fishes are triacylglycerol (specific gravity 0.93), alkyl diacylglycerol (0.91), wax esters (0.86), aqualene (0.86) and pristine (0.78) (Pelster, 1997). As the density is generally between 0.8 and 1.0, large volumes are therefore necessary to generate sufficient upthrust by displacement. Figure 3.9 shows that over 6 litres of squalene are necessary to produce 1 kg of buoyancy. In some sharks the liver in which squalene is accumulated may account for more than 25 per cent of the body weight. Generally sharks tend to be slightly negatively buoyant (Bone and Roberts, 1969) but Nakamura et al. (2015) suggest that some deep-sea Squalomorph sharks, *Hexanchus griseus* and *Echinorhinus cookei* are positively buoyant, enabling them to glide upward to surprise pelagic prey from below or to aid upward migration at dusk after spending the day in cold deep waters.

Lipids have the advantage over gas bladder buoyancy in that the change in volume with pressure (depth) is almost negligible compared with water, so no adjustment is necessary during ascent and descent. However density may change with temperature by as much as 8 per cent, and lipids can melt or solidify,

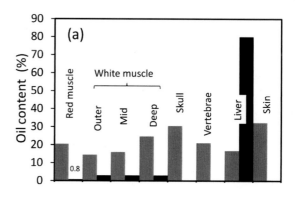

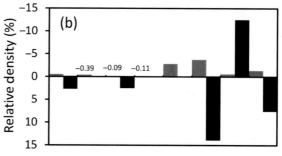

Figure 3.10 Comparison of different tissues of a teleost, *Ruvettus pretiosus* (oilfish), (grey histogram bars) and a shark, *Cetorhinus maximus* (basking shark), (black histogram bars) that both use oil for buoyancy. (a) Oil content, % weight. (b) Relative densities, negative values indicate the tissues are buoyant (Data from Bone, 1972).

influencing buoyancy equilibrium (Phleger and Grigor, 1990). Lipids represent an energy store that can be utilised either for metabolism or be mobilised into maturation of eggs during reproduction. However squalene, an intermediate in the synthesis pathway to cholesterol, is relatively inert compared to other lipids and tends to be accumulated by chondrichthyes in the liver purely for buoyancy. Figure 3.10 shows the density of different tissues of the basking shark (*Cetorhinus maximus*), a species that is close to neutrally buoyant. It can be seen that muscles, skin and the cartilaginous skeleton are all negatively buoyant, and the fish achieves neutral buoyancy by virtue of the 80 per cent oil content of the liver. By contrast *Ruvettus pretiosus* (Oilfish), a teleost with no swim bladder, achieves neutral buoyancy by the storage of oils (wax esters) in all its tissues. Even the bones have a high enough oil content to make them

Table 3.1 **Comparison of the buoyancy balance sheet for a shallow-water fish with a swim bladder** (*Ctenolabrus rupestris*) **and a bathypelagic fish with no swim bladder** (*Gonostoma elongatum*), **100 g** **wet body weight in air. Negative weight in seawater indicates the tissue is buoyant. (After Denton &** **Marshall, 1958).**

	Ctenolabrus rupestris		*Gonostoma elongatum*	
Depth range	1–50 m		785–4740 m	
	Weight (g)		Weight (g)	
Component	In air	In seawater	In air	In seawater
Fat	0.5	−0.1	3.7	−0.5
Protein	16.6	+3.8	5.0	+1.1
Body fluids	73.3.	−0.9	87.6	−1.2
Skeleton, etc.	9.2	+2.6	3.2	+1.1
Swim bladder	0.4	−5.4	–	–
Total	100	0	100	+0.5

buoyant (Bone, 1972). *Hoplostethus atlanticus* (orange roughy) also has wax esters in the bones of the skull and vertebrae (20 per cent by weight) as well as in the viscera, swim bladder, skin and flesh. Orange roughy are neutrally buoyant at depths of around 1000 m, but in warm surface waters the lipid melts and they become positively buoyant (Phleger and Grigor, 1990). Both in *Ruvettus* and *Hoplostethus* the distribution of buoyancy tends to give a 45° head up posture when not moving, which may be advantageous for stalking prey in the water column above.

3.5.3 Weight Reduction

For fishes in the upper layers of the ocean the metabolic cost of maintaining buoyancy or creating lift by continuous swimming is easily met by the rich food supply available. In the bathypelagic zone with very sparse feeding opportunities this is not possible, and it has been found that most bathypelagic fishes without a swim bladder have a much reduced skeleton and high water content. Denton and Marshall (1958) compared the composition and buoyancy arrangements of the NE Atlantic inshore fish

Ctenolabrus rupestris (Goldsinny-wrasse, 50 m_{fbd}) with the bathypelagic Stomiiform species *Gonostoma elongatum* (4750 m_{fbd}) (Table 3.1). In the wrasse the weight of a robust skeleton and muscle system is offset by the buoyancy from the swim bladder. In the bathypelagic species the skeleton and protein content is greatly reduced, with the watery tissues providing buoyancy. The scales and otoliths are also reduced to save weight but the jaws retain full function. Although near-neutral buoyancy is achieved, swimming performance is much impaired.

In addition to a general reduction in protein and skeletal components, some species have large deposits of gelatinous material composed of hygroscopic polysaccharide material known as glycosaminoglycans. In nonmigrating bathypelagic species with no swim bladder, these deposits are found between the myotomes and around the axial skeleton of the Stomiidae, e.g. *Tactostoma macropus* and *Chauliodus macouni*. The Bathylagidae, *Bathylagus pacificus* and *B. milleri* have an additional subcutaneous layer. The gelatinous material has very high water (96 per cent), low protein (3 per cent) and low ion contents, so it is positively buoyant whilst providing some structural rigidity

(Yancey et al., 1989). Snailfishes (family Liparidae) do not have a swim bladder, and yet the Antarctic species (*Paraliparis devriesi*) has been found to be neutrally buoyant (Eastman et al., 1994). The skeleton is very light with minimal bone density and extensive cartilage. The main morphological adaptation is an extensive subdermal extracellular matrix (SECM) occupying at least one-third of the body volume filled with low-density glycosaminoglycans. Liparids are observed in hadal trenches beyond the depths at which swim bladders can function (Linley et al., 2016). It is likely that these species use similar means to achieve neutral buoyancy.

3.6 Muscle Adaptations

Fishes generally have two kinds of muscle fibres, red and white. The red muscle functions aerobically, has a rich blood supply, myoglobin pigment that gives the dark colour, many large mitochondria, small-diameter fibres that contract relatively slowly and is used for endurance swimming. The white muscle system functions anaerobically, has a poor blood supply, little myglobin (hence the white colour), few mitochondria, large-diameter twitch fibres that contract fast and is used for short-duration sprinting (Bone and Moore, 2008). In the myotomes of the trunk the red muscle typically forms a strip on either side under the skin, and the white muscle makes up most of the bulk of the fish cross-section (Fig. 3.11).

Epipelagic fishes such as the Scombridae (mackerels) and Clupeidae (herrings) have a well-developed red muscle system, whereas sedentary bottom-living fish may have little or no red muscle in the body cross section. Analysis of available data shows that demersal fishes generally have less red muscle the deeper they live (Fig. 3.11) and that benthic fishes have less red muscle than benthopelagic species (Drazen et al., 2013). This implies that the cruising speed and endurance-swimming capabilities of deeper-living fishes are less than in shallow-water species.

The maximum speed of fish is determined by the twitch speed of the white muscle. One to-and-fro

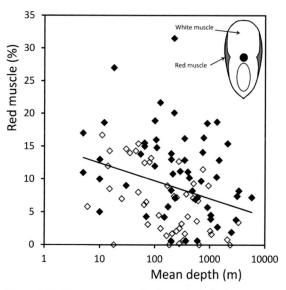

Figure 3.11 The percentage of red muscle in body cross sections of demersal fishes living at different depths. (Data from Drazen et al., 2013). Black symbols – benthopelagic species, open symbols – benthic species. The line is fitted to the pooled data set: $y = -1.179 \ln(x) + 15.166$ $R^2 = 0.0942$ ($P = 0.002$). The inset diagram shows a fish cross section and area of red muscle measured.

beat of the tail, two times the twitch duration, drives the fish forward 0.6 of a body length (Wardle, 1975). Wardle et al. (1987) measured the twitch durations in muscle samples taken from the deep-sea eel (*Histiobranchus bathybius*) captured at 4000 m depth. At atmospheric pressure the muscles had a very weak response but repressurising to 200 to 400 atmospheres restored function and twitch duration to 200 ms. This gives a predicted maximum tail beat frequency of 2.5 Hz and maximum sprint speed 1.05 m.s^{-1} or 1.5 body lengths.s^{-1}; very much slower than the 8 body lengths.s^{-1} that could be expected in shallow-water species of the same size. The European eel (*Anguilla anguilla*) in the same experiment at atmospheric pressure had a twitch duration of 70 ms giving a maximum tail beat frequency of 7.1 Hz, and a speed of 2.6 m.s^{-1} or 4.26 body lengths.s^{-1}. In both eels, *Histiobranchus bathybius* and *Anguilla anguilla*, increase in pressure increased the contraction time of the muscles.

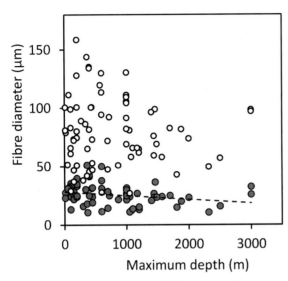

Figure 3.12 Diameters of Red and White muscle fibres. Each point represents a species mean diameter (data from Greer-Walker and Pull, 1975). Species maximum depths are from FishBase (Froese and Pauly, 2016).

Bailey et al. (2003) measured the muscle contraction time of Blue hake (*Antimora rostrata*), in situ at 2500 m depth by exposing fish attracted to bait within view of a video camera to an electrical stimulus. The mean contraction time was 170 ms, which for a 50 cm fish predicts a theoretical maximum sprint speed of 0.88 m.s^{-1} or 1.76 body lengths.s^{-1}. Actual average maximum speed observed was 0.7 m.s^{-1}, and white muscle power was estimated as only 17.0 W.kg^{-1} compared with values of over 100 W.kg^{-1} achieved in shallow-water species.

The observed contraction times of the deep-sea fish muscles are much slower than the 55–60 ms predicted by Wardle (1975) for shallow-water fish living at 0°C. It seems that deep-sea fish muscles have intrinsically lower power output, probably reflecting their high water content and low enzyme activity levels. However, differences between shallow and deep benthic fishes may not be so great because many have sedentary lifestyles even at shallow depths. Bailey et al. (2003) point out that the contraction times for *Antimora rostrata* are similar to values recorded for shallow-water Antarctic ice fish of the genus *Notothenia*.

Zero values for red muscle percentage in Figure 3.11 need to be interpreted with care because fishes with no red muscle in the trunk region may have well-developed red muscles in the pelvic or caudal fins, which are adapted to swimming with a sculling motion rather than oscillation of the entire body (Greer-Walker and Pull, 1975). White fibres vary greatly in diameter, the largest and presumably the fastest being found in shallow-dwelling species (Fig. 3.12). Red fibres are smaller and less variable because of the need to ensure efficient diffusion of oxygen from the blood supply. Fibres tend to be smaller in deeper fish.

3.7 Gill Area

Ultimately the capacity for sustained metabolic activity is defined by the oxygen uptake capacity of the gills and the ability of the circulatory system to deliver that oxygen to the tissues. In addition to respiratory gas exchange, the gills are also important in salt balance and excretion of the nitrogenous end products of metabolism. The gills on either side of the head are typically made up of four gill arches, each of which has two rows of filaments. There are hundreds of filaments in each row, and on each filament there are 10–30 lamellae per mm, which provide the large surface area over which gas exchange takes place. In teleost fishes the osmotic concentration of blood plasma is approximately 250 mOsm compared with 1000 mOsm of sea water. This means that there is continuous loss of water across the gills by osmosis, which has to be compensated for by drinking. The excess salt taken up in the imbibed water must then be excreted through the gill chloride cells, a process that is energetically costly. Deep-sea fishes live in an energy-limited environment and therefore cannot afford to maintain gills that are any larger than absolutely necessary for normal function.

Hughes and Iwai (1978) surveyed the gill morphology of six different species of deep-sea fishes captured at depths of 1255–1520 m in the vicinity of seamounts in the mid-Pacific. Owing to the problems of retrieving specimens in good

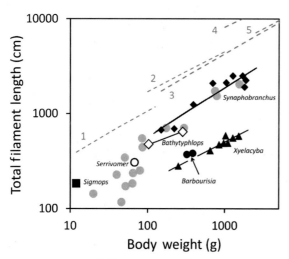

Figure 3.13 Total gill filament lengths (index of gill area) of deep-sea fishes (black) compared with surface and shallow water species (grey). *Synaphobranchus affinis* (Grey cutthroat, 2400 m$_{fbd}$) Anguilliformes, *Serrivomer sector* (Sawtooth eel, 3243 m$_{fbd}$) Anguilliformes, *Sigmops elongatus* (Elongated bristlemouth fish, 4750 m$_{fbd}$) Stomiatiformes, *Bathytyphlops marionae* (Marion's spiderfish, 1920 m$_{fbd}$) Aulopiformes, *Xyelacyba myersi* (Gargoyle cusk, 2500 m$_{fbd}$) Ophidiiformes, *Barbourisia rufa* (Velvet whalefish, 2000 m$_{fbd}$) Cetomimiformes (data from Hughes and Iwai, 1978). Round grey symbols – Shallow marine species from Hughes (1966). Lines, high speed epi-pelagic species: 1–*Trachurus* (Atlantic horse mackerel) (Hughes 1966), 2 – *Scomber japonicus* (Pacific chub mackerel), 3 – *Sarda chiliensis* (Eastern Pacific Bonito), 4 – *Katsuwonus pelamis* (Skipjack tuna), 5 – *Acanthocybium solandri* (Wahoo), data from Wegner et al. (2010).

condition it was not possible to measure the area of the secondary lamellae, but the number of gill filaments was counted, and their total length was estimated as an index of gill area (Fig. 3.13). The size of gills in fishes varies considerably according to lifestyle, e.g. from 8200 cm in the Skipjack tuna to 503 cm in *Xyelacyba myersi*, the deep-sea Gargoyle cusk eel (at one kilogram body weight). Some deep-sea fishes such as *Synaphobranchus affinis*, (1866 cm in a 1 kg fish) have gill sizes similar to shallow-water benthic fishes. Bailey et al. (2005) found that the metabolic rate and muscle metabolic enzyme activities of *Synaphobranchus*

kaupii were an order of magnitude higher than those of other deep-sea fishes. This genus appears to be characterised by high activity levels and represents the upper limits of metabolic rates found in deep-sea species. Generally deep-sea fishes have small gill sizes reflecting low aerobic metabolic capacities. Friedman et al. (2012) found increased gill areas in flatfishes and marcrourids living in the oxygen-minimum zone at depths of 500–1000 m in the NE Pacific, presumably to enable sustained oxygen uptake at low partial pressures.

3.8 Heart and Circulatory System

Based on measurements from 21 specimens, Greer-Walker et al. (1985) determined the heart weight of the round-nose grenadier (*Coryphaenoides rupestris*) to be 0.059 per cent of body weight compared with 0.44 per cent in the skipjack tuna (*Katsuwonus pelamis*; Hughes, 1979), 0.2 per cent in rainbow trout (*Oncorhynchus mykiss*; Denton and Yousef, 1976) and 0.13 per cent in the cod (*Gadus morhua*), the small size clearly reflecting the relatively low metabolic rate of the deep-sea species. Furthermore all 29 species of demersal macrourid, captured at depths of 500–4000 m, examined by Greer-Walker et al. (1985) lacked the outer compact layer of ventricular mycocardium invested with coronary arteries that is found in fast-swimming fishes such as *Salmo*, *Scomber*, *Xiphias* and *Thunnus* (Tota, 1978). Instead the macrourid ventricular is spongy with the myocardial fibres bathed in venous blood with no oxygenated arterial supply, a condition typical of relatively sluggish fish species. The spongy structure provides a large surface area for the myocardium to take up oxygen at low partial pressures.

In a study of pelagic species, Blaxter et al. (1971) found there is a clear inverse relationship between heart weight and body water content (Fig. 3.14) and recognised three main categories:

i) Deep-sea species with a reduced skeleton, high water content and small heart reflecting low metabolic capacity.

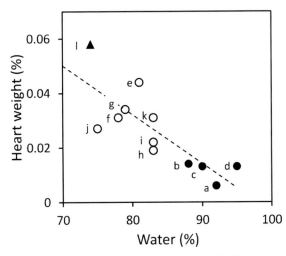

Figure 3.14 Heart weights in relation to total body water content. Open symbols – species with swim bladders. Closed symbols – species with no swim bladder. Black circles – Category 1 species: a. *Xenodermichthys copei* (Bluntsnout smooth-head, 2650 m_{fbd}) Alepocephalidae, b. *Chauliodus sloani* (Sloane's viperfish, 4700 m_{fbd}) Stomiidae, c. *Sigmops elongatum* (Elongated bristlemouth, 4740 m_{fbd}) Gonostomatidae, d. *Melanocoetus johnsoni* (Humpback anglerfish, 4500 m_{fbd}) Melanocetidae. Open circles – Category 2 species: e. *Chiasmodon niger* (Black swallower, 2745 m_{fbd}) Chiasmodintidae, f. *Gonostoma denudatum* (700 m_{fbd}) Gonostomatidae, g. *Margrethia obtusirostra* (Bighead portholefish, 1739 m_{fbd}), Gonostomatidae, h. *Poromitra capito* (1000 m_{fbd}) Melamphaidae, i. *Diaphus rafinesquii* (White-spotted lantern fish, 2173 m_{fbd}) Myctophidae, j. *Hygophum hygomi* (Bermuda lantern fish, 1485 m_{fbd}) Myctophidae, k. *Argyropelecus aculeatus* (Lovely hatchetfish, 2056 m_{fbd}) Sternoptychidae. Triangle – Category 3 species: l. *Scomber japonicus* (Atlantic chub mackerel). Heart weights are dry weights expressed as percentage of wet body weight.
(Data from Blaxter et al., 1971)

ii) Deep-sea species with a swim bladder, robust skeletons, well-developed muscles and greater metabolic capacities.

iii) Epipelagic species such as *Exocoetus* (flying fish) and *Scomber* (mackerel) that are fast swimmers and have no swim bladder. It is interesting to compare two closely related species in the Gonostomidae, *Gonostoma denudatum* (700 m_{fbd}) with an air-filled swim bladder, 78 per cent water content and heart weight 0.031 per cent, whereas

the deeper-living *Sigmops elongatum* (4740 m_{fbd}) has an oil-filled swim bladder, 90 per cent water content and heart weight 0.013 per cent.

Heart size in deep-sea fishes is therefore reduced reflecting their generally lower metabolic capacities. Blaxter et al. (1971) showed that when the dry weight of fish hearts is expressed as a percentage of dry body weight the trends in Figure 3.14 become insignificant and the heart size is constant at *ca.* 0.15 per cent of body weight, albeit with considerable variation about the mean (range 0.05–0.27 per cent).

The red blood cell content (haematocrit) is reduced in mesopelagic and bathypelagic fishes, with typical values below 10 per cent in category 1 fishes, around 20 per cent in category 2 fishes compared with a mean of 52 per cent in category 3 fishes, the mackerel and flying fish. Haematocrit is clearly inversely correlated with body water content (Blaxter et al., 1971).

Blaxter et al. (1971) also point out the large size of the neural lymph canal in category 1 watery-bodied fishes compared with smaller canals in category 2 fishes with swim bladders. They suggest that the lymph system increases in importance as the heart weight, haematocrit and body dry weight decrease. It is now recognised that there is a secondary circulatory system in fishes with its own network of arteries, capillaries and veins that are in parallel with the primary blood circulatory system (Olsen, 1996). The secondary vessels carry no erythrocytes and were first described in the anglerfish (*Lophius piscatorius*) by Burne (1926), who associated them with the lymphatic system. In the Atlantic cod (*Gadus morhua*) Skov and Steffensen (2003) estimate that the plasma volume in the secondary circulatory system is about 50 per cent of the plasma volume in the primary circulatory system. It seems likely that the secondary circulatory system may be particularly well developed in deep-sea species.

3.9 The Art of Slow Swimming and the Optimum Speed Fallacy

In order to swim, a fish must produce a force sufficient to overcome the drag force, which increases in proportion to the square of velocity:

$$D \propto C_d \cdot v^2 \tag{1}$$

Then the power (force x distance per second) required for swimming is proportional to the cube of velocity.

$$P_s = C_d v^3 \tag{2}$$

The cube law indicates that swimming at high speeds is very costly in terms of energy, but the converse is also true, that low-speed swimming can be very economical, which is more important for deep-sea fishes.

For swimming fish Alexander (1999) shows that the preceding equation can be written as:

$$P_s = 0.5 \, C_d \, \rho^{0.33} m^{0.67} v^3 \tag{3}$$

where P_s - power in watts (W) for swimming, C_d – drag coefficient (based on volume) here we assume a value of 0.02, which is typical for streamlined bodies, m – body mass in kilograms (kg), v – velocity metres per second (m.s^{-1}). We can use this to estimate energy expenditure during swimming in the abyssal grenadier fish (*Coryphaenoides armatus*). The oxygen consumption of this species has been measured in situ by Smith (1978) and Bailey et al. (2002) with values ranging from 2.7–3.7 ml O$_2$.kg^{-1}.h^{-1}. Though the activity levels of these fish could not be controlled in the respirometers, Bailey et al. (2002) made the comment that their fish swam continuously around the respirometer chamber at a speed of approximately 0.1 m.s^{-1}. Here we assume that the standard metabolic rate (SMR) or resting rate is 1.4 ml O$_2$.kg^{-1}.h^{-1}, and that the active metabolic rate (AMR) or maximum aerobic rate is 4.0 ml O$_2$.kg^{-1}.h^{-1} in a 1 kg fish. Using an oxycalorific conversion factor of 13.6 J per mg of oxygen (Priede, 1985), these can be expressed in Watts (W) as SMR = 0.08 W (P_{rm}) and AMR = 0.23 W (Fig.3.15a). The metabolic cost of swimming (P_{sm}) is calculated using equation (3) assuming 10 per cent efficiency in transformation of metabolic energy into useful propulsive work. The metabolic rate of the fish at any given speed is given by:

$$P = P_{rm} + P_{sm}$$

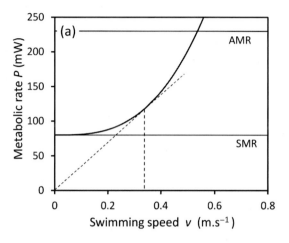

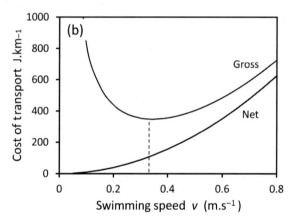

Figure 3.15 Metabolic cost of swimming in the abyssal grenadier (*Coryphaenoides armatus*).

(a) Metabolic rate in relation to swimming speed. Solid curved line – calculated metabolic power (milliwatts). AMR – Active (maximum) metabolic rate. SMR – Standard (resting) metabolic rate. The tangent to the curve (dashed line) indicates optimum or maximum range speed.

(b) Cost of transport per km. Gross – is SMR plus the cost of swimming, note the minimum at 0.34 m.s^{-1}. Net – is the cost of swimming only.

It can be seen that as the speed of the fish increases so the metabolic rate increases until the line intersects the AMR line, the point of maximum sustainable speed at 0.54 m.s^{-1}. The optimum or maximum range speed is derived from the tangent through the origin of the graph, representing the point where gross cost of transport is minimal. The gross cost of transport is calculated as

$$T_g = (P_{rm} + P_{sm})/v$$

The speed of minimum cost of transport calculated in this way is often cited in standard texts (Videler, 1993) as the optimum speed that the fish should use during migrations. In Figure 3.15b it can be seen that swimming at slow speeds is very costly. However, if the net cost of the transport is calculated then swimming at slow speeds appears to be more efficient. The net cost of transport is then:

$$T_n = P_{sm}/v$$

Shown in the lower curve in Figure 3.15b.

Using gross cost of transport, if our 1 kg *C. armatus* swims 1000 km at its optimum speed of 0.34 m.s^{-1} the total energy cost is 348,263 J and the journey takes 34 days. The same fish swimming at 0.1m.s^{-1} would take 116 days and expend 809,772 J, using more double the energy and taking more than three times longer. The assumption is, that having accomplished the journey quickly and 'economically', the optimally swimming fish can then proceed to its next appointment. However, we can compare the two strategies in another way. Over the 116 days the faster-swimming fish has 34 days of swimming plus 82 days of waiting at resting metabolic rate so that its total energy expenditure becomes 348,263 J + (82 days @ 0.08 W) = 912,969 J. Thus during 116 days the optimally swimming fish consumes 13 per cent more energy than the continuously swimming slower fish.

For a neutrally buoyant fish the energy cost of swimming at very slow speeds is almost negligible; small undulations of the body or the fins can set the fish in motion at 0.1 m.s^{-1} with the metabolic rate *ca.* 1 per cent above resting. A slow-swimming nomadic lifestyle is readily supported. A speed of 0.1 m.s^{-1} enables a fish to travel over 3000 km per year. For negatively buoyant fishes this is not true; there is a significant penalty of swimming at slow speed, analogous to hovering flight in birds, and for them the classical optimum speed calculations are generally applicable (Videler, 1993).

These calculations are rather sensitive to the assumptions regarding energetic efficiency of

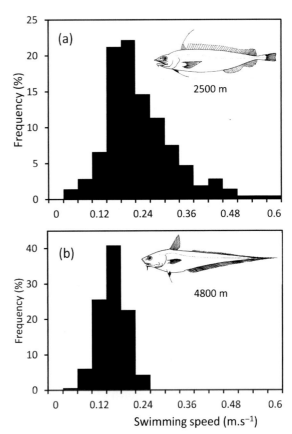

Figure 3.16 Swimming speeds of bathyal and abyssal demersal gadiform fishes tracked using ingestible acoustic transponders. (Data from Collins et al., 1999).
(a) *Antimora rostrata* (blue antimora, blue hake) Moridae. 2500 m depth
(b) *Coryphaenoides armatus* (Abyssal grenadier) Macrouridae. 4800 m depth.

swimming, although the fundamental arguments are not altered. Webb (1971) estimated swimming efficiency of rainbow trout *Oncorhynchus mykiss* to be 15 per cent at 0.6 m.s^{-1} falling to less than 5 per cent at speeds less than 0.2 m.s^{-1}. Van Ginneken (2005) maintained freshwater eels *Anguilla* in a water tunnel swimming continuously for 173 days to mimic their deep-sea transoceanic migration and found that they were four to six times more efficient than rainbow trout. It seems likely that deep-sea species with elongated bodies may be able to swim much more efficiently than expected. Fish swimming efficiency remains poorly understood (Ellerby, 2010).

Tracking studies show that deeper-living demersal fish species tend to swim more slowly (Fig. 3.16; Collins et al., 1999), and at 4000 m depth the Ophidiid, cusk eel *Barathrites iris* swims at 0.11 m.s^{-1} (Henriques, 2004). However, within species, the abyssal grenadier *Coryphaenoides armatus* shows an increase in speed with depth from 0.05 m.s^{-1} at 2500 m to 0.11 m.s^{-1} at 4000 m depth. This is partially explained by increase in size of the fish with depth (Bagley et al., 1994). Pop-up, satellite, archival, transmitting tags (PSAT) are beginning to provide information on movements of deep-sea sharks. Leaf scale gulper sharks (*Centrophorus squamosus*) migrate at an average speed of up to 20 nautical miles per day (0.42 m.s^{-1}) at depths of 200 to 1500 m, including possible transits in mid-water above an abyssal plain with a bottom depth of 4000 m (Rodríguez-Cabello et al., 2016).

3.10 Feeding and Growth

Amongst the diversity of fishes in the deep sea there is a corresponding diversity of feeding strategies. Three main feeding modes found in shallow or surface-living species, herbivory, filter feeding and coral polyp feeding are absent or rare in the deep sea. In the absence of primary production, herbivory is in principle not possible in the deep sea. However, Jeffreys et al. (2010) found that macrourids (*Coryphaenoides mediterraneus*) and ophidiids *Spectrunculus* sp. can ingest artificial parcels of spinach deployed on the seafloor and suggest they may utilise macrophyte detritus such as sea grasses that reaches the seafloor in large quantities in some regions. Also Robison (1984) found that in North Pacific gyre the myctophid, *Ceratoscopelus warmingii* ascends at night and feeds on floating mats of the diatom *Rhizosolenia* spp. and presents evidence of adaptation of the gut for such opportunistic herbivory. There is insufficient particulate organic matter in the deep sea to support filter feeding (Herring, 2002), a commonly used strategy in coastal

waters. However, dense concentrations of zooplankton can occur seasonally at depths of over 600 m, e.g. *Calanus* spp. at high latitudes (Rabindranath et al., 2011), providing feeding opportunities for deep filter-feeding fishes. Coral reefs in shallow tropical seas provide numerous niches for visual feeders such as butterfly fishes, which is clearly not possible in the deep sea although there are associations between fishes and deep cold-water corals (Ross and Quatrini, 2007).

For deep-sea fishes Gartner et al. (1997) proposed ten trophic guilds amongst the demersal species and three guilds in the pelagic species. These have been further revised by Drazen and Sutton (2016) and are presented with slight modification in Table 3.2. The piscivores are divided into two guilds: (1) ambush predators that sit and wait for their prey and (2) active predators. Generally it is the largest individuals within a species that consume fish, switching to this diet when they grow to a suitable size. The deep-sea lizard fish *Bathysaurus* is often seen motionless on the seafloor resting on its pectoral fins ready to snatch passing fishes in its jaws, which have hinged teeth that ratchet the prey into its mouth (Sulak et al., 1985). Other ambush predators are equipped with lures as in the Lophiiformes; *Lophius* spp. with the lure on top of the head or *Thaumatichthys* with a bioluminescent lure inside the mouth. The active predators comprise a range of squaloid sharks, rays and large, more active teleost fishes with swim bladders that pursue their prey continuously, moving to encounter feeding opportunities often up into the water column.

The three most important deep-demersal guilds are (3) Micronektivores, (4) hyperbenthic feeders and (5) epifaunal browsers. The Micronektivores were originally termed 'Macronekton foragers' by Gartner et al. (1997), but here, following the definition of Brodeur et al. (2005), micronekton includes a wide range of free-swimming animals in the size range 2–10 cm. This form of feeding is predominant at bathyal depths where concentrations of pelagic nekton, such as in the Deep Scattering Layer, impinge on the slopes of the continental margins, islands, seamounts, canyons and ridges and may represent

Table 3.2 **Deep-sea fish trophic guilds. (After Gartner et al., 1997 and Drazen and Sutton, 2016).**

Habitat	Guild	Prey categories	Common biological characteristics	Examples
Demersal	1. Piscivores ambush	Fish, large cephalopods and crustaceans	Large, muscular, sedentary, large gape, no gas bladder, often large eyes, large teeth	*Bathysaurus*, Large *Helicolenus*, *Lophius*, *Reinhardtius*.
	2. Piscivores active	Fish, large cephalopods and crustaceans and frequently carrion	As above but with a gas bladder, active swimmers, generally smaller sharp teeth (for teleosts)	*Synaphobranchus*, *Diastrobranchus*, *Antimora rostrata*, Large *Coryphaenoides armatus*, *Amblyraja hyperborea*, Medium to large squaloid sharks
	3. Micronektonivores	Small midwater fishes, mysids, euphausids, decapods, cephalopods, often with some epibenthic invertebrates	Active swimmers, some are schooling, many demersal taxa make forays into the pelagic, moderate to large size, moderate gapes, well-developed gill rakers, large eyes.	Many medium- to large-sized macrourids (*Coryphaenoides*, *Malacocephalus*, *Albatrossia*) *Alepocephalus*, *Bassozetus Hoplostethus*, *Bothrocara brunneum*, *Antigonia*, *Zenopsis Beryx splendens*; Many skates (genus *Bathyraja*) and small to medium squaloid sharks
	4. Hyperbenthic Crustacean feeders	Small hyperbenthic crustaceans and some epibenthic invertebrates such as polychaetes and crustaceans	Small to medium in size, terminal but often subterminal mouths, often small eyes	Small- to medium-sized morids Many small macrourids (e.g. *Coelorinchus*, *Coryphaenoides*, *Hymenocephalus*, *Nezumia*), *Halosauropsis*, *Aldrovandia*, *Bathypterois*, *Cataetyx alleni*, *Antigonia*, deep water notothenioids
	5. Epifaunal browsers	Polychaetes, isopods, crabs, gastropods with some	Small to medium in size, terminal but often subterminal	*Hydrolagus*, *Polyacanthonotus*, *Notacanthus*

Table 3.2 (*cont.*)

Habitat	Guild	Prey categories	Common biological characteristics	Examples
Demersal (continued)		echinoderms and hyperbenthic prey	mouths, often small eyes	
	6. Infaunal predators	Bivalves, polychaetes, tanaids, gastropods, often ingest sediment	Small- to medium-sized fishes, various body forms, some have crushing palatine teeth or beak-like jaws	Zoarcidae, *Glyptocephalus*, *Microstomus*, *Laemonema barbatulum*, several chimaerids
	7. Megafaunal croppers	Sponges, anemones, corals, echinoderms	No distinct unifying characters	*Notocanthus chemnitzi*, several chimaerids, *Barathrites parri*
	8. Necrophages/ scavengers	Dead fish, elasmobranchs, whales, jellies	Many species facultative. Specialists, mostly eel-like with small mouths full of stout teeth or rasping surfaces	Hagfishes (*Myxine* and *Eptatretus*), Sharks, Synaphobranchidae, *Antimora rostrata*, *Coryphaenoides armatus*, *Barathrites iris*
	9. Necrophagivores	Mostly amphipods that are attracted to carrion	Variety of bait attending fishes, large to small size, variety of body forms	*Paraliparis bathybius*, *Notoliparis kermadecensis*, *Pachycara* spp., *Bassozetus* sp.
Demersal/ Pelagic	10. Gelativores	Medusa, ctenophores, salps, pyrosomes, often supplemented by micronekton prey	Often large eyes, *Alepocephalidae*, with triturating crumenal organ at back of throat and Stromateoids with a pharyngeal mill	*Helicolenus percoides*, *Neocyttus rhomboidalis*, *Alepocephalus*, *Conocara*, large *Melamphaidae*, *Bathylagidae*, most *Stromateoidei*
Pelagic	11. Zooplanktivores	Diverse zooplankton – nauplii, copepods, ostracods, euphausiids, etc.	Small meso- and bathypelagic fishes with small to moderate-sized mouths and teeth, often vertically migrating	Most larval teleosts, Myctophidae, Bathylagidae, Sternoptychidae, Phosichthyidae, Many smaller Gonostomatids, *Cyclothone* spp.

Table 3.2 (*cont.*)

Habitat	Guild	Prey categories	Common biological characteristics	Examples
Pelagic (cont.)	12. Micronektonivores	Many focus on fishes but also shrimps, larger mysids, and occasionally cephalopods	Many small to moderate-sized fishes, often with bioluminescent lures, large mouths and fanglike teeth	Stomiidae, Ceratioid anglerfishes; Nemichthyid and Serrivomerid eels; Evermannellidae, Paralepididae
	13. Generalists	Diversity of zooplankton and micronekton	Mostly bathypelagic fishes with enormous gape, reduced musculature and reduced visual predation (small eyes)	Saccopharyngiformes – gulper eels; *Anoplogaster cornuta*; smaller Ceratioidei; Giganturidae; smaller Stomiidae

one of the major pathways for transfer of biomass in the oceans (Mauchline and Gordon, 1991). Several important commercial species specialise in this form of feeding, the macrourid, *Corypyhaenoides rupestris*, orange roughy, *Hoplostethus atlanticus* and Alfonsino, *Beryx splendens,* all equipped with well-developed eyes for spotting prey in the dimly lit twilight depths. They may venture considerable distances above the bottom either horizontally away from the slope or vertically into the mesopelagic and bathypelagic.

In most analyses of deep-sea fish diets, crustacea feature as a major component, and feeding on mobile prey in the benthic boundary layer is a common strategy. As with the piscivores amongst the hyperbenthic feeders there are two alternative strategies, either sit-and-wait or active foraging. The Ipnopidae (deep-sea tripod fishes) stand on the seafloor supported by elongated pelvic and caudal fins while the pectoral fins are extended into the water flow to detect prey items. Gill rakers help extract prey from the water flow through the mouth. Conversely many macrourids, ophidiids notacanths and morids browse continuously above the seafloor picking up a variety of swimming crustacea, including copepods, amphipods and mysids.

The epifaunal browsers overlap considerably with hyperbenthic guild but feed on truly benthic prey, such as polychaetes, isopods and crabs. The spiny eels (Notacanthidae) have very small ventral mouths and pick small benthic crustaceans and polychaetes off the seafloor without ingesting any sediment (Crabtree et al., 1985).

Infaunal predators are relatively rare in the deep sea. Digging holes and extracting prey from muddy sediments is a common strategy in shallow waters, and Marshall (1965) suggested that some deep-sea macrourids may use the armoured projecting snout found in some species to turn over the surface of bottom oozes to access to invertebrate prey. There is no evidence that they use their snouts in this way, and it seems that there is not sufficient infaunal biomass to support fish predation except for some Zoarcids, flatfish, the morids and chimaerids that feed on polychaetes and brittle stars.

Megafauna are by definition a conspicuous feature of the deep-sea ocean floor. However, most have a low energy content and are rendered unpalatable by mechanical and chemical defences. Sessile organisms such as sponges, corals and anemones, as well as the most abundant mobile megafaunal invertebrates, holuthurians, are rarely consumed by fishes.

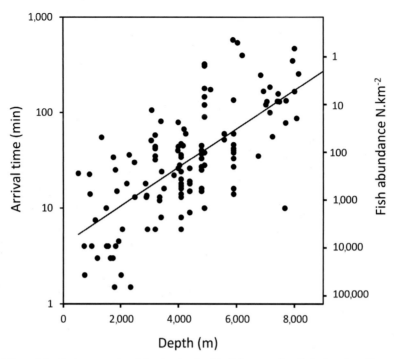

Figure 3.17 Arrival time of the first scavenging fish at a standard bait (one mackerel) deployed on the seafloor at different depths. Data from Jones et al. (2003), Jamieson et al. (2009) and Linley et al. (2016).

Ophiuroids, sea urchins, molluscs and crustacea are eaten by some fishes.

Necrophages, scavengers and necrophagivores can be grouped together as 'bait-attending fishes', which have received much attention because they can be readily observed by baited camera systems. Priede et al. (2010) found that 20 per cent of deep demersal fish species are attracted to bait on the continental slopes of the NE Atlantic, whereas on the Mid-Atlantic Ridge 30 per cent were attracted to bait (King et al., 2006). Carrion falls reaching the seafloor are quickly intercepted with a clear trend of longer delay with increasing depth reflecting the decreasing abundance of fishes (Fig. 3.17). Small food falls (e.g. single mackerel) are consumed and dispersed within a few hours (Priede et al. 1991), whereas larger carcasses such as cetaceans result in a persistent aggregation of fishes over weeks or months until the soft flesh is consumed (Smith et al., 2015). Using ingestible transmitters Jones (1999) showed that individual abyssal grenadiers *Coryphaenoides armatus* remain within 50 to 100 m of a white-sided dolphin (*Lagenorhynchus acutus*) carcase where they had been feeding at 4800 m depth in the NE Atlantic. They depart after 2 h to 30 h presumably when satiated and did not return. The fish aggregation was therefore sustained by a continuous turnover of new fish arriving, with possibly a thousand or more visiting the feeding location before it was exhausted.

The necrophages, which feed directly on the carrion fall, include hagfishes (*Myxine*, *Eptatretus*), sharks (*Hexanchus*, *Centroscymnus*, *Etmopterus*), Synaphobranchid eels, *Antimora rostrata*, *Coryphaenoides armatus*, *C. yaquinae*, *Barathrites iris*, and *Spectrunculus*. Lampitt et al. (1983) first noted that the Liparid *Paraliparis bathybius* (Black seasnail), although attracted to a bait at 4009 m depth, did not feed on the carcase but consumed amphipods feeding on the carcase. This form of indirect feeding on carrion, defined as Necrophagivory by Gartner et al. (1997), is found more widely in the Liparidae and is characteristic of hadal trench endemic species.

Linley et al. (2016) show that at great depths in the Pacific Ocean there is a trend amongst fishes of changing from necrophagy at abyssal depths to necrophagivory at hadal depths greater than 6000 m. Ruxton and Houston (2004) show that although there are sufficient food falls to theoretically support obligate necrophagous fishes in the abyss, most scavengers are also predators and necrophagivores. Drazen et al. (2008) found that 69 per cent of the diet of the abyssal grenadier (*Coryphaenoides armatus*) is carrion, and isotopic signatures showed that carrion was the major food source for both *C. armatus* and *C. yaquinae*, indicating these fish largely bypass the abyssal food web by feeding on material arriving direct from the sea surface.

Since 2000 there has been an increasing awareness that largely transparent gelatinous zooplankton is an important component of deep-sea pelagic biomass, possibly as much as 25 per cent. Robison (2004) coined the term 'jelly web' to describe the role of these organisms that can dominate the second and third trophic levels of mid-water food webs. Because they are usually destroyed by traditional sampling nets and also because their remains are very difficult to identify in fish stomach contents, they had previously been widely ignored. Sutton et al. (2008) discovered a deep maximum of bathypelagic fish biomass at 1500 and 2300 m depth over the Mid-Atlantic Ridge, and it now seems likely that this is supported by predation on gelatinous zooplankton (Sutton, 2013). Gelatinous prey are utilised by pelagic fishes such as *Bathylagus euryops* (Bathylagidae) and *Scopelogadus beanii* (Melamphaidae), as well as the benthopelagic slickheads of the family Alepocephalidae. The Alepocephaliformes (Platytroctidae, Bathylaconidae and Alepocephalidae) and Argentiniformes (Argentinidae, Microstomatidae, Opisthoproctidae and Bathylagidae) have a pharyngeal crumenal organ that enables them to triturate gelatinous prey. Episodic mass deposition of moribund jelly fish on the deep-sea floor also provides feeding opportunities for demersal predatory or scavenging fishes (Billett et al., 2006; Sweetman et al., 2014).

There are three main pelagic feeding guilds, zooplanktivores, micronektivores and generalists.

Most larvae of teleosts feed on zooplankton, and they are found in the surface layers of the ocean. There is a great diversity of specialised small mesopelagic and bathypelagic fishes often associated with the deep-scattering layers that feed on zooplankton. This assemblage including myctophids and gonostamatids represents a large proportion of the fish biomass in the oceans. The micronektivores feed on larger prey, mainly fishes. The Paralepididae (barracudinas) are probably agile swift swimmers and able to pursue prey, which they capture in their large jaws. In the sparse deep-pelagic environment where nearest-neighbour distances are long, sit-and-wait foraging is used by long-bodied fishes with well-developed lateral line sense organs such as the Nemichthyid and Serrivomerid eels aided by lures to attract prey such as in the Stomiidae (barbeled dragon fishes) and Ceratioid anglerfishes. Large teeth are characteristic as in the Evermannellidae (sabretooth fishes). The effectiveness of these feeding specialisations is exemplified by the fact that dragonfishes can consume 53–230 per cent of the annual standing stock of their fish prey (Clarke, 1982; Davison et al., 2013; Sutton and Hopkins, 1996).

The concept that in a very food-sparse deep-sea environment fish should be generalist feeders, opportunists taking anything available is superficially persuasive. However as knowledge of fish diets has improved, so it has become evident that there is specialisation even at great depths. Nevertheless the notion of broad-spectrum opportunism in the deep bathypelagic has some support. The Pelican eel, *Eurypharynx pelecanoides* (Pelican eel), with very large weak jaws, is presumed to behave like a living net engulfing a wide variety of prey including shrimps, fishes, copepods and benthic organisms (Gartner et al., 1997). The fangtooth, *Anoplogaster cornuta*, also takes a wide spectrum of prey.

It is extraordinarily difficult to estimate the food consumption rates of deep-sea fishes; however, two approaches have been used. Stomach fullness data from fishes captured in sampling programmes can be converted to daily ration intake by applying estimates of gastric evacuation rates. MacPherson (1985) used different calculations according to whether the species

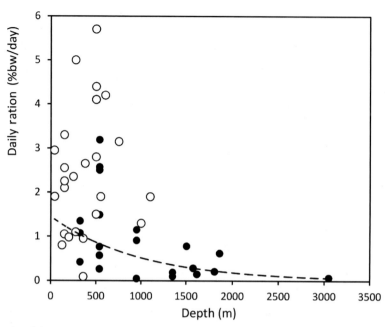

Figure 3.18 Food intake of deep-sea fishes in relation to depth. Daily ration as a percentage of body weight. Open circles – pelagic species. Closed circles – demersal species. Values plotted are medians of the ranges given in Drazen and Sutton (2016) from which the data are taken.

was a continuous feeder (*Coelorhinchus fasciatus,* banded whiptail 1086 m$_{fbd}$ Macrouridae) or a discontinuous feeder that does not take new prey until the stomach has been evacuated (*Helicolenus dactylopterus* Blackbelly Rosefish 600 m$_{fbd}$: Sebastidae, and *Lophius upsicephalus* Devil anglerfish 400 m$_{fbd}$ Lophiidae). The latter case greatly simplifies the calculation, but obtaining reliable estimates of gastric evacuation rates remains problematic. The alternative method is based on construction of an energy budget model as applied to the macrourids *Coryphaenoides acrolepis* and *C. armatus* by Drazen (2002) using data on respiration, growth and reproduction. Despite these caveats, plotting of available data (Fig. 3.18) shows that feeding rates of pelagic species are generally higher than in demersal species and within the demersal species food intake decreases exponentially with increasing depth. These ration rates are low compared with shallow-water species, e.g. in the cod *Gadus morhua,* the satiation ration is 3.67–5.21 per cent body weight per day (Soofiani and Hawkins, 1982).

3.11 Life History Strategies

Deep-sea fishes are characterised as slow-growing and exhibiting 'K-selected' life history properties (Pianka, 1970) with low fecundity and low intrinsic rate of population increase. This has been further emphasised by maximum ages of over one century reported in several species (Table 3.3). However, only a few deep-sea species live to over 100 years, and there is wide variation, as in any diverse fauna. More important than maximum age is the age when individuals reach sexual maturity, as this defines the generation time and hence is a limit on rate of population growth. The age of maturity can be measured as the age at which 50 per cent of the population is sexually mature (A_{50}). Figure 3.19a shows that with increasing depth the A_{50} of demersal teleost fishes increases from an average of 3–4 years at continental shelf depths to 20 years at depths of over 700 m. There is a continuum across the continental slope towards slower growth and longer delay of sexual maturity with increasing

Table 3.3 **Maximum ages of teleost fishes with minimum and maximum depths of occurrence. (After Drazen and Haedrich, 2012).**

Order/Common Name	Species	Age Max	Depth (m) Min	Depth (m) Max
ORDER ARGENTINIFORMES				
Greater silver smelt	*Argentina silus*	35	150	550
ORDER ALEPOCEPHALIFORMES.				
Baird's slickhead	*Alepocephalus bairdi*	38	650	1700
ORDER ZEIFORMES				
John dory	*Zeus faber*	12	50	150
Black oreo	*Allocyttus niger*	153	600	1200
Smooth oreo	*Pseudocyttus maculatus*	86	700	1400
ORDER GADIFORMES				
Giant grenadier	*Albatrossia pectoralis*	58	500	1300
Pacific grenadier	*Coryphaenoides acrolepis*	73	700	2000
Popeye grenadier	*Coryphaenoides cinerus*	15	500	1400
Roundnose grenadier	*Coryphaenoides rupestris*	72	400	1500
Roughhead grenadier	*Macrourus berglax*	25	300	1000
Ridge-scaled grenadier	*Macrourus carinatus*	37	500	1000
Bigeye grenadier	*Macrourus holotrachys*	27	670	1600
Whitson's grenadier	*Macrourus whitsoni*	55	600	1500
Hoki	*Macrouronus novaezelandiae*	25	250	700
Pacific hake	*Merluccius productus*	16	45	500
Atlantic cod	*Gadus morhua*	25	50	200
Greater forkbeard	*Phycis blennoides*	20	150	300
Blue ling	*Molva dypterygia*	30	350	1200
ORDER BERYCIFORMES				
Orange roughy	*Hoplostethus atlanticus*	149	700	1400
Alfonsino	*Beryx splendens*	23	400	800
ORDER SCORPAENIFORMES				
Pacific ocean perch	*Sebastes alutus*	100	150	825
Splitnose rockfish	*Sebastes diploproa*	84	200	600
Widow rockfish	*Sebastes entomelas*	60	23	200
Acadian redfish	*Sebastes fasciatus*	50	130	500
Yellowtail rockfish	*Sebastes flavidus*	64	50	250
Golden redfish	*Sebastes marinus*	60	100	1000
Deepwater redfish	*Sebastes mentella*	75	300	900
Shortspine thornyhead	*Sebastoloblus alascanus*	80	300	1200
Longspine thornyhead	*Sebastoloblus altivelis*	45	500	1400
Sablefish	*Anoplopoma fimbria*	114	200	1000
ORDER PERCIFORMES				
Black cardinalfish	*Epigonus telescopus*	104	450	1100

Table 3.3 (*cont.*)

Order/Common Name	Species	Age	Depth (m)	
		Max	Min	Max
Pelagic armorhead	*Pseudopentaceros wheeleri*	7	220	500
Patagonian toothfish	*Dissostichus eliginoides*	53	600	2100
Black scabbardfish	*Aphanopus carbo*	12	200	1600
Largehead hairtail	*Trichiurus lepturus*	15	100	350
ORDER PLEURONECTIFORMES				
Rex sole	*Errex zachirus*	29	60	500
Witch flounder	*Glyptocephalus cynoglossus*	25	50	500
Pacific halibut	*Hippoglossus stenolepis*	55	25	275
Dover sole	*Microstomus pacificus*	58	370	1000
Greenland halibut	*Rheinhardtius hippoglossoides*	36	500	1200

depth (Drazen and Headrich, 2012). The trend is also significant within each of the major taxa, Gadiformes and Scorpaeniformes indicating the phenomenon is driven by environment rather than presence of different taxa at different depths. Some deep-sea teleosts produce relatively few large eggs, e.g. Alepocephaliformes and Liparidae that may deposit eggs on the seafloor or show parental care with no larval stage. Excluding such species and only considering broadcast spawners, it is evident that fecundity (number of eggs per female per season) generally decreases with depth, a 10-fold difference between 50 m and 700 m depth (Fig. 3.19b). Reproductive output appears to be limited by lack of available food energy in the deep sea. Fernandez-Arcaya et al. (2016) also found a trend of decrease in fecundity with depth in the Western Mediterranean Sea, but egg size increased probably resulting in larvae hatching at a more advanced stage of development and better able to survive. Drazen (2002) compared the energy budgets of two macrourids within the genus *Coryphaenoides*, the slope-dwelling *C. acrolepis* (Pacific grenadier) and the abyssal *C. armatus*. Sustaining annual reproduction (iteroparity) in *C. acrolepis* requires more than half the total energy expenditure, whereas *C. armatus* saves energy and grows to a larger size by reproducing just once

in a lifetime (semelparous). Combining the growth and fecundity information, Drazen and Haedrich (2012) found a significant trend of decrease with depth in r^1, a measure of potential rate of population increase:

$$r^1 = \log_n\left(\frac{F_{50}}{A_{50}}\right)$$

where F_{50} is the fecundity at 50 per cent maturity (Jennings et al., 1998). The trend is from $r^1 = 2.3$ at 50 m depth to 0.6, at 700 m depth. Clarke et al. (2003) derived an r^1 value of 0.95 for the roundnose grenadier *Coryphaenoides rupestris* compared with 6.86 for the Atlantic Cod *Gadus morhua*. Values of r^1 for deep-sea sharks are much lower; 0.05 in the Leafscale gulper shark, *Centrophorus squamosus* and 0.01 in Birdbeak dogfish *Deania calcea*. In general in the deep-living Chondrichthyes maximum ages are not much greater than in shallow representatives but maturation is delayed with the A_{50} value, approximately doubled compared with shallow-water species (Garcia et al., 2008). The finding that the Greenland shark, *Somniosus microcephalus*, becomes sexually mature at 156 ± 22 years and lives to over 250 years (Nielsen et al., 2016) further reinforces the concept of slow growth in deep-water Chondrichthyes.

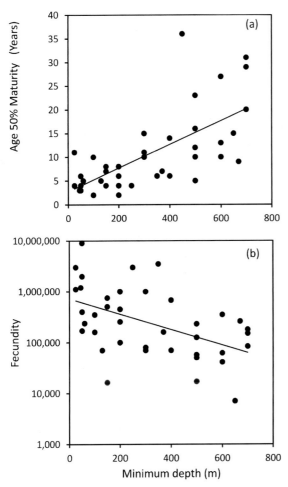

Figure 3.19 Age and fecundity of broadcast-spawning teleosts in relation to depth. The species are listed in Table 3.3. (a) Age at 50 per cent maturity. (b) Maximum fecundity, egg number in the largest females. (Data from Drazen and Haedrich (2012).

Childress et al. (1980) investigated the growth and reproduction of small pelagic fishes off southern California. Three groups were considered: the epipelagic *Sardinops sagax* (South American pilchard, Clupeidae); mesopelagic vertical migrators that feed near the surface at night; *Leuroglossus stilbius* (California smoothtongue, Bathylagidae) and three Myctophids, *Stenobrachius leucopsaurus* (Northern lampfish), *Triphoturus mexicanus* (Mexican lampfish), *Lampanyctus ritteri* (Broadfin lampfish) and deeper-living nonmigrating bathypelagic species: *Poromitra*

crassiceps, (Crested bigscale, Melamphaidae), *Borostomias panamensis* (Panama snaggletooth, Stomiidae), *Lampanyctus regalis* (Pinpoint lampfish, Myctophidae), *Bathylagus milleri* (Stout blacksmelt, Bathylagidae) and *Bajacalifornia burragei* (Sharpchin slickhead, Alepocephalidae). The life spans of the epipelagic and bathypelagic species ranged from four to eight years and the mesopelagic species five to eight years. *Sardinops* grows rapidly to a large size (maximum length 240 mm) with early maturity and repeated reproduction throughout the life span. The mesopelagic species are smaller (mean maximum length 98 mm), have slow growth, also mature early and have repeated reproduction. The bathypelagic species have fast growth to a large size (mean maximum length 201 mm) but at the expense of low reproductive output delayed to a single reproductive event late in the life span. Growth efficiency is high in the bathypelagic species because of low metabolic rate and low energy content of the tissues. The bathypelagic species appear to prioritise growth to a large size, reaching a possible minimum size necessary to feed and survive.

In both pelagic and demersal species, therefore, there is a trend towards a decrease in reproductive effort with depth, lower fecundity and reduced frequency of spawning to the extent that there is often only a single spawning event at the end of the life span. Allocation of the entire life span reproduction to one burst might be considered risky (all the eggs in one basket), but Childress et al. (1980) argue that this can be selected for because of the relative stability of the deep-sea environment. Semelparity has a consequence for scientific research in that mature individuals are exceedingly rare in most samples, and it is thus very difficult to study such life histories.

The Myxiniformes (Hagfishes) generally spawn in deep water, and there is no evidence that hagfish sampled at different depths show differences in their reproductive characteristics, but information is sparse (Powell et al., 2005). Mead et al. (1964) stated that 'abyssal Chondrichthyes differ reproductively in no known way from their epipelagic or coastal relatives'. However, several studies have since shown that maturation is delayed in deep-sea species, e.g. *Bathyraja minispinosa* (Whitebrow skate) reaches a maximum age of 37 years and sexually maturity (A_{50}) at 23 years, more than halfway through its maximum life span (Ainsley et al., 2011). In the NE Atlantic Stehmann and Merrett (2001) retrieved egg capsules of the deep-sea ray *Bathyraja richardsoni* with developing embryos from 1541 m depth in the Porcupine Seabight, and Henry et al. (2016) observed egg capsules of the same species on a sandy seabed on the flanks of the Hebrides Terrace seamount at similar depths 1489–1580 m. These depths of egg deposition are at the shallow end of the normal depth distribution range for this species. There is a conservation concern that these egg-laying habitats may be vulnerable to destruction by trawls, although for this species the depth is beyond the limits of normal commercial fishing activity.

In the teleosts, many species are broadcast spawners; the larvae generally develop in the surface layers and are collected in routine ichthyoplankton surveys (Moser, 1996). Marshall (1953) argued that eggs of deep-sea species tend to be relatively larger than in shallow-water species, thus giving rise to an advantage in that larvae hatch at a relatively advanced stage of development and probably begin their feeding on small copepods. Large larvae also have lower relative food requirements and can swim faster to obtain food of a suitable size. These factors would tend to offset the disadvantages of lower fecundity in deeper-living species. For broadcast-spawning fishes the long distance between the seafloor and surface waters where the larvae can feed is a potential problem. Robertson (1981) first suggested that complex sculpturing of the surface of the egg in some species may influence the ascent rate of eggs from a deep-sea spawning site to the sea surface. In particular the eggs of deep-water myctophid *Maurolicus muelleri* (Silvery lightfish, 1524 m_{fbd}) have a pattern of hexagonal ridges that apparently slow the ascent through the water column to 41 per cent of the rate expected of a smooth-surfaced egg of the same diameter. Merrett and Barnes (1996) found similar ornamentation in the eggs of six genera of macrourids, *Coelorhincus*, *Coryphaenoides*, *Macrurus*, *Hymenocephalus*, *Malacocephalus* and *Nezumia*, and suggested these ensure that development occurs at depth remaining within the narrow bathymetric strata on the bathyal slopes where the adults live. Deep-sea fishes display a wide variety of reproductive traits, direct and indirect development, parental care, sexual dimorphism, demersal or pelagic eggs, oviparity, ovoviviparity and viviparity, simultaneous hermaphoroditism, shoaling, and spawning migration, all of which are known in shallow-water and surface-species (Mead et al., 1964). The two main groups of deep-sea fishes that show distinctive reproductive adaptations are the Aulopiformes (lizardfishes), which evolved synchronous hermaphrodism as they invaded the deep sea (Davis and Fielitz, 2010), and the superfamily Ceratioidea (deep-sea anglerfishes: Lopiiformes) with extreme sexual dimorphism characterised by dwarf males, which are parasitic on adult females (see Chapter 4). These can be considered to be extreme solutions to maintaining reproduction in a food-sparse environment with low probability of encounter of conspecifics for mating.

3.12 Excretion

Marine teleost fishes continuously lose water by osmosis through the gills, and this loss is compensated for by drinking. The excess sodium (Na^+) and chloride (Cl^-) ions taken up by drinking are excreted through the chloride cells in the gills, a process that requires energy so the chloride cells are richly endowed with mitochondria. The main metabolic end product of protein metabolism is ammonia (NH_3), which readily dissolves in water, passively diffusing across the gill membrane into the surrounding seawater. Parallel processes coupled to NH_4^+ ion transport may be involved in blood pH regulation (Eddy and Handy, 2012). Because the gills are mainly responsible for water, ammonia and salt excretion, the kidneys play a secondary and minor role in excretion. The main function of the kidney is to excrete divalent ions, Mg^{++} and SO_4^-. The primary urine in most species is formed by ultrafiltration in the glomeruli, and then re-absorption occurs in the tubule to retain water and other essentials. These processes are energetically costly, and therefore it is not surprising that some marine teleosts have lost the function of the glomeruli, and the kidneys are aglomerular. The kidneys of Antarctic Notonthenioids and eelpouts (e.g. *Lycodichthys dearborni*) are aglomerular to avoid loss of low-molecular-weight antifreeze glycopeptides (Eastman et al., 1979). Noting that the kidney of the *Ateleopus japonicus* (Pacific jellynose fish: Ateleopodidae) is aglomerular, Ozaka et al. (2009) suggest this condition may be advantageous to deep-sea fishes that depend on high water content in the tissues to maintain neutral buoyancy. Aglomerular kidneys have also been found in the deep-sea species *Eurypharynx pelecanoides* (Pelican eel: Saccopharyngiformes) and *Pogonophryne scotti* (Saddleback plunderfish, Artedidraconidae), and Ozaka et al. (2009) further propose that species such as *Chauliodus* and *Bathylagus*, which do not have swim bladders, should also be aglomerular. Lophiiformes that have been examined hitherto, *Lophius americanus, Lophius piscatorius* (Brull et al., 1953), *Histrio histrio, Antennarius striatus, Halieutaea stellata*, all have aglomerular or pseudo-glomerular kidneys, suggesting the entire order may have no glomeruli. Histological analysis is needed on many more species to properly test the Osaka et al. (2009) hypothesis.

3.13 Sensory and Communication Systems

Much of the strange appearance of deep-sea fishes, absence of colours, large or small eyes, presence of barbels, etc. is related to sensory function. They live in an environment that is totally different to human experience. Newly captured mesopelagic or bathypelagic fishes placed in a shipboard aquarium typically swim downward repeatedly crashing against the bottom of the tank until they die of their injuries.

Living in an essentially infinite three-dimensional volume, neither they nor their ancestors for many generations have ever encountered an extended solid surface. Natural selection has provided them with no means of detecting, interpreting or responding to such a phenomenon. Benthic and benthopelagic species can respond appropriately to surfaces, but living in total darkness they are unlikely to perceive these visually. Understanding how sensory systems work in deep-sea fishes is a fascinating field of research questioning many human assumptions about perception. Here there is only space for a cursory examination of broad trends in sensory adaptation of fishes to the deep-sea environment.

3.13.1 Diversity of Brain Morphology

The morphology of animal brains reflects the functional demands placed on them. In relation to the sensory biology of deep-sea fishes Wagner (2001a, b) recognised four main areas of the brain that receive primary projections from the senses in accordance with classical vertebrate anatomy (Romer, 1970); from anterior to posterior they are (Fig. 3.20a, b):

1. Olfactory bulb. At the anterior end of the brain this receives inputs from chemo-sensory receptors of the olfactory epithelium in the nasal capsules on either side of the nose.
2. Optic tectum. This receives visual inputs from the eyes via the optic nerves.
3. Trigeminal/octavo lateral. This region of the brain combines sound and vibration inputs from external sensory cells of the lateral line along the length of the body and head region with acoustic information from the inner ear.
4. Gustatory area. This combines chemo-sensory inputs from taste buds on the body surface, oral cavity and pharynx.

In most deep-sea fishes the optic tectum is the largest area of the brain, on average occupying 60 per cent of the volume of the sensory areas. This should be interpreted with caution because it may simply mean that visual information requires more processing

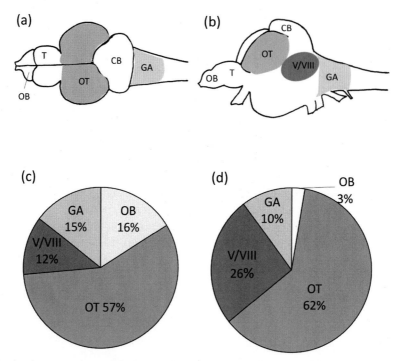

Figure 3.20 The four main sensory areas of the brains of deep-sea fishes: Olfactory bulb (OB), Optic Tectum (OT), Trigeminal/ octavolateral area (V/VIII), Gustatory Area (GA).

(a) *Bathytroctes microlepis* (Smallscale smooth-head, 4900 m$_{fbd}$) Alepocephalidae. Dorsal view of the brain.

(b) Left lateral view of the same brain. (After Wagner, 2001a). CB – cerebellum, T – telencephalon.

(c) Abyssal demersal fishes. Mean proportions of volumes of the four main sensory areas in 33 species. (Data from Wagner, 2001a).

(d) Mesopelagic and bathypelagic fishes. Mean proportions of volumes of the four main sensory areas in 67 species. (Data from Wagner, 2001b).

capacity than other senses. The ranking of brain areas is radically different in pelagic and demersal species: Optic, Trigeminal/octavo lateral, Gustatory and Olfactory in order of decreasing volume in pelagics and Optic, Olfactory, Gustatory and Trigeminal/ Octavo lateral in demersal species. Chemosensory function is much more important in demersal than in pelagic species (Fig. 3.20c, d).

Within the demersal species, the Alepocephalids (slick heads) stand out as exceptional with a very large optic tectum (mean 81 per cent) and small olfactory bulb (mean 3.6 per cent) regardless of depth (Fig. 3.21a) (Wagner, 2002). In the remaining demersal families there is a significant trend of decrease in optic tectum proportion with depth, i.e. vision becomes less important with increasing depth. The data form two clusters: a main group including the Halosauridae, Notacanthidae and Macrouridae with the optic tectum volume mostly larger than 50 per cent and a group with small optic tecta: Ipnopidae, Synaphobranchidae and Ophidiidae. The species with the smallest optic tectum is the Ophidiid cusk eel, *Barathrites iris* with brain area percentages as follows: olfactory bulb 81.34 per cent, optic tectum 13.77 per cent, Trigeminal/ Octavolateral 2.3 per cent and gustatory 2.59 per cent. In the Synaphobranchidae (cutthroat eels) the mean olfactory bulb volume is *ca.* 50 per cent. Wagner (2001a) describes these species as 'swimming noses' that locate their prey largely through olfaction. The ophidiids can also be included in this description as

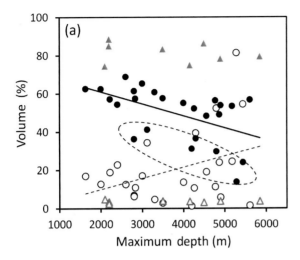

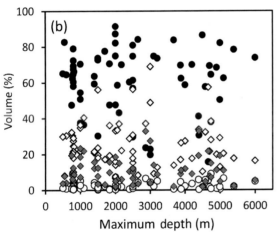

Figure 3.21 Sensory areas of the brains of deep-sea fishes. The proportions of the four main sensory areas in relation to maximum depth of occurrence.

(a) Abyssal demersal species (data from Wagner, 2001a). Black circles – Optic tectum volume for all species except Alepocephalidae. Solid trend line. $y = -0.0063x + 73.535$ $P = 0.006$. Dashed Ellipse encloses data for Ipnopidae, Synaphobranchidae and Ophiidae with small optic tecta. Open circles – Olfactory bulb volume for all species except Alepocephalidae. Dashed trend line NS.

(b) Mesopelagic and bathypelagic species (data from Wagner, 2001b). Black circles – Optic tectum, Open circles – Olfactory bulb, Open diamonds – Trigeminal/octavalateral area (V/VIII), Grey diamonds – Gustatory area.

more extreme examples. In the Ipnopidae (tripod fishes), instead of large olfactory bulbs, it is the gustatory area that is highly developed (32–42 per cent) together with the Trigeminal/octavo lateral (10–25 per cent) so skin taste buds and lateral line are important for these sit-and-wait predators. The Bathysauridae (Lizard fishes), which are also 'sit-and-wait' predators, have similarly large Trigeminal/octavo lateral and gustatory areas.

In the pelagic species there is no trend with depth (Fig. 3.21b) for any of the brain areas. The olfactory bulbs are universally small (maximum 11.5 per cent), but there are some interesting exceptions to the rule that vision is dominant in the pelagic. In the Alephocephalid, *Photostylus pycnopterus* (Starry smooth-head), the largest brain area is the Trigeminal/octavo lateral (37 per cent) followed by the gustatory (33 per cent). In the Gonostomatidae (Bristlemouths), which have small eyes, the Trigeminal/octavo lateral is often the largest area (34–37 per cent), and the gustatory area is also well developed (15–27 per cent). The largest Trigeminal/octavo lateral volume (69 per cent) is found in the *Melanonus zugmayeri* (Arrow tail, 3000 m_{fbd}), which probably has the most sensitive known system of neuromast sensory cells on the head and the lateral line. *Scopelogadus beanii* (Bean's bigscale, 2500 m, Melamphaidae), which has a system of sensory canals on the head and has a correspondingly large Trigeminal/octavo lateral volume (56 per cent). In the Ceratioidea (deep-sea anglerfishes) ratios of the brain areas are olfactory, optic, Trigeminal/octavo lateral, gustatory: 0.25 per cent, 30.5 per cent, 33.79 per cent, 32.8 per cent in *Ceratias holboelli* and 1.4 per cent, 30.27 per cent, 56.07 per cent, 12.26 per cent in *Melanocetus johnsonii*. In these Ceratioid adult female fishes all the senses other than olfaction are used to detect prey in the bathypelagic.

The importance of different brain areas can shift during development. Wagner (2003) showed that when the abyssal grenadier (*Coryphaenoides armatus*) reaches a size of about 40 cm total length there is a decrease in relative volume of the optic tectum and an increase in the olfactory bulbs as the fish switches from pelagic visually oriented feeding as a juvenile to feeding on benthic prey as an adult.

3.13.2 Vision

For deep-sea fishes the light regime in the oceans can be divided into two layers, mesopelagic and bathypelagic. In the mesopelagic zone down to 1000 m, sunlight becomes attenuated and monochromatic; short wavelengths are scattered, and long wavelengths are absorbed, leaving only dim blue-green light of 470–480 nm wavelength. As a result of scattering, the angle of the Sun at any given time becomes irrelevant, lighting is the same from all azimuth directions in the horizontal plane but the down-welling intensity is 200 times greater than the upwelling illumination (Herring, 2002; Fig. 3.22). Below 1000 m depth in the bathypelagic, bathyal and abyssal zones there is no sunlight, and the only light is from bioluminescence produced by fishes and other organisms. Most bioluminescence is in the blue-green part of the spectrum that penetrates furthest through sea water and is made up of mainly point source flashes from animals (Craig et al., 2015). The quantity of bioluminescent targets is greatest near the surface and decreases with depth from 80 m^{-3} at 600 m depth to < 0.5 m^{-3} at depths over 4000 m in the North Atlantic (Gillbrand et al., 2007) and 0.01 m^{-3} at 5000 m depth in the eastern Mediterranean (Craig et al., 2010). There are potential visual stimuli at all depths and all fish therefore have eyes, including the deepest-living fishes in hadal trenches (Jamieson, 2015), although in some cases the eyes are degenerate and may be of limited visual function.

Lockett (1977) points out that in the deep sea there is an absence of reference marks of known size or other cues that can enable an animal to judge distance using monocular vision, as is common for terrestrial creatures including man. This means that for many predators the only solution is effective binocular vision. A good example is *Stylephorus chordatus* with forward-facing tubular eyes (see Chapter 4) that enable it to locate and pounce on its prey with its high-speed suction apparatus. Binocular vision is remarkably

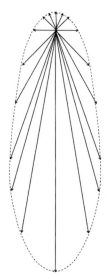

Figure 3.22 Diagram of the light field at mesopelagic depths. The relative radiance is indicated by the lengths of the arrows; low levels of upwelling radiance and strongest radiance in the down-welling direction, the ratio between these two should be 1:200 but the scale is compressed for clarity. (After Denton, 1970).

well developed in this and other species such as the barreleye *Macropinna microstoma*, which can rotate its eyes in tandem whilst keeping the binocular field of view correctly lined up (Robison and Reisenbichler, 2008).

In deep-living mesopelagic fishes there is a trend towards larger eyes in order to capture more light. Because the lens of a fish eye is spherical, the focal length is 2.55 times the lens radius (Matthiessen's ratio), i.e. the diameter of the eye is *ca.* 2.55 the lens diameter. Such large eye diameters cannot be accommodated in the head of the fish, particularly if it is laterally flattened. The evolutionary solution has been the development of tubular or telescope eyes, which typically look upward. This can be regarded as a large spherical eye that has been cut away to leave a central part with the retina set at the Matthiessen's ratio distance to provide good visual acuity in one direction (Fig. 3.23). Fishes with upward-directed eyes can perceive the silhouette of prey against the background of down-welling light, estimate distance and pounce upward to capture the item.

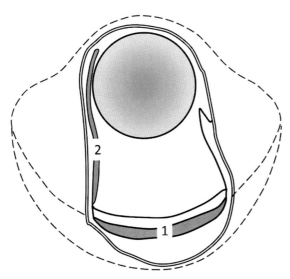

Figure 3.23 Diagram of the outline of a tubular eye superimposed on the outline of a normal eye (dashed line) with a spherical lens of the same diameter. 1 – Main retina of the tubular eye. 2 – Accessory retina present in some species with tubular eyes forms an unfocused image from a wider field of view.
(After Franz, 1907).

A disadvantage of tubular eyes is that the fish is blind in other directions and may be very vulnerable to attack from below. The field of view of such fishes can be extended in various ways. In *Opisthoproctus*, *Rhynchohyalus*, *Winteria*, *Dolichopteryx* and *Macropinna* there are accessory retinas that collect light from different directions, not necessarily producing a well-focused image but sufficient to detect movement. *Dolycopteryx* is noted for the use of reflective optics to focus the image on accessory retinas (Wagner et al., 2009). In the Scopelarchids (pearl eyes) there is a transparent lens pad that has a layered structure that guides light from the lateral ventral aspect into the lens. This provides a field of view downward to within 20° of the vertical and gives the fish its 'pearl eye' appearance (Collin et al., 1998). The Evermannelids (sabretooth fishes) have a similar structure, although of different anatomical origin, known as an optical fold that directs upwelling light into the lens of the eye (Locket, 1977). Possibly the most extreme example of extension of the visual field is the

Bathylychnops exilis (Javelin spookfish), which has an accessory globe with separate lens and retina giving rise to the name 'four-eyed' fish (Pearcy et al., 1965).

In fishes with spherical eyes the field of view is nevertheless restricted by the pupil so that the full aperture of the lens can only be used from one direction. This can be overcome by enlarging the pupil so that it is larger than the lens, allowing light to pass from a wider range of angles to the retina (Fig. 3.24). This creates a space between the lens and the iris, known as the aphakic gap. Munk and Frederiksen (1974) show that in the Gonostomatidae (Bristlemouths) the eye becomes smaller but relative pupil diameter and the aphakic gap increase in size with depth. In threshold visibility conditions the aphakic gap greatly increases the retinal illumination for peripheral targets. The large aperture reduces contrast in a normal illuminated visual field near the surface but for spotting bioluminescent flashes against a dark background performance is excellent. Small eyes should not be dismissed as necessarily of poor performance as demonstrated by the tiny cameras in modern cell phones and portable computers. In some species, e.g. *Scopelosaurus* (waryfishes, Notosudidae) and *Bajacalifornia megalops* (Bigeye smooth-head, Alepocephalidae) the aphakic gap is only in the forward direction, improving image illumination in the binocular field of view with a corresponding fovea for high resolution of targets(Lockett, 1977).

The retinas of deep-sea fishes are also modified to maximise sensitivity. Because the light is monochromatic there are generally no cones, although larvae and juveniles of some Aulopiformes, e.g. *Bathysauroides* and *Chlorophthalmus* (see Chapter 4), do have cones. The rods have very long outer segments to maximise the length of light path intercepted. Furthermore there may be multiple banks of rods, in two or more layers. Wagner et al. (1998) found 12 banks of rods in the *Epigonus telescopus* (black cardinal fish: Epigonidae) and four–six banks in the macrourids they sampled. Because a typical rod intercepts 95 per cent of the incident light, there is some doubt as to whether additional layers of rods intercepting the remaining 5 per cent can be of

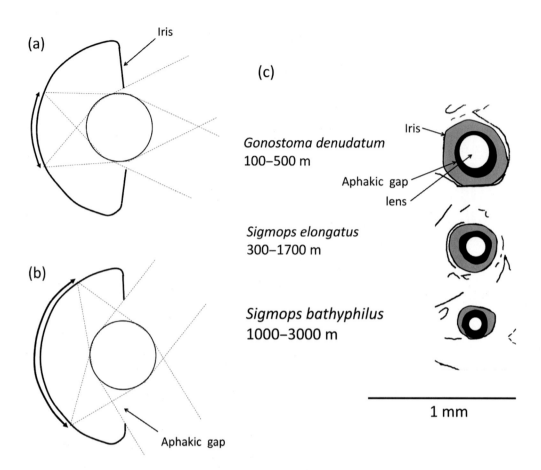

Figure 3.24 The aphakic gap in deep mesopelagic and bathypelagic fish eyes. (a) Diagram of a fish eye with a normal iris. The field of view over which the full diameter of the lens is used is limited to the arc shown by the arrow. (b) Diagram of a fish eye with an aphakic gap; the pupil aperture is larger than the lens diameter. This allows light from a much wider arc to enter the eye. (c) The aphakic gap in the Gonostomatidae, scale diagrams of the eyes of three species living at different depths. Note the very small eye and large aphakic gap in the deepest living *Sigmops bathyphilus* adapted to detect bioluminescent flashes. (After Munk and Frederiksen, 1974).

any value. Adult demersal species such as *Antimora rostrata*, *Coryphaenoides guentheri* and *C. armatus* have around 200×10^3 rods.mm^{-2}, about half the number found in typical nocturnal mammals such as the cat, but in some species parts of the retina have much higher densities: *Winteria telescopa* (429×10^3.mm^{-2}), *Bathylagus benedicti* (765×10^3.mm^{-2}) and *Lamanyctus ater* (500×10^3. mm^{-2}). A reflective tapetum in the outer layer of the eye redirects light back through the retina, presumably to further enhance sensitivity. The tapetum in Chondrichthyes is generally composed of

guanine crystals in the choroid layer providing mirror-like reflection, whereas in teleosts there is usually a more diffuse reflector in the retinal pigment epithelium (Douglas et al., 1998). The myctophids and aulopiformes have guanine-based mirror-like tapeta. Across the teleosts there can be interesting colour variations such as blue, green, white, silver and gold as well as the red tapetum of *Malacosteus niger*, probably linked to the red bioluminescence in this species.

The pigments in the rods have peak spectral absorbance values in the range 468 to 494 nm overlapping the predicted range for down-welling light of 450–475 nm but more closely matching the blue colour of bioluminescence (470–494 nm). Douglas et al. (1998) conclude that deep-sea fish eyes are primarily adapted to the detection of bioluminescent flashes. However, there is an interaction with eye size, small fish with small eyes, which can only detect bioluminescence at short ranges and need relatively long wavelength pigments in their rods, and fish with larger eyes, which need visual pigments with shorter wavelength values for maximum sensitivity.

The photoreceptor cells are connected into functional groups by ganglion cells that perform the first stage of conversion of the optical image into a neural image. The density of ganglions cells defines the limit of spatial resolving power of the eye, and variation across the retina gives some insight into the importance of different visual directions.

Many pelagic species from a range of families have relatively uniform retinal ganglion cell densities: *Serrivomer beani* (Serrivomeridae), *Scopelogadus beanii* (Melamphaidae), *Diaphus raphinesquii, Myctophum punctatum, Nannobrachium atrum* (Myctophidae) and *Melanonus zugmayeri* (Melanonidae) (Collin and Partridge, 1996). These can be regarded as unspecialised retinas with high sensitivity in all directions. Amongst the demersal species *Notacanthus bonapartei* (Notacanthidae) has an unspecialised retina with uniform ganglion cell densities, but *Synaphobranchus kaupii* (Synaphobranchidae) has a ventral *area retinae* with high ganglion cell densities conferring resolving power in the upward direction for capture of mobile prey in the water column above the seafloor. Heger et al. (2007) recorded *S. kaupii* feeding on luminescent crustacea (ostracods); behaviour presumably aided by such an arrangement. The tripod fish *Bathypterois dubius* (Ipnopidae), although it has small eyes previously regarded as degenerate, has two *areae retinae*, one for forward vision (probably to intercept bioluminescent copepod prey) and one for rearward vision to avoid stalking predators. The mesopelagic hatchetfish *Sternoptyx diaphana*, with nontubular eyes has a pronounced ventral *area*

retinae providing good upward visual resolution for stalking prey in the mesopelagic. The Alepocephalidae (slick-heads) have highly specialised retinas with very high ganglion cell densities in a fovea providing high visual acuity in the forward-looking area of binocular vision. The tubular eyes of *Argyropelecus affinia* (Sternoptychidae), *Stylephorus chordatus* (Stylephoridae) and *Scopelarchus michaelsarsi* (Scopelarchidae) all have specialised *areae retinae* with enhanced ganglion cell densities.

Certain species that live at depths greater than 900 m and that do not themselves have luminous organs have been recorded as having degenerate eyes with varying degrees of loss of function (Munk, 1965; Lockett, 1977): the tripod fishes *Benthosaurus grallator, Bathymicrops regis* and *Bathypterois longipes* (Ipnopidae) the grenadiers *Echinomacrurus mollis, Macrouroides inflaticeps* (Macrouridae), the cusk-eels *Leucicorus lusciosus, Holcomycteronus profundissimus, Typhlonus* spp. (Ophidiidae), the blind cusk-eels *Sciadonus pedicellaris, Nybelinella erikssoni* (Aphyonidae) and the hadal snail fish *Notoliparis kermadecensis* (Liparidae). The three species of *Ipnops* (Ipnopidae, Chapter 4) and some flabby whale fishes (*Ditropichthys storeri, Gyronomimus* sp. (Cetomimidae)) have the eye reduced to a flat photosensitive area on the surface of the cranium. *Ipnops* eyes have a reflective tapetum and may be a unique photosensing organ with unknown capabilities. Also Collin and Partridge (1996) have demonstrated functional eyes in *Bathypterois dubius* suggesting that assumptions regarding loss of visual function in some genera may have to be revised.

3.13.3 Camouflage in the Deep

The low levels of light in the deep sea present opportunities for animals to be invisible to their potential predators or prey. Camouflage adaptations can either be aimed towards remaining unseen or towards appearing to be something different.

An obvious means of becoming invisible is to be transparent so that light passes straight through the body as if it were not there (Fig. 3.25 a, b). Many

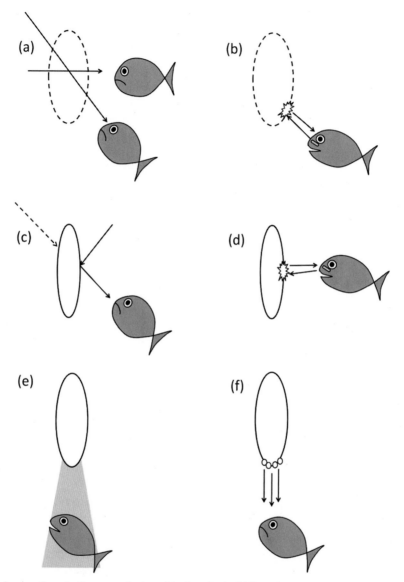

Figure 3.25 Optical camouflage in the mesopelagic and bathypelagic. (a) Transparent prey, solar light passes through unimpeded. (b) Transparent prey can be detected by fish with headlights detecting reflections from parts of the body. (c) Silvery prey with vertical mirror reflectors is invisible because the predator sees reflected light that matches the scattered sunlight. (d) Silvery prey are easily detected by fish with headlights particularly in the bathypelagic. (e) In typical down-welling mesopelagic light the dark silhouette of prey is very conspicuous to a predator hunting from below. (f) Ventral photophores provide counterillumination that eliminates or reduces the contrast of the silhouette. (After Denton, 1970 and Zylinski and Johnsen, 2011).

marine invertebrates and most fish larvae are transparent. Leptocephalus larvae of the Elopomorpha (bonefishes, Halosoaurs and eels) are transparent and can grow to over one metre long in some species. Transparency is aided by a flat thin body form. At metamorphosis the body becomes pigmented, and transparency is lost. No adult fishes are completely transparent. Haemoglobin pigment in the blood makes the circulatory system and gills visible, the eyes with pigment in the retina cannot be transparent, the stomach contents and the liver are also opaque. These parts can be disguised by silvery reflectors. The nearest approach to transparency in fishes are some of the Antarctic crocodile icefishes (Channichthydae), which have no haemoglobin or myglobin. Some deep-living fishes including the hadal snail fishes are unpigmented and the viscera can been seen through the translucent body wall. In the mesopelagic some species of mesopelagic hatchetfishes (Sternoptychidae) are partially transparent mainly in the tail region, whilst the rest of the body is silvery, e.g. *Sternoptyx diaphana,* *S. pseudobscura* and *Argyropelecus hemigymnus.* Transparent organisms are susceptible to detection by predators equipped with searchlight photophores and eyes that can detect the scattered light (Zylinski and Johnsen, 2011).

In the mesopelagic with a symmetrical light environment around a vertical axis, a flat vertical mirror is invisible because it reflects light into the observer's eye at exactly the angle and intensity that is expected (Fig. 3.25). Many mesopelagic fishes have silvery reflectors in the skin giving a chromium-plated appearance to freshly caught myctophids and hatchet fishes. In the hatchet fishes (Sternoptychidae), which are laterally flattened, there is a close approach to the ideal vertical mirror. The sharp ventral keel or hatchet blade also helps to make the fish invisible to upward-looking predators. Dark blue or black on the dorsal surface helps to camouflage the fish against the low levels of background upwelling light when viewed from above. There are limits to the use of mirrors, and they are no defence against a predator hunting from directly vertically below. The only option is then to use bioluminescent counterillumination from the ventral surface of the body to mimic the down-welling light as described below in Section 3.13.4.

In the bathypelagic the vertical mirror camouflage system no longer works because of the absence of the symmetrical scattered solar light field. A mirror would simply reflect bioluminescence from other organisms potentially making it more conspicuous to other creatures (Fig. 3.25d). From about 600 m depth silveriness in fishes progressively disappears, becoming more bronze-coloured, and below 1000 m pelagic fishes are generally dark matte black or brown-coloured. At greater depths the intense black colour of bathypelagic species such as the *Saccopharynx,* *Eurypharynx* and Ceratioid anglerfishes is a stark contrast to the silvery fishes from mesopelagic depths. The fish is invisible provided that it does not reflect the predominantly blue light. Near the abyssal seafloor colours return with the grey, brown or albino appearance of macrourids and ophidiids, but these can have no significance for the living fish because they are only revealed by the lights of a submersible or if the fish is retrieved to the surface (Herring, 2002).

3.13.4 Bioluminescence

Bioluminescence is often regarded as a defining feature of deep-sea fishes. It is therefore a little surprising to find that bioluminescence is confined to only 33 of more than 200 families that are considered in the present volume (Herring, 1987; Haygood, 1993; Suntsov et al., 2008) in Table 3.4. None of the hagfishes (Class Myxine) have bioluminescence. Within the Chondrichthyes, no Chimaeriformes have bioluminescence, and only two families of squaliform sharks have light organs. Marshall (1965) points out that apart from macrourids and a few morids, no benthic deep-sea fish have light organs. Suntsov et al. (2008) comment that the total absence of bioluminescence in deep-sea anguilliform eels is perplexing. In fact there are no light organs in any of the Elopomorpha except for two aberrant bathypelagic genera, *Saccopharynx* (swallowers, 10 species) and *Eurypharynx* (pelican eel, 1 species), which have caudal light organs plus 'white line organs' that run down the length of the body from behind the head on each side of the body, the function of which is poorly understood. In the Alepocephaliformes, characteristic shoulder luminescent organs are found in the Platytroctidae as well as ventral photophores in the mesopelagic

Table 3.4 Occurrence of Bioluminescence in Deep-Sea Fishes

I -Intrinsic luminescence B -Bacterial luminescence. (After Herring, 1987; Haygood, 1993; Suntsov et al., 2008).

Class, ORDER	Family	Subfamily	Genus	Type
Chondrichthyes				
SQUALIFORMES	Dalatiidae		*Dalatias, Euprotomicroides, Heteroscymnoides, Isistius, Mollisquama, Squaliolus*	I
	Etmopteridae		*Centroscylium, Etmopterus*	I
Actinopterygii				
ANGUILLIFORMES II	Saccopharyngidae		*Saccopharynx*	I
	Eurypharyngidae		*Eurypharynx*	I
ALEPOCEPHALIFORMES	Platytroctidae		*Persparsia*, Holtbyrnia*, Sagamichthys*, Searsia*, Searsioides*, Maulisia*, Barbantus, Mendotus, Normichthys, Pectinantus, Platytroctes*	I
	Alepocephalidae		*Microphotolepis, Photostylus, Rouleina, Xenodermichthys*	I
ARGENTINIFORMES	Opisthoproctidae		*Dolichopteryx, Opisthoproctus, Rhynchohyalus, Winteria*	B
STOMIIFORMES	Gonostomatidae		*Bonapartia, Cyclothone, Gonostoma, Margrethia, Sigmops, Diplophos, Manducus, Triplophos*	I
	Sternoptychidae	Maurolicinae	*Araiophos, Argyripnus, Danaphos, Maurolicus, Sonoda, Thorophos Valenciennellus*	I
		Sternoptychinae	*Argyropelecus, Sternoptyx, Polyipnus*	I
	Phosichthyidae		*Ichthyococcus, Phosichthys, Pollichthys, Polymetme, Vinciguerria, Woodsia, Yarrella,*	I
	Stomiidae	Astronesthinae	*Astronesthes, Borostomias, Eupogonesthes, Heterophotus, Neonesthes, Rhadinesthes*	I
		Stomiinae	*Stomias, Chauliodus*	I
		Melanostomiinae	*Bathophilus, Chirostomias, Echiostoma, Eustomias, Flagellostomias, Grammatostomias, Leptostomias, Melanostomias,*	I

Table 3.4 (*cont.*)

Class, ORDER	Family	Subfamily	Genus	Type
STOMIIFORMES (continued)			*Odontostomias, Opostomias, Pachystomias, Photonectes, Tactostoma, Thysanactis, Trigonolampa*	
		Idiacanthinae	*Idiacanthus*	I
		Malacosteinae	*Aristostomias, Malacosteus, Photostomias*	I
AULOPIFORMES	Chlorophthalmidae		*Chlorophthalmus*	B
	Scopelarchidae		*Benthalbella, Scoperarchoides*	I
	Evermannellidae		*Coccorella, Evermannella?*	I
	Paralepididae		*Lestidium, Lestrolepis*	I
MYCTOPHIFORMES	Neoscopelidae		*Neoscopelus*	I
	Myctophidae		*Benthosema, Bolinichthys, Centrobranchus, Ceratoscopelus, Diaphus, Diogenichthys, Electrona, Gonichthys, Gymnoscopelus, Hintonia, Hygophum, Idiolychnus, Krefftichthys, Lampadena, Lampanyctodes, Lampanyctus, Lampichthys, Lepidophanes, Lobianchia, Loweina, Metelectrona, Myctophum, Nannobrachium, Notolychnus, Notoscopelus, Parvilux, Protomyctophum, Scopelopsis, Stenobrachius, Symbolophorus, Taaningichthys, Tarletonbeania, Triphoturus*	I
GADIFORMES	Macrouridae		*Cetonurus, Coelorinchus, Haplomacrourus, Hymenocephalus, Lepidorhynchus, Malacocephalus, Mesobius, Nezumia, Odontomacrurus, Sphagemacrurus, Trachonurus, Ventrifossa*	B
	Steindachneriidae		*Steindachneria*	B
	Trachyrinchidae		*Idiolophorhynchus*	B
	Moridae		*Antimora? Gadella, Lotella? Physiculus, Tripterophycis*	B

Table 3.4 (*cont.*)

Class, ORDER	Family	Subfamily	Genus	Type
BATRACHOIDIFORMES	Batrachoididae		*Porichthys*	I
BERYCIFORMES	Anomalopidae		*Parmops*	B
	Trachichthyidae		*Aulotrachichthys*	B
PERCIFORMES	Howellidae		*Howella*	I
	Acropomatidae		*Acropoma*	B
	Epigonidae		*Epigonus, Florenciella*	I
	Chiasmodontidae		*Pseudoscopelus*	I
LOPHIIFORMES	Ogcocephalidae		*Dibranchus*	I
LOPHIIFORMES (superfamily Ceratioidea)	Melanocetidae		*Melanocetus*	B
	Himantolophidae		*Himantolophus*	B
	Diceratiidae		*Bufoceratias, Diceratias*	B
	Oneirodidae		*Bertella, Chaenophryne, Chirophryne, Ctenochirichthys, Danaphryne, Dermatias, Dolopichthys, Leptacanthichthys, Lophodolos, Microlophichthys, Oneirodes, Pentherichthys, Phyllorhinichthys, Puck, Spiniphryne, Tyrannophryne*	B
	Thaumatichthyidae		*Lasiognathus, Thaumatichthys*	B
	Centrophrynidae		*Centrophryne*	B
	Ceratiidae		*Ceratias, Cryptopsaras*	B
	Gigantactinidae		*Gigantactis, Rhynchactis*	B
	Linophrynidae		*Acentrophryne, Borophryne, Haplophryne, Linophryne, Photocorynus*	B

* Mesopelagic platytroctidae with ventral photophores as well as the shoulder organ

species, and 4 of the 18 genera of Alepocephalidae have photophores. In the Argentiformes the barreleye Opisthoprocts have light organs that produce diffuse light through the ventral sole plate.

However, the Stomiiformes with over 400 species is one of the great bioluminescent orders of deep-sea fishes. The predominantly zooplanktivorous families, Gonostomatidae (bristlemouths),

Sternoptychidae (hatchetfishes) and Phosichthyidae (lightfishes) have primary light organs on the head and the ventral surface of the body. The family Stomiidae (barbelled dragonfishes, 287 species) has a greater diversity of light organs with the piscivorous species having bioluminescent barbels and secondary photophores scattered over the body. The Malacosteinae, *Aristostomias*, *Malacosteus* and

Photostomias have long-wavelength-emitting red photophores.

In the Aulopiformes (lizardfishes) bioluminescence is relatively rare, occurring in 4 of the 11 deep-sea families with differing morphologies suggesting that bioluminescence has evolved several times. *Chlorophthalmus* has a perianal light organ, *Benthalbella* and *Scoperarchoides* have ventral light organs, *Coccorella* has modified pyloric caecae and *Lestidium* and *Lestrolepis* have abdominal luminescent tissue. The Myctophiformes (lanternfishes) with 255 species is the second major order of bioluminescent fishes. They form an important component of pelagic biomass in the oceans, and all except one species are bioluminescent. They have ventral, lateral and circumorbital photophores.

In the Gadiformes members of 4 out of 11 families considered in this volume are bioluminescent, with 12 out of 29 macrourid genera possessing light organs. The gadiform light organs are associated with the intestine and provide ventral light sources, either shining through one or more windows in the body wall or possibly emitting luminescent fluid through the anus. In some macrourid species (*Coelorinchus* and *Hymenocephalus*) there is evidence that light organs are confined to the younger individuals, and it is suggested that the light organ is important during the pelagic early life of the fish and regresses when the fish adopts its adult benthic lifestyle (Marshall, 1965). There is no sexual dimorphism in the macrourid light organs, and their function remains obscure. Batrachoidiformes (toadfishes) of the genus *Porichthys* (midshipmen) have ventral photophores but are mostly shallow-water species with only one that has been found deeper than 200 m (*P. bathoiketes*). In the Bercyiformes there are few deep-sea bioluminescent species: one species of flashlight fish (Anomalopidae) *Parmops echinatus* is recorded in the deep sea, and there are four of species of luminous roughies in the genus *Aulotrachichthys* (Trachichthyidae).

Considering the vast diversity of Perciformes with over 7500 species, bioluminescence is remarkably rare, confined to five genera in four families. *Howella* has modified pyloric caecae, *Acropoma* a ventral bacterial light organ connected to the gut, the Epigonidae have an oesophageal light organ that can exude bioluminescent fluid through the gills and *Pseudoscopelus* has photophores distributed over the body.

In the Lophiiformes bioluminescence is confined to the bathypelagic Ceratioid anglerfishes except for one report of luminescence in the batfish (*Dibranchus atlanticus*, 1300 m$_{fbd}$). Crane (1968) elicited bright flashes of light from postmortem samples of the skin of *D. atlanticus* by chemical stimulation with peroxidase. There has been no confirmation of such luminescence in living specimens of this, or related species. All families of the Ceratioidea have luminescent organs except for the Caulophrynidae and Neoceratiidae. There are symbiotic luminous bacteria in the esca and some other light organs, whereas *Linophryne* is unique in also having intrinsically luminescent photophores in the barbel.

Bioluminescence in fishes can either be derived from symbiotic bacteria or intrinsic self-luminescence. Luminous bacteria occur in several marine ecological niches, either planktonic, in the guts of marine animals (enteric bacteria), or in decomposing organic matter (saprophytic). The light organ bacterial symbionts are derived from these within the Vibrionaceae, the genera *Vibrio* and *Photobacterium*. Fishes can acquire their bacteria from the environment in the case of facultative symbionts or by intergenerational transmission in the case of obligate symbionts (Anomalopidae, Ceratioidea). Individual bacteria do not produce light; luminescence can only occur when a certain critical population size has been reached; through a process known as quorum sensing the culture then begins to emit a steady light as long as it is provided with sufficient nutrients. This means that flashes of light are not possible from a bacterial light organ, and fishes require various shutter, reflector and filter systems to modify the light output (Haygood, 1993).

Most luminescent fishes have intrinsic light organs. The light is produced when a substrate known as a luciferin is oxidised in the presence of luciferase to create an unstable excited molecule that decays to oxyluciferin, releasing the excess energy as a photon. Two kinds of luciferins are found in fishes, the coelenterazine and crustacean *Vargula* types. The *Vargula* type is only found in coastal fishes, including *Porichthys* in Table 3.4. In all other deep-sea fishes, including the Alepocephaliformes, Stomiiformes and Myctophiformes, intrinsic luminescence is entirely of the coelenterazine type. It is likely that many species acquire the coelenterazine in their diet in a similar way to vitamins (Thompson and Rees, 1995). Coelenterazine is widespread in marine organisms, and fishes feeding particularly on bioluminescent invertebrate prey would have no difficulty in acquiring the necessary quantities. Like vitamin C, coelenterazine has antioxidative properties, and Rees et al. (1998) suggested that with reduced oxidative stress in the darkness of the deep sea, animals living there have been able to harness the light-producing properties of coelenterazine while maintaining beneficial antioxidative properties in other tissues.

Bioluminescence in deep-sea fishes fulfils multiple functions. Flashes of light or emission of luminous material can act as defence mechanisms. Dragonfishes *Astronesthes* emit trains of light flashes at up to 5 Hz from patches of luminous tissues on the head that may distract a predator. In the Platytroctidae the shoulder organ can emit scintillating groups of cells with luminous granules. Shallow-water flashlight fish often use a 'blink and run' escape response in which they suddenly turn off their lights and dash off in a different direction. The postorbital photophores of many dragonfishes and other deep-living species could be used to mislead predators before an escape response (Herring, 2002). Most fishes in the upper mesopelagic have ventral counterilluminating photophores for camouflage, and Myctophids have been shown to alter the intensity of light emitted from photophores to match the down-welling light.

Photophores on the head and around the eyes can be used as searchlights to illuminate potential prey. The mytophids such as *Diaphus effulgens* (Headlight fish) use blue lights around the eye to visualise transparent prey in the mesopelagic (Land and Osorio, 2011), and the dragon fishes *Malacosteus, Pachystomias* and *Aristostomias* (family Stomiidae) have large red-emitting photophores below the eye that emit light that cannot be detected by prey organisms (Herring and Cope, 2005). The use of lures by shallow-water anglerfishes (*Lophius* spp.) to attract prey is well established, and it is reasonable to assume that deep-sea Ceratioid anglerfishes use their luminescent lures in a similar way, although direct evidence is difficult to obtain. The Stomiid dragonfishes are remarkably effective predators on other fishes, and this success is attributed to a large extent to their elaborate luminescent barbels that are used as lures. The cookie cutter shark *Isistius* uses an unusual pattern of ventral photophores with a dark unilluminated decoy area behind the head to attract potential prey within range for attack (Widder, 1998, see Chapter 4).

It is likely that bioluminescence is used for intraspecific communication. The strongest evidence probably comes from the observation that the distribution of photophores is species specific in the stomiidae (dragonfishes) and sexually dimorphic as well as species specific in many myctophid (lanternfishes) species. This implies that bioluminescence can be used to recognise other members of the same species for schooling purposes with mating and sexual behaviour mediated through bioluminescence. Davis et al. (2014) show that the speciation rate in the stomiidae and myctophidae is much higher than in related families with undifferentiated photophore patterns (Gonostomatidae). They propose that intraspecific communication provides a mechanism for reproductive isolation leading to exceptional species richness in the Myctophids. There remains the caveat that human taxonomists do use photophore patterns as a convenient means of distinguishing species thus increasing the recognised species richness of a fauna, which might otherwise be comprised of numerous cryptic species.

3.13.5 The Olfactory and Gustatory Systems

The olfactory and gustatory systems of deep-sea fishes are generally similar to those found in shallow-water species but with increasing importance of olfaction and taste below the mesopelagic-bathypelagic transition at 1000 m depth (Kotrschal et al., 1998). Marshall (1967) estimated that more than 80 per cent of the fish species living at depths greater than 1000 m exhibit sexual dimorphism in the olfactory system. Typically the olfactory receptors of males are more complex and greatly enlarged in comparison with those of females as exemplified by bathypelagic species such as *Cyclothone* and *Gonostoma* and demersal species, including *Notacanthus*, *Halosaurus*, *Aldrovandia* and the Zoarcids (Mead et al., 1964). Sexual dimorphism reaches its extreme in the Ceratioid anglerfishes in which the juvenile males have very highly developed olfactory bulbs to aid their search for females, which are presumed to release a sexual pheromone. Mesopelagic species generally have less well-developed olfactory systems and no sexual dimorphism. However, Baird et al. (1990) reported enlarged olfactory organs in male Sternoptychids, *Valenciennellus tripunctulatus* (Subfamily Maurolicinae) and *Argyropelecus hemigymnus* (Subfamily Sternoptychinae), whereas the closely related *Sternoptyx diaphana* has much smaller nasal rosettes and no sexual dimorphism, the more typical mesopelagic condition (Baird and Jumper, 1993). Jumper and Baird (1991) modelled the olfactory search behaviour of *Argyropelecus hemigymnus* in an open ocean environment and predicted that a male should be able to detect pheromones released by a female at ranges of 30–80 m within about one hour compared with eight days required to find a female in the absence of olfactory cues. There is no direct information on pheromones of deep-sea fishes, but pheromonal communication is well established in the homing and mating behaviour of salmon (*Salmo*, *Oncorhynchus*; Sorensen and Wisenden, 2015), and it is highly likely that similar principles apply in deep-sea species.

An obvious application of olfaction and gustatory sensing is in the search for food (Montgomery & Pankhurst, 1997), and this is evident in benthic baited camera observations. Fishes approach the bait from downstream and change their position to remain within odour plume during tidal changes in current direction (Priede and Bagley, 2000). Bailey et al. (2007) observed that abyssal grenadiers *Coryphaenoides armatus* adopt a head-down swimming posture with the barbel touching the seafloor. On arrival at the bait source they search using the barbel across the bait and surrounding surfaces. Histological investigations showed that there are about 450 taste buds per mm^2 on the barbel, which is richly innervated with 20,000 nerve axons (compared with 50,000 axons in the optic nerve). They propose that olfactory searching enables the fish to initially locate the odour source, but that this is followed by tactile and gustatory searching using the barbel to locate the food source. Additional taste buds in the skin of the head, fins and in the oropharyngeal cavity are also probably implicated.

3.13.6 Underwater Sound and the Octavo Lateral System

In the quiescent regime of the abyss, remote from surface sea noise, waves and strong currents, the acoustic domain has the potential to be particularly effective for the detection of prey and for communication. Marshall (1954) first proposed that sound may be an important means of communication for deep-sea fishes based on the observation that the males of most species of the dominant Macrourinae subfamily of the macrourids have drumming muscles on either side of the fore part of the swim bladder (Marshall, 1965). These are homologous to similar muscles found in shallow-water Gadoids and Scianids, which are known to produce knocking or croaking sounds as part of their courtship, agonistic and defensive displays (Hawkins and Amorim, 2000). The sounds are generated by fast contractions of the drumming muscles that vibrate the swim bladder at 100–300 Hz. Drumming muscles have also been described in the Ophidiidae (cusk eels) in the two major subfamilies Ophidiinae and Neobythitinae. In the Neobythitinae from the upper slopes in the W Pacific; *Hoplobrotula armata* (Armoured cusk, 350 m_{fbd}), *Neobythites longipes* (Longray cusk 480 m_{fbd}) and *N. unimaculatus* (Onespot cusk, 567 m_{fbd}) Ali et al. (2016) found larger medial drumming muscles in males than in females and suggested that males produce calls that attract females. Because the Ophidiidae and Macrouridae make up more than half the deep-sea benthic fish fauna, it might be supposed that the deep sea would be rich in sound produced by the fishes living there. Several attempts have been made to test Marshall's hypothesis by deploying hydrophones in the deep sea, but convincing recordings of fish sounds have proved elusive. Mann and Jarvis (2004) located a probable deep-sea fish sound at 548–696 m depth off Bermuda, and Rountree et al. (2012) recorded a few candidate fish sounds at 682 m depth described as 'drumming, duck-like or unknown' in the frequency range expected from fishes below 1.2 kHz. Attempts both at a cabled observatory in the Barkley Canyon at 985 m depth in NE Pacific Ocean (Wall et al., 2014) and using a lander at multiple locations across the Mid-Atlantic Ridge at 2500 m depth (Cousins et al., 2013, acoustic results unpublished) have failed to obtain unequivocal evidence of deep-sea fish sounds. Marshall (1965) points out that the abyssal species that he examined (*Coryphaenoides carapinus*, 5610 m_{fbd}, and *C. armatus*, 5180 m_{fbd}) have no drumming muscles. It is possible that only slope-dwelling bathyal species have sound-producing capabilities. Deep-sea fishes also may only use sound production very sparingly, for example, during very rare, possibly once in a lifetime, spawning events, so that the concept of a continuous acoustic chatter between individuals may be erroneous. Compared with acoustic recordings from marine mammals with echolocation clicks up to over 100 kHz, the low-frequency sounds of fishes < 2 kHz can be remarkably difficult to discern against background noise from the sea, ships and other human activities.

The Octavo lateral system of fishes, also known as the Acoustico-lateralis system, comprises the hair-cell-based sensory systems of the ear and the lateral line (Popper and Fay, 1993). Deng et al. (2011) described the inner ear morphology in the deep-sea Morid, *Antimora rostrata*. It follows the typical plan for a teleost fish (Bone and Moore, 2008) with three semicircular canals in each labyrinth for detection of angular accelerations with hair cells in the ampullae. Three chambers at the base of the labyrinth each contain an otolith, in order of size the sagitta, which is the largest lies in the sacculus, the asteriscus in the lagena and the lapillus in the utriculus. The otoliths are attached to the hair cells of the sensory epithelium (or macula) by a gelatinous otolith membrane. The hair cells of the maculae constitute the hearing organ but also with the otoliths as a reference mass, detect linear accelerations. The pattern of hair cell orientation in the saccular macula of *Antimora* is more complex than in any other vertebrate, and Deng et al. (2011) speculate that this may be linked to exceptionally well-developed hearing capability. There is a direct connection between the swim bladder and the saccule, which would further enhance hearing sensitivity because compression of the swim bladder converts sound pressure waves to fluid displacements.

The sagitta otoliths are larger in *Antimora* than in most other fishes, and Marshall (1965) also notes that sagittas are larger in those macrourids that have sound-producing mechanisms in the swim bladder. The deeper-living *Coryphaenoides* species with no drumming muscles have smaller sagitta otoliths. Lombarte and Cruz (2007) found in the NW Mediterranean that epipelagic fishes have very small otoliths, and amongst demersal species the sagitta size increases with depth until 750 m. Fishes living deeper than 1000 m, were found to have smaller otoliths. This is suggestive of a peak in sound production and hearing capability at around 750–1000 m depth with sound becoming less important in the abyss.

The inner ear also may be implicated in pressure sensing in fishes. Tytler and Blaxter (1973) show that gadoids can detect pressure changes of 0.4–1 per cent of the ambient pressure and suggested that this is mediated through pressure stretch receptors in the swim bladder wall. However, although the swim bladder improves sensitivity, it is not essential; the threshold of sensitivity in the dab (*Limanda limanda*) with no swim bladder is 0.01–0.02 bar (Blaxter, 1979). Using direct recordings from the auditory nerve of dogfish *Scyliorhinus canicula* Fraser et al. (2003) found that the nerve spike frequency tracked cyclical pressure changes of 0.3 bar amplitude with a threshold of detection by hair cells in the labyrinth of around 0.03 bar. The precise mechanism of pressure transduction by hair cells remains unclear.

The lateral line system extends from the head of the fish to tail and is made of rows of neuromast organs with hair cells that can detect movement of water relative to the body of the fish. The neuromasts can be free-standing on the surface of the skin or buried inside canals below the surface of the skin. Free-standing neuromasts are sensitive to slow flows and low frequencies (10–60 Hz), whereas those in lateral line canals are most sensitive to higher frequencies, 50–200 Hz). Proliferation of free-standing neuromasts, widening of the canal system on the head and enlargement of the canal organs are perceived as adaptations to the still-water conditions of the deep sea (Marshall, 1996). The most extreme examples of free-standing neuromasts are probably in the deep-sea Ceratioid anglerfishes (Plate 8b), *Bufoceratias wedli* (Diceratiidae), which has no canal system, and the neuromasts are raised up on papillae in rows around the head, lower jaw and along the sides of the body. *Neoceratias spinifer* (Neoceratiidae) has neuromasts on long stalks. Such exposed neuromasts are extremely sensitive but must also be vulnerable to noise from the fish's own movements and water currents. They can only work if the fish is neutrally buoyant and motionless (Montgomery and Pankhurst, 1997). Marshall (1996) also describes in *Poromitra capito* (Melamphaidae) the presence of an array of specialised superficial neuromasts on the front of the head between the nares, which may work in tandem with the olfactory system to detect and evaluate potential prey.

A further means by which sensitivity of the lateral line can be enhanced is by making it as long as possible, which may partially account for the elongated shape of many deep-sea fishes. Marshall (1966) attributed this hypothesis to Dr. P. Orkin of the University of Aberdeen citing Wynne-Edwards (1962). The beam width (β) of a sonar is inversely proportional to the length or aperture of the array (L).

$$\beta \approx \frac{\lambda}{L}$$

where λ is the wavelength. Thus a long lateral line enables more precise localisation of disturbances in the water. Wynne-Edwards (1962) lists the scabbard fishes Family Trichiuridae (*Lepidopus* and *Trichiurus*), the eels including the deep-sea gulpers (*Saccopharynx*), snipe eels (*Nemichthys*), the deal fish (*Trachypterus*), oarfish (*Regalecus*), Macrourids, the frilled shark (*Chlamydoselachus*) and the chimaeras as examples of deep-sea fishes with elongated lateral lines. In *Harriotta raleighana* (Pacific longnose chimaera) the lateral line is extended not only along the tail but also anteriorly along the elongated snout. Wynne-Edwards (1962) shows that a functional lateral line extends to the tip of the whip-like tail in these species, innervated by well-developed branches of the vagus nerve. He further comments that few of the long-bodied species have bioluminescent organs and that most luminescent species are small short fish and suggests these represent two alternative adaptions to maintaining social contacts in the deep sea.

An intriguing possibility is that with appropriate neural processing, fishes might be able to use a synthetic aperture sonar technique whereby the fish moves forward slowly integrating the lateral line signal into a long baseline exceeding the body length of the fish. This would provide a very narrow effective beam width and high spatial resolution at long ranges.

3.14 Time in the Deep Sea

For shallow-water fishes it is well established that activity, feeding, growth and reproduction are under control of a circadian (≈24 h) clock (Kulczykowska et al., 2010). Mesopelagic fishes of the deep scattering layer and those that feed on them are clearly coupled to the circadian cycle but the question arises as to whether fishes living deeper than 1000 m beyond the reach of solar light have a biological clock. Priede et al. (1999) investigated a series of demersal fishes from increasing depths *Phycis blennoides* (1200 m$_{fbd}$) Phycidae, *Trachyrinchus murrayi* (1630 m$_{fbd}$) Trachyrincidae, *Nezumia aequalis* (2320 m$_{fbd}$), *Coryphaenoides rupestris* (2600 m$_{fbd}$), *Coryphaenoides guentheri* (2830 m$_{fbd}$) *Coryphaenoides armatus* (5180 m$_{fbd}$) Macrouridae and found that despite variation in brain morphology all species had melatonin-binding sites and melatonin receptor gene expression in the optic tectum and specific brain regions similar to shallow-water fishes. Melatonin is the mediator hormone that transmits biological rhythms generated in the central nervous system to peripheral tissues and is secreted by photoreceptor-related cells in the pineal and the retina. Wagner et al. (2007) detected release of melatonin from freshly isolated pineals and retinas of *Synaphobranchus kaupii* and *Coryphaenoides armatus* and for the former found a significant cycle of spontaneous release related to the local tidal/lunar cycle. As this showed that there is a functional biological clock in these deep-sea species, they hypothesised that it has become adapted to use the tidal cycle as a *zeitgeber* instead of the solar cycle. The fish would be aware of the tidal cycle from changes in current flow and possibly pressure transduction. Biological rhythms of activity, growth and reproduction are therefore probably sustained in the deep sea but do not depend on the solar light as the environmental cue.

3.15 Diseases and the Immune System

Individual deep-sea fishes are often found with physical injuries such as damage to the skin or the tip of the tail missing, a common phenomenon in macrourids. It is also well established that deep-sea fishes have parasites (Klimpel et al., 2009). In a video survey of demersal fishes of the NW Atlantic, Quattrini and Demopoulos (2016) found that 9 per cent of individuals were carrying ectoparasites. The diversity of parasites decreased with depth from 494–3262 m. Campbell (1980) showed that in the abyssal grenadier *Coryphaenoides armatus* the intensity of infections with helminth parasites decreases with depth and showed a correlation with macrofaunal abundance. Bray (2004) surveyed the depth range of digenean parasites in fishes and found 17 families that occur in the deep sea with a clear decrease in diversity with depth and only four families occurring at depths greater than 4000 m. Haematozoan parasite infections have been detected in blood smears taken from a variety of deep-sea demersal species including a trypanosome found in the chimaera *Harriotta raleighana*, which hints at a possible ancient origin of such host–parasite relationships (Davies et al., 2012). Davies and Merrett (1998) also report putative viral erythrocytic necrosis (VEN) in the abyssal grenadier *Coryphaenoides armatus*. It seems that deep-sea fishes are exposed to the full spectrum of potential pathogens but that there is a general trend of decrease in prevalence and intensity of infections with depth as potential hosts and sources of infection decrease in abundance.

The immune system has not been properly described in any deep-sea fish. It can be assumed that they possess the generalised adaptive immune system that evolved in the lower vertebrates at the same time as the origin of jaws (Matsunaga and Rahman, 1998). Fishes exhibit a full range of specific defence mechanisms, both humoral and cell-mediated with an immunological memory that enables vaccination of species kept in fish farms (Secombes and Ellis, 2012). However sequencing of the genome of cod (*Gadus morhua*) has revealed that an important link that activates the cellular immune response, MHC II (major histocompatibility complex II), found in all other vertebrates is missing (Star et al., 2011). Star and Jentoft (2012) propose that gadiform fishes lost the MHC II genes during a deep-sea phase in their evolutionary history (Howes, 1991) when the adaptive value was insufficient to compensate for the metabolic cost of maintaining such a system in the face of reduced challenges from pathogens and low energy availability. Those lineages like the modern cod that have invaded shallow seas have acquired compensatory changes in the immune system that make them no more susceptible to disease than most other vertebrates. There is no information available as to whether deep-sea gadiforms retain a reduced immune system or whether other deep-sea lineages have similar deficiencies.

Table 3.5 **The Maximum Depth Limits of Fishes**

Depth (m)	Name	Description	Reference
2000	Lethal limit for shallow species	At this depth shallow-water fish cannot survive compression experiments. Many shallow-water invertebrates can survive much higher pressures (8000 m)	Sébert (1997)
3003	Class Myxine Deepest species	Deepest record of a hagfish, *Eptatretus carlhubbsi* (Myxinidae)	Yeh and Drazen (2009)
3000	Class Chondrichthyes deepest species	General maximum depth limit of Chondrichthyes; only 3 species with deeper records.	Priede et al. (2006)
3000	Class Chondrichthyes deepest holocephalan	Deepest Holocephalan; *Hydrolagus affinis* (Chimaeridae)	Krefft (1990)
3700	Class Chondrichthyes deepest shark	*Centroscymnus coelolepis* reported to 3700 m	Forster (1973)
4156	Class Chondrichthyes deepest species	Deepest recorded Chondrichthyan, *Rajella bigelowi* (Rajidae)	Stehmann (1990)
4699	Class Chondrichthyes Limit of muscle TMAO/Urea ratio	Muscle urea content theoretically reaches zero, it is not possible to further increase TMAO concentration to protect protein function against pressure whilst retaining osmoconformity.	Laxson et al. (2011)
8145	Deepest Fish observation	*In situ* video. 'Ethereal snailfish' Mariana Trench.	Linley et al. (2016)
8370	Deepest Fish capture	*Abyssobrotula galatheae* single trawled specimen from the Puerto Rico Trench. Further hadal occurrence of this species has never been confirmed	Nielsen (1977)
8400	Ultimate depth limit for teleost fishes	Body fluids are isosmotic with seawater; the concentration of the piezolyte TMAO cannot be increased further. Enyme function impaired at greater depths.	Yancey et al. (2014)
11000	Full ocean depth	Depth of the Challenger deep in the Mariana Trench	Jamieson (2015)

3.16 The Maximum Depth Limit for Fishes

As shown in the present chapter, fishes have adapted in various ways to increasing constraints of living at greater and greater depths (Table 3.5). The deepest observation of a living fish is a Liparid, the ethereal snailfish at 8145 m depth in the Mariana Trench (Linley et al., 2016), and the deepest captured specimen is *Abyssobrotula galatheae* reported to have been trawled from 8370 m in the Puerto Rico Trench (Nielsen, 1977). No fishes have been seen or captured from greater depths, hence the deepest 25 per cent of the ocean depth range 8400–11000 m is devoid of fish. Yancey et al. (2014) show that in teleost fishes, the concentration of the piezolyte TMAO increases in body fluids with increasing depth, successfully protecting their protein function at progressively higher pressures. A limit is reached when the body fluids are isosmotic with the surrounding water and the TMAO concentration can be increased no further. Yancey et al. (2014) predicted that the maximum depth limit for teleost fishes is reached at a depth of 8200–8400 m, which is remarkably close to the maximum observed depths of fishes. Buoyancy from swim bladders and lipid accumulation are probably too energetically costly and hadal fishes appear to depend on watery tissues and reduced skeletons to attain neutral buoyancy.

The reasons for exclusion of chondrichthyans from the abyss are not so clearcut. Musick and Cotton (2015) show that sharks and rays are large fishes that occupy the highest trophic levels (> 4) in bathydemersal communities and argue that they are therefore ecologically excluded from the food-sparse niches of the abyss and are probably unable to accumulate sufficient high energy lipids in the liver to maintain neutral buoyancy. Because chondrichthyans accumulate TMAO in their tissues it might be supposed that they are physiologically preadapted to the deep-sea by virtue of the protein-protecting function of TMAO. This may be true because Laxson et al. (2011) show that muscle TMAO concentration does increase with depth but in order to maintain osmotic balance the urea concentration correspondingly decreases. Extrapolating the urea data, zero concentration is reached at 4699 m which Laxson et al. (2011) propose may represent the physiological maximum depth for Chondrichthyes because no further increase in TMAO would be possible. Treberg and Speers-Roesch (2016) review this and other potential physiological limits to the depth distribution of chondrichthyans. They identify three major physiological hypotheses. The first is essentially the Laxson et al. (2011) hypothesis, that the chondrichthyan osmoregulatory system, using urea and TMAO, is impossible to reconcile with the effects of changes in pressure and temperature and the need to maintain stability of proteins and membranes. The second extends the hypothesis of Priede et al. (2006) regarding the high cost of lipid buoyancy and shows that accumulating the necessary energy for buoyancy compromises the growth rate and reproductive output in the oligotrophic conditions of the abyss in comparison with teleost fishes. The third suggests that chondrichthyans may be unusually nutrient-limited through their osmoregulatory strategy, which entails constant loss of nitrogen across the gills, which must be replaced through the diet. They conclude that these three mechanisms probably act in concert. Together with the ecological hypothesis of Musick and Cotton (2015) it is probable that multiple factors lead to the general exclusion of Chondrichthyes from the abyss.

4 Systematic Description of Deep-Sea Fishes

The deep sea has been independently colonised by fishes numerous times, resulting in a diverse assemblage of unrelated species, genera, families and higher taxa that can be regarded to varying degrees as deep-sea specialists. It is therefore not possible to describe a single pathway representing the evolution of deep-sea fishes. Rather, in this section deep-sea species found on the branches and twigs of the evolutionary tree of fishes are described in relationship one to the other. Broadly the classification of fishes in Nelson (2006), Catalog of Fishes (Eschmeyer, 2014) and FishBase (Froese and Pauly, 2016) is followed, in particular at family level, but higher-level arrangement of orders takes into account recent evidence from molecular phylogeny (Betancur-R et al., 2013; Janvier, 2010; Licht et al., 2012; Near et al., 2012, 2013; Sorensen et al., 2014). The historical origin of these groups, over geological time, is described in Chapter 2.

4.1 Class Myxini

4.1.1 Order Myxiniformes (Hagfishes)

4.1.1.1 Family Myxinidae (1) (Hagfishes)

Within the Myxiniformes there is one family of hagfishes, the Myxinidae, with 78 species divided between two subfamilies, the Eptatretinae, including *Eptatretus, Paramyxine* and *Rubicundus* (51 species), and Mxyininae, including *Myxine, Nemamyxine, Neomyxine* and *Notomyxine* (27 species). The Myxininae tend to occupy deeper waters than the Eptatretinae, with over 75 per cent of species with maximum depths > 500 m. Nevertheless the deepest recorded species of hagfish is *Eptatretus carlhubbsi* at

3003 m from the Hawaiian region of the North Pacific Ocean (Yeh and Drazen, 2009). Species have regional distributions on bathyal slopes around the ocean basins, for example *Myxine glutinosa* and *Myxine ios* are found in the NE Atlantic and are replaced by *Myxine limosa* in NW Atlantic with an overlap zone off Canada and Northern USA between *M. glutinosa* and *M. limosa*. Hagfish occur in all oceans except the Arctic Ocean (albeit *M. glutinosa*, however, does extend from the North Atlantic as far as Murmansk), the Southern Ocean and the Red Sea. They are absent from shallow tropical seas, remaining in deep water at low latitudes. Hagfishes are nonmigratory, considered not to move more than 100 km during their life span. They generally adopt a burrowing habit, so are most abundant on soft sediment-covered sea beds, emerging when there are feeding opportunities or remaining within the sediment to forage on benthic infauna (Jørgensen et al., 1998). Møller and Jones (2007) described a putative vent-endemic species, *Eptatretus strickrotti*, from a hydrothermal vent site at 2211 m depth on the East Pacific Rise.

The hagfish body form is simple; elongated and eel-like with no fins other than the caudal fin, which has no rays and extends onto dorsal and ventral surfaces. Though the giant hagfish, *Eptatretus carlhubbsi*, has been recorded at 116 cm long, the average maximum length attained is ca. 50 cm. The Eptatretinae are considered to be the more primitive of the two subfamilies and retain the ancestral multiple pairs of external gill openings on either side of the pharynx. In the Myxininae there is a duct on either side taking the outflow water from the gills to a single pair of external gill openings (Fig. 4.1). In addition all hagfish have a pharyngocutaneous duct on the left side, which acts as a shunt that can expel

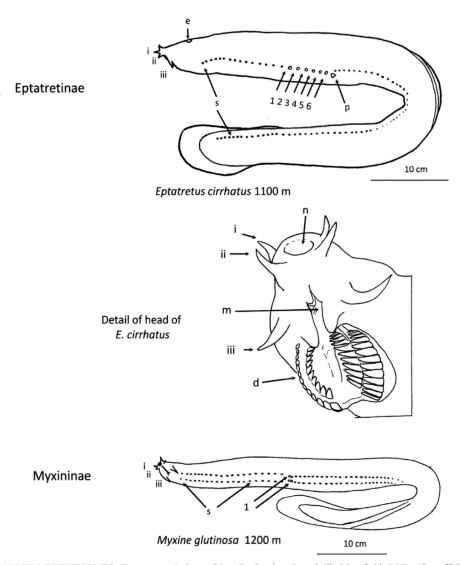

Eptatretinae

Eptatretus cirrhatus 1100 m

Detail of head of
E. cirrhatus

Myxininae

Myxine glutinosa 1200 m

Figure 4.1 ORDER MYXINIFORMES. *Eptatretus cirrhatus* (New Zealand or Broadgilled hagfish), W Pacific, off S and E Australia and New Zealand. Note the multiple paired gill openings and dorsal eye spot. Detail of the ventral view of the head with the dental plate everted after Zintzen et al. (2011). *Myxine glutinosa* (Atlantic hagfish) N Atlantic and W Mediterranean. Note only one pair of gill openings and no eye spots. Key: d – dental plate with two rows of teeth, anterior and posterior on each side, e – eye, i – first pair of barbels, ii – second pair of barbels, iii – third pair of barbels, m – mouth, n- nostril, p – pharyngocutaneous duct aperture, s - slime pores, 1,2,3,4,5,6 - paired gill apertures. The depths given are the maximum depths for each species recorded in FishBase.

excess water and debris directly from the pharynx to the outside without passing through the gills. In Myxininae the pharyngocutaneous duct connects to the left common branchial duct and therefore has no separate external opening. Around the head there are three pairs of barbels, the first two pairs surrounding a single nostril on the snout, which is the main route for the intake of water for respiration (McNeil Alexander, 1975). In the Eptatretinae there are conspicuous eye spots on the top of the head, whereas the Myxininae, considered to have become specialised for deep-water conditions, have no eye spots. Fernholm and Holmberg (1975) found rudimentary eyes in *Myxine* and concluded that there has been an evolutionary trend towards reduction of visual capability and that deep-water occurrence of *Eptatretus* species with functional eyes is the result of recent diversification.

Although hagfish have no jaws, they have a remarkable cartilaginous toothplate that can be everted from the mouth onto the ventral surface of the head. This is equipped with keratinous teeth that can exert a strong biting action from side to side. Hagfish are well known as carrion feeders (Plate 2a) arriving in vast numbers at whale falls (Smith et al., 2015), but they are also able to actively pursue and consume live prey (Zintzen et al., 2011). Stomach content analyses indicate consumption of a wide variety of prey including polychaetes, shrimps, hermit crabs, cephalopods, brittlestars, bony fishes, sharks, birds and whale flesh. Stable isotope analysis suggests selective feeding by the three different species found in New Zealand waters (Zintzen et al., 2013). They are very agile, can burrow into prey carcases and knot the body to create an anchor or gain purchase.

Hagfishes are notable for their ability to produce mucus or slime; the name Myxine is derived from the Greek μύξα (=mucus), and they are widely known in 'slime eels'. The skin is naked, devoid of scales so the ability to produce slime is an important defence mechanism. The slime has a remarkably low concentration of mucin, 700 times less than human saliva, so cannot be regarded as a mucus (Fudge et al., 2005). The most important constituent of the slime is tapered threads of protein, 10–17 cm long and 1–3 μm in diameter. When expelled, these form a fine mesh-work held together by mucin, which acts as a spot adhesive between fibres, trapping sea water to produce a viscous mass when agitated. The body of the hagfish is not generally coated with slime, the slime is produced from 70 to 200 pairs of pores distributed along the ventral surface of the body from each of which slime is expelled as a jet with a velocity of 0.18 ms^{-1} in the part of the body that is stimulated (Lim et al., 2006). A fish predator attempting to bite on a hagfish quickly finds its mouth and gills clogged with fibrous slime, effectively deterring further pursuit of the prey

(Zintzen et al., 2011). The stored slime represents about 3–4 per cent of the hagfish body mass, and a typical Pacific hagfish *Eptatretus stoutii* weighing 60 g can generate 24 l of hydrated slime (Fudge et al., 2005), representing a highly efficient bulk defence from a small quantity of protein.

Hagfish produce up to 30 large (ca. 35 mm diameter) yolky eggs, which are equipped with hooks so that they form bunches on the seafloor. In general there is no evidence of a breeding season with a very small proportion of males and females being in the mature state throughout the year (Martini and Beulig, 2013). There is no larval stage; young hatch from the eggs as miniature adults.

Hagfish have been traditionally fished in Japan and Korea in small quantities for food and bait but since the 1940s there has been increased exploitation for food and for leather, so-called eelskin, largely processed in South Korea (Ellis et al., 2015).

4.1.2 Order Petromyzontiformes (Lampreys)

4.1.2.1 Family Petromyzontidae (2) (Lampreys)

The lampreys are anadromous and freshwater fishes of the Northern Hemisphere and are included here mainly because of a single capture of *Petromyzon marinus* (sea lamprey) at 4099 m depth in the NW Atlantic (Haedrich, 1977). They are well known as ectoparasites on other fishes, and freshwater populations of *P. marinus* have had a devastating effect on trout populations in the Great Lakes of North America. In the oceans *P. marinus* are often seen attached to fishes, sharks and marine mammals, leaving characteristic circular wounds from the teeth, tongue and sucker-like oral disc. When not attached to a host the oral disc is folded to give a streamlined shape to the snout, and they can swim fast enough to attach to an adult mackerel (*Scomber scombrus*, personal observation). Sea lampreys grow to over 1 m in length and are capable of travelling large distances during several years at sea before ascending rivers in springtime to spawn. They are certainly common at upper bathyal depths, and the abyssal capture gives this species one of the largest depth ranges known for any fish.

4.2 Class Chondrichthyes

Subclass Holocephali

This subclass has a rich fossil history with seven orders, only one of which remains extant, the Chimaeriformes.

4.2.1 Order Chimaeriformes

This order comprises three extant families, Callorhinchidae (plownose chimaeras), which are shallow-water species, and the two families found in the deep sea, the Rhinochimaeridae (longnose chimaeras, eight species) and the Chimaeridae (Shortnose chimaeras,) with mean maximum depths of 1390 m and 1200 m, respectively. They are variously known as rabbitfish, ratfish, spook-fish or ghost sharks. The deepest Rhinochimaerids are *Harriotta haeckeli*, (Smallspine spookfish) and *Harriotta raleighana* (Pacific longnose chimaera) both recorded at 2600 m and the deepest Chimaerid is *Hydrolagus affinis* (Smalleyed rabbitfish) recorded at 3000 m (Plate 2b). *H. haeckeli* and *H. raleighana* both have global distributions from the North Atlantic to the SW Pacific off Australia and New Zealand, whereas *H. affinis* is confined to the North Atlantic Ocean. The Rhinochaemaeridae has relatively few worldwide species, whereas the Chimaeridae has more species but with regional distributions, such as *Hydrolagus deani* (Philippine chimaera, 770 m$_{fbd}$) known only from the Philippines in the Central West Pacific. The average maximum length of Chimaeras is ca. 1m, though some species can reach 1.4 m. They are demersal fishes, found on bathyal slopes of continental margins, ocean ridges, islands and seamounts in all the world's oceans from the Arctic to the sub-Antarctic. The chimaeras are true cartilaginous fishes with no bony skeleton, fins, rays or scales; the only mineralised parts are their highly resistant tooth plates and fin spines which account for the rich fossil record for these fishes. There are three pairs of tooth plates, two pairs in the upper jaw and one pair in the lower jaw, the anterior plates protrude from the mouth to resemble the incisors of rodents giving rise to their names, rat-fish, rabbitfish, or

Hydrolagus (=water hare in Greek). Unlike sharks, which can replace their teeth, the tooth plates grow continuously throughout the life of the fish and are not replaced. The head is large with conspicuous eyes, and the body tapers to a long whip-like tail. There are paired ventral nostrils connected to the mouth via grooves. The lateral line system is very conspicuous on the naked skin with a branching pattern of canals and pores over the head region. There is no spiracle, and the four gill slits on each side of the head are covered by a fleshy operculum with a single exhalent opening on either side of the head anterior to the pectoral fin. The pectoral fins are large and propel the animal forward with a wing-like flapping motion. There are two dorsal fins, the first of which is reinforced with a sharp anterior spine, which can be mildly toxic and has been the cause of significant injuries to deep-sea fishermen handling *Chimaera monstrosa* in trawl bycatches in the NE Atlantic (Hayes and Sim, 2011).

Mature male chimaeras are equipped with remarkable copulatory organs. First, there is a pair of claspers attached to the pelvic fins, as in sharks, but each of these divides into two branches. Second, there is a pair of prepelvic tenaculae armed with teeth, which retract into pouches anterior to the pelvic fins. Third, on top of the head there is a club-like frontal tenaculum also equipped with teeth to grasp the female. Males have been observed to hold onto the pectoral fin of the female using the frontal tenaculum during mating (Tozer and Dagit, 2004). All chimaeras are oviparous, the female usually extruding one pair of eggs at a time, one from each oviduct. Each egg is contained in a horny case, which has a central longitudinal spindle-shaped egg chamber and wing-like lateral webs that prevent rolling and help anchor the case to the surface of the sediment. Chembian (2007) observed egg cases of *Rhinochimaera atlantica*, ca. 25 cm long at 281–301 m depth in the Northern Indian Ocean. Incubation time can be up to one year, and the hatchlings resemble small adults so that advanced embryos can be identified to species. There is generally no targeted fishery for chimaeras; however, the flesh is edible, and livers can be processed for oil, which is valued so they are taken as a bycatch in some fisheries together with deep-water

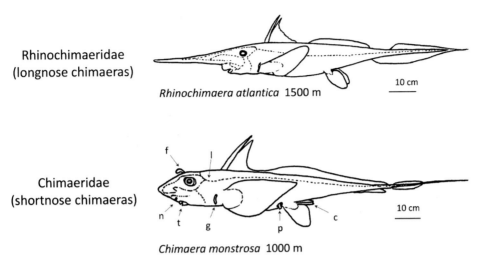

Rhinochimaeridae
(longnose chimaeras)

Rhinochimaera atlantica 1500 m

10 cm

Chimaeridae
(shortnose chimaeras)

Chimaera monstrosa 1000 m

Figure 4.2 ORDER CHIMAERIFORMES. *Rhinochimaera atlantica* (Straightnose rabbitfish) N Atlantic. *Chimaera monstrosa* (Rabbitfish) NE Atlantic and Mediterranean. Mature male: c – claspers, f – frontal tenaculum, g – gill aperture, l – lateral line canals, n – nostril, p – prepelvic tenaculum, t – tooth plates.

sharks. In some deep-sea trawl fisheries in the NE Atlantic *Chimaera monstrosa* may constitute up to 15 per cent of the discards (Dagit et al., 2007). Two species of the family Callorhinchidae, *Callorhinchus capensis* (Cape elephantfish, 374 m_{fbd}) from off South Africa and *C. milii* (Ghost shark, 227 m_{fbd}) off Australia and New Zealand are listed deep-sea sea commercial species by Tandstad et al. (2011).

4.2.1.1 Family Rhinochimaeridae (3) (Longnose chimaeras)

This family comprises eight species in three genera (Fig. 4.2). The genus *Harriotta* is almost circumglobal with an antitropical distribution. *H. haeckeli* (Smallspine spookfish, 2600 m_{fbd}) occurs in the N Atlantic, Indian Ocean and SW Pacific and *H. raleighana* (Pacific longnose chimaera, 2600 m_{fbd}) in the Atlantic, Pacific and W Indian Oceans. *Neoharriotta* is confined to lower latitudes, *N. carri* (600 m_{fbd}) in the W central Atlantic, *N. pinnata* (Sicklefin chimaera, 500 m_{fbd}) in the E Atlantic and W Indian Oceans and *N. pumila* (380 m_{fbd}) W Indian Ocean. *Rhinochimaera atlantic* (Straightnose rabbitfish, 1500 m_{fbd}) is antitropical in the N and SE Atlantic and *R. pacifica* (Pacific spookfish, 1490 m_{fbd}) in the Pacific off Japan, Australasia and Peru.

4.2.1.2 Family Chimaeridae (4) (Shortnose chimaeras)

The family Chimaeridae (Fig. 4.2) comprises 39 species in two genera, *Chimaera* (14 species) and *Hydrolagus* (25 species). The genus *Chimaera* has five species endemic to the Atlantic Ocean (N and SE), one in the Indian Ocean, three Indo-West Pacific species around Australia and six species from the W Pacific from Japan to New Zealand but appears to be absent from the E Pacific. The genus *Hydrolagus* (Plate 2b) is circumglobal with six species in the Atlantic Ocean, 14 species endemic to different parts of the Pacific Ocean, two species endemic to the Indian Ocean, two Indo-W Pacific species around Australia and a Southern Ocean species *H. homonycteris* (Black ghostshark, 1447 m_{fbd}) off Australia and New Zealand. *H. novaezealandiae* (Dark ghostshark, 950 m_{fbd}) endemic to New Zealand and *C. monstrosa* (rabbitfish, 1500 m_{fbd}) in the NE Atlantic are listed by FAO as exploited deep-sea species (Tandstad et al., (2011, 2005).

Subclass Elasmobranchii
Subdivision Selachii (Sharks)

The sharks are divided into two clades, here ranked as Superorders according to Nelson (2006), the Squalomorphi and Galeomorphi.

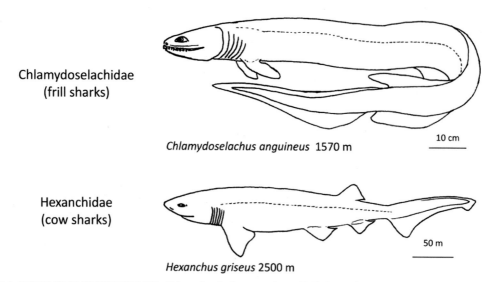

Chlamydoselachidae (frill sharks)

Chlamydoselachus anguineus 1570 m

10 cm

Hexanchidae (cow sharks)

Hexanchus griseus 2500 m

50 m

Figure 4.3 ORDER HEXACANTHIFORMES. *Chlamydoselachus anguineus* (frilled shark) Almost circumglobal, patchy distribution74°N–58°S. *Hexanchus griseus* (bluntnose sixgill shark) Circumglobal 65°N–48°S.

Superorder Squalomorphi

The Squalomorphi comprises four orders, Hexanchiformes, Pristiophoriformes, Squatiniformes and Squaliformes.

4.2.2 Order Hexanchiformes

The Hexanchiformes is a small (six species) but distinctive group of sharks often regarded as primitive owing to the six or seven gill slits compared with the five pairs that are normal in neoselachii. There are two families.

4.2.2.1 Family Chlamydoselachidae (5) (Frilled sharks)

The family Chlamydoselachidae has two species: *Chlamydoselachus anguineus* (Frilled shark; Fig. 4.3), which has a patchy worldwide distribution, and *Chlamydoselachus africana* (African frilled shark), found only in the SE Atlantic off Angola and Namibia. They grow up to 2 m long, the females larger than the males, and are demersal or benthopelagic down to 1500 m depth. There are six pairs of gills with characteristic frills, with the first gill slits almost completely encircling the body giving a cutthroat appearance. The body is elongated with dorsal and anal

fins set far back. The teeth, each with three inward-curving needle-like cusps, protrude slightly beyond the edge of the jaws, aiding capture of squid that comprise up to 60 per cent of the diet in specimens captured in Suruga Bay off Japan (Kubota et al., 1991). *C. anguineus* is ovoviviparous, producing young 40–60 cm long after a gestation period of possibly three years.

4.2.2.2 Family Hexanchidae (6) (Cow sharks)

The family Hexanchidae are also ovoviviparous and the deepest-living are *Heptranchias perlo* (Sharpnose sevengill shark), 1000 m_{fbd}, and *Hexanchus griseus* (Bluntnose sixgill shark), 2500 m_{fbd} (Fig. 4.3). *H. perlo* is a relatively small shark, maximum length 137 cm but is considered an aggressive active top predator consuming other sharks, teleost fishes and squids (Braccini, 2008). *H. griseus* is a large shark that grows to 4.8 m long and is often attracted to carrion falls.

4.2.3 Order Pristiophoriformes

4.2.3.1 Family Pristiophoridae (7) (Saw sharks)

This order comprises one family, the Pristiophoridae, with seven species in two genera. The monotypic *Pliotrema warreni* (Sixgill sawshark, 430 m_{fbd}) is found in the SW Indian Ocean off South Africa and Madagascar. Of the six species of the genus *Pristiophorus* three species are found around Australia, *P. japonicus* (Japanese sawshark, 500 m_{fbd}) is found in the NW Pacific, *P. nancyae* (African dwarf sawshark, 500 m_{fbd}) and the only really deep-living species is *Pristiophorus schroederi* (Bahamas sawshark, 1000 m_{fbd}) from the NW Atlantic.

4.2.4 Order Squatiniformes

4.2.4.1 Family Squatinidae (8) (Angel sharks)

This order comprises 23 species in one family Squatinidae and one genus found only in the Atlantic and Pacific Oceans. Their mean maximum depth of occurrence is 368 m, but *Squatina dumeril* (Atlantic angelshark, 1375 m_{fbd}) occurs on the upper bathyal slopes of the tropical Western Atlantic, and *Squatina tergocellata* (Ornate angelshark, 400 m_{fbd}) is considered bathydemersal off Southern Australia.

4.2.5 Order Squaliformes

The order Squaliformes comprises over 130 species, of which 80 per cent can be considered deep-sea (Fig. 4.4) representing the main ancient diversification of neoselachii into the deep sea (Fig. 4.5) with a remarkable diversity in body form and size, including some of the smallest, e.g. *Etmopterus perryi* (dwarf lantern shark) with 16 cm long adult males (Springer and Burgess, 1985) and also one of the largest known sharks, e.g. *Somniosus microcephalus* (Greenland shark), which grows to 7.3 m length. There are seven families, two of which have light organs, Dalatiidae and Etmopteridae, implying that bioluminescence evolved at least twice in the Squaliformes; however, Klug and Kriwet (2010) propose that bioluminescence emerged just once in a common ancestral group. All squaliformes produce live young by ovoviviparity, the eggs being retained inside the female.

4.2.5.1 Family Echinorhinidae (9) (Bramble sharks)

There is just one genus, *Echinorhinus*, with two species, *E. brucus* (Bramble shark, 900 m_{fbd}) found

mainly in the Atlantic Ocean, the Mediterranean Sea and the Indian Ocean, with *E. cookei* (Prickly shark, 1100 m_{fbd}) confined to the Pacific Ocean notably around New Zealand (Fig. 4.4a). They are large demersal sharks, (3.1 and 4.0 m maximum length, respectively) found down to 900 m depth. Both species are uncommon, occurring as an occasional bycatch in commercial and recreational fisheries.

4.2.5.2 Family Dalatiidae (10) (Kitefin sharks)

This family comprises 10 species in six genera of generally small bathypelagic sharks less than 50 cm long. An exception to this is the genus *Dalatias* with one species *D. licha* (the kitefin shark, 1800 m_{fbd}) that grows to 182 cm length, with an almost worldwide distribution, bathydemersal mainly at depths 200–600 m but a maximum reported depth of 1800 m. It is often captured in deep-sea shark fisheries off Portugal and Japan but is absent from the Eastern Pacific Ocean and the Northern Indian Ocean. This black-bodied dogfish has photophores distributed over the ventral surface interspersed between the dermal denticles (Reif, 1985). The genera *Euprotomicroides*, *Heteroscymnoides*, *Mollisquama*, and *Squaliolus* are all bathypelagic pygmy or pocket sharks with simple photophores on the ventral surface providing visual camouflage in the mesopelagic environment (Claes et al., 2012). *Squaliolus laticaudus* (spined pygmy shark, 1200 m_{fbd}) and *S. aliae* (Smalleye pygmy shark, 2000 m_{fbd}) with maximum recorded lengths of 22 cm are amongst the smallest known sharks. They exhibit diel vertical migrations in pursuit of fish and squid prey in the deep scattering layer of the oceans. Some of these sharks are very poorly known from only a few specimens. In addition to photophores scattered over the body, *Euprotomicroides zantedeschi* (tail light shark) has an abdominal pouch from which blue bioluminescent fluid can be extruded (Munk and Jørgensen, 2010) as well as enlarged pectoral fins, which apparently provide propulsion by a wing-like flapping motion. Much attention has been directed to the genus *Isistius* (cookiecutter sharks), which feed on larger fishes and marine mammals by removing discs

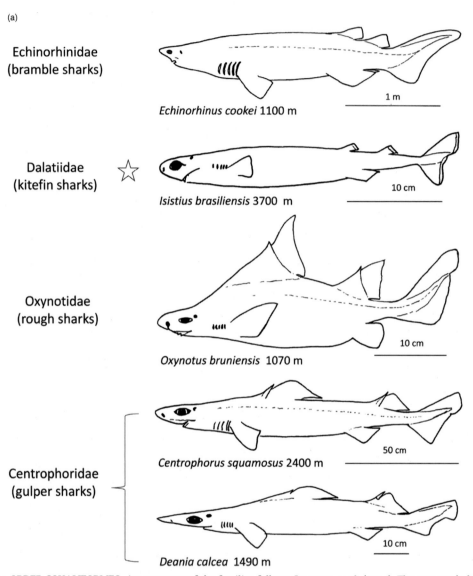

(a)

Echinorhinidae
(bramble sharks)

Echinorhinus cookei 1100 m

1 m

Dalatiidae
(kitefin sharks)

Isistius brasiliensis 3700 m

10 cm

Oxynotidae
(rough sharks)

Oxynotus bruniensis 1070 m

10 cm

Centrophoridae
(gulper sharks)

Centrophorus squamosus 2400 m

50 cm

Deania calcea 1490 m

10 cm

Figure 4.4a ORDER SQUALIFORMES. Arrangement of the families follows Sorensen et al. (2014). The star symbols indicate families with bioluminescent species. *Echinorhinus cookei* (prickly shark) Pacific Ocean 47°N–55°S, *Isistius brasiliensis* (cookiecutter shark) Circumglobal 35°N–40°S. *Oxynotus bruniensis* (prickly dogfish) SW Pacific off S Australia and New Zealand. *Centrophorus squamosus* (leafscale gulper shark) E Atlantic, Indo-W Pacific 69°N–54°S. *Deania calcea* (birdbeak dogfish) E Atlantic and Pacific 70°N–56°S.

of superficial flesh using their sharp mandibular teeth creating a crater wound. The cause of the mysterious wounds was first elucidated by Jones (1971), who demonstrated that they were the result of attacks by *I. brasiliensis* (Fig. 4.4a). Best and Photoplouou (2016) show that this genus is the cause of most of the crater-like wounds seen on large whales in South African waters, typically 69 unhealed wounds per adult Sei Whale (*Balaenoptera borealis*). Widder (1998) proposed that a pigmented collar behind the head, devoid of ventral photophores could appear as an attractive dark patch in the down-welling mesopelagic light field luring potential visual predators, which would then be in turn attacked by

(b)

Squalidae
(dogfish sharks)

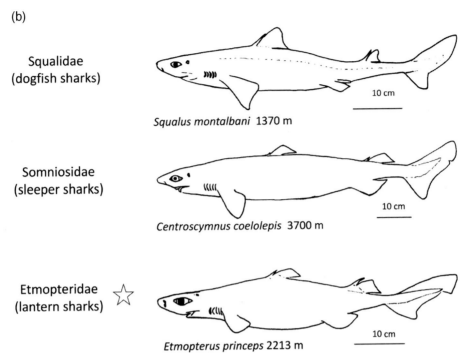

Squalus montalbani 1370 m

Somniosidae
(sleeper sharks)

Centroscymnus coelolepis 3700 m

Etmopteridae
(lantern sharks)

Etmopterus princeps 2213 m

Figure 4.4b ORDER SQUALIFORMES. *Squalus montalbani* (Indonesian greeneye spurdog), Philippines, Indonesia to Australia. *Centroscymnus coelolepis* (Portuguese dogfish), Atlantic, W Mediterranean, Indo-W Pacific. *Etmopterus princeps* (great lanternshark), North Atlantic.

the cookiecutter. *I. brasiliensis* is distributed worldwide around the tropics between 20°N and 20°S whereas *I. labialis* is known from a single specimen in the South China Sea and *I. plutodus* (largetooth cookiecutter shark) from scattered locations in the Atlantic and Pacific Oceans. The cookiecutters feed on a range of deep oceanic prey, squid, fish and crustacea as well as the eponymous ectoparasitic feeding.

4.2.5.3 Family Oxynotidae (11) (Rough sharks)

These sharks have a very unusual appearance with large dermal denticles, a triangular body cross section with a flat ventral surface and two exaggerated high dorsal fins, each with a spine mounted on a large dorsal hump and a trailing fin membrane. They are bathydemersal on upper slopes, and there are five species, each with a regional distribution: *Oxynotus centrina* (Angular roughshark, 777 m_{fbd}) throughout the Eastern Atlantic Ocean and the Mediterranean Sea,

O. paradoxus (Sailfin roughshark, 720 m_{fbd}) in the North East Atlantic 41–11°N only, *O. caribbaeus* (Caribbean roughshark, 457 m_{fbd}) in the Caribbean, *O. bruniensis* (Prickly dogfish, 1070 m_{fbd}) in the Southwest Pacific Ocean notably around southern Australia and New Zealand (Fig. 4.4a) and *O. japonicus* (Japanese roughshark, 270 m_{fbd}) in the NW Pacific Ocean. Maximum length ranges from 49 cm (*O. caribbaeus*) to 1.5 m (*O. centrina*). They are widely reported to possess light organs (e.g. Nelson, 2006), but Ebert (2013) states there are no photophores on the body, and there are no detailed reports of their presence.

4.2.5.4 Family Centrophoridae (12) (Gulper sharks)

These are deep demersal spiny dogfish of maximum length from 80 to 160 cm found at depths down to 2400 m. They have an almost global distribution along bathyal continental margins, mid-ocean ridges and seamounts but are absent from the NE Pacific

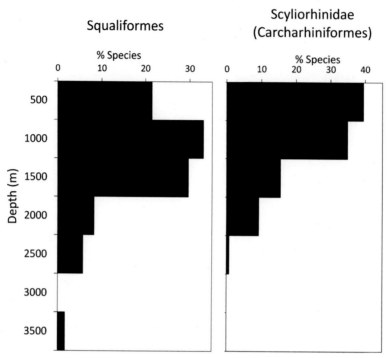

Figure 4.5 Comparison of the maximum depth distribution of the Squaliformes and Scylliorhnidae.

Ocean and polar seas. *Centrophorus squamosus* (leafscale gulper shark, 2400 m$_{fbd}$) is found in the Eastern and Mid-Atlantic Ocean from north to south, the Western Indian Ocean and Western Pacific Ocean and is an important component of deep-water fisheries off Ireland, Spain, Portugal and France (Fig. 4.4a). Bathypelagic capture of one specimen at 1250 m depth in water 4000 m deep suggests it may be able to move freely in mid-water between different areas of fragmented bathyal benthic habitat. There are two genera, *Centrophorus* (15 species) with a flattened rounded snout and *Deania* (bird beak sharks, four species) (Fig. 4.4a) with a characteristic flattened elongated snout (bird beak or spatulate).

4.2.5.5 Family Squalidae (13) (Dogfish sharks)
This family comprises two genera, *Cirrhigaleus* (3 species) and *Squalus* (26 species) of dogfish with spines on the two dorsal fins 50–160 cm total length. They have a global distribution and occur from inshore to the upper slope, overlapping in their depth

distribution with deep demersal Centrophoridae, for example, the benthopelagic *Squalus acanthias* (picked or spiny dogfish) has been recorded down to 1460 m. Several species are considered deep-sea specialists, for example, *Squalus montalbani* (Indonesian greeneye spurdog, 1370 m$_{fbd}$; Fig. 4.4b), *Squalus chloroculus* (Greeneye spurdog, 1360 m$_{fbd}$) and *Squalus griffini* (Northern spiny dogfish, 950 m$_{fbd}$) and are exploited in deep-water fisheries in the Western and SW Pacific regions.

4.2.5.6 Family Somniosidae (14) (Sleeper sharks)
The name of this family reflects their passive habits, for example *Somniosus microcephalus* (Greenland shark) when captured by baited hooks through holes in winter ice show no fight as they are winched up from the depths of coastal fjords. However there are reports of voracious predatory behaviour. They are large, predominantly demersal sharks occurring globally at bathyal depths from polar to tropical seas and are taken by deep-sea fisheries, trawl and

longlines in several regions. The maximum depth attained is 3700 m by *Centroscymnus coelolepis* (Portuguese dogfish) (Fig. 4.4b, Plate 3a) but most species do not extend beyond 1000–1500 m. There are five genera: *Centroscymnus* (six species), *Scymnodalatias* (six species), *Somniosus* (five species), *Zameus* (two species) and the monotypic *Scymnodon*. The life span of *Somniosus microcephalus* (Greenland shark, 2200 m_{fbd}) has been estimated to be 392 ± 120 year, making it the longest-lived vertebrate (Nielsen et al., 2016).

4.2.5.7 Family Etmopteridae (15) (Lantern sharks)

This family comprises 47 species of predominantly small mesopelagic and bathypelagic sharks of average maximum length 46 cm, which are found at depths down to 2490 m. They have photophores arranged predominantly on the ventral surface to provide counterillumination camouflage in the down-welling light of the mesopelagic. The photophores are larger and more elaborate than those of the Dalatiidae, leading Claes et al. (2014a) to speculate that the Dalatiidae represents a more primitive condition from which the Etmopterid bioluminescence evolved. The photophore number is related to depth of occurrence, so shallower-dwelling species have more photophores allowing matching to higher intensities of down-welling light. The theoretical upper limit for shark counterillumination in daytime is about 200 m depth. In addition to the photophores needed for counterillumination, Etmopterids have complex patterns of shading and photophores on the side of the body that may be used for intraspecific signalling, which can be detected by appropriately adapted eyes (Claes et al., 2014b). There is no sexual dimorphism in photophore distribution, and it is suggested that signalling may be used in schooling behaviour (Reif, 1985). There are five genera: *Aculeola* (one species), *Centroscyllium* (seven species) *Etmopterus* (37 species; Fig. 4.4b), *Miroscyllium* (one species) and *Trigonognathus* (one species). The diet is made up of mid-water fishes (e.g. Clupeidae, Myctophidae and Stomiidae), cephalopods (including cuttlefish and histioteuthid squids),

crustacea (decapod crabs, penaeid and euphausiid shrimp) and jellyfish.

Superorder Galeomorphi

The Galeomorphi is the shallow-dwelling clade of Selachii generally found in coastal waters or in the epipelagic zone of the open ocean. The orders Heterodontiformes (bullhead sharks) and Orectolobiformes (carpet sharks) occur mostly at depths less than 100 m, the deepest-living being in the family Parascylliidae (collared carpet sharks) being found down to 290 m. However, the tropical epipelagic planktivorous shark, *Rhincodon typus* (whale shark, Rhincodontidae, Orectolobiformes), the world's biggest fish with total length up to over 12 m, has been tracked down to a depth of 750–1000 m in the Indian Ocean, although spending most time at depths less than 100 m (Rowat and Gore, 2007).

4.2.6 Order Lamniformes

This order comprises 16 species of mainly epipelagic oceanic sharks but some species do occur in the deep sea.

4.2.6.1 Family Odontaspididae (16) (Sand tiger sharks)

In the genus *Odontaspis; O. ferox* (Smalltooth sand tiger, 2000 m_{fbd}) is benthopelagic in tropical to temperate slopes down to bathyal depths and *O. noronhai* (Bigeye sand tiger shark, 1000 m_{fbd}) is deep pelagic in the East Pacific and Atlantic Oceans. They grow to over 3 m in length. The genus *Carcharias* are inshore species.

4.2.6.2 Family Mitsukurinidae (17) (Goblin shark)

The comparatively rare *Mitsukurina owstoni* (Goblin shark) is found on bathyal slopes 300–1000 m deep with a patchy worldwide distribution (Fig. 4.6). Classified within its own family, it has a unique body form with a dorsally extended snout and highly protrusible jaws. It appears to be more closely related to the fossil Mesozoic shark *Scapanorhynchus* than

Mitsukurinidae
(goblin sharks)

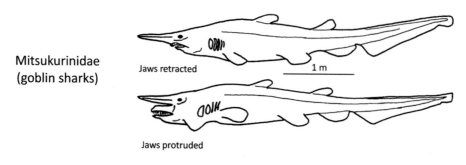

Jaws retracted   1 m

Jaws protruded

Mitsukurina owstoni 1300 m

Figure 4.6 ORDER LAMNIFORMES. *Mitsukurina owstoni* (goblin shark), Circumglobal 48°N–55°S. Note the change in appearance when the jaws and branchial apparatus are protruded for feeding.

other extant Lamniformes. Growing to a maximum length of over 6 m, the body is slender and flabby, and it is probably not a fast swimmer. The diet includes mesopelagic teleosts, squid and bioluminescent ostracods, and the snout may act as a forward-projecting detector of prey (Duffy, 1997). *M. oswtoni* probably float motionless in midwater capturing prey by rapid jaw extrusion and buccal suction.

4.2.6.3 Family Megachasmidae (18) (Megamouth sharks)

The single species in this family, the rare *Megachasma pelagios* (Megamouth shark, 600 m_{fbd}) is often popularly referred to as a deep-sea species but it is a large (total length up to 5.5 m) tropical oceanic epipelagic planktivor usually found at 120–170 m depth with no evidence of a deep-sea habit (Leighton et al., 1983).

4.2.6.4 Family Alopiidae (19) (Thresher sharks)

With their characteristic long and curving caudal fin these are oceanic sharks known to dive occasionally to mesopelagic depths, e.g. *Alopias superciliosus* (Bigeye thresher, 730 m_{fbd}),

4.2.6.5 Family Cetorhinidae (20) (Basking sharks)

The planktivorous *Cetorhinus maximus* (Basking shark) follows the seasonal vertical migration of its copepod prey and spends the autumn and winter months beyond

the continental shelf edge feeding at depths down to 1000 m (Sims et al., 2003; Francis, and Duffy, 2002). One individual tracked during a summertime trans-Atlantic migration spent most time at depths >200 m and reached a maximum depth of 1264 m (Gore et al., 2008). Births of young have also been rarely observed in coastal surface waters (Priede, 1984) suggesting important stages of the lifecycle may normally occur in the deep sea. Although well known in coastal areas on the surface in the summer months, the basking shark might be considered a half-time deep-sea fish.

4.2.6.6 Family Lamnidae (21) (Mackerel sharks)

This family comprises mainly epipelagic sharks such as the predatory *Carcharodon carcharias* (Great white shark, 1280 m_{fbd}). Diving to great depths is observed in this and several other species including *Lamna ditropis* (Salmon shark, 650 m_{fbd}) and *Lamna nasus* (Porbeagle, 715 m_{fbd}), but they are not generally considered to be deep-sea fishes. Such dives are also undertaken by large pelagic teleosts such as blue marlin and Bluefin tuna (Block et al., 1992, 2001) as well as the whale shark and may aid orientation in the open ocean by integrating information from above and below the thermocline.

4.2.7 Order Carcharhiniformes

This order includes a wide range of epipelagic and shallow-water sharks, some of which make deep

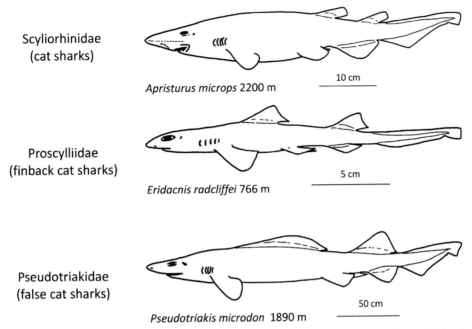

Scyliorhinidae
(cat sharks)

Apristurus microps 2200 m   10 cm

Proscylliidae
(finback cat sharks)

Eridacnis radcliffei 766 m   5 cm

Pseudotriakidae
(false cat sharks)

Pseudotriakis microdon 1890 m   50 cm

Figure 4.7 ORDER CARCHARHINIFORMES. *Apristurus microps* (Smalleye catshark) N and SE Atlantic. *Eridacnis radcliffei* (pygmy ribbontail catshark) Indo-W Pacific. *Pseudotriakis microdon* (false catshark) circumglobal.

dives, but a few species are considered to be bathydemersal.

4.2.7.1 Family Scyliorhinidae (22) (Cat sharks)

This is the main radiation of Carcharhiniformes into the deep sea. The Scyliorhinidae have an average maximum length of 59 cm and are mostly demersal species distinguished from the superficially similar Squalidae by the absence of dorsal fin spines. They are oviparous. The depth distribution (Fig. 4.5) shows that 40 per cent of species live at depths less than 500 m, average maximum depth is 718 m and there is a decrease in species number with depth, characteristic of an invasive or secondary deep-sea family. The genus *Apristurus* (ghost or demon catsharks) is the main deep-sea family with 31 species, which are bathydemersal with a mean maximum depth of 1264 m, and the deepest recorded is *Apristurus microps* (Smalleye catshark, 2200 m$_{fbd}$; Fig. 4.7). Each *Apristurus* species has a regional distribution, for example *A. microps* is found in the North and South East Atlantic Ocean, *A. indicus* (smallbelly catshark,

1840 m$_{fbd}$) in the Western Indian Ocean and *A. garricki* is a new species endemic to New Zealand waters. As new discoveries are made, the number of described species is expected to increase, making this the most specious genus of extant sharks (Sato et al., 2013). Although they are primarily demersal species, pelagic prey are taken. It is thought that younger stages adopt a pelagic lifestyle before descending and adopting the benthic habit. The species are generally poorly known, not specifically targeted by fisheries but are taken as bycatch by trawls and longlines. Nine other genera have ranges extending into the deep sea. The deepest recorded species are *Asymbolus galacticus* (Starry catshark, 550 m$_{fbd}$), *Cephaloscyllium signouru* (Flagtail swellshark, 700 m$_{fbd}$), *Figaro boardmani* (Australian sawtail catshark, 823 m$_{fbd}$), *Galeus melastomus* (Blackmouth catshark, 1200 m$_{fbd}$), *Holohalaelurus regain* (Izak catshark, 1075 m$_{fbd}$), *Parmaturus macmillani* (McMillan's cat shark, 1500 m$_{fbd}$), *Pentanchus profundicolus* (Onefin catshark, 1071 m$_{fbd}$), *Schroederichthys saurisqualus* (Lizard catshark, 435 m$_{fbd}$) and *Scyliorhinus meadi* (Blotched catshark, 750 m$_{fbd}$).

4.2.7.2 Family Proscylliidae (23) (Finback cat sharks)

The three species in the genus *Eridacnis* (ribbon-tail catsharks) are bathydemersal down to 766 m depth.

4.2.7.3 Family Pseudotriakidae (24) (False catsharks)

This family comprises three demersal species in two genera including *Gollum attenuatus* (Slender smooth-hound, 660 m_{fbd}) and *Pseudotriakis microdon* (False catshark, 1890 m_{fbd}; Fig. 4.7).

4.2.7.4 Family Triakidae, (25) (Houndsharks)

Comprising 46 species in nine genera these are mostly demersal coastal species but the two species in the genus *Iago* are considered bathydemersal: *Iago garricki* (Longnose houndshark, 475 m_{fbd}) and *Iago omanensis* (Bigeye houndshark, 2195 m_{fbd}).

Subdivision Batoidea

The Batoidea, skates and rays are characterised by the flattened body form with pectoral fins extended into 'wings', ventral openings of the gills slits and dorsal eyes and spiracles. There are four orders, Torpediformes (electric rays), Pristiformes (sawfishes), Rajiformes (skates) and Myliobatiformes (stingrays).

4.2.8 Order Torpediformes (Electric Rays)

There are over 60 species of electric rays or numbfishes, which are generally demersal in shallow waters and shelf seas. However, a few species are found at upper slope depths.

4.2.8.1 Family Narkidae (26) (Sleeper rays)

Typhlonarke aysoni (Blind electric ray, 900 m_{fbd}) occurs typically at 300–400 m depth off New Zealand.

4.2.8.2 Family Narcinidae (27) (Numbfishes)

Benthobatis marcida (Deepsea Blind Ray, 923 m_{fbd}) occurs on upper bathyal slopes of the W Central Atlantic and *Narcine tasmaniensis* (Tasmanian numbfish, 640 m_{fbd}) occurs off South Australia.

4.2.8.3 Family Torpedinidae (28) (Torpedo electric rays)

Torpedo fairchildi (New Zealand torpedo, 1153 m_{fbd}.) is found in New Zealand waters.

4.2.9 Order Pristiformes (Sawfishes)

These are all shallow-water fishes except one species, *Rhinobatos variegatus* (Stripenose guitarfish), which has been recorded at 366 m depth in the Gulf of Mannar, India.

4.2.10 Order Myliobatiformes (Stingrays)

The Myliobatiformes are predominantly shallow-water rays that displaced the Rajiformes into deeper waters following the end-Cretaceous extinction events. A few species are recorded at upper slope depths.

4.2.10.1 Family Hexatrygonidae (29) (Sixgill stingrays)

The single species in this family *Hexatrygon bickelli* (Sixgill stingray) occurs at depths of 350 to 1120 m on slopes in the Indo-Pacific area from South Africa to Hawaii (Fig. 4.8). Maximum length is 1.43 m, and it is viviparous, giving birth to litters of two to five live young.

4.2.10.2 Family Plesiobatidae (30) (Deepwater stingrays)

In this monotypic family (Fig. 4.8) *Plesiobatis daviesi* (Deepwater stingray, 780 m_{fbd}) is distributed in the Indo-Pacific area from Southern Africa to Hawaii, where it has been seen at baits deployed at 350 m depth in submarine canyons.

4.2.10.3 Family Urolophidae (31) (Round stingrays)

Several species of this family are endemic to restricted geographical ranges at bathyal upper slope depths, *Urolophus expansus* (Wide stingaree, 420 m_{fbd}) off SW Australia, *U. deforgesi* (Chesterfield Island stingaree,

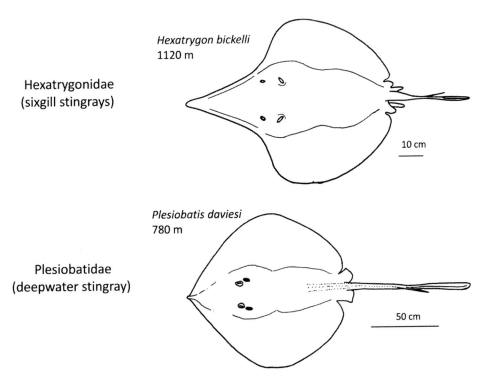

Hexatrygonidae
(sixgill stingrays)

Hexatrygon bickelli
1120 m

10 cm

Plesiobatidae
(deepwater stingray)

Plesiobatis daviesi
780 m

50 cm

Figure 4.8 ORDER MYLIOBATIFORMES. *Hexatrygon bickelli* (Sixgill stingray) Indo-Pacific. *Plesiobatis daviesi* (Deepwater stingray) Indo-Pacific.

300 m$_{fbd}$) around Chesterfield Island, *U. neocaledoniensis* (New Caledonian stingaree, 428 m$_{fbd}$) found off New Caledonia and adjacent islands and *U. piperatus* (Coral sea stingaree, 370 m$_{fbd}$) off northern Queensland. These demersal species from Australia and the SW Pacific are known only from a few specimens collected during research cruises.

4.2.10.4 Family Myliobatidae (32) (Eagle rays)

These are tropical coastal fishes: the largest species is *Manta birostris* (Giant manta) up to 9.1 m span, a tropical epipelagic zooplanktivor, which like the whale shark and basking shark, is capable of diving to over 1000 m depth (Marshall et al., 2011).

4.2.11 Order Rajiformes

Here we only consider one family, the Rajidae, which constitutes the main radiation of batoidea into deep-sea environments.

4.2.11.1 Family Rajidae (33) (Skates)

This family exhibits high species diversity but with remarkable morphological conservatism (McEachran and Dunn, 1998; Fig. 4.9). There are over 280 species, which occur from inshore locations to the deep sea. All produce eggs in characteristic rectangular 'mermaid's purse' capsules with a point at each corner. Whilst Myliobatiformes have diversified into pelagic open ocean environments, the Rajidae have remained benthic fishes specialising in soft substrates. There have been multiple radiations into the deep sea so that most genera have deep-sea representatives. High species diversity arises partly from regional endemism, e.g. *Bathyraja abyssicola* is found in the North Pacific, *B. richardsoni* in the North Atlantic, *B. smithii* in the South East Atlantic. Two major clades are recognised within the Rajidae ranked by Nelson (2006) as subfamilies: Rajinae and Arhychobatinae.

Subfamily Rajinae (Hardnose skates) comprises 187 species in 14 genera, the deepest species in each

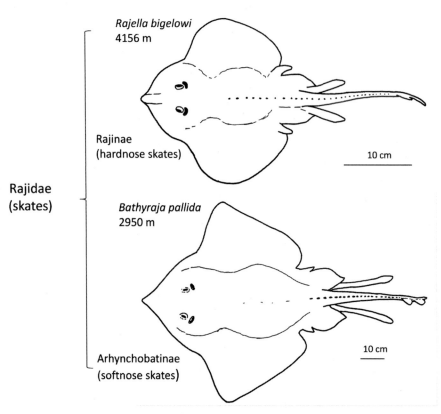

Figure 4.9 ORDER RAJIFORMES. *Rajella bigelowi* (Bigelow's ray) N Atlantic. *Bathyraja pallida* (Pale ray) NE Atlantic.

being *Amblyraja hyperborea* (Arctic skate, 2500 m$_{fbd}$), *Breviraja marklei* (Nova Scotia skate, 988 m$_{fbd}$), *Cruriraja rugosa* (Rough leg skate, 1007 m$_{fbd}$), *Dactylobatus clarkia* (Hookskate, 1000 m$_{fbd}$), *Dipturus innominatus* (New Zealand smooth skate, 1310 m$_{fbd}$), *Fenestraja mamillidens*, (Prickly skate, 1093 m$_{fbd}$), *Gurgesiella dorsalifera*, (Onefin skate, 800 m$_{fbd}$), *Leucoraja melitensis* (Maltese ray, 600 m m$_{fbd}$), *Malacoraja kreffti* (Krefft's ray, 1200 m$_{fbd}$), *Neoraja africana* (West African pygmy skate, 1640 m$_{fbd}$), *Okamejei leptoura* (Thintail skate, 735 m$_{fbd}$), *Raja rhina* (Longnose skate, 1069 m$_{fbd}$), *Rajella bigelowi* (Bigelow's ray, 4156 m$_{fbd}$) and *Zearaja nasuta* (New Zealand rough skate, 1500 m$_{fbd}$). The mean maximum depth of all species is 724 m, and the deepest living genus is *Rajella* with a mean maximum depth of 1472 m. The mean maximum size of the Rajinae is 73 cm (total length) with the largest being *Dipturus innominatus* (New Zealand smooth skate) at 2.4 m

long. *R. bigelowi*, the deepest living Chondrichthyan is a rather small ray, maximum total length 55 cm (Fig. 4.9).

Subfamily Arhynchobatinae (Softnose skates) with a mean maximum depth of 1011 m are predominantly deeper-living than the Rajinae. The soft nose with a reduced rostral cartilage is considered advantageous for life in deeper water because the resulting flexible snout can be manipulated by radial muscles of the pectoral fins for grubbing in soft sediments. The nasal capsules are also enlarged, indicating large nasal rosettes increasing chemosensitivity in an environment devoid of solar light (McEachran and Dunn, 1998). There are 89 species in 8 deep-sea genera, the deepest species in each being *Arhynchobatis asperrimus* (Longtail skate, 1070 m$_{fbd}$), *Bathyraja pallida* (Pale ray, 2950 m$_{fbd}$; Fig. 4.9), *Brochiraja spinifera* (Prickly deep-sea skate, 1460 m$_{fbd}$),

Insentiraja subtilispinosa (Velvet skate, 1100 m$_{fbd}$), *Notoraja sapphira* (Sapphire skate, 1313 m$_{fbd}$), *Pavoraja arenaria* (Sandy skate, 712 m$_{fbd}$), *Psammobatis scobina* (Raspthorn sand skate, 450 m$_{fbd}$) and *Rhinoraja longicauda* (White-bellied softnose skate, 1000 m$_{fbd}$).

Skates are widely targeted in fisheries around the world, but the deepest-living species such as *Bathyraja pallida* (depth range 1,879–2,952 m) and *Rajella bigelowi* (depth range 367–4,156 m) live mainly at depths beyond their reach. *Bathyraja spinicauda* (Spinetail ray) that lives on the upper slopes down to 1463 m has been depleted by fisheries in the NW Atlantic to the point of danger of extinction (Devine et al., 2006). *Bathyraja meridionalis* (Dark-belly skate, 2240 m) is thought to be circumglobal in the Southern Ocean and appears as a bycatch in fisheries for toothfish (*Dissostichus* spp.) in the SW Atlantic (Stehman and Pompert, 2009).

4.3 Class Actinopterygii

The deep-sea ray-finned fishes (Actinopterygii) all belong to the Division teleostei, which is the main group of fishes that has come to dominate the world's freshwaters and oceans. This is the most species-rich and diversified group of vertebrates with about 30,000 species of which 10–14 per cent occur in the deep sea. Within the teleostei there have been multiple diversifications into the deep sea.

Subdivision Elopomorpha

The Elopomorpha is a distinctive subdivision of the teleostei characterised by a unique marine larval stage known as a leptocephalus. This is a transparent, laterally flattened leaf-like stage that feeds on plankton and particulate matter in the upper layers of the ocean. There are no paired, pectoral or pelvic, fins. Leptocephali hatch at less than 5 mm length and typically grow to ca. 5 cm but can reach 2 m in some species, before metamorphosis to the adult, which often involves decrease in length. The anus typically

moves forward along the body during metamorphosis from an almost terminal caudal position in the larva to about halfway from the head in adults. Also the simple 'V' shaped myotomes of the larva are replaced by adult 'W' shaped muscle blocks. Metamorphosis involves major reorganisation of the body plan, which makes morphological identification of leptocephali to adult species extraordinarily difficult. Positive identification is only possible from large collections including late stages of metamorphosis or by molecular taxonomy. During metamorphosis the larva does not feed, and the gut is blind ending with no anal connection. Compared with other teleosts the larval life span is very long, often up to several years (Harrison, 1966). Fecundity is very high, with adult females producing vast numbers, often in excess of 1 million small eggs.

Order Elopiformes

This order comprises shallow-water and estuarine fishes that enter freshwater with no deep-sea representatives.

4.3.1 Order Albuliformes (Bonefishes)

4.3.1.1 Family Albulidae (34) (Bonefishes)

The only family in this order comprises superficially herring-like, mainly tropical, coastal fishes that enter brackish and freshwaters although one genus, *Pterothrissus*, occurs in the deep sea (Fig. 4.10). There are two species, both bathydemersal: *P. gissu* (Japanese gissu) considered rare, found at depths down to 1000 m$_{fbd}$ in the NW Pacific, and *P. belloci* (Longfin bone fish) found in the tropical SW Atlantic off the coast of Africa down to 500 m$_{fbd}$ and landed as bycatch in fisheries. *P. belloci* is recorded as feeding on benthic copepods and polychaetes. The leptocephalus is recognised by the presence of a well-developed caudal fin (Fig. 4.10) and in *P. gissu* grows to a length of 18 cm before metamorphosis, when it shrinks to 8 cm length to attain the adult form (Tsukamoto, 2002). Adult *P. gissu* ultimately reach a maximum length of 50 cm and have been seen close to the seafloor in benthic surveys by ROVs.

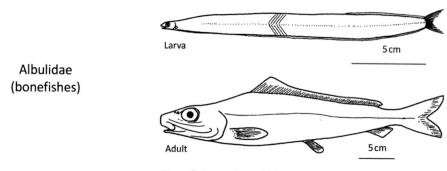

Albulidae
(bonefishes)

Larva   5 cm

Adult   5 cm

Pterothrissus gissu 1000 m

Figure 4.10 ORDER ALBULIFORMES. *Pterothrissus gissu* (Japanese gissu) NW Pacific. Larva and Adult. Leptocephalus larva image after Tsukamoto (2002).

4.3.2 Order Notacanthiformes

This order has two important deep-sea families, the Halosauridae (halosaurs) and Notacanthidae (spiny eels; Fig. 4.11).

4.3.2.1 Family Halosauridae (35) (Halosaurs)

With 16 species in three genera this is a truly endemic deep-sea family of fishes with a mean maximum depth of 2012 m. The deepest species in each genus are *Aldrovandia rostrata* (5026 m_{fbd}), *Halosauropsis macrochir* (Abyssal halosaur, 3300 m_{fbd}; Fig. 4.11) and *Halosaurus attenuatus* (2500 m_{fbd}). They are benthic species distributed worldwide at bathyal to abyssal depths on continental slopes, mid-ocean ridges, around islands and seamounts.

A leptocephalus larva identified as possibly of the genus *Aldrovandia* was captured at 1100 m depth in the NE Atlantic. Metamorphosis commences at about 19 cm in length (Harrisson, 1966) with forward displacement of the anus and development of fins. The adult body form is elongated, with relatively large scales and a long tapering tail. *Halosauropsis macrochir* (*Aldrovandia macrochir* in some sources) grows to a total length of up to 90 cm and is found throughout the Atlantic Ocean at depths from 1000 to 3500 m, also in the Western Pacific, off Australia, New Zealand and Japan, and the Western Indian Ocean. On the Mid-Atlantic Ridge the minimum size is 10 cm and minimum age 2 years, possibly reflecting

the size and age of metamorphosis. There is no obvious depth-related size trend with the peak of size distribution at 45 to 65 cm total length. The diet comprises mainly crustacea, followed by Teleostei, Polychaeta and Cephalopoda. With up to 700 individuals.km^{-2}, *H. macrochir* is one of the most abundant species captured in deep-sea trawls at 1700–3500 m on the Mid-Atlantic Ridge (Bergstad et al., 2012; Cousins et al., 2013a), but it is not attracted to baited cameras or longlines (Cousins et al., 2013b). In ROV video transects, individuals are seen holding station with a slow tail beat (0.3 Hz) or drifting with the current just above the sediment (Linley et al., 2013), the pectoral fins held forward possibly with a sensory function.

4.3.2.2 Family Notacanthidae (36) (Spiny eels)

The Notacanthidae occupy a depth range from 125 m to over 4000 m, and the distribution of maximum depths shows this is a deep-sea endemic family with no shallow-water species. Mean maximum depth is 2092 m. There are 11 species in three genera with the deepest being *Lipogenys gillii* (2000 m_{fbd}), *Notacanthus chemnitzii* (Snubnosed spiny eel, 3285 m_{fbd}; Fig. 4.11) and *Polyacanthonotus challengeri* (Longnose tapirfish, 4560 m_{fbd}). The average maximum length is 50 cm. The general body form is elongated cylindrical with a tapering tail and a rounded snout. The dorsal fin has characteristic

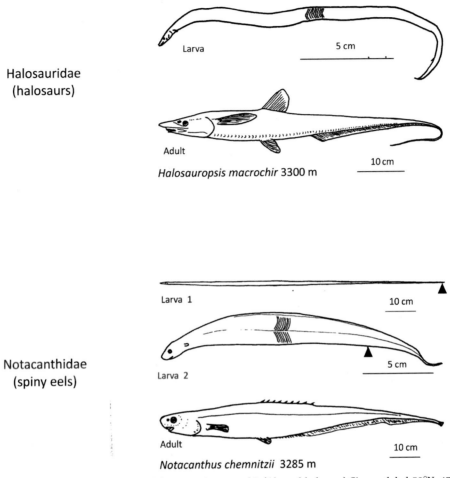

Figure 4.11 ORDER NOTACANTHIFORMES. *Halosauropsis macrochir* (Abyssal halosaur) Circumglobal 50°N–47°S, except E Pacific. *Notacanthus chemnitzii* (Snubnosed spiny eel) Circumglobal 61°N–50°S except tropics. Putative larval stage images *Halosauroropsis* based on Harrisson (1966), *Notacanthus* early stage upper image after *Leptocephalus giganteus* (Castle, 1959) and late stage of metamorphosis after Merrett (1981). The triangle symbol indicates the position of the vent.

isolated spines. Although some species, for example *P. challengeri*, are listed as bathypelagic, the diet is benthic amphipods, polychaetes and mysids. In ROV surveys on the Mid-Atlantic Ridge this species was observed just above the seafloor sediment, undulating the tip of its tail at 2.6 Hz, holding station against the bottom current (Linley et al., 2013). They are not attracted to baited gears. Individual species are very widespread; *P. challengeri* is found in the Pacific, Atlantic and Southern Oceans, and *Notacanthus chemnitzii* occurs worldwide except for tropical waters usually on the upper slopes down to 1000 m.

A late-stage metamorphosing leptocephalus, total length 197 mm, was captured at *ca.* 650 m depth at an altitude of 6–31 m above the seafloor in the NE Atlantic (Merrett, 1981). The head had developed a typical notacanthid shape with a rounded snout (Fig. 4.11). *Leptocephalus giganteus*, first described from a specimen 893 mm long captured off New Zealand (Castle, 1959) is now thought to be the larva of a Notacanthid and may grow to 1 to 2 m in total length. Moser and Charter (1996) describe a developmental series of leptocephali from 22 mm to 560 mm total length captured off Southern California,

which they tentatively identified as larvae of
N. chemnitzii. Metamorphosis presumably entails a
considerable shrinkage in length to the size observed
by Merrett (1981). Although not exploited by
fisheries, *N. chemnitzii* lives at depths at which
commercial trawlers operate and has been impacted
sufficiently in the NW Atlantic to be regarded as
endangered (Devine et al., 2006).

4.3.3 Order Anguilliformes (Eels)

Molecular phylogeny of the Elopomorpha shows that the
Saccopharyngiformes nest within the Anguilliformes
alongside the morphologically and ecologically similar
bathypelagic eel families Nemichthyidae and
Serrivomeridae (Inoue et al., 2010). These two orders are
therefore here considered together but split into two
basic divisions: Anguilliformes I (demersal) and
Anguilliformes II (bathypelagic). Within this grouping
there are 889 species of marine eel ranging from shallow
water to abyssal depths in 17 families (Fig. 4.12).

Anguilliformes I (Demersal)

This comprises the demersal radiation of relatively
robust Anguilliformes into the deep sea at bathyal and
down to abyssal depths. A few species have
secondarily adopted a pelagic lifestyle.

4.3.3.1 Family Synaphobranchidae (37) (Cutthroat eels)

The Synaphobranchidae are often a very conspicuous
and abundant component of the deep-sea demersal
fish assemblages from the upper slopes down to the
abyss. They are small to medium-sized eels, mostly
less than 1 m long (average maximum length 66 cm),
but some species can grow to 1.8 m in length. They
are rather generalised eels with few distinguishing
features, and the anus is anterior to the mid-body.
Their depth distribution indicates (Priede and Froese,
2013) they are deep-sea invasive with 30 per cent of
species at <500 m but extend down over 5000 m
depth on the abyssal plains. There are 38 species
distributed between three subfamilies.

Subfamily Simenchelyinae has a single species,
Simenchelys parasitica (Snubnosed eel, 2620 m_{fbd}),
which is readily distinguished by its short jaws and
snub nose. It has well-developed fins and is an active
forager on bathyal slopes in the Atlantic Ocean and
the West Pacific, often attracted to and caught in
large numbers in baited traps. The name is derived
from the reported habit of burrowing into the flesh of
larger fishes including a remarkable observation of
two living individuals, 21 cm and 24 cm total length
found inside the heart of a moribund shortfin mako
shark, *Isurus oxyrinchus*, hauled up from 1000 m
depth in the Atlantic Ocean off New York (Caira et al.,
1997). The authors concluded that *S. parasitica* are
facultative parasites and probably enter hosts trapped
by fishing gears just prior to or after death.

Subfamily Ilyophinae with 27 species in nine
genera is the most morphologically diverse and
speciose of the Synaphobranchid subfamilies. There is
variation in the body shape, dentition, fins, eye size
and the snout. They are generally distinguished from
the Synathobranchinae by the lower jaw being shorter
than the upper. Many of the species are known from a
small number of specimens caught in restricted areas,
and it is likely that many more species will be
discovered. The deepest in each genus are
Atractodenchelys robinsorum (710 m_{fbd}) in the SW
Pacific, *Diastobranchus capensis* (Basketwork eel,
2000 m_{fbd}) in the Southern Hemisphere except the
E. Pacific, *Dysomma fuscoventralis* (1425 m_{fbd}) in the
Red Sea and NW Indian Ocean, *Dysommina rugosa*
(775 m_{fbd}) in the W Atlantic and E Central Pacific to
Hawaii, including the 'eel city' location on the
Nafanua hydrothermal vent site off Samoa (Staudigel
et al., 2006), *Ilyophis robinsae* (4800 m_{fbd}) in the
Central W Pacific including a record at 6068 m the
Kermadec Trench (Linley et al., 2016), *Linkenchelys
multipora* (237 m_{fbd}) in the W Atlantic Bahamas area
and *Meadia roseni* (1020 m_{fbd}) in the NW Pacific off
Taiwan. *Ilyophis saldanhai* is reported from
hydrothermal vents sites on the Mid-Atlantic Ridge
and East Pacific Rise and depths of 2838–3100 m
(Cause et al., 2005) and *Thermobiotes mytilogeiton*
(1750 m_{fbd}) is endemic to the Vau Lili vent site in the
W Pacific (Geistdoerfer, 1991).

Subfamily Synaphobranchinae has 10 species in three genera. These are archetypical eels with few morphological peculiarities; at various times the two main genera, *Synaphobranchus* and *Histiobranchus*, have been synonymised under *Synaphobranchus* and within these genera the species distinctions have been questioned (Svendsen and Byrkjedal, 2013). *Histiobranchus* is the deepest living genus with three species distributed in different parts of the world: *H. bathybius* (5440 m_{fbd}) in the North Atlantic and Pacific Oceans (Fig. 4.12a), *H. australis* (3001 m_{fbd}) in the Southern Atlantic, Indian and Pacific Oceans and *H. bruuni* (4974 m_{fbd}) in the Southwest Pacific, Tasman Sea and off SE New Zealand. From otter trawl surveys in the NE Atlantic *H. bathybius* is most abundant around 2400 m depth (up to 300 fish. km^{-2}) but its relative importance increases with depth so that at 4800 m it comprises up to 30 per cent of the catch (Karmovskaya and Merrett, 1998). The diet consists of fish, crustacea and molluscs with evidence of benthopelagic foraging, but *H. bathybius* can also be a scavenger as evidenced by arrival at baits deployed on the seafloor of the Madeira and Porcupine Abyssal Plains (Armstrong et al., 1992). Species in the genus *Synaphobranchus* are generally found shallower than *Histiobranchus* on the mid to upper slopes around continents, islands, seamounts and mid-ocean ridges. *S. kaupii* (4800 m_{fbd}) occurs worldwide except in the E. Pacific, *S. affinis* is (2400 m_{fbd}) circumglobal except in the NE Pacific, *S. brevidorsalis* (3000 m_{fbd}) is circumtropical except in the NE Pacific, *S. oregoni* is more narrowly circumtropical except in the E. Pacific and *S. calvus* (2000 m_{fbd}) only in the SW Atlantic off Brazil. The absence of the genus *Synaphobranchus* in the NE Pacific Ocean is notable. In surveys of the upper slopes in the NE Atlantic at depths <2250 m *S. kaupii* comprises over 33 per cent of the demersal trawl catch and is the dominant species at baited cameras deployed on the seafloor (Priede et al., 1994) with a clear increase in size with depth from a mean length of 17 cm at 600 m to 42 cm at 2000 m (Priede and Bagley, 2000). Using the submersible *Nautile*, Uiblein et al. (2002) observed large numbers of *S. kaupii* swimming forward, holding station or

drifting just above the seafloor at depths of 933–1822 m on slopes around the Bay of Biscay and estimated that the density was up to 13140 fish.km^{-2}. *S. kaupii* has a relatively high metabolic rate for deep-sea fishes, reflecting their continuous locomotor activity (Bailey et al., 2005). Heger et al. (2007) observed bursts of bioluminescence associated with swarms of small *S. kaupii* attracted to bait at ca. 900 m depth in the NE Atlantic. The luminescence was produced by the Ostracod, *Vargula norvegica*, apparently as a defence response to attacks by the eels. In the areas where it occurs, *S. kaupii* is often an abundant, conspicuous and highly active part of the demersal fish assemblage. The species *S. dolichorhynchus* is known only from leptocephali collected from the North Atlantic, and the adult species has not been identified. *Haptenchelys texis* (4086 m_{fbd}) occurs across the Atlantic Ocean from the Bahamas in the west to NW Africa in the east including at hydrothermal vent sites. Synaphobranchid eels live either too deep or are too small to be of interest to commercial deep-water fisheries; however, there is evidence that the population density of *S. kaupii* has decreased following the onset of deep-water fisheries in the Porcupine Seabight area of the NE Atlantic (Bailey et al., 2009).

4.3.3.2 Family Myrocongridae (38) (Myroconger eels)

This is an extremely rare family of demersal eels known only from a few specimens. *Myroconger gracilis* was collected by bottom trawl on the Kyushu-Palau Ridge in the NW Pacific Ocean towed at depths of 640–320 m (Castle, 1991).

4.3.3.3 Family Muraenidae (39) (Moray eels)

There are 197 species of moray eel widely distributed in coastal and shelf seas habitats around the world with one potentially deep-sea species, *Gymnothorax bacalladoi* (Canary Moray, 640 m_{fbd}), known from a few specimens found on slopes around the Canary and Madeira Islands.

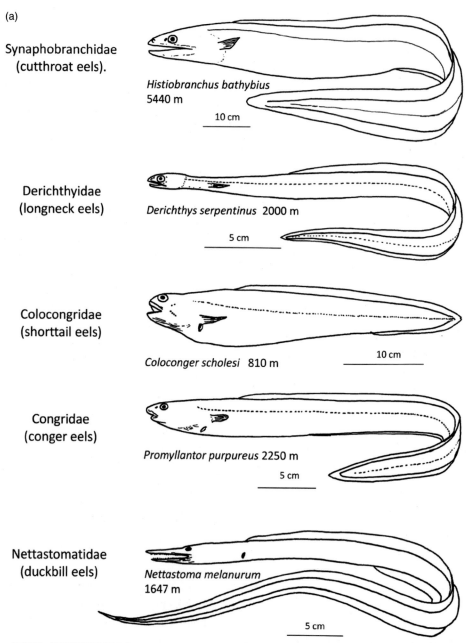

(a)

Synaphobranchidae
(cutthroat eels).

Histiobranchus bathybius
5440 m

10 cm

Derichthyidae
(longneck eels)

Derichthys serpentinus 2000 m

5 cm

Colocongridae
(shorttail eels)

Coloconger scholesi 810 m

10 cm

Congridae
(conger eels)

Promyllantor purpureus 2250 m

5 cm

Nettastomatidae
(duckbill eels)

Nettastoma melanurum
1647 m

5 cm

Figure 4.12a ORDER ANGUILLIFORMES. Demersal families. *Histiobranchus bathybius* (deep-water arrowtooth eel), circumglobal 70°N–56°S. *Derichthys serpentinus* (narrownecked oceanic eel), circumglobal. *Coloconger scholesi* (Indo-Pacific shorttail conger), Indo-Pacific. *Promyllantor purpureus*, Indo-West Pacific. *Nettastoma melanurum* (blackfin sorcerer), Atlantic 45°N–30°S.

4.3.3.4 Family Chlopsidae (40) (False morays)

These are shallow-water tropical and subtropical eels, but one species (*Chlopsis bidentatus*) has been described from 300 to 503 m depth off New Caledonia and Fiji, in the Western Central Pacific Ocean (Tighe and McCosker, 2003).

4.3.3.5 Family Derichthyidae (41) (Longneck eels)

The Derichthyidae comprises three species of mesopelagic to bathypelagic eels recorded down to 2000 m depth. They feed on small fishes and crustacea and grow to 30–60 cm total length. *Derichthys serpentinus* (Narrownecked oceanic eel) has a snub nose (Fig. 4.12a), whereas the two *Nessorhamphus* species have long snouts. They are widespread circumglobally from tropical to temperate latitudes. The branchial region is reduced in diameter creating the characteristic long-neck appearance, unusual in fishes. Castonguay and McCleave (1987) captured leptocephalus larvae of *D. serpentinus* in the Sargasso Sea in the same area as those of *Anguilla* in relatively high numbers compared with other deep-sea species.

4.3.3.6 Family Colocongridae (42) (Shorttail eels)

The Colocongridae have the shortest body proportions of all the eels, with the total length typically less than 10 times the body height, converging with the shape of some gadiformes. The maximum length range is 38–90 cm. The genus *Thalassenchelys* is known only from deep-bodied leptocephali found in the Indian and Pacific Oceans (Shimokawa et al., 1995), molecular evidence confirming affinity to the Colocongridae (Lopez et al., 2007). The genus *Coloconger* comprises demersal species distributed circumglobally at upper bathyal depths from 300 m to over 1000 m throughout the tropics but extending as far north as Japan and south to Australia. There are five species known as adults, each with a regional distribution: *C. cadenati* (600 m$_{fbd}$) in the E Atlantic, *C. japonicus* (760 m$_{fbd}$) in the NW Pacific, *C. meadi* (925 m$_{fbd}$) in the W Atlantic and the Gulf of Mexico, *C. raniceps* (Froghead eel, 1134 m$_{fbd}$) in the Indian Ocean and the NW Pacific and *C. scholesi* (Shorttail conger, 810 m$_{fbd}$) in the Indian and Pacific Oceans (Fig. 4.12a).

4.3.3.7 Family Congridae (43) (Conger eels)

The Congridae comprises over 190 species of demersal eels in 32 genera. The subfamily Heterocongrinae (garden eels) contains 35 entirely shallow-water species, although the larvae may be bathypelagic. Of the remaining Congridae, the type species *Conger conger* (European conger, 1171 m$_{fbd}$) is the largest known eel with a maximum recorded length of 3 m but typically not more than 1.5 m. It generally lives inshore at continental shelf depths, and the maximum reported depth reflects the onset of migration to deep-sea offshore spawning locations. Shelf sea commercial fisheries capture entirely immature individuals (Campillo, 1992). Suggested spawning in the Sargasso Sea after a transatlantic migration analogous to that of *Anguilla anguilla* has been rejected by McCleave and Miller (1994), and there appears to be no evidence for often-cited spawning at great depths (3000–4000 m) in the NE Atlantic between Gibraltar and the Azores. The only validated source of larvae is at 600–800 m depth, south of the island of Sardinia in the Western Mediterranean Sea (Correia et al., 2006). A consensus view is emerging that *C. conger* do not make long spawning migrations and that there are multiple hitherto unknown spawning locations in the Mediterranean Sea and the NE Atlantic Ocean with larvae from different sources mixing in the pelagic realm (Correia et al., 2012).

Most Congridae probably have a deep-water component to their lifecycle similar to that of *Conger conger,* and there are some bathydemersal species in most genera. Six genera can be regarded as deep-sea endemic, predominantly occurring on slopes at depths of 200–1000 m: *Bassanago, Bathycongrus, Japonoconger, Promyllantor, Pseudophichthys* and *Xenomystax*. The genus *Bassanago* comprises four species of Southern Hemisphere deep-water Conger eels: *B. albescens* (1700 m$_{fbd}$) around S America and SW Africa, *B. bulbiceps* (1100 m$_{fbd}$) off New Zealand, *B. hirsutus* off Eastern Australia and *B. nielseni* (340 m$_{fbd}$) discovered on SE Pacific seamounts. The genus *Bathycongrus* is circumglobal at low latitudes with 22 species, the most widespread being *B. guttulatus* (1270 m$_{fbd}$) throughout the Indo-Pacific region from East Africa to Hawaii and *B. varidens* (935 m$_{fbd}$) from southern Canada to Chile in the East Pacific. Six species are endemic to the Atlantic Ocean (some extending into the Gulf of Mexico), two species to the Indian Ocean and 13 species, including

B. varidens to different parts of the Pacific Ocean. *B. odontostomus* (886 m$_{fbd}$) and *B. wallacei* (500 m$_{fbd}$) straddle the boundary between the Eastern Indian and the West Pacific. The genus *Japonoconger* comprises three species, *J. africanus* (650 m$_{fbd}$) from the E. Atlantic off West Africa, *J. caribbeus* (576 m$_{fbd}$) in the W. Atlantic and *J. sivicolus* (535 m$_{fbd}$) in the N.W. Pacific off Japan. The genus *Promyllantor* has the deepest Congridae: *P. adenesis* (2325 m$_{fbd}$) in the NW Indian Ocean and *P. purpureus* (2250 m$_{fbd}$) from the coasts of SW India and Sulawesi (Fig. 4.12a). The single species in the genus *Pseudophichthys splendens* (Purple-mouthed conger, 1647 m$_{fbd}$) is distributed across the Atlantic Ocean between the coast of NW Africa in the east and the Gulf of Mexico to Brazil in the West. The genus *Xenomystax* has four species distributed on either side of the Isthmus of Panama: *X. atrarius* (935 m$_{fbd}$) in the East Pacific from Canada to Chile and three species in the Central West Atlantic Ocean, *X. austrinus* (732 m$_{fbd}$), *X. bidentatus* (604 m$_{fbd}$) and *X. congroides* (825 m$_{fbd}$). *X. trucidans* (1316 m$_{fbd}$) is an outlier collected off the Maldives in the Indian Ocean.

4.3.3.8 Family Nettastomatidae (44) (Duckbill eels)

The Nettastomatidae are predominantly tropical and warm temperate slender-bodied demersal eels of moderate size, average maximum length 57 cm. They are characterised by an elongated snout and an absence of pectoral fins in most species. There are 42 species in six genera. In most genera some species are recorded at upper bathyal depths, but *Nettastoma* and *Venefica* can be regarded as deep-sea endemic. *Nettastoma melanurum* (Blackfin sorcerer, 1647 m$_{fbd}$; Fig. 4.12a) is found in both the Eastern and Western Atlantic Ocean, from Portugal to the Gulf of Guinea including the Mediterranean and Gulf of Mexico to southern Brazil, respectively, benthopelagic on slopes generally between 300 and 900 m depth. Their diet is mainly crustacea. *N. parviceps* (Duck-billed eel, 1190 m$_{fbd}$) is found in the Indo-Pacific region from SE Africa to Eastern Australia, Japan, Hawaii and Chile; with a bi-temperate distribution it is rare or absent in equatorial waters.

N. solitarium (Solitary duckbill eel, 610 m$_{fbd}$) occurs in the Indo-Pacific region from the Western Indian Ocean to Hawaii. *N. syntresis* (641 m$_{fbd}$) is found in the Western Central Atlantic and in the North Eastern Gulf of Mexico. In the genus *Venefica* the geographic range is dominated by *V. proboscidea* (Whipsnout sorcerer, 2200 m), which is distributed on the lower slopes worldwide in tropical, subtropical and temperate waters. Other species in the genus have regional distributions: *V. multiporosa* (1300 m$_{fbd}$) Indo-West Pacific from Madagascar to Australia and the Philippines, *V. ocella* (1953 m$_{fbd}$) East Pacific, *V. procera* (2304 m$_{fbd}$) West Central Atlantic and Gulf of Mexico and *V. tentaculata* (500 m$_{fbd}$) NW Pacific off Japan. Trawl surveys of *Nettastoma melanurum* in the Mediterranean Sea at 580–1598 m depth show a total length range 30.2–75.3 cm with females about 10 cm larger than the males. Spawning appears to be synchronous with a single batch of 8000–19000 eggs each year between September and January (Porcu et al., 2013). Miller and McLeave (2007) observed that in the NW Atlantic, Nettastomatid eels probably do not migrate offshore but spawn on the slopes, and the leptocephali are carried by currents offshore to the Sargasso Sea area.

4.3.3.9 Family Ophichthidae (45) (Snake eels and Worm eels)

This is a family of over 300 species of mainly tropical coastal demersal eels. There are a few deep-sea species. The genus *Benthenchelys* comprises three species of rare small (10 cm total length) bathypelagic eels from the tropical E Indian Ocean and Pacific Ocean recorded down to over 1000 m depth. *Ophichthus tetratrema* is a demersal eel up to 56 cm long from the E Pacific Ocean off Costa Rica and Ecuador at depths of 700 to 1,000 m.

4.3.3.10 Family Muraenesocidae (46) (Pike congers)

Within the family Muraenesocidae the genus *Gavialiceps* comprises five bathydemersal species found on bathyal slopes down to depths of over 1000 m in the Indo-Pacific Area: *G. javanicus*

(Duckbill conger, 600 m_{fbd}) in the Indo-West
Pacific, *G. arabicus* (496 m_{fbd}) in the W Indian
Ocean, *G. bertelseni* (1200 m_{fbd}) in the W Indian
Ocean, *G. taeniola* (1046 m_{fbd}) in the Indian Ocean
and *G. taiwanensis* (750 m_{fbd}) in the NW Pacific.

Anguilliformes II (Bathypelagic)

This group encompasses the Order
Saccopharyngiformes as defined by Nelson (2006) and
those families of the Anguilliformes that have
radiated into the deep pelagic environment with
rather elongated body forms. Molecular phylogeny
groups these together with the freshwater eels of the
family *Anguilla* that have retained the deep-sea
spawning habit (Inoue et al., 2010).

This group of fishes has greatly reduced body
morphology lacking several bones of the skull,
branchiostegal rays, scales, pelvic fins, ribs, pyloric
caecae and swim bladder. These are considered extreme
adaptations to the bathypelagic environment, minimising
energy requirements and conferring neutral buoyancy.
The jaws are greatly expanded to enable ingestion of large
prey. There are 34 species in six families (Fig. 4.12b).

4.3.3.11 Family Saccopharyngidae (47) (Swallowers)

There is one genus, *Saccopharynx*, with 10 species
distributed in the Atlantic and Pacific Oceans. The
largest species, *S. ampullaceus*, can grow to over 1.5
total length, including a long whip-like tail; the main
part of the body is typically 30 cm long (Fig. 4.12b). The
large mouth and distensible abdomen enables the
ingestion of large prey items, which comprise mainly
fishes. A luminous organ on the tail may act as a lure
(Smith, 2002). They appear to be a solitary species with
just occasional catches of single specimens from deep-
water net tows. There are seven species recorded from the
Atlantic Ocean: *S. ampullaceus* (Gulper eel, 3000 m_{fbd}),
S. thalassala (1000 m_{fbd}), *S. trilobatus* (1305 m_{fbd}),
S. harrisoni, *S. hjorti* (1400 m_{fbd}), *S. paucovertebratis*
and (1700 m_{fbd}), *S. ramosus* (1500 m_{fbd}) and three from
the Pacific, *S. lavenbergi* (3000 m_{fbd}), *S. schmidti*
(Whiptail gulper) and *S. berteli* (1100 m_{fbd}).

4.3.3.12 Family Eurypharyngidae (48) (Gulper or Pelican eels)

There is one species, *Eurypharynx pelecanoides* (Pelican
eel; Fig. 4.12b, Plate 3b), that is found worldwide in
tropical and temperature waters, e.g. in the Atlantic
Ocean from Iceland to 48°S. It can grow to a total length
of 1 m but a more typical length is around 55 cm. The
tail ends with an expanded luminous organ. As the
stomach is not as distensible as in *Saccopharynx*, it
seems to take smaller prey, mainly crustacea as well as
some fishes, cephalopods, and other invertebrates. The
large mouth enables it to enclose large volumes of
water, which are then expelled through the gills
retaining the food items, which are then swallowed. The
maximum recorded depth at the 7625 m in the hadal
zone is almost certainly erroneous (Jamieson, 2015); the
species is usually reported from 1200 to 1400 m depth.

4.3.3.13 Family Cyematidae (49) (Bobtail snipe eels)

There are two species in two different genera. *Cyema
atrum* (Fig. 4.12b) occurs circumglobally in all the
oceans at depths from 330 to 5100 m. Maximum size
is 15 cm. *Neocyema erythrosoma* is known only from
a few deep pelagic specimens taken in the SE and NW
Atlantic Ocean (DeVaney et al., 2009) and over the
northern Mid-Atlantic Ridge. When freshly caught,
they appear bright orange or red and are about 9 cm
long. The general appearance is of a part-
metamorphosed leptocephalus with the head of an
adult and tail of a larva but with mature gonads
evident, indicating paedomorphosis.

4.3.3.14 Family Monognathidae (50) (One jaw gulpers)

In this family, simplification of the basic vertebrate
body plan has reached an extreme, with loss of the
upper jaw and skeletal supports of the median fins.
The eyes and other sense organs are greatly reduced.
The lower jaw closes directly against the
neurocranium. Feeding is probably aided by a
venomous fang on the front of the skull. Stomach
contents have been reported as mainly composed of
shrimps. Monognathids are small fish, mostly less

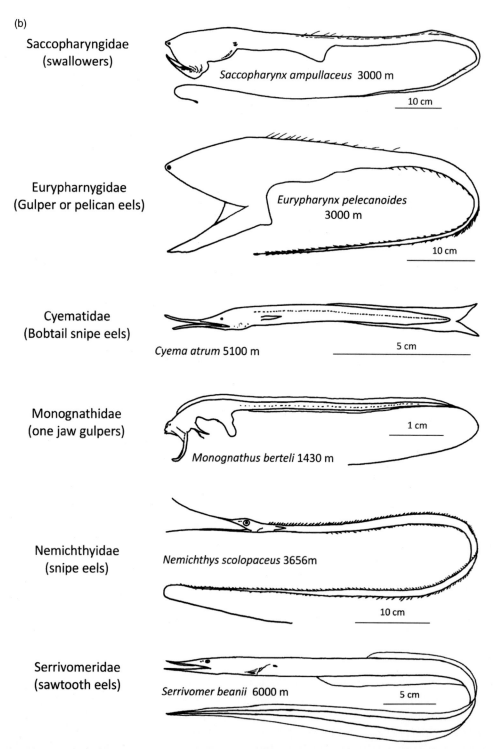

(b)

Saccopharyngidae (swallowers)

Saccopharynx ampullaceus 3000 m

10 cm

Eurypharyngidae (Gulper or pelican eels)

Eurypharynx pelecanoides 3000 m

10 cm

Cyematidae (Bobtail snipe eels)

Cyema atrum 5100 m

5 cm

Monognathidae (one jaw gulpers)

Monognathus berteli 1430 m

1 cm

Nemichthyidae (snipe eels)

Nemichthys scolopaceus 3656m

10 cm

Serrivomeridae (sawtooth eels)

Serrivomer beanii 6000 m

5 cm

Figure 4.12b ORDER ANGUILLIFORMES. Pelagic families (Saccopharyngiformes) *Saccopharynx ampullaceus* (gulper eel), Atlantic. *Eurypharynx pelecanoides* (Pelican eel), circumglobal. *Cyema atrum* (bobtail eel), circumglobal. *Monognathus berteli*, W Indian Ocean. *Nemichthys scolopaceus* (Slender snipe eel), circumglobal. *Serrivomer beanii* (Stout sawpalate), circumglobal.

than 10 cm total length and generally taken at depths greater than 2000 m. One captured mature male had greatly reduced feeding apparatus, the lower jaw and rostral fang were rudimentary but the olfactory organs were enlarged, suggesting location of the female by pheromone trails. 14 species are recognised, known mostly from a few specimens of each species, two species from the Indian Ocean; *Monognathus berteli* (1430 m$_{fbd}$; Fig. 4.12b) and *M. rajui*, six from the Atlantic Ocean; *M. bertini, M. boehlke* (5430 m$_{fbd}$), *M. herringi* (5400 m$_{fbd}$) *M. nigeli* (3910 m$_{fbd}$), *M. jesperseni* (3000 m$_{fbd}$) and *M. taningi* (1000 m$_{fbd}$) and five from the Pacific Ocean: *M. bruuni* (3000 m$_{fbd}$), *M. isaacsi* (900 m$_{fbd}$) *M. ozawai, M. rosenblatti* (5266 m$_{fbd}$) and *M. smithi* (5800 m$_{fbd}$; Nielsen and Hartel, 1996).

4.3.3.15 Family Nemichthyidae (51) (Snipe eels)

The Nemicththyidae comprises nine species in three genera, very widespread throughout the oceans at tropical and temperate latitudes. They are highly specialised mesopelagic and bathypelagic eels (depths down to 4580 m) with a thin, black-coloured, scaleless, extremely elongated body up to over 1.5 m long. *Nemichthys scolopaceus* (Slender snipe eel, 3656 m$_{fbd}$) has the highest number of vertebrae (>750) of any vertebrate (Fig. 4.12b). The genus *Avocettina* has three species, *A. acuticeps* (Southern snipe eel, 2000 m$_{fbd}$) has a Southern Hemisphere distribution in the Atlantic, Indian and Pacific Oceans, except in the East Pacific, *A. infans* (Avocet snipe eel, 4580 m$_{fbd}$) has a Northern Hemisphere distribution north of 20°S, except the Mediterranean Sea and East Pacific Ocean, and *Avocettina bowersii* (641 m$_{fbd}$) is confined to East Central Pacific Ocean. The genus *Nemichthys* (Plate 4a) also has three species: *N. curvirostris* (Boxer snipe eel, 2000 m$_{fbd}$) occurring worldwide 40°N to 40°S, *N. scolopaceus* (Slender snipe eel, 3656 m$_{fbd}$) occurring worldwide 55°N–42°S including the Mediterranean Sea (Bilecenoglou et al., 2006) and *N. larseni* (1000 m$_{fbd}$) confined to the East Pacific Ocean. *Labichthys carinatus* (2000 m$_{fbd}$) is found in the W Indian Ocean, the Atlantic and East Pacific Oceans between 40°N and 20°S and *L. yanoi* (Yano's snipe eel) from the SW Pacific Ocean. The jaws of the adults are long, thin

and curving. They feed on crustacea with evidence of vertical migration in pursuit of prey. In pelagic net tows in the NW Atlantic at depths of 450–670 m off Georges Bank *Nemichthys scolopaceus* was the most important fish species by biomass, feeding on predominantly on euphausiids and decapods, indicating that these that eels form a major trophic pathway in the ocean rim ecosystem (Feagans-Bartow and Sutton, 2014). At sexual maturity, strong sexual dimorphism appears with shortening of the male jaws, which together with degenerative changes suggests semelparity, a single burst of reproduction at the end of the life span (Charter, 1996). Recently hatched larvae of *N. scolopaceus* have been captured in the Sargasso Sea and Bahamas area of the North West Atlantic Ocean (Wippelhauser et al., 1996). Spawning occurs every year over a protracted season possibly over most of the year with larvae concentrating in the vicinity of fronts associated with the subtropical convergence zone. *L. carinatus* spawns over a more restricted period during the summer months. Leptocephali occupy the upper 200 m of the ocean before descending to adult depths to metamorphose.

4.3.3.16 Family Serrivomeridae (52) (Sawtooth eels)

The Serrivomeridae are most closely related to the Nemichthyidae with which they share many similarities, being deep pelagic elongated eels with long slender jaws; however, these adaptations are less extreme, and the average maximum length is 57 cm. There are eight species in two genera. *Stemonidium hypomelas* (Black serrivomerid eel, 1229 m$_{fbd}$) is found mesopelagically throughout the Pacific Ocean and South Atlantic Ocean. It is reported ascending to 175 m depth at nighttime in waters off Brazil. The genus *Serrivomer* has seven species, *S. beanii* (Bean's sawtooth eel, 5998 m$_{fbd}$) circumglobal between 60°N and 20°S but absent from the NW Pacific (Fig. 4.12b), *S. bertini* (Thread eel, 1750 m$_{fbd}$) in the Indian and West Pacific Oceans and Southeast Pacific to Chile, *S. lanceolatoides* (Short-tooth sawpalate, 1000 m$_{fbd}$) in the North Atlantic and the Mediterranean Sea, *S. garmani* (2250 m$_{fbd}$) in the Indian Ocean and *S. jesperseni* (Crossthroat sawpalate, 825 m$_{fbd}$), *S. samoensis* (2250 m$_{fbd}$) and *S. sector* (Sawtooth eel,

3243 m$_{fbd}$) in the Pacific Ocean. *Serrivomer* species are predominantly mesopelagic, migrating upward at night, occasionally reported near the surface. They feed mainly on crustacea but also on fish and cephalopods. In the NW Atlantic larvae of *S. lanceolatoides* are found in the spawning area of freshwater eels (*Anguilla* spp.), in the Sargasso Sea (Miller and McLeave, 2007). In the NW Pacific Ocean there is a similar overlap in the spawning areas of *Serrivomer* and *Anguilla* species. The eggs of *Serrivomer* are larger (*ca.* 2 mm diameter) than the eggs of *Anguilla* (1.5 mm; Yoshinaga et al., 2011).

4.3.3.17 Family Anguillidae (53) (Freshwater eels)

Freshwater eels are mentioned in this text because they spend part of their lifecycle in the deep sea, and there is strong evidence from molecular genetics (Inoue et al., 2010) that they fall within the Anguilliformes II (bathypelagic) clade. There are 16 species of freshwater eel distributed around the world. The European eel (*Anguilla anguilla*) spawns in the Sargasso Sea area of the NW Atlantic where newly-hatched leptocephali are found in the plankton at 50–300 m depth. Leptocephali of deep-sea bathypelagic eels, *Nemichthys scolopaceus*, *Serrivomer beani*, *S. brevidentatus*, *Avocettina infans* and *Nessorhamphus ingolfianus*, also occur in the same area (Castonguay and McCleave, 1987). The leptocephali of *A. anguilla* drift towards Europe, and metamorphosed glass eels ascend into rivers and streams throughout the Atlantic and Mediterranean coasts. Tracking of the return migration of adults using satellite pop-up tags reveals that they swim predominantly at depths between 200 and 1000 m in the Atlantic Ocean ascending during the night and descending during the day progressing at about 13.8 km day^{-1} towards the Sargasso Sea (Aarestrup et al., 2009). Wahlberg et al. (2014) recorded one such returning adult, *A. anguilla*, being consumed by predatory toothed whales while swimming at depths between 470 and 700 m.

Subdivision: Otocephala

This major subdivision of the Teleostei includes a vast array of freshwater fishes Cypriniformes (carps) and Siluriformes (catfishes) together with some marine or euryhaline orders, the Clupeiformes (herrings), Gonorhynchiformes (milkfishes) and one deep-sea order, the Alepocephaliformes (Near et al., 2102, Bentancur-R et al., 2013).

4.3.4 Order Alepocephaliformes

Comprising 137 species in three families, Platytroctidae, Bathylaconidae and Alepocephalidae, the Alepocephaliformes is an ancient order of endemic deep-sea fishes that originated in the mid-Triassic (Fig. 4.13). Their classification has been changed several times; in most older texts they were placed in the Salmoniformes (Matsui and Rosenblatt, 1987; Nelson, 2006) included them in the marine smelts (Argentiniformes) with which they share the crumenal or eipibranchial organ, but Ambrose (1996) points out that the Alepocephaliformes have larger eggs than the smelts, and there is direct development with no distinctive larval stage. Most recent molecular genetic studies (Bentancur-R et al., 2013; Near et al., 2012) place the Alepocephaliformes in a separate group of their own, most closely related to the shallow-water Clupeiformes (herrings) and Ostarophysii (freshwater superorder of teleosts) from which they diverged over 200 Ma. This placing is in agreement with earlier proposals by Marshall (1966).

The Alepocephaliformes are small to medium-sized fishes, mean maximum length (28 cm) that are mainly meso- or bathypelagic down to a mean maximum depth of 2015 m with some species found at over 5000 m. The name derives from the lack of scales on the head. The head and eyes are large, and many species have bioluminescent organs. The swim bladder is absent, the skeleton often only partially ossified, the flesh is watery and in some species there are subdermal fluid spaces (Best and Bone, 1976). The median fins are on the posterior half of the body. The adults generally have black or dark-coloured skin, with the mouth, branchial cavity and peritoneum also black. Whilst most species are bathypelagic, some are often captured in demersal trawls, suggesting a benthopelagic habit.

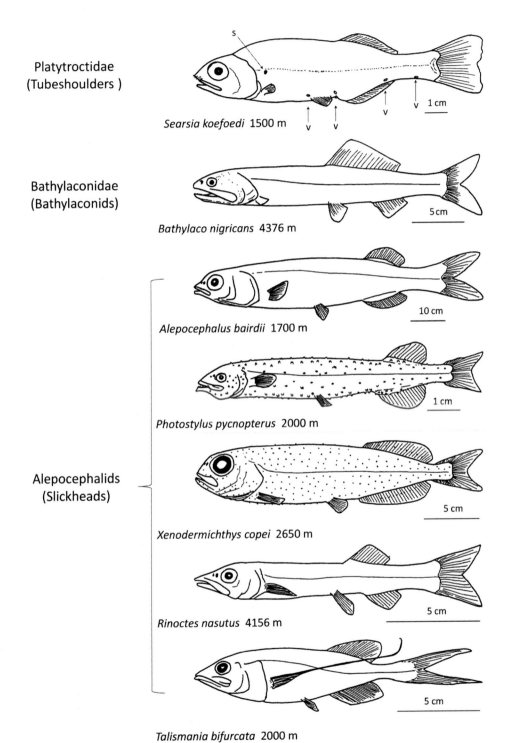

Platytroctidae
(Tubeshoulders)

Searsia koefoedi 1500 m

Bathylaconidae
(Bathylaconids)

Bathylaco nigricans 4376 m

Alepocephalids
(Slickheads)

Alepocephalus bairdii 1700 m

Photostylus pycnopterus 2000 m

Xenodermichthys copei 2650 m

Rinoctes nasutus 4156 m

Talismania bifurcata 2000 m

Figure 4.13 ORDER ALEPOCEPHALIFORMES *Searsia koefoedi* (Koefoed's searsid) circumglobal tropics. s – opening of the tube of the shoulder or postclavicular organ that can exude bioluminescent material, v = ventral light organs. *Bathylaco nigricans* (black warrior) probably circumglobal 40°N–40°S. *Alepocephalus bairdii* (Baird's slickhead) N Atlantic. *Photostylus pycnopterus* (starry smooth-head) circumglobal. *Xenodermichthys copei* (bluntsnout smooth-head) probably circumglobal. *Rinoctes nasutus* (abyssal smooth-head) probably circumglobal, *Talismania bifurcata* (threadfin slickhead) Pacific.

4.3.4.1 Family Platytroctidae (54) (Tubeshoulders)

The diagnostic feature of the Platytroctidae (formerly known as Searsiidae) that separates them from the other Alepocephaliformes is a unique pair of shoulder luminescent organs comprising a sac on each side that opens via a black tube through a hole in a modified scale. This can discharge a bright luminous fluid into the water, which produces numerous bright points of blue-green light (Nicol, 1958). Regular photophores mainly on the ventral surface of the body, similar to those found in myctophids, are also found in six genera that can be considered mesopelagic: *Persparsia* (one species), *Holtbyrnia* (eight of the nine species), *Sagamichthys* (three species), *Searsia* (one species; Fig. 4.13) and *Searsioides* (two species) and *Maulisia* (two out of five species; Sazonov, 1996). The deeper-living species such as the bathypelagic genus *Mentodus* tend to have no ventral photophores (Matsui and Rosenblatt, 1987). There are 39 species recognised in 13 genera. The deepest species in the bathypelagic genera are *Barbantus curvifrons* (Palebelly searsid, 4500 m_{fbd}) widespread in the Atlantic, Pacific and Indian Oceans, *Maulisia microlepis* (Smallscale searsid, 2000 m_{fbd}) in the Atlantic Ocean, *Mentodus rostratus* (2100 m_{fbd}) in the tropical Atlantic, Pacific and Indian Oceans, *Normichthys yahganorum* (2000 m_{fbd}) circumglobal in the Southern Hemisphere subtropical convergence zone between 30°S and 50°S, *Pectinantus parini* (1750 m_{fbd}) in the Pacific and Indian Oceans and *Platytroctes apus* (Legless searsid, 5393 m_{fbd}) in the deep waters of the Atlantic and Indian Oceans 67°N–21°S. Matsui and Rosenblatt (1987) comment that whilst Platytroctidae are globally distributed in tropical to temperate waters, they tend to be absent from oligtrophic gyre regions and that most specimens are caught in relatively narrow bands of highly productive equatorial waters and in the eastern boundary currents on the western side of continents. There is also a strong association with seafloor features such as ridges, fracture zones, seamounts and oceanic islands. The deeper-living species are regarded as more advanced.

4.3.4.2 Family Bathylaconidae (55) (Bathylaconids)

The Bathylaconidae comprise four species of bathypelagic to abyssopelagic, deep-living smoothheads with a worldwide distribution in warm and temperate waters. Nelson (1994) considered them a subfamily of the Alepocephalidae. Photophores, and other luminescent organs are absent; earlier reports of luminous organs (Parr, 1948) appear to have been erroneous. There are four species in two genera: *Bathylaco macrophthalmus* (3570 m_{fbd}) in the East Pacific, *B. nielseni* (2000 m_{fbd}) in the Eastern Central Atlantic and Indian Oceans, *B. nigricans* (Black warrior, 4375 m_{fbd}) circumglobal in warmer waters (Fig. 4.13) and *Herwigia kreffti* (Krefft's smooth-head, 3200 m_{fbd}) global except for the western Atlantic. Examining specimens of mature female *B. nigricans* from the Indian Ocean, Nielsen and Larson (1968) found that they appear to produce batches of 450–1000 eggs approximately 3 mm diameter with no suggestion of a distinct breeding season. The larvae are pelagic (Ambrose, 1996). It is suggested that well-developed visual and muscular systems may enable *B. nigricans* to hunt by stalking, and fast-forward sprinting to capture the prey in the large mouth.

4.3.4.3 Family Alepocephalidae (56) (Slickheads or Smoothheads)

The Alepocephalidae comprise 97 species in 18 genera that have a worldwide distribution in the mesopelagic and bathypelagic realms. All are deep-sea species, and the mean maximum depth of occurrence is 2037 m. Diversity ranges from small (*ca.* 10 cm total length) delicate pelagic bioluminescent species such as *Photostylus pycnopterus* (Starry smooth-head) to large robust (up to 100 cm total length) nonbioluminescent species such *Alepocephalus bairdii* (Baird's slickhead) that are caught in quantity in deep-sea demersal trawl fisheries (Fig. 4.13). Photophores on the body are confined to four genera: *Microphotolepis, Photostylus, Rouleina* and *Xenodermichthys. Microphotolepis schmidti* has two kinds of photophores: large regular photophores (0.8 mm diameter) distributed in rows on the body and the head, and a much larger number of small secondary photophores on top of almost every scale of the body (Sazanov, 1996). In its monotypic genus *Photostylus pycnopterus* (Starry smooth-head, 2000 m$_{fbd}$; Fig. 4.13) has a unique arrangement of photophores on small stalks sparsely set over the entire head and body. It is a small fish, 11 cm total length, naked with no scales, distributed worldwide at 65°N–35°S. *Xenodermichthys* has a scale-less dark skin covered almost entirely with small nodular photophores (0.5 mm diameter). *X. copei* (Bluntnose smooth-head, 2650 m$_{fbd}$) occurs almost worldwide 70°N–56°S (Fig. 4.13) but in the Central West Pacific it is replaced by *X. nodulosus*, with a diet mainly of pelagic crustacea. The photophores of the genus *Rouleina* are generally similar to those of *Xenodermichthys*, and 7 of the 10 species are known to be bioluminescent. Sazanov (1996) also describes putative light organs on the lower jaw of *Bathyprion danae* (Fangtooth smooth-head, 3200 m) and *Mirognathus normani* (Norman's smooth-head, 3000 m). There is no evidence that the rare *Aulastomatomorpha phospherops* (Luminous slickhead, 2200 m$_{fbd}$) from the Indian Ocean has bioluminescent capability despite its name (Markle and Sazonov, 1996).

The genus *Leptochilichthys* is often placed in its own family Leptochilichthyidae but, following Nelson (2006), here it is included in the Alepocephalidae. There are three species up to 30 cm total length, all of them rare with scattered records. *L. agassizii* (Agassiz' smooth-head, 3100 m) and *L. pinguis* (Vaillant's smooth-head, 1400 m) are bathypelagic from the Atlantic and Indo-Pacific Oceans. *L. microlepis* (Smallscale smoothhead, 726 m) was discovered off Japan (Machida and Shoiogaki, 1988).

Most Alepocephalids are benthopelagic, living within a few metres of the seafloor and forming a conspicuous component of the bathyal to abyssal fish fauna throughout the world (Crabtree and Sulak, 1986). Demersal trawl surveys in the Porcupine Seabight area of the NE Atlantic captured 15 species in 7 genera ranging from *Alepocephalus bairdii* at the shallowest depths (706–2500 m) to *Conocara salmoneum* (Salmon smooth-head) on the abyssal plain at 3639–4842 m (Priede et al., 2010). A similar pattern is observed in different regions with Alepocephalids often a dominant element of the demersal fish fauna. They can be considered to consist of 10 benthopelagic genera: *Alepocephalus, Asquamiceps, Bajacalifornia, Bathytroctes, Conocara, Einara, Leptoderma, Narcetes, Rinoctes* and *Talismania*. The most speciose is *Alepocephalus* with *A. australis* (Small-scaled brown slickhead, 2600 m$_{fbd}$) distributed worldwide at 55°N–56°S and a series of regionally localised species. *Alepocephalus agassizii* (Agassiz' slickhead, 2500 m$_{fbd}$), *A. bairdi* (Baird's slickhead, 2500 m$_{fbd}$), *A. productus* (Smalleye smooth-head, 3500 m$_{fbd}$) and *A. rostratus* (Risso's smooth-head, 2250 m$_{fbd}$) occur in the Atlantic Ocean, *A. andersoni* (726 m$_{fbd}$), *A. asperifrons* (1867 m$_{fbd}$), *A. fundulus* (3060 m$_{fbd}$), *A. longiceps* (Longfin slickhead, 1300 m$_{fbd}$), *A. longirostri* (Longsnout slickhead, 1140 m$_{fbd}$), *A. melas, A. owstoni* (Owston's slickhead, 1000 m$_{fbd}$), *A. planifrons* (750 m$_{fbd}$), *A. tenebrosus* (Owston's slickhead, 5500 m$_{fbd}$), *A. triangularis* (Triangulate slickhead, 1140 m$_{fbd}$) and *A. umbriceps* (Slickhead, 2000 m$_{fbd}$) in the Pacific Ocean, *A. bicolor* (bicolor slickhead, 1080 m$_{fbd}$) and *A. antipodianus* (Antipodean slickhead, 1160 m$_{fbd}$) in the Indo-Pacific, *A. blanfordii* (2000 m$_{fbd}$) and *A. dentifer* (1760 m$_{fbd}$) are exclusive to the Indian Ocean. *Alepocephalus rostratus* captured at 1000 to 2250 m on the slopes of the Catalan Sea in the Western Mediterranean have a predominantly gelatinous macroplanktonic diet (*Pyrosoma* and siphonophores), but they also take some benthic prey (Carrassón and Matallana, 1998). It is presumed that spawning is demersal with eggs deposited on the seafloor. *A. rostratus* in the Mediterranean has an extended spawning season with peak activity from September to October but in the Eastern Atlantic Ocean spawning is concentrated more in the winter months (Folessa et al., 2007). In the NE Atlantic Ocean, *A. bairdii* and *A. rostratus* were taken as a bycatch in mixed-species deep demersal fisheries with declared landings reaching a peak of over 13, 000 tonnes for *A. bairdii* in 2002 (ICES, 2012). *A. antipodianus* and *A. australis* have similarly been captured in fisheries on the Chatham Rise off New Zealand, though they are generally discarded (Jones and Breen, 2013).

The genus *Asquamiceps* consists of four species: *A. caeruleus* (2740 m$_{fbd}$) 32°N–22°S in the Atlantic and Indian Oceans, *A. hjorti* (Barethroat slickhead, 2865 m$_{fbd}$) in the E Atlantic and Indo-West Pacific Oceans, *A. longmani* from the West Pacific Ocean and *A. velaris* (3660 m$_{fbd}$), which is possibly circumglobal. *Bajacalifornia* comprises five species with a combined circumglobal distribution though rather few specimens have been examined (Miya and Markle, 1993). The genus *Bathytroctes* comprises 11 deep-living species with a mean maximum depth of 3087 m, including the deepest-living Alepocephalid, *B. macrolepis* (Koefoed's smooth-head, 5850 m$_{fbd}$) from the Atlantic, Indian and Pacific Oceans. In the NE Atlantic, *B. macrolepis*, *B. michaelsarsi* (5075 m$_{fbd}$) and *B. microlepis* (4900 m$_{fbd}$) have all been captured on slopes at depths >1800 m beyond the reach of commercial fisheries (Priede et al., 2010). The genus *Conocara*, similarly distributed with 10 species at tropical and temperate latitudes, generally occurs at shallower depths than *Bathytroctus* and is absent from most of the Pacific Ocean except in the SW around Australia and New Zealand. The most abundant species are *C. fiolenti* (Fiolenti's smooth-head, 1600 m$_{fbd}$), *C. kreffti* (Wrinkled slickhead, 1700 m$_{fbd}$), *C. macropterum* (Longfin smooth-head, 2200 m$_{fbd}$) and *C. murrayi* (Murray's smooth-head, 2603 m$_{fbd}$). In the Bahamas region *C. macropterum* is one of the most frequently captured demersal fish species at bathyal depths. Adult length modes centred on 30 cm for males and 33 cm for females with a gap between these and juveniles at <100 mm in the length-frequency distribution, suggesting fast growth to the adult reproductive size (Crabtree and Sulak, 1986). Small *C. macropterum* consume mostly polychaetes, copepods, and ostracods, whereas large ones feed on decapod crustaceans and teleost fishes with salps in the diet of all size classes. In common with other Alepocephalids, ingestion of sediment is frequently observed, providing evidence of benthic feeding. The ova diameters of *Conocara macropterum* and *C. fiolenti*, 6.0 and 5.0 mm, respectively, are the largest of any Alepocephalid species. It is suggested that they are buried in the seafloor in a way similar to the deposition of salmonid eggs in redds that have

been excavated by the parents (Crabtree and Sulak, 1986). The two species of the genus *Einara* are known from scattered specimens at 700–2000 m depth taken in the Atlantic, Indian and Pacific Oceans at tropical to temperate latitudes.

Leptoderma comprises five species of thin eel-like fishes with naked scale-free skin recorded at depths down to over 2000 m. The type specimen of *L. macrophalum* is 151 mm long had a fresh weight of 16 g (Byrkjedal et al., 2011). Two species are found in the Atlantic Ocean and three in the Indo-West Pacific at tropical to temperate latitudes. The genus *Narcetes* has two main species, *N. stomias* (Blackhead salmon, 2300 m) found between 54°N–28°S in the Atlantic and East Pacific Oceans and *N. lloydi* (Lloyd's slickhead, 1350 m) 34°N–34°S, in the Indo-West Pacific. Three other species are more rarely caught. The common name, Blackhead salmon, is an apt description of these conspicuous fishes that generally occur at depths greater than 1000 m. In its monotypic genus *Rinoctes nasutus* (Abyssal smooth-head, 4256 m) is a small fish, maximum length 19 cm, apparently highly adapted to its abyssal habitat with reduced ossification, no scales, fatty subcutaneous deposits and an unpigmented body. One mature female was found to contain 20 ripe eggs, 4–4.5 mm diameter (Markle and Merrett, 1980). The species was thought to be confined to the deep basins of the North Atlantic and the Gulf of Mexico but it may also occur in the West Central Pacific and the Indian Oceans. Catches of groups of individuals suggest that *R. nasutus* is gregarious and feeds on epibenthic organisms. The genus *Talismania* is currently considered to have 11 species, *T. antillarum* (Antillean smooth-head) distributed circumglobally between 49°N and 35°S and 455 and 1460 m depth, *T. bifurcata* (Threadfin slickhead, 2000 m) in the Pacific Ocean (Fig. 4.13), *T. homoptera* (Hairfin smooth-head, 1690 m) 31°N–4°S in the tropical Atlantic Ocean, *T. longifilis* (Longtail slickhead) 23°N–46°S in the East Atlantic and the Indo-West Pacific Oceans, *T. mekistonema* (Threadfin smooth-head) 40°N–24°N in the Atlantic and Indo-West Pacific Oceans and the other species are rarities. The characteristic features of *Talismania* are the 'threadfins' or greatly elongated pectoral fins with a very long first ray extending more than half the body length. The tail fins are often also extended and forked. Specimens are generally caught on or near the bottom, and the elongate pectoral rays may be a sensory adaptation for the detection of prey near the seafloor analogous to the specialised fin rays in the Bathyteroidae (Sulak, 1975). The eggs of *T. bifurcata* in the NE Pacific Ocean are reported to be pelagic (Ambrose, 1996) in contrast to the more usual demersal egg deposition found in most Alepocephaliformes.

Subdivision Euteleostei

This taxon includes all the remaining teleost fishes after the Elopomorpha, Osteoglossomorpha and Otocephala.

4.3.5 Order Argentiniformes (Marine Smelts)

This order comprises 87 species in four families in order of increasing depth of occurrence (Fig. 4.14); Argentinidae (argentines or herring smelts) mean maximum depth 458 m, Microstomatidae (pencil smelts) 1044 m, Opisthoproctidae (barreleyes of spookfishes) 1424 m and Bathylagidae (deep-sea smelts) 1442 m. A characteristic feature they share with the Alepocephaliformes is the crumenal organ, a specially adapted posterior branchial structure or epibranchial organ that grinds up or triturates planktonic prey notably gelatinous organisms.

4.3.5.1 Family Argentinidae (57) (Argentines or Herring smelts)

The argentines are small (mean maximum length 19 cm) shoaling silvery herring-like fishes. There are 27 species in two genera occurring widely around the world. *Argentina silus* (Greater argentine, 1440 m_{fbd}) is the deepest and largest recorded species, with a (Fig. 4.14a), maximum length up to 70 cm distributed on the continental shelves and slopes around the North Atlantic Ocean. In Norwegian waters it is

exploited by demersal trawling at depths between 200 and 600 m depth with total catches having reached a maximum of 22, 000 tonnes in 1983. It is often taken as a bycatch in the fishery for blue whiting *Micromesistius poutassou* and also occurs together with *Sebastes* sp. It reaches sexual maturity from 4 to 12 years (mean age of 7 for males and 6 for females (Johannessen and Monstad, 2003). Other species, such as *A. brucei* (400 m_{fbd}) *A. georgei* (457 m_{fbd}) *A. stewarti* (750 m_{fbd}) in the West Central Atlantic, *A. euchus* (590 m_{fbd}) in the W Indian Ocean, *A. sphyraena* (Argentine, 700 m_{fbd}) in the E Atlantic Ocean and the Mediterranean Sea and *Glossanodon semifasciatus* (Deep-sea smelt, 1017 m_{fbd}) in the NW Pacific are considered to be deep-sea fishes. The eggs and larvae are pelagic, and young stages are usually found at shallower depths on the continental shelves. The argentines are generally fishes of the outer shelf and upper slopes.

4.3.5.2 Family Microstomatidae (58) (Pencil smelts)

The Microstomatidae are slender-bodied silvery mesopelagic fishes with 19 species in three genera. In its monotypic genus *Microstoma microstoma* (Slender argentine) is distributed worldwide 55°N–49°S including the Mediterranean Sea. Generally solitary, it feeds on zooplankton. The genus *Nansenia* contains 17 species, some of which, such as *N. ardesiaca* (553 m), are found on the slopes of islands, seamounts and continental shelves, indicating a pseudo-oceanic distribution, *N. crassa*, *N. oblita* and *N. candida* are intermediate, whereas the others are truly oceanic. The genus is distributed from sub-Arctic to sub-Antarctic latitudes. Cold-water species are *N. candida*, found in the sub-Arctic North East Pacific, *N. antarctica*, which is circumglobal along the Antarctic Convergence, and *N. tenera* and *N. groenlandica*, which have a bipolar distribution in the Northern and Southern Hemispheres (Fig. 4.14a). Most species are from warm oceanic waters, for example, *N. atlantica* is associated with the Mauritanian upwelling (Kawaguchi and Butler, 1984). Average maximum length is 16 cm for the genus. *Xenophthalmichthys*

(a)

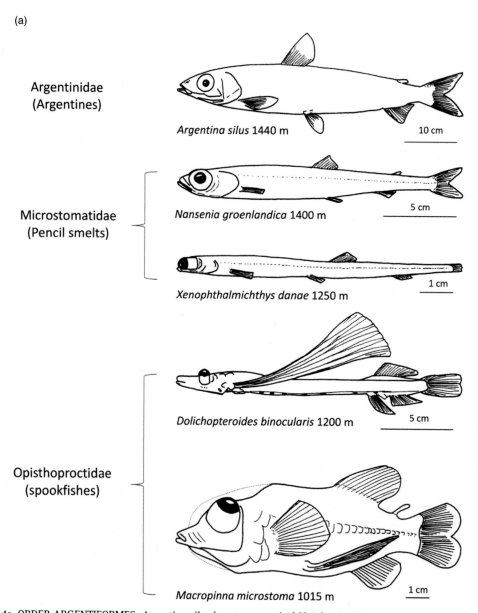

Argentinidae
(Argentines)

Argentina silus 1440 m   10 cm

Microstomatidae
(Pencil smelts)

Nansenia groenlandica 1400 m   5 cm

Xenophthalmichthys danae 1250 m   1 cm

Dolichopteroides binocularis 1200 m   5 cm

Opisthoproctidae
(spookfishes)

Macropinna microstoma 1015 m   1 cm

Figure 4.14a ORDER ARGENTIFORMES. *Argentina silus* (greater argentine) N Atlantic. *Nansenia groenlandica* (Greenland argentine) N Atlantic and Indian Ocean. *Xenophthalmichthys danae*, probably circumtropical. *Dolichopteroides binocularis*, Atlantic and SW Pacific 34°N–40°S. *Macropinna microstoma* (barreleye) N Pacific, the dashed line around the head indicates the position of the transparent cowl in living specimens.

danae (1250 m) has been sporadically caught in the Atlantic and the Pacific Oceans but is probably circumtropical in its distribution. It has remarkable forward-looking telescope eyes that occupy most of the anterior elevation of the head.

4.3.5.3 Family Opisthoproctidae (59) (Barrel eyes or Spookfishes)

The Opisthoproctidae comprise 19 species of relatively rare small fishes (mean length 15 cm) in eight genera highly adapted for life in the mesopelagic zone. They appear to be solitary, living at the limits of solar light penetration in the ocean. Opisthoproctidae can be divided into two kinds: four genera of long-bodied slender species (spookfishes) with a high vertebral count, *Bathylychnops, Dolichopteroides, Dolichopteryx* and *Ioichthys,* and four genera of short deep-bodied species (barreleyes) with low vertebral count, *Macropinna, Opisthoproctus, Rhynchohyalus* and *Winteria* (Parin et al., 2009). Here we first consider the long-bodied genera. *Bathylychnops* contains three species: *B. brachyrhynchus,* possibly distributed in all the oceans, *B. chilensis* recorded from the SE Pacific and *B. exilis* (Javelin spookfish) from the North Pacific and Atlantic Oceans. These are the longest of the Opisthoproctidae; *B. exilis* grows to 50 cm length, *B. brachyrhynchus* to at least 30 cm, whereas *B. chilensis* is known from only one juvenile specimen. They are oviparous with pelagic eggs and larvae. The eyes are large, globular, placed dorsally providing an upward-looking binocular visual field, and much of the head is transparent. *B. chilensis* has an additional accessory lens on each eye directed anteriorly (Parin et al., 2009), *B. exilis* has an accessory bulb and lens directed ventrally (Pearcy et al., 1965) and a similar arrangement is reported in *B. brachyrhynchus* (Parin, 2004). *B. exilis* is a strong swimmer and is likely to be an active predator in the mesopelagic. The monotypic genus *Dolichopteroides* has one species, *D. binocularis,* which occurs in all the oceans, though mainly in the tropical zone (Fig. 4.14a). The eyes are telescopic (tubular), upward-looking with accessory lenses directed laterally and ventrally. The body is transparent and white. The genus *Dolichopteryx* comprises nine species: *D. longipes* is probably circumglobal in the Northern Hemisphere 45°N–27°N, whereas *D. trunovi* is probably circumglobal in the Southern Hemisphere (Parin et al., 2009), *D. anascopa* and *D. minuscula* occur in the Indo-West Pacific (Misuzawa et al., 2015); *D. andriashevi, D. parini, D. pseudolongipes* and *D. vityazi* are from the Pacific Ocean, and *D. rostrata* is described from the NE Atlantic (Fukui and Kitagawa, 2006). All species have upward-looking tubular eyes, with accessory diverticula to provide lateral and ventral visual sensitivity.

Wagner et al. (2009) found in addition to the main lens, *D. longipes* has concave Fresnel-type reflective mirrors that focus light onto accessory retinas. Ventral light organs are reported in *D. andriashevi* by Parin et al. (2009). Descriptions of the morphology are hampered by the effects of capture and preservation. Fresh specimens seem to be largely transparent with white axial musculature and skeleton (Wagner et al., 2009) and in life the skin forms a streamlined transparent envelope over the body and head enclosing the eyes. The pectoral and pelvic fins, often damaged in trawled specimens, are greatly extended like wings, apparently enabling the fish, which has no swim bladder, to hover in mid-water (Fig. 4.14a). *Dolichopteryx* species are oviparous, *D. minuscula* produces batches of about 100 eggs (Fukui and Kitagawa, 2006), and *D. rostrata* shows a similar pattern with a total egg number in the ovary of 473 with a batch of 68 developed eggs (0.9–1.3 mm diameter) observed in a 6.6 cm long adult female. The monotypic *Ioichthys kashkini* from the NW Indian Ocean differs from *Dolichopteryx* in having nontubular spheroid eyes and from *Bathylychnops* in having no accessory lens or retina; the eyes are simple and conventional although with an upward-looking aspect (Parin, 2004).

The deep-bodied barreleye Opisthoproctidae (Fig. 4.14a,b) are distributed worldwide mainly in tropical and temperate seas: *Macropinna microstoma* (Barreleye, 700 m_{fbd}) in the North Pacific and the Bering Sea (Fig. 4.14a), *Opisthoproctus grimaldii* (Mirrorbelly, 4750 m_{fbd}, usually 300–400 m) in the Atlantic and W Pacific 43°N–49°S (Fig. 4.14b), *O. soleatus* (Barreleye, 800 m_{fbd}) circumglobal in tropical to temperate waters (Fig. 4.14b), *Rhynchohyalus natalensis* (Glasshead barreleye, 549 m_{fbd}) in the tropical Atlantic and Pacific 35°N–49°S (Fig. 4.17b), and *Winteria telescopa* (Binocular fish, 2500 m_{fbd}) in the East Atlantic and Indo-West Pacific 28°N–49°S (Fig. 4.14b). All species have the eponymous barreleyes, tubular eyes that are directed upward in all species except *W. telescopa* in which they face forward. However, observations of living *M. microstoma,* both in the field at *ca.* 700 m depth and in a ship-board aquarium, show that it can rotate

(b)

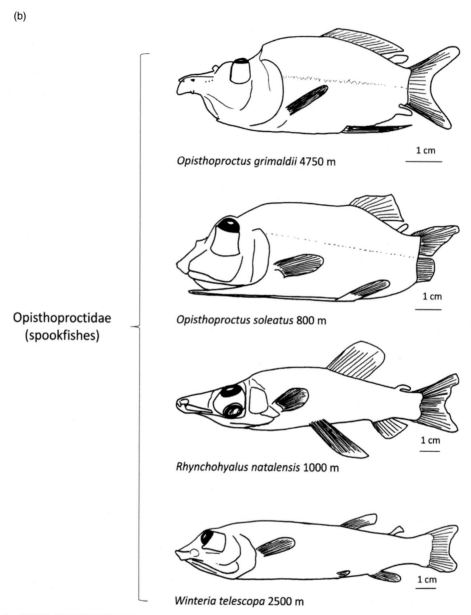

Opisthoproctidae
(spookfishes)

Opisthoproctus grimaldii 4750 m

1 cm

Opisthoproctus soleatus 800 m

1 cm

Rhynchohyalus natalensis 1000 m

1 cm

Winteria telescopa 2500 m

1 cm

Figure 4.14b ORDER ARGENTIFORMES. *Opisthoproctus grimaldii* (mirrorbelly) Subtropical Atlantic and W Pacific. *Opisthoproctus soleatus* (barreleye) circumglobal. *Rhynchohyalus natalensis* (glasshead barreleye) Atlantic and Pacific. *Winteria telescopa* (binocular fish) circumglobal except E Pacific.

its eyes from an upward-looking to a forward-looking direction through an arc of 75°, both eyes moving in tandem (Robison and Reisenbichler, 2008). It is not known if other species can rotate the eyes. The top of the head, including the eyes, is enclosed in a transparent cowl, which is lost in dead specimens. The eyes of *R. natalensis* have diverticula with accessory retinas and mirror system similar to *Dolichopteryx longipes,* providing visual sensitivity towards the ventro-lateral aspect (Partridge et al., 2014). It was originally thought that the ventral flat sole plate of *Opithroproctus* might be the swim bladder (in which case the fish would have to swim upside-down) but Bertelsen (1958) showed that it is a reflector organ that transmits light from a bacterial light organ associated with the rectum directing light downward from the belly of the fish, which swims with a normal upright posture. Haedrich and Craddock (1968) suggested that *W. telescopa* feeds on gelatinous plankton such as siphonophores, medusa and salps. Robison and Reisenbichler (2008) found cnidarian remains in the stomachs of *M. microstoma* and suggest that with the small mouth they browse on the tentacles, zooiids and captured prey of large colonial siphonophores of the genus *Apolemia*. The fish adopts an almost horizontal posture with the head tilted up about 10° when searching for prey. Presumably as the prey is approached the eyes can be rotated so that items can be visually identified and targeted for ingestion. Reproduction and fecundity seems to similar to that of other Opisthoproctidae with maturing batches of 59–171 eggs observed in mature female *W. telescopa* (Haedrich and Craddock, 1968).

4.3.5.4 Family Bathylagidae (60) (Deepsea smelts)

The Bathylagidae are planktivorous fishes of the mesopelagic or upper bathypelagic. They are dark-coloured in appearance and show no overt adaptations to the deep sea other than large eyes and small mouth for planktivory. Their morphology is similar to that of the salmonids, often including an adipose fin but with some variation in shape, for example *Bathylagus euryops* is short and fat, whereas other species are relatively slender. They are often captured along with typical mesopelagic fishes such as myctophids though not in large numbers, usually in small aggregations. There are 21 species in 8 genera. The genus *Bathylagichthys* has a predominantly southern distribution, *B. australis* in the Indian and Pacific Oceans, *B. longipinnis* and *B. problematicus* off Southern Africa, *B. parini* in the SE Pacific, and *B. greyae* in the SW Pacific and in the Atlantic up to 32°N. The three species of *Bathylagoides* are tropical in the Indo-Pacific region. *Bathylagus* occurs beneath the highly productive polar and subpolar waters in the Northern and Southern Hemispheres. *B. pacificus* is found in the North Pacific and *B. euryops* in the North Atlantic. *B. antarcticus* occurs around the Southern Ocean between 45°S and 74°S, consuming Euphausid krill. *B. tenuis* has a similar distribution showing diurnal migration, feeding mainly on copepods, appendicularians, euphausiids, pteropods, chaetognaths and cephalopods. Other new species have been identified in the Scotia Sea and off Chile (Kobylianskii, 2006). *Dolicholagus longirostris* is circumglobal between 39°N and 21°S in tropical waters, also exhibiting daily vertical migrations. The genus *Leuroglossus* is confined to the North Pacific, where *L. schmidti* (northern smoothtongue) is an important species. The monotypic *Lipolagus ochotensis* is also confined to the North Pacific. *Pseudobathylagus milleri* (Stout blacksmelt) is also a monotypic genus from the North Pacific but deeper-living with a minimum adult depth of 550 m. *Melanolagus bericoides* is a widespread global species occurring in all the oceans 71°N–58°S (Fig. 4.14c). Particularly in the Southern Ocean, the Bathylagidae have been identified as a significant link in the krill-based pelagic food chain leading to top predators such as birds and marine mammals (Stowasser et al., 2009). Bathylagidae are all oviparous with planktonic eggs and larvae, the pelagic stages of which are highly evolved (Moser and Ahlstrom, 1996). Several genera have stylophthalmine larvae with eyes on stalks: *Leuroglossus* has short eye stalks in early larvae, *Lipolagus ochotensis* has stalks of moderate length, *Bathylagus pacificus* has longer eye stalks, and *Melanolagus bericoides* and *Dolicholagus longirostris* have extremely long eyestalks (Fig. 4.14c). The stalked eyes increase the volume of water searched visually for individual items of plankton.

4.3.6 Order Stomiiformes

The Stomiiformes or Stomiatiformes (Fig. 4.15) comprises over 400 species of small (average maximum length 15 cm) deep-sea pelagic fishes living at mesopelagic and bathypelagic depths and characterised by the possession of nonbacterial photophores or light organs unique to this group. The photophores are globular structures with an outer

(c)

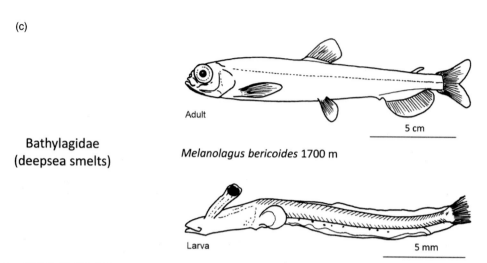

Bathylagidae
(deepsea smelts)

Melanolagus bericoides 1700 m

Adult
5 cm

Larva
5 mm

Figure 4.14c ORDER ARGENTIFORMES. *Melanolagus bericoides* (bigscale deepsea smelt) circumglobal 71°N–58°S.

layer of flat cells, rich in guanine, which serves as a reflector and an inner layer containing the photocytes or light-producing cells, which are connected to the nervous system. Bassot (1966) recognised three basic types of photophores, the occurrence of which can be related to taxonomic position within the stomatiatiformes. There are four main families, Gonostomatidae, Sternoptychidae, Phosichthyidae and Stomiidae. The Stomiiformes are regarded as close to basal teleosts having a wide jaw gape, which includes the maxilla and premaxilla.

4.3.6.1 Family Gonostomatidae (61) (Bristlemouths)

With 31 species in eight genera, this family is not particularly diverse but through their widespread occurrence down to an average maximum depth of over 2500 m in the bathypelagic these fishes occupy the largest living space on the planet and may be numerically the most abundant of vertebrates (Plate 4b). The common name arises from the numerous uniform-sized teeth in the jaws. They are generally dark-coloured or black fish with an elongated cylindrical body and without enlarged eyes. The single species in its genus *Bonapartia pedaliota* (Longray fangjaw, 1200 m_{fbd}) occurs in the Atlantic and Indian Oceans, mainly in tropical waters. The

genus *Cyclothone* has seven species with circumglobal distributions living at different typical depth ranges or different latitudinal distributions: *C. alba* (300–800 m), *C. braueri* (200–900 m), *C. pseudopallida* (300–1400 m) 61°N–68°S, *C. acclinidens* (300–1500 m), *C. pallida* (600–1800 m) 61°N–40°S (Plate 4b), *C. obscura* (2000–2600 m) 36°N–45°S, *C. microdon* (500–2700 m; Fig. 4.15a). *C. kobayashii* is circumglobal in the Southern Ocean, 38°S–67°S (Miya, 1994). Other species are more restricted in their distributions, *C. pygmaea*, only 2–2.9 cm long, is confined to the Mediterranean Sea. *C. livida* to the E Atlantic, *C. parapallida* is patchily distributed in the Atlantic and Pacific south of 25°N, *C. atraria* in the North Pacific and *C. signata* in the Pacific and Indian Oceans. *Cyclothone* species are oviparous with planktonic eggs and larvae with protandrous hermaphroditism in some species. They tend not to undertake diel vertical migrations. The genus *Gonostoma* has three species, two of which are circumglobal, *G. atlanticum* between 40°N and 40°S and *G. elongatum* over a wider latitudinal range whereas *G. denudatum* is found only in the Atlantic and Indian Oceans. This genus shows extensive vertical migration between mesopelagic and bathypelagic depths. The genus *Margrethia* has two more or less circumglobal species of which *M. obtusirostra* (Bighead portholefish, 1739 m_{fbd}) is

(a)

Gonostomatidae
(bristlemouths)

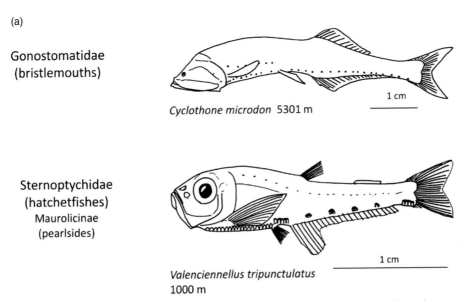

Cyclothone microdon 5301 m

1 cm

Sternoptychidae
(hatchetfishes)
Maurolicinae
(pearlsides)

Valenciennellus tripunctulatus
1000 m

1 cm

Figure 4.15a ORDER STOMIIFORMES. *Cyclothone microdon* (Veiled anglemouth) circumglobal 67°N–45°S. *Valenciennellus tripunctulatus* (Constellationfish) probably circumglobal including the Mediterranean.

associated with island and continental slopes. The genus *Sigmops* has four species: *S. bathyphilus* (Spark anglemouth, 3000 m_{fbd}) does not show vertical migrations and is most widespread at 66°N–35°S in the Atlantic and Pacific Oceans. They are protandrous hermaphrodites, though some males do not change sex but become super males. The genera *Diplophos* (five species), *Manducus* (two species) and *Triplophos* (one species) are placed provisionally in a separate family, Diplophidae, by Nelson (2006). *Diplophos taenia* (Pacific portholefish, 1594 m_{fbd}) is circumglobal at latitudes 40°N–30°S. It exhibits diel vertical migrations with juveniles and adults found at 300–800 m by day, sometimes reaching the surface at night. *Manducus maderensis* (850 m_{fbd}) is confined to the Atlantic Ocean and has characteristic photophores on the lateral line as well as the typical ventral photophores of the Gonostomatidae.

4.3.6.2 Family Sternoptychidae (62) (Hatchetfishes)
The Sternoptichidae is divided into two subfamilies: the Sternoptychinae, which includes three genera of laterally flattened, deep-bodied highly silvered hatchetfishes, and the Maurolicinae, a more diverse

assemblage of seven genera of photophore-bearing fishes but without the characteristic 'hatchet' body form.

Subfamily Maurolicinae (Pearlsides)
The Maurolicinae have an elongated typical fish body shape quite unlike the hatchet shapes of the Sternoptychinae, but they have the composite light organs characteristic of hatchetfishes. There are 30 species in seven genera distributed at temperate or tropical latitudes. Most species are benthopelagic on upper slopes of continents, islands or seamounts down to a depth of no more than 800 m and are captured either in bottom trawls or mid-water trawls towed close to the seafloor (Badcock and Merrett, 1972; Harold and Lancaster, 2003; Parin and Kobyliansky, 1996). The genus *Argyripnus* is found in all the oceans but the seven species are quite rare, represented by few specimens in collections; *A. brocki* is endemic to the Hawaiian Islands, *A. pharos* is widespread in the Indo-Pacific region and *A. atlanticus* in the Atlantic Ocean and East Central Pacific. The genus *Sonoda* (two species) is confined to the central West Atlantic. The circumglobal genus

Maurolicus comprises 15 species, each with a restricted geographical range, for example *M. amethystinopunctatus* in the Mediterranean Sea and adjacent NE Atlantic and *M. mucronatus* in the Red Sea (Parin and Kobyliansky, 1996). In the Bay of Biscay area of the NE Atlantic *M. muelleri* is the dominant diet component of juvenile Albacore tuna (*Thunnus alalunga*; Pusineri et al., 2005). The two species of *Thorophos* are confined to the West Central Pacific, and *Araiophos* is a rarity from the Pacific Ocean. In contrast to the general benthopelagic habit of the Maurolicinae, *Danaphos oculatus* is mesopelagic in the Indo-Pacific region beyond the continental shelf primarily at 183–914 m depth (Watson 1996). *Valenciennellus tripunctulatus* (Constellationfish), the most abundant global representative of the Maurolicinae (Fig. 4.15a), is truly pelagic with a worldwide distribution in tropical to temperate waters 67°N–34°S, and it plays a major role in deep-sea food chains. Highest densities occur between 290 and 460 m depth in the centres of gyres, and there is little upward movement at night. Feeding follows a diurnal pattern with most activity at 1200 h to 2200 h. The diet is made up of copepods, ostracods, euphausids, amphipods and chaeognaths (Hopkins and Baird, 1981). They are small fishes, maximum length 3.1 cm and have a sparse nonaggregated distribution; each individual must search a large volume of water per day to survive (Baird and Hopkins, 1981).

Subfamily Sternoptychinae (Hatchetfishes)

The hatchetfishes are a very widespread and conspicuous component of the mesopelagic ichthyofauna. Mean maximum length is 6 cm, and they are typically highly silvered, appearing as if chromium-plated when freshly caught (Plate 5a). The body is deep and laterally flattened, tapering towards a sharp ventral keel and giving rise to the miniature hatchet shape with the tail considered to represent the handle. The eyes are large and sometimes tube-like, looking upward. The photophores are arranged to emit light mainly downward through transparent windows along the keel and ventral sides of the body

and tail. Some species are partially transparent. The body shape, silvering and bioluminescent counterillumination make hatchetfishes almost invisible in the mesopelagic layers of the oceans. Although some species, e.g. *Argyropelecus affinis* (Pacific hatchetfish, 3872 m_{fbd}) have been recorded very deep, the mean maximum depth for the subfamily of 985 m more accurately reflects the normal depth distribution of hatchetfishes in the mesopelagic where they exhibit diel vertical migrations, with some occasionally appearing at the surface at night. They feed on small pelagic prey, including ostracods, various crustacea including copepods and decapod larvae, and fish larvae. There are 43 species in three genera. The genus *Argyropelecus* has seven species of which five are circumglobal: the *A. aculeatus* (Lovely hatchetfish, 600 m_{fbd}; Fig. 4.15b), *A. affinis*, *A. gigas* (Hatchetfish, 1000 m_{fbd}; Plate 5a), *A. hemigymnus* (Half-naked hatchetfish, 2400 m_{fbd}) and *A. sladeni* (Sladen's hatchetfish, 2926 m_{fbd}) All show migrations from 300–600 m depth during day, moving upward at night, except *A. gigas*, which remains at depths of 400–600 m. The genus *Sternoptyx* has four species: *S. diaphana* (Fig. 4.15b), *S. pseudobscura* and *S. pseudodiaphana*, which are circumglobal, and *S. obscura*, which is confined to the Indo-Pacific. They generally occur deeper than *Argyropelecus* with fewer marked vertical migrations. Compared with *Argyropelecus*, the body shape of *Sternoptyx* has a characteristic backwards-tilted appearance. The diet of *S. diaphana* is very diverse, mainly amphipods, euphausids, chaetognaths and some fishes, varying between ocean regions, which leads Hopkins and Baird (1973) to suggest that this species is a solitary predator with limited pursuit capability that takes the nearest available prey items over a short detection distance. The genus *Polyipnus* has a generally more rounded, less angular body profile and is benthopelagic, occurring on continental slopes, around oceanic islands and seamounts at depths down to over 1000 m, being caught more often in bottom trawls than in mid-water tows (Harold et al., 2002). The genus *Polyipnus* is more specious than the oceanic genera with 32 species in total, five species in

(b)

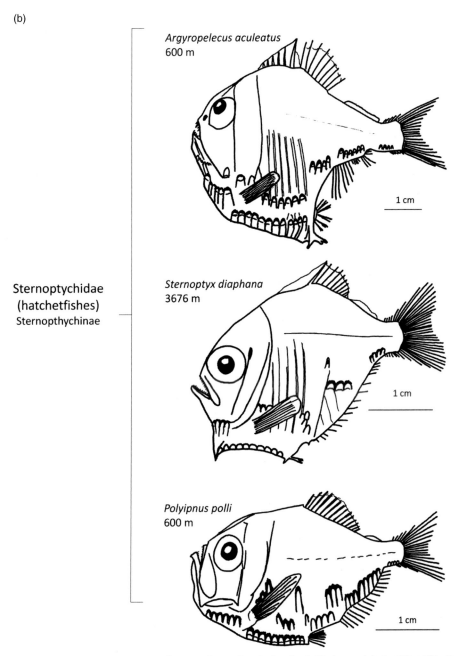

Sternoptychidae
(hatchetfishes)
Sternopthychinae

Argyropelecus aculeatus
600 m

1 cm

Sternoptyx diaphana
3676 m

1 cm

Polyipnus polli
600 m

1 cm

Figure 4.15b ORDER STOMIIFORMES. *Argyropelecus aculeatus* (lovely hatchetfish) circumglobal 40°N–35°S. *Sternoptyx diaphana* (Diaphanous hatchetfish) circumglobal mainly tropical. *Polyipnus polli* (round hatchetfish) E Atlantic.

the Atlantic Ocean (Fig. 4.15b), three endemic to the Indian Ocean, seven Indo-Pacific species and 17 species in different parts of the Pacific Ocean (Harold, 1994). Some species, e.g. *P. spinosus* and *P. clarus*, may be found on the continental shelf, but the latter is most often captured at 300–400 m depth. *Polyipnus* is planktivorous, feeding on copepods, euphausids and mysids where the pelagic fauna impinges on the slopes. Spawning occurs through much of the year with a possible peak in winter.

4.3.6.3 Family Phosichthyidae (63) (Lightfishes)

These deep pelagic fishes are very similar to the Gonostomidae in external appearance but have distinctive light organs with a lumen and a duct. There are 24 species in seven genera. The genus *Vinciguerria* comprises highly abundant small fishes less than 5 cm in length, silvery and transparent in contrast to the black of *Cyclothone*, with which it often appears abundant in mid-water catches. Three species are circumglobal with differing diel vertical migrations in the mesopelagic: *V. attenuata* (Fig. 4.15c; 48°N–57°S) 250–600 m by day ascending to 100–500 m at night, *V. nimbaria* 200–400 m by day and 0–100 m at night and *V. poweriae* 300–600 m during the day and at 50–350 at night. *V. lucetia* is confined to the Pacific and *V. mabahiss* to the Red Sea. *Ichthyococcus ovatus* is circumglobal, 46°N–20°S (Fig. 4.15c) usually found at 200–500 m with no evidence of diel vertical migrations. Six other species in the genus *Ichthyococcus* have regional distributions, for example *I. polli* only in the tropical Atlantic Ocean usually at 550–750 m depth. Other pelagic genera are *Pollichthys* and *Woodsia*. Three genera are considered benthopelagic associated with slope environments of continental margins, islands and seamounts: *Phosichthys argenteus* (Silver lightfish) in its monotypic genus occurs widely at 6°S–49°S, *Yarrella argenteola* in the East Central Pacific, *Y. blackfordi* in the tropical Atlantic, probably schooling at 500–700 m depth and the genus *Polymetme* has one circumglobal species *P. corythaeola* distributed at 66°N–48°S, plus five regional species, for example *P. surugaensis* from the West Pacific and *P. thaeocoryla* associated with Atlantic seamounts.

4.3.6.4 Family Stomiidae (64) (Barbeled dragonfishes)

The Stomiidae is a family of 287 species of specialised deep-sea predatory fishes, mean maximum length 18 cm, mean maximum depth 2303 m that inhabit the mesopelagic and bathypelagic zones of the oceans. The name is derived from the first description of *Stomias boa boa* (boa dragonfish) as possessing 'the head of a reptile on the body of an esocid' (Risso, 1810). Stomiidae usually have an elongated, dark-coloured body, photophores in rows on the ventral and latero-ventral aspects, modest-sized eyes and often large teeth fang-like teeth. All species have a characteristic prominent postorbital photophore on each side of the head that emits blue light (Nicol, 1960). Also characteristic is the mental barbel on the lower jaw, present in almost all species (Plate 5b). The barbel is often equipped with complex photophores and is presumed to be used as a lure to attract prey. It is likely to be most effective in attracting fishes that are relatively sparsely distributed in the mesopelagic. Young stages, in which the barbel is not fully developed, feed on small crustacea, whereas the adults and larger species tend to be exclusively piscivorous. The Stomiidae are upper trophic level predators in the mesopelagic and upper bathypelagic. Clarke (1982) suggested they consume a large proportion of the standing crop of planktivorous fishes each year off Hawaii.

In common with most predators, Stomiidae occur in relatively small numbers compared with their prey. Furthermore, their sensory and locomotor capabilities enable them to avoid capture by nets. Consequently they are caught much more rarely than zooplankton consumers, and knowledge of this group is often confined to a few specimens of each species (Sutton and Hopkins, 1996). Here we follow Nelson (2006) in recognising five subfamilies: Astronesthinae, Stomiinae, Melanostomiinae, Idiacanthinae and Malacosteinae. The skin is scale-less except for the Stomiinae.

Subfamily Astronesthinae (Snaggletooths)

This subfamily has 59 species in six genera. The genus *Astronesthes* is the most diverse with 48 species, the two most widespread being *A. gemmifer* (2400 m$_{fbd}$) and *A. indicus*. *A. gemmifer* is almost circumglobal between 62°N and 32°S. *A. indicus* (Black snaggletooth, 2000 m$_{fbd}$) occurs in all the oceans,

(c)

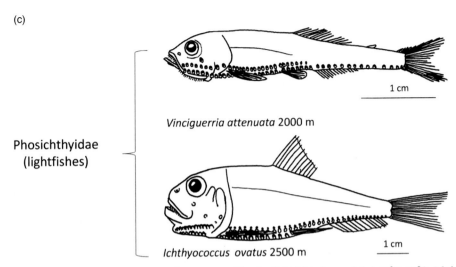

Phosichthyidae
(lightfishes)

Vinciguerria attenuata 2000 m

1 cm

Ichthyococcus ovatus 2500 m

1 cm

Figure 4.15c ORDER STOMIIFORMES. *Vinciguerria attenuata* (Slender lightfish) circumglobal 48°N–57°S. *Ichthyococcus ovatus* (Lightfish) circumglobal 46°N–20°S except N Pacific.

especially in the equatorial zones, usually deeper than 500 m during the day, feeding on mid-water fishes and crustacea. Four species are circumglobal in the Southern Hemisphere: *A. kreffti* (850 m$_{fbd}$) 30°S–50°S, *A. boulengeri* (4000 m$_{fbd}$) 30°–40°S, *A. psychrolutes* (1500 m$_{fbd}$) and probably *A. illuminatus* (2000 m$_{fbd}$) at 32°S–42°S. In the Atlantic Ocean there are two relatively common endemic species: *A. leucopogon* 41°N–30°S and *A. richardsoni* 20°N–26°S, both of which are usually deeper than 500 m during the day, though smaller individuals may migrate to near the surface at night. Twelve other *Astronesthes* species have been recorded as endemic to the Atlantic Ocean. Only two species are potentially endemic to the Indian Ocean, *A. lamellosus* from the Arabian Sea and the Bay of Bengal, whilst *A. exsul* is known only from its type locality in the Western Indian Ocean. 12 species are distributed across the Indo-Pacific Region with *A. cyaneus* widespread 36°N–20°S in the Indo-West Pacific and *A. lucifer* throughout the Indo-pacific 36°N–33°S benthopelagic associated with seamounts, slopes and ridges. The other 10 Indo-Pacific species have patchy or localised distributions. A further 12 species are endemic to the Pacific Ocean, some of which are rare, whereas others are confined to defined

regions, for example, *A. lupine* in Australian waters and *A. galapagensis* found in the Peruvian Current. *A. tanibe* is found in the Southeast Atlantic and Southeast Pacific and *A. niger* in the Atlantic and Indian Oceans 43°N–36°S. Seven species of *Astronesthes* occur in the Eastern Gulf of Mexico. Six of these, *A. cyclophotus, A. indicus, A. macropogon, A. micropogon, A. niger* and *A. richardsoni* are generalised zooplankton/ micronekton consumers with euphausids comprising the dominant component of the diet, whereas *Astronesthes similis* is a piscivore (Sutton and Hopkins, 1996).

There are six species in the genus *Borostomias*, three of which are very widespread: *B. antarcticus* (Snaggletooth, 2630 m$_{fbd}$) from the North Atlantic to the Southern Ocean 66°N–66°S, *B. elucens* in the Atlantic, Indian and parts of the Pacific 36°N–18°S and *B. monomena* almost circumglobal 35°N–50°S. They are usually deeper than 500 m during the day. *B. antarcticus* feeds predominantly on fishes but also takes cnidaria (Gasket et al., 2001) and crustacea, but Stowasser et al. (2009) suggest that these latter items might be derived from the diet of the prey fish. *B. monomena* appears to feed exclusively on fish (Sutton and Hopkins, 1996). In its own monotypic genus, *Eupogonesthes xenicus* has been described from the Eastern Indian Ocean on the continental slopes of Western Australia at 200–600 m depth. The genus *Heterophotus* also has just one species, *H. ophistoma* (Wingfin snaggletooth), which is distributed almost worldwide 55°N–40°S at 790–1420 m depth (Fig. 4.15d). It is unusual amongst stomiidae for its wing-like pectoral fins and deep body shape. Stomach contents of the few available specimens suggest they may specialise in hunting for squid (Clarke, 1982; Sutton and Hopkins, 1996). The genus *Neonesthes* has two species, which are both widespread, *N. capensis* (Cape snaggletooth) circumglobal 40°N–55°S and *N. microcephalus* (Smallhead snaggletooth) 3°N–40°S in the Atlantic and at scattered locations elsewhere. *Rhadinesthes decimus* (slender snaggletooth, 4900 m$_{fbd}$), the single species in its genus, is the longest of the Astronesthinae, up to 41 cm total length with a large head and slender body reported from scattered occurrences worldwide 63°N–28°S.

Subfamily Stomiinae (Dragonfishes and Viperfishes)

This subfamily comprises two distinctive genera considered to be in separate tribes by Nelson (2006): *Stomias* (Scaly dragonfishes) and *Chauliodus* (Viperfishes). The genus *Stomias* has 11 species, four of which are circumglobal, *S. affinis* (Günther's boafish) 35°N–39°S, *S. boa boa* (Boa dragonfish) deep-water, 45°N–50°S including the Mediterranean (Fig. 4.15d, Plate 5b), *S. longibarbatus* (Longbarb scaly dragonfish) 41°N–40°S and *S. gracilis*, confined to the Southern Hemisphere 40°S–62°S in the subtropical convergence and sub-Antarctic waters. Three species are endemic to the Atlantic Ocean, *S. boa ferox* north of 30°N, *S. brevibarbatus* across the North Atlantic from Portugal and Mauretania to Canada and *S. lampropeltis* in the Eastern Atlantic off tropical West Africa. Two species are endemic to the Pacific Ocean, *S. atriventer* in the East Pacific and *S. danae* occurring widely throughout the ocean. *S. nebulosus* occurs in tropical and subtropical waters of the Indo-Pacific, and *S. boa colubrinus* is found in the East Pacific and Atlantic Oceans. In the Eastern Gulf

of Mexico. *S. affinis* was found to feed mainly on fishes (*ca.* 90 per cent by weight) especially myctophids but also some crustacea (Sutton and Hopkins, 1996), and in the Pacific Ocean *S. danae* is also considered to be piscivorous (Clarke, 1982).

The genus *Chauliodus* comprises eight species, *C. sloani,* which is global 70°N–56°S (Fig. 4.15d) and seven regional species, three in the Atlantic Ocean, *C. danae* 42°N–14°N, *C. minimus* 5°S–40°S and *C. schmidti* East Atlantic 20°N–20°S, four in the Pacific Ocean, *C. macouni* 66°N–23°N, *C. barbatus* and *C. vasnetzovi* in the SE Pacific and *C. dentatus* in the South Pacific, one in the NW Indian Ocean *C. pammelas* 25°N–0°N. In the Atlantic Ocean, *C. schmidti* is tolerant of the lowest dissolved oxygen concentrations 17–28 per cent saturation at 50–500 m depth where *S. danae* and *S. sloani* are absent. *S. danae* is most prevalent at 53–75 per cent oxygen saturation. *S. sloani* can tolerate a wide range of oxygen concentrations, from 22 to 100 per cent saturation but is competitively excluded where the other two species are present (Haffner, 1952). *Chauliodus* has no scales but there are hexagonal markings on the skin showing the pattern of scales. Unlike most stomiidae the chin barbel in *Chauliodus* is either short, albeit with photophores, or absent. Its function as a lure is replaced by a greatly elongated mobile first dorsal fin ray on which are mounted photophores. *Chauliodus* also has small light organs scattered over the body, large bell-shaped light organs arranged in four longitudinal rows along the lateral and ventral surfaces of the body and large suborbital light organs near the eye. There are also bell-shaped light organs inside the lower jaw. *Chauliodus* is noted for its very long fang-like teeth, the longest being the front teeth of the lower jaw, which extend in front of the head to above the eyes when the mouth is closed. The teeth are firmly attached to the upper and lower jaws, with no hinge as in the venomous fangs of the viper snakes (Greven et al., 2009). Because the teeth are fixed, the jaw gape must open very wide when feeding, the skull is hinged and can rotate whilst the absence of a floor to the mouth enables the ingestion of large prey (Tchernavin, 1953) up to 63 per cent of its body length (Clarke, 1982). *Chauliodus* are predominantly

(d)

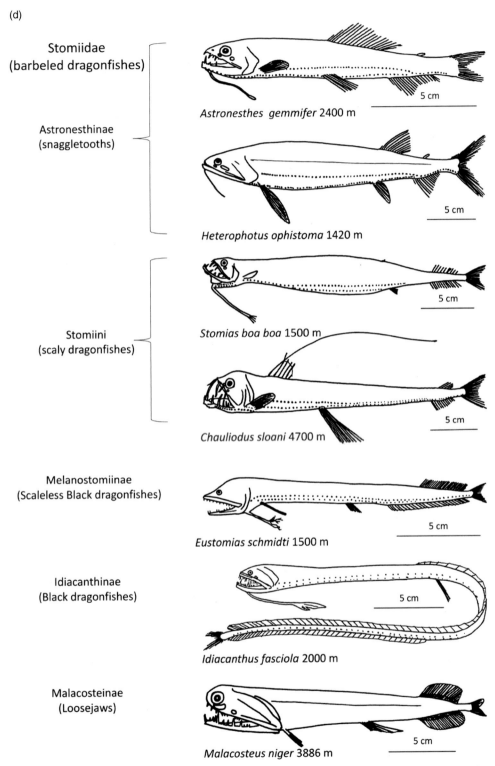

Stomiidae
(barbeled dragonfishes)

Astronesthinae
(snaggletooths)

Astronesthes gemmifer 2400 m

5 cm

Heterophotus ophistoma 1420 m

5 cm

Stomiini
(scaly dragonfishes)

Stomias boa boa 1500 m

5 cm

Chauliodus sloani 4700 m

5 cm

Melanostomiinae
(Scaleless Black dragonfishes)

Eustomias schmidti 1500 m

5 cm

Idiacanthinae
(Black dragonfishes)

5 cm

Idiacanthus fasciola 2000 m

Malacosteinae
(Loosejaws)

5 cm

Malacosteus niger 3886 m

Figure 4.15d ORDER STOMIIFORMES. *Astronesthes gemmifer*. Probably circumglobal. *Heterophotus ophistoma* (Wingfin snaggletooth) Atlantic, Indian and Pacific 55°N–40°S. *Stomias boa* (Boa dragonfish) anti-tropical N Atlantic, W Mediterranean, Sub-Antarctic SE Pacific, SW Atlantic and S Indian Ocean. *Chauliodus sloani* (Sloane's viperfish) circumglobal 70°N–56°S. *Eustomias schmidti* (Schmidt's dragonfish) Atlantic and Pacific 35–40°N and 30–35°S. *Idiacanthus fasciola* (Ribbon sawtail fish) Atlantic, Indian and W Pacific 40°N–54°S. *Malacosteus niger* (stoplight loosejaw, red flashlight fish) circumglobal 66°N–30°S except Mediterranean.

piscivorous with myctophids comprising 72 per cent of prey numbers in the eastern Gulf of Mexico (Sutton and Hopkins, 1996). Despite being recorded down to 4700 m_{fbd} *C. sloani* is usually found at 500–1500 m depth with smaller individuals occasionally found at the surface at night (Haffner, 1952). The eggs of *Chauliodus* are relatively large compared with those of other Stomiidae, 2.7–3.1 mm diameter in *C. macouni*, the Pacific viperfish (Moser, 1996).

Subfamily Melanostomiinae (Scaleless black dragonfishes)

With 191 species in 16 genera the Melanostomiinae account for over 65 per cent of stomiidae species. The genera *Bathophilus*, *Eustomias* and *Grammatostomias* are considered to be a clade within the subfamily (Nelson, 2006) and hence are here considered together.

The genus *Bathophilus* comprises 16 species, four of which are probably circumglobal; *B. digitatus* 28°N–36°S, *B. longipinnis* 31°N–50°S, *B. nigerrimus* 45°N–57 and *B. schizochirus* 28°N–17°S; four species are confined to the Pacific Ocean: *B. abarbatus*, *B. filifer*, *B. flemingi* and *B. kingi*. *B. indicus* is endemic to the Indian Ocean. *B. proximus* from the NW Atlantic is the only species endemic to the Atlantic Ocean; six other species found in the Atlantic have also been recorded from other Oceans: *B. brevis*, *B. altipinnis*, *B. ater*, and *B. vaillanti* in the Pacific Ocean, *B. irregularis* and *B. pawneei* from the Indian and Pacific Oceans. Body form is quite variable from the short deep-bodied *B. brevis* maximum standard length 5.5 cm to the more typical Stomiid elongated shape of *B. digitatus* length 17 cm. In the eastern Gulf of Mexico at least six species of *Bathophilus* coexist, *B. digitatus*, *B. longipinnis*, *B. nigerrimus*, *B. pawneei* and *B. proximus* all of which feed mainly on myctophids (Sutton and Hopkins, 1996).

Fishes of the genus *Eustomias* are black, elongate, with large fangs, serial photophores and chin barbels typical of many Stomiidae. They are distinguished from other Melanostomiids by a protrusible upper jaw. They occur in tropical and subtropical waters around the globe and undertake vertical migrations from depths greater than 300–400 m during the day to less than 200–250 m at night (Gibbs et al., 1983). With 116 described species the genus is extraordinarily species-rich, though most can only be reliably identified from the distinctive pattern of bulbs and branches of the barbel. Because the barbel is often lost, it is not always possible to identify to species and consequently the validity of many described species has been questioned (Sutton and Hartel, 2004). However, Gibbs et al. (1983) argue that distinguishing species by minor differences in the barbel is mostly correct, although the diagnostic characters may not be evident until the fish has grown to about 8 cm length. There is no sexual dimorphism in barbels; species show remarkable consistency in their barbel morphology, and attempts to reduce the number of recognised species are rejected by Gibbs et al. (1983), who propose 9 subgenera, some of which are further split into groups. Three species are probably circumglobal, *E. trewavasae* in the Southern Hemisphere 33°S–40°S, *E. braueri* 36°N–34°S, and *E. satterleei* 40°N–20°S. At least a further 15 species including *E. schmidti* (Fig. 4.15d) from equatorial boundary currents, *E. enbarbatus*, *E. fissibarbis* and *E. melanostigma* are reported from more than one ocean. Approximately 50 species are endemic to the Atlantic Ocean, 40 in the Pacific Ocean and two in the Indian Ocean, probably a reflection of relative sampling effort as much as real differences between the oceans. *E. trewavasae* in the south 40°S and *E. borealis* in the Atlantic north of 30°N define the latitudinal limits of distribution of the genus. In the Eastern Gulf of Mexico at least 11 species co-occur and seven species in samples off Hawaii (Clarke, 1982). Their diet consists almost entirely of fishes, especially myctophids of the genus *Diaphus* (Sutton and Hopkins, 1996).

The genus *Grammatostomias*, characterised by an enigmatic luminous loop on each flank behind the head, comprises three species largely endemic to the Atlantic Ocean: *G. circularis* 36°N–24°S, *G. dentatus* 38°N–30°S and *G. flagelibarba* from the NE Atlantic. In its monotypic genus, *Chirostomias pliopterus* with an adipose fin and long pectoral fins with branching tips is also endemic to the Atlantic

45°N–20°N. *Echiostoma barbatum* (Threadfin dragonfish) is circumglobal 48°N–40°S, and although recorded to 4200 m$_{fbd}$ depth has been captured off the Mississippi Delta at depths of 70–500 m. There is sexual dimorphism in the postocular light organs, and females are larger than males, typically 29 cm and 25 cm, respectively (Kruegger and Gibbs, 1966). Also in a monotypic genus *Flagellostomias boureei* (Longbarb dragonfish) is known from all oceans 45°N–49°S.

Leptostomias is a genus of slender-bodied dragonfishes growing up to 38 cm total length. 12 species are currently recognised, but there have long been doubts about the validity of more than half of these. Three species are widespread: *L. bilobatus* across the tropical Atlantic, *L. gladiator* 51°N–41°S in the Atlantic and elsewhere, and *L. gracilis,* probably circumglobal 30°N–34°S if putative synonymous species are included. Four species co-occur in the Gulf of Mexico, all feeding on fishes, mainly myctophids (Sutton and Hopkins, 1996). The genus *Melanostomias* comprises 16 species, the most widespread and possibly circumglobal being *M. bartonbeani* 56°N–40°S, *M. macrophotus* 42°N–29°S Atlantic and Indo-West Pacific, *M. melanops* 36°N–49°S except NE Pacific, *M. tentaculatus* 45°N–49°S, and *M. valdiviae* 41°N–49°S. *M. niger* occurs around the Southern Hemisphere at 20°S–50°S. Three species are endemic to the Atlantic Ocean, five to the Pacific Ocean and there are two Indo-Pacific species. At least four species co-occur in the Gulf of Mexico feeding on myctophids (Sutton and Hopkins, 1996). *Odontostomias* has two little-known species described from the tropical Atlantic Ocean close to the continental shelf of West Africa. There are two species in the genus *Opostomias*. *O. micripnus* (Obese dragonfish) is the largest of the Stomiidae, reported up to 55 cm total length with a fat rounded body and occurring circumglobally in the Southern Hemisphere including multiple records off S. Australia. *O. mitsuii* (Pitgum lanternfish) is the complementary species in the Northern Hemisphere found in the Pacific Ocean 55°N–32°N and grows to 36 cm length.

In its monotypic genus *Pachystomias microdon* is almost circumglobal 50°N–37°S and is best known as one of the few species of fish capable of emitting red light. The suborbital and postorbital photophores around the eye are red light emitters (Herring and Cope, 2005), and the eyes have corresponding red light sensitivity (Kenaley et al., 2013). Bone and Moore (2007) suggest that *Pachystomias* may use these adaptations to illuminate and detect prey, which consists entirely of fishes (Sutton and Hopkins, 1996).

The genus *Photonectes* is distinguished from other melanostomiinae by a lower jaw longer than the upper jaw, a large postorbital light organ with no suborbital light organ and other characters (Klepadlo, 2011). There are 17 species of which *P. margarita*, *P. braueri* and *P. parvimanus* are reported from all the major oceans at tropical and subtropical latitudes. Three species are endemic to the Atlantic Ocean, and five are reported only from the Pacific Ocean. Two species are Indo-Pacific, two species Atlanto-Pacific and three are from scattered collections in several oceans. In the study by Sutton and Hopkins (1996) in the Gulf of Mexico, *P. margarita* and *P. leucospilus* were found to ingest fishes and decapod crustacea. The three remaining genera in the subfamily Melanostomiinae are all monotypic: *Tactostoma macropus* from the North Pacific, *Thysanactis dentex* 18°N–30°S in the Pacific and Atlantic Oceans and *Trigonolampa miriceps,* which is circumglobal in the southern subtropical convergence but extends also into the North Atlantic. In the Pacific Ocean off Hawaii *Thysanactis dentex* was found to take fishes, including *Bregmaceros* sp. up to 48 per cent of its own body length in size during the day and more crustacea with some fishes during the night (Clarke, 1982).

Subfamily Idiacanthinae (Black dragonfishes)

The single genus in this subfamily comprises three species worldwide of long slender eel-like dragonfishes. *Idiacanthus fasciola* (Fig. 4.15d) occurs throughout all oceans 40°N–54°S except in the Eastern Pacific Ocean, where it is replaced by *I. antrostomus*. *I. atlanticus* is circumglobal in the Southern Hemisphere 25°S–60°S. There is marked sexual dimorphism, the females of *I. fasciola* can grow up to 48 cm total length compared with just 7 cm for males.

Adult males lack teeth, paired fins and barbel. They have relatively large eyes and a large postorbital photophore. The gut is poorly developed, and the gonads extend the entire length of the body cavity. Males are bathypelagic, unable to feed and presumed to be short-lived, whereas the females are active predators making extensive vertical migrations in pursuit of prey and may have longer life spans. Spawning seems to occur in deep water, and the eyes of larvae are mounted on long stalks with length up to 25 per cent of the body length (Moser, 1996). Both in the Pacific Ocean off Hawaii and in the Gulf of Mexico *I. fasciola* were found to feed on a variety of Myctophid fishes (Clarke, 1981, Sutton and Hopkins, 1996).

Subfamily Malacosteinae (Loosejaws)

This subfamily comprises 14 species in three genera, *Aristostomias* (six species), *Malacosteus* (two species) and *Photostomias* (six species). They are typically less than 25 cm long and have jaws longer than the skull, 15 – 30 per cent standard body length enabling a very large gape. The absence of a floor membrane to the mouth appears to further facilitate ingestion of large prey. Paradoxically, despite the large gape, the main prey of *Malacosteus niger* (Fig. 4.15d) is copepods, and Sutton (2005) suggests that this species does not vertically migrate but subsists on small prey in between opportunistic large meals. *M. niger,* also known as the stoplight fish, has red-emitting photophores, which together with red visual sensitivity may enable efficient visual detection of copepods. Furthermore a diet of copepods provides a source of red-sensitive pigment for the retina. *Aristostomias* also has red-emitting photophores (Herring and Cope, 2005) and red visual sensitivity (Kenaley et al., 2013) but retains a piscivorous diet typical of many Stomiidae. *Photostomias*, which has conventional blue-emitting photophores and blue-sensitive eyes, feeds predominantly on crustacea, in particular penaeidean shrimps (Sutton and Hopkins, 1996). Interpretation of feeding adaptations in the Malacosteinae is more complex than is at first apparent.

In the genus *Aristostomias* there are four species, which are almost circumglobal: *A. grimaldii* 35°N–35°S, *A. lunifer* 50°N–33°S, *A. polydactylus* 52°N–57°S and *A. xenostoma* 30°N–34°S but are replaced in the NE Pacific Ocean by *A. scintillans,* which is the only species off California (Moser, 1996) and extends north to the Bering Sea. *A. tittmanni* is found in the Atlantic and SE Pacific. *Malacosteus niger* is circumglobal 66°N–30°S and is replaced by *M. australis* in the Southern Hemisphere 25°–45°S and in the Indian Ocean (Kenaley, 2007). In the genus *Photostomias*, *P. atrox* is circumtropical, *P. guernei* is circumglobal 40°N–56°S, *P. goodyeari* endemic to the Atlantic Ocean, *P. tantillux* endemic to the Pacific Ocean, *P. liemi* is Indo-Pacific and *P. lucingens* is recorded from the SE Atlantic and Pacific Oceans (Kenaley, 2009).

4.3.7 Order Ateleopodiformes

This small group comprising one family occupies an interesting position in the evolutionary tree of fishes (Fig. 4.16). Morphological characters differentiate them from the Stomiatiformes. They clearly belong to the neoteleosts but do not share key characters with lower neoteleost orders such as the Aulopiformes and Lampridiformes, leading Olney et al. (1993) to an unresolved trichotomy between the Ateleopodiformes, Stomiatiformes and the eurypterygian neoteleosts. Since then, molecular studies (Betancur-R et al., 2013; Near et al., 2012) place the Ateleopodiformes on their own as a separate branch of the neoteleosts.

4.3.7.1 Family Ateleopodidae (65) (Jellynose fishes)

This family is almost circumglobal at tropical and subtropical latitudes comprising 13 demersal species that live on the upper slopes down to an average maximum depth of 750 m. The body is rather gelatinous with a poorly ossified, mainly cartilaginous skeleton, and mean maximum total length is over 1 m. The large bulbous head with a tapering tail and reduced caudal fin give a shape superficially similar to a chimaera or macrourid, also suggesting the alternate common name, tadpole

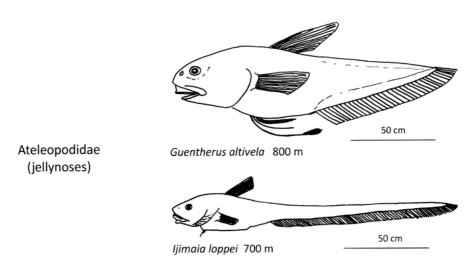

Ateleopodidae
(jellynoses)

Guentherus altivela 800 m

50 cm

Ijimaia loppei 700 m

50 cm

Figure 4.16 ORDER ATELOPODIFORMES. *Guentherus altivela* (Jellynose) Atlantic and E Central Pacific. *Ijimaia loppei* (Loppe's tadpole fish) E Atlantic off Africa.

fishes. There are four genera. The genus *Ateleopus* is confined to the Indo-Pacific region with three species from the West Pacific, including Japanese waters, and two species from the Indian Ocean. *A. natalensis* is found from the Red Sea to the Cape of Southern Africa. The genus *Ijimaia* occurs in the Atlantic Ocean, *I. antillarum* in the West around Cuba, *I. loppei* in the East from Morocco to South Africa (Fig. 4.16), and there are three species in the Pacific Ocean. In its own genus, *Parateleopus microstomus* is reported from the Western Pacific and around Hawaii. The genus *Guentherus* has two species, *G. katoi* from the NW Pacific off Japan and *G. altivela* from the Eastern Atlantic and Eastern Central Pacific (Fig. 4.16). Bussing and López (1977) suggest that *G. altivela* is a transatlantic species that expanded its range into the Eastern Pacific before the last emergence of the Central American land bridge and that the species might eventually be found in the Western Atlantic. The *Guentherus* species are rather deep-bodied compared with the slender shape of other Ateleopodidae. *G. atlivela* grows up to 2 m in length, off the coasts of Costa Rica and Panamá, and is taken as a bycatch in deep-water shrimp fisheries. It has been observed by manned submersible off Cocos Island in the Pacific Ocean (Starr and Green, 2010). Ateleopodidae tend to be rare, and their biology is

poorly understood. They are oviparous, producing eggs about 3 mm diameter (Amaoka, 2005). Their diet is reported as consisting of decapods (*A. japonicus*) ophiuroids and fishes (*I. loppei*).

4.3.8 Order Aulopiformes (Lizardfishes)

The Aulopiformes is the first order of advanced Eurypterygian teleosts, but they retain primitive features such as the adipose fin and posterior location of the pelvic fins (Fig. 4.17). A diagnostic feature is a unique specialisation of the gill arches including greatly elongated pharyngobranchial cartilages. There are 258 species, which range from inshore to a remarkable variety of specialisations to the deep sea both in pelagic and benthic habitats. Tubular eyes have evolved three times in deep-pelagic families (Scopelarchidae, Evermannelidae and Giganturidae), and deep-sea tripod fishes (Ipnopidae) are uniquely adapted for sit-and-wait foraging in the benthic boundary layer. Aulopiformes originated in the early Cretaceous and shortly thereafter evolved synchronous hermaphrodism, which remains a characteristic feature of the group (Davis and Fielitz, 2010). The swim bladder is absent. Here we follow Nelson (2006) in recognising four suborders.

Suborder: Synodontoidei

The Synodontoidei live mainly in continental shelf and inshore waters including brackish estuaries. Exceptions are several species of *Paraulopus* (cucumber fishes, Paraulopidae), which are bathydemersal on upper slopes at 200–400 m depth in the NW and SW Pacific and *Harpadon erythraeus* (Synodontidae, Lizardfishes) of the Bombay Duck genus, which is apparently endemic to the deep warm basins of the Red Sea at 779–1150 m depth.

Suborder: Chlorophthalmoidei

The Chlorophthalmoidei comprises 65 species in four main families, the Notosudidae that are mainly pelagic down to 1160 m mean maximum depth and three benthic families living at progressively greater depths, Chlorophthalmidae (518 m), Bathysauropsidae (1860 m) and Ipnopidae (3228 m).

4.3.8.1 Family Chlorophthalmidae (66) (Greeneyes)

The family Chlorophthalmidae are typically 10 cm long (mean maximum length 25 cm) shoaling demersal fishes, the adult populations of which are concentrated at upper slope depths of 200–500 m They are distributed circumglobally in tropical to temperate regions. The genus *Chlorophthalmus* has three species in the Atlantic Ocean, two in the Western Indian Ocean, two in the Indo-West Pacific and eight species endemic to different parts of the Pacific Ocean. *Chlorophthalmus agassizi* (shortnose greeneye; Fig. 4.17a) is the largest species in the family and is found worldwide between 45°N and 19°S at depths down to 1000 m, in water temperatures of 5°C–13°C, most commonly at 350–450 m depth, and is one of the most abundant upper bathyal fishes in the Eastern Mediterranean Sea (D'Onghia et al., 2006). Size increases with depth, and diet is extremely varied indicating a mixed feeding strategy including decapod crustacea, copepods, cephalopods and fishes (Anastasopoulou and Kapiris, 2008). Sexual maturity as a hermaphrodite with maturing eggs and sperm in different parts of the ovo-testis occurs at about three years old, total length 12 cm. *C. agassizi* spawn offshore with several batches of eggs between spring and autumn each year with a peak in August. Larvae are pelagic, and juveniles settle at shelf depths before moving into deeper water (D'Onghia et al., 2006). In the Mediterranean Sea *C. agassizi* are regularly caught in deep-water commercial trawls but are generally discarded (Stergiou et al., 2003) though they may be retained for utilisation in other regions. The genus *Parasudis* comprises two species confined to the Atlantic Ocean, *P. fraserbrunneri* in the east and *P. truculenta* in the west. *Chlorophthalmus* species have a small perianal light organ, which produces continuous blue-green luminescence (510 nm wavelength) generated by symbiotic bacteria. The eyes have twin cones in part of the retina providing high-resolution detection of this light from the upper anterior direction with a yellow lens to aid wavelength discrimination (Somiya, 1977). It is hypothesised that this enables schooling orientation and intraspecific sexual attraction avoiding self-fertilization (Anastasopoulou et al., 2006).

4.3.8.2 Family Bathysauropsidae (67) (Lizardfishes)
4.3.8.3 Family Bathysauroididae (68) (Lizardfishes)

The family Bathysauropsidae comprises two species, *Bathysauropsis gracilis* (Black lizardfish, 2835 m$_{fbd}$), which is circumglobal in the southern oceans (Fig. 4.17a) and *Bathysauropsis malayanus* (886 m$_{fbd}$) from the West Pacific. They are rare demersal predatory fish with elongated cylindrical bodies up to about 30 cm in length with the eponymous lizard-like head with gaping mouth and many teeth. *Bathysauroides gigas* (Pale deepsea lizardfish) also from the West Pacific was formerly included in the Bathysauropsidae but is now placed in its own monotypic family (Baldwin and Johnson, 1996). Unusually for deep-sea fish *Bathysauroides* has twin cones in the retina similar to those of the *Chlorophthalmus*, but unlike *Chlorophthalmus* rods are entirely absent. There is also a peculiar process in the iris that is not seen in *Bathysauropsis*. *Bathysauroides* lives at 200–400 m, much shallower than *Bathysauropsis* and is presumably adapted to a quite different photic environment (Somiya et al., 1996).

(a)

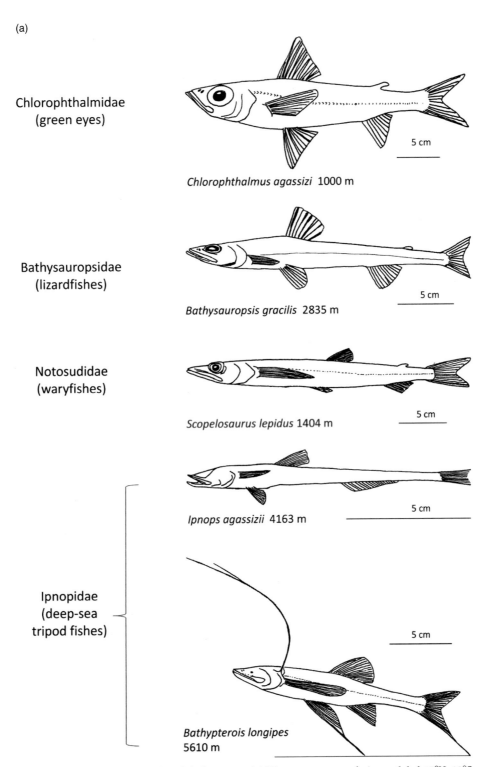

Chlorophthalmidae
(green eyes)

5 cm

Chlorophthalmus agassizi 1000 m

Bathysauropsidae
(lizardfishes)

5 cm

Bathysauropsis gracilis 2835 m

Notosudidae
(waryfishes)

5 cm

Scopelosaurus lepidus 1404 m

5 cm

Ipnops agassizii 4163 m

Ipnopidae
(deep-sea
tripod fishes)

5 cm

Bathypterois longipes
5610 m

Figure 4.17a ORDER AULOPIFORMES. *Chlorophthalmus agassizi* (Shortnose greeneye) circumglobal 45°N–19°S. *Bathysauropsis gracilis* (black lizardfish) Southern Hemisphere circumglobal. *Scopelosaurus lepidus* (blackfin waryfish) N Atlantic 71°N–15°S, *Ipnops agassizii* (grideye fish) patchy circumglobal 6°N–22°S. *Bathypterois longipes* (Abyssal spiderfish) Atlantic and Pacific.

4.3.8.4 Family Notosudidae (69) (Waryfishes)

The Notosudidae comprises 19 species of mesopelagic and bathypelagic fishes often associated with island, seamount and continental slopes. The body form is slender with a long head, large mouth and large eyes, maximum total length 50 cm. There is no swim bladder and no light organs (Watson and Sandkop, 1996). They are circumglobal from the sub-Arctic (Bering Sea) to the sub-Antarctic and feed on small prey such as copepods, euphausids and small fishes. In the genus *Ahliesaurus*, *A. berryi* (2000 m_{fbd}) is circumglobal between 40°N–40°S and *A. brevis* (3000 m_{fbd}) is confined to the NE Pacific Ocean. In its monotypic genus *Luciosudis normani* (Paperbones, 800 m_{fbd}) is mesopelagic and distributed in temperate and subtropical waters of the Southern Hemisphere. The genus *Scopelosaurus* has 16 species including two North Pacific species, two South Pacific Species, two North Atlantic species (*S. lepidus*, Blackfin waryfish, 1404 m_{fbd}, recorded up to 71°N; Fig. 4.17a) and one SE Atlantic species, *S. smithii* (815 m_{fbd}), which is circumtropical 40°N–30°S. There is a series of more or less circumglobal Southern Hemisphere species: *S. ahlstromi* 12°–45°S, *S. meadi* (Blackring waryfish, 760 m_{fbd}) 20°–40°S and *S. hamiltoni* (Smallscale waryfish 1000 m_{fbd}) in the sub-Antarctic 30°–61S°. Like all Aulopiformes, Notosudidae are synchronous hermaphrodites producing planktonic eggs and larvae. Adults may migrate towards the surface to spawn. The eyes of Notosudids are adapted for high-resolution forward-looking binocular vision with a fovea in the retina equipped with cones. The rest of the retina has rods grouped with a reflective tapetum, which would impart high sensitivity. It is suggested that these adaptations allow *S. lepidus* to visually fixate small prey animals during hunting (Munk, 1977).

4.3.8.5 Family Ipnopidae (70) (Deep-Sea tripod fishes)

The Ipnopidae includes 29 species of slender-bodied demersal fishes in five genera that occur at depths from 476 to 6000 m worldwide in tropical and temperate seas. The head and ventral surface of the body are flattened, and the gill slits are large. The eyes are very much reduced; there are no luminous organs or swim bladder, leading Nielsen (1966) to comment that ipnopids are not provided with many means of communication. However, the lateral line system is fairly well developed in some species, and patterns of catches indicate they may live in aggregations on the seafloor. Franco et al. (2009) found 9 species of Ipnopidae at bathyal depths (580–2271 m) on the continental slopes off Brazil in the SW Atlantic Ocean.

The genus *Bathymicrops* comprises four species of quite rare small abyssal fishes (minimum reported depth 3300 m, standard length 9–12 cm). The entire body and head are covered with scales including over the eyes, which have degenerated to less than 0.8 mm diameter with no lens, residual retinal pigment and doubtful function (Munk, 1965). In addition, the lateral line system is poorly developed (Nielsen, 1966). The mouth is large, and elongated gill rakers give the potential to capture both large and small prey items. Food items found in stomachs are epibenthic crustacea including amphipods. *B. regis* is found in the Atlantic Ocean 36°N–37°S, *B. brevianalis* in the SW Pacific and Indian Oceans and *B. belyaninae* and *B. multispinis* in the West Pacific.

The genus *Bathypterois* comprises 19 species of the iconic tripod fishes, which are familiar from deep-sea images captured by submersibles since the 1960s. The tripod is made up of elongated rays of the two pelvic fins and the ventral lobe of the caudal fin. The fish stands on the seafloor facing into the current with umbrella-like pectoral fins expanded above the head to intercept prey in the benthic boundary layer flow. Chave and Mundy (1994) observed that these tripod fin rays could alternate between being flexible during swimming and rigid when standing, presumably controlled by muscle tension on the tendons that connect the bead-like stack of bones comprising the fin ray. Bathypteroids feed on benthopelagic crustacea, small fishes and cephalopods from the benthic boundary layer. They vary in size from the 37 cm total length for *B. grallator* (Tripodfish, 4720 m_{fbd}) found at abyssal depths to 13 cm for *B. bigelowi* (986 m_{fbd}), relatively shallow-living species from slope depths. They can stand up to about one body length or more above the sediment. Two species are almost circumglobal: *B. atricolor* (Attenuated spider fish, 5150 m$_{fbd}$) 32°N–22°S except in the Western Atlantic, *B. grallator* 45°N–27°S except in the Eastern Indian Ocean and *B. longipes* (Abyssal spiderfish, 5610 m$_{fbd}$) is found in the Atlantic Ocean 40°N–35°S and also in the Pacific Ocean (Fig. 4.17a). Five species are endemic to the Pacific Ocean, six to the Atlantic Ocean, two to the Indian Ocean and two to the Indo-West Pacific. *B. preceptor* (1600 m$_{fbd}$) is found in the Western Indian Ocean and Southwest Atlantic Ocean. Although the eyes of *B. grallator* and *B. longipes* are small and degenerate, they may retain some residual functionality (Munk, 1965) and in *B. dubius* there are retinal specialisations giving visual sensitivity both in the forward and lateral visual fields (Collin and Partridge, 1996).

The genus *Bathytyphlops* has two species both recorded from the Atlantic and Indian Oceans: a slope-dwelling species *B. marionae* (1920 m$_{fbd}$) and a deeper species *B. sewelli* (4200 m$_{fbd}$). They are comparable to *Bathymicrops* but occur at shallower depths (Franco et al., 2009), are larger, over 30 cm maximum length, the mouth is bigger, teeth are stronger and gill rakers are reduced, suggesting adaptation to take larger prey. The lateral line is well developed (Nielsen, 1966)

Discoverichthys praecox is the single species of a genus from the abyssal NE Atlantic (5440 m depth); it is a small fish 12.5 cm long with a poorly ossified skeleton, vestigial scales, flaccid musculature and minute eyes hidden beneath the expanded frontal bones of the skull (Merrett and Nielsen, 1987). The gill rakers are well developed, suggesting an ability to feed on small prey. A megapterygium larva of *D. praecox* with remarkably large fins has been described from the tropical West Pacific (Okiyama et al., 2007).

The three species of *Ipnops* have strangely modified eyes, which are flattened and located on top of the head beneath transparent cranial bones. They appear as bright iridescent patches contrasting with the general dark colour of the fish. For some time it was thought that these cephalic organs may be luminescent organs but Munk (1959) showed there is no luminescent function, and they are in fact degenerate eyes. The retina has typical rods but there

is no lens, so it is concluded that the cephalic organ can detect down-welling light but has no image-forming capability. They are small fishes, less than 15 cm long, *I. agassizii* (4163 m$_{fbd}$; Fig. 4.17a), *I. murrayi* (3477 m$_{fbd}$) and *I. meadi* (4940 m$_{fbd}$) occurring at widely scattered locations at tropical subtropical latitudes. The diet includes benthic crustacea and polychaetes (Nielsen, 1966).

Suborder: Giganturoidei

This suborder erected by Nelson (2006) comprises just two families, each with two species: the Bathysauridae, which are bathydemersal, and the Giganturidae, which are bathypelagic. Molecular genetic studies place these two families in a clade together with the Ipnopidae (suborder Chlorophthalmoidei; Davis and Fielitz, 2010). Here the suborder Giganturoidei is retained but moved adjacent to the Chlorophthalmoidei in recognition of this close relationship.

4.3.8.6 Family Bathysauridae (71) (Deep-sea lizardfishes)

These fishes are often caught in deep demersal trawls, albeit in small numbers (Bailey et al., 2006) and are readily recognised by the flattened head with large gape, pointed jaws, upward-looking eyes and numerous teeth giving a lizard-like appearance, with a body length of up to 60 or 80 cm. They are generally solitary and in submersible images are seen resting on the sediment waiting in ambush for prey (Linley et al., 2013). Both species are circumglobal in their distribution, *Bathysaurus ferox* (Fig. 4.17b) between 70°N and 46°S, typically at bathyal depths on continental slopes, mid-ocean ridges, island and seamount slopes at 600–3500 m depth; *B. mollis* (4903 m$_{fbd}$) is the deeper species generally >2000 m depth between 50°N and 20°S at abyssal depths. The diet comprises mainly fishes and decapod crustacea, live prey, which are probably detected by a combination of visual and lateral line sensory capabilities (Wagner, 2001a). *Bathysaurus* are rarely seen at baited cameras in which olfactory cues are important. In the NW

(b)

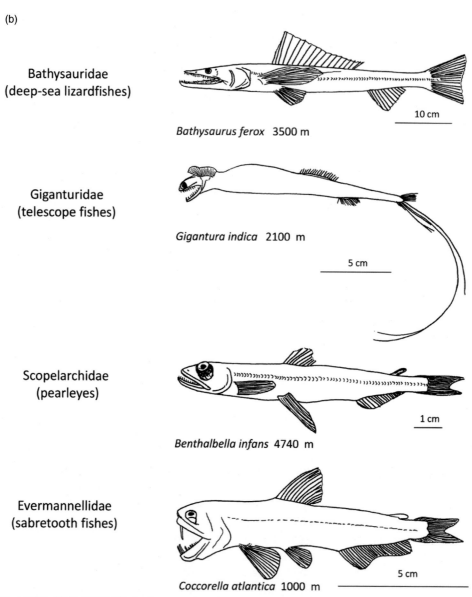

Bathysauridae
(deep-sea lizardfishes)

Bathysaurus ferox 3500 m

10 cm

Giganturidae
(telescope fishes)

Gigantura indica 2100 m

5 cm

Scopelarchidae
(pearleyes)

Benthalbella infans 4740 m

1 cm

Evermannellidae
(sabretooth fishes)

Coccorella atlantica 1000 m

5 cm

Figure 4.17b ORDER AULOPIFORMES. *Bathysaurus ferox* (Deep-sea lizardfish) circumglobal 70°N–50°S. *Gigantura indica* (Telescopefish) circumglobal tropical to subtropical. *Benthalbella infans* (Zugmayer's pearleye) circumglobal except Mediterranean. *Coccorella atlantica* (Atlantic sabretooth) circumglobal.

Atlantic *B. ferox* have a very unusual parasite fauna dominated by nematodes and the absence of any helminth parasites. They are known to bite tails off macrourids and by not eating the body and viscera of the prey, avoid exposure to infection from larval helminths (Campbell et al., 1980).

4.3.8.7 Family Giganturidae (72) (Telescope fishes)

The two species of *Gigantura* (Johnson and Bertelsen, 1991) have a highly distinctive appearance with forward-looking telescope eyes, large jaw gape, pectoral fins high on the body dorsal to the operculi, no pelvic fins and the eponymous giant lower lobe of the caudal fin. They are small to medium-sized bathypelagic fishes; maximum length for *G. chuni* is 15 cm and 20 cm for *G. indica* (Fig. 4.17b), which is more slender-bodied than *G. chuni*. Both species are circumglobal in their distribution, never caught in large numbers, and *G. indica* was recorded above and below 1000 m during intensive sampling to 5000 m depth in the Sargasso Sea (Sutton et al., 2010). *Gigantura* metamorphose to the adult form at quite a large size (25–34 mm) whereupon they lose their pelvic and adipose fins, the jaw gape extends, the tubular eyes develop and the head decreases in size relative to the rest of the body. The maximum density of larvae of *G. chuni* in the water column is at around 100 m (Jackson, 2002), and the adults can be regarded as neotenous lizardfish in that they fail to descend to the seafloor but retain the larval pelagic habitat, albeit deeper and with some larval anatomical features. The adults have forward-looking high-resolution binocular vision and Robison (2008) reports that they hang vertically, looking upward, sculling the caudal fin slowly with the pectorals fanned out to stabilize the head. The stomach is distensible enabling ingestion of large prey.

Suborder: Alepisauroidei

The Alepisauroidei are predominantly small to medium-sized (mean length 26 cm) predatory mesopelagic and bathypelagic fishes with characteristic fang-like barbed teeth (Kriwet, 2003), found from surface and inshore waters down to abyssal depths. Like most Aulopiformes they are synchronous hermaphrodites and have no swim bladder. There are four extant families.

4.3.8.8 Family Scopelarchidae (73) (Pearleyes)

The Scopelarchidae have distinctive upward-looking cylindrical eyes with a conspicuous 'pearl organ' from which they derive their common name. The pearl organ is a modified part of the cornea on the lateral aspect of the eye that is thought to transmit light originating from the side and below the fish onto a retina (Collin et al., 1998). This compensates for the otherwise very narrow field of view and gives this group of fishes their distinctive appearance. The upward-looking sector of the eye can resolve small mobile point sources of light enabling the hunting of bioluminescent prey from below, and the pearl organ confers an ability to detect movements from below. Of the 18 species of Scopelarchidae only two are known to be bioluminescent, *Benthalbella infans* and *Scopelarchoides kreffti* in which ventral photophores are uniquely derived from modified muscle tissue (Johnston and Herring, 1985). There are 18 species in four genera (Johnson, 1974). The greatest diversity is in central-tropical waters, the region within which pearleyes are thought to have originated (Davis, 2015). The genus *Benthabella* with four species is considered basal and is most widespread with two species at high latitudes in the North Pacific: *B. dentata* and *B. linguidens*, one species circumglobally in the Southern Hemisphere, *B. elongata* from 40°S to Antarctica and *B. infans* (Fig. 4.17b), which is found globally throughout tropical and subtropical seas but not in the Mediterranean Sea or the East Pacific Ocean. *B. infans* (Zugmayer's pearleye, 4750 m_{fbd}) grows to 14 cm maximum length. *Lagiacrusichthys macropinnis* constitutes a secondary invasion into the Southern Ocean alongside *B. elongata* (Davis, 2015). The genus *Rosenblattichthys* (four species) is confined to low latitudes in the Indo-Pacific region with one species, *R. hubbsi* extending into the Atlantic Ocean. The genus *Scopelarchoides* has four more or less circumglobal tropical and subtropical species: *S. danae*, *S. analis*, *S. guentheri* and *S. michaelsarsi*; *S. siginifer* is Indo-Pacific; *S. kreffti* is confined to the high latitudes in the SW Atlantic and SW Pacific, and there are three species endemic to the Pacific Ocean. The young stages are generally found between the surface and 200 m and adults between 300–800 m depth.

4.3.8.9 Family Evermannellidae (74) (Sabretooth fishes)

The Evermannellidae are mesopelagic predators with large fang-like teeth in the robust lower jaw and palate, the eyes are upward-looking tubular or semitubular, the body is elongated and laterally flattened, maximum length is 12–19 cm. There are eight species in three genera that live mainly in tropical and subtropical waters, usually at 200–1000 m depth. *Coccorella atlantica* (Atlantic sabretooth) is found in central waters of the Atlantic (Fig. 4.17b), Indian and Pacific Oceans, whereas *C. atrata* is confined to the Indo-Pacific region. The genus *Evermannella* has five species: *E. ahlstromi* and *E. megalops* in the Pacific Ocean, *E. indica* in the Indo-Pacific, *E. melanoderma* and *E. balbo* in the Atlantic with the latter extending into the Southern Ocean to 40°S. In its monotypic genus *Odontostomops normalops* occurs in the Atlantic, Indian and Pacific Oceans but not in the Mediterranean Sea nor the E Pacific Ocean. *C. atrata* has ventrally located light organs that emit blue light downward from under the lower jaw, the abdomen and anus. The light originates from modified areas of the intestine and pyloric caecae and is not of bacterial origin (Herring, 1971). Luminescence has been observed only in *C. atrata* but Swinney (1994) suggested that bioluminescence may occur in the genus *Evermanella,* which also has evidence of modified pyloric caecae. The stomach is highly distensible enabling ingestion of large prey items (Bray and Gomon, 2011).

4.3.8.10 Family Alepisauridae (75) (Lancetfishes)

The three species of Alepisauridae consist of medium to large slender-bodied pelagic predatory fishes distributed circumglobally. *Omosudis lowii* (4000 m_{fbd}), the smallest, maximum length 23 cm, occurs at 40°N–40°S typically between 100–1000 m depth and feeds on squid and other pelagic fishes. It is sometimes placed in its own family Omosudidae (Hammerjaws). *Alepisaurus brevirostris* is found at depths of 640–1591 m, over a wide latitudinal range 68°N–72°S, grows to 96 cm maximum length and the diet includes euphausid shrimps as well as fishes. *A. ferox* is the largest (Fig. 4.17c), maximum length over 2 m and occurs between 84°N and 57°S from the surface down to 1830 m, feeding on fishes, cephalopods, tunicates, and crustaceans. Analysis of stomach contents of *A. ferox* and *A. brevirostris* captured by longline in the North Atlantic found a complete absence of meso- and bathypelagic prey; the diet

mostly comprises epipelagic species including smaller *Alepisaurus* (Haedrich, 1964). *A. ferox* are often a bycatch in tuna and sports fisheries and probably should not be considered a deep-sea species. Both species of *Alepisaurus* have characteristic long sail-like dorsal fins. The eyes are large, round and laterally directed, a minimal adaptation to deep-sea conditions compared with the highly adapted eyes of other Aulopiformes (Davis and Fielitz, 2010).

4.3.8.11 Family Paralepididae (76) (Barracudinas)

The Paralepididae are elongated, large-eyed species with long jaws and a superficial similarity to the barracudas (Sphyraenidae) but differ in having no spiny fin rays, no swim bladder, generally smaller body size (mean maximum length 26 cm) and a deep pelagic habit (mean maximum depth 1322 m). There are 62 species in 13 genera making this the largest family in the Alepisauroidei. *Arctozenus risso* (2200 m_{fbd}) in its monotypic genus is a widespread circumglobal species ranging 71°N–55°, generally found at 200–1000 m and associated with continental, island and seamount slopes (Fig. 4.17c). *Dolichosudis fuliginosa* (1200 m_{fbd}) is a relatively rare species confined to the Atlantic Ocean. The genus *Lestidiops* comprises 16 species. Two species occur in the tropical and subtropical regions of all three major oceans: *L. jayakar* (2000 m_{fbd}) and *L. mirabilis* (825 m_{fbd}); six species are endemic to the Atlantic Ocean including *L. sphyrenoides* recorded by Risso (1820) in the Mediterranean Sea; seven species in the Pacific Ocean and one, *L. indopacifica,* in the Indo-Pacific region. *Lestidium* has four species: *L.atlanticum* (1000 m_{fbd}) which is circumglobal 46°N–37°S, two species, *L. nudum* and *L. prolixum* in the Pacific Ocean and *L. bigelowi* in the Indo-Pacific. *L. bigelowi* has three mid-ventral light organs, one on the isthmus between the operculi and the others between the pectoral and pelvic fins (Graae, 1967). *L. atlanticum* and *L. nudum* have a single mid-ventral band of luminous tissue from the isthmus to the pelvic fins different in structure from that in *L. bigelowi*. There are three species of *Lestrolepis* and whilst they are considered mesopelagic they do not occur very deep and are often

(c)

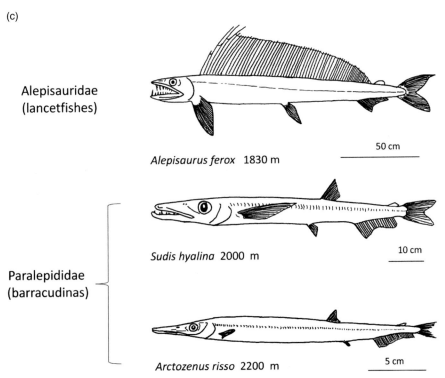

Alepisauridae
(lancetfishes)

Alepisaurus ferox   1830 m

50 cm

Paralepididae
(barracudinas)

Sudis hyalina 2000 m

10 cm

Arctozenus risso 2200 m

5 cm

Figure 4.17c ORDER AULOPIFORMES. *Alepisaurus ferox* (Long snouted lancetfish) circumglobal 84°N–57°S. *Sudis hyalina*, circumglobal 50°N–40°S. *Arctozenus risso* (Spotted barracudina) circumglobal 71°N–55°S.

found inshore. They have two parallel bands of ventral luminous tissue. The genus *Macroparalepis* tends to have an antitropical distribution with *M. affinis* (2000 m$_{fbd}$) and *M. brevis* (650 m$_{fbd}$) in the North and South Atlantic; *M longilateralis* and *M. macrogeneion* are confined to the South Atlantic, the latter extending into the Southern Indian and Pacific Oceans. There are two species endemic to the Pacific Ocean and *M. nigra* in the Eastern Atlantic. Diet is mostly zooplankton rather than fishes. *Magnisudis* comprises three quite large species (up to 55 cm); the most widespread is *M. atlantica* (Duckbill barracudina, 4750 m m$_{fbd}$) circumglobal 69°N–61°S, *M. prionosa* (Southern barracudina, 1000 m m$_{fbd}$) is confined to between 20°S and the Antarctic Convergence, and *M. indica* is found in the Indo-West Pacific. *Notolepis* is a Southern Hemisphere genus with two species, *N. annulata* (2050 m$_{fbd}$) and *N. coatsi* (2000 m$_{fbd}$), which are both probably circumglobal in Antarctic waters with the former

extending into the SW Atlantic Ocean. *Paralepis* has four mainly tropical species: *P. speciosa* is endemic to the Mediterranean Sea, *P. coregonoides* to the North Atlantic, *P. elongata* is found in the Atlantic and Indian Oceans and *P. brevirostris* in the tropical Atlantic and West Pacific. *Stemonosudis* has 11 species, three Atlantic, two Pacific, one Indian Ocean, three Indo-Pacific and four species found in several of the main oceans. *Sudis hyalina* (2000 m$_{fbd}$) grows to over 1 m in length (Fig. 4.17c) and is distributed circumglobally at 50°N–40°S. It has been observed in the Mediterranean Sea in underwater videos at 774 m depth slowly swimming in the water column above the sedimentary seafloor (D'Onghia et al., 2011). Lacerations or 'fish bites' at 600–800 m depth on oceanographic cables on continental slopes of the NW Atlantic have been attributed to adult *S. hyalina* (Haedrich, 1965). The incidence of bites over short time periods suggested high numbers of *S. hyalina* may have been present. The other species in the genus,

S. atrox, is reported from the Atlantic and Pacific Oceans. The genus *Uncisudis* has four species: *U. advena* W Atlantic, *U. quadrimaculata* E Atlantic, *U. longirostra* in the E Atlantic and Indian Oceans and *U. posteropelvis* from the W Pacific Ocean. Formerly placed within its own family, the genus *Anotopterus* (daggertooths) comprises three species found at high latitudes: *A. nikparini* in the North Pacific, *A. pharao* in the North Atlantic and *A. vorax* circumglobal 55°S–72°S in the Southern Ocean. They are predators on fishes, including other Paralepidae, and the stomach is distensible allowing ingestion of remarkably large amounts of large prey. In the North Pacific *A. nikparini* preys on molluscs, crustaceans, marine worms, coelenterates and salps as well as fishes. Larger adults migrate to colder subpolar waters, whereas younger adults inhabit more temperate regions. Although recorded at up to 147 cm long the body is slender and the largest specimen weighed only 1.6 kg (Ambrose, 1996). The Paralepididae are generally not well known, with many species known only from the juvenile stages. Without a swim bladder they are not good sound scatterers and therefore do not appear conspicuous in echograms. In their survey of mesopelagic fishes over the Nova Scota slope of the NW Atlantic, Themelis and Halliday (2012) recorded 14 species of paralepidids, the most abundant being *Arctozenus risso* but most others were represented by just a few individuals per net tow. Paralepidae are agile, swift swimmers, and it is likely that they are able to avoid nets. By analogy with deep-sea oceanic squids (Clarke, 2003) they are probably under-represented in net samples but nevertheless occur widely in stomach contents of marine predators including *Alepisaurus* (Haedrich, 1964) and marine mammals (Pusineri et al., 2007).

4.3.9 Order Myctophiformes (Lanternfishes)

The Myctophiformes comprises 255 species of predominantly small deep-sea pelagic and benthopelagic fishes in two families, Neoscopelidae and Myctophidae (Fig. 4.18). They often occur in high abundance in the mesopelagic and usually have light organs.

4.3.9.1 Family Neoscopelidae (77) (Blackchins)

The Neoscopelidae has six species in three genera. The genus *Neoscopelus* has large eyes, ventro-lateral rows of photophores on the body and also light organs under the tip of the tongue. The bioluminescent organs of *Neoscopelus* are different in structure from those of the Myctophidae and probably evolved independently. *Neoscopelus* are benthopelagic, generally associated with continental, island and seamount slopes at depths from 250 to over 1000 m. There are three species: *N. macrolepidotus* (large-scaled lantern fish) in the Atlantic and Pacific Oceans from 55°N to 49°S; *N. microchir* 35°N–42°S in the Atlantic, Indian and West Pacific (Fig. 4.18a) and *N. porosus* in the Indo-Pacific. *N. microchir* is the largest, growing to 30 cm length. They have a swim bladder, silvery appearance, well-ossified skeleton and firm musculature (Moser, 1996); they are robust fishes benefitting from relatively rich benthic feeding opportunities. The genus *Scopelengys* by contrast is bathypelagic, lacks photophores and swim bladder, the eyes are small, the skeleton is weakly ossified and the muscles are flabby. *S. tristis* (Pacific blackchin, maximum length 20cm) is more or less circumglobal (Fig. 4.18a) between 35°N and 35°S at depths greater than 1000 m for the adults and 500–800 m for the juveniles. *S. clarkei* is known only from the West Central Pacific (Butler and Ahlstrom, 1976). The sole species in its genus *Solivomer arenidens* (1413 m_{fbd}) is endemic to the Sulu Sea Basin in the Philippines, has no light organs but has a swim bladder and shows few obvious adaptations to life in the deep sea despite being captured at great depths (Miller, 1947); it is probably benthopelagic. The Neoscopelidae represent the kinds of fishes from which the great diversity of the Myctophidae probably originated (Poulsen et al., 2013).

4.3.9.2 Family Myctophidae (78) (Lanternfishes)

This is arguably the most important family of all deep-sea fishes with 249 species in 32 genera and is found throughout the world's oceans often in massive abundances. They are small fishes, mean maximum length 9.2 cm, and are dominant in epipelagic, mesopelagic and bathypelagic environments, leading

(a)

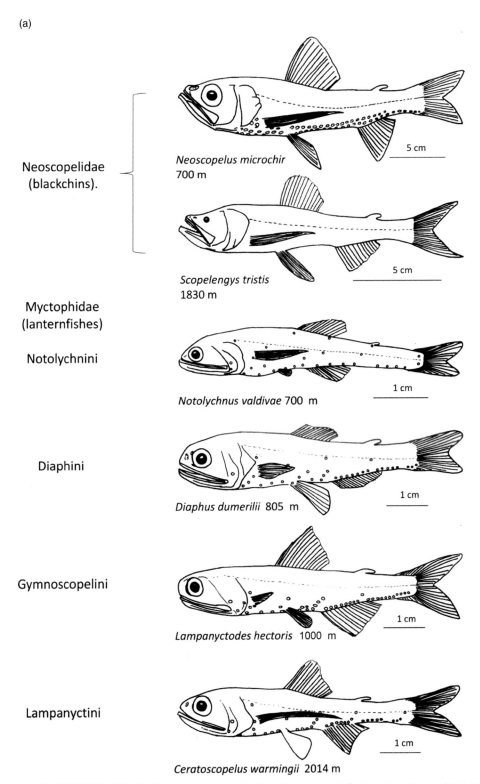

Neoscopelidae (blackchins).

Neoscopelus microchir
700 m

5 cm

Scopelengys tristis
1830 m

5 cm

Myctophidae (lanternfishes)

Notolychnini

Notolychnus valdivae 700 m

1 cm

Diaphini

Diaphus dumerilii 805 m

1 cm

Gymnoscopelini

Lampanyctodes hectoris 1000 m

1 cm

Lampanyctini

Ceratoscopelus warmingii 2014 m

1 cm

Figure 4.18a ORDER MYCTOPHIFORMES *Neoscopelus microchir* (Shortfin neoscopelid) Atlantic, Indian and West Pacific 35°N–42°S. *Scopelengys tristis* (Pacific blackchin) probably circumglobal 35°N–35°S. *Notolychnus valdiviae* (Topside lampfish) circumglobal 56°N–40°S. *Diaphus dumerilii*, mainly Atlantic 51°N–40°S. *Lampanyctodes hectoris* (Hector's lanternfish) E Atlantic and Pacific 13°S–43°S. *Ceratoscopelus warmingii* (Warming's lantern fish) circumglobal 42°N–65°S.

to estimates that they may account for half of all fish biomass in the oceans (Poulsen et al., 2013). They occupy an important position in oceanic food chains: they consume zooplankton, transport organic matter vertically through their migrations, provide forage for larger predatory species and their faeces contribute to the rain of particulate organic matter descending towards the seafloor. Myctophids are characterised by the presence of photophores (Plate 6a), which are found in all species except one: *Taaningichthys paurolychnus,* a highly oceanic very deep-living species never found shallower than 900 m. Myctophid photophores are distinctive round eye-like iridescent spots distributed over the body of the fish. A typical adult individual has 80–100 photophores mainly on the ventral surface but also around the eye and on the lateral aspects of the body. Even if damaged and stripped of scales when recovered from nets, the pattern of photophores often remains intact, making them a valuable diagnostic feature in species identification. In addition, there are supra- and infracaudal luminescent glands as well as luminous patches elsewhere on the body. The unique pattern of photophores in each species has led to the suggestion that they may function as a means of intraspecific recognition or communication. However as typical abundances in the ocean distances between individuals are likely to be greater than 10 m, the eyes could not resolve the details of individual photophore patterns at such a range and would only be useful in tandem with other sensory mechanisms (such as chemoreception) to bring individuals close together (Herring, 2000). Light emission and retinal sensitivity are essentially monochromatic in the blue/green wavelength range 480–492 nm maximising the effective range through seawater. Maximum range of detection of a conspecific under ideal conditions is estimated to be 30 m (Turner et al., 2009). Two species (*Myctophum nitidulum* and *Bolinichthys longipes*) have retinas sensitive to longer wavelengths and could theoretically detect red bioluminescence such as emitted by some of the Stomiidae. Davis et al. (2014) distinguish the functions of ventral photophores for camouflage and lateral photophores for communication and suggest that there has been an

evolutionary elaboration of the latter providing a mechanism for genetic isolation and increased speciation rate in the Myctophids compared with other pelagic deep-sea taxa. The hypothesis is compelling, but direct evidence that the eyes of these fishes can indeed detect small differences in pattern of photophores on the body remains lacking.

The larvae of myctophids are remarkably varied, much more so than the adults, indicating they occupy a wide range of distinctive pelagic niches, which may account for the high species diversity in this group. Sabates et al. (2003) compared the larvae of *Benthosema glaciale* and *Myctophum punctatum* that coexist in the same part of the Western Mediterranean Sea. *M. puncatum* has elongated eyes set on stalks and feeds on a more diversified diet at greater depths than *B. glaciale,* which specialises in feeding on copepod eggs and larval stages. Most species show daily vertical migration with some coming to the surface at night; mean maximum depth reached during the day is 1121 m.

The myctophidae are classified into two subfamilies, Lampanyctinae and Myctophinae. Early taxonomic studies were primarily based on photophore patterns. Based on osteological evidence as well as photophores, Paxton (1972) further divided the families into seven tribes. Moser and Ahlstrom (1974) proposed that in view of their distinctiveness in myctophids, larval stages could be also used to resolve taxonomic and evolutionary relationships. Combining photophore patterns, osteology, larval stages and an increasing numbers of characters has resulted in successive revisions of myctophid relationships (Stiassny, 1996, Yamaguchi, 2000). Poulsen et al. (2013) found mitochondrial DNA gene sequences unique to the Myctophidae supporting their division into seven tribes. This arrangement is followed here, but within each tribe the genera are listed alphabetically. Most genera occur across all three major oceans, Atlantic, Indian and Pacific.

Tribe Notolychnini

The single species (Fig. 4.18a) in this tribe *Notolychnus valdiviae* (Topside lampfish) is recognised as a sister taxon to all the other Myctophidae and represents a basal clade standing alone outside the two major subfamilies (Poulsen et al., 2013). Uniquely, in addition to the usual ventral photophores, *Notolychnus* has a series of three dorsally located photophores on each side of the body between the leading edges of the dorsal and adipose fins. *Notolychnus* is an oceanic species, distributed worldwide between 56°N and 40°S at a depth of 375–700 m during the day and 25–350 m at night, feeding on copepods, ostracods and euphausiids. It occurs in sufficient abundance in the Gulf of Mexico to be considered a marginal candidate for fishery exploitation (Gjøsaeter and Kawaguchi, 1980).

Subfamily Lampanyctinae

With 170 species this is the larger of the two subfamilies within the Myctophidae.

Tribe Diaphini

This tribe is dominated by the genus *Diaphus*, the most diverse of the myctophid genera with 78 species: 19 species occur in all three major oceans, five species are endemic to the Atlantic Ocean, six to the Indian Ocean and 29 to the Pacific Ocean. 13 species are Indo-Pacific, one species is found in the Atlantic and Indian Oceans and four species occur in the Atlantic and Pacific. The latitudinal range is from 62°N to 58°S (Moser et al., 1983) occupying a range of habitats from abundant oceanic pelagic species such as *D. coeruleus* to benthopelagic or demersal species such as *D. burtoni, D. confusus, D. mascarensis* and *D. watasei*, which are often caught in bottom trawls on slopes. Eight species are dominant in upper ocean layers at night in their respective areas and are potential candidates for commercial exploitation: *D. dumerilii* (Atlantic SW, Mauretania and Caribbean), *D. garmani* (Equatorial Pacific; Fig. 4.18a), *D. theta* (N Pacific sub-Arctic), *D. kuroshio* (NW Pacific), *D. luetkeni* (E Indian Ocean), *D. malayanus, D. regani* and *D. suborbitalis* (W Central Pacific; Gjøsaeter and Kawaguchi, 1980).

Idiolychnus urolampus in its monotypic genus is an Indo-Pacific benthopelagic species found in inshore areas or at seamounts at depths down to over 500 m. There are two species in the genus *Lobianchia*: *L. dofleini* is an oceanic species found in the Atlantic Ocean from 50°N to 40°S and circumglobally in the South, whereas *L. gemelarii* is one of the most widespread myctophid species occurring circumglobally from 60°N to 55°S, including the Mediterranean Sea.

Tribe Gymnoscopelini

There are 18 species in six genera, all of which are confined to the Southern Hemisphere apart from *Notoscopelus*. There are six species of *Notoscopelus*: *N. resplendens* is global between 55°N and 61°S reaching a depth of 2000 m during the day but the upper 300 m of the water column at night. *N. caudispinosus* occurs in the Atlantic and Indo-West Pacific (54°N–58°S), *N. elongatus* is endemic to the Western Mediterranean, *N. boloni* to the NE Atlantic and the Mediterranean, *N. kroyeri* to the North Atlantic and *N. japonicus* to the NW Pacific off Japan. *Gymnoscopelus* comprises eight species found between 24°S and the coast of Antarctica, mostly associated with the Antarctic polar front. They form an important part of the food chain in the Southern Ocean at depths down to 1000 m, feeding on amphipods, copepods, euphausids and salps, the diet differing according to species (Shreeve et al., 2009). *G. piabilis* tends to be benthopelagic and is consumed by the demersal Patagonian toothfish, *Dissostichus eleginoides*, whereas *G. hintonoides* is oceanic, generally deeper than 800 m. The remaining genera are all monotypic with Southern Hemisphere distributions. *Lampanyctodes hectoris* is tropical (13°S–43°S; Fig. 4.18a) associated with coastal upwelling areas where it is sufficiently abundant to have been fished commercially off South Africa by trawling at nighttime in the upper 100 m of the water column (Gjøsaeter and Kawaguchi, 1980). Spawning aggregations of *L. hectoris* have been encountered at 80 m depth off Northern New Zealand and eggs with developing embryos captured in plankton samples

(Robertson, 1977). *Scopelopsis multipunctatus* is circumglobal between 25°S and 42°S, *Lampichthys procerus* at the Subtropical Convergence (23°S–46°S) and *Hintonia candens* in the Southern Ocean (40°S–50°S).

Tribe Lampanyctini

This tribe has 71 species in 10 genera. The genus *Bolinchthys* comprises seven widespread oceanic species: *B. photothorax* is circumglobal between 48°N–48°S, *B. indicus* and *B. supralateralis* are in the Atlantic and Indo-West Pacific Oceans, *B. longipes* and *B. pyrsobolus* occur in the Indo-Pacific, *B. distofax* in the Atlantic and West Pacific Oceans and *B. nikolayi* is confined to the SW Pacific. *Ceratoscopelus* has two circumglobal species, *C. townsendi* and *C. warmingii*, as well as (Fig. 4.18a) *C. maderensis,* which is confined to the North Atlantic. The nine species of the genus *Lampadena* are relatively large (mean maximum length 14 cm) and deep-living compared with most myctophidae and consequently are less well known (Niass and Ozawa, 2000). Five species are probably circumglobal including *L. luminosa* (44°N–37°S) and *L. notialis,* two species are endemic to the Atlantic, two to the Pacific, and *L. anomala* is distributed around Africa in the E. Atlantic and W. Indian Ocean. *Lampanyctus* comprises 20 species, five of which occur in all three major oceans; four are circumglobal in the Southern Ocean with range extensions into the major oceans with *L. macdonaldi* bipolarly distributed in the North and South Atlantic Oceans. Seven species are endemic to the Pacific Ocean, one is endemic to the Atlantic, one to the Atlanto-Pacific and two are Indo-Pacific, including *L. steinbecki,* which occurs in high abundance in the Eastern North Pacific (Gjøsaeter and Kawaguchi, 1980). The two species of *Lepidophanes* are confined to the Atlantic Ocean. The genus *Nannobrachium* comprises 17 species of deep-living myctophids characterised either by rudimentary or absent pectoral fins, weak flaccid musculature, absent or undeveloped swim bladder and a 'pinched' concave body profile. They are distributed worldwide from the tropics to sub-Arctic regions (Zahuranec, 2000).

N. atrum (Fig. 4.18b) and *N. lineatum* are most widely distributed almost circumglobally, 58°N–46°S and 42°N–37°S, respectively. There are two southern species: *N. wisneri* in the Subtropical Convergence and furthest south is *N. achirus* from the Subtropical Convergence to the Antarctic Polar Front, where it can occur as a dominant species in catches (Shreeve et al., 2009). Ten species are endemic to different parts of the Pacific Ocean. There are two species endemic to the Atlantic Ocean: *N. cuprarium* in subtropical waters and *N. isaacsi,* which is found only in the Mauritanian upwelling off West Africa. *N. indicum* is confined to the equatorial Indian Ocean. The genera *Parvilux* and *Stenobrachius* each have two species endemic to the Pacific Ocean, the former only in the East and the latter in the North, including sub-Arctic waters. The deep-living genus *Taaningichthys* comprises three more or less circumglobal species, which are nonmigratory; *T. paurolychnus* is noted for the absence of photophores (Fig. 4.18b). Two species of *Triphoturus* are confined to the Eastern Pacific Ocean, whereas *T. nigrescens* is found in the tropical waters of all the major oceans.

Subfamily Myctophinae

The Myctophinae comprises 78 species in three tribes.

Tribe Electronini

The Electronini are predominantly Southern Hemisphere myctophids that originated during the Early Miocene (16–21 million years ago) following the opening of the Drake Passage and the formation of the Antarctic Circumpolar Current (Chapter 1). Of the three genera, a few species have colonised the Atlantic, Indian and Pacific Oceans (Gordeeva, 2013). *Electrona* has four southern species distributed between 35° and 78°S and one global species, *E. risso,* which is widespread in all the oceans including the Mediterranean Sea (Fig. 4.18b). Its monotypic genus *Krefftichthys anderssoni* is found only in the south between 32°S and 78°S. Eleven species of the genus *Protomyctophum* are confined to the Southern Ocean but one species, *P. arcticum,* is endemic to the North

(b)

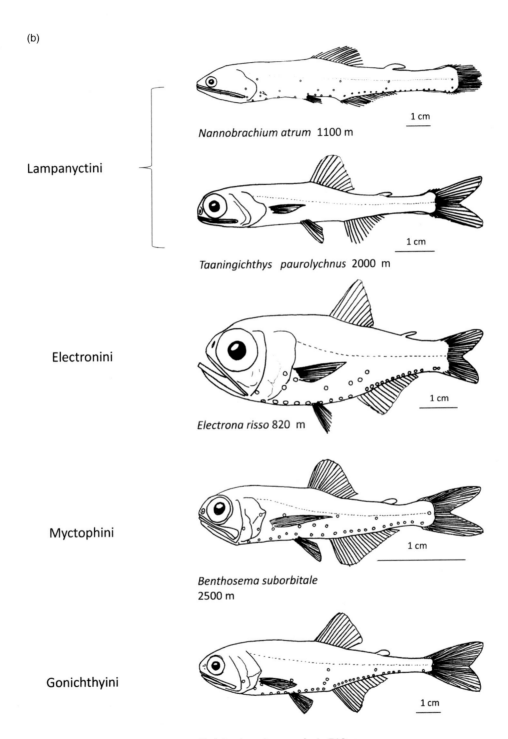

Lampanyctini

Nannobrachium atrum 1100 m

1 cm

Taaningichthys paurolychnus 2000 m

Electronini

Electrona risso 820 m

1 cm

Myctophini

Benthosema suborbitale
2500 m

1 cm

Gonichthyini

Tarletonbeania crenularis 710 m

1 cm

Figure 4.18b ORDER MYCTOPHIFORMES *Nannobrachium atrum* (Dusky lanternfish) Bipolar N Atlantic, S Atlantic, S Indian, SW Pacific and Southern Ocean. *Taaningichthys paurolychnus*, circumglobal 40°N–20°S. *Electrona risso* (Electric lantern fish) circumglobal 55°N–46°S including the Mediterranean. *Benthosema suborbitale* (Smallfin lanternfish) circumglobal 50°N–50°S. *Tarletonbeania crenularis* (Blue lanternfish) NE Pacific.

(c)

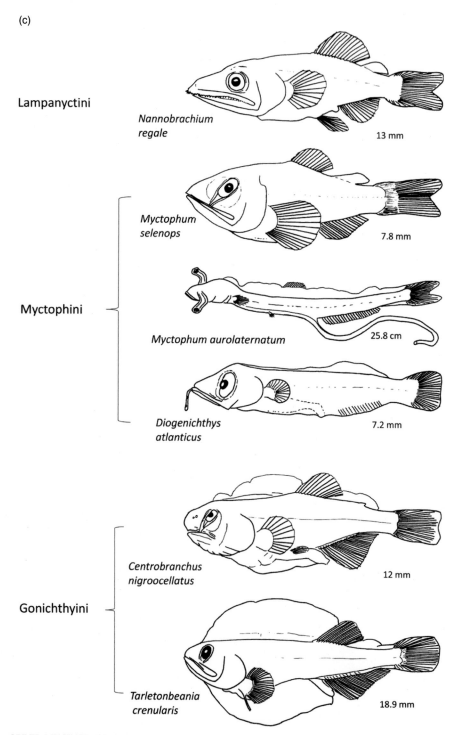

Lampanyctini

Nannobrachium regale

13 mm

Myctophini

Myctophum selenops

7.8 mm

Myctophum aurolaternatum

25.8 cm

Diogenichthys atlanticus

7.2 mm

Gonichthyini

Centrobranchus nigroocellatus

12 mm

Tarletonbeania crenularis

18.9 mm

Figure 4.18c ORDER MYCTOPHIFORMES. Myctophiform larvae showing the extreme variation in morphology in different species. After Moser (1996) and Stiassny (1996). Standard lengths are indicated.

Atlantic Ocean to 70°N, and three species to the Pacific Ocean: *P. thompsoni* and *P. crockeri* in the North Pacific and *P. beckeri* in the East Central Pacific. *P. crokeri* is a dominant species at the surface at night where it occurs in the Pacific Ocean (Gjøsaeter and Kawaguchi, 1980). Five species of the Electronini featured in catches in the northern Scotia Sea in the vicinity of the Antarctic Polar Front where they feed on a variety of crustacea; however, *E. carlsebergi* was noted for consuming salps (Shreeve et al., 2009).

Tribe Myctophini

The Myctophini is a worldwide tribe of short-jawed myctophids including some of the most dominant species in terms of biomass near the surface at night (Gjøsaeter and Kawaguchi, 1980). There are four genera. *Benthosema* comprises five species: *B. suborbitale* is circumglobal in the tropical and temperate waters and is abundant in the Kuroshio current system of the NW Pacific, *B. glaciale* is an abundant species found in the Atlantic Ocean from Baffin Bay to the Mediterranean Sea, *B. fibulatum* and *B. pterotum* occur in the Indo-Pacific where the latter is particularly important in the Arabian Sea; *B. panamense* is confined to the NE Pacific. The genus *Diogenichthys* comprises three species, which all occur in the three major oceans with high abundances of *D. atlanticus* in the West Central Atlantic and parts of the NW Pacific, and *D. laternatus* in the Equatorial East Pacific. The genus *Hygophum* has nine species: *H. hygomii* and *H. reinhardtii* are found in all three major oceans and *H. hanseni* is circumglobal in the Southern Hemisphere 30°S–43°S in the Antarctic Convergence. Two species are endemic to the Pacific and two species to the Atlantic. *H. macrochir* occurs in the tropical Atlantic and SW Pacific. *H. proximum* is a tropical and subtropical Indo-Pacific species, which is abundant in the equatorial East Pacific. *Metelectrona* is a Southern Hemisphere genus comprising two circumglobal species, *M. herwigi* (35°S–41°S) and *M. ventralis* (48°–50°S) with *M. ahlstromi* confined to the SE Pacific. The genus *Myctophum* is relatively well known because of its nocturnal surface migratory habit though its maximum depth during the day can

be down to 1000 m. There are 16 species distributed between 48°N and 57°S: six occur in all three major oceans, notably *M. asperum* and *M. nitidulum,* which are circumglobal, two species are endemic to the Atlantic with *M. punctatum* (Plate 6a) also in the Mediterranean, two are endemic to the Pacific Ocean, two to the Indian Ocean and four species are Indo-Pacific.

Tribe Gonichthyini

First recognised by Paxton (1972), the Gonichthyini comprises four genera with similar body forms, which feature a projecting snout, subterminal mouth, slender tail and a relatively well-ossified skeleton.

The genus *Centrobranchus* comprises one circumglobal species, *C. nigroocellatus,* which is found in all oceans at 40°N–45°S, two Pacific species and *C. andreae,* distributed throughout the Indo-Pacific. *Gonichthys barnesi* occurs at 30°S–40°S in the southern Subtropical Convergence zone of all the oceans, whereas other species in the genus are confined either to the Atlantic Ocean (*G. cocco*) or to the Pacific Ocean (*G. tenuiculus* and *G. venetus*). The genus *Loweina* has three species: *L. interrupta* found in all three major oceans, *L. rara* in the Atlantic and Pacific Oceans and *L. terminata* in the subtropical North Pacific. The two species of the genus *Tarletonbeania* are endemic to the North Pacific Ocean, where *T. crenularis* is a dominant species in the sub-Arctic North East extending into the Bering Sea (Fig. 4.18b). The genus *Symbolophorus* has eight species: *S. barnardi* and *S. boops* are circumglobal Southern Hemisphere species, *S. rufinus* is tropical and subtropical in all three major oceans, two species are endemic to the Atlantic Ocean, two to the Pacific and *S. evermanni* is widespread in the Indo-Pacific region.

4.3.10 Order Lampriformes

The Lampriformes are generally large laterally compressed fishes of the open ocean (Fig. 4.19) varying in morphology from the disc-shaped Opah, (*Lampris guttatus*), which grows up to 2 m diameter, to the ribbon-like King of Herrings (*Regalecus glesne*),

(a)

Lampridae
(Opahs)

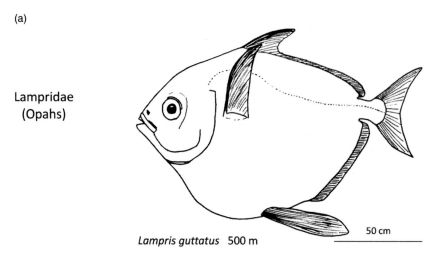

Lampris guttatus 500 m

50 cm

Figure 4.19a ORDER LAMPRIFORMES *Lampris guttatus* (opah) circumglobal 70°N–45°S.

which is the longest teleost fish recorded at 11 m length. They typically occur between the surface and 1000 m depth and have a silvery body and red-coloured fins. There are 24 species in seven families. They are a conspicuous species of the epipelagic and mesopelagic and can be considered, at least partly, to be deep-sea fishes. There are two suborders: Bathysomi, with deep short-bodied forms, presumed to be ancestral, and represented mostly by extinct species such as *Bathysoma* from the late Cretaceous onward (Olney et al., 1993; Davesne et al., 2014) and Taenosomi, ribbon-like elongated species. They are all oviparous, with planktonic eggs and larvae.

Suborder Bathysomi
4.3.10.1 Family Veliferidae (79) (Velifers)
This family comprises two relatively small demersal species < 40 cm long found on the upper slopes and seamount summits of the Indo-West Pacific region at depths less than 200 m. They are a sister group to all the lampridiforms. They cannot be considered to be deep-sea fishes.

4.3.10.2 Family Lampridae (80) (Opahs)
There are two species in this family, *Lampris guttatus* the Opah, which occurs in all the world's oceans

between 70°N and 45°S (Fig. 4.19a) and *L. immaculatus* in the Southern Opah, which is circumglobal in the Southern Hemisphere between 34°S and the Antarctic Polar front. *L. immaculatus* (maximum length 1.1 m) is somewhat smaller than *L. guttatus*, which grows to a weight of 250 kg. They are never caught in large numbers but if encountered as a bycatch, for example, in tuna fisheries, they are considered to be valuable food fish. They swim by means of a wing-like flapping motion of the pectoral fins; *L. guttatus* has been shown to maintain an elevated body temperature, up to 6°C above ambient water temperature, by virtue of a heat exchanger network of blood vessels between the heart and the gills. This species appears to be unique amongst fishes in thus attaining whole-body endothermy including a warm heart, which enables it to spend most of its time at depths of 50–400 m in cold waters below the thermocline pursuing their prey, which comprises mid-water fishes, invertebrates and squids (Wegner et al., 2015).

Suborder Taenosomi
This suborder of elongated ribbon-like fishes encompasses most of the extant diversity within the lampiformes. They have long been known from occasional strandings on shores around the major oceans but are primarily mesopelagic and epipelagic

predators that adopt a head-up vertical posture in the water column. Camouflaged by their silvery skin, they can be very inconspicuous in the mesopelagic and take their prey from below by accelerating upward whilst exerting suction with their protrusible jaws and buccal apparatus.

4.3.10.3 Family Lophotodidae (81) (Crestfishes)

The Lophotodidae are represented by four species in two genera. In its own monotypic genus *Eumecichthys fiski* (Unicorn crestfish) probably has a worldwide distribution outwith polar seas at depths of around 1000 m (Fig. 4.19b). It has a ribbon-like body up to 150 cm long and characteristic extension of the skull forward of the eyes with a crest extension of the dorsal fin giving the unicorn appearance. The genus *Lophotus* has three species, *L. guntheri* and *L. lacepede*, which are circumglobal, and *L. capellei*, which is confined to the Atlantic and Pacific Oceans. These fish grow up to 2 m long; *L. lacepede*, known to feed on fishes and squids, does not generally occur as deep as *Eumecichthys*. The Lophotodidae have an ink sac that can eject dark-coloured fluid from the cloaca that forms a cloud in the water similar to squid or cuttlefish ink providing a means of avoiding predators or confusing prey.

4.3.10.4 Family Radiicephalidae (82) (Tapertails)

This family is represented by a single species, *Radiicephalus elongatus*, (tapertail; Fig. 4.19b), known from the tropical Eastern Atlantic Ocean and East Pacific Ocean. Very few specimens have been examined (Harrison and Palmer, 1968). The body length is up to 76 cm, and it has a large head, large eyes and an eponymous tapering tail. There is an ink sac as in the Lophotodidae. Stomach contents indicate that in the Eastern Atlantic *Radiicephalus* feeds on myctophidae of the genus *Lampanyctus* during the night at depths less than 700 m.

4.3.10.5 Family Trachipteridae (83) (Ribbon fishes)

The Trachipteridae comprises 10 species in three genera. The genus *Desmodema* has two species:

D. polystictum occurs globally (42°N–52°S) between 0 and 500 m depth in the epipelagic and mesopelagic, whereas *D. lorum* is confined to the Eastern Central Pacific. *D. polystictum* grows up to 110 cm total length and feeds on fish, squids, octopus and crustacea. It has been suggested that the young are epipelagic but that at metamorphosis the adults descend to occupy the twilight zone at a few hundred metres, depth, hovering with a head-up posture while maintaining position by undulating the dorsal fin but accelerating upward by rapid bursts of anguilliform swimming to capture their prey (Rosenblatt and Butler, 1977). The lateral line runs the length of the tapering tail greatly increasing sensitivity to vibrations in the water. The genus *Trachipterus* comprises five species, three with regional distributions in the Northern Hemisphere: *T. altivelis* (King-of-the-salmon) in the E. Pacific, *T. ishikawae* (Slender ribbonfish) in the NW pacific and *T. arcticus* (Dealfish) in the N Atlantic. (Fig. 4.19b); *T. jacksonensis* (Blackflash ribbonfish) is found in all three major oceans and *T. trachypterus* (Mediterranean dealfish) occurs in the Mediterranean and in the Atlantic and Pacific Oceans. Maximum length ranges from 1.5 to 3 m. The genus *Zu* comprises two species, which are probably circumglobal. *Z. cristatus* is a mesobathypelagic species occurring in the Mediterranean Sea, Atlantic, Pacific and Indian Oceans in tropical and temperate regions (Dulčić, 2002), whereas *Z. elongatus* occurs in the Southern Hemisphere including the SE Atlantic and SW Pacific off New Zealand.

4.3.10.6 Family Regalicidae (84) (Oarfishes)

There are two genera, *Agrostichthys* and *Regalecus*. *Agrostichthys parkeri*, the sole species in its genus, is known only from rather few specimens taken in the Southern Hemisphere 25–59°S. Most records are from around New Zealand, where they have been trawled offshore, stranded on beaches or speared by scuba divers in inshore waters. Maximum reported length is 3 m. Personnel handling a live specimen have experienced mild electric shocks pulsed at

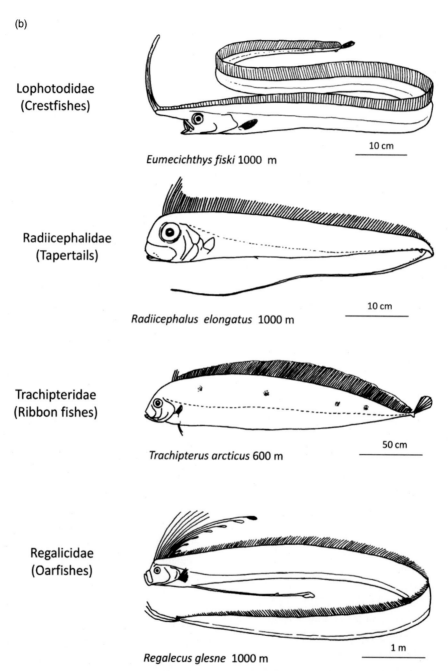

(b)

**Lophotodidae
(Crestfishes)**

10 cm

Eumecichthys fiski 1000 m

**Radiicephalidae
(Tapertails)**

10 cm

Radiicephalus elongatus 1000 m

**Trachipteridae
(Ribbon fishes)**

50 cm

Trachipterus arcticus 600 m

**Regalicidae
(Oarfishes)**

1 m

Regalecus glesne 1000 m

Figure 4.19b ORDER LAMPRIFORMES *Eumecichthys fiski* (unicorn crestfish) probably circumglobal but rare. *Radiicephalus elongatus* (tapertail) E Atlantic and E Pacific. *Trachipterus arcticus* (dealfish) N Atlantic, *Regalecus glesne* (king of herrings) circumglobal 72°N–52°S.

approximately 10s intervals, suggestive of functioning electric organs (McDowall and Stewart,1999). *Agrostichthys* probably feeds offshore in deep pelagic waters while the morphology of the head and mouth suggests that it adopts an oblique rather than a vertical posture in the water column. It is

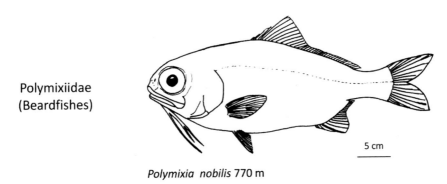

Polymixiidae
(Beardfishes)

Polymixia nobilis 770 m

Figure 4.20 ORDER POLYMIXIIFORMES. *Polymixia nobilis* (Stout beardfish) Atlantic.

also suggested that this species may be able to autotomize its tail. A juvenile specimen has been described from near the Walvis Ridge in the Southern Atlantic between 0 and 550 m depth (Evseenko and Bol'shakova, 2014). The genus *Regalecus* is represented by two species, *R. glesne* (King of herrings) which is distributed worldwide between 72°N and 52°S (Fig. 4.19b) and *R. russelii*, which is confined to the tropical Pacific Ocean. These are large fishes, usually encountered alone or in groups of no more than two or three together simultaneously. They have been recorded down to 1000 m, and a live specimen was filmed at 467 m depth in the Gulf of Mexico. Most specimens of *Regalecus* that have been retrieved are incomplete, and it is considered that this may be because of an ability to autotomize the posterior part of the body (T. Roberts in Schmitter-Soto, 2008). They feed by adopting the typical vertical head-up posture of the Taenosomi with the rays of the dorsal fin crests and pelvic fins spread out, sensing the water column and stabilising the fish with the body straightened out. Protrusible jaws permit suction feeding, which is aided by complex gill rakers able to retain krill, small fishes and squid that comprise its diet.

4.3.11 Order Polymixiiformes (Beardfishes)

4.3.11.1 Family Polymixiidae (85) (Beardfishes)
This order comprises a single genus, *Polymixia*, (Fig. 4.20), formerly placed within the Beryciformes but now recognised together with the Percopsiformes as representing the earliest diverging lineage of acanthomorph teleosts (Near et al., 2013). There are 10 species of *Polymixia* distributed in tropical and subtropical latitudes of the Atlantic, Indian and West Pacific Oceans. They are medium-sized (mean maximum length 27 cm) deep-bodied bathydemersal fishes found on the upper slopes of continental margins, seamounts, oceanic islands and ridges down to a maximum depth of 800 m. There are two species in the Atlantic Ocean: *P. lowei* (600 m_{fbd}) in the West including the Gulf of Mexico (Powell et al., 2003) and the more widespread *P. nobilis* (770 m_{fbd}), which also extends into the Western Indian Ocean (Fig. 4.20). *P. fusca* (435 m_{fbd}) is endemic to the Indian Ocean, and there are two Indo-Pacific species, *P. berndti* (585 m_{fbd}) and *P. busakhini* (660 m_{fbd}). Five species are endemic to different parts of the Pacific Ocean. In the Eastern Atlantic, *P. nobilis* is associated with the slopes of volcanic islands and seamounts and has been captured off the Canary Islands at depths of 270–700 m as part of a small-scale longline fishery mainly targeted at Alfonsinos; *Beryx splendens*, *Beryx decadactylus* and large specimens of *Trachurus*. *P. nobilis* in this fishery grows to over 45 cm in length and a maximum age of 14 years; the smaller individuals feed on crustacea and fish and the largest specimens consume fish (e.g. *Chlorophthalmus* and Myctophidae) and cephalopods. Spawning is continuous throughout the year with peaks in April and June (Garcia-Mederos, 2010).

4.3.12 Order Zeiformes (Dories)

The Zeiformes are deep-bodied, laterally flattened fishes with large eyes and large protrusible jaw gape enabling them to capture a wide variety of crustacean and fish prey (Fig. 4.21) They generally occur at upper slope or mesopelagic depths (200–1000 m) and even if pelagic, are often associated with submarine features such as summits, ridges or canyons. Most species show no evidence of vertical migration, although they may feed on migratory species of the Deep Scattering Layer. There are 33 species in six families. Owing to their depth of occurrence, they are often caught in mixed deep-water fisheries, and those with sufficient edible flesh are landed commercially. They generally do not occur in great enough abundance to sustain targeted fisheries, although a few species are considered highly desirable.

4.3.12.1 Family Cyttidae (86) (Lookdown dories)

This family is confined to the Southern Hemisphere, notably around South Australia and New Zealand. There are three species: *Cyttus australis* (Silver dory, 350 m$_{fbd}$) endemic to Australian waters, *C. novaezealandiae* (New Zealand dory, 450 m$_{fbd}$) and *C. traversi* (King dory, 978 m$_{fbd}$) off Australia, New Zealand and South Africa (Fig. 4.21a). All are taken as bycatch in mixed fisheries. In New Zealand waters *C. traversi* is most frequently captured in trawls between 200 and 800 m depth with a total annual catch limit catch limit of 783 tonnes in 2009 under a quota management system (Forman and Dunn, 2010). *C. traversi* is considered a middle-depth species by New Zealand authorities (Ballara, 2014).

4.3.12.2 Family Oreosomatidae (87) (Oreos)

With a mean maximum depth of 1330 m this is the deepest-living Zoarciform family. There are four genera. *Allocyttus* comprises four species: *A. folletti* (Oxeye oreo, 732 m$_{fbd}$) in the North Pacific, *A. guineensis* (Guinea oreo, 1900 m$_{fbd}$) in the tropical East Atlantic, *A. niger* (Black oreo, 1300 m$_{fbd}$) 43°–55°S off New Zealand and Australia and *A. verrucosus* (Warty dory, 1800 m$_{fbd}$), which is

circumglobal (28°N–58°S) with the exception of the North Pacific (Fig. 4.21a). The latter two species are landed as bycatch in commercial fisheries. The genus *Neocyttus* also comprises four species: *N. helgae* (False boarfish, 1210 m$_{fbd}$) in the NE Atlantic, *N. acanthorhynchus* (900 m$_{fbd}$) in W Indian Ocean, *N. psilorhynchus* (Rough oreodory, 1170 m$_{fbd}$) and the largest species, maximum length 40 cm, *N. rhomboidalis* (Spiky oreo, 1240 m$_{fbd}$) is circumglobal in the Southern Hemisphere (24°S–48°S) and of minor commercial interest. The remaining two genera are both monotypic: *Oreosoma atlanticum* (Ox-eyed oreo, 1550 m$_{fbd}$) is a small fish < 20 cm from the South Atlantic, Indian and SW Pacific Oceans whereas *Pseudocyttus maculatus* (Smooth oreo dory, 1500 m$_{fbd}$) grows to 68 cm, is circumglobal (23°S–67°S) and is taken in fisheries.

4.3.12.3 Family Parazenidae (88) (Smooth dories)

This family comprises four species of deep tropical fishes in three genera. *Cyttopsis rosea* (Rosy Dory, 750 m$_{fbd}$; Fig. 4.21a) is the most important, found in warm waters around the world except for the East Pacific and is landed in commercial fisheries. In the NE Atlantic its distribution has extended northward at upper slope depths (average 320 m) from 37°N in the 1960s to 52°N in the 1990s, apparently associated with the warming of waters off Western Europe (Quero et al., 1998). The second species in the genus, *C. cypho*, is confined to the Central West Pacific. *Parazen pacificus* (Parazen, 500 m$_{fbd}$) is a small silvery fish (maximum length 14 cm) found on slopes of the tropical West Atlantic and Indo-west Pacific and *Stethopristes eos* (686 m$_{fbd}$) is confined to the East Pacific.

4.3.12.4 Family Zenionidae (89) (Armoreye dories)

This family comprises seven species of small fishes generally less than 10 cm long, found at upper slope depths down to 700 m. *Capromimus abbreviatus* (Capro dory, 500 m$_{fbd}$) is endemic to New Zealand waters, and two species of *Cyttomimus* are confined to the Pacific Ocean. The genus *Zenion* has four species: *Z. hololepis*

(a)

Cyttidae
(Lookdown dories)

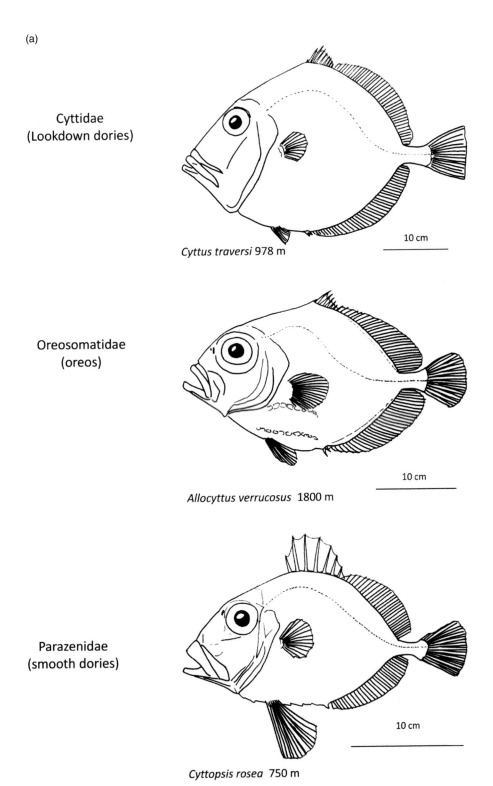

Cyttus traversi 978 m

10 cm

Oreosomatidae
(oreos)

Allocyttus verrucosus 1800 m

10 cm

Parazenidae
(smooth dories)

Cyttopsis rosea 750 m

10 cm

Figure 4.21a ZEIFORMES *Cyttus traversi* (king dory) SE Atlantic and Indo-W Pacific. *Allocyttus verrucosus* (warty dory) circumglobal 28°N–58°S except North Pacific. *Cyttopsis rosea* (rosy dory) Atlantic and Indo-W Pacific.

(Fig. 4.21b) is widespread between 44°N and 28°S in the Atlantic, Pacific and Indian Oceans, *Z. japonicum* is confined to the Pacific Ocean, *Z. leptolepis* to the Indian Ocean and off New Zealand and *Z. longipinnis* is found in the Atlantic and Indian Oceans.

4.3.12.5 Family Grammicolepididae (90) (Tinselfishes)

This family has three species in three different genera. *Grammicolepis brachiusculus* (Thorny tinselfish, 1026 m_{fbd}) is found in scattered locations around oceanic islands, seamounts and continental slopes in the Atlantic Ocean and North Pacific usually at 500–700 m depth (Fig. 4.21b). Barreiros et al. (2011) speculate that islands and seamounts act as stepping-stones that have enabled this species to extend its range in the Atlantic Ocean. *Grammicolepis* is a relatively large fish (up to 64 cm) but is too rare to be targeted by fisheries. *Macrurocyttus acanthopodus* (dwarf dory, 878 m_{fbd}) has been described from the Western Central Pacific. *Xenolepidichthys dalgleishi* (Spotted tinselfish, 885 m_{fbd}) less than 15 cm in length with a silvery body, is circumglobal between latitudes 38°N and 38°S, and benthopelagic on slopes.

4.3.12.6 Family Zeidae (91) (Dories)

These are the largest of the Zeiformes ranging in maximum length from 40 to 90 cm and are not particularly deep-living species. The genus *Zenopis* comprises four species of which *Z. conchifer* (Silvery John Dory, 600 m_{fbd}) is the most widespread, circumglobal between 58°N and 28°S and often captured in fisheries (Fig. 4.21b). There are two Indo-Pacific species, and *Z. oblonga* is endemic to the East Pacific. In the NE Atlantic *Z. conchifer* is captured at 100–500 m depth with a mean depth of occurrence of 280 m on the upper slope. It has extended its range northward over the later decades of the twentieth century, possibly because of ocean warming (Quero et al., 1998). The genus *Zeus* comprises two predominantly shelf-dwelling species including *Z. faber* (John Dory, 400 m_{fbd}).

4.3.13 Order Stylephoriformes

The monotypic family Stylephoridae was formerly considered to be a member of the Lampriformes, alongside the ribbonfishes but recent evidence places *Stylephorus chordatus* in its own order, most closely related to the Gadiformes (Miya et al., 2007; Fig. 4.22).

4.3.13.1 Family Stylephoridae (92) (Tube eyes or Thread tails)

Stylephorus chordatus has a ribbon-like body up to 28 cm long with a thread-like extension to the ventral lobe of the caudal fin. It has forward-looking tubular eyes and a unique protrusible tubular jaw mechanism with a small mouth opening, which enables it to suck in very rapidly a large volume of water at speeds up to 3 $m.s^{-1}$ into the buccal cavity (Pietsch, 1978). It adopts a vertical head-up posture in the water column at depths of 300–600 m at night and 600–800 m during the day, feeding on zooplankton (mainly copepods) using the slurp action of its jaw and buccal apparatus (Fig. 4.22). It is considered a rare species but has been reported occasionally in large numbers in the Florida current near the surface and is found in all the oceans between 45°N and 37°S (Olney, 2005).

4.3.14 Order Gadiformes

Gadiforms are primarily cold-water fish, occurring worldwide from the surface to the deep sea in high latitudes, whereas in tropical waters they are mainly confined to cooler deep layers (Fig. 4.25). They include some of the most important commercially exploited fishes in the world such as cod, haddock and hake. Here we divide the gadiforms into three suborders: Muraenolepidoidei (nine species), Macrouroidei (394 species) and Gadoidei (210 species) according to Rao-Varón and Orti (2009).

Suborder Muraenolepidoidei

This is considered to be the most primitive branch of the Gadiformes comprising just one family with nine species.

(b)

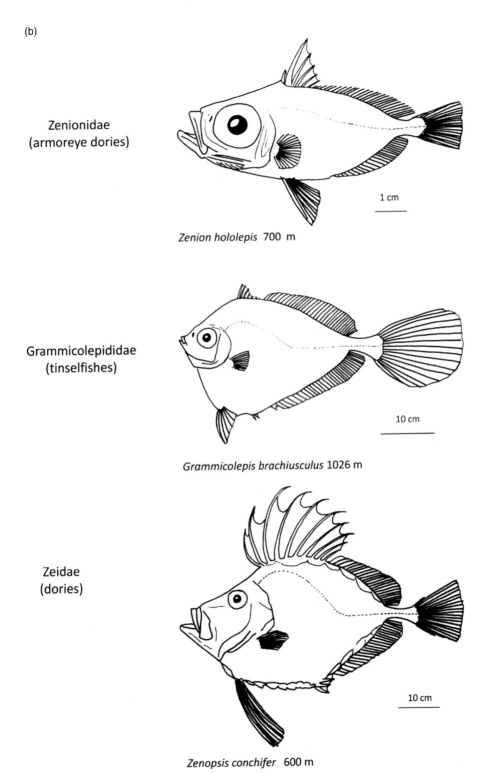

Zenionidae
(armoreye dories)

Zenion hololepis 700 m

1 cm

Grammicolepididae
(tinselfishes)

Grammicolepis brachiusculus 1026 m

10 cm

Zeidae
(dories)

Zenopsis conchifer 600 m

10 cm

Figure 4.21b ZEIFORMES *Zenion hololepis*, Atlantic, W Indian, W and SE Pacific 44°N–28°S, *Grammicolepis brachiusculus* (thorny tinselfish) Atlantic 42°N–37°S and NW Pacific. *Zenopsis conchifer* (silvery John Dory) Atlantic, Mediterranean and Indian.

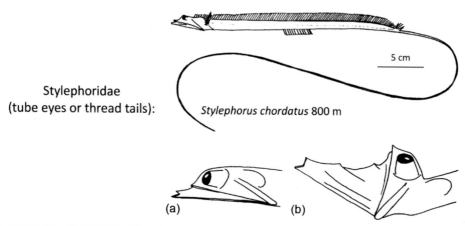

Stylephoridae
(tube eyes or thread tails):

Stylephorus chordatus 800 m

5 cm

(a)   (b)

Figure 4.22 ORDER STYLEPHORIFORMES *Stylephorus chordatus* (tube-eye) probably circumglobal 45°N–37°S. Lower panel, close up of the head region showing (a) jaws in the resting position. (b) jaw mechanism protruded and head tilted back during suction feeding (after Pietsch, 1978).

4.3.14.1 Family Muraenolepididae (93) (Eel cods or Moray cods)

This family with two genera is confined to the cold waters of the Southern Ocean. Until recently it was considered that the genus *Muraenolepis* comprised just three species. *M. microps* (Smalleye Moray cod, 1600 m_{fbd}) is possibly circumglobal from 50°S to 78°S (Fig. 4.25a), *M. marmorata* (Marbled moray cod, 1600 m_{fbd}) described by Günther (1880) in the Challenger report is confined to the slopes of the Crozet, Kerguelen and Heard Islands and *M. orangiensis* (Patagonian moray cod, 860 m_{fbd}) found off Argentina. In a series of reports Balushkin and Prirodina (2010) have described five new species, each of which is confined to a limited range on continental, island, seamount or ridge slopes in the Southern Ocean, and they suggest that these were often previously identified as *M. microps*. *Muraenolepis* are captured by trawls and longlines around the Antarctic Peninsula and in the Scotia Sea with some fishery exploitation around island slopes. Low levels of landings of *Muraenolepis* are reported from mixed fisheries around Bouvet Island in the South Atlantic (Padilla et al., 2015). In its monotypic genus *Notomuraenobathys microcephalus* (Smallhead moray cod, 3040 m_{fbd}) is found mainly in the Atlantic sector of the Southern Ocean including the Scotia Sea and off the Antarctic Peninsula. The Muraenolepididae are

medium-sized fish (mean maximum length 32 cm) that feed on zooplankton, are synchronous spawners and have a demersal or benthopelagic habit.

Suborder Macrouroidei

This is the most speciose of the gadiform suborders comprising three families: Macrouridae (389 species), Macrouronidae (four species) and Steindachneriidae (one species).

4.3.14.2 Family Macrouridae (94) (Grenadiers, Rattails or Whiptails)

The Macrouridae encompasses a diversity of gadiform fishes that are unequivocally residents of the deep sea. Their mean minimum depth of occurrence is 639 m, and they are associated with bathyal and abyssal depths on the slopes of continents, islands, seamounts and mid-ocean ridge systems (Iwamoto, 2008a). They are dominant demersal fish species on abyssal plains with *Coryphaenoides yaquinae* reported down to 7012 m depth in the Mariana Trench (Linley et al., 2016) A few species are found, exceptionally, at shallow continental shelf depths particularly during cold-water periods (Iwamoto, 2008). Although generally caught in bottom trawls, most species have swim bladders, are neutrally buoyant, can be observed

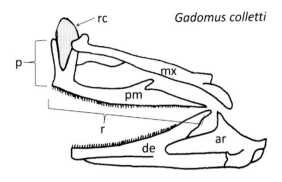

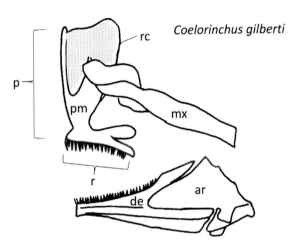

Figure 4.23 ORDER GADIFORMES. Family Macrouridae, jaw protrusion index. Lateral view of the left side of the jaws of two species of Macrourids with contrasting premaxillary protrusion indices, *Gadomus colletti* (28 per cent) and *Coelorinchus gilberti* (150 per cent). Ar – articular, de – dentary, mx – maxillary, pm – premaxillary, rc – rostral cartilage. p – premaxillary pedicel length, r – premaxillary ramus length. Protrusion index % = 100 × p/r. After Okamura (1970).

swimming at some altitude above the seafloor (Priede et al., 1990) and should be regarded as benthopelagic rather than purely demersal. The subfamilies Trachyrincinae and Macrouroidinae, considered by Nelson (2006) to be within the Macrouridae, are here placed within the suborder Gadoidei as proposed by Roa-Varón and Ortí (2009). This leaves 30 genera within the Macrouridae. Further changes in classification are highly likely including revision of

the genera in the light of detailed morphological and molecular studies (Iwamoto, 2008a).

Macrourids are generally medium-sized fish (mean maximum length 40 cm) characterised by a large mouth and head with the body tapering to a thin tail (Plate 6b). Despite the high species diversity, the general body shape is rather uniform; adaptations to different environments involve subtle modifications of the different sense capabilities (Wagner, 2001a), anatomical variations and physiological adaptations to different depths (Morita, 2008). Diversity is further enhanced by the fragmented nature of bathyal habitat resulting in the isolation of populations and evolution of regional species. In a study of the macrourid fishes of Japan, Okamura (1970) compared the jaw morphology of different species in relation to feeding habit. In the macrourids the premaxilla forms the tooth-bearing part of the upper jaw and is supported by the maxillary bone (Fig. 4.23). At the anterior end there is an upward extension, the premaxillary pedicel giving the bone an L-shape appearance from the lateral aspect. The premaxillary pedicel slides into a large cavity in the anterior cranium between the nasals when the mouth is closed. When the mouth opens the premaxilla can be protruded downward by an amount determined by the length of the pedicel. Okamura (1970) proposed that the ratio of the lengths of the two segments of the premaxilla, pedicel height (p) and jaw ramus (r) can be termed the Protrusion Index expressed as a percentage:

Protrusion Index = $100 \times p/r$

Figure 4.24 shows the Protrusion Index for the Japanese macrourids studied by Okamura (1970). There is a general increase in Protrusion Index from less than 30 per cent in the more basal genera, *Gadomus* and *Bathgadus* (subfamily Bathygadinae), to over 100 per cent in the genus *Coelorinchus* (subfamily Macrourinae) Okamura (1970) proposed an evolutionary trend of decrease in mouth size and increase in Protrusion Index correlated with in an increase in the fraction of benthic food items in the stomach contents. Merrett et al. (1983) used the Protrusion Index to predict the likely dietary habit of a

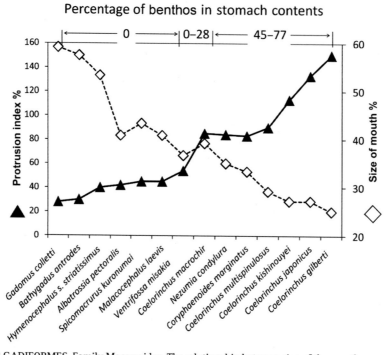

Percentage of benthos in stomach contents

Figure 4.24 ORDER GADIFORMES. Family Macrouridae, The relationship between size of the mouth, upper jaw (premaxillary) protrusion index and stomach contents in Japanese macrourids. (Size of mouth = 100 × upper jaw length/head length). Species with a small mouth and highly protrusible jaws feed predominantly on benthic prey. After Okamura (1970).

rare newly discovered species of Macrourid, *Macrosmia phalacra*.

Here we divide the Macrouridae into two subfamilies: Macrourinae, which contains most of the species, and Bathygadinae, which has fewer species but is a very distinct clade within the Suborder Macrouroidei (Roa-Varón and Orti, 2009).

Subfamily Macrourinae

Comprising over 360 species this subfamily contains most of the diversity of the family Macrouridae. There are 28 genera, several of which have been split into subgenera by different authors, and future revisions are likely (Iwamoto, 2008).

In its monotypic genus, *Albatrossia pectoralis* (Giant grenadier, 3500 m$_{fbd}$) is the largest known marcourid, growing to over 2 m in length and 80 kg weight. It occurs around the slopes of the North Pacific Ocean from the north of Japan in the west to Northern Mexico in the east, including the Okhotsk

and Bering Seas and the Gulf of Alaska, usually between 700 and 1100 m depth. It is one of the most abundant deep-water species in the NW Pacific. Densest aggregations are usually within 5 m of the seafloor. Although considered a candidate for fishery exploitation the water content of the flesh is too high for direct human consumption (Tuponogov et al., 2008). In Alaskan waters it is estimated that the annual bycatch of Giant grenadier in longline fisheries for sable fish (*Anoplopoma fimbria*) and Greenland halibut (*Reinhardtius hippoglossoides*) was 16,000 tonnes per annum during 1997 to 2005, all of which was discarded (Clausen, 2008). Nuclear and mitochondrial DNA studies indicate that *Albatrossia* should probably be classified in the genus *Coryphaenoides* alongside slope-dwelling species such as *C. acrolepis*, which is also found in the North Pacific (Roa-Varón and Ortí, 2009).

The genus *Asthenomacrurus* has two species. *A. victoris* (Victory whiptail, 3530 m$_{fbd}$) is known from less than ten specimens captured worldwide

(a)

Muraenolepididae
(eel cods or moray cods)

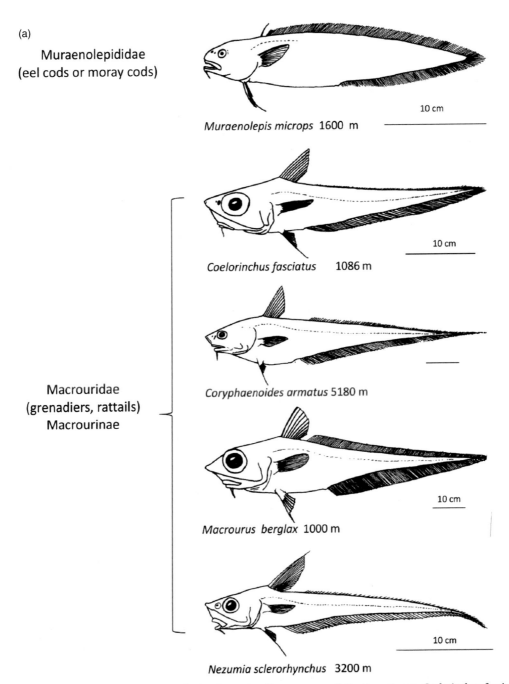

Muraenolepis microps 1600 m

10 cm

Coelorinchus fasciatus 1086 m

10 cm

Coryphaenoides armatus 5180 m

Macrouridae
(grenadiers, rattails)
Macrourinae

Macrourus berglax 1000 m

10 cm

Nezumia sclerorhynchus 3200 m

10 cm

Figure 4.25a ORDER GADIFORMES. *Muraenolepis microps* (smalleye moray cod) Southern Ocean. *Coelorinchus fasciatus* (Banded whiptail) Southern Hemisphere 20°S–59°S. C*oryphaenoides armatus* (Abyssal grenadier) circumglobal 65°N–61°S. *Macrourus berglax* (roughhead grenadier) N Atlantic 82°N–37°N. *Nezumia sclerorhynchus* (Roughtip grenadier) Atlantic and Mediterranean 43°N–4°N.

from the Indo-West Pacific, the Indian Ocean and the North Atlantic Mid-Ocean Ridge at lower bathyal and upper abyssal depths (Iwamoto and Orlov, 2008). Its congener *A. fragilis* (fragile grenadier, 3336 m_{fbd}) is recorded at similar depths in the Eastern Pacific Ocean. They are small fish less than 25 cm long, and their rarity is probably related to difficulties experienced in sampling in rough terrain at upper abyssal depths. The genus *Cetonurus* has two species, both found in the Pacific and Atlantic Oceans: *C. crassiceps* (Globosehead rattail, 1490 m_{fbd}) at upper bathyal depths and *C. globiceps* (Globehead grenadier, 4621 m_{fbd}) at abyssal depths. *C. globiceps* grows up to 50 cm long and feeds on small fishes and crustacea. The monotypic *Cetonurichthys subinflatus* (Smallpore whiptail, 1700 m_{fbd}) is found in the Indian Ocean on the slopes off Western Australia (Williams et al., 1996).

Coelorinchus is the most specious genus of macrourids with 122 species at the time of writing and more yet to be described. Spelling is variable in the literature, alternatives being *Caelorinchus* and *Coelorhynchus;* here we adopt the recommendation of Iwamoto (2008b). *Coelorinchus* species are grenadiers of the upper slope, mean maximum depth of occurrence is 774 m, and the deepest species are *C. labiatus* (Spearsnouted grenadier) and *C. occa* (Swordsnout grenadier), both recorded from 2220 m_{fbd} in the North Atlantic Ocean. Mean minimum depth of occurrence is 430 m but a few species have been recorded at shelf depths less than 200 m, *C. denticulatus* (Filesnout grenadier, Indian Ocean), *C. fasciatus* (Banded whiptail, Southern Ocean), *C. kermadecus* (Kermadec rattail, Southwest Pacific) and *C. oliverianus* (Hawknose grenadier, New Zealand waters). Many species are rare, known only from a few specimens. Three species are circumpolar in the Southern Hemisphere: *C. acanthiger* (Surgeon grenadier, 1500 m_{fbd}) between 6°S–44°S in the Atlantic, Indian and Pacific Oceans including off Australia and New Zealand, *C. fasciatus* (Banded whiptail, 1086 m_{fbd}) between 20°S–59°S (Fig. 4.25a), and *C. kaiyomaru* (Campbell whiptail, 1150 m_{fbd}) between 40°S–56°S. All are taken as bycatch in regional fisheries. *C. fasciatus* (banded whiptail, 1100

m_{fbd}) is the most abundant small grenadier in fishery samples around the Falkland/Malvinas Islands in the SW Atlantic, mainly captured between 200 and 600 m, showing a pattern of seasonal spawning in springtime and maximum age 19 years (Laptikhovsky et al., 2008). 82 species of *Coelorinchus* are endemic to the Pacific Ocean, and local diversity can be very high; 21 species have been recorded off Taiwan at depths from 100 to 1211 m (Shao et al., 2008), McMillan and Paulin (1993) described nine new species from New Zealand waters and nine species are known off Chile including *C. chilensis* (1480 m_{fbd}) and *C. fasciatus,* recognised as bycatch in commercial fisheries (Pequeño, 2008). This diversity, however, does not extend into the NE Pacific with only one species, *C. scaphopsis,* known in waters off California (Ambrose, 1996).

12 species are endemic to the Indian Ocean, including *C. amydrozosterus* (Faintbanded whiptail, 717 m_{fbd}), *C. charius* (Graceful whiptail, 981 m_{fbd}) and *C. hoangi* (700 m_{fbd}) on the slopes of Western Australia. Off the slopes of East Africa the dominant species are *C. braueri* (Shovelnose grenadier, 1200 m_{fbd}) *C. denticulatus* (Filesnout grenadier, 333 m_{fbd}) and *C. trunovi* (552 m_{fbd}). The Atlantic Ocean also has 12 endemic species including *C. caelorhincus* (Hollowsnout grenadier, 1250 m_{fbd}), which was first described from the Mediterranean by Risso (1810) and is now known to extend around the margins of the North Atlantic Ocean from NW Canada to Cape Verde off West Africa and on the Mid-Atlantic Ridge (Bergstad et al., 2008). *C. mediterraneus* (1200 m_{fbd}) is endemic to the Western and Central basins of the Mediterranean Sea (Iwamoto and Ungaro, 2002), which together with *C. caelorhincus* means there are only two species of *Coelorinchus* in the Mediterranean Sea. In general, species numbers in the North Atlantic are lower than in the Pacific Ocean with two species observed in the Porcupine Seabight area, *C. caelorhincus* and *C. labiatus* (Priede et al., 2010) and two off NW Africa, *C. caelorhincus* and *C. occa* (Merrett and Marshall, 1981). Eight species span the Indo-Pacific divide but are restricted to small ranges such as *C. australis* (Javelin, 300 m_{fbd}) and *C. mirus* (Gargoyle fish, 400 m_{fbd}), which are endemic to

Australian waters. In total, Australia has 38 species of *Coelorinchus* in its faunal records. Two species occur in both the South Atlantic and the Indian Oceans, *C. vityazae* (920 m$_{fbd}$) and *C. karrerae* (Karrer's whiptail, 1150 m$_{fbd}$), and two species in the Atlantic and Pacific Oceans, *C. matamua* (Mahia whiptail, 1000 m$_{fbd}$) and *C. aconcagua* (Aconcagua grenadier, 450 m$_{fbd}$). In their description of macrourid fishes of the Philippine Islands and adjacent areas to the south, Gilbert and Hubbs (1928) point out that this region appears to be the centre of distribution of the genus *Coelorinchus* containing 45 per cent of the 51 species known in the world at the time. Despite the discovery of numerous new species, it remains true that the West Pacific and seas between Australia and SE Asia have the highest diversity of *Coelorinchus* species. Most species of *Coelorinchus* have a ventral abdominal light organ between the pelvic fin bases comprising sacs filled with luminescent bacteria (Dunlap et al., 2014).

Coryphaenoides, the deepest-living genus of macrourids, occurs at a mean minimum depth of 1103 m and a mean maximum depth in 2380 m. They are primarily demersal fishes of the lower bathyal and abyssal regions of the world's oceans. *Coryphaenoides armatus* (Abyssal grenadier, 5180 m$_{fbd}$) occurs between 65°N and 61°S throughout the world's oceans except for the Arctic Ocean, the Southern Ocean and the Indian Ocean north of 45°S; it (Fig. 1.52, Plate 6b) is the dominant fish species appearing at artificial food falls deployed at abyssal depths (King and Priede, 2008). It has a ventral light organ between the anal-fin origin and the pelvic-fin bases. At depths greater than 4700 m in the Pacific Ocean, *C. armatus* is replaced by *C. yaquinae* down to over 7000 m depth in the upper parts of hadal trenches. The two species overlap in their distribution, are morphologically very similar but in underwater images can be distinguished from one another by the greater silvery reflectance of the skin of *C. armatus* (Jamieson et al., 2012). Wilson and Waples (1984) recognised a subspecies, *C. armatus variabilis* found in the NE Pacific. Morita (1999) showed that *C. armatus* and *C. yaquinae* occupy a basal position in the *Coryphaenoides* phylogenetic tree and diverged from the shallower nonabyssal species 5.2 million years ago.

In the Southern Ocean *C. ferrieri* (3931 m$_{fbd}$) and *C. lecointei* (3931 m$_{fbd}$) 47°S–73°S are more or less endemic species at depths greater than 2000 m. *C. filicauda* (5000 m$_{fbd}$) is found between 34°S and 60°S in the Southern Ocean but extends into the SW Atlantic, the Southern Indian Ocean and the SW Pacific off Australia and New Zealand. *C. grahami* (Graham's whiptail, 1280 m$_{fbd}$) and *C. mcmillani* (McMillan's whiptail, 1400 m$_{fbd}$) are shallower-living species found off South Africa, Australia and the SW Pacific Ocean. *C. rudis* (Bighead grenadier, Rudis rattail, 2380 m$_{fbd}$) is circumglobal at tropical and subtropical latitudes (35°N–8°N), bathypelagic, can grow to over 1 m total length and the largest individuals feed on cephalopods. The remaining species in the genus have more restricted ranges; 34 species are endemic to the Pacific Ocean, 23 to the Atlantic Ocean and five to the Indian Ocean. Three species can be considered Indo-Pacific in their distribution: *C. murrayi* (2350 m$_{fbd}$), *C. asprellus* (2290 m$_{fbd}$) and *C. serrulatus* (2070 m$_{fbd}$), which is best known off South Australia and New Zealand, where it is trawled as a bycatch. *C. carapinus* (Carapine grenadier, 5610 m$_{fbd}$) occurs widely in the Atlantic Ocean and also in the Southern Indian Ocean. *Coryphaenoides leptolepis* (Ghostly grenadier, 4000 m$_{fbd}$) is found in the North Atlantic and the NE Pacific.

Many of the larger shallower-living *Coryphaenoides* probably appear as bycatch in longline and trawl fisheries but may not be correctly identified. A few species, however, are targeted by fisheries, notably *C. rupestris* (roundnose grenadier, 2600 m$_{fbd}$) in the North Atlantic Ocean, *C. acrolepis* (Pacific grenadier, 3700 m$_{fbd}$) and *C. cinerus* (Popeye grenadier, 2600 m$_{fbd}$) in the North Pacific Ocean. *C. rupestris* was exploited most heavily on both sides of the North Atlantic where landings peaked at ca. 80,000 t in 1971; as a result, by 2000 the stock was depleted and considered to be endangered (Devine et al., 2006). *C. rupestris* has large eyes, is a benthopelagic visual feeder and consequently is not attracted to baits on longlines or traps, and therefore can only be captured by trawl.

The monotypic *Cynomacrurus piriei* (dogtooth grenadier, 3800 m$_{fbd}$) is endemic to the Southern

Ocean, circumpolar (36°S–79°S), occurring usually between 1000 and 2000 m depth. The genus *Echinomacrurus* has two deep abyssal species *E. mollis* (6000 m$_{fbd}$) from the NE Atlantic and *E. occidentalis* (4337 m$_{fbd}$) from Eastern Pacific Ocean. The peculiar monotypic species *Haplomacrourus nudirostris* (Naked snout rattail, 1590 m$_{fbd}$) is found in the Southern Atlantic, Indian and Pacific Oceans associated with continental slopes, for example, off Australia. The genus *Hymenocephalus* comprises 23 small (mean maximum length 16 cm) fragile species found at upper bathyal depths, typically 400–850 m, in tropical waters. The majority (18 species) are endemic to different areas of the Pacific Ocean, and two species are endemic to the Indo-West Pacific region north of Australia. *H. gracilis* (Graceful grenadier, 345 m$_{fbd}$) is circumglobal occurring on both sides of the North Atlantic including the Mediterranean Sea, the Indian and Pacific Oceans and feeds on planktonic crustacea. *H. italicus* (Glasshead grenadier, 1400 m$_{fbd}$) also occurs in the Mediterranean Sea, the Atlantic and Indian Oceans but is absent from the Pacific Ocean. *H. billsam* (900 m$_{fbd}$) is endemic to the Western Central Atlantic Ocean. The genus *Kumba* comprises eight species of rare bathypelagic grenadiers, mean maximum depth 1115 m, with large heads, the snout free of scales and a ventral light organ with a single gland and lens anterior to the rectum. *Kumba* is found in widely scattered localities around the world (Iwamoto and Sazonov, 1994): five species from the Pacific Ocean, one from the Indian Ocean and two from the Atlantic. They are probably related to the genus *Nezumia*. The genus *Kuronezumia* comprises six species. *K. bubonis* (Bulbous rattail, 1300 m$_{fbd}$) is most widespread, found in the Atlantic, Indian and Pacific Oceans, including on the seamounts of New Zealand (Tracey et al., 2004). Three species are endemic to the Pacific Ocean, one in the Indian Ocean and *K. leonis* (Snubnose whiptail, 850 m$_{fbd}$) on the upper slopes of the South Atlantic. The monotypic *Lepidorhynchus denticulatus* (Thorntooth grenadier, toothed whiptail, Javelin fish, 1000 m$_{fbd}$) is the most abundant macrourid on upper slopes off SE Australia (Bray, 2011) and occurs on 40 per cent of seamount

complexes in New Zealand waters (Tracey et al., 2004). It is landed as bycatch with some being utilised as bait. The genus *Lucigadus* is mainly found in the Pacific Ocean, where five species are endemic, but *L. nigromaculatus* (Blackspotted grenadier, 1463 m$_{fbd}$) extends into the Indian Ocean off Southern Australia, and *L. ori* (Bronze whiptail, 1153 m$_{fbd}$) is found in the South Atlantic and SW Indian Oceans. The monotypic *Macrosmia phalacra* (1660 m$_{fbd}$) may be widespread in tropical seas but is known only from two specimens, one from the Eastern Indian Ocean Ninety East Ridge and a second from the Eastern North Atlantic off Morocco (Merrett et al., 1983).

The five species of *Macrourus* are all cold-water fishes of polar and subpolar regions. *M. berglax* (Roughhead grenadier, 1000 m$_{fbd}$), the only northern species (Fig. 4.25a), is endemic to the slopes and ridges of the North Atlantic basin from the Norfolk Canyon off Chesapeake Bay in the West, to the Irish Atlantic slope in the East, including the waters of Labrador, Davis Strait, Greenland, Iceland, Norway, Spitzbergen, and the Barents Sea. *M. berglax* is found at temperatures from below 0C (Bullough et al., 1998) to 5°C. *M. berglax* landings from commercial fisheries in the NW Atlantic peaked at 7,700 t in 1998, taken mainly as a bycatch in the fishery for Greenland halibut, *Reinhardtius hippoglossoides* (González-Costas and Muria, 2008). They have also been reported as a minor bycatch in deep-sea fisheries of the Barents Sea (Dolgov et al., 2008) and NE Atlantic (Lorance et al., 2008). In the Southern Hemisphere *M. whitsoni* (Whitson's grenadier, 3185 m$_{fbd}$) is circumpolar, generally south of the Antarctic Convergence zone, where *M. caml* (Caml grenadier, 2080 m$_{fbd}$) is also widespread, including off South Georgia. *M. carinatus* (Ridge scaled rattail, 1200 m$_{fbd}$) occurs further north in sub-antarctic to temperate waters off South America, South Africa, Tasmania and New Zealand and *M. holotrachys* (Bigeye grenadier, 1400 m$_{fbd}$) is known from the Southwest Atlantic Patagonian slope, off South Georgia and the Ross Sea. All these species occur as bycatch in fisheries for toothfish (*Dissostichus* spp.) in the Southern Ocean (Macmillan et al., 2012) and there is a targeted fishery for *M. carinatus* in waters off

Argentina reaching over 12000 t per annum in 2008 (Laptikhovsky, 2010). According to an analysis of nuclear and mitochondrial DNA, *Macrourus* spp. can be considered as polar seas representatives of the genus *Coelorinchus* (Roa-Varón and Ortí, 2009).

The genus *Malacephalus* has seven species, one of which, *M. laevis* (Softhead grenadier, 1000 m_{fbd}), occurs in all three major oceans 62°N–43°S, two species are endemic to the Atlantic Ocean and four to the Pacific Ocean. They are demersal fishes of the upper slopes and have well-developed ventral light organs with two windows (Iwamoto and Arai, 1987). They are taken as bycatch in fisheries but are nowhere sufficiently abundant to be targeted. In the French fishery for *Coryphaenoides rupestris* in the NE Atlantic, the *M. laevis* bycatch amounts to only 0.015 per cent of the weight of target species caught weight and are all discarded (Lorance et al., 2008). Longline fishermen in the NW Atlantic are reputed to have used the luminescent tissues of *M. laevis* to enhance the effectiveness of baits used for cod.

The genus *Mataeocephalus* comprises five species of relatively small pointed-snout grenadiers of the tropical Indo-Pacific region mainly at depths 800–1350 m. *M. acipenserinus* (Sturgeon grenadier, 1300 m_{fbd}) ranges from the Indian Ocean to Hawaii and Sala y Gómez in the Pacific Ocean, whereas *M. tenuicauda* (Slendertail grenadier, 1600 m_{fbd}) is endemic to the East Pacific off Panama, Columbia and Ecuador. These species are locally abundant off Hawaii and Ecuador, respectively, but are too small and too deep to be of interest to fisheries. The other species are all endemic to the West Pacific.

The genus *Mesobius* comprises two species of bathypelagic grenadiers generally caught in mid-water nets. *M. antipodum* (Bathypelagic rattail, 1300 m_{fbd}) is found from the SE Atlantic to Australia and New Zealand, whereas *M. berryi* (Berry's grenadier, 2700 m_{fbd}) is confined to the East Pacific Ocean. The skeleton is poorly ossified, and the single ventral light organ is well developed (Hubbs and Iwamoto, 1977).

The genus *Nezumia*, first described by Jordan and Starks (1904) in their account of Macrouridae caught off the coast of Japan by the expedition of the steamer *Albatross*, was given the Japanese name for a rat or mouse. Recent nuclear and mitochondrial DNA analysis indicates this to be the most advanced recently derived genus of Macrourids (Roa-Varón and Ortí, 2009). *Nezumia* (Fig. 4.25a) currently comprises 51 species distributed worldwide at bathyal depths of 580–1100 m. All have ventral light organs. On the basis of depth of occurrence and abundance Merrett and Haedrich (1997) speculated that the genus *Nezumia* might provide likely candidates for exploitation by deep-water fisheries, but although widely taken as bycatch, its small size (<32 cm total length) and slender body form mean that they are of little commercial value and thus are usually discarded (Lorance et al., 2008). The genus is made up of a diversity of species with localised distributions. There are 12 species endemic to the Atlantic Ocean, 10 in the Indian Ocean and 27 species endemic to different parts of the Pacific Ocean. Two species are considered Indo-Pacific in their distribution. Two species occur in the Mediterranean Sea: *N. aequalis* (Common Atlantic grenadier, 2320 m_{fbd}) and *N. sclerorhynchus* (Roughtip grenadier, 3200 m_{fbd}; Fig. 4.25a), where they appear to spawn asynchronously throughout the year (D'Onghia et al., 2008).

Odontomacrurus murrayi (Roundhead grenadier, 2500 m_{fbd}) is a monotypic bathypelagic macrourid that grows to over 60 cm in length, including the long whip-like tail. It feeds on fish and is generally dark in coloration with a ventral light organ and occurs in the Atlantic, Indian and Pacific Oceans.

The monotypic *Pseudocetonurus septifer* (950 m_{fbd}) occurs across the Pacific Ocean from Taiwan to Hawaii and the SE Pacific. *Paracetonurus flagellicauda* (4500 m_{fbd}) is a rather rare deep-living fish known from the Northeast Atlantic and a single specimen from the Western Indian Ocean. Unusually abundant captures of *P. flagellicauda* on the Mid-Atlantic Ridge in the Azores area suggest that they may migrate along the oceanic ridge system between widely spaced localities where they have been recorded (Iwamoto and Orlov, 2006). The genus *Pseudonezumia* comprises four species from the Western Pacific Ocean where *P. parvipes* (2308 m_{fbd}) is the most widespread; the distribution of *P. pusilla* (Tiny whiptail, 2000 m_{fbd}) also extends into the Indian Ocean on the slopes off

Western Australia (Williams et al., 1996). There are six species in the genus *Sphagemacrurus*, which are small (generally <20 cm total length) fishes living at 800–1500 m depth in the tropics: *S. decimalis* is endemic to the W. Pacific, *S. gibber* in the E. Pacific, *S. grenadae* in the W Atlantic, *S. hirundo* in the Mid and E Atlantic and two species that are Indo-Pacific in their range. The genus *Spicomacrurus* comprises four rare species of small grenadiers from the Western Pacific Ocean at around 600 m depth from Southern Japan to the north of New Zealand. *Trachonurus* comprises five species with characteristic bristly or spiny scales giving rise to common names such as furry, velvet and shaggy. *T. villosus* (Furry whiptail, 1590 m$_{fbd}$) is worldwide in its distribution mainly in warmer waters; three species are endemic to the Pacific Ocean, and *T. yiwardaus* (Yiwarda whiptail, 1293 m$_{fbd}$) has been recorded from the Indian Ocean slopes of West Australia.

The genus *Ventrifossa* has 24 species with a generally silvery appearance and a ventral light organ occurring predominantly in tropical waters between 400 and 800 m depth on the upper slopes. 16 species are endemic to different parts of the Pacific Ocean, three species to the Indian Ocean and two species to the tropical Western Atlantic Ocean. Three species are Indo-Pacific in their distribution, from the east coast of Africa to the West Pacific. *V. occidentalis* recorded off West Africa by Merrett and Marshall (1981) is now classified in the genus *Malacocephalus* so the genus *Ventrifossa* can be considered to be absent from the Eastern Atlantic and the Mediterranean Sea.

Subfamily Bathygadinae

The Bathygadinae comprises 26 species in two genera, which are distributed worldwide in tropical and subtropical waters mainly within the depth range 650–1350 m. In the genus *Bathygadus* four species are endemic to the Atlantic Ocean where *B. melanobranchus* (Vaillant's grenadier, 2600 m$_{fbd}$) occurs most widely between 53°N and 34°S, feeding on mysids, pelagic copepods, chaetognaths and small shrimps (Fig. 4.25b). The Indian Ocean has just one species, *B. furvescens* (Blackfin rat tail, 1295 m$_{fbd}$), and there are eight species in the Pacific Ocean. The

genus *Gadomus* has three species endemic to the Atlantic Ocean, seven species in the Pacific Ocean and one species endemic to the Indian Ocean. *G. capensis* (1480 m$_{fbd}$) occurs around the Cape in the SE Atlantic and SW Indian Oceans, and *G. pepperi* (Blacktongue rat tail, 1500 m$_{fbd}$) is found around Australia.

4.3.14.3 Family Macruronidae (95) (Southern hakes)

This family comprises four species in one genus: *Macruronus*, which were formerly considered to be hakes within the family Merlucciidae (Nelson, 2006) but DNA analysis supports their placement as a separate family within the suborder Macrouroidei, most closely related to the basal macrourids (Roa-Varón and Orti, 2009). *M. maderensis* is known only from a few specimens collected in the NE Atlantic off the island of Madeira. The other three species all occur in the Southern Hemisphere: *M. capensis* (Cape grenadier, 658 m$_{fbd}$) off the Cape of Good Hope, South Africa, *M. magellanicus* (Patagonian grenadier, 500 m$_{fbd}$) in the SE Pacific and SW Atlantic off South America and *M. novaezelandiae* (blue grenadier, hoki, 1000 m$_{fbd}$) off New Zealand and southern Australia (Fig. 4.25b). *M. capensis* is not abundant and is of no interest to fisheries. However, *M. magellanicus* and *M. novaezelandiae* support major fisheries; they are robust fish that grow to over 1 m in length, and the meat produces a white fillet for which there is considerable demand. They are often considered as two subspecies, *Macruronus novaezelandiae magellanicus* and *Macruronus novaezelandiae novaezelandiae*, respectively. Off New Zealand *M. novaezelandiae*, where it is known as Hoki, is the main fishery. Together with landings from Australian fishing grounds the total catch of *M. novaezelandiae* reached 300,000 tonnes in 1999 (Lloris et al., 2005). In the SE Pacific off Chile the annual catch of *M. magellanicus* peaked at over 350,000 tonnes, and in the SW Atlantic over 100,000 tonnes were taken in 1998 mainly by Argentinean vessels. Blue grenadiers are captured by bottom trawl predominantly at depths of 200–800 m. Juveniles may occur inshore, but adults are offshore in water depths over 400 m, where

(b)

Macrouridae
(grenadiers, rattails)
Bathygadinae

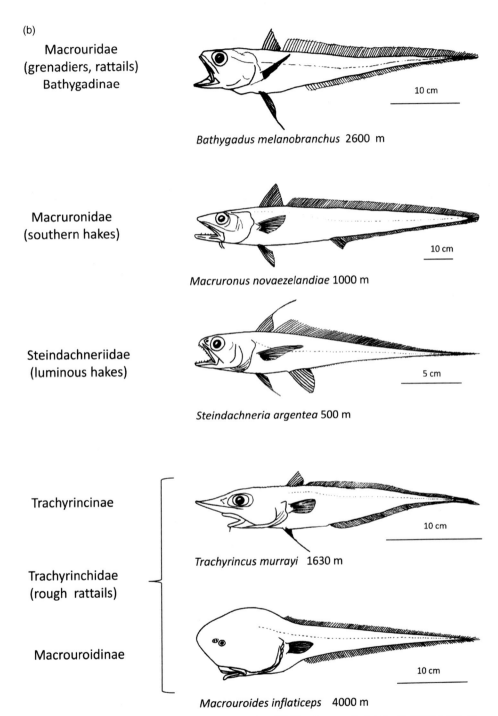

Bathygadus melanobranchus 2600 m

Macruronidae
(southern hakes)

Macruronus novaezelandiae 1000 m

Steindachneriidae
(luminous hakes)

Steindachneria argentea 500 m

Trachyrincinae

Trachyrinchidae
(rough rattails)

Macrouroidinae

Trachyrincus murrayi 1630 m

Macrouroides inflaticeps 4000 m

Figure 4.25b ORDER GADIFORMES. *Bathygadus melanobranchus* (Vaillant's grenadier) Atlantic 53°N–34°S. *Macruronus novaezelandiae* (Blue grenadier, hoki) SW Pacific off New Zealand and S Australia. *Steindachneria argentea* (Luminous hake) W Atlantic 31°N–8°N. *Trachyrincus murrayi* (roughnose grenadier) N Atlantic. *Macrouroides inflaticeps* (inflated whiptail) circumglobal.

they feed benthopelagically on small fishes, including myctophids, crustacea and squid.

4.3.14.4 Family Steindachneriidae (96) (Luminous hakes)

This family comprises just one species *Steindachneria argentea* (Luminous hake, 500 m$_{fbd}$) with unique morphological features including separate anal and urogenital pores (Fig. 4.25b). It inhabits the outer shelf and slopes (400–500 m depth) of the Western Atlantic Ocean from Florida, around the Gulf of Mexico to Venezuela, where it has been recorded descending during the day to the top of the oxygen minimum layer at 300–440 m depth over the Cariaco Trench (Love et al., 2004). It has a silvery appearance with a light organ on the ventral half of the body and sides of the head. Maximum total length is 30 cm.

Suborder Gadoidei

The Gadoidei comprises eight families, Trachyrinchidae, Moridae, Merluciidae, Melanonidae, Euclichthyidae, Gadidae, Ranicipitidae and Bregmacerotidae.

4.3.14.5 Family Trachyrinchidae (97) (Roughnose grenadiers)

These were previously considered as subfamilies within the Macrouridae (Nelson, 2006), but following Roa-Varón and Orti (2009) based on DNA evidence here we place them within the Gadoidei as a separate family. There are nine species in two subfamilies.

Subfamily Trachyrincinae

The Trachyrincinae comprises seven species found in temperate waters of the Northern and Southern Hemispheres, all characterised by a pointed rough snout and ventral mouth. They feed on small crustacea including copepods and occupy slope depths typically 450–1400 m. The monotypic *Idiolophorhynchus andriashevi* lacks a chin barbel and is found off West and South Australia and also off New Zealand. The six species in the genus

Trachyrincus all have a chin barbel. There are two species in the SE Pacific, *T. villegai* (Grey grenadier, 980 m$_{fbd}$) and *T. helolepis* (Armorhead grenadier, 960 m$_{fbd}$) and two species in the SW Pacific *T. aphyodes* (800 m$_{fbd}$) and *T. longirostris* (Slender unicorn rattail, 1280 m$_{fbd}$). *T. scabrus* (Roughsnout grenadier, 1700 m$_{fbd}$) is confined to the NE Atlantic Ocean and the Mediterranean Sea and is taken as bycatch in deep-water fisheries throughout the area. *T. murrayi* (Roughnose grenadier, 1630 m$_{fbd}$) has a bipolar distribution and is found around the Northern Atlantic basin and in the SW Pacific off New Zealand (Fig. 4.25b).

Subfamily Macrouroidinae

There are two species that are both bathypelagic fishes found more or less circumglobally in tropical waters. They have a large spherical head, ventral mouth, small eyes and a tapering tail with no caudal fin. There is no light organ, and the swim bladder is reduced. They are placed in two monotypic genera. *Macrouroides inflaticeps* (Inflated whiptail, 4000 m$_{fbd}$) has no pelvic fins, is circumglobal in the Atlantic, Indian and Pacific Oceans and grows to 48 cm maximum length (Fig. 4.25b). *Squalogadus modificatus* (Tadpole whiptail, 1740 m$_{fbd}$) has pelvic fins, ranges from Japan to New Zealand in the West Pacific, also occurs in the Atlantic Ocean, but is possibly absent from the Indian Ocean. It grows to 35 cm. Both species are associated with continental slopes. There is no fishery significance.

4.3.14.6 Family Moridae (98) (Deep-sea cods)

The Moridae comprises 110 species in 18 genera including a number of species that are very well known in deep-water catches and video surveys. They generally occur on the upper slopes of continents, islands, ridges and seamount down to a maximum depth of 3000 m.

The genus *Antimora* has two species. *A. rostrata* (Blue antimora, Blue hake, 3000 m$_{fbd}$) is found throughout the world's oceans between 62°N and 62°S (Fig. 4.25c), except for the North Pacific Ocean,

(c)

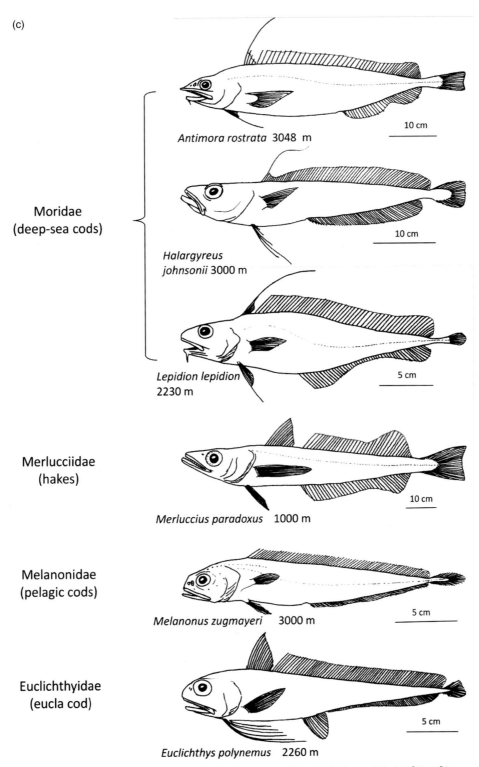

Moridae
(deep-sea cods)

Antimora rostrata 3048 m

10 cm

Halargyreus johnsonii 3000 m

10 cm

Lepidion lepidion 2230 m

5 cm

Merlucciidae
(hakes)

Merluccius paradoxus 1000 m

10 cm

Melanonidae
(pelagic cods)

Melanonus zugmayeri 3000 m

5 cm

Euclichthyidae
(eucla cod)

Euclichthys polynemus 2260 m

5 cm

Figure 4.25c ORDER GADIFORMES. *Antimora rostrata* (blue antimora, blue hake) circumglobal 62°N–62°S except the N Pacific. *Halargyreus johnsonii* (slender codling) Subpolar and temperate waters of the Northern and Southern Hemispheres. *Lepidion lepidion* (Mediterranean codling) NW Mediterranean. *Merluccius paradoxus* (deep-water Cape hake) off South Africa. *Melanonus zugmayeri* (Arrowtail) almost circumglobal 60°N–49°S. *Euclichthys polynemus* (Eucla cod) SW Pacific: off New Zealand and Australia.

where it is replaced by *A. microlepis*. They have a two-chambered swim bladder, which enables them to hover over the seafloor and probe the substrate for prey using long tactile pelvic fin rays and the barbel (Iwamoto, 1975). Both species form aggregations and are a frequent bycatch in trawl and longline fisheries but generally are not targeted commercially. *A. rostrata*, the dominant fish species, is attracted to baits at depths of 1500–2500 m in the NE Atlantic Ocean (Priede et al., 1994) with a general trend of larger individuals occurring deeper. On the Mid-Atlantic Ridge *A. rostrata* are captured between 1200 and 2700 m depth both in trawls and longlines. They reach a maximum age of 25 years, maximum length of 75 cm and circumstantial evidence indicates possible ocean basin scale migrations between spawning, nursery and adult feeding areas. There is sexual and size segregation with females dominant in most catches (Fossen and Bergstad, 2006).

The monotypic *Eretmophorus kleinenbergi* has been regarded as a small epipelagic morid endemic to the NW Mediterranean Sea. However, bottom-trawl capture of a 101 mm long specimen at 950 m depth suggests the pelagic stages may be juveniles; also the range possibly extends into the Atlantic Ocean (Lloris et al., 1994).

Gadella comprises 13 species of medium-size (mean maximum length 25 cm) morids found on upper slopes generally in tropical and subtropical waters. They have a small ventral light organ on the belly anterior to the anus. There are seven species in the Pacific Ocean, three in the Indian Ocean, two in the Atlantic Ocean and one, *G. norops* (750 m_{fbd}) around Australia and New Zealand. *G. imberbis* (Beardless codling, 800 m_{fbd}) is reported as bycatch in some deep-water shrimp fisheries in the Atlantic Ocean. *Guttigadus* comprises eight species generally less than 15 cm in length occurring on slopes mainly at 600–1200 m depth in the Southern Hemisphere. *G. latifrons* (1875 m_{fbd}) is exceptional for the genus in that it occurs in the NE Atlantic and the Mediterranean Sea as well as in the SW Atlantic and Indian Oceans. *G. globosus* (Tadpole cod, 1600 m_{fbd}) is confined to waters around New Zealand and the South Atlantic islands reflecting the origins of the genus in the Southern Ocean north of the Antarctic Convergence (Meléndez and Markle, 1997).

The monotypic *Halargyreus johnsonii* (Slender codling, 3000 m_{fbd}) is a an elongated silvery fish that grows up to 56 cm in length (Fig. 4.25c) and is found over a wide latitudinal range (63°N–60°S) from the sub-Arctic to the slopes of Antarctica in the Atlantic and Pacific Oceans. In the NE Atlantic off the west coast of Scotland, Mauchline and Gordon (1984) captured *H. johnsonii* in bottom trawls principally between 750 m and 1000 m depth; the stomach contents consisted almost entirely of crustacea with evidence of change in diet during growth. *Laemonema*, the sister genus of *Guttigadus*, are larger fishes, commonly up to 30 cm length with a more northerly circumglobal distribution (60°N–40°S), though absent from the NE Pacific. There are nine species in the Pacific Ocean, one in the Indian Ocean and seven North Atlantic species. Closely related species pairs are distributed allopatrically, so *L. barbatulum* (Shortbeard codling, 1620 m_{fbd}) occurs on the slopes of the western Atlantic and *L. laureysi* (Guinean codling, 690 m_{fbd}) in the eastern Atlantic Ocean. *L. melanurum* (Blackfin codling, 644 m_{fbd}) is commonly taken as bycatch in NW Atlantic deep-water shrimp trawls but is not utilised (Meléndez and Markle, 1997). The genus *Lepidion* comprises nine species that occur mainly between 500 and 1300 m on slopes; they are benthopelagic in their habits and grow to 40–85 cm length depending on species. None are commercially exploited, although they are conspicuous members of the demersal fish assemblage where they occur. *L. schmidti* (Schmidt's cod, 1520 m_{fbd}) is the most widespread, recorded from Australia, New Zealand, the Northwest Pacific off Japan, the Southern Indian and North Atlantic Oceans. Two species are endemic to the Pacific Ocean, two in the North Atlantic and two species are recorded around South Africa. *L. lepidion* (Mediterranean codling, 2230 m_{fbd}) is endemic to the Mediterranean Sea (Fig. 4.25c). *L. ensiferus* (Patagonian codling, 1000 m_{fbd}) is confined to 39°S–55°S in the South Atlantic and the sub-Antarctic islands of the Indian Ocean. *L. eques* (North Atlantic codling, 1880 m_{fbd}) is abundant on the slopes of the NE Atlantic (Mauchline and Gordon, 1984a), where it occurs together with *Halargyreus johnsonii,* but the diets of the two species differ. The monotypic *Mora moro* (Common mora, 2500 m$_{fbd}$) is found on slopes from 64°N to 51°S in the Eastern Atlantic, Indian and SW Pacific Oceans. It has no light organs but has well-developed eyes and grows to 80 cm length. *Notophycis marginata* (Dwarf codling, 1200 m$_{fbd}$) is a benthopelagic species found from 32°S to 61°S off the south of New Zealand, Australia and South America in the Atlantic and Pacific Oceans. A second species in *N. fitchi* (600 m$_{fbd}$) has been described from the NE Pacific. *Physiculus* is the most speciose of the morid genera with 42 species found in tropical waters and subtropical waters on the upper slopes typically between 260 and 480 m depth. This genus has a ventral light organ that appears as a dark transparent window between the pelvic fins anterior to the anus. There are 18 species endemic to the Pacific Ocean, 10 in the Indian Ocean and 9 in the Atlantic Ocean with *P. dalwigki* (Black codling, 738 m$_{fbd}$) extending into the Western basin of the Mediterranean Sea. There are three Indo-Pacific species of which *P. rhodopinnis* (540 m$_{fbd}$) from Madagascar to Hawaii and *P. roseus* (Rosy cod, 550 m$_{fbd}$) from the Bay of Bengal to New Caledonia are the most widespread.

The monotypic *Rhynchogadus hepaticus* (700 m$_{fbd}$) is a rare bathypelagic morid, maximum length 10 cm, known only from the Western Mediterranean basin. A specimen captured in the Catalan Sea was found to have been feeding predominantly on *Calanus helgolandicus,* a copepod that occurs predominantly between 500 and 1000 m depth in this area (Lloris et al., 1994).

Tripterophycis gilchristi (Grenadier cod, 1000 m$_{fbd}$) is a Southern Hemisphere species distributed 0°S–59°S from the Mid Atlantic across the Indian Ocean to Australia and New Zealand. The second species in the genus, *T. svetovidovi* (Brown grenadier cod, 950 m$_{fbd}$) is more pelagic over a similar range. They have ventral light organs.

4.3.14.7 Family Merlucciidae (99) (Hakes)

There are two genera: the genus *Lyconodes* comprises three rare species known only from a few specimens,

L. argenteus from one specimen of the Cape of Good Hope, *L. brachycolus* (997 m$_{fbd}$) has been recorded from muddy slopes of the North Atlantic and *L. pinnatus* (700 m$_{fbd}$) may be circumglobal in the Southern Ocean. *Merluccius* comprises 16 species, many of which are very important commercially in their respective regions. The total world catch of *Merluccius* spp. reached 2.1 million tonnes in 1973 and since then has fluctuated between 1 and 2 million tonnes per annum (Alheit and Pitcher, 1995). The mean maximum depth of occurrence is 730 m, and although six species have been recorded to 1000 m depth or more, they occur widely in shelf seas and would not necessarily be considered to be deep-sea species. An exception is *M. paradoxus* (Deep-water Cape hake, 1000 m$_{fbd}$), which never occurs shallower than 200 m (Fig. 4.25c) and is taken in a mixed fishery off South Africa together with *M. capensis* (Shallow-water Cape hake, 1000 m$_{fbd}$). *M. paradoxus* is dominant off the west coast of South Africa, where it can account for up to 90 per cent of the total hake catch. Hakes are cannibalistic, and *M. capensis* feeds extensively on *M. paradoxus,* but the deep-water species never feeds on shallow-water hake. Species in upwelling areas such as *M. albidus* (offshore silver hake, 1170 m$_{fbd}$) in the W Atlantic, *M. gayi* (South Pacific hake, 500 m$_{fbd}$) and *M. gayi peruanus* (Peruvian hake, 500 m$_{fbd}$) from the SE Pacific and *M. polli* (Benguela hake, 910 m$_{fbd}$) from the E Atlantic are commercial species captured in deep water.

4.3.14.8 Family Melanonidae (100) (Pelagic cods)

This family comprises two species in a single genus, *Melanonus.* There are divergent views on their relationship to other Gadiformes (Nelson, 2006). Here, based on nuclear and mitochondrial DNA data (Roa-Varón and Orti, 2009), they are placed as a separate family most closely related to the merluciid hakes. They are a very distinctive, widely occurring though uncommon bathypelagic species that grow up to 18 cm total length. In a survey of the pelagic fishes of the Sargasso Sea, *Melanonus* comprised less 0.1 per cent of the total catch (Sutton et al., 2010). The general coloration is black, the head is covered with naked neuromast sensory cells and the lateral system is well developed, suggesting sit-and-wait type of foraging in the bathypelagic. *M. zugmayeri* (Arrowtail, tropical pelagic cod, 3000 m$_{fbd}$) is found at latitudes 60°N–49°S in the Atlantic, Indian and Pacific Oceans (Fig. 4.25c), whereas *M. gracilis* (Pelagic cod, 3813 m$_{fbd}$) is circumpolar in the Southern Hemisphere between Antarctica and the Subtropical Convergence.

4.3.14.9 Family Euclichthyidae (101) (Eucla cod)

There is only one species *Euclichthys polynemus* (Eucla cod, bearded cod, 920 m$_{fbd}$), which is endemic to the slopes off Australia and New Zealand, mainly between 250 and 800 m depth (Fig. 4.25c). A distinctive long 'beard' is made up of extended free rays of the pelvic fins, four on each side, that reach from the head as far back as the anus. Maximum total length is 35 cm.

4.3.14.10 Family Gadidae (102) (Cods)

The family Gadidae comprises 69 species including many of the important commercially exploited fishes of the shelf seas. Following Roa-Varón and Orti (2009) we here recognise four subfamilies.

Subfamily Gadinae

This subfamily includes well-known species such as cod, haddock, whiting and pollock, which are commercially exploited in demersal shelf seas fisheries. Several species are recorded down to great depths, for example *Arctogadus glacialis* (Arctic cod, 1000 m$_{fbd}$), *Gadiculus argenteus* (Silvery pout, 1000 m$_{fbd}$) and *Theragra chalcogramma* (Alaska Pollock, 1280 m$_{fbd}$); however, because they most commonly occur at shelf depths they cannot be considered as deep-sea species. *Micromesistius poutassou* (Blue whiting, 3000 m$_{fbd}$) is an exception (Fig. 4.25d); it is a pelagic species most commonly found at 300–600 m depth around the North Atlantic Ocean from the NE coast of the USA to the

Mediterranean Sea and north to the Norwegian and Barents Seas. *M. poutassou* spawn along the edge of the continental shelf or around offshore banks and seamounts and have a shoaling habit, feeding on zooplankton and small fish. Maximum length is 50 cm. Blue whiting are extensively exploited by pelagic fisheries with landings in the NE Atlantic reaching over 2 million tonnes per annum during 2003–2006. In the Southern Hemisphere the larger (90 cm maximum length) *M. australis* (Southern blue whiting, 900 m$_{fbd}$) is also very abundant 37°S–65°S, off Chile, Argentina, the South Atlantic Islands and southern New Zealand. It is considered that there are four separate stocks in New Zealand waters with total landings recorded as 20,000–40,000 tonnes per annum during 1990–2010 (New Zealand, 2011).

Subfamily Lotinae

Excluding *Lota lota*, the freshwater burbot, there are two marine genera, *Brosme* and *Molva* that occur around the N Atlantic Ocean down to depths of 1000 m. They are demersal fishes with an elongated body form that grow to over 1 m long. The monotypic *Brosme brosme* (Tusk, 1000 m$_{fbd}$), *Molva dypterygia* (Blue ling, 1000 m$_{fbd}$) and *Molva molva* (Ling, 1000 m$_{fbd}$) are all commercially exploited, whereas *M. macrophthalma* (Spanish ling, 754 m$_{fbd}$) of lesser importance is often landed with the other species.

Subfamily Gaidropsarinae (Rocklings)

This subfamily comprises 18 species of mainly shallow-water fishes. One species has a genuinely deep-water distribution at high latitudes; *Gaidropsarus argentatus* (Arctic rockling, 2260 m$_{fbd}$) is endemic to the slopes of the North Atlantic off East Greenland, Iceland, Norway and the Faroe-Shetland Ridge in waters at or below 0°C and is commercially exploited (Fig. 4.25d). The congener *G. ensis* (Threadfin rockling, 2000 m$_{fbd}$) is found in deep waters off the east coast of North America but also occurs inshore.

(d)

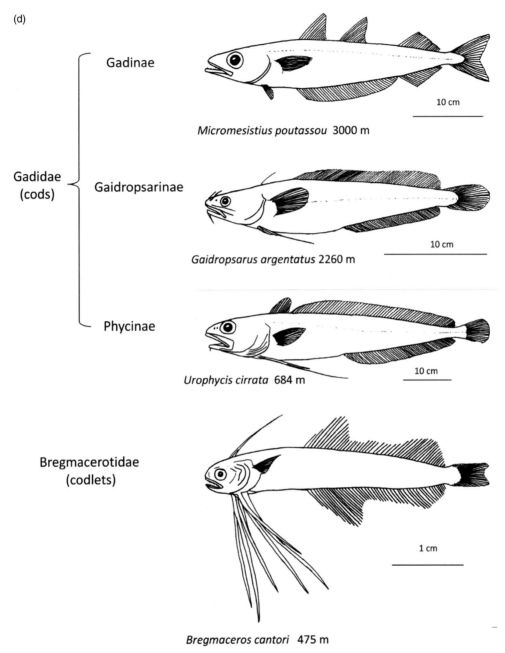

Gadinae

Micromesistius poutassou 3000 m

10 cm

Gadidae (cods)

Gaidropsarinae

Gaidropsarus argentatus 2260 m

10 cm

Phycinae

Urophycis cirrata 684 m

10 cm

Bregmacerotidae (codlets)

Bregmaceros cantori 475 m

1 cm

Figure 4.25d ORDER GADIFORMES. *Micromesistius poutassou* (blue whiting) N Atlantic. *Gaidropsarus argentatus* (Arctic rockling) N Atlantic 72°N–45°N. *Urophycis cirrata* (Gulf hake) W Atlantic 31°N–38°S and Gulf of Mexico. *Bregmaceros cantori* (striped codlet) W Atlantic and Gulf of Mexico.

Subfamily Phycinae (Phycid hakes)

This family comprises 11 species in two genera, *Phycis* and *Urophysis* commonly known as forkbeards, hakes and codlings with a mainly shelf-depth distribution with some species occasionally recorded down to over 1000 m. *Urophycis cirrata* (Gulf hake, 684 m$_{fbd}$)

usually occurs beyond the shelf edge at 360–470 m depth on muddy slopes in the Western Atlantic Ocean from Florida to Brazil and forms part of the upper slope (315–875 m depth) deep-sea demersal fish assemblage in the Northern Gulf of Mexico (Powell et al., 2003).

4.3.14.11 Family Bregmacerotidae (103) (Codlets)

The phylogeny of the Bregmacerotidae is uncertain with mitochondrial DNA suggesting a close relationship with the Macrourinae but nuclear DNA indicating placement in a separate clade together with the shallow-water monotypic family Rancipitidae as a sister group to the Gadidae. Roa-Varón and Orti (2009) suggest that there may have been rapid evolutionary change in this family creating these inconsistencies. They are small pelagic fishes generally less than 10 cm long, mostly found inshore and on the continental shelf in tropical and subtropical waters. There are 14 species in a single genus, and the main interest in the present context are those species that range offshore forming part of the deep scattering layer in several parts of the world. *Bregmaceros atlanticus* (Antenna codlet) is widespread in the Atlantic and Indo-Pacific, *B. cantori* (Striped codlet) occurs in the Western Atlantic on the slopes from the Gulf of Mexico to southern Brazil (Fig. 4.25c), *B. mcclellandi* (Unicorn cod, 2000 m_{fbd}) occurs in the Eastern Indian Ocean and *B. japonicus* (Japanese codlet, 905 m_{fbd}) is found across the Indo-Pacific from the central Indian Ocean to Hawaii. Baird et al. (1973) found that *Bregmaceros* spp. descend to 850 m depth during the day into anoxic hydrogen sulphide-rich waters of the Cariaco Trench off Venezuela. Love et al. (2004) showed that whilst *Steindachneria argentea* remain above the oxic-anoxic interface, their prey *B. cantori* (Milliken and Houde, 1984) descend into the anoxic layers and apparently survive during the day by utilising oxygen stored in the swim bladder from before dawn. *B. mcclellandi* is captured in a seasonal fishery during inshore migrations from October to March in the NW Indian Ocean (James, 1981).

4.3.15 Order Ophidiiformes

The Ophidiiformes are primarily marine benthic fishes divided into two suborders: Ophiodiodei (284 species), which are oviparous, and Bythitoidei (226 species), which are viviparous (Nelson, 2006; Fig. 4.26). They occur at all depths from inshore coastal settings to the abyss where they constitute a conspicuous component of the deep-sea ichthyofauna and extend into the upper hadal zone at depths greater than 6000 m (Jamieson, 2015). The viviparous Bythitoidei predominate in shallow waters but some have nevertheless diversified into the deep sea (Suarez, 1975). The body is elongated and tapering. The pelvic fins, if present, are located far forward in a jugular position. The dorsal and anal fins have long bases and are often connected with the caudal fin forming a continuous structure from behind the head on the dorsal surface round the tail to the anus on the ventral surface.

Suborder Ophidioidei

This suborder comprises all the oviparous Ophidiiform species, which are distinguished by the lack of an external intromittent organ in the males and anterior nostrils located well above the upper lip. It is assumed that all species follow the same reproductive pattern with pelagic eggs produced in a buoyant gelatinous mass that floats near the sea surface.

4.3.15.1 Family Carapidae (104) (Pearlfishes)

These are highly specialised fishes of tropical and temperate outer shelf and upper slopes that mostly occur in the body cavities of various invertebrates such as bivalves, holothurians and asteroids (Ambrose, 1996). They have a unique planktonic larval stage known as the vexillifer characterised by a long thread-like process anterior to the dorsal fin known as the vexillum. This is retained in some adult species. Govoni et al. (1984) showed that the vexillum is innervated by a cranial nerve, has an arterial and venous blood circulation and is supported by collagenous shafts. Living larvae can change the shape and size of the vexillum, apparently by control of blood flow, and it is speculated that it probably serves multiple functions: sensory, predatory deception and hydrodynamic. The vexillifer stage is followed by metamorphosis into benthic tenuis larval stage (Ambrose, 1996).

Subfamiily Pyramodontinae

This subfamily comprises five species that are all free-living with a benthopelagic or pelagic habit (Parmentier and Vandewalle, 2003). They are uncommon or rare with an average maximum length 28 cm and occur on the outer shelf and upper slope. The four species of the genus *Pyramodon* are confined to the tropical Indo-West Pacific including Japan, Indonesia, Australia and New Zealand at depths from 180 to 500 m. The monotypic *Snyderidia canina* (1762 m_{fbd}) is circumglobal around the tropics but is absent from the Eastern Pacific (Fig. 4.26a). The swim bladder is small, and their jaws have conspicuous canine-like teeth. *S. canina* is reported to feed on shrimps (Parmentier and Vandewalle, 2003).

Subfamily Carapinae

This subfamily comprises 30 species of which six can be considered to be deep-sea fishes. *Echiodon cryomargarites* (Messmate) is circumglobal in the Southern Hemisphere (Fig. 4.26a) on slopes down to 2000 m depth including off New Zealand, Australia and as far south as South Georgia. It is uncommon and assumed to be a facultative commensal of sponges and tube-dwelling polychaetes (Nielsen et al., 1999). *E. dentatus* (3250 m_{fbd}) occurs on slopes of the Eastern Atlantic from Spain to the Equator including the Mediterranean Sea. *E. drummondii* (545 m_{fbd}) is found further north off Norway, with records from deep-water demersal trawl surveys in the west of Ireland (Priede et al., 2010). Two further deep-water species of *Echiodon* are recorded from the SW Pacific off New Zealand including *E. pukaki* (860 m_{fbd}) and *E. neotes* (8300 m_{fbd}) from the Kermadec Trench. The reported demersal hadal occurrence at over 8000 m depth of *E. neotes* is now considered unlikely (Jamieson, 2015). The monotypic *Eurypleuron*

owasianum (Eel pearlfish, 445 m_{fbd}) is endemic to the Northwest Pacific off Japan.

4.3.15.2 Family Ophidiidae (105) (Cusk–eels)

This family comprises approximately 250 species, the dorsal and anal fins are continuous with the caudal fin. There are four subfamilies.

Subfamily Brotulinae

There is one genus, *Brotula* distributed around the tropics. The distribution extends down to 650 m depth, but because they usually occur at less than 100 m they are not considered to be deep-sea fishes.

Subfamily Brotulotaeniinae

This subfamily comprises one genus, *Brotulotaenia*, with four species characterised by the absence of barbels and scales reduced to small prickles. They are uncommon, caught in midwater and bottom trawls at mesopelagic and bathyal depths down to over 2500 m beneath tropical and subtropical fishes (Nielsen et al., 1999). *B. brevicauda* (2650 m_{fbd}; Fig. 4.26a) and *B. crassa* (Violet cuskeel, 1100 m_{fbd}) occur widely in the Atlantic and Indian Oceans, *B. nigra* (Dark cusk, 1100 m_{fbd}) has a discontinuous distribution in the Atlantic Ocean , and *B. nielseni* (900 m_{fbd}) is widespread in the Indian and Pacific Oceans.

Subfamily Ophidiinae

This family comprises 56 species usually subdivided into two tribes, the Ophidiini, which are all shallow-water species, and the Lepophidiini, with seven species in three genera that can be considered deep-sea species. *Cherublemma emmelas* (Black brotula, 740 m_{fbd}) occurs on slopes of the Eastern Pacific Ocean (32°N–16°S) from Baja California to Northern Chile. The genus *Lepophidium* has three uncommon or rare species found beyond the shelf edge on the upper slopes of the West Central Atlantic from the Bahamas to the Caribbean at depths down to 500 m. The genus *Genypterus* has three abundant deep-water

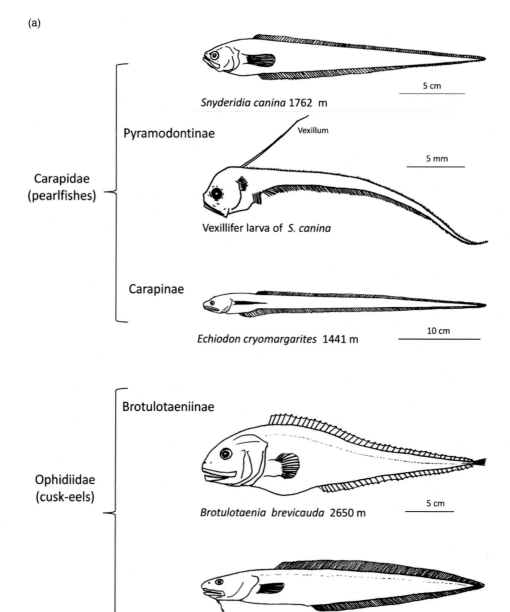

(a)

Snyderidia canina 1762 m

5 cm

Pyramodontinae

Vexillum

Carapidae
(pearlfishes)

Vexillifer larva of *S. canina*

5 mm

Carapinae

Echiodon cryomargarites 1441 m

10 cm

Brotulotaeniinae

Ophidiidae
(cusk-eels)

Brotulotaenia brevicauda 2650 m

5 cm

Ophidiinae

Genypterus blacodes 1000 m

50 cm

Figure 4.26a ORDER OPHIDIIFORMES. *Snyderidia canina*, circumglobal in the tropics except E Pacific. *Echiodon cryomargarites* (messmate) circumglobal in Southern Hemisphere subtropical and temperate waters. *Brotulotaenia brevicauda*, Atlantic and Indian Oceans. *Genypterus blacodes* (pink cusk-eel) Southern Hemisphere 17°S–57°S, Australia, New Zealand, Chile, SW Atlantic.

species in the Southern Hemisphere. *G. blacodes* (Pink cusk-eel, 1000 m$_{fbd}$) occurs in the SW Pacific off Australia and New Zealand, the SE Pacific off Chile and the SW Atlantic off Brazil (Fig. 4.26a). It is a highly commercial species taken by longline and trawl throughout middle depths of 250–800 m in New

Zealand waters and marketed as 'New Zealand Ling'. The total allowable catch was over 20,000 tonnes in 2008–2009 (Marine Stewardship Council, 2015). *G. capensis* (Kingklip, 500 m$_{fbd}$) is the subject fishery off South Africa usually at depths of 350–450 m (Badenhorst, 1988) with landings of up to 10,000 tonnes per annum. *G. chilensis* (Red cusk-eel) occurs on slopes from Peru to southern Chile and is also commercially exploited with declared landings of ca. 1000 tonnes per annum (Nielsen et al., 1999). Smaller quantities of the shallower-living *G. maculatus* (Black cusk-eel) are also caught in Chilean waters.

Subfamily Neobythitinae

This subfamily comprises 183 species in 37 genera representing a remarkable diversification into the deep sea including the hydrothermal vent endemic species *Ventichthys biospeedoi*. Mean minimum depth of occurrence is 1030 m, and the eyes are generally small.

The monotypic *Abyssobrotula galatheae* is an uncommon fish that occurs circumglobally at abyssal depths below tropical and subtropical areas 46°N–48°S (Nielsen et al., 1999). It has also been recorded in the Japan Trench at 6480 to 6640 m (Jamieson, 2015) and most famously in the Puerto Rico Trench at 8370 m (Nielsen, 1977), which is the maximum depth for any fish, an observation now considered doubtful in the absence of subsequent confirmation. It is a small fish, maximum length 16.5 cm. *Acanthonus armatus* (Bony-eared assfish, 4415 m) is also monotypic, occurring benthopelagically at bathyal and abyssal depths below tropical and subtropical oceans of all the oceans. It is most abundant in the Central Western Atlantic, where it is one of the dominant species in demersal trawl surveys of the Northern Gulf of Mexico at depths greater than 1533 m (Powell et al., 2003) and was also captured on slopes off West Africa (Merrett and Marshall, 1981). It grows to 40 cm length and has a large bony head. *Alcockia rostrata* (4040 m$_{fbd}$) is a rare monotypic species of the Indo-Pacific region with an elongated body and depressed head. The genus *Apagesoma*

comprises three abyssal species: *A. australis* (4275 m$_{fbd}$) recorded from around the Crozet Islands in the Southern Indian Ocean, *A. delosommatus* (4150 m$_{fbd}$) from the lower slopes of the Central Atlantic from the Bahamas to Angola and *A. edentatum* (5082 m$_{fbd}$) endemic to the deep basins of the NW Atlantic off the Bahamas. There are two species of *Barathrites*, the abyssal *B. iris* (5285 m$_{fbd}$) from the NE Atlantic and bathyal *B. parri* (3000 m$_{fbd}$) from the NW Atlantic, where it can be locally abundant. *B. iris* may have a circumglobal distribution beneath tropical and subtropical waters (Nielsen et al., 1999). The genus *Barathrodemus* comprises two bathyal species, *B. manatinus* (1318 m$_{fbd}$) from the Western Central Atlantic and *B. nasutus* (1998 m$_{fbd}$) from the Western Central Pacific and Eastern Indian Oceans. There is sexual dimorphism, and males may be capable of sound production as in some shallow-water Ophidiidae. The male *B. manatinus* has modified ribs and drumming muscles attached to the swim bladder (Carter and Musick, 1985). *Bassogigas* comprises two rare species of cusk-eel from bathyal depths, *B. gillii* (2239 m$_{fbd}$), which occurs in the Atlantic, Indian and W. Pacific, and *B. walkeri* (2590 m$_{fbd}$), described from the W. Pacific off Guam (Nielsen and Møller, 2011). The genus *Bassozetus* is a cosmopolitan deep-sea genus (Nielsen and Merrett, 2000) with 13 species, two of which are circumglobal in their distribution (Fig. 4.26b); *B. levistomatus* (5200 m$_{fbd}$) is rare but widespread at abyssal depths, and *B. robustus* (Robust assfish, 4420 m$_{fbd}$) usually occurs at 1000–2750 m depth on the slopes (Nielsen et al., 1999). However, Jamieson (2015) argues that Ophidiids photographed around the edges of the Mariana and Peru–Chile Trenches down to 6173 m may be *B. robustus*. Three species of *Bassozetus* are Indo-West Pacific in their distribution, three endemic to the Pacific Ocean, three in the Atlantic Ocean, one in the Indian Ocean and *B. compressus* (Abyssal assfish, 5456 m$_{fbd}$) occurs in the Atlantic and Pacific Oceans. One of the Pacific species, *B. zenkevitchi*, is recognised as pelagic from rare captures in the North Pacific Ocean (Nielsen et al., 1999). The genus *Bathyonus* comprises three species: *B caudalis* (4040 m$_{fbd}$) from the Indo-Pacific and is probably circumglobal, *B. laticeps* (4775 m$_{fbd}$) from

(b)

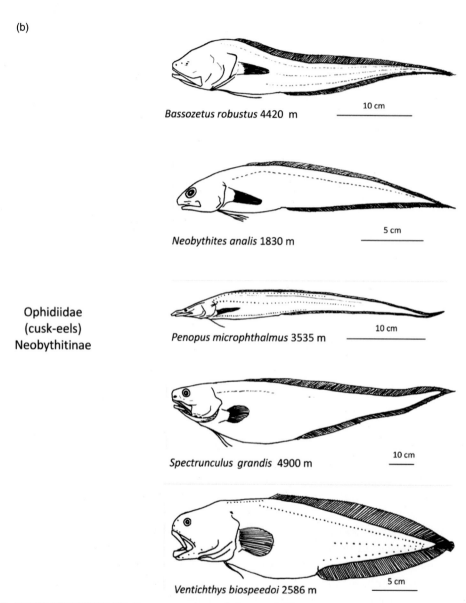

Ophidiidae
(cusk-eels)
Neobythitinae

Bassozetus robustus 4420 m 10 cm

Neobythites analis 1830 m 5 cm

Penopus microphthalmus 3535 m 10 cm

Spectrunculus grandis 4900 m 10 cm

Ventichthys biospeedoi 2586 m 5 cm

Figure 4.26b ORDER OPHIDIIFORMES. *Bassozetus robustus* (robust assfish) Circumglobal in tropical to temperate waters. *Neobythites analis* (Black-edged cusk-eel) around southern Africa from SE Atlantic to Mozambique. *Penopus microphthalmus*, Atlantic. *Spectrunculus grandis* (Pudgy cusk-eel) Atlantic and Pacific. *Ventichthys biospeedoi* (East-Pacific vent brotula) endemic to hydrothermal vents of the SE Pacific Rise.

the Atlantic Ocean and *B. pectoralis* (4600 m$_{fbd}$) from the Atlantic and SW Indian Oceans. The two species of *Benthocometes* have a short stubby-headed appearance and occur on upper slopes, *B. australiensis* off NW Australia and *B. robustus* (1000 m$_{fbd}$) in the Atlantic and Mediterranean including the Ionian Sea (Mytilineou et al., 2005). The genus *Dicrolene* comprises 15 species of bathyal cusk-eel that occur over an average depth range between 1000 and 1600 m. In the Atlantic Ocean two species are relatively common, *D. introniger* (Digitate cusk-eel, 1785 m$_{fbd}$), which occurs on both sides of the

Atlantic, is trawled at 690–1806 m depth off West Africa (Merrett and Marshall, 1981) and *D. kanazawai* (2342 m$_{fbd}$) is a dominant constituent of the demersal fish assemblage at 1533–3075 m depth in the Northern Gulf of Mexico (Powell et al., 2003). Elsewhere *Dicrolene* spp. are rare or uncommon, five species in the Pacific Ocean, three in the Indo-Pacific, four in the Indian Ocean and one rare species off SW Africa.

The genus *Enchelybrotula* comprises two rare benthopelagic elongated slim species from the Indo-Pacific region found between 2000 and 3200 m depth. The monotypic *Epetriodus freddyi* (Needletooth cusk, 1750 m$_{fbd}$) is a short-bodied uncommon species from the Indo-West Pacific. *Eretmichthys pinnatus* (2820 m$_{fbd}$) occurs below tropical areas of all the oceans and grows to over 40 cm length. The genus *Glyptophidium* comprises seven species with large eyes found on the upper slopes at 300–650 m depth in the Indo-West Pacific region. *G. argenteum* (815 m$_{fbd}$), *G. longipes* (Bigeye brotula, 825 m$_{fbd}$) and *G. oceanium* (565 m$_{fbd}$) are landed occasionally in regional fisheries.

The genus *Holcomycteronus* comprises six rare species of cusk-eel with an average minimum depth of occurrence 2400 m generally in tropical and subtropical areas of all the oceans. They are demersal fishes of the lower slopes and abyssal plains with very small eyes. *H. profundissimus* is the deepest-living (5160–7160 m) species found in all the major oceans and has been recorded at its maximum depth in the Java Trench in the Eastern Indian Ocean. *H. squamosus* (5055 m$_{fbd}$) is found in the Atlantic Ocean, *H. pterotus* (3195 m$_{fbd}$) in the Indian Ocean, *H. digittatus* (4080 m$_{fbd}$) in the Eastern Pacific, *H. brucei* (4575 m$_{fbd}$) in the Southern Ocean including the Weddell Abyssal Plain and *H. aequatoris* (4030 m$_{fbd}$) in the Indo-West Pacific region. The genus *Hoplobrotula* comprises three species found at upper bathyal depths, 180–550 m across the Indo-Pacific region from SE Africa to Japan and Eastern Australia that are occasionally landed in fisheries. The monotypic *Homostolus acer* (Filament cusk, 1000 m$_{fbd}$) with long pelvic fin rays occurs in the central West Pacific and is also occasionally landed in Japanese markets. The genus *Lamprogrammus*

comprises five deep oceanic species distributed around the tropical and subtropical regions of the world. The lateral line has special conspicuous scales on which are mounted spindle-shaped neuromasts organs. *L. niger* (2000 m$_{fbd}$) occurs circumglobally and can be locally abundant. The other species are uncommon or rare. The giant *L. shcherbachevi* (Scaleline cusk, 1000 m$_{fbd}$), which grows up to almost 2 m long, is known from only a few specimens but is probably also distributed circumglobally (Cohen and Rohr, 1993). Smaller individuals are generally meso- or bathypelagic whilst larger specimens are captured in bottom trawls. However, the 1.93 m long holotype of *L. shcherbachevi* was captured by mid-water trawl. The rare monotypic *Leptobrotula breviventralis* (780 m$_{fbd}$) is benthopelagic at bathyal depths in the Indo-Pacific region. The genus *Leucicorus* comprises two species, *L. atlanticus* (6800 m$_{fbd}$) from the tropical Western Atlantic and *L. lusciosus* (3435 m$_{fbd}$) from the East Pacific on opposite sides of the Isthmus of Panama. *L. atlanticus* has been captured at depths of 5800–6800 m in the Cayman Trench and 4590 to 4600 m in the Yucatan basin (Rass et al., 1982) and is one of few fishes recorded at hadal depths (Jamieson, 2015). The eyes are degenerate with a rudimentary or absent lens, although the orbit is large (Nielsen et al., 1999). The six species of *Luciobrotula* are found at depths of 500–1200 m at tropical latitudes. *L. bartschi* is most widely distributed in the Indo-Pacific from South Africa to Hawaii and is replaced by *L. corethromycter* in the Atlantic Ocean. Although they grow to over 60 cm, they are of no interest to fisheries. The monotypic *Mastigopterus imperator* (2365 m$_{fbd}$) is a rare deep benthopelagic cusk-eel found at bathyal depths from Madagascar to the Philippines. The genus *Monomitopus* comprises 14 species of bathyal cusk-eel, generally distributed between 500 and 1200 m depth throughout tropical and subtropical regions. Three species groups are recognised. The *M. nigripinnis* (1510 m$_{fbd}$) group has a down-turned head and has seven species distributed in the Atlantic, Indian and Pacific Oceans. The *M. pallidus* (1211 m$_{fbd}$) group with a slender straight head has three species from the Atlantic (two species) and the Central W Pacific (one species). The *M. torvus* (1260 m$_{fbd}$)

group with four species is confined to the Pacific Ocean (Nielsen et al., 1999).

Neobythites is the largest genus of the Ophidiiformes with over 50 upper bathyal demersal and benthopelagic species found beyond the shelf edge at an average depth range between 280 and 570 m. They are distributed circumglobally at tropical and subtropical latitudes but are absent from the Eastern Atlantic. There are 24 species endemic to the Pacific Ocean, 14 in the Indian Ocean, nine in the Atlantic Ocean, three Indo-Pacific species and *N. analis* (Black-edged cusk-eel, 1830 m_{fbd}) found in the Indian and Atlantic Oceans. Shallow-occurring species from the NW Pacific *N. sivicola* (249 m_{fbd}) and *N. stigmosus* (300 m_{fbd}) are recorded in fish markets in Japan but common deeper-living species such as *N. analis* (Fig. 4.26b) and *N. elongatus* (1270 m_{fbd}) from the Caribbean are not exploited. A new monotypic genus and species *Neobythitoides serratus* (950 m_{fbd}) is described from the East China Sea by Nielsen and Machida (2006).

Penopus comprises two rare species of deep bathyal benthopelagic species up to 35 cm long with flattened heads and reduced eyes. *P. microphthalmus* (3535 m_{fbd}) is known from both the Western and Eastern Atlantic Ocean (Fig. 4.26b) and *P. japonicus* from the NW Pacific. The genus *Porogadus* comprises 13 species of benthopelagic bathyal and abyssal cusk-eels with an elongated slender body, total length up to 30 cm with spines on head and unusual patterns of lateral lines varying from three lateral lines in *P. miles* (Slender cuskeel, 5055 m_{fbd}) to no lateral line in *P. catena* (3500 m_{fbd}). *P. miles* is the most common and widespread, distributed circumglobally at tropical and subtropical latitudes, four species are endemic to the Pacific Ocean, four to the Indian Ocean and three in the Atlantic. *P. catena* occurs on both sides of the Isthmus of Panama and is captured in deep demersal fish surveys in the Northern Gulf of Mexico at depths greater than 1500 m (Powell et al., 2003). The genus *Pycnocraspedum* comprises five rare or uncommon species, three from the Pacific Ocean, one from the Atlantic and *P. squamipinne* (pelagic cusk, 500 m_{fbd}), which has an Indo-Pacific distribution. The monotypic *Selachophidium guentheri* (980 m_{fbd}) is

found on upper slopes around Southern Africa from Angola to Mozambique.

The genus *Spectrunculus* was for some time considered to comprise a single species, *S. grandis*, which is very widespread at depths of 800–4900 m in the Atlantic and Pacific Oceans (Fig. 4.26b). They are often attracted to baited cameras and are a conspicuous component of the ichthyofauna of slopes (Priede et al., 1994), abyssal plains (Armstrong et al., 1992) and mid-ocean ridge systems (King et al., 2006). There are two colour morphs, pale and dark brown with various shades of mottling in between, which it was suggested may reflect sexual dimorphism. Following a comprehensive review of available fresh and preserved specimens Uiblein et al. (2008) concluded that coloration is not related to sex but that there are two species, the smaller and often darker *S. crassus* (3677 m_{fbd}), which grows up to 60 cm standard length, and *S. grandis*, which grows up to 130 cm and is generally pale in colour. Colour, however, should not be taken as a reliable characteristic for species discrimination. Stomach contents indicate a varied diet including detritus, foraminifera, polychaetes, crustacea, gastropods, bivalves and ophiurids.

The monotypic *Tauredophidium hextii* (2260 m_{fbd}) is a small (<10 cm), rare benthopelagic cusk-eel with very reduced eyes and a large head with prominent spines found in the Indo-West Pacific. *Typhlonus nasus* (5100 m) is a deeper-living (>4000 m) species of the Indo-Pacific with a soft rounded head with no spines, vestigial eyes and total length up to 28 cm. *Xyelacyba myersi* (2500 m_{fbd}) is a rare but possibly circumglobal bathyal benthopelagic species with a relatively short deep body length up to 57 cm.

The type specimen of *Ventichthys biospeedoi* (2586 m_{fbd}) was collected on 29 April 2004 by the submersible *Nautile* in a baited trap deployed at 2586 m during investigations of a hydrothermal vent on the East Pacific Rise as part of the BIOSPEEDO cruise of the R.V. *Atlantis* (Fig. 4.26b). It appears to be an obligatory hydrothermal vent species that was observed swimming in shimmering warm vent fluids at temperatures between 2 and 7°C at three different sites. It is morphologically distinctive because of its posteriorly located large kidneys, broad head and pattern of four lateral lines on each side and therefore is placed in its own monotypic genus (Nielsen et al., 2006).

Suborder Bythitoidei

This suborder comprises the viviparous Ophidiiformes, which are distinguished by the presence of an intromittent organ in the male and the anterior nostril located immediately above the upper lip. The copulatory apparatus of the males is normally hidden beneath a fleshy hood and includes a median penis, which is formed by modification of an anal fin ray and pairs of pseudoclaspers on either side. The female also has a copulatory apparatus with voluntary muscles that may aid insemination. Embryos are retained and develop within the ovary, and sperm may also be stored within the ovary to fertilize successive broods of eggs (Suarez, 1975). There are three families within the suborder, Bythitidae, Aphyonidae and Parabrotulidae. Suarez (1975) points out that whereas oviparous Opidiidae predominate in the deep sea the viviparous brotulids predominate in shallow waters, and the Aphyonidae represents the main radiation of viviparous species into the deep sea.

4.3.15.3 Family Bythitidae (106) (Viviparous brotulas)

This family comprises 200 species divided into two subfamilies. They are predominantly shallow-water species.

Subfamily Bythitinae

This is the smaller of the two subfamilies of Bythitidae with 79 species, 35 of which are found in the deep sea, including the hydrothermal vent fish *Thermichthys hollisi*.

The genus *Bellottia* comprises five species of small (<7 cm) fishes found at depths down to 580 m, either demersal or benthopelagic. *B. apoda* (569 m) occurs around the Mediterranean Sea and in the subtropical NE Atlantic Ocean. It is regularly caught in low numbers at upper slope depths in deep demersal trawl

surveys in the Ionian Sea (Mytilineou et al., 2005) and feeds on planktonic and benthic prey. The other species are rare: two described from the Western Atlantic associated with deep-water coral habitats (Nielsen et al., 2009) and two from the West Pacific. The genus *Bythites* comprises three rare species from the North Atlantic Ocean, *B. fuscus* (526 m_{fbd}) from West Greenland, *B. gerdae* (832 m_{fbd}) from the Florida Strait and *B. islandicus* (285 m_{fbd}) off SE Iceland. The genus *Cataetyx* is distributed circumglobally in temperate to tropical seas generally deeper than 500 m. *C. alleni* (1000 m_{fbd}) and *C. laticeps* (2400 m_{fbd}) are both found in the NE Atlantic and Western Mediterranean Sea and can be locally abundant (Priede et al., 2010). *C. messieri* (1649 m_{fbd}) occurs on slopes around the southern tip of South America, and *C. rubrirostris* (Rubynose brotula, 1000 m_{fbd}) is locally abundant in the NE Pacific off the coast of the USA to Baja California. The larvae of *C. rubirostris* are born at <7 mm length and are planktonic (Ambrose, 1996c). The other *Cataetyx* species, three from the Atlantic and five from the Pacific, are all rare (Fig. 4.26c). *Diplacanthopoma brachysoma* (1670 m_{fbd}) is locally abundant on slopes in tropical waters of the Western Atlantic including the Gulf of Mexico and grows to a standard length of 10 cm. The other species in the genus, four from the Indian Ocean, two Indo-Pacific and two from the Pacific Ocean, are all rare.

Hephthocara comprises two rare bathydemersal species with large heads, deep tapering bodies and no pelvic fin rays: *H. crassiceps* (1510 m_{fbd}) from the West Tropical Pacific and *H. simum* (1650 m_{fbd}) from the Eastern Indian Ocean, which may prove to be the same species. The genus *Pseudonus* has two rare small bathydemersal species, *P. acutus* (1620 m_{fbd}) from the tropical E. Pacific and *P. squamiceps* (1270 m_{fbd}) from the Indo-West Pacific. The genus *Saccogaster* includes six rare small brotulas generally <10 cm long, found down to a maximum depth of 834 m circumglobally at tropical and subtropical latitudes but apparently absent from the Eastern Atlantic.

Thalassobathia pelagica (1000 m_{fbd}) is an unusual mesopelagic ophidiiform (Fig. 4.26c) captured at 700 m depth in the Mid-North Atlantic, where it lives apparently in association with the giant jellyfish

(c)

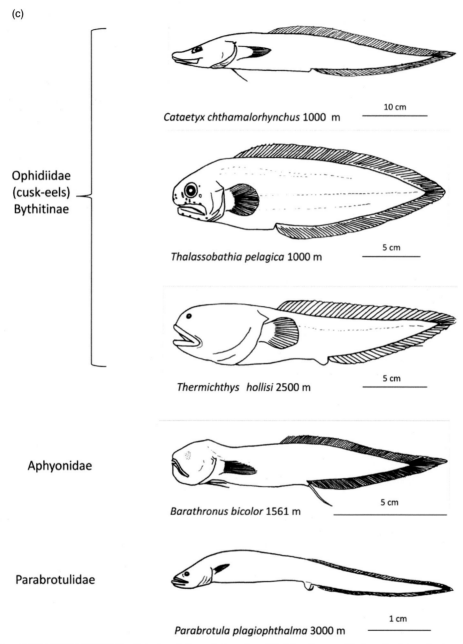

Ophidiidae (cusk-eels) Bythitinae

Cataetyx chthamalorhynchus 1000 m   10 cm

Thalassobathia pelagica 1000 m   5 cm

Thermichthys hollisi 2500 m   5 cm

Aphyonidae

Barathronus bicolor 1561 m   5 cm

Parabrotulidae

Parabrotula plagiophthalma 3000 m   1 cm

Figure 4.26c ORDER OPHIDIIFORMES. *Cataetyx chthamalorhynchus*, SE Atlantic rare. *Thalassobathia pelagica*, Atlantic and Pacific associated with giant jellyfish. *Thermichthys hollisi*, endemic to hydrothermal vents in the E Pacific. *Barathronus bicolor*, W Central Atlantic and Gulf of Mexico. *Parabrotula plagiophthalma* (false cusk) Atlantic and W Pacific.

Stygiomedusa fabulosa (Harbison et al., 1973). The body is relatively short, the skin is thick, scales are largely absent and specimens are about 20 cm long. More recently *T. pelagica* has been observed by ROV in the Pacific Ocean in association with *Stygiomedusa gigantea* at 1300 m depth below the oxygen minimum zone off the Gulf of California (Drazen and Robison, 2004). A second species, *T. nelsoni* (2000 m$_{\mathrm{fbd}}$), has been described from the Pacific Ocean off northern Chile, but there is no evidence as to whether there is a jellyfish association.

Thermichthys hollisi (2595 m_{fbd}) was first described as *Bythites hollisi* by Cohen et al. (1990) based on a gravid female 304 mm long standard length, captured from a hydrothermal vent on the Galapagos Rise in the tropical eastern Pacific using a net attached to the manipulator arm of the DSRV *Alvin* operated by the pilot Ralph Hollis (Fig. 4.26c). The species has since been recorded at three vent sites on the East Pacific Rise (Biscoito et al., 2002) where a second specimen, an adult male, was captured at the Hobb site in a baited trap by the French BIOSPEEDO expedition. The replacement generic name was then proposed by Nielsen and Cohen (2005). The original gravid female contained approximately 10,000 larvae with mean length of 5.1 mm, which are presumed to be pelagic. The species evidently disperses quite widely in the East Pacific region as indicated by the 4000 km distance between the sites at which the two known specimens were collected.

The monotypic *Timorichthys disjunctus* (392 m_{fbd}) is a small fish living on the deep northwestern Australian Plateau. The genus *Tuamotuichthys* comprises three small species of brotulas found on different island slopes down to 536 m depth in the Western Pacific.

Subfamily Brosmophycinae

The Brosmophycinae comprises 120 shallow-water species of brotulas distributed worldwide. The only exception is *Melodichthys hadrocephalus* (400 m_{fbd}), known from a single specimen taken in deep water off the coast of Brittany in the NE Atlantic Ocean.

4.3.15.4 Family Aphyonidae (107) (Aphyonids, Blind cusk-eels)

The Aphyonidae represents the main diversification of the viviparous brotulas into the deep sea. There are 24 species in six genera spanning a depth range down to over 5000 m. They produce relatively few large young, which with the exception of *Barathronus*

remain near the bottom. Lacking a swim bladder, they probably live close to the seafloor and are generally small fragile fishes with degenerate eyes in the adults (Plate 7a).

The genus *Aphyonus* comprises four species found at tropical and subtropical latitudes, *A. bolini* (1300 m_{fbd}) in the Indo-Pacific, *A. brevidorsalis* (1500 m_{fbd}) Indian Ocean, *A. gelatinosus* (2500 m_{fbd}) circumglobal (Plate 7a) and *A. rassi* (4412 m_{fbd}) in the Caribbean. Maximum length is 25 cm, and the eyes are either small or not externally visible. *Barathronus* has 11 species, all of which are rare except for *B. bicolor* (1561 m_{hbd}), which is relatively common on the slopes of the Western Atlantic Ocean from the Gulf of Mexico to Southern Brazil. Standard length range is from 6 to 14 cm. There is a suggestion of a disjunctive distribution between the Northern and Southern Hemispheres with *B. linsi* (2045 m_{fbd}) found off northern Brazil (Nielsen et al., 2015). *Barathronus* are characterised by a scaleless loose transparent unpigmented skin. In the adults the eyes are degenerate and probably nonfunctional (Lockett, 1977). In addition to *B. bicolor* (Fig. 4.26c) four species are recognised in the Atlantic Ocean, two in the Indian Ocean, two in the Pacific and two Indo-Pacific species. *Meteoria erythrops* (5320 m_{fbd}) is a rare fragile abyssal demersal species from the North Atlantic Ocean (28–42°N), less than 8 cm long with residual eyes. The genus *Nybelinella* comprises two rare species less than 10 cm long: *N. brevidorsalis* (5160 m_{fbd}) from the SE Indian Ocean and *N. erikssoni* (5440 m_{fbd}) from the Atlantic Ocean between 45°N and 27°S. The genus *Parasciadonus* comprises two species less than 8 cm long collected off New Caledonia in the Western Pacific Ocean at depths of 4640 to 5440 m. *Sciadonus* probably occurs worldwide below tropical and subtropical seas. There are four species that are benthopelagic between 1785 and 5610 m depth, *S. cryptophthalmus* (5000 m) is known from one specimen caught in the NE Atlantic off Spain, *S. galatheae* (5440 m_{fbd}) from two specimens in the NE Atlantic and samples from the Kermadec Trench in the SW Pacific, *S. jonassoni* (5610 m) from the Eastern Atlantic and *S. pedicellaris* (4880 m_{fbd}) from the North Atlantic and SE Pacific off

Chile. These are small slim fishes less than 10 cm long with a protruding lower jaw, almost horizontal mouth and indistinct eyes (Nielsen et al., 1999).

4.3.15.5 Family Parabrotulidae (108) (False brotulas)

The taxonomic position of these fishes is doubtful. Here, following Nelson (2006), they are considered within the Ophidiiformes as a separate family. However, they have been variously placed within the Blennoidei, Gadiformes, Ophidiiformes within the family Brotulidae or within the Zoarcidae (Anderson, 1986; Kimura, 1990). They were subsequently raised to family status Parabrotulidae within the Zoarciformes, a scheme followed by Anderson (1986), although he points out that they may be derived from the Aphyonidae.

The Parabortulas are small, *ca.* 5 cm long, viviparous, bathypelagic eel-like fishes with a small mouth, protruding lower jaw, loose thin skin that also covers the eyes, no pelvic fins, no gas bladder and no scales. There are free neuromasts located on the head, body and tail with no sensory canals or pores. Two genera are recognised, *Parabrotula* (two species), which has a black skin, slender head and pointed tail, and *Leucobrotula* (one species), which has a pale colour, robust head and square tail (Nielsen et al., 1990). *Parabrotula plagiophthalma* (3000 m_{fbd}) is best known from the NE Atlantic but has a worldwide distribution including off Japan and Australia. Most catches have been by mid-water trawl at depths of 760–1500 m at water temperatures of 5.0–9.0°C in the meso- to bathypelagic with no evidence of diel vertical migration. There is sexual dimorphism with mature females (49 mm) larger than the males (41 mm). At maturity, the male penis extends beyond the genital hood indicating ripeness of the testes similar to the situation in *Barathronus bicolor* (Aphyonidae). A gravid female of *P. plagiophthalma* has been reported to contain 23 embryos in one lobe of the ovary, and partial extrusion of embryos has been observed. The breeding season appears to be extended and occurs throughout the geographic range of the species. A second species, *Parabrotula tanseimaru* (1300 m_{fbd}), has been described from Sagami Bay, Japan at bathypelagic depths. It is found to feed primarily on copepods (Miya and Nielsen, 1991) and dominates over *P. plagiophthalma* in Sagami Bay, whereas only *P. plagiophthalma* is found in offshore open-ocean samples (Fig. 4.26c). All specimens of *Leucobrotula adipata* (1290 m_{fbd}) are from the Northeast Atlantic (60°N–35°N) at depths between 610 and 1290 m with the majority of catches close to the seafloor with the net 10–60 m off the bottom.

4.3.16 Order Batrachoidiformes

This order comprises just one family with approximately 75 species.

4.3.16.1 Family Batrachoididae (109) (Toadfishes)

The Batracoididae are benthic fishes with a large frog-like head and eyes occurring predominantly in coastal waters with some species found in brackish and fresh water. Only one species, *Porichthys bathoiketes* (320 m_{fbd}, maximum length 16 cm), is considered to be a deep-sea species, found at depths greater than 180 m on slopes in the Western Atlantic from Honduras to the Guianas. Like the other species in this genus, it has rows of ventral photophores.

4.3.17 Order Gobiesociformes

This order comprises one family with approximately 150 species.

4.3.17.1 Family Gobiesocidae (110) (Clingfishes)

The Gobiesocidae are primarily shallow-water or intertidal fishes with a few in freshwater. They are characterised by pelvic fins modified to form a sucking disc enabling them to cling to hard surfaces. Two species occur in deep water: *Alabes bathys* (348 m_{fbd}) on slopes of southern Australia and *Gymnoscyphus ascitus* (258 m_{fbd}) off the island of St Vincent in the W Atlantic. Both are small benthic fishes less than 4 cm long.

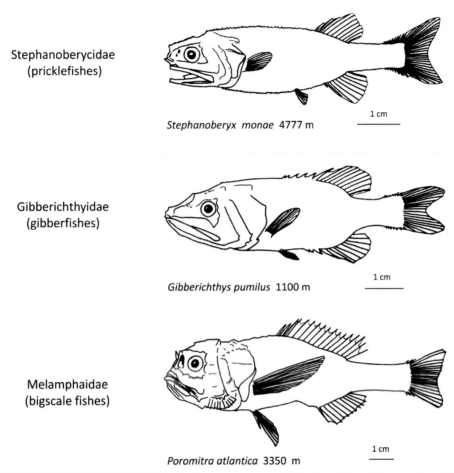

Stephanoberycidae
(pricklefishes)

Stephanoberyx monae 4777 m

1 cm

Gibberichthyidae
(gibberfishes)

Gibberichthys pumilus 1100 m

1 cm

Melamphaidae
(bigscale fishes)

Poromitra atlantica 3350 m

1 cm

Figure 4.27 ORDER STEPHANOBERYCIFORMES. *Stephanoberyx monae*, W Atlantic. *Gibberichthys pumilus* (gibberfish) W Atlantic. *Poromitra atlantica*, circumglobal 30–60° S.

ORDER ATHERINIFORMES: No deep-sea representatives.
ORDER CYPRINODONTIFORMES: No deep-sea representatives.
ORDER BELONIFORMES: No deep-sea representatives.

4.3.18 Order Stephanoberyciformes

This order comprises 68 species in four families of predominantly bathypelagic fishes with a short roundish body, mean maximum length 9 cm and mean maximum depth approximately 2000 m (Fig. 4.27). These are generally rare deep-sea fishes distributed at depth throughout the Atlantic, Indian and Pacific Oceans. In pelagic surveys of the Sargasso Sea area of the North Atlantic Ocean seven species were captured, comprising less than 2 per cent of the catch in the 1000 to 5000 m depth stratum (Sutton et al., 2010).

4.3.18.1 Family Stephanoberycidae (111) (Pricklefishes)

This family comprises four species in four monotypic genera (Fig. 4.27). They are all rare fishes occasionally caught in deep-water trawls. *Abyssoberyx levisquamosus* (4565 m_{fbd}) was retrieved by demersal trawl in the abyssal tropical North Atlantic Ocean

(20°N) off West Africa. Maturing females were 152 to 168 mm long containing multiple batches of developing eggs ca. 1 mm diameter. Stomachs contained chitinous remains. There is evidence that *Abyssoberyx* also occurs in the Pacific Ocean (Merrett and Moore, 2005). *Acanthochaenus luetkenii* (5308 m$_{fbd}$) is found in the Atlantic, Indian and Pacific Oceans beneath subtropical and temperate waters (40°N–57°S). They become sexually mature at 8 cm and reach a maximum length of 14 cm. The diet includes benthic crustacea. *Malacosarcus macrostoma* (4438 m$_{fbd}$) is known from the Western Central Pacific and possibly the NE Atlantic (Merrett and Moore, 2005). *Stephanoberyx monae* (4777 m$_{fbd}$) is recorded on abyssal slopes in the Western Atlantic from New England to the Caribbean.

4.3.18.2 Family Gibberichthyidae (112) (Gibberfishes)

There is one genus, *Gibberichthys*, that comprises two benthopelagic species typically found at 320–1100 m depth, usually captured near continental or island slopes where they feed on small crustacea especially pelagic amphipods. Females reach maturity at 8 cm length, and maximum length is 13 cm. *G. latifrons* (2000 m$_{fbd}$) occurs throughout the Indo-West Pacific though *G. pumilus* (1100 m$_{fbd}$) is confined to the Western Atlantic (Fig. 4.27). Larvae with characteristic filamentous pelvic fins are found in surface waters.

4.3.18.3 Family Melamphaidae (113) (Bigscale fishes)

The Melamphaidae is the most speciose of the Stephanoberyciform families with over 60 species in five genera distributed widely at depths of 200–2000 m in all oceans except the Mediterranean Sea and the Arctic. They are meso- to bathypelagic fishes that feed primarily on gelatinous zooplankton and small crustacea. The head is large with an elaborate network of sensory canals. The larvae feed in the surface layers, and there is a downward migration during development with juveniles at 100–200 m depth before descent to the respective adult depths of each species. The genus *Melamphaes*

comprises 25 species of which 11 are endemic to the Pacific Ocean, seven in the Atlantic Ocean, *M. danae* (1275 m$_{fbd}$) is Indo-Pacific and three species, *M. suborbitalis* (Shoulderspine bigscale, 1000 m$_{fbd}$), *M. longivelis* (Eyebrow bigscale, 1500 m$_{fbd}$) and *M. microps* (2516 m$_{fbd}$), are recorded from the Atlantic and Pacific Oceans. Three species are more or less circumglobal in their distribution: *M. polylepis* (2250 m$_{fbd}$) 34°N–31°S, *M. simus* (Ridgehead, 800 m$_{fbd}$) and *M. eulepis* (200 m$_{fbd}$). The genus *Poromitra* comprises 22 species including *P. atlantica* (3350 m$_{fbd}$), which is circumglobal in the Southern Hemisphere between 30° and 60° S (Kotlyar, 2009). Four species are endemic to the Atlantic Ocean (Fig. 4.27), four to the Indian Ocean and nine to the Pacific Ocean. *P. macrophthalma* (3125 m$_{fbd}$) and *P. oscitans* (Yawning, 5320 m$_{fbd}$) are distributed beneath the tropical and subtropical waters of the Indo-Pacific. *P. capito* (1000 m$_{fbd}$) has a possible bipolar distribution in the North Atlantic and South West Pacific, and *P. coronata* (1647 m$_{fbd}$) is distributed around Southern Africa from the SE Atlantic to SW Indian Ocean. Recorded occurrence of *P. curilensis* from hadal depth (8000 m) in the Kuril Trench is based on a misinterpretation of results from the use of a nonclosing net towed above the trench. The normal depth of occurrence is much shallower, and indeed one specimen was caught near the surface using a dip net (Kotlyar, 2008). The genus *Scopeloberyx* comprises nine species. *S. opisthopterus* (3000 m$_{fbd}$) and *S. robustus* (Longjaw bigscale, 4740 m$_{fbd}$) are circumglobal and tropical and subtropical latitudes. Two species are endemic to the Atlantic Ocean, three to the Pacific Ocean and two species are Indo-Pacific in their distribution, including *S. microlepis* (Southern bigscale, 1717 m$_{fbd}$) confined to latitudes south of 30°S. The genus *Scopelogadus* comprises two species that are worldwide in their distribution, *S. beanii* (Bean's bigscale, 2500 m$_{fbd}$) 65°N–40°S and *S. mizolepis* (Ragged bigscale, 3385 m$_{fbd}$) 40°N–30°S plus two Pacific species. The monotypic *Sio nordenskjoldii* (Nordenskjold's bigscale, 3000 m$_{fbd}$) is circumglobal in the Southern Hemisphere between 20°S and 68°S with the young distributed

(a)

Anoplogastridae
(fangtooths)

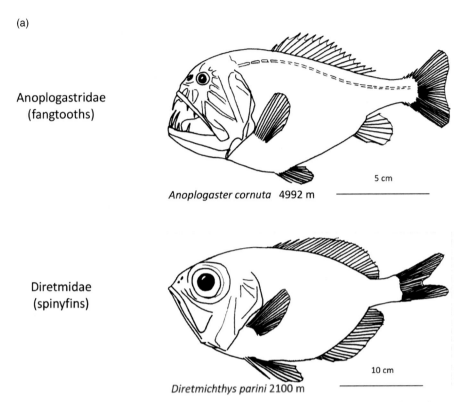

Anoplogaster cornuta 4992 m

5 cm

Diretmidae
(spinyfins)

Diretmichthys parini 2100 m

10 cm

Figure 4.28a ORDER BERYCIFORMES. *Anoplogaster cornuta* (common fangtooth) circumglobal 65°N–46°S. *Diretmichthys parini* (Parin's spinyfish) Atlantic and Indo-W Pacific 40°N–40°S.

in warmer waters of the northern part of the species distribution.

4.3.18.4 Family Hispidoberycidae (114) (Spiny-scale pricklefishes)

This family comprises one rare species, *Hispidoberyx ambagiosus* (1019 m_{fbd}) known from the Eastern Indian Ocean and NW Pacific.

4.3.19 Order Beryciformes

The Berciformes comprises over 160 species, many of which live in shallow water but all tend to avoid bright light by nocturnal behaviour or remaining at depth during the day (Fig. 4.28). Some are true deep-sea species including commercially exploited species such as the orange roughy (*Hoplostethus atlanticus*). Following Nelson (2006) here we recognise three

suborders: Trachichthyoidei, Berycoidei and Holocentroidei.

Suborder Trachichthyoidei

There are five families, Anoplogastridae, Diretmidae, Anomalopidae, Monocentridae and Trachichthyidae.

4.3.19.1 Family Anoplogastridae (115) (Fangtooth)

This family comprises two species in a single genus, the very distinctive *Anoplogaster*, which are small, short-bodied bathypelagic fishes with the eponymous fang teeth. *A. cornuta* (Common fangtooth, 4992 m_{fbd}) occurs worldwide between 65°N and 46°S (Fig. 4.28a). Adults are usually between 500 and 2000 m depth, and they grow to 18 cm total length. They feed on crustacea when young and on fish as adults and are in turn consumed by deep-diving predatory pelagic fishes such as tuna, albacore and marlin. *A. brachycera* (Shorthorn fangtooth, 1500 m$_{fbd}$) is a smaller fish (6 cm maximum length) found in tropical waters of the Western Atlantic and the Western Pacific Oceans.

4.3.19.2 Family Diretmidae (116) (Spinyfins)

Often grouped together with the Anoplogastridae in a superfamily, the Diretmidae are also short-bodied bathypelagic fishes but rather larger in size, with a maximum length 24–40 cm. There are four species in three genera. *Diretmichthys parini* (Parin's spinyfish, 2100 m$_{fbd}$) is widely distributed around the globe between 40°N and 40°S (Fig. 4.28a). *Diretmoides pauciradiatus* (Longwing spinyfin, 600 m$_{fbd}$) is a mesopelagic species in the Atlantic, Indian and West Pacific Oceans, whereas its congener *D. veriginae* (1300 m$_{fbd}$) has a more restricted distribution in the E. Indian Ocean and W. Pacific. *Diretmus argenteus* (Silver spinyfin, 2000 m$_{fbd}$), in contrast to the dark coloration of the previous species, is silvery in appearance and is distributed circumglobally between 65°N and 40°S but has not been recorded from the Mediterranean.

4.3.19.3 Family Anomalopidae (117) (Flashlight fishes)

These are predominantly reef-associated shallow-water species with a characteristic luminous organ beneath each eye with a shutter mechanism for controlling light emission. Of the eight recognised species, only one, *Parmops echinatus* (458 m$_{fbd}$), is considered as a possible deep-sea species described from deep -water off the island of Fiji in the Pacific Ocean.

Family Monocentridae (Pinecone fishes)

These are all shallow-water species.

4.3.19.4 Family Trachichthyidae (118) (Roughies)

This family comprises 50 species in eight genera. Six genera contain deep- sea species that range down to a maximum depth of 1500 m. The body is laterally compressed, oval-shaped with a large head covered with characteristic ridges between which there are mucus-filled cavities beneath the skin, leading to the alternate common name 'slimeheads'. Some species with well- developed tooth-like scales or scutes on the ventral median line are known as 'sawbellies'.

The genus *Aulotrachichthys* comprises eight species of luminous roughies, which have light organs populated by symbiotic luminous bacteria (Ast and Dunlap, 2005). Four species are considered to be bathypelagic deep-sea fishes: *A. argyrophanus* (Western luminous roughy, 229 m$_{fbd}$) from the SW Atlantic off the mouth of the Amazon River, *A. latus* (Philippine luminous roughy, 723 m$_{fbd}$), *A. rosthemius* (West Pacific luminous roughy, 198 m$_{fbd}$) from the West Pacific and *A. sajademalensis* (Saya de Malha luminous roughy, 274 m$_{fbd}$) from the Indian Ocean and the Mediterranean Sea.

The two species of *Gephyroberyx* are benthopelagic, generally found between 200 and 500 m depth on the upper slopes. *G. darwinii* (1210 m$_{fbd}$) is circumglobal between 43°N and 35°S (Fig. 4.28b), and *G. japonicus* (Big roughy, 660 m$_{fbd}$) is confined to the NW Pacific. Despite its common name, *G. japonicus* is the smaller of the two species, which grow to 60 and 20 cm length, respectively. Both are exploited in commercial fisheries.

Hoplostethus is the most diverse genus of Trachichthyidae with 28 species varying in size from 10 to 75 cm total length. Most species are of no commercial interest. They are distributed circumglobally in tropical to temperate waters generally associated with upper slopes and summits down to 1000 m depth. The centre of species distribution is in the Indo-Pacific region with nine species endemic to the Indian Ocean, ten to the Pacific Ocean and two to the Atlantic Ocean. Four species are Indo-Pacific in their distribution including *H. gigas* (Giant sawbelly, 522 m$_{fbd}$) found off South Australia on the upper slopes around canyon edges (Roberts and Gomon, 2012). Three species are distributed in the Atlantic and Indian oceans: *H. mikhailini* (525 m$_{fbd}$) around the Cape of Good Hope, *H. cadenati* (Black slimehead, 1000 m$_{fbd}$) between 28°N and 27°S in the

(b)

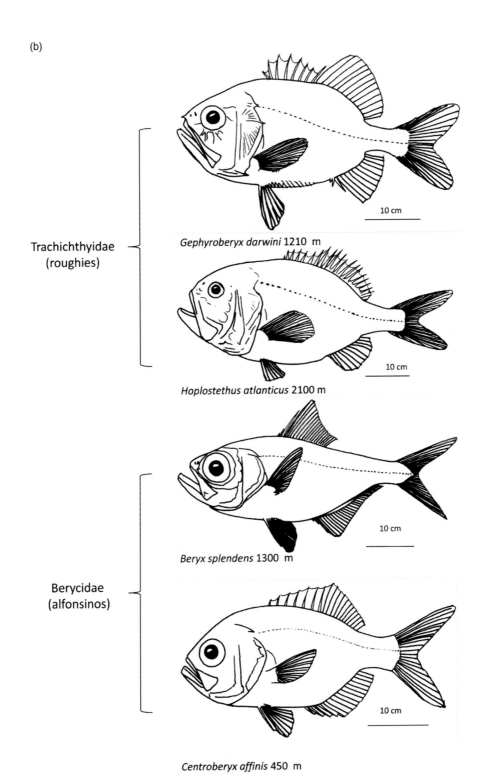

Trachichthyidae
(roughies)

Gephyroberyx darwini 1210 m

Hoplostethus atlanticus 2100 m

Berycidae
(alfonsinos)

Beryx splendens 1300 m

Centroberyx affinis 450 m

Figure 4.28b ORDER BERYCIFORMES. *Gephyroberyx darwini* (Darwin's slimehead) Atlantic and Indo-W Pacific 43°N–35°S. *Hoplostethus atlanticus* (orange roughy) Atlantic, Indian Ocean, SW and SE Pacific 65°N–56°S. *Beryx splendens* (splendid alfonsino) circumglobal 45°N–43°S except the Mediterranean and NE Pacific. *Centroberyx affinis* (redfish) W Pacific, Australia and New Zealand.

Eastern Atlantic and Western Indian Oceans and *H. mediterraneus* (Silver roughy,1175 m$_{fbd}$), which is cosmopolitan throughout the Atlantic Ocean from Iceland to Southern Brazil and South Africa including the Mediterranean Sea, also in the Western Indian Ocean and the Red Sea. *H. mediterraneus* is often divided into regional subspecies. The most widespread and largest species (up to 75 cm total length) is *H. atlanticus* (orange roughy,1809 m$_{fbd}$) that occurs in most oceans between 65°N and 56°S including the North Western Atlantic, the Eastern Atlantic from Iceland to Morocco, Namibia, South Africa, South-Central Indian Ocean, Australia and New Zealand and the Eastern Pacific off Chile (Fig. 4.28b). Only two species of *Hoplostethus* are commercially exploited: *H. cadenati* is commonly caught in trawls around the coasts of Africa including off Namibia at depths of 200–700 m (Bianchi et al., 1999). The orange roughy (*H. atlanticus*) is a most valuable fish occurring in large spawning aggregations around seamounts and on slopes where it is accessible to trawlers. From the late 1970s it was heavily exploited on fishing grounds around the world reaching a peak global annual catch of 90,000 tonnes in 1990 since when production has declined. The eggs of *H. atlanticus* are relatively large (2.25 mm diameter) compared with most pelagic fishes, and hence fecundity, is low *ca.* 22, 000 eggs per kg of female body weight. This, together with late age of maturation, 20 years and life span of over 100 years leads to a low intrinsic rate of population increase so the species is not resistant to high rates of exploitation. Most stocks around the world have crashed, and for example, in the NE Atlantic since 2010 scientific advice has been that there should be no directed fisheries for this species and bycatch should be minimised (ICES, 2012). In New Zealand with much larger initial stock sizes it may be possible to pursue sustainable exploitation albeit at much lower levels than during the boom years when fisheries for the species first developed. *H. atlanticus* feeds on a wide range of pelagic crustacea, fish and squid and is not attracted to carrion, hence they cannot be caught on baited longlines or in traps. *H. mediterraneus* is widely caught as a bycatch in bottom trawls but is of no interest to commercial

fisheries. It is caught in trawl fisheries for deep-sea shrimps around the Mediterranean Sea and is generally discarded. This has resulted in a gradual depletion and distortion of the age structure of the *H. mediterraneus* stocks leading to absence of large female mega-spawners and a population dominated by juveniles; so-called longevity overfishing even though the species is not being targeted (Vitale et al., 2014).

Paratrachichthys trailli (Sandpaper fish, 500 m$_{fbd}$) is reported as bathypelagic and bioluminescent in New Zealand waters and *Trachichthys australis* (southern roughy, 1557 m$_{fbd}$) endemic to southern Australia and although reported to great depths, is generally found in coastal shallow waters.

Suborder Berycoidei
4.3.19.5 Family Berycidae (119) (Alfonsinos)
This family comprises 10 species in two genera. The genus *Beryx* has three species. *B. decadactylus* (Alfonsino, 1000 m$_{fbd}$) occurs worldwide in temperate to tropical latitudes 70°N–48°S) except in the Eastern Pacific. It can grow to 1 m long but the more usual maximum size is 35 cm. *B. splendens* (Splendid alfonsino, 1300 m$_{fbd}$) is also worldwide (Fig. 4.28b) but over a narrower latitudinal range (45°N–43°S) and is absent from the Mediterranean Sea and the NE Pacific. *B. mollis* (500 m$_{fbd}$) is from the NW Pacific off Japan and the W Indian Ocean. The alfonsinos are benthopelagic, often abundant around the summits of seamounts and upper slopes. Following depletion of other seamount species, commercially exploitation increased from a nominal world annual catch of *B. splendens* of 2,000 tonnes in 1990 to 15,000 tonnes in 2003 (Tandstad et al., 2011). Fishing is by mid-water trawl between depths of 300 and 500 m. The genus *Centroberyx* comprises seven species from the Indo-Pacific region especially around Australia occurring down to a maximum depth of 500 m (Fig. 4.28b). *C. affinis* (redfish or nannygai, 450 m$_{fbd}$) is exploited by commercial and recreational fisheries around Australia and New Zealand, reaching a maximum age of 30 years and a size of 1 kg. *C. affinis* is listed by FAO in global deep-water fishery landings (Tandstad et al., 2011). *C. gerrardi* (Bight redfish, 500 m$_{fbd}$) is

(a)

Rondeletiidae
(redmouth whalefishes)

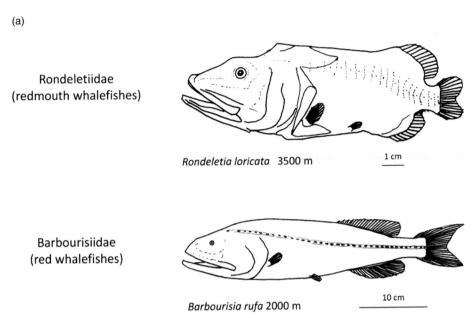

Rondeletia loricata   3500 m

1 cm

Barbourisiidae
(red whalefishes)

Barbourisia rufa 2000 m

10 cm

Figure 4.29a CETOMIMIFORMES. *Rondeletia loricata* (Redmouth whalefish) circumglobal 67°N–42°S. *Barbourisia rufa* (Velvet whalefish) circumglobal in tropical and temperate waters.

endemic to South Australia and is also taken in commercial fisheries.

Suborder Holocentroidei
4.3.19.6 Family Holocentridae (120) (Soldierfish)

This suborder comprises one family, the Holocentridae with over 80 species of mainly reef-associated shallow-water fishes known as squirrelfish and soldierfish. *Ostichthys kaianus* (Kai soldierfish, 640 m_{fbd}) is a deep-water Indo-Pacific species that is commercially exploited.

4.3.20 Order Cetomimiformes

This order comprises 33 species of small to medium-size bathypelagic fishes that occur worldwide, generally deeper than 1000 m with some species down to depths over 5000 m (Fig. 4.29). Paxton (1989) speculates that they may be the most abundant bathypelagic species at depths greater than 1800 m. There are three families and 19 genera. They are collectively known as whalefishes owing to the general whale-like shape and large gape of the mouth. Some species when freshly captured are conspicuously coloured red to orange-brown; at the depths they live, this bright colour is of no significance.

4.3.20.1 Family Rondeletiidae (121) (Redmouth whalefishes)

There is a single genus, *Rondeletia,* with two species, both about 11 cm long. *R. loricata* (Redmouth whalefish, 3500 m_{fbd}) occurs worldwide between 67°N and 42°S and feeds on various crustacea including amphipods (Fig. 4.29a). *R. bicolor* (3003 m_{fbd}) is confined to the tropical and subtropical Western Atlantic Ocean.

4.3.20.2 Family Barbourisiidae (122) (Red whalefishes)

This family is represented by a single species, *Barbourisia rufa* (Velvet whalefish, 4218 m_{fbd}), which is found throughout the world oceans in tropical and temperate latitudes (Fig. 4.29a). At a maximum length of 39 cm this is the largest of the Cetomimiformes. The body is covered with minute spinules giving the surface a velvety touch; the colour of fresh specimens is red except for darker patches, where the viscera show through the semitransparent body wall.

4.3.20.3 Family Cetomimidae (123) (Flabby whalefishes)

This family encompasses all the remaining species of Cetomimiformes and is one of the most speciose of bathypelagic fish families. A major revision occurred when Johnson et al. (2009) demonstrated that fishes from related families as listed by Nelson (2006), with very differing morphologies, the Mirapinnidae (tapetails) and Megalomycteridae (bignose fishes) are respectively the larvae and males of the Cetomimidae (Fig. 4.29b). Sexual dimorphism is extreme; the females have the definitive whalefish shape with the large mouth, tiny eyes, large cavernous lateral line canals, no pelvic fins or external scales and may grow to more than 40 cm long. In contrast the males are much smaller (3-7 cm long) with an elongated body shape, huge nasal organs and small mouth with immobile upper jaws. The stomach and oesophagus degenerate, and the males apparently do not feed during adult life, surviving on energy stored in a massive liver that occupies the peritoneal cavity together with the paired testes. Late-stage juveniles metamorphosing into males have stomachs distended with large quantities of copepods presumably to sustain them during adult life. The juveniles are distinct from both adult forms; they are elongated, have a large vertical mouth and pelvic fins. The most striking feature of the larva is a long ribbon-like extension of the tail, which may extend up to nine times the body length. This streamer extension is often shed during capture (Fig. 4.29b). The Mirapinna or hairyfish stage in which the body is covered with short hair-like pile appears to be a postlarva: Johnson et al. (2009) assigned the species *Mirapinna esau* as the postlarva of *Procetichthys kreffti*. The species of Cetomimidae are therefore known variously from larvae, postlarvae, males, females

and intermediate forms, the proper relationships between which are yet to be determined.

The genus *Ataxolepis* comprises two species, *A. apus* (1464 m_{fbd}), which is known from the West Central Pacific Ocean solely from male specimens (Fig. 4.29b), and *A. henactis,* known from the East Pacific off Panama. *Cetichthys* also has two species but described from maturing females (Paxton, 1989), *C. indagator* (3620 m_{fbd}) from the Indian, Pacific and Atlantic Oceans between 30°and 40°S and *C. parini* (5000 m_{fbd}), which is probably the deepest-living whalefish from the Indo-Pacific. The monotypic *Cetomimoides parri* is from the Western Central Pacific, and the only known specimens are males. The largest genus with seven species is *Cetomimus*, with five species from the Atlantic Ocean, one from the Pacific Ocean (Fig. 4.29b) and *C. gillii* (2300 m_{fbd}) from all three major oceans. *Cetostoma regani* (2250 m_{fbd}) has been caught in all three oceans from 50°N to 40°; out of over 150 specimens all were females and very few were captured in depths shallower than 600 m. *Danacetichthys galathenus* (2200 m_{fbd}) is known from a few specimens taken in tropical regions of the Atlantic, Indian, and Pacific Oceans and all were females with immature ovaries. *Ditropichthys storeri* (3400 m_{fbd}) occurs in all three major oceans between 45°N and 45°S and all taken were females. *Eutaeniophorus festivus* (Festive ribbonfish, 200 m_{fbd}) is circumglobal in tropical to subtropical waters except in the E. Pacific Ocean and is known only from the tapetail larvae found down to 200 m depth (Fig. 4.29b). The genus *Gyrinomimus* comprises five species distributed between 52°N and 72°S, three of which are circumglobal and all are represented by developing or adult females. *Megalomycter teevani* (1647 m_{fbd}) is known only from large-nose male specimens taken off Bermuda in the Western Atlantic Ocean and previously placed in the family Megalomyceteridae. The single specimen of *Mirapinna esau* (Hairyfish), 5.5 cm long, was captured in the Atlantic Ocean north of the Azores and is now regarded as a postlarval stage of an unknown cetomimid. *Notocetichthys trunovi* (Trunov's southern cetomimid, 1360 m_{fbd}) is described from two female specimens taken in waters off Antarctica within

(b)

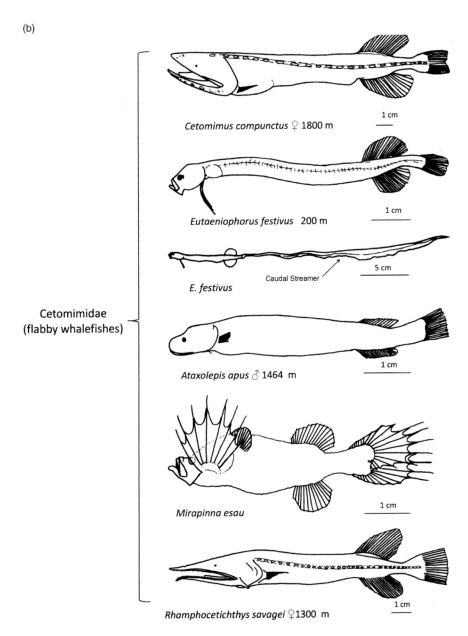

Cetomimus compunctus ♀ 1800 m   1 cm

Eutaeniophorus festivus   200 m   1 cm

Caudal Streamer   5 cm

E. festivus

Cetomimidae
(flabby whalefishes)

Ataxolepis apus ♂ 1464 m   1 cm

Mirapinna esau   1 cm

Rhamphocetichthys savagei ♀1300 m   1 cm

Figure 4.29b CETOMIMIFORMES. *Cetomimus compunctus*, NW Pacific, Japan adult female. *Eutaeniophorus festivus* (Festive ribbonfish) circumglobal 35°N–25°S except E Pacific. This fish with its extraordinary caudal streamer is the larva of a hitherto unknown adult cetomimid. *Ataxolepis apus*, W central Pacific, male of which the corresponding female remains unknown. *Mirapinna esau* (hairyfish) E Atlantic a postlarval stage of *Procetichthys kreffti*? (Johnson et al., 2009). *Rhamphocetichthys savagei* (Savage's birdsnouted whalefish) Pacific and E Central Atlantic adult female.

100 m of the seafloor. The genus *Parataeniophorus* comprises three species described from tapetail juveniles, *P. bertelseni* (1000 m$_{fbd}$) from the Central Atlantic, *P. brevis* (Short tapetail, 1400 m$_{fbd}$) from the Indo-Pacific and *P. gulosus* (1400 m$_{fbd}$) from the Atlantic and Indian Oceans. The monotypic

Centriscidae
(snipefishes &
shrimpfishes)

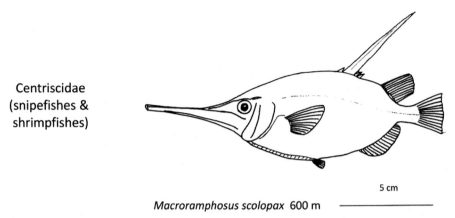

5 cm

Macroramphosus scolopax 600 m

Figure 4.30 ORDER SYNGNATHIFORMES. *Macroramphosus scolopax* (longspine snipefish) Atlantic and Indo-W Pacific 66°N–43°S.

Procetichthys kreffti (2200 m_{fbd}) is known from a female specimen taken in the SE Atlantic Ocean and is regarded as a primitive sister group to the other whalefishes (Paxton, 1989). *Rhamphocetichthys savagei* (Savage's birdsnouted whalefish, 2100 m_{fbd}) has remarkable long beak-like jaws and specimens, all immature presumably female, have originated from the tropical Pacific and Atlantic Oceans (Fig. 4.96b). *Vitiaziella cubiceps* (6200 m_{fbd}) has been taken from the Northwest Pacific in the vicinity of the Kuril-Kamchatka Trench; the maximum depth recorded is probably the depth of the seafloor at the location of capture rather than the true depth at which this species lives.

Order Gasterosteiformes

This order comprises mainly fresh and brackish water fishes, about 12 species including some sticklebacks (Gasterosteidae) and seamoths (Pegasidae) are wholly marine but they are shallow-water and coastal fishes.

4.3.21 Order Syngnathiformes

The Syngnathiformes comprises over 270 species including pipefishes and seahorses, most of which occur in shallow water. Of the five families only two have deep-water representatives, the Centriscidae and Syngnathidae (Fig. 4.30).

4.3.21.1 Family Centriscidae (124) (Snipefishes and Shrimpfishes)

Three of the five genera occur in the deep sea. The monotypic *Centriscops humerosus* (Banded yellowfish, 1000 m_{fbd}) is found circumglobally in southern temperate latitudes usually at 400–750 m depth. It is benthopelagic on continental, island, seamount and ridge slopes and grows to a length of 30 cm. *Macroramphosus scolopax* (Longspine snipefish, 600 m_{fbd}) is circumglobal in the Northern and Southern Hemispheres but occurs mainly in temperate latitudes between 20° and 40° (Fig. 4.30). It is particularly abundant off the coast of Portugal with an estimated biomass of up to 500,000 tonnes along the shelf edge distributed in depth between sardines (*Sardina pilchardus*) that occur further inshore and Blue whiting (*Micromesistius poutassou*) found in deeper waters (Marques et al., 2005). They are captured as bycatch, although efforts were made to develop a targeted fishery, and stocks have apparently declined owing to fishing pressure. Distribution is patchy with a tendency towards shoaling behaviour. Genetic evidence fails to distinguish between different putative species of *Macroramphosus* (Robalo et al., 2009) in the North Atlantic, but there may be other species elsewhere in the world. Juveniles feed mainly on pelagic invertebrates including copepods, while adults feed on bottom invertebrates. The species is listed by FAO as an exploited deep-water fish

(Tandstad et al., 2011) although it is probably best regarded as an outer shelf species. The genus *Notopogon* has four demersal species found in the southern tropical and subtropical zones predominantly at depths of 200–600 m: *N. fernandezianus* (Orange bellowfish, 580 m$_{fbd}$) in the South Pacific and SW Atlantic, *N. lilliei* (Crested bellowfish, 600 m$_{fbd}$) around S Australia and New Zealand, *N. macrosolen* (Longsnout bellowfish, 500 m$_{fbd}$) in the SE Atlantic off Namibia and South Africa and *N. xenosoma* (Longspine bellowfish, 610 m$_{fbd}$) throughout the southern Indo-Pacific from South Africa to New Zealand and New Calendonia (Duhamel, 1995). They swim with a characteristic nose-down vertical posture and the body is razor-shaped with a sharp ventral edge, maximum lengths are 17–33 cm.

4.3.21.2 Family Syngnathidae (125) (Pipefishes and Seahorses)

This is a diverse family with over 250 species, but only one can be considered to have deep-sea affinities, *Cosmocampus profundus* (Deepwater pipefish, 270 m$_{fbd}$), which occurs on the upper slopes of the Western Atlantic Ocean from Florida to the Western Caribbean. It has the typical reproductive strategy of this family in which the male retains the eggs in a brood pouch on the ventral surface of the body.

4.3.22 Order Scorpaeniformes

The Scorpaeniformes comprises 1500 species termed collectively the mail-cheeked fishes, which all have a suborbital stay bone on each side of the head connecting the suborbital bones around the eye to the preopercular bone (Fig. 4.31). Most species live in shallow waters, and Shinohara and Imamura (2007) recognise two main lineages, the Scorpeanoid lineage, including the suborders Scorpaenoidei (scorpionfishes) and Platycephaloidei (gurnards and flatheads), and the Cottoid lineage, including the suborders Anoplopomatoidei (sablefishes), Hexagrammoidei (greenlings and combfishes), Normanichthyoidei and Cottoidei (sculpins, lumpsuckers and snailfishes). Here with the addition

of suborder Dactylopteroidei (flying gurnads), we follow the classification of Nelson (2006) in recognising seven suborders. There are 33 families, 18 of which include some deep-sea species.

Suborder Dactylopteroidei

The relationship of this group to the Scorpaeniformes is questionable, and) it was provisionally ranked as a separate order, Dactylopteriformes by Shinohara and Imamura (2007).

4.3.22.1 Family Dactylopteridae (126) (Flying gurnards)

This family with seven species derives its name from the large wing-like pectoral fins; they are demersal fishes with no flying propensities. One species has been recorded in deep water: *Dactyloptena tiltoni* (Plain helmet gurnard, 565 m$_{fbd}$) in the Indo-Pacific off Western Australia and the Philippines.

Suborder Scorpaenoidei

This order with over 500 species comprises 13 families of which only five have deep-sea representatives.

4.3.22.2 Family Sebastidae (127) (Rockfishes)

The nine genera in this family all have deep-sea representatives amounting to ca. 20 per cent of the 137 described species. All have venom glands associated with the dorsal, ventral or pelvic fin spines (Smith and Wheeler, 2006). Viviparity is the dominant form of reproduction, but in some species, after internal fertilization the eggs are released before they hatch (so-called zygoparity).

The genus *Helicolenus* generally occurs between 190 and 600 m depth; four species are endemic to different parts of the Pacific Ocean, one species in the SW Atlantic and one around southern Africa in the SE Atlantic and the SW Indian Oceans. *H. dactylopterus* (Blackbelly rosefish, Bluemouth, 1100 m$_{fbd}$) is the most widespread species (Fig. 4.31a) and is found in the Atlantic Ocean (70°N–46°S) as well as the

(a)

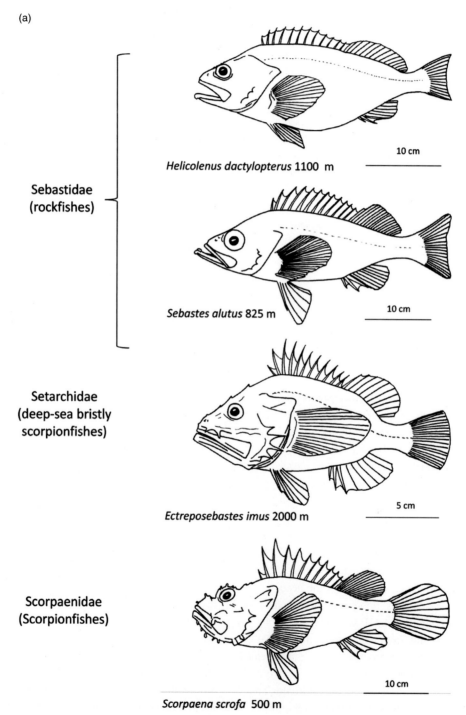

Sebastidae
(rockfishes)

Helicolenus dactylopterus 1100 m

10 cm

Sebastes alutus 825 m

10 cm

Setarchidae
(deep-sea bristly
scorpionfishes)

Ectreposebastes imus 2000 m

5 cm

Scorpaenidae
(Scorpionfishes)

Scorpaena scrofa 500 m

10 cm

Figure 4.31a ORDER SCORPAENIFORMES. *Helicolenus dactylopterus* (blackbelly rosefish) Atlantic. *Sebastes alutus* (Pacific ocean perch) N Pacific. *Ectreposebastes imus* (Midwater scorpionfish) circumglobal in tropical and temperate waters. *Scorpaena scrofa* (Red scorpionfish) E Atlantic.

Mediterranean Sea, where the greatest biomass occurs between 200 and 500 m depth (Pirrera et al., 2009). They reach a maximum age of 30 years and a length of 36 cm with the males larger and older than the females. In the Porcupine Seabight area of the NE Atlantic it is recorded at 247–853 m depth (Priede et al., 2010). Reproduction is by zygoparity, in which the embryos develop within the ovary and are extruded in a gelatinous mass from which the larvae escape. *H. dactylopterus* is exploited commercially around the Atlantic Ocean, appearing as a bycatch in deep-sea prawn fisheries in the Mediterranean Sea and in targeted fisheries with 50,000 tonnes per annum landed in fish markets in Catalonia, SE Spain (Ribas et al., 2006). Submersible observations show that it is a cryptic sit-and-wait predator on the upper slopes (Uiblein et al., 2003) often associated with reefs and coral mounds. *H. percoides* (Red gurnard perch, 750 m_{fbd}) and *H. barathri* (Bigeye sea perch, 800 m) occur off South Australia and New Zealand. In New South Wales they are collectively known as 'Ocean Perch'. *H. percoides* or inshore ocean perch is captured at 80–350 m depth, whereas *H. barathri*, the offshore ocean perch, occurs on the upper slope at depths 250–800 m. Ocean perch are regarded as being fished within acceptable limits. Maximum age has been recorded as 60 years but most landings are in the 25–35 cm size class, less than 15 years old (Rowling et al., 2010). The genus *Hozukius* comprises two species of rockfish described from the NW Pacific around seamounts and off Japan down to 1000 m depth.

The genus *Sebastes* has over 100 species more than 90 of which are found in the North Pacific Ocean. Some of these are abundant and form the basis for major fisheries. The mean maximum depth of occurrence is 400 m; here we shall consider only those species that occur in deeper waters. In the Atlantic Ocean there is one species of interest, *S. mentella* (Beaked redfish, 1444 m_{fbd}), also known as the deep-sea redfish to distinguish it from the shallower-occurring *S. norvegicus* (Golden redfish, 1000 m_{fbd}, formerly *S. atlanticus*). *S. mentella* is found around the continental margins of the North Atlantic from Nova Scota in the West, to the Irminger Sea,

Greenland, Iceland, Reykjanes Ridge, the Norwegian Sea and Barents Sea at depths of 300–1000 m. This viviparous species releases larvae during April to June and occurs in shoals feeding on euphausids, hyperiids, cephalopods, chaetognaths and small fishes, reaching a maximum size of 55 cm. *S. mentella* are exploited by pelagic trawlers fishing down to 1000 m depth with annual landings of 120–130,000 tonnes from the Irminger Sea and adjacent areas by vessels of 15 different nations (Sigurðsson, 2006). All the other deep-water *Sebastes* species are from the North Pacific, where there are major fisheries for *S. aleutianus* (Rougheye rockfish, 900 m_{fbd}) and *S. alutus* (Pacific ocean perch, 825 m_{fbd}; Fig. 4.31a), which are both distributed from Japan to the Bering Sea, Aleutian Islands and off California, as well as *S. aurora* (Aurora rockfish, 769 m_{fbd}) and *A. diploproa* (Splitnose rockfish, 800 m_{fbd}) in the NE Pacific and *S. flammeus* (500 m_{fbd}) in the NW Pacific off Japan. *S. levis* (Cowcod, 500 m_{fbd}) is of interest as a gamefish off California. There are 12 further bathydemersal species around the northern Pacific Ocean that are not significantly targeted by fisheries. Catches of Pacific ocean perch (*S. alutus*) expanded during the 1960s reaching 480,000 tonnes in 1965, followed by a decline. From 2010 management permits an annual catch of 30–50,000 tonnes.

The monotypic *Adelosebastes latens* (1200 m_{fbd}) is described from the Emperor Seamounts in the NW Pacific Ocean. The genus *Sebastolobus* comprises three North Pacific species, *S. alascanus* (Shortspine thornyhead, 1600 m_{fbd}) from Japan to California, *S. altivelis* (Longspine thornyhead, 1757 m_{fbd}) from the Aleutian Islands to California and *S. macrochir* (Broadbanded thornyhead, 1504 m_{fbd}) from the NW Pacific. They produce eggs in gelatinous egg masses. *S. alascanus* and *S. altivelis* occur at oxygen minimum zone depths between 430 and 1010 m depth off California; they have low aerobic metabolic capacities and small gills, which reflects more sedentary behaviour than most Scorpaeniformes (Friedman et al., 2012). Life span may be up to 100 years in *S. alascanus* for which there is a minor fishery interest. The genus *Trachyscorpia* comprises seven demersal species. *T. cristulata* has two subspecies in the North Atlantic, *T. cristulata* (Atlantic thornyhead, 1100 m$_{fbd}$) from the W. Atlantic, Massachusetts to the Gulf of Mexico and *T. cristulata echinata* (Spiny scorpionfish, 2500 m$_{fbd}$) from the E. Atlantic, Ireland to Senegal and also the Western Mediterranean Sea. *T. cristulata echinata,* which grows up to 50 cm in length, is landed commercially by deep-sea trawl fisheries in the NE Atlantic (Clarke et al., 2015). Three species of *Trachyscorpia* have been described from the Pacific Ocean, one from the Indian Ocean and *T. eschmeyeri* (Cape rockfish, 1025 m$_{fbd}$) known from the SE Atlantic and off New Zealand.

4.3.22.3 Family Setarchidae (128) (Deep-sea bristly scorpionfishes)

This family comprises seven species in three genera. *Ectreposebastes imus* (Midwater scorpionfish, 2000 m$_{fbd}$) is the Scorpaeniform species most adapted to a bathypelagic lifestyle (Fig. 4.31a), the skeleton is poorly ossified, the scales are thin and the muscles are soft with a high water content; however, the venom glands in at least the anal fin spines are retained. It occurs in all the major oceans usually at 500–850 m depth beneath tropical and subtropical waters. The genus *Lioscorpius* comprises two species of small demersal scorpionfish from the upper slopes of the W. Pacific and Western Australia at 200–400 m depth. *Setarches guentheri* (Channeled rockfish, 780 m$_{fbd}$) has a very wide distribution in all three major oceans at tropical latitudes and appears as a bycatch in some bottom trawl fisheries. Its range has extended as far north as Portugal in the NE Atlantic (Velasco et al., 2010). Its congener, *S. longimanus* (Red deepwater scorpionfish, 704 m$_{fbd}$), is only found in the Indo-West Pacific.

Family Neosebastidae has no deep-sea species

4.3.22.4 Family Scorpaenidae (129) (Scorpionfishes)

This family comprises over 200 species of mainly shallow-water demersal species, of which about 10 per cent are found in the deep water down to a maximum recorded depth of 920 m. The genus

Idiastion comprises three species, two of which are found in the Pacific Ocean, *I. pacificum* (Flame humpback scorpionfish, 430 m$_{fbd}$), a benthopelagic species from the Central North Pacific seamounts and ridges and *I. hageyi* (522 m$_{fbd}$), known from the steep slopes of the Galápagos Islands. *I. kyphos* (622 m$_{fbd}$) is from the slopes of the central Western Atlantic and the Caribbean Sea. These are small fishes less than 13 cm long. The genus *Neomerinthe* comprises 11 species of scorpionfish distributed at average depths of 100–330 m in the tropical Indo-West Pacific, the Red Sea and the Atlantic Ocean ranging down to 920 m. The genus *Phenacoscorpius* with five species occurs over an average depth range of 330–570 m in the tropical Indo-Pacific and the Western Atlantic. *Pontinus* comprises 20 tropical and subtropical species, some of which are known from upper slope depths such as *P. kuhlii* (Offshore rockfish, 600 m$_{fbd}$), which occurs between 42°N and 8°S in the Eastern Atlantic and the Mediterranean. There are over 60 species of *Scorpaena* distributed worldwide, of which only two relatively large species (maximum length 50 cm) occurring in the Eastern Atlantic Ocean might be considered to be deep-sea fishes. *S. scrofa* (Red scorpionfish or large-scaled scorpionfish, 500 m$_{fbd}$) extends from south of the British Isles to Senegal, including Madeira, the Canary Islands, Cape Verde and throughout the Mediterranean Sea (Fig. 4.31a). It is solitary and sedentary over rocky, sandy or muddy bottoms, where it feeds on fishes, crustaceans and molluscs. *S. scrofa* can constitute 30 per cent of the catch by some longliners in Greek waters (Katsanevakis et al., 2010), averages ca. 6 per cent of the total fish catch in the Azores (Pham et al., 2013) and is listed as an exploited deep-sea species by FAO (Tandstad et al., 2011). The deeper-living *S. elongata* (Slender rockfish, 800 m$_{fbd}$) has a similar geographic range but is of lesser commercial interest. The genus *Scorpaenodes* comprises 29 species of mainly reef-associated shallow-water scorpionfishes and only *S. rubrivinctus* (412 m$_{fbd}$) observed on rocky reefs with sponges at 200–400 m depth off the Galápagos and Cocos Islands in the Eastern Pacific is found in deep water.

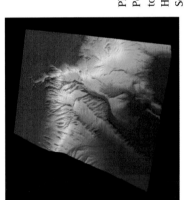

(a)

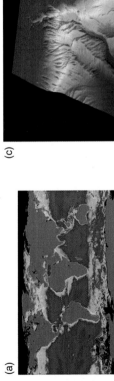

(c)

Plate 1(a) Global sea surface chlorophyl pigment concentration averaged over operational duration of the Coastal Zone Color Scanner (CZCS) between 1978 and 1986. The image is in false colour with red denoting 10 mg.m⁻³ and violet <0.01 mg.m⁻³. Image from NASA/GSFC-DAAC (Feldman et al., 1989). (*A black and white version of this figure will appear in some formats.*)

Plate 1(c) The Nazare Canyon off the coast of Portugal. The depth range is from <100 m inshore to >5000 m on the abyssal plain. (Courtesy Veerle Huvenne, National Oceanography Centre, Southampton, 3 D bathymetry from the EU FP6 HERMES project). (*A black and white version of this figure will appear in some formats.*)

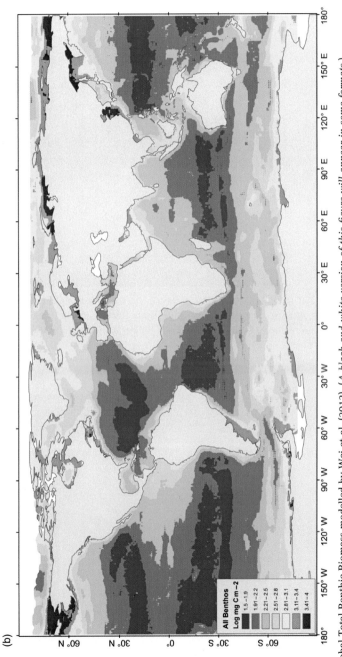

Plate 1(b) Global Total Benthic Biomass modelled by Wei et al. (2013). (*A black and white version of this figure will appear in some formats.*)

(a)

Plate 2(a) ORDER MYXINIFORMES. Myxinidae (Hagfishes) *Eptatretus deani* (Black hagfish) at a gray whale (*Eschrichtius robustus*) carcass at 1675 m depth in the Santa Cruz Basin, NE Pacific Ocean. After 18 months most of the whale's soft tissue has been removed. (Courtesy. Craig Smith, University of Hawaii). (*A black and white version of this figure will appear in some formats.*)

(b)

Plate 2(b) ORDER CHIMAERIFORMES. Chimaeridae (Chimaeras). *Hydrolagus affinis* (Smalleyed rabbitfish) at 2000 m depth in the SW Atlantic Ocean off Angola (Courtesy. Alan Jamieson, University of Aberdeen). This is the world's deepest recorded Chimaerid at 3000 m. (*A black and white version of this figure will appear in some formats.*)

(a)

Plate 3(a) ORDER SQUALIFORMES. Somniosidae (Sleeper sharks) *Centroscymnus coelolepis* (Portuguese dogfish) at 2000 m depth in the South West Atlantic Ocean off Angola. Recorded down to 3000 m depth this is the world's deepest living shark. (Courtesy. Alan Jamieson, University of Aberdeen). (*A black and white version of this figure will appear in some formats.*)

(b)

Plate 3(b) ORDER ANGUILLIFORMES. Eurypharyngidae, *Eurypharynx pelecanoides* (Gulper or Pelican eel) pelagic capture over the Mid-Atlantic Ridge, North Atlantic Ocean. (Photo courtesy David Shale). (*A black and white version of this figure will appear in some formats.*)

(a)

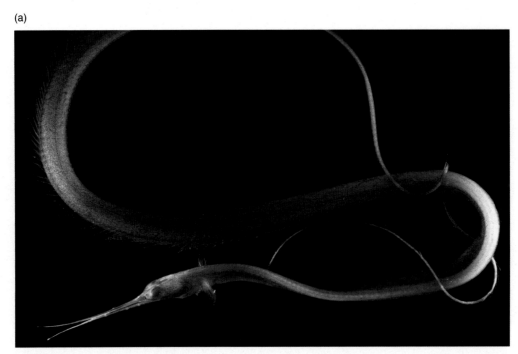

Plate 4(a) ORDER ANGUILLIFORMES. Nemichthyidae. *Nemichthys* sp. (snipe eel) pelagic capture over the Mid-Atlantic Ridge, North Atlantic Ocean. (© David Shale / naturepl.com). (*A black and white version of this figure will appear in some formats.*)

(b)

Plate 4(b) ORDER STOMIIFORMES. Gonostomatidae. *Cyclothone pallida* (Tan bristlemouth). Pelagic capture over the Mid-Atlantic Ridge, North Atlantic Ocean. (© David Shale / naturepl.com). (*A black and white version of this figure will appear in some formats.*)

(a)

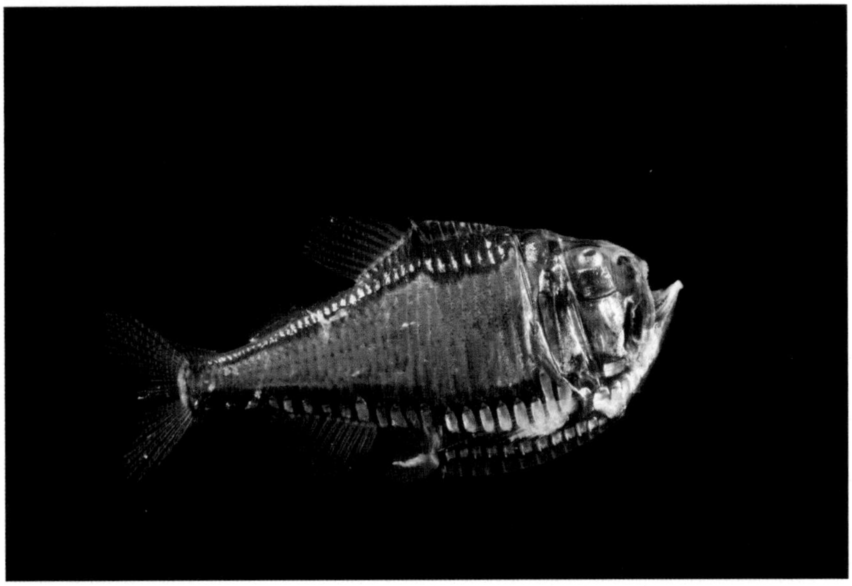

Plate 5(a) ORDER STOMIIFORMES. Sternoptychidae. *Argyropelecus gigas* (Hatchetfish) pelagic capture over the Mid-Atlantic Ridge, North Atlantic Ocean. (© David Shale / naturepl.com). (*A black and white version of this figure will appear in some formats.*)

(b)

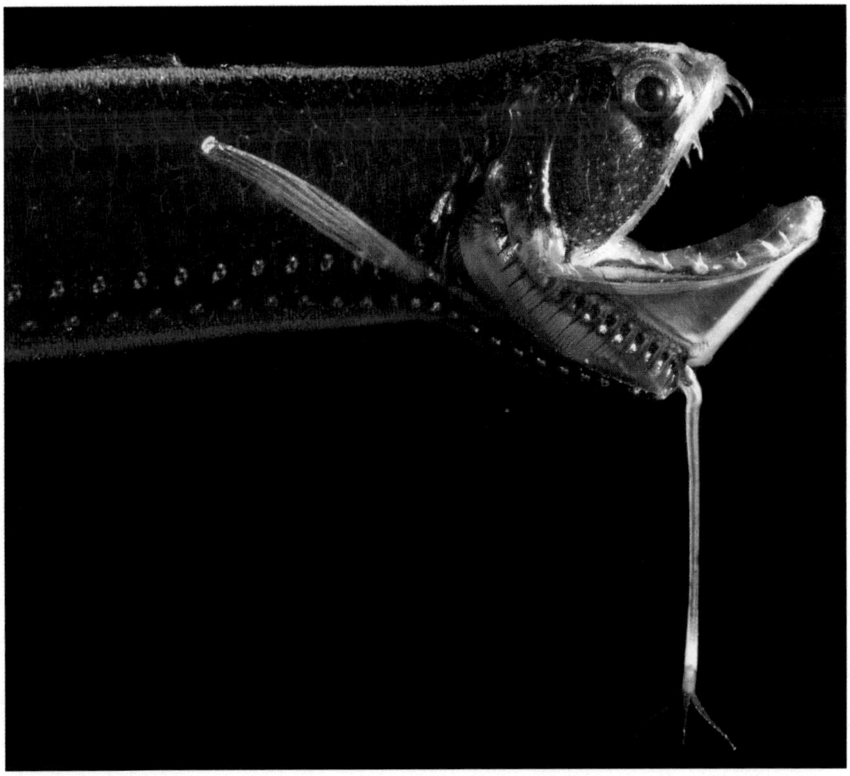

Plate 5(b) ORDER STOMIIFORMES. Stomiidae. *Stomias boa* (Boa dragonfish) detail of head showing the barbel lure. Pelagic capture over the Mid-Atlantic Ridge, North Atlantic Ocean. (© David Shale / naturepl.com). (*A black and white version of this figure will appear in some formats.*)

(a)

Plate 6(a) ORDER MYCTOPHIFORMES. Myctophidae. *Myctophum punctatum* (Spotted lanternfish) pelagic capture over the Mid-Atlantic Ridge at 100–150 m depth during the night. (© David Shale / naturepl.com). (*A black and white version of this figure will appear in some formats.*)

(b)

Plate 6(b) ORDER GADIFORMES. Macrouridae *Coryphaenoides armatus* (Abyssal grenadier). Attracted to baited camera at abyssal depth in the NE Atlantic. (Photo Alan Jamieson). (*A black and white version of this figure will appear in some formats.*)

(a)

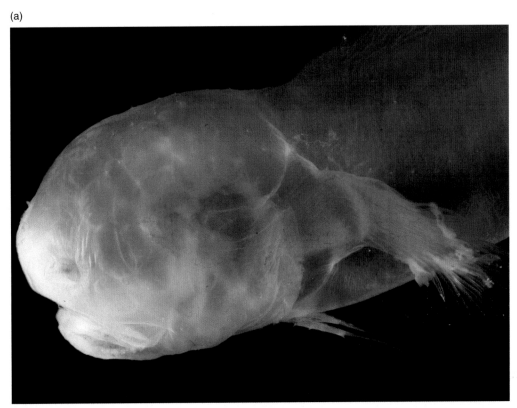

Plate 7(a) ORDER OPHIDIIFORMES (Cusk eels). Aphyonidae (Blind cusk eels). *Aphyonus gelatinosus* (Gelatinous blindfish). Benthic Atlantic Ocean (© David Shale / naturepl.com). (*A black and white version of this figure will appear in some formats.*)

(b)

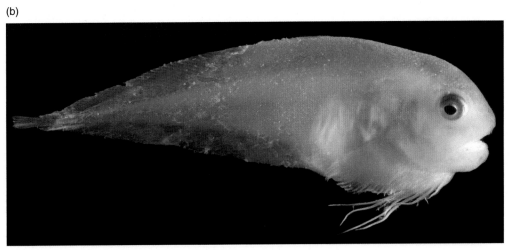

Plate 7(b) ORDER SCORPAENIFORMES Liparidae (Snail fishes) *Careproctus* sp. 2362 m depth in the Barents Sea, Atlantic Arctic region. (© David Shale / naturepl.com). (*A black and white version of this figure will appear in some formats.*)

(a)

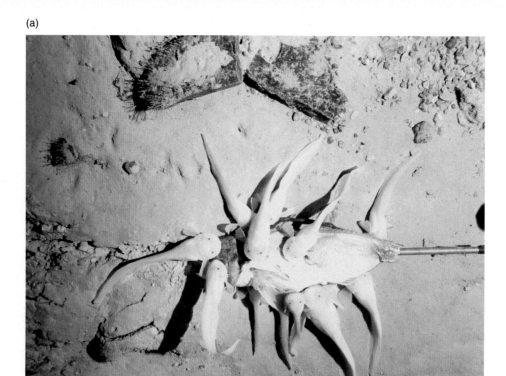

Plate 8(a) ORDER SCORPAENIFORMES Liparidae (Snailfishes) *Notoliparis kermadecensis* Hadal snailfishes at 7561 m depth in the Kermadec trench SW Pacific Ocean. One of the world's deepest living fishes. (Courtesy, Alan Jamieson) (*A black and white version of this figure will appear in some formats.*)

(b)

Plate 8(b) ORDER LOPHIIFORMES Himantolophidae (footballfishes). *Himantolophus* sp. Juvenile bathypelagic female anglerfish with lure, Atlantic Ocean (© David Shale / naturepl.com). (*A black and white version of this figure will appear in some formats.*)

4.3.22.5 Family Tetarogidae (130) (Waspfishes)

Of 25 species, only one of these venomous fishes occurs at depths greater than 200 m, *Neocetropogon mesedai* (383 m_{fbd}) from the Western Indian Ocean and the Red Sea.

4.3.22.6 Family Plectrogeniidae (131) (Stinger flatheads)

With one genus, *Plectrogenium,* this family comprises two species of small (<7 cm) fishes from the Indo-Pacific region found at depths greater than 250 m, *P. nanum* (Dwarf Thornyhead, 650 m_{fbd}), widespread in the Indo-Pacific (Fig. 4.31b) and *P. barsukovi* (310 m_{fbd}) from the SE Pacific Nasca Ridge. They should probably be classified with the other flatheads in the Platycephaloidei (Nelson, 2006).

4.3.22.7 Family Congiopodidae (132) (Pigfishes, Horsefishes or Racehorses)

This family comprises nine species from the Southern Hemisphere found in temperate and sub-Antarctic regions with a characteristic pig or horse-like projecting snout with a small mouth. Three of these species are considered bathydemersal, *Alertichthys blacki* (549 m_{fbd}) and *Congiopodus coriaceus* (deep-sea pigfish, 385 m_{fbd}) from New Zealand waters of the SW Pacific (Fig. 4.31b), and *C. kieneri* (500 m_{fbd}) from the SE Pacific off Peru and Chile. The scales are minute, and the skin of *C. coriaceus* is described as thick, smooth and leathery (Paulin and Moreland, 1979). The common name of *C. coriacus* serves to distinguish it from a shallower-living congener rather than to indicate that it lives at great depths.

Suborder Platycephaloidei

This suborder comprises over 260 species in six families, none of which have a maximum depth greater than 1000 m.

4.3.22.8 Family Triglidae (133) (Searobins, Gurnards).

These are demersal fishes with characteristic prehensile anteriorly-placed free pelvic fin rays on which they can 'walk' on the seafloor detecting benthic prey items. This specifically adapts them to feeding on soft sediments. The swim bladder is well developed with drumming muscles for producing sound. Most of the ca. 120 species live at continental shelf depths. The genus *Lepidotrigla* has nine species, which are described as bathydemersal, living offshore at depths down to 350 m in the Western Pacific Ocean, mainly around islands. Most of the deeper-living Triglidae are in the genus *Pterygotrigla*. *P. andertoni* (Painted latchet, 500 m_{fbd}) is generally caught between 200 and 500 m depth off New Zealand and Australia with small amounts landed commercially. The closely related *P. picta* (Spotted gurnard, 500 m_{fbd}), which occurs at similar depths (Fig. 4.31b) around islands and banks in the S Pacific and off Chile, is listed by FAO as a deep-water fish resource (Tandstad et al., 2011). Five uncommon or rare species of *Pterygotrigla* occur in the W Pacific with depth ranges of 250–500 m. *Trigla lyra* (Piper gurnard, 700 m_{fbd}) from the Eastern Atlantic and the Mediterranean is of minor commercial interest and mostly caught at the shallow end of its depth range.

4.3.22.9 Family Peristediidae (134) (Armoured searobins, Armoured gurnards)

These fishes are rather similar to the Triglidae though the body is encased by four rows of spiny plates on each side and they occupy a deeper average depth range of 250–450 m. They have multiple barbels on the lower jaw, and the snout is flattened, often with a pair of anterior projections. With a total length averaging 24 cm, they are caught as bycatch, although the armouring reduces their commercial value. *Heminodus philippinus* (500 m_{fbd}) and *Paraheminodus murrayi* (Murray's armoured gurnard, 710 m_{fbd}) are endemic to the Indo-West Pacific. The genus *Peristedion* has 10 deep-water species endemic to the Western Central Atlantic with varying latitudinal and depth distributions from the Georges Bank to Brazil though mostly clustered in tropical latitudes. In the Gulf of Mexico *Peristedion* species (Fig. 4.31b) are found in shelf and upper slope fish assemblages at 188–785 m depth (Powell et al., 2003).

(b)

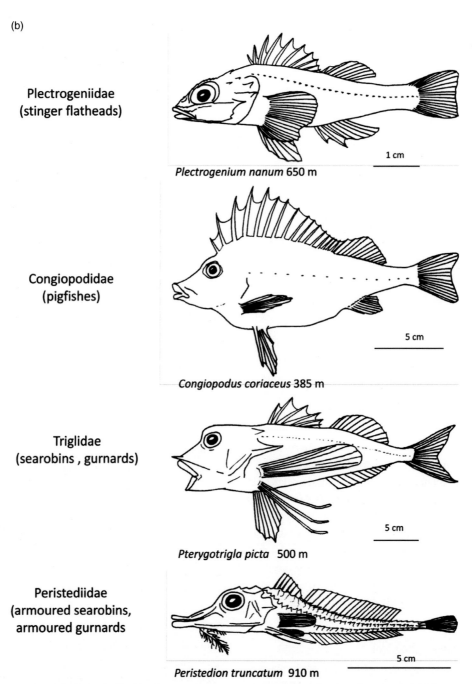

Plectrogeniidae
(stinger flatheads)

Plectrogenium nanum 650 m

Congiopodidae
(pigfishes)

Congiopodus coriaceus 385 m

Triglidae
(searobins , gurnards)

Pterygotrigla picta   500 m

Peristediidae
(armoured searobins,
armoured gurnards

Peristedion truncatum 910 m

Figure 4.31b ORDER SCORPAENIFORMES. *Plectrogenium nanum*, Indo-Pacific. *Congiopodus coriaceus* (deep-sea pigfish) SW Pacific New Zealand. *Pterygotrigla picta* (Spotted gurnard) S Pacific. *Peristedion truncatum* (black armoured searobin) W Central Atlantic and Gulf of Mexico.

Four deep-water *Peristedion* species are endemic to the Pacific Ocean, one to the Indian Ocean and two having an Indo-West Pacific distribution. The genus *Satyrichthys* is exclusively Indo-Pacific in its distribution with nine deep-water species at tropical and subtropical latitudes from Costa Rica in the east to the Arabian Sea in the west at depths down to 858 m (Kawai, 2013). Some species grow to over 50 cm in length and are exploited in small-scale fisheries. The monotypic *Scalicus orientalis* (510 m_{fbd}) is confined to the tropical West Pacific.

4.3.22.10 Family Bembridae (135) (Deep-water flatheads)

The Bembridae are considered more primitive than the true flatheads Platycephalidae and are represented by eight species in three genera that occur in the tropical Indo-Pacific region. The term *deep-water* refers to their habitat on upper slopes. The genus *Bembradium* has two species endemic to the Pacific Ocean with *B. roseum* (800 m_{fbd}) the deepest and most widespread (Fig. 4.31c) from Japan, Indonesia to the Hawaiian Islands and the coast of Chile. *Bembras* comprises four species known from the West Pacific Ocean and the Eastern Indian Ocean north of Australia and one species known only from the Gulf of Aden. The monotypic *Brachybembras aschemeieri* (583 m_{fbd}) is from the Western Pacific.

4.3.22.11 Family Parabembridae (136) (Sprat-like flatheads)

The two species in this family are often included in the Bembridae (Nelson, 2006). The deeper-living species is *Parabembras robinsoni* (African deep-water flathead, 600 m_{fbd}) found from the Western Indian Ocean between southern Mozambique and Durban, South Africa.

Family Platycephalidae (Flatheads)

These are shallow-water species predominantly of the Indo-Pacific Region.

4.3.22.12 Family Hoplichthyidae (137) (Ghost flatheads)

This family comprises about eight species of benthic fishes known only from the Indo-West Pacific in tropical to temperate latitudes. They have an extremely broad flattened head with a slim round cross-section tail region. *Hoplichthys hasswelli* (Armoured flathead or Deep-Sea flathead, 1500 m) is found on the outer shelf and continental slopes in the SW Pacific around Australia south of 26°S. It is common in New Zealand waters, where it is usually caught at 25–30 cm length (maximum 45 cm (Fig. 4.31c). Although often caught in fisheries the meat yield is too low to warrant significant exploitation (CSIRO, 2011).

Suborder Anoplopomatoidei

4.3.22.13 Family Anoplopomatidae (138) (Sablefishes)

This family comprises two species both endemic to the North Pacific Ocean. *Anoplopoma fimbria* (Sablefish, 2740 m_{fbd}) is found around the North Pacific from southern Japan to central Baja California including the Bering Sea, off Kamchatka and Alaska 64°N–23°N (Fig. 4.31c); *Erilepis zonifer* (Skilfish, 680 m_{fbd}) has a similar distribution over a smaller latitudinal range 62°N–36°N. These are large fishes that grow to 120 and 183 cm length, respectively, and are both valued in fisheries. *A. fimbria* is superficially a salmon-shaped fish living on the muddy seafloor at 300–600 m. Spawning occurs on the bottom, and eggs rise to 200–500 m depth, where they hatch. The larvae then sink deeper utilising their yolk sac reserves before a swift ascent to the surface, where they feed and grow rapidly in the neuston. Late-stage juveniles descend to the seafloor at shelf depths and thereafter habitat depth increases with age and size (Moser, 1996). The fishery for *A. fimbria* is based on trawls, traps and longlines. The peak annual reported catch was 64,900 tonnes in 1972, and since 2010 landings have been around 20,000 tonnes per annum. *E. zonifer* is caught in much smaller amounts and is targeted by sport fishermen. Russian longliners have focussed on catching the giant *E. zonifer* on the Emperor Seamounts (Anon, 2014).

Suborder Hexagrammoidei

This order comprises about 12 demersal species known as greenlings endemic to the coastal waters of the North Pacific and occasionally recorded to over 500 m depth.

(c)

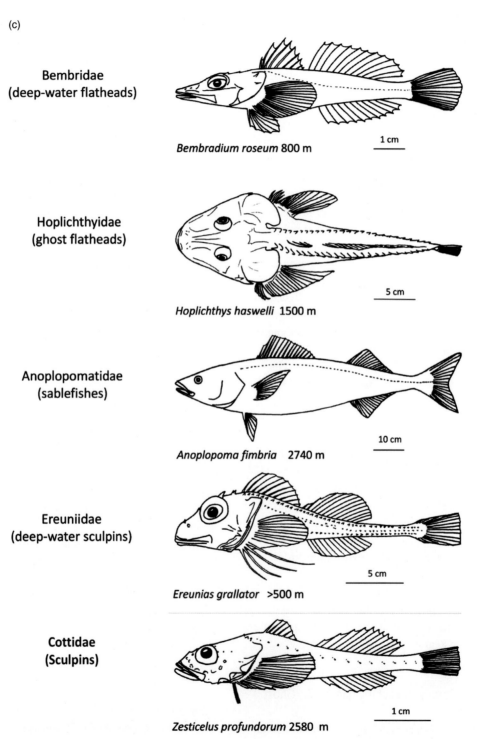

Bembridae
(deep-water flatheads)

Bembradium roseum 800 m

1 cm

Hoplichthyidae
(ghost flatheads)

Hoplichthys haswelli 1500 m

5 cm

Anoplopomatidae
(sablefishes)

Anoplopoma fimbria 2740 m

10 cm

Ereuniidae
(deep-water sculpins)

Ereunias grallator >500 m

5 cm

Cottidae
(Sculpins)

Zesticelus profundorum 2580 m

1 cm

Figure 4.31c ORDER SCORPAENIFORMES. *Bembradium roseum*, Pacific. *Hoplichthys haswelli* (Armoured flathead) SW Pacific. *Anoplopoma fimbria* (Sablefish) N Pacific. *Ereunias grallator*, NW Pacific Japan. *Zesticelus profundorum* (flabby sculpin) N Pacific.

Suborder Normanichthyoidei

This suborder comprises just one shallow-water species.

Suborder Cottoidei

The Cottoidei includes approximately 700 marine species in nine families that occupy a vast diversity of demersal habitats throughout the world's oceans from inshore to depths of over 7000 m in hadal trench environments of the North and South Pacific Ocean. There are also numerous freshwater species emphasising the remarkable habitat versatility of this group of fishes.

4.3.22.14 Family Ereuniidae (139) (Deep-water sculpins)

This little-known family with three species is distinguished from typical sculpins of the family Cottidae by the presence of free pectoral fin rays similar to those in triglids. They probably live at a minimum depth of 500 m in the North West Pacific, *Ereunias grallator* (Fig. 4.31c) and *Marukawichthys ambulator* off Japan and *M. pacificus* found on the Emperor Seamount Chain.

4.3.22.15 Family Cottidae (140) (Sculpins)

Members of this large family occur in freshwater, brackish and marine environments. Over 180 species are marine, mostly at continental shelf depths, and they are absent from the Indian Ocean. There are a few deep-sea species in the Pacific Ocean. The genus *Antipodocottus* has two species found at depths of 400–600 m in the SW Pacific, *A. galatheae* (Galathea sculpin, 594 m_{fbd}) off Australia and *A. megalops* (600 m_{fbd}) off New Zealand. Several species are endemic to deep water of the North Pacific. The monotypic *Artediellina antelope* (615 m_{fbd}) is endemic to the Sea of Okhotsk, and *Artediellus* spp. occur at similar depths in the Sea of Japan and off the Kuril islands. The monotypic *Bolinia euryptera* (410 m_{fbd}) is found in at depths greater than 200 m in the Bering Sea (Yabe, 1991). The genus *Icelus* has approximately six species living in deep waters of the marginal seas of the NW Pacific down to 900 m depth. The monotypic *Rastrinus scutiger* (740 m_{fbd}) occurs in the Bering Sea and the Gulf of Alaska, *Stlengis* has two deep-water species off Japan, and the monotypic *Triglops scepticus* (Spectacled sculpin, 925 m_{fbd}) is found from the Sea of Japan to Alaska. The genus *Zesiticulus* has the three deepest-living cottid species, *Zesticelus bathybius* (1270 m_{fbd}) minimum depth 700 m off Japan, *Z. ochotensis* (1845 m_{fbd}) minimum depth 1000 m in the Sea of Okhotsk and *Z. profundorum* (Flabby sculpin, 2580 m_{fbd}), which though rare is the most wide-ranging from the Bering Sea to Baja California, Mexico (Fig. 4.31c). Cottids spawn adhesive demersal eggs, and the larvae are pelagic.

4.3.22.16 Family Hemitripteridae (141) (Sea ravens)

This family comprises eight generally shallow-water species. There is one deep-water species, *Hemitripterus bolini* (Bigmouth sculpin, 925 m_{fbd}) that usually occurs at 200–300 m depth around the North Pacific from the Bering Sea to northern California; grows to over 70 cm in length (Fig. 4.31d).

4.3.22.17 Family Agonidae (142) (Poachers)

These rather elongated Scorpaeniform fishes, of average maximum length 20 cm, with a body covered with bony plates and large fan-like pelvic fins, range from intertidal depths to over 1000 m depth. Few can be considered true deep-sea species; *Bathyagonus nigripinnis* (Blackfin poacher, 1290 m_{fbd}) usually occurs at 400–700 m in the North Pacific from the Bering Sea to Northern California (Fig. 4.31d), two *Percis* spp. are found down to 500 m in the NW Pacific and two *Sarritor* in the N Pacific from Japan to Alaska. The genus *Xeneretmus* comprises four species from the NE Pacific with an average depth range of 80–400 m.

4.3.22.18 Family Psychrolutidae (143) (Fathead sculpins and Blobfish)

This family comprises 30 species with an average depth range of 430–970 m occurring in the Atlantic,

(d)

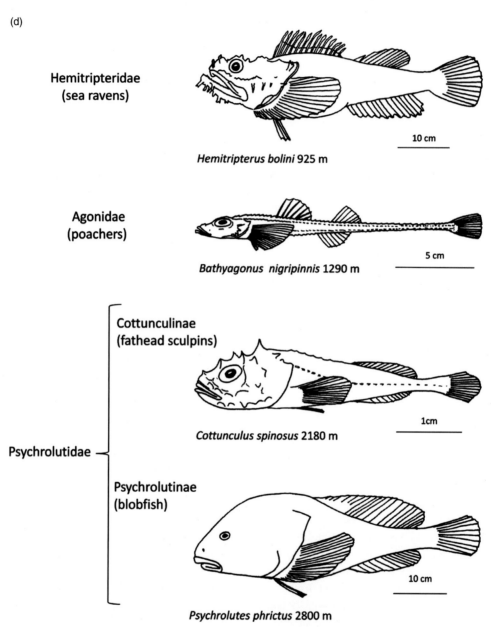

Hemitripteridae
(sea ravens)

10 cm

Hemitripterus bolini 925 m

Agonidae
(poachers)

5 cm

Bathyagonus nigripinnis 1290 m

Cottunculinae
(fathead sculpins)

1cm

Cottunculus spinosus 2180 m

Psychrolutidae

Psychrolutinae
(blobfish)

10 cm

Psychrolutes phrictus 2800 m

Figure 4.31d ORDER SCORPAENIFORMES. *Hemitripterus bolini* (Bigmouth sculpin) N Pacific. *Bathyagonus nigripinnis* (blackfin poacher) N Pacific. *Cottunculus spinosus*, SE Atlantic. *Psychrolutes phrictus* (blob sculpin) N Pacific.

Indian, Pacific and Arctic Oceans with an antitropical distribution with only a few species in warm-temperate environments. They can be divided into two almost equal subfamilies, the Cottunculinae (fathead sculpins), which have a well-ossified large head usually with spines, and the Psychrolutinae (blobfish), which have a poorly ossified head and no spines (Nelson, 2006). While as in the Cottidae the eggs are demersal, these eggs are large, 4.5 mm diameter, which suggests that they may be able to hatch directly into a demersal lifestyle. Fecundity in view of the large egg size is low, *ca.* 200 eggs in adults about 30 cm long (Hare and Maranncik, 2005).

Subfamily Cottunculinae (Fathead sculpins)

The genus *Ambophthalmos* comprises three species endemic to SW Pacific waters around New Zealand at depths of 200–600 m. The genus *Cottunculus* has five species endemic to the Atlantic Ocean. Two species range across the N Atlantic: *C. thomsonii* (Pallid sculpin, 1600 m_{fbd}) from the east coast of North America to N W Africa and *C. microps* (Polar sculpin, 1342 m_{fbd}) further north from New Jersey in the west to SE Greenland, Spitzbergen, Bear Island and the English Channel in the east. *C. sadko* (700 m_{fbd}) occurs in the Arctic Ocean off Northern Greenland, the Kara Sea and the Norwegian Sea area of the North Atlantic. *C. tubulosus* (1950 m_{fbd}) is described from the northern Mid-Atlantic Ridge. There are three species of *Cottunculus* in the Southern Hemisphere, *C. spinosus* (2180 m_{fbd}) from the S E Atlantic at depths greater than 1400 m (Fig. 4.31d), *C. granulosus* (Fathead, 1250 m_{fbd}) around South America from southern Brazil to Chile and *C. nudus* (Bonyskull toadfish, 1123 m_{fbd}) in the SW Pacific off New Zealand. The monotypic *Dasycottus setiger* (Spinyhead sculpin, 850 m_{fbd}) is found around the North Pacific slopes from the Sea of Japan to Washington USA. The genus *Eurymen* comprises two species from the North Pacific that reach a maximum depth of 400 m. The four species of *Malacocottus* inhabit the upper slopes of the North Pacific: *M. aleuticus* (Whitetail sculpin, 1000 m_{fbd}) usually occurs at 300–600 m depth from the Sea of Okhotsk to Prince William Sound Alaska, *M. gibber* (1200 m_{fbd}) from Northern Japan to South Korea, *M. kincaidi* (Blackfin sculpin, 1025 m_{fbd}) and *M. zonurus* (Darkfin sculpin, 1980 m_{fbd}) both range from Northern Japan in the west to Washington State USA in the east.

Subfamily Psychrolutinae (Blobfish)

The genus *Ebinania* has two deep-living species in the Pacific Ocean, *E. vermiculata* (1000 m_{fbd}) off Japan, and *E. gyrinoides* (794 m_{fbd}) in the W. Central Pacific,

one in the Indian Ocean off southern Australia, *E. australiae* (1170 m_{fbd}), one in the Eastern Atlantic, *E. costaecanariae* (700 m_{fbd}). There are two species in the Southern Ocean: *E. macquariensis* (Macquarie blobfish, 714 m_{fbd}) around Macquarie Kerguellen Islands and *E. malacocephala* (603 m_{fbd}) in the S. Pacific sector. *Gilbertidia pustulosa* (700 m_{fbd}) is found in the Sea of Okhotsk. *Neophrynichthys heterospilos* (290 m_{fbd}), although not yet recorded at great depth, is thought to occur in deep water around New Zealand. The genus *Psychrolutes* comprises 11 species, 10 of which occur in the deep sea. Four species have been recorded in the Atlantic Ocean, *P. subspinosus* (1750 m_{fbd}) in the NE Atlantic including around Iceland, *P. inermis* (1550 m_{fbd}) in the Eastern Atlantic from Mauritania to South Africa and also the Western Indian Ocean, *P. macrocephalus* (1012 m_{fbd}) in the SE Atlantic and *P. marmoratus* in the SW Atlantic but also the SE Pacific off South America. Four species are endemic to the Pacific Ocean: *P. paradoxus* (Tadpole sculpin, 1100 m_{fbd}) and *P. phrictus* (Blob sculpin, 2800 m_{fbd}) in the North Pacific (Fig. 4.31d), *P. marcidus* (Blobfish, 1200 m_{fbd}) and *P. microporos* (986 m_{fbd}) in the Southwest Pacific off South Australia and New Zealand, respectively. *P. occidentalis* (Western Australian sculpin, 704 m_{fbd}) is found in the SE Indian Ocean and *P. sio* (967 m_{fbd}) in the SE Pacific. *Psychrolutes* are benthic feeders on polychaetes, crustaceans and molluscs.

4.3.22.19 Family Bathylutichthyidae (144)

This family comprises a single species, *Bathylutichthys taranetzi* (1650 m_{fbd}), known from one specimen 10 cm long discovered in deep water in the Southern Ocean near South Georgia. It is similar to *Psychrolutes* though with a conspicuous barbel on each lower jaw at the corner of the mouth and the caudal fin joined to the dorsal and ventral fins (Nelson, 2006).

4.3.22.20 Family Cyclopteridae (145) (Lumpsuckers)

These are mainly shelf-dwelling species in temperate to Arctic waters of the Northern Hemisphere with a characteristic globose shape and pelvic fins modified to form a ventral sucking disc. *Cyclopterus lumpus* (Lumpfish, 868 m_{fbd}) of the North Atlantic and Arctic makes seasonal migrations into deep water and feeds pelagically on gelatinous prey such as ctenophores and medusas. They are close to neutrally buoyant enabling efficient pelagic foraging but the sucking disc is necessary to hold station in strong currents at shelf depths where they guard their egg masses on the seafloor during the breeding season (Davenport and Kjørsvik, 1986). Other species such as *Aptocyclus ventricosus* (Smooth lumpsucker, 1700 m_{fbd}) from the North Pacific have also been recorded in deep water.

4.3.22.21 Family Liparidae (146) (Snailfishes)

Together with the Cyclopteridae the Liparidae comprise a monophyletic group recognised by Nelson (2006) as forming a super-family, the Cyclopteroidea. Like the lumpsuckers, Liparidae have a ventral sucking disc, and they produce demersal eggs, which are often guarded by the parents. The body is elongated with a scaleless jellylike skin. The dorsal and ventral fins are long, extending to the tail and almost joining the caudal fin. This is a remarkably successful group ranging geographically from the Arctic to the Antarctic and from the intertidal zone to hadal depths of over 7000 m. They are thought to have originated in the North Pacific from where they have spread northward into the Arctic and North Atlantic and southward into the Southern Ocean and the deep sea (Knudsen et al., 2007). The present-day distribution is most species-rich in the Pacific, Arctic and Antarctic Oceans with 30 genera and over 380 species, *ca.* 250 of which are found in the deep sea. The deep-sea genera are regarded as most derived and have secondarily lost the ventral sucking disc: *Acantholiparis*, some *Careproctus*, *Edentoliparis*, *Genioliparis*, *Lipariscus*, *Nectoliparis*, *Odontoliparis*, *Paraliparis*, *Psednos*, *Rhinoliparis* and *Rhodichthys*. The sucker is largest in shallow-water species to enable holding station in strong current regimes but at greater depths with weaker currents the sucker becomes smaller and is eventually lost in the abyssal genera. Bathyal species generally have a distinct

spawning season, whereas abyssal species spawn throughout the year releasing small batches (<10) of large eggs *ca.* 5 mm diameter, which hatch as benthic juveniles (Stein, 1980). Here the genera are arranged according to the molecular phylogeny of Knudsen et al. (2007) with those they did not analyse inserted according to the phylogeny of Balushkin (1996). The general arrangement is therefore from basal shallow-water genera with a large sucker to the more derived deep-water forms with no ventral sucker.

The monotypic *Nectoliparis pelagicus* (tadpole snailfish, 3383 m_{fbd}) occupies the most basal position in liparid phylogeny and is a bathypelagic species up to 6.5 cm long (Fig. 4.31e) found around the North Pacific Ocean from Northern Japan to California. Although basal, it has secondarily lost the ancestral Cyclopteroid ventral disc (Knudsen et al., 2007) presumably because it serves no function in a pelagic environment. *N. pelagicus* commonly occurs in the Monterey Canyon and has been observed by ROV at 419–541 m depth actively swimming in midwater or floating static in a curled posture (Stein et al., 2006). The genus *Liparis* occurs worldwide and comprises *ca.* 60 generally demersal shallow-water species with large ventral discs 7–15 per cent of standard length in diameter. A few species are found in deep water including *L. fabricii* (Gelatinous snailfish, 1800 m_{fbd}) in the Arctic Ocean and North Atlantic 84°N–42°N and *L. zonatus* (910 m_{fbd}) in the Northwest Pacific. The genus *Crystallichthys* comprises three demersal species recorded at depths down to 830 m in the North Pacific in the Kamchatka, Bering Sea, Aleutian and Gulf of Alaska regions. The monotypic *Crystallias matsushimae* (700 m_{fbd}) is found from the Sea of Japan to the Bering Sea. Also monotypic, *Palmoliparis beckeri* (Gloved snailfish, 800 m_{fbd}) is described from the northern Kuril Islands in the NW Pacific and is noted for extensions of the anterior pelvic fin rays with terminations that look remarkably like a pair of hands. The ventral sucking disc diameter is ca. 10 per cent of standard length, and individuals can grow to 1 kg weight (Balushkin, 1996).

The genus *Careproctus* (Plate 7b) comprises 112 species of predominantly bathyal snailfishes with a reduced ventral sucker (2–7 per cent of

(e)

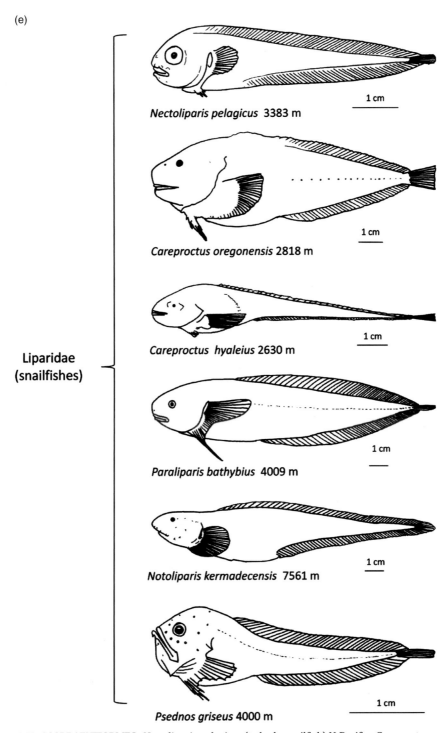

Liparidae (snailfishes)

Nectoliparis pelagicus 3383 m

1 cm

Careproctus oregonensis 2818 m

1 cm

Careproctus hyaleius 2630 m

1 cm

Paraliparis bathybius 4009 m

1 cm

Notoliparis kermadecensis 7561 m

1 cm

Psednos griseus 4000 m

1 cm

Figure 4.31e ORDER SCORPAENIFORMES. *Nectoliparis pelagicus* (tadpole snailfish) N Pacific. *Careproctus oregonensis* (smallfin snailfish) NE Pacific. *Careproctus hyaleius,* endemic to hydrothermal vents of the E Pacific Ridge. *Paraliparis bathybius* (black seasnail) Arctic and NE Atlantic. *Notoliparis kermadecensis,* SW Pacific endemic to the Kermadec Trench. *Psednos griseus,* E Pacific Mexico.

standard length). There are 48 species endemic to the North Pacific Ocean mostly occurring along the ocean margin from Japan to the Bering Sea, Alaska and the west coast of North America (Fig. 4.31e). There are three species in the Arctic Ocean, two of which extend also into the North Atlantic, and 11 species in the North Atlantic. The North Atlantic species can be divided into two kinds: relatively shallow forms originating from the Arctic Ocean with affinities to North Pacific species, and recently discovered abyssal species *C. merretti* (Merret's snailfish, 3990 m_{fbd}) and *C. aciculipunctatus* (4100 m_{fbd}), which are assumed to have colonised the abyssal Atlantic from the Southern Ocean (Andriashev, 1998). There are 45 species of *Careproctus* in the Southern Hemisphere, 25 of which are endemic to the Southern Ocean. Of seven S Pacific species two are in the SW Pacific off New Zealand and Australia and four in the SE Pacific off Chile and extending into the Southern Ocean. 11 of the 12 South Atlantic species are in the SW Atlantic in the Patagonian and Scotia Arc region. There is only one species in the SE Atlantic off South Africa, *C. albescens* (1460 m_{fbd}), and one species recorded in the Indian Ocean off the Crozet islands, *C. crozetensis* (4290 m_{fbd}; Duhamel and King, 2007). As recently as 1990 there was thought to be only one species of *Careproctus* in the Southern Ocean, and the discovery of the high diversity and importance of these fishes in the south has been a remarkable development in ichthyological research. Andriashev (1998) points out that *Careproctus* in the Southern Hemisphere are relatively rare on the shelf and are common in the abyss down to 5500 m depth, whereas the genus *Paraliparis* is found on the shelf and down to 2500 m, the reverse of the pattern in the Northern Hemisphere, where *Paraliparis* is the deeper-living genus.

The deepest-living *Careproctus* species is *C. sandwichensis* (5459 m_{fbd}), apparently endemic to the South Sandwich Trench. It is not yet clear whether the relative absence of *Careproctus* and other Liparid genera from the Indian Ocean is because of lack of sampling effort or a true zoogeographic phenomenon. *Careproctus hyaleius* (2630 m_{fbd}) was discovered on hydrothermal vent sites along the East Pacific Rise,

swimming actively like a tadpole amongst *Riftia* worms and on blocks of basalt surrounding the smoker chimneys. They are slim and transparent in appearance, up to over 10 cm in length (Fig. 4.31e) and females were observed with clutches of eggs 3.5 to 5.5. mm diameter (Geistdoerfer, 1994). Fishes of the genus *Liparis* are known to lay eggs on various substrates including living bivalve shells and polychaete tubes (Overdick et al., 2014); however, several species of *Careproctus* lay their eggs in the gill cavities of living lithoid crabs (stone crabs), where they develop and hatch. Parrish (1972) found over 200 newly hatched larvae of *C. melanurus* (Blacktail snailfish, 2286 m_{fbd}) on box crab (*Lopholithoides foraminatus*) taken at 100 m depth in Monterey Bay off California. The larvae were free-swimming but apparently remain with the crab, suggesting a symbiotic relationship. Stein (2006) observed *Careproctus*, possibly *C. melanurus*, at 1432 m depth in the Monterey Canyon attaching to the legs of *Paralomis* crab. In the Southern Ocean *Careproctus* 20–90 mm long have been observed hitching rides on crabs attached by the sucker on the back or the legs of *Paralomis formasa* at 625–1525 m depth (Yau et al., 2000). Approximately 6 per cent of crabs had such passengers and up to three fish were seen on a single crab. Larger *Careproctus* could move on and off crabs easily and would opportunistically feed when the crab arrived at bait. Fish were also seen on *Lithodes santolla* but not the spiny *Paralomis spinosissim*. In the North Pacific and the Sea of Okhotsk, eggs of *Careproctus* spp. were found on the gills of three species of stone crabs in 1–12 per cent of individuals at depths of 800–1200 m (Poltev, 2013). *C. abbreviatus* (1143 m_{fbd}) are known to deposit eggs in lithoid crabs in this region. To aid egg deposition either on substrates or within the branchial cavities of crabs female *Careproctus* have an evertable ovipositor organ.

The genus *Volodichthys* includes three species previously considered within *Careproctus*, possibly representing a group of earlier colonisers of the Southern Hemisphere. They are distinguished by relatively large ventral suckers and fewer anal and dorsal fins rays than in the *Careproctus* southern species (Balushkin, 2012). *V. solovjevae* (2012 m_{fbd}) is a new species captured on longlines together with Antarctic toothfish *Dissostichus mawsoni* in the Southern Ocean off Eastern Antarctica in the Cooperation Sea area. Other species in the genus are all from off S. America: *V. herwigii* (1250 m$_{fbd}$) and *V. smirnovi* (1570 m$_{fbd}$) and *V. parini* (1570 m$_{fbd}$) further south off the South Shetland Islands.

Gyrinichthys minytremus (640 m$_{fbd}$), *Temnocora candida* (Bigeye snailfish, 400 m$_{fbd}$), *Lopholiparis flerxi* (Hardhead snailfish, 284 m$_{fbd}$), *Squaloliparis flerxi* (900 m$_{fbd}$), *Pseudonotoliparis rassi* (2200 m$_{fbd}$) and *Prognatholiparis ptychomandibularis* (Wrinkle-jaw snailfish, 455 m$_{fbd}$; Orr and Busby, 2001) are all from monotypic genera endemic to the North Pacific Rim. *Allocareproctus* comprises five species from the same area with an average depth range of 200–500 m. *A. jordani* (Cherry snailfish, 631 m$_{fbd}$) grows to 34 cm length at the age of eight years and has a most varied diet, including polychaetes, amphipods, decapods, gastropods, octopods, ophiuroidea, fishes and discarded fish offal (Orlov and Tokranov, 2011). The monotypic *Eknomoliparis chirichignoae* (920 m$_{fbd}$) is from the SE Pacific off Chile.

The genera *Pseudoliparis* from the Northern Hemisphere and *Notoliparis* from the Southern Hemisphere are endemic to deep-sea trench systems. They are distinguished from other liparids by the presence of characteristic pores on the cephalic canals, skeletal features, reduced eyes and lack of pigmentation, appearing pale pink and transparent in life. *P. amblystomopsis* has been captured from depths of 6156 to 7587 m in the Japan and Kuril-Kamchatka Trenches of the North West Pacific Ocean but aggregations of up to 20 individuals have been seen attracted to baited cameras deployed at 7703 m in the Japan Trench (Fujii et al., 2010). *P. belyaevi* is recorded from 6380 to 7587 m in the Japan Trench and was recognised by Chernova et al. (2004) as the deepest known living fish. The genus *Notoliparis* comprises four species: *N. antonbruuni* (6150 m$_{fbd}$), described from one rather damaged specimen taken in the Peru Trench in the SE Pacific Ocean (Stein, 2005), *N. kermadecensis* (Fig. 4.31e, Plate 8a) from 6474 to 7561 m depth in the Kermadec Trench of the SW Pacific (Jamieson, 2015), *N. kurchatovi* (5474 m$_{fbd}$),

described from one specimen captured in the South Orkney Trench in the Southern Ocean and *N. macquariensis* (5410 m$_{fbd}$) from the Macquarie-Hjort Trench, in the Southern Ocean. All these species are endemic to their respective trenches or trench systems and have not been observed in adjacent abyssal plain regions nor in the deepest parts of the trenches. During a dive of the bathscaphe *Archimède* at 7300 m depth in the Puerto-Rico Trench in the NW Atlantic Pérès (1965) observed numerous liparids, which were very similar in general appearance and behaviour to those from the Pacific Trenches, but because the bathyscaphe had no means of photographing or capturing specimens, the species remains unidentified. This depth is similar to the maximum observed depths of *Pseudoliparis* and *Notoliparis* in the Pacific Ocean. Jamieson (2015) shows an image of a possible further new species of hadal liparid from the Peru–Chile Trench at 7050 m depth and Linley et al. (2016) observed two putative new liparid species from the Mariana Trench at over 8000 m depth

Menziesichthys bacescui (1317 m$_{fbd}$) in its monotypic genus is known only from its type locality, the Peru–Chile Trench, but Chernova et al. (2004) state that it is probably an Ophidiid.

The genus *Osteodiscus* comprises two species of deep-living but large-disked liparids from the North Pacific, *O. andriashevi* (1950 m$_{fbd}$) from the Sea of Okhotsk in the NW and *O. cascadiae* (Bigtail snailfish, 3000 m$_{fbd}$) from the NE Pacific off British Columbia and Oregon. The genus *Elassodiscus* comprises three species from the North Pacific with reduced or vestigial ventral discs: *E. caudatus* (Blackbelly snailfish, 1040 m$_{fbd}$) from Alaska to California, *E. obscurus* (1773 m$_{fbd}$) from the NW including the Sea of Okhotsk and *E. tremebundus* (1800 m$_{fbd}$) from Japan to the Aleutian Islands. The genus *Rhinoliparis* comprises two small slim snailfishes with no sucker disc occurring around the North Pacific Rim down to 2189 m depth.

Paraliparis is the most speciose genus of liparids with over 130 species, all of which have no ventral disc. They are deep-water bathyal demersal species with an average minimum depth of occurrence of

1027 m, though a few species have been reported from abyssal depths greater than 3000 m. Two species are particularly associated with the Arctic Ocean, *P. bathybius* (Black seasnail, 4009 m$_{fbd}$), which is found in the Norwegian and Greenland Seas, Baffin Bay as well as the Central Polar Basin (Fig. 4.31e) and *P. violaceus* (2365 m$_{fbd}$) from the Nansen Basin. 31 species are endemic to the North Pacific Ocean but the greatest diversity is found in the Southern Ocean from which 55 species have been described, many of them recent discoveries. In addition to these Southern Ocean species, 23 species have been described from the SE Indian Ocean off Australia, 10 from the South Pacific and three from the SW Atlantic making a total of over 94 Antarctic and sub-Antarctic species currently described. There are 10 species in the North Atlantic with *P. copei copei* (Blacksnout seasnail, 1976 m$_{fbd}$), possibly the most widespread from Canadian waters in the NW Atlantic to Cape Point in the SE Atlantic, it is epibenthic, feeding on ctenophores and cnidaria. *P. murieli* (600 m$_{fbd}$) is described from one specimen taken in the Western Mediterranean Sea off Spain.

Eutelichthys leptochirus (700 m$_{fbd}$) in its monotypic genus is known from the Western Mediterranean basin off the coast of Spain, has no ventral sucker, is benthopelagic at bathyal depths and is distinguished from *Paraliparis* by the presence of only five branchiostegal rays (Mattallanas, 2000). Also monotypic, *Lipariscus nanus* (Pygmy snail fish, 910 m$_{fbd}$) is a small slim fish distributed around the North Pacific from Japan to California. *Rhodichthys regina* (Threadfin seasnail, 2365 m$_{fbd}$) is a large (up to 31 cm) liparid from the Central Arctic Basin that is also found in deep cold waters of the NE Atlantic north of the Wyville-Thomson Ridge. *Genioliparis* comprises three species of large-headed fishes with massive lower jaws, one from the NE Pacific, *G. ferox* (2884 m$_{fbd}$) and two from the Southern Ocean. Balushkin and Voskoboinikiva (2008) suggest the genus originated in the North Pacific and penetrated to the south along the western coasts of North and South America in parallel with *Careproctus* and *Paraliparis*. *G. kafanovi* (1727 m$_{fbd}$) is a deep-bodied fish up to 34 cm long, and the holotype was extracted from the stomach of

an Antarctic toothfish (*Dissostichus mawsoni*) caught by longline in the Ross Sea. *Edentoliparis terraenovae* (550 m$_{fbd}$) is a very unusual purely pelagic liparid (length up to 6.3 cm) with no teeth on the jaws or pharyngeal apparatus, distributed around the Antarctic (Andriashev, 1998).

Until 1978 the genus *Psednos* was known from only one species, *P. micrurus* (Barnard's dwarf snailfish, 1208 m$_{fbd}$) that had been described in 1927; a small pelagic fish 4.5 cm long from the SE Atlantic Ocean off South Africa. In 1992 a second species, *Psednos christinae* (1500 m$_{fbd}$) was discovered in the NE Atlantic and active searches since then have revealed *Psednos* to have a worldwide distribution (Chernova and Stein, 2004) with 35 species described at the time of writing. Named the dwarf snailfishes (Andriashev, 1998), they are small mesopelagic or bathypelagic fishes with an elaborate sensory canal system on the head. They become sexually mature at a length of just 3–4 cm and although pelagic are often associated with deep slopes. Specimens are small, fragile and easily damaged, compounding the problems of studying this elusive group of fishes (Chernova and Stein, 2002). The currently known distribution of the genus is 15 species in the Pacific Ocean, 15 species in the Atlantic Ocean and five from the Indian Ocean. *Psednos* are absent from the North Pacific Rim, where liparids originated, possibly reflecting the highly derived character of this genus. The northernmost species in the Pacific are *P. cathetostomus* (2030 m$_{fbd}$) and *P. griseus* (4000 m$_{fbd}$) from 32°N off California and Northern Mexico (Fig. 4.31e). Most Pacific species are from the Southern Hemisphere including off New Zealand, Australia and Chile. In the North Atlantic the distribution is from 65°N in the Davis Strait off W. Greenland, *P. groenlandicus* (Greenland dwarf snailfish, 1055 m$_{fbd}$) to 35°S off Cape Point South Africa where *P. micrurus* is found.

4.3.23 Order Perciformes

The order Perciformes is the largest order of fishes and in fact is the largest order of vertebrates occurring in all major aquatic habitats, including lakes, rivers and oceans, with over 7500 species recorded in the marine environment (Fig. 4.32). Their classification is to some extent due for revision in the light of new evidence from molecular genetics (Near et al., 2013). Following Nelson (2006) we recognise 19 suborders, of which three occur almost exclusively in freshwater: Percoidei, Elassomatoidei, Zoarcoidei, Notothenioidei, Trachinoidei, Pholidichthyoidei, Blennioidei, Icosteoidei, Callionymoidei, Gobioidei, Kurtoidei (Freshwater), Acanthuroidei, Scombrolabracoidei, Scombroidei, Stromateoidei, Anabantoidei (Freshwater), Channoidei (Freshwater) and Caproidei. About 500 species, 7 per cent of the marine perciformes, are recognised as deep-sea fishes, and these are mostly concentrated in certain families.

Suborder Percoidei

This suborder includes about 3000 marine species of which about 125 are deep-sea species. Here we consider only the families with deep-sea representatives.

4.3.23.1 Family Howellidae (147) (Oceanic basslets)

This family comprises eight species in three genera (Prokofiev, 2007). They are small, less than 10 cm long pelagic or benthopelagic fishes with large eyes, often with black scaly skin, that are found circumglobally in tropical and subtropical regions. *Bathysphyraenops simplex* (500 m_{fbd}) occurs in tropical regions of all the major oceans, whereas *B. declivifrons* is confined to the West Pacific. The genus *Howella* has five species: the two most widespread are *H. brodiei* (Pelagic basslet, 1829 m_{fbd}) and *H. sherborni* (Sherborn's pelagic basslet, 950 m_{fbd}). *H. brodiei* (Fig. 4.32a) is circumglobal in tropical to subtropical regions and the Atlantic Ocean and is found from 20°S to 65°N with the adults furthest north. In the Atlantic Ocean *H. sherborni* is confined to the Southern Hemisphere and also occurs in the South Pacific and south of Australia (Post and Quèro, 1991). It appeared to be a Southern Hemisphere species until it was discovered in the North Pacific off Alaska by Busby and Orr (1999). Post and Quèro (1991) found that some juveniles ascend to near the surface at night whilst larger specimens generally

occur deeper. Adult *H. brodiei* are found between 700 m and 1800 m depth during the day and 400–1000 m during the night, whereas *H. sherborni* adults occur between 1500 and 2700 m during the day and 100 to 2000 m during the night. In the SE Atlantic *S. sherborni* is recorded as feeding on pelagic crustacea (Macpherson, 1989). The other three *Howella* species are all found in the Pacific Ocean. Herring (1992) identified the pyloric caeca and posterior intestine of *H. brodiei* as bioluminescent organs with six or seven caeca emitting blue light through the ventral body wall anterior to the pelvic fins. The monotypic *Pseudohowella intermedia* (700 m_{fbd}) is known only from a few specimens found in the Western Central Pacific off Papua New Guinea and from the Hawaiian Islands.

4.3.23.2 Family Acropomatidae (148) (Lanternbellies)

The Acropomatidae, also known as temperate ocean basses, reach an average maximum length of 19 cm, are usually silvery in colour and occur down to a mean maximum depth of 460 m. Although described as a 'motley and ill-defined assemblage of fishes' (Fisher and Bianchi, 1984) they are the deepest-living oceanic representatives of a diverse group formerly known as the Percichthyidae (Yamanoue and Yoseda, 2001). Only one genus, *Acropoma*, has light organs, which are ventral between the pelvic fins. The light is produced by bacteria contained in a U-shaped duct alongside the gut connected to the pyloric caecae and a small pore near the anus. Dorsally there is a reflector membrane, and the ventral muscles are transparent, allowing light to penetrate to the exterior (Thacker and Roje, 2009). There are five species. *A. japonicum* (Glowbelly, 500 m_{fbd}) occurs in the Indo-West Pacific from East Africa to Japan and Australia on sandy and muddy slopes (Fig. 4.32a). The other *Acropoma* species are all confined to the West Pacific. All species are exploited to varying degrees by bottom trawl fisheries.

The monotypic *Apogonops anomalus* (Three-spined cardinalfish, 600 m_{fbd}) is endemic to Australia and occurs usually between 100 and 400 m depth on the outer shelf and slope. Also monotypic *Doederleinia*

(a)

Howellidae
(oceanic basslets)

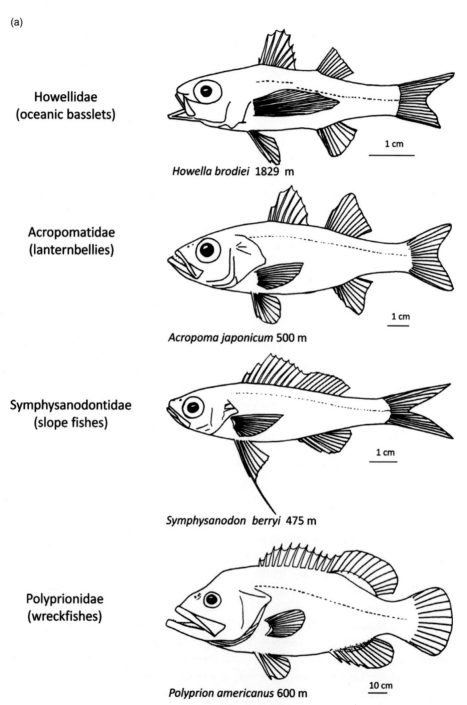

Howella brodiei 1829 m

Acropomatidae
(lanternbellies)

Acropoma japonicum 500 m

Symphysanodontidae
(slope fishes)

Symphysanodon berryi 475 m

Polyprionidae
(wreckfishes)

Polyprion americanus 600 m

Figure 4.32a ORDER PERCIFORMES. *Howella brodiei* (pelagic basslet) circumglobal. *Acropoma japonicum* (glowbelly) Indo-W Pacific. *Symphysanodon berryi* (slope bass) W Atlantic. *Polyprion americanus* (wreckfish) Atlantic, Mediterranean, Indian and SW Pacific.

berycoides (Rosy seabass, blackthroat seaperch, 600 m$_{fbd}$), the largest of the Acropomatidae reaching 40 cm in length, is found in the Eastern Indian Ocean and the Western Pacific off Japan and Australia. It feeds demersally on large crustacea and molluscs and is highly valued in commercial fisheries. The genus *Malakichthys* comprises seven small pelagic seabass species often associated with the outer shelf and upper slope regions in the Indo-West Pacific, for example *M. barbatus* (600 m$_{fbd}$), which grows to 18 cm length and off the coast of Japan is distributed between 200 and 600 m depth (Yamanoue and Yoseda, 2001). *Neoscombrops* comprises three species: *N. atlanticus* (192 m$_{fbd}$) from the tropical Western Atlantic, *N. cynodon* (Silver splitfin, 550 m$_{fbd}$) from the Indian Ocean and *N. pacificus* (500 m$_{fbd}$) from the Pacific Ocean associated with islands and seamounts. The genus *Synagrops* comprising 11 species is circumglobal at temperate and tropical latitudes. *S. adeni* (Aden splitfin, 600 m$_{fbd}$) is fished commercially at depths of 280–600 m in the NW Indian Ocean and is much smaller (11 cm maximum length) than *S. japonicus* (Blackmouth splitfin, 800 m$_{fbd}$), which grows to 35 cm length, is also commercially exploited and occurs throughout the Indo-Pacific from East Africa to Hawaii. Four species are endemic to the West Pacific Ocean and three species to the Atlantic Ocean, where *S. microlepis* (Thinlip splitfin, 500 m$_{fbd}$) is reported to feed on pelagic crustacea including euphausiids, mysids and decapods. Two species, *S. bellus* (Blackmouth bass, 910 m$_{fbd}$) and *S. spinosus* (Keelcheek bass, 544 m$_{fbd}$), have an Atlanto-Pacific distribution.

4.3.23.3 Family Symphysanodontidae (149) (Slope fishes)
This family comprises 12 species in a single genus *Symphysanodon* of small to medium-sized fishes that occur on the outer edge of the continental shelf and upper slope at an average depth range of 165–380 m, also around islands and on submarine ridges in tropical regions 25° N–25° S. (Anderson and Springer, 2005). There are three species endemic to the W. Atlantic including *S. berryi* (Slope bass, 476 m$_{fbd}$; Fig. 4.32a), five species in the Indian Ocean and four

species in the Pacific Ocean. They occur on offshore trawl grounds and are of minor commercial interest.

4.3.23.4 Family Polyprionidae (150) (Wreckfishes)
This family comprises four species of large archetypal perciform fishes that grow to 1.5–2.5 m length and over 100 kg weight, all of which are exploited both for game and commercial fishing. There are two genera, *Polyprion* and *Stereolepis*, each with two species. Owing to their size they are very conspicuous where they occur in deep-water catches or surveys. *P. americanus* (Wreckfish, 600 m$_{fbd}$) occurs from 70°N–55°S in the Atlantic Ocean, as well as the Mediterranean Sea, the Western Indian Ocean and the SW Pacific off New Zealand (Fig. 4.32a). It is a demersal species found on rough seafloors, in caves and around shipwrecks, feeding on crustaceans, cephalopods and benthic fishes. *P. oxygeneios* (Hapuku wreckfish, 854 m$_{fbd}$) is circumglobal in the Southern Hemisphere (28°S–50°S). *S. doederleini* (600 m$_{fbd}$) occurs at depths greater than 400 m in the NW Pacific off Japan, Russia and South Korea. The largest species, *S. gigas* (Giant seabass), is found no deeper than 50 m.

4.3.23.5 Family Serranidae (151) (Sea basses and Groupers)
This family comprises over 500 species, most of which are reef-associated shallow-water species. A few genera contain bathydemersal species found down to a maximum depth of just over 600 m. The genus *Anthias* comprises colourful species of the tropical and subtropical Atlantic Ocean, some of which occur at depths greater than 200 m: *A. woodsi* (Swallowtail bass, 421 m$_{fbd}$) in the Western Atlantic and *A. cyprinoides* and *A. helenensis* (460 m$_{fbd}$) in the Eastern Atlantic on deep island slopes. *Bathyanthias* comprises three species from the tropical Western Atlantic found down to 400 m depth. *Epinephelus suborbitalis* (Seamount grouper, 570 m$_{fbd}$) is known only from the Minami-Koto Seamount in the NW Pacific. Eight of the 14 grouper species in the genus *Hyporthodus* are reported from depths greater than 200 m: *H. mystacinus* (Misty grouper, 400 m$_{fbd}$) from

the Western Atlantic and the Eastern Pacific and *H. octofasciatus* (Eightbar grouper, 300 m$_{fbd}$) are considered to be deep-water specialists. *Lepidoperca aurantia* (New Zealand orange perch, 604 m$_{fbd}$) is a minor commercial species endemic to waters around New Zealand commonly found at 200–400 m depth that feeds mainly on euphausiids and grows to 26 cm length. Catches of *L. aurantia* in trawl surveys on the Chatham Rise indicate a patchy distribution with locally high abundance (Horn et al., 2013). *Pronotogrammus eos* (Bigeye bass, 325 m$_{fbd}$) occurs on both sides of the Isthmus of Panama on upper slopes of the tropical East Pacific and Western Atlantic and is of minor commercial interest. *Pseudanthias rubrolineatus* (400 m$_{fbd}$) is reported from deep water off New Caledonia in the Pacific Ocean.

4.3.23.6 Family Ostracoberycidae (152) (Shellskin alfonsinos)

This family comprises three species of distinctive silvery-coloured demersal fishes with a large head and eyes that occur at depths of 250–700 m in the Indo-West Pacific region. There is only one genus, *Ostracoberyx*, comprising three species: *O. dorygenys* (711 m$_{fbd}$), the largest, which grows to 23 cm length distributed from Madagascar to the Philippines (Fig. 4.32b), *O. fowleri* off Japan and *O. paxtoni* (Spinycheek seabass, 512 m$_{fbd}$) off SE Australia (Quéro and Ozouf-Costaz, 1991).

4.3.23.7 Family Callanthiidae (153) (Gropos)

Also known as the 'splendid perches', these are highly colourful fishes generally found at depths beyond the reach of conventional scuba diving. There are 14 species in two genera. *Callanthias* comprises five species endemic to the Pacific Ocean, one species from the Indian Ocean and SW Pacific and two species from the Atlantic Ocean including the largest Callanthiid, *C. ruber* (Parrot seaperch, 500 m$_{fbd}$), which grows to 60 cm long, is found throughout the Mediterranean Sea and parts of the NE Atlantic and is exploited in small-scale fisheries. *Grammatonotus* is

represented by six species found in different parts of the Pacific Ocean, the deepest-living being *G. macrophthalmus* (500 m$_{fbd}$) found off southern Japan and Hawaii.

4.3.23.8 Family Grammatidae (154) (Basslets)

These are small, average maximum length 5 cm, colourful fishes popular in marine aquaria; however a few species in the genus *Lipogramma* from the Bahamas to the Caribbean region of the Western Atlantic occur beyond scuba diving depth: *L. evides* (Banded basslet, 365 m$_{fbd}$), *L. flavescens* (Yellow basslet, 290 m$_{fbd}$) and *L. robinisi* (Yellowbar basslet, 290 m$_{fbd}$), some of which have been collected by means of the manned submersible, *Johnson Sea Link* (Gilmore, 1997).

4.3.23.9 Family Opistognathidae (155) (Jawfishes)

These are elongated mouthbrooding fishes, average length 15 cm that live in burrows and occur circumglobally in tropical regions. There are 80 mainly shallow-water species, but a few occur in deep water, e.g. *Lonchopisthus lemur* (275 m$_{fbd}$) and *Opistognathus leprocarus* (Roughcheek jawfish, 308 m$_{fbd}$).

4.3.23.10 Family Epigonidae (156) (Deep-water cardinalfishes)

The Epigonidae are distinguished from the well-known aquarium Apogonidae (shallow-water cardinalfishes) by the presence of one extra vertebra, scales on the nonspiny dorsal and anal fins, other anatomical features as well as their much deeper occurrence. They are distributed in tropical to temperate waters around the world at an average depth range of 300–860 m and reach a maximum depth of 3000 m. Several species have bioluminescent organs developed from different parts of the intestine. They are mostly small fishes, average length 17 cm, but an exception is *Epigonus telescopus* (Black cardinal fish, 1200 m$_{fbd}$), which grows to 75 cm (Fig. 4.32b). *E. telescopus* occurs widely in the

(b)

Ostracoberycidae
(shellskin alfonsinos)

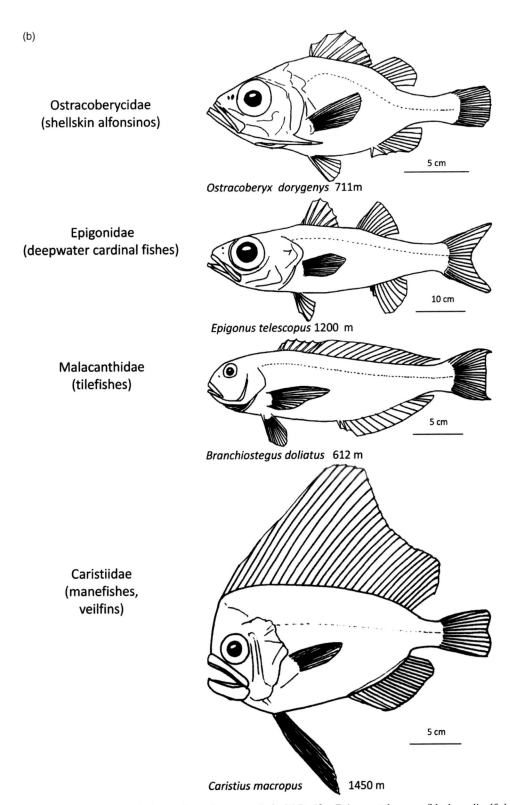

Ostracoberyx dorygenys 711m

5 cm

Epigonidae
(deepwater cardinal fishes)

Epigonus telescopus 1200 m

10 cm

Malacanthidae
(tilefishes)

Branchiostegus doliatus 612 m

5 cm

Caristiidae
(manefishes,
veilfins)

Caristius macropus 1450 m

5 cm

Figure 4.32b ORDER PERCIFORMES. *Ostracoberyx dorygenys*, Indo-W Pacific. *Epigonus telescopus* (black cardinalfish) Atlantic, Indian and SW Pacific, New Zealand. *Branchiostegus doliatus* (ribbed tilefish) W Indian Ocean. *Caristius macropus* (manefish) N Pacific.

Atlantic Ocean (65°N–43°S), the Indian Ocean and the SW Pacific and is exploited in commercial fisheries (Tandstad et al., 2011). In New Zealand waters *E. telescopus* is captured both in targeted fisheries and as a bycatch with other deep-water species such as the orange roughy (*Hoplostethus atlanticus*), alfonsino (*Beryx splendens*) and bluenose (*Hyperoglyphe antarctica*). Trawlers target schools of Black cardinals at 600–900 m depth.

There are three monotypic genera of small bathydemersal Epigonids, *Brephostoma carpenter* (2872 m_{fbd}) from the Pacific Ocean off Papua New Guinea and Hawaii, *Brinkmannella elongata* from the Western Central Atlantic and *Florenciella lugubris* (800 m_{fbd}) from the Indian Ocean. Herring (1992) showed that *F. lugubris* has bioluminescent diverticulae on the oesophagus and is able to eject pulses of luminescent material from beneath the operculum. The genus *Epigonus* comprises over 30 species, 6 of which are endemic to the Atlantic Ocean, 5 in the Indian Ocean and 14 in the Pacific Ocean. *E. lenimen* (Big-eyed cardinalfish, 820 m_{fbd}) is distributed across the Indo-Pacific region from Madagascar to New Zealand. *E. pectinifer* (1200 m_{fbd}) is found in the Pacific and Atlantic Oceans from Japan and Tasmania to the Caribbean, although it may be made up of more than one species. Four species can be considered almost circumglobal: *E. denticulatus* (Pencil cardinal, 830 m_{fbd}) from the Gulf of Mexico to West Africa, the Western Mediterranean, the Adriatic Sea, the Indian Ocean, Japan, Australia, New Zealand and the Hawaiian Ridge; *E. macrops* (Luminous deep-sea cardinalfish, 1300 m_{fbd}) from the Western Atlantic off Florida to the Indian Ocean, western Australia, Indonesia and Vietnam; *E. robustus* (Robust cardinalfish, 3000 m_{fbd}) circumglobal in the Southern Hemisphere (30°S–50°S) and *E. telescopus,* which has already been discussed. Bioluminescence in *E. macrops* is probably produced by a specialised pylorica caeca separated from the others and associated with reflective structures (Herring, 1992).

The genus *Microichthys* comprises two species: *M. coccoi* (549 m_{fbd}) maximum length 3 cm, found throughout the Mediterranean Sea and in the Atlantic Ocean off the Azores, and *M. sanzoi,* maximum length 6 cm, known only from the Straits of Messina off Sicily (Pollard, 2014). *M. coccoi* appears in low numbers in the 500–700 m depth zone of demersal trawl surveys of the Eastern Ionian Sea (Mytilineou et al., 2005).

There are two monotypic genera of bathypelagic Epigonidae, *Rosenblattia* and *Sphyraenops*. *R. robusta* (Stout cardinalfish, 2000 m_{fbd}) is circumglobal in southern temperate waters, has a short body shape, maximum length 10 cm, is mesopelagic and probably has oesophageal bioluminescence similar to that of *Florenciella* (Herring, 1992). *S. bairdianus* (Triplespine deep-water cardinalfish, 1750 m_{fbd}) occurs in the NW Atlantic, across the Pacific Ocean and in the Eastern Indian Ocean.

4.3.23.11 Family Malacanthidae (157) (Tilefishes)

This family comprises over 40 species in five genera with an average length 40 cm distributed in the Atlantic, Indian and Pacific Oceans down to a maximum depth of over 600 m. They have an elongated body with a long continuous dorsal fin. Only a few species can be considered to be deep-sea fishes. *Branchiostegus doliatus* (Ribbed tilefish, 612 m_{fbd}) is from the W Indian Ocean, where it is caught mainly between 200 and 400 m over muddy bottoms where it may inhabit burrows (Fig. 4.32b). Some species of *Caulolatilus* in the W Atlantic reach down to almost 500 m. *Lopholatilus chamaeleonticeps* (Great northern tilefish, 540 m_{fbd}) is widespread in the W Atlantic from Nova Scotia to the Caribbean, grows to over 1 m length and is captured in commercial and game fisheries most commonly around 200 m depth.

4.3.23.12 Family Scombropidae (158) (Gnomefishes)

This family comprises a single genus *Scombrops* with three species that grow to 1–1.5 m length in the Indo-West Pacific and Atlantic Oceans. The deepest-living is *Scombrops oculatus* (Atlantic scombrops, 610 m_{fbd}), usually caught by anglers at depths between 200 and 600 m in the subtropical W Atlantic.

4.3.23.13 Family Carangidae (159) (Jacks and Pompanos)

This is a very important family comprising over 140 species including oceanic species, which may dive to great depths, for example *Seriola lalandi* (Yellowtail amberjack, 825 m$_{fbd}$), *Trachurus capensis* (Cape horse mackerel, 500 m$_{fbd}$) and *Trachurus* (Atlantic horse mackerel, 1050 m$_{fbd}$), but because they all occur mostly near the surface, they cannot be considered to be deep-sea species.

Family Bramidae (160) (Pomfrets)

These are very widespread oceanic fishes that range down to an average depth of 550 m and maximum depth of over 1200 m. There are 20 species in six genera. Despite having been recorded relatively deep, *Brama brama* (Atlantic pomfret, 1000 m$_{fbd}$) spends most time at depths less than 200 m. However, *Eumegistus brevorti* (Tropical pomfret, 1317 m$_{fbd}$) has a mesopelagic or bathypelagic habit distributed in the Western Atlantic from the Bahamas to Brazil, avoiding warm surface waters while remaining at depths greater than 384 m.

4.3.23.14 Family Caristiidae (161) (Manefishes, Veilfins)

The Caristiidae are mesopelagic fishes that occur worldwide in tropical to temperate waters down to an average maximum depth of 1350 m. They have a laterally compressed deep body with elaborate veil-like fins, steep frontal profile to the head and an average maximum length of 18 cm. There are 19 species in two subfamilies, the Caristiinae, large-mouthed species and Paracaristiinae, the small-mouthed species (Kukuev et al., 2013).

Subfamily Caristiinae

The genus *Caristius* comprises eight species. Most widespread are *C. fasciatus* (420 m$_{fbd}$), which is reported from the North Atlantic off Greenland, the South Atlantic, the W Indian Ocean and the NE Pacific, and *C. meridionalis* (1335 m$_{fbd}$), which is probably circumglobal at southern temperate latitudes including off Argentina, Australia and New Zealand (Stevenson and Kenaley, 2013). Three species have been described from the SE Atlantic, two species are endemic to the Pacific Ocean (Fig. 4.32b) and *C. digitus* (1260 m$_{fbd}$) is recorded from the Indian and SW Pacific Oceans. *Platyberyx* comprises six species, *P. andriashevi* (200 m$_{fbd}$), which is circumglobal at tropical latitudes 57°N–5°S, two species endemic to the Atlantic Ocean and three from the Pacific Ocean.

Subfamily Paracaristiinae

The genus *Paracaristius* comprises four species: *P. nudarcus* and *P. maderensis* (2000 m$_{fbd}$) recorded in all three major oceans plus two further species described from the Atlantic Ocean, *P. nemorosus* from the tropics (20°N–10°S) and *P. aquilus* off Angola (Stevenson and Kenaley, 2011). The monoptypic *Neocaristius heemstrai* (1360 m$_{fbd}$) is found in the south of the Indian, Pacific and Atlantic Oceans.

The unusual body form of the Caristiidae seems ill-adapted for the pursuit of pelagic prey, and it seems they likely derive most of their food from a symbiosis with mid-water colonial Hydroza of the order Siphonophorae. Jannsen et al. (1989) first observed from the HOV *Johnson Sea Link*, in the NW Atlantic, a *Caristius* sp. associated with the vertical string of a siphonophore *Bathyphysa conifera* manoeuvring with its mouth close to the zooids, often swimming upside down. Stomach contents of Caristiids were found to comprise fragments of amphipods, mesopelagic and bathypelagic fishes, various crustacea, chaetognaths and pieces of siphonophores. *Caristius* appears to steal prey from siphonophores, eat fragments of the zooids and may also ingest amphipods that attack the siphonophore resulting in a mutually beneficial symbiosis. There was evidence of *Caristius* exploiting several species of siphonophore. Lindsay et al. (2001) observed that the relationship between a Caristiid and *Praya* sp. siphonophore may be a moderate association with the fish occasionally venturing free into the water column. Benfield et al. (2009) recorded a presumed *Paracaristius* sp. associated with a physonect siphonophore (family Apolemiidae) at

(c)

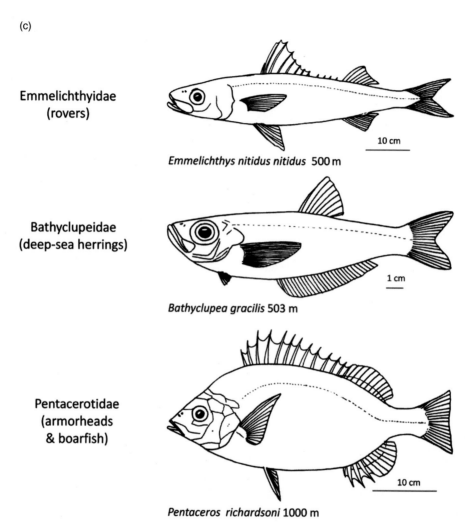

Emmelichthyidae
(rovers)

Emmelichthys nitidus nitidus 500 m

Bathyclupeidae
(deep-sea herrings)

Bathyclupea gracilis 503 m

Pentacerotidae
(armorheads
& boarfish)

Pentaceros richardsoni 1000 m

Figure 4.32c ORDER PERCIFORMES. *Emmelichthys nitidus* (Cape bonnetmouth) South Africa, Australia and New Zealand 16°N–43°S. *Bathyclupea gracilis* (slender deep-sea herring) central W Pacific. *Pentaceros richardsoni* (pelagic armorhead) Southern Hemisphere 23°S–56°S, South Africa, Australia, New Zealand and Cape Horn.

495 m depth in the northern Gulf of Mexico, and Angulo et al. (2014) found a juvenile *Paracaristius* sp. associated with siphonophore *Praya reticulata* in the Eastern Pacific at a depth of 100 m off Costa Rica. There remains great uncertainty regarding the specificity of the fish–siphonophore symbiosis, although Angulo et al. (2014) attempted to link the geographic distribution of *Paracristius maderensis* and the siphonophore *Praya reticulata*. The eggs and larvae of Caristiids are planktonic and eggs can be quite large, up to 2 mm diameter (Moser, 1996).

4.3.23.15 Family Emmelichthyidae (162) (Rovers)
The Emmelichthyidae are elongated round-bodied fishes with toothless or near toothless jaws and forked tails that grow to an average maximum length of 35 cm and are widely distributed in tropical to warm temperate seas over an average depth range of 146–342 m. There are 19 species in three genera. The genus *Emmelichthys* comprises seven species (Fig. 4.32c). *E. nitidus* (Cape bonnetmouth, 500 m$_{fbd}$) occurs across the Indo-Pacific from the western Cape coast of South Africa to Australia, New Zealand and

the coast of Chile, where the SE Pacific stocks have often been recognised as a subspecies, *E. nitidus cyansescens*. *E. nitidus* is commercially exploited, known as redbait off Australia and Red Pearl Fish off New Zealand. Off Tasmania it is captured in mid-water trawls with a total catch of *ca.* 8000 tonnes in 2004/5, mostly utilised as feed for farmed bluefin tuna. Batches of pelagic eggs are released every three days during a short spawning season (Neira and Lyle, 2011). *E. struhsakeri* (Golden redbait, 428 m_{fbd}) occurs in the Pacific Ocean from southern Japan to Hawaii and Australia and is also fished commercially on the shelf edge and on seamounts. Two species have been described from the Pacific Ocean, one from the Red Sea and one in the Atlantic. The genus *Erythrocles* comprises six species, of which three are endemic to the Pacific Ocean, including *E. scintillans* (Golden kali kali, 606 m_{fbd}) that is found from Japan to Hawaii and is exploited in minor fisheries. *E. schlegelii* (Japanese rubyfish, 300 m_{fbd}) is known from the NW Pacific and the Western Indian Ocean, is the largest of the Emmelichthyidae, growing to 72 cm and featuring in some commercial catches. There is one species endemic to the Atlantic Ocean and in the Indian Ocean. The genus *Plagiogeneion* comprises five species found mainly in the southern Indo-Pacific region. *P. macrolepis* (Bigscale rubyfish, 390 m_{fbd}) is endemic to the Great Australian Bight, where it occurs as a bycatch in fisheries. *P. rubiginosum* (Rubyfish, 600 m_{fbd}) is found from off Cape Town to Sri Lanka, S Australia and New Zealand, where it forms part of the deep-water fishery resource. *P. rubiginosum* feeds on large zooplankton, salps and myctophids growing to a maximum of 60 cm length with a life span of over 40 years.

4.3.23.16 Family Lutjanidae (163) (Snappers)

These are well-known mainly shallow tropical and subtropical fishes. There are about 100 truly marine species of which a few occur in deep water just beyond the shelf-break depth. The genus *Etelis* comprises four such species that inhabit rocky seafloors: *E. carbunculus* (Deep-water red snapper, 400 m_{fbd}), *E. coruscans* (Deep-water longtail red

snapper, 400 m_{fbd}) and *E. radiosus* (Pale snapper, 360 m_{fbd}) from the Indo-West Pacific, and *E. oculatus* (Queen snapper, 450 m_{fbd}) from the West Central Atlantic. All are highly valued as food fish.

4.3.23.17 Family Nemipteridae (164) (Threadfin breams)

Belonging to the sparoid lineage of the Percoids, these are demersal and reef-associated fishes of mostly shallow tropical waters but most recent discoveries of new species in the genus *Parascolopsis* (monocle breams) have all been in deep water, *P. baranesi* (500 m_{fbd}) from the northern Red Sea, *P. melanophrys* (Dwarf monocle bream, 500 m_{fbd}) of the Indo-Malayan Archipelago and *P. rufomaculatus* (Red-spot dwarf monocle bream, 321 m_{fbd}) from NW Australia.

4.3.23.18 Family Sparidae (165) (Sea breams, Porgies)

Occurring in coastal tropical and subtropical regions of all the major oceans, Sparids are demersal or benthopelagic fishes ranging down to 700 m depth. There are 128 species of which a few occur in deep water. *Dentex fourmanoiri* (290 m_{fbd}) has been described from deep water off New Caledonia in the SW Pacific. *Pagellus bogaraveo* (Red Sea Bream or Blackspot seabream, 700 m_{fbd}) is the deepest-living species and supported an important fishery in the Bay of Biscay area of the NE Atlantic. This stock collapsed during the 1980s, and the centre of activity of the fishing fleets has moved further south to the Strait of Gibraltar area with longliners and artisanal vessels involved (ICES, 2102).

4.3.23.19 Family Sciaenidae (166) (Drums, Croakers)

Comprising over 120 species the Sciaenidae occur in tropical and subtropical coastal waters of all the major oceans and range down to 800 m depth *Atrobucca genaie* (700 m_{fbd}) from the Red Sea, *A. trewavasae* (minimum depth 300 m) from the Bay of Bengal and *Protosciaena bathytatos* (Deep-water drum, 600 m_{fbd}) from central W Atlantic are found at upper slope depths.

4.3.23.20 Family Mullidae (167) (Goatfishes)

This family of 80 species is found circumglobally in warm shallow waters, but *Upeneus davidaromi* (Arom's goatfish, 500 m_{fbd}) is reported with a usual depth range of 250–300 m in the Red Sea.

4.3.23.21 Family Bathyclupeidae (168) (Deep-sea herrings)

This is a small group of benthopelagic schooling fishes that grow up to ca. 20 cm length inhabiting the upper bathyal depth zone of the tropical Atlantic, Indian and West Pacific Oceans. They are notably absent from the Northern and Central waters of the Eastern Atlantic. Bathyclupeids are rather poorly known despite often being quite abundant, comprising 20 per cent of the catch at depths of 600–800 m in trawl surveys off Western Australia, twice as many as the most abundant macrourid (Williams et al., 1996). It is a highly specialised group that probably diverged independently from ancestral Percoids and shows a combination of primitive and specialised traits; for example, they have lost the spiny dorsal fin characteristic of Perciformes while retaining the supporting skeletal elements. Although Prokofiev (2014) proposes two genera, here we retain the single genus *Bathyclupea* with seven species: *B. hoskynii* (Indian Deep-sea Herring, 550 m_{fbd}), from the NW Indian Ocean; *B. schroederi* (Schroeder's Deep-sea Herring, 750 m_{fbd}), from the Gulf of Mexico and the Caribbean; *B. argentea* (West Atlantic deep-sea herring, 677 m_{fbd}), W Atlantic from New England to Suriname including the Caribbean; *B. elongata* (Southern deep-sea herring, 785 m_{fbd}), Southern Hemisphere (18–34°S), SE Atlantic to E Australia and New Zealand; *B. gracilis* (West Pacific dwarf deep-sea herring, 500 m_{fbd}), W Pacific Taiwan to NE Australia (Fig. 4.32c); *B. malayana* (Malaysian deep-sea herring, 930 m_{fbd}), Indo-West Pacific from Madagascar to Java and the Coral Sea; and *B. megaceps* (Big-headed deep-sea herring, 765 m_{fbd}), Philippines. Bathyclupeids must play an important role in upper bathyal food chains, as specimens have been taken from the stomachs of predatory fishes, and there is a clear trend that larger individuals of *B. elongata* and *B. malayana* occur deeper. Janssen (2004) observed an extraordinary putative foraging behaviour from the *Johnson Sea Link* submersible. Bathyclupeids descended motionless, head down vertically and body rigid until the snout touched the seafloor and then ascended tail first with the body remaining vertical propelled by undulations of the pectoral fins. It is presumed that the fish were slightly negatively buoyant; several individuals were observed with chalky snouts, suggesting that bottom contact was frequent. Sinking motionless in this way would allow maximum lateral line sensitivity to detect any prey movements in the water column.

4.3.23.22 Family Chaetodontidae (169) (Butterfly fishes)

These are well-known colourful members of inshore tropical reef environments, but one species, *Prognathodes guyotensis* (320 m_{fbd}), appears to be confined to guyot summits at depths greater than 280 m in the NW Pacific and Indian Oceans.

4.3.23.23 Family Pentacerotidae (170) (Armorheads)

The Pentacerotidae are laterally compressed deep-bodied fish with the head encased in bone and conspicuous spines on the dorsal and anal fins. They are presumed to have originated from Southern Australia and are distributed around the Southern Hemisphere in the Indo-Pacific with range extensions into the North Pacific and SE Atlantic (Kim, 2012). There are 13 species in seven genera divided into two subfamilies, the Pentacerotinae and shallower-living Histiopterinae with average maximum depths of 600 and 270 m, respectively.

Subfamily Pentacerotinae

This subfamily comprises one genus, *Pentaceros*, and six species. *P. capensis* (Cape armorhead, 300 m_{fbd}), found around South Africa from the SE Atlantic to Reunion in the W Indian Ocean, is of minor fishery interest, *P. quinquespinis* (210 m_{fbd}) is described from the SE Pacific off Chile, *P. decacanthus* (Bigspined boarfish, 460 m_{fbd}) is usually captured at depths greater than 300 m off southern Australia and New

Zealand and *P. japonicus* (Japanese armorhead, 830 m$_{fbd}$) occurs in the W Pacific from Japan to Australia and New Zealand and is commercially exploited. There are two species known as the Pelagic Armorheads (Fig. 4.32c): *P. richardsoni* (Southern Pelagic armorhead, southern boarfish, 1000 m$_{fbd}$) maximum length 56 cm, distributed in the Southern Hemisphere from around South Africa to southern Australia, New Zealand and Cape Horn, and *P. wheeleri* (North Pacific Pelagic Armorhead, Slender armorhead, 800 m$_{fbd}$) maximum length 44 cm, found across the North Pacific from Japan to Alaska and Central California, including seamounts of the Central North Pacific. There is some confusion in the literature regarding the naming of these species; they were placed in a separate genus *Pseudopentaceros* prior to a revision by Kim (2012). For some time the North Pacific armorheads were considered to comprise two separate species, *Pseudopentaceros wheeleri* and *Pseudopentaceros pectoralis* (Humphreys et al., 1989). Both northern and southern species of pelagic armorhead are commercially exploited, and in the North Pacific the fishery for *P. wheeleri* on seamounts of the Emperor Chain and northern Hawaiian Ridge removed ca. 1,000,000 tonnes between 1968 and 1975, a classic example of boom and bust in deep-sea fisheries with only low levels of catches sustained since that time. In view of its commercial importance, the life history of *P. wheeleri* in the North Pacific has been better studied than the other species. *P. wheeleri* has an extended epipelagic and pelagic subadult stage lasting 1.5–2.5 years in the sub-Arctic waters of the NE Pacific and the Gulf of Alaska, where they build up enormous fat reserves. They return to the seamounts where they were spawned at a size of 26–33 cm fork length. The large quantity of fat is then metabolised over the remaining 3–5 years of its life span, during which time the fish spawn, lose weight, become emaciated and finally moribund. The deep-living demersal population of adults on the seamounts therefore comprises a mixture of different morphotypes from fat to lean giving rise to the erroneous recognition of more than one species, which was finally resolved by Humphreys et al. (1989). Nevertheless *P. wheeleri* continue to feed on the seamounts, particularly exploiting vertically migrating prey from the deep-sea scattering layer that impinges on the summit and slopes of the seamounts. Although the diet comprises a large variety of oceanic prey and seamount resident species, it is dominated by a large proportion of gelatinous tunicates (*Pyrosoma*) and salps. Older fishes have a higher food intake than recent arrivals. The gelatinous prey has a very high water content, and it is estimated that throughout most of their adult phase food intake remains below maintenance requirement, hence the loss in weight (Seki and Somerton, 1994). Despite the limited food intake *P. wheeleri* show a conventional annual spawning cycle with a peak of egg production during December and January, possibly spawning up to four or five times before they die. Previous hypotheses that these fish may be semelparous, spawning once in their life span, are rejected (Yanagimoro and Humphries, 2005).

Subfamily Histiopterinae

This subfamily comprises seven species in six genera, including the giant boarfishes *Paristiopterus* from around Australia and New Zealand that grow to 1 m in length but are not found at beyond 260 m depth. Only two species, *Histiopterus typus* (Sailfin armorhead, 421 m$_{fbd}$), which is widespread in the Indo-West Pacific from the Red Sea to Japan and Australia, and *Zanclistius elevatus* (Longfin boarfish, 500 m$_{fbd}$) from Southern Australia and New Zealand, extend into deeper waters of the upper slope. Both species are commercially exploited.

4.3.23.24 Family Cepolidae (171) (Bandfishes)

The Cepolidae are elongated medium-size fishes that occur in the Eastern Atlantic, Mediterranean and the Indo-West Pacific down to a maximum depth of 500 m. There are 23 species in five genera. Most are found at shelf depths where some live in burrows, and only a few can be regarded as deep-sea species. *Cepola haasti* is endemic to plateau depths (ca. 400 m) off New Zealand. *Owstonia dorypterus* (320 m$_{fbd}$) and *O. sarmiento* (307 m$_{fbd}$) are recorded from the

Philippines. *O. weberi* (350 m$_{fbd}$), *Sphenanthias simoterus* (500 m$_{fbd}$) and *S. whiteheadi* (Indian bandfish, 350 m$_{fbd}$) are from the Western Indian Ocean, where the latter is a bycatch in the commercial deep-sea shrimp trawl fishery.

Suborder Elassomatoidei

The Elassomatoidei comprises over 1000 marine species, including well-known shallow-water tropical and subtropical families such as the Pomacentridae (damselfishes), Labridae (wrasses) and Scaridae (parrotfishes). Only in the genus *Bodianus* (hogfishes, Labridae) have a few species have been recorded at over 200 m depth with the deepest down to 510 m depth.

Suborder Zoarcoidei

The Zoarcoidei comprises ca. 400 species that occur primarily in the North Pacific Ocean, though they range from the Arctic to the Antarctic and represent one of the main diversifications of Perciformes into the deep sea, extending from inshore down the slopes to abyssal depths of over 4000 m and on hydrothermal vents of the Mid-Atlantic Ridge (Geistdorfer, 1994). Zoarcoids are mainly demersal, and reproduction is by oviparity with relatively large eggs produced in a mass on the seafloor, which is often guarded by the parents (the genus *Zoarces* is ovoviparous).

4.3.23.25 Family Bathymasteridae (172) (Ronquils)

This family comprises seven species from the North Pacific Ocean margin that are distributed between the intertidal zone and the outer shelf. Only *Bathymaster signatus* (825 m$_{fbd}$) found in the Arctic, the North Pacific and the Sea of Okhotsk ranges significantly deeper onto upper slope depths, usually being found at 35–380 m depth.

4.3.23.26 Family Zoarcidae (173) (Eelpouts)

There are approximately 300 species in this family of which over 60 per cent occur in the deep sea. Here we

recognise five subfamilies: Zoarcinae, Gymnelinae, Lycodinae, Lycozoarcinae (Anderson and Fedorov, 2004) and Neozoarcinae, the latter having formerly been considered in the Stichaeidae (Pricklebacks) until recent revision based on molecular genetic evidence (Radchenko, 2015; Radchenko et al., 2015).

Subfamily Zoarcinae

This subfamily comprises one genus with six species that occur at high northern latitudes in the Atlantic and Pacific Oceans; only one species ranges deep, *Zoarces americanus* (Ocean pout, 388 m$_{fbd}$) from the NW Atlantic, where it grows to over 1 m long and is fished commercially, mostly in coastal waters.

Subfamily Gymnelinae

The Gymnelinae comprises 48 species in 14 genera that are mostly found in the North Pacific Ocean. They are elongated eel-like mostly demersal fishes. The monotypic *Andriashevia aptera* (1025 m$_{fbd}$) is from the Pacific coast of Japan associated with gorgonian corals, where it feeds on crustacea and encrusting invertebrates. The genus *Bilabria* comprises two shallow-water species from the NW Pacific. *Davidijordania* comprises five species from the Sea of Japan and the Sea of Okhotsk, some of which range down to 300 m depth. *Ericandersonia sagamia* (930 m$_{fbd}$) is monotypic and recorded from depths greater than 880 m off Japan. *Gymnelopsis* comprises five species recorded from the Sea of Okhotsk, the deepest-living being *G. brevifenestrata* (783 m$_{fbd}$) and *G. ochotensis* (780 m$_{fbd}$).

The monotypic *Nalbantichthys elongatus* (520 m$_{fbd}$) and *Opaeophacus acrogeneius* (800 m$_{fbd}$) are both recorded from the Bering Sea. The genus *Puzanovia* comprises two demersal species from the NW Pacific, *P. rubra* (800 m$_{fbd}$) and *P. virgata* (600 m$_{fbd}$). The genus *Gymnelus* is confined to high northern latitudes, with five species found in the Arctic Ocean, one in the North East Atlantic Ocean and six species from the North Pacific. They are mainly shelf-dwelling fishes; only *G. andersoni* (300 m$_{fbd}$) found off Spitzbergen is regarded as a deep-sea species, although other species have been recorded down to 480 m depth. *Hadropareia, Krusensterniella* and *Magadanichthys* are mainly shelf-dwelling genera from the North Pacific. *Melanostigma* comprising seven species is a circumglobal deep-water genus occurring in both the Northern and Southern Hemispheres. They are bathypelagic medium-sized fish that grow to an average maximum length of 16 cm and are generally found at depths greater than 250 m. *M. atlanticum* (Atlantic soft pout, 1853 m_{fbd}) occurs in the North Atlantic (74°N–19°N; Fig. 4.32d) typically off slopes in midwater, feeding on foraminifera, copepods and ostracods (Mauchline and Gordon, 1984). In the NW Atlantic both male and female *M. atlanticum* have been found burrowing 25–35 cm below the sediment surface together with eggs, 2.8–3.5 mm diameter within the abdomen of females and 3.9 mm diameter in homogenous egg masses in the burrows. Water depth was 350 m. Burrowing appears to be an important part of the reproductive behaviour of the species, and they are apparently able to keep their burrows oxygenated despite the surrounding anoxic sediment (Silverberg et al., 1987). *M. orientale* (600 m_{fbd}) is from the NW Pacific, *M. pammelas* (Midwater eelpout, >699 m_{fbd}) NE Pacific, *M. bathium* (2635 m_{fbd}) SE Pacific, *M. inexpectatum* W. Central Pacific, *M. vitiazi* (890 m_{fbd}) across the Indo-West Pacific (6°S–50°S) and *M. gelatinosum* (Limp eelpout, 2561 m_{fbd}) is circumglobal in the Southern Hemisphere south of 30°S. The genus *Seleniolycus,* a sister group to *Melanostigma,* may be circumglobal in the Southern Ocean (latitudes 57°S–63°S) with three species known so far: *S. laevifasciatus* (1500 m_{fbd}) recorded from South Shetland and South Sandwich Islands and *S. pectoralis* (2594 m_{fbd}) and *S. robertsi* (2290 m_{fbd}) from the Pacific-Antarctic Ridge. *Seleniolycus* have been captured by baited longlines set on the seafloor so they are probably benthic rather than benthopelagic. Their flesh is gelatinous, the skin fragile and there are no pelvic fins (Møller and Stewart, 2006). *Barbapellis pterygalces* (Moose eelpout, 1136 m_{fbd}) is a new genus described from the continental slope of eastern Antarctica with unique skin folds on the head presumed to be an aid to benthic foraging on the muddy seafloor. The type specimen was a 204 mm long female carrying ca. 20 eggs 2.4–3 mm in diameter (Iglésias et al., 2012).

Subfamily Lycodinae

This is the most diverse of the Zoarcid subfamilies spread through all the oceans from the Arctic to the Antarctic with 233 species, over 70 per cent of which are resident in the deep sea. Here we consider the genera in order of their latitudinal range from North to South. The genus *Lycodes* is northernmost, comprising 64 demersal species, about half of which occur at deep-sea depths. They grow to an average maximum length of 40 cm Three species are endemic to the Arctic Ocean, including *L. mcallisteri* (McAllister's eelpout, 668 m_{fb}) that feeds on fishes and *L. sagittarius* (Archer eelpout, 600 m_{fbd}) that feeds on benthic invertebrates. There are 18 more species that occur in the Arctic, 5 with range extensions into the North Pacific and 13 into the North Atlantic. Amongst the Arcto-Pacific species only one can be considered to be deep-sea, *L. palearis* (Wattled eelpout, 925m_{fbd}). The Arcto-Atlantic species occur much deeper, for example *L. adolfi* (Adolf's eelpout, 1880 m_{fbd}), *L. frigidus* (Glacial eelpout, 3000 m_{fbd}), *L. luetkenii* (Lütken's eelpout, 1463 m_{fbd}), *L. paamiuti* (Paamiut eelpout, 1337 m_{fbd}), *L. squamiventer* (Scalebelly eelpout, 1808 m_{fbd}), and *L. vahlii* (Vahl's eelpout, 1200 m_{fbd}). 41 species of *Lycodes* are endemic to the North Pacific Ocean, mostly shallow-water species, the deepest being *L. concolor* (Ebony eelpout, 1025 m_{fbd}) from the Sea of Okhotsk to the Gulf of Alaska and *L. tanakae* (1100 m_{fbd}) from the Seas of Japan and Okhotsk. Two species are endemic to the Atlantic Ocean, *L. esmarkii* (Greater eelpout, 1090 m_{fbd}) from the cold waters north of the Iceland–Faroe–Shetland ridge and *L. terraenovae* (2604 m_{fbd}), which ranges from the Davis Strait in the NW Atlantic to off South Western Africa, the only *Lycodes* species found in the Southern Hemisphere. Both species are benthic feeders; *L. esmarkii* has a diet almost exclusively comprised of ophiuroids, and *L. terraenovae* feeds on sponges, worms, molluscs, pycnogonids and crustaceans as well as ophiuroids. The genus

(d)

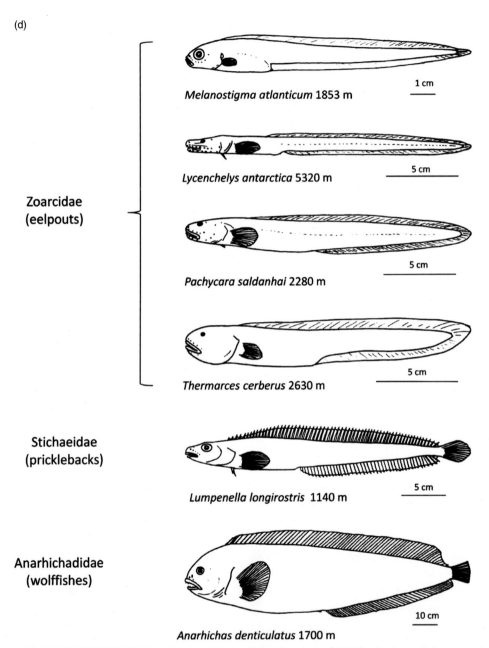

Zoarcidae (eelpouts)

Melanostigma atlanticum 1853 m

1 cm

Lycenchelys antarctica 5320 m

5 cm

Pachycara saldanhai 2280 m

5 cm

Thermarces cerberus 2630 m

5 cm

Stichaeidae (pricklebacks)

Lumpenella longirostris 1140 m

5 cm

Anarhichadidae (wolffishes)

Anarhichas denticulatus 1700 m

10 cm

Figure 4.32d ORDER PERCIFORMES. *Melanostigma atlanticum* (Atlantic soft pout) N Atlantic. *Lycenchelys antarctica*, Southern Ocean 23°S–63°S. *Pachycara saldanhai*, endemic to hydrothermal vents on the Mid-Atlantic Ridge. *Thermarces cerberus*, endemic to hydrothermal vents of the E Pacific Rise and Galapagos Rift. *Lumpenella longirostris* (longsnout prickleback) N Pacific. *Anarhichas denticulatus* (northern wolffish) N Atlantic.

Lycodonus comprises four species found in the Atlantic Ocean, *A. flagellicauda* (1993 m_{fbd}) in the Arctic and North Atlantic Oceans, *L. mirabilis* (Chevron scutepout, 2394 m_{fbd}) in the NW Atlantic, *L. malvinensis* (2044 m_{fbd}) in the SW Atlantic and *L. vermiformis* (1152 m_{fbd}) in the SE Atlantic. The monotypic *Exechodontes daidaleus* (1004 m$_{fbd}$) is endemic to the NW Atlantic. Eight genera are endemic to the North Pacific: *Bothrocarina, Derepodichthys, Eucryphycus, Hadropogonichthy, Japonolycodes, Lycogrammoides, Lyconema* and *Taranetzella*. *T. lyoderma* (Looseskin eelpout, 3000 m$_{fbd}$) occurs around the ocean margin from Japan to Mexico. *Bothrocara* comprises eight predominantly deep-sea species, six of which are endemic to the North Pacific, whereas *B. elongatum* (1271 m$_{fbd}$) is from the Gulf of Panama and *B. molle* (Soft eelpout, 2688 m$_{fbd}$) ranges around the Pacific Ocean from Japan in the West to Mexico in the East and to the Southern Ocean. *Lycodapus* has 10 deep-sea species endemic to the North Pacific as well as *L. australis* from the coasts of Chile, *L. pachysoma* (Stout slipskin, 2600 m$_{fbd}$) found off Oregon and in the Southern Ocean and *L. antarcticus* (1200 m$_{fbd}$) found in the Southern Ocean and around sub-Antarctic islands and seamounts.

The genus *Lycenchelys* comprises 62 species found in all oceans from the Arctic to the Antarctic. All are demersal deep-sea species, the average minimum depth of occurrence is 1180 m, and includes the deepest known Zoarcid *Lycenchelys antarctica* (5320 m$_{fbd}$) found in the SE Pacific and off Antarctica (Fig. 4.32d). Three species are found in the Arctic Ocean, two are endemic and one, *L. sarsii* (Sar's wolf eel, 600 m$_{fbd}$), also occurs in the N Atlantic, three species are endemic to the N Atlantic and three to the SW Atlantic. 36 species are endemic to the N Pacific Ocean and nine to the S Pacific, including three found in the SW Pacific off New Zealand and six in the SE Pacific off South America. One species, *L. hureaui* (937 m$_{fbd}$), has been found in the Indian Ocean off the Kerguelen Islands. Six species have been recorded in the Southern Ocean around Western Antarctica, the deepest being *L. nanospinata* (2815 m$_{fbd}$) from the Scotia Ridge. The genus *Pachycara* with 24 species has a more southerly distribution; only two species are found in the N Pacific and eight in the S and E Pacific including two from off New Zealand. Five species are recorded from the Indian Ocean and one species, *P. brachycephalum* (1810 m$_{fbd}$), is circumglobal in

the Southern Ocean around Antarctica. Eight species are recorded from the Atlantic Ocean, including *P. bulbiceps* (4780 m$_{fbd}$), an abyssal species also found in the NE Pacific.

Three genera of Lycodinae are associated with deep-sea hydrothermal vent fields. Amongst the Atlantic species of *Pachycara,* two are endemic to hydrothermal vent fields on the Mid-Atlantic Ridge (Fig. 4.32d), *P. thermophilum* (3480 m$_{fbd}$) and *P. saldanhai* (2280 m$_{fbd}$; Geistdoerfer, 1994; Biscoito and Almeida, 2004). *Pachycara caribbaeum* has been described from hydrothermal vents and a methane seep in the Mid-Cayman Spreading Centre of the Caribbean Sea (Anderson et al., 2016). Two Pacific species of *Pachycara* have been collected from hydrothermal vent sites: *P. rimae* (2500 m$_{fbd}$) from the Galapagos rift zone and *P. gymninium* (3225 m$_{fbd}$) from the NE Pacific, although the latter is not endemic to the vents as it is found in a variety of deep-water habitats. The genus *Pyrolycus* comprises two species of vent endemic eelpouts, *P. manusanus* (1971m$_{fbd}$) from the Bismarck Sea off Papua New Guinea and *P. moelleri* (1336 m$_{fbd}$) from the Kermadec Ridge off New Zealand (Anderson, 2006). The genus *Thermarces* comprises three species, *T. andersoni* (2620 m$_{fbd}$) and *T. cerberus* (2630 m$_{fbd}$) collected from hydrothermal vent sites on the East Pacific Rise (Fig. 4.32d), and *T. pelophilum* (1947 m$_{fbd}$) from cold fluid seeps in the W Atlantic Ocean on the Barbados accretionary complex (Geistdoerfer, 1999).

A feature of the Lycodinae is their diversification into the Southern Hemisphere. The monotypic *Leucogrammolycus brychios* (632 m$_{fbd}$) has been described from slope depths in the SW Atlantic off Brazil, and *Oidophorus* comprises two species, *O. brevis* (900 m$_{fbd}$) from off Argentina and *O. mcallisteri* (3038 m$_{fbd}$), found off South Georgia. *Dieidolycus* comprises three species, *D. adocetus* (1957 m$_{fbd}$) from off Papua New Guinea in the W Pacific, *D. gosztonyii* (2165 m$_{fbd}$) from the SE Pacific off Chile and *D. leptodermatus* (3040 m$_{fbd}$) distributed around the Southern Ocean (53°S–69°S). *Ophthalmolycus* comprises nine species, three of which occur in the SE Pacific from the Gulf of Panama to Chile, three are confined to seas around Western Antarctica and three species are circum-Antarctic. In the Magellanic province of the SW Atlantic and the SE Pacific around the tip of South America, 19 species in 14 genera, *Aiakas*, *Argentinolycus*, *Austrolycus*, *Crossostomus*, *Dadyanos*, *Iluocoetes*, *Letholycus*, *Maynea*, *Notolycode*, *Patagolycus*, *Phucocoetes*, *Piedrabuenia*, *Plesienchelys* and *Pogonolycus*, make up a radiation of endemic species of Lycodinae onto the slopes down to a maximum depth of 900 m (Corbella, 2013). The suffix 'lycus' from the Greek 'λύκος', meaning 'wolf', is commonly used by taxonomists when naming Zoarcids of the Southern Hemisphere.

Five genera of Lycodinae are endemic to the Southern Ocean: *Bellingshausenia*, *Bentartia* and *Gosztonyia* are monotypic, *Lycodichthys* has two species and *Santelmoa* four species. All are endemic to Western Antarctica except *L. antarcticus* (540 m$_{fbd}$), which is probably circum-Antarctic in its distribution 60°S–74°S.

Subfamily Lycozoarcinae

This subfamily comprises one species from the NW Pacific, Sea of Okhotsk region *Lycozoarces regani* (300 m$_{fbd}$).

Subfamily Neozoarcinae

The Neozoarcinae comprises three genera, *Neozoarces*, *Eulophias* and *Zoarchias* (formerly classed within the family Stichaeidae), which are all confined to intertidal and shallow waters.

4.3.23.27 Family Stichaeidae (174) (Pricklebacks)

This family comprises about 70 species of shallow-water fishes mainly found in the North Pacific Ocean with an elongated body and a long dorsal fin with multiple spines. Two genera have species that occur at upper slope depths, *Bryozoichthys* comprises two species found down to *ca.* 400 m, and *Lumpenella longirostris* (Longsnout prickleback, 1140 m$_{fbd}$) usually occurs at 300–600 m depth (Fig. 4.32d) from Japan to British Columbia in the Pacific Ocean, the Bering Sea and in the Arctic Ocean off Greenland.

4.3.23.28 Family Anarhichadidae (175) (Wolffishes)

The Anarhichadidae comprises three species from the North Atlantic and two species from the North Pacific. The North Atlantic species of genus *Anarhichas* occur on upper slopes and *A. denticulatus* (Northern wolffish, 1700 m$_{fbd}$) in particular ranges deep and is commonly found at 100–900 m depth in midwater offshore feeding on ctenophores, medusas, small fishes, echinoderms, crustaceans and soft-bodied molluscs (Fig. 4.32d). It grows to 1.8 m length and individuals or possible breeding pairs were very conspicuous at 900–1100 m depth in a baited camera survey of the Mid-Atlantic Ridge (King et al., 2006). Godø (1997) describes *A. denticulatus* intercepting and defending bait at 250 m in the Barents Sea. *A. lupus* (Atlantic wolffish, 600 m$_{fbd}$) shows extended courtship behaviour, pairing, internal fertilisation and produces large eggs 6 mm diameter, which are guarded by the male. Larvae hatch at 20 mm long and fecundity is low. Reproductive behaviour has not been directly observed in *A. denticulatus*, but they are presumed to spawn deep, and the eggs and larvae are larger than in *A. lupus*. Both species have been depleted by fisheries exploitation in the NW Atlantic (Kulka et al., 2007).

Suborder Notothenioidei

The Notothenoidei dominate the diversity, biomass and abundance of teleost fishes in the Southern Ocean (Near et al., 2015). Many are adapted to living in waters with subzero temperatures and about 30 per cent of the *ca.* 150 species are associated with the deep sea. There are eight families, five of which represent the major Antarctic diversification and three, Bovichtidae (temperate icefishes, from Australia, New Zealand and South America), Pseudaphritidae (catadromous icefishes, from Australia) and Eleginopsidae (Patogonian blennies, from South America), represent relict fresh- and shallow-water species from continental land masses surrounding the Southern Ocean reflecting the evolutionary origin of the group around the southern supercontinent of Gondwana. Only the families with deep-water representatives are considered below.

4.3.23.29 Family Nototheniidae (176) (Cod icefishes)

The Nototheniidae comprises 57 species in 14 genera including rockcods and toothfishes, which are the major fishery resources of the Antarctic region.

The genus *Aethotaxis* has one bathypelagic species found around Antarctica at depths from 390 to 850 m. Two subspecies are recognised: *A. mitopteryx pawsoni* from West Antarctica and the more widespread *A. mitopteryx* (Longfin icedevil, 850 m$_{fbd}$). There is no swim bladder so neutral buoyancy is attained by reduction of the bones, weakly ossified scales and extensive lipid deposits. Although apparently adapted to the pelagic lifestyle most specimens are caught in bottom fishing gears suggesting a benthopelagic habit. Maximum length is 42 cm, blood erythrocyte count is low and it is assumed the species has a sluggish mode of life, using a sit-and-wait foraging strategy to feed on zooplankton. Reproductive effort is high with numerous larvae seen in shelf areas of the Weddell Sea (Kunzmann and Zimmerman, 1992). *Cryothenia* comprises two species of pelagic oceanic fishes less than 17 cm long, found down to a maximum depth of 400 m in the Bellinghausen and Ross Seas around West Antarctica.

Dissostichus comprises two species of benthopelagic fishes that grow to over 2 m in length and are probably the most valuable commercial fishery resource in Antarctic and sub-Antarctic waters. *D. eleginoides* (Patagonian toothfish, 3850 m$_{fbd}$) occurs at latitudes of 33°S–66°S in the Southern Ocean, SE Pacific, South Atlantic and Southern Indian Oceans around sub-Antarctic islands (Fig. 4.32e). *D. mawsoni* (Antarctic toothfish, 2200 m$_{fbd}$) occurs at higher latitudes (45°S–78°S) around the entire Antarctic continent. They are found over a very wide depth range from < 50 m to their respective maximum depths. They are exploited throughout their geographical range by both legal and illegal fisheries using trawls, traps and mainly baited bottom-set longlines working at 1200–1800 m depth in the Southern Ocean. In the Antarctic area total catch of *D. eleginoides* has been from 11,000 to 17,000 tonnes since 2000 and over the same time period *D. mawsoni* landings have increased from 750 tonnes to over 4000 tonnes (CCAMLR, 2015). Including landings from South America, the Indian Ocean and Australia and illegal landings, it is estimated that the world catch of *D. eleginoides* peaked at 100,000 tonnes in 1997. Generally the size of *D. eleginoides* increases with depth with the females larger than the males. Consequently depth-related sex segregation develops in the population, and at spawning time during July and August it is hypothesised that the females move up slope and males down slope to meet in the spawning area at 800–1200 m depth (Agnew et al., 1999). Eggs and larvae are dispersed widely on oceanic currents. The monotypic *Lepidonotothen squamifrons* (Grey rockcod, grey notothen, 900 m$_{fbd}$) is benthopelagic with an opportunistic diet dominated by gelatinous salps and tunicates. It is patchily distributed but often abundant around Antarctic and sub-Antarctic islands and seamounts with highest catches occurring at depths greater than 200 m, somewhat shallower than the *Dissostichus* species. *L. squamifrons* grows to a maximum length of 55 cm with sexual maturity attained at eight years and 37–38 cm length. Populations have been depleted by fisheries and directed fishing for this species has been prohibited since 1989 off South Georgia (Gregory et al., 2013).

The remaining genera of Nototheniidae, *Gobionotothen, Gvozdarus, Lindbergichthys, Notothenia, Nototheniops, Pagothenia, Paranotothenia, Patagonotothen* and *Pleuragramma* are primarily shelf-depth or epipelagic fishes often known as rockcods, although maximum depths of over 500 m are recorded. An exception is the genus *Trematomus* with 11 demersal species occurring down to an average maximum depth of 650 m including *T. lepidorhinus* (Slender scalyhead, 800 m$_{fbd}$), *T. loennbergii* (Scaly rockcod, 1191 m$_{fbd}$) and *T. tokarev* (Bigeye notothen, 700 m$_{fbd}$), all three of which occur predominantly at depths greater than 300 m around Antarctica. As they are less than 30 cm in length they are of no interest to fisheries.

4.3.23.30 Family Artedidraconidae (177) (Barbelled plunderfishes)

The Artedidraconidae are distinguished from all the other Notothenioidei and their shallower-dwelling

(e)

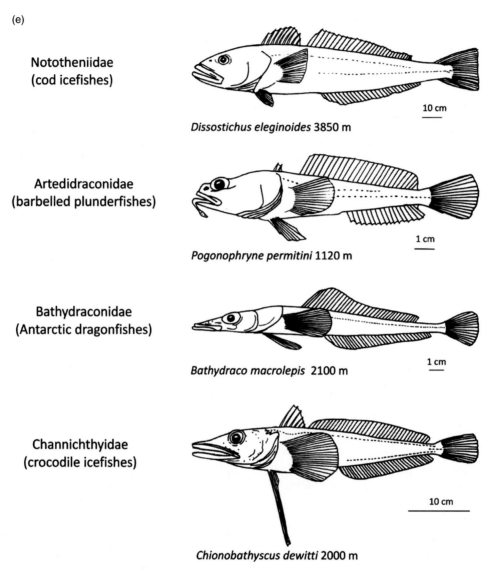

Nototheniidae
(cod icefishes)

Dissostichus eleginoides 3850 m

10 cm

Artedidraconidae
(barbelled plunderfishes)

Pogonophryne permitini 1120 m

1 cm

Bathydraconidae
(Antarctic dragonfishes)

Bathydraco macrolepis 2100 m

1 cm

Channichthyidae
(crocodile icefishes)

Chionobathyscus dewitti 2000 m

10 cm

Figure 4.32e ORDER PERCIFORMES. *Dissostichus eleginoides* (Patagonian toothfish) Southern Ocean 33°S–66°S. *Pogonophryne permitini* (finespotted plunderfish) Southern Ocean 60°S–78°S. *Bathydraco macrolepis*, Southern Ocean 68°S–79°S. *Chionobathyscus dewitti*, Southern Ocean 60°S–79°S.

sister family Harpagiferidae (spiny plunderfishes) by the presence of a barbel on the tip of the lower jaw. The shape of the barbel may be unique to each species. They are small demersal fishes, 10–34 cm maximum length, living at shelf depths and on slopes around Antarctica and the Southern Ocean islands. They have a large head, large mouth with protrusible jaws, and several species are successfully captured on baited longlines. They are probably sit-and-wait predators that can ambush mobile prey such as shrimps as well as carrion falling to the seafloor. There are 28 species in four genera (Lecointre et al., 2011). The genus *Artedidraco* comprises six species mainly found at sublittoral to slope depths down to 800 m. *Dolloidraco longedorsali* (1145 m_{fbd}) and *Histiodraco veliferi* (667 m_{fbd}) are two monotypic species that live at depths greater than 200 m. The genus *Pogonophryne* is the deepest and most diverse genus with 21 species distributed around the Antarctic at an average depth range of 660–960 m (Fig. 4.32e). The deepest recorded species is *P. immaculata* (2542 m_{fbd}) known only from the holotype collected from the South Shetland Islands. Some species have a vivid colour patterning on the body and there is evidence of sexual dimorphism with a higher second dorsal fin in the male than in the female *P. brevibarbata* (1157 m_{fbd}; Shanidikov et al., 2013). *P. ventrimaculata* (460 m_{fbd}) produces a small number of large eggs, and it is suggested that there may be parental nesting behaviour.

4.3.23.31 Family Bathydraconidae (178) (Antarctic dragonfishes)

The Bathydacrocnidae are small to medium-sized fishes, average maximum length 27 cm, with nonprotrusible jaws and no spiny dorsal fin (Fig. 4.32e). They range from inshore under ice cover to over 2000 m depth, are relatively uncommon and of no interest to fisheries. There are 17 species in 11 genera, the deepest-living are in the genus *Bathydraco*, which comprises five species, all always found at depths greater than 300 m. *B. macrolepis* (2100 m_{fbd}) and *B. scotiae* (2950 m_{fbd}) are probably circum-Antarctic, whereas *B. antarcticus* (2400 m_{fbd}), *B. joannae* (1800 m_{fbd}) and *B. marri* (1250 m_{fbd}) are more restricted in their ranges centred on the Scotia Sea area. Juvenile *B. marri* captured at 340 m depth in the Ross Sea have been found to feed exclusively on crustacea; the smallest individuals feed on copepods and other small pelagic prey, shifting to larger benthopelagic prey such as amphipods and mysids as they grow (Le Mesa, 2007). The nonprotrusible jaws appear to preclude truly benthic feeding, and *B. macrolepis* has been observed at 1200 m depth hovering a few metres above the seafloor with the body horizontal, sculling against the current with the pectoral fins to hold position apparently waiting to intercept mobile prey (Near et al., 2009). Other relatively deep-living Bathydacrocnidae are the monotypic *Akarotaxis nudiceps* (915 m_{fbd}), *Gerlachea australis* (670 m_{fbd}),

Vomeridens infuscipinnis (813 m_{fbd}) and the two species of *Racovitzia*, the circum-Antarctic *R. glacialis* (610 m_{fbd}) and *R. harrissoni* (600 m_{fbd}) found in the Southeast Pacific off Chile. The other genera and species all occur on the shelf or inshore.

4.3.23.32 Family Channichthyidae (179) (Crocodile icefishes)

The Channichthyidae have a crocodile-like flattened snout with nonprotrusible jaws, a spiny dorsal fin and an average maximum total length of 43 cm with some species growing to over 70 cm. They are remarkable for their colourless blood with no erythrocytes or haemoglobin and appear to be able to survive by virtue of high concentrations of dissolved oxygen at low temperatures and high cardiac output. These species have compensated for lack of haemoglobin and myoglobin by development of large hearts and other circulatory adaptations (Sidell and O'Brien, 2006). There are 25 species in 11 genera, mostly demersal, all found in cold waters of the Southern Ocean at shelf and slope depths. Several species in the genera *Chaenocephalus*, *Chaenodraco*, *Champsocephalus*, *Channichthys*, *Chionodraco* and *Pseudochaenichthys* are of interest to fisheries and although distributed down to 800 m depth are mostly exploited at shelf depths. The monotypic *Chionobathyscus dewitti* (2000 m_{fbd}) is a truly deep-sea species probably circum-Antarctic at depths greater than 500 m (Fig. 4.32e). Specimens retrieved at 976–1838 m depth from the stomachs of Antarctic toothfish were found to contain fishes, krill and squid. Lacking a strong jaw apparatus *C. dewitti* appears to swallow large prey whole; the stomach is distensible, enabling the weight of food in it to exceed 30 per cent of the body weight (Petrov, 2011). Species in other genera are distributed to upper slope depths: *Cryodraco* (600 m_{fbd}), *Dacodraco* (800 m_{fbd}), *Neopagetopsis* (900 m_{fbd}) and *Pagetopsis* (800 m_{fbd}).

Suborder Trachinoidei

This suborder comprises 11 families and about 300 species, the majority of which are tropical shallow-water species including the Trachinidae, the weever fishes. Only the five families that include deep-sea species are considered here.

4.3.23.33 Family Chiasmodontidae (180) (Swallowers)

This is a deep-sea endemic family (Priede and Froese, 2013) with a mean depth range of 538 to 2136 m. It comprises about 30 species in four genera distributed worldwide at tropical to subpolar latitudes. They are bathypelagic fishes with a cylindrical body, average maximum length 16 cm, dark skin, large jaws with long fang-like teeth and the stomach and lateral muscles greatly distensible enabling very large prey to be ingested. Some species have photophores.

The genus *Chiasmodon* comprises seven species. *C. pluriradiatus* (1600 m$_{fbd}$) found in the tropical Atlantic, Indian Ocean and S Pacific, and *C. asper* (5691 m$_{fbd}$) confined to the Pacific Ocean, form a species group (Melo, 2009). The remaining species form the *niger* group, *C. microcephalus* (2500 m$_{fbd}$), which is circumglobal 32°S- 54°S, *C. niger* (Black swallower, 2745 m$_{fbd}$) and *C. harteli* (1417 m$_{fbd}$) confined to the Atlantic Ocean, *C. braueri* (2326 m$_{fbd}$) from the Indian and Atlantic Oceans and *C. subniger* (4568 m$_{fbd}$) from the E Pacific. Although the species is widespread, capture is rare with only one *C. niger* having been caught in an extensive survey of the Sargasso Sea down to 5000 m depth (Sutton et al., 2010). The genus *Dysalotus* comprises two rare species distinguished from other Chiasmodontidae by the presence of enlarged prickles on the posterior flank and multiple rows of needle-like teeth. Both species are circumglobal in their distribution at tropical and subtropical latitudes at depths down to 2100 m. The genus *Kali* is distinctive for its long recurved teeth, fragile cranial bones and enlarged fangs in some species; they are found worldwide in the Atlantic, Indian, Pacific and Southern Oceans (53°N-60°) mostly at depths of 1000 m to 3000 m. There are seven species, which are all circumglobal except for *K. parri* (1900 m$_{fbd}$), which is known only from a few samples in the S Atlantic. *K. colubrina* (2270 m$_{fbd}$) is found 28°N-40°S, *K. falx* (2870 m$_{fbd}$) 29°N-8°N, *K. indica* (3300 m$_{fbd}$) 40°N-60°S, *K. kerberti*

(4465 m$_{fbd}$) 38°N-33°S, *K. macrodon* (2500 m$_{fbd}$) 24°N-19°S (Fig. 4.32f) and *K. macrura* (1000 m minimum) 28°N-15°S.

Most of the species on the genus *Pseudoscopelus* have photophores distributed over the body, and little seems to be known about their structure and function. Indeed Lütken (1892) designated them as pores, having failed to recognise them as functional bioluminescent organs. Melo (2009) recognises 11 possible photophore locations on the head region, including on the maxilla, around the orbit and opercular bones and 16 locations on the body on the isthmus, trunk and around the various fins. The pattern of photophores on the body is unique to each species and can be used as a diagnostic character. Two species, *P. aphos* and *P. parini,* lack photophores, although the latter has white lines on the ventral part of the body where some photophore groups might occur. Three species are circumglobal in their distribution: *P. altipinnis* (2390 m$_{fbd}$; Fig. 4.32f) is found in the tropics and subtropics (46°N-27°S), *P. australis* (1250 m$_{fbd}$) occurs in the Southern Ocean (31°S-60°S) and *P. cordilluminatus* (1164 m$_{fbd}$) is known from the Atlantic and Indo-West Pacific. Six species are endemic to the Pacific Ocean, five to the Atlantic Ocean and there are two Indo-Pacific species.

4.3.23.34 Family Champsodontidae (181) (Gapers)

There is only one genus in this family, *Champsodon*, comprising 13 species of small (average maximum length 11 cm) tropical demersal fishes of the Indo-Pacific region, characterised by a large head and mouth and elongated laterally flattened body. They are silvery-coloured fishes often associated with myctophids in acoustic scattering layers impinging on the upper slopes. Six species are endemic to the Pacific Ocean, four to the Indian Ocean and three species are Indo-Pacific in their distribution. Two of the Indo-Pacific species have been recorded in the Eastern Mediterranean. Several species can be regarded in varying degrees as deep-sea species, such as *C. capensis* (gaper, 552 m$_{fbd}$) that can form great shoals in the Western Indian Ocean, which move

(f)

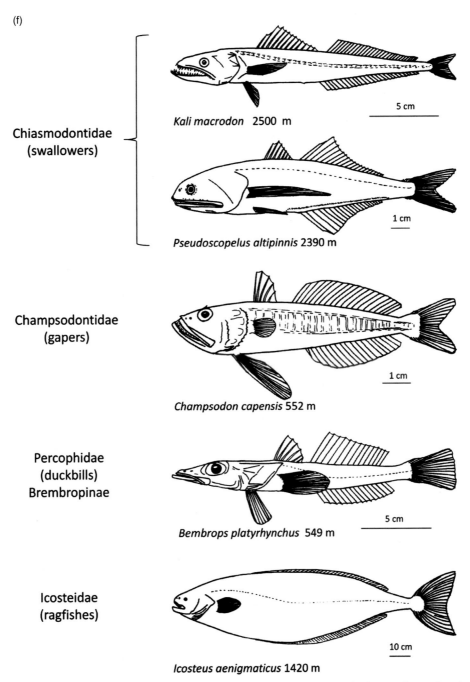

Chiasmodontidae (swallowers)

Kali macrodon 2500 m

5 cm

Pseudoscopelus altipinnis 2390 m

1 cm

Champsodontidae (gapers)

Champsodon capensis 552 m

1 cm

Percophidae (duckbills) Brembropinae

Bembrops platyrhynchus 549 m

5 cm

Icosteidae (ragfishes)

Icosteus aenigmaticus 1420 m

10 cm

Figure 4.32f ORDER PERCIFORMES. *Kali macrodon*, probably circumglobal tropical and subtropical seas. *Pseudoscopelus altipinnis*, probably circumglobal 46°N–36°S. *Champsodon capensis* (gaper) Indian Ocean. *Bembrops platyrhynchus* (Natal duckbill) Indo-W Pacific. *Icosteus aenigmaticus* (ragfish) N Pacific.

towards the surface at night (Fig. 4.32f), *C. omanensis* (1120 m_{fbd}) may form aggregations above and below the oxygen-minimum zone in the NW Indian Ocean and *C. guentheri* (1020 m_{fbd}) in the Western Pacific from Japan to Australia. *Champsodon* species have two lateral lines on each side with vertical rows of sensory papillae in between (Nemeth, 1994).

4.3.23.35 Family Percophidae (182) (Duckbills)

The Percophidae comprises 47 species of demersal fishes divided into three subfamilies, one of which, the Brembropinae, includes species occurring on the upper slopes at depths down to over 900 m.

Subfamily Brembropinae

This subfamily comprises two genera and 23 species with an average depth range of 219–471 m distributed around the tropical and warm temperate regions of the world. They have a dorsal fin with six spines, a flattened head with the lower jaw projecting beyond the upper jaw and a cylindrical elongated body, and average maximum total length is 21 cm. The genus *Bembrops* (Fig. 4.32f) has nine species endemic to the Atlantic Ocean, one species in the Indian Ocean, three Indo-West Pacific species and four Pacific species distributed in the west from Japan to New Zealand. There are no *Bembrops* in the East Pacific. The deepest-living species is *B. quadrisella* (914 m_{fbd}) found at depths greater than 350 m in the Caribbean Sea and W Atlantic off Suriname. *Bembrops* feed primarily on other smaller fishes and crustacea (Das and Nelson, 1996). The genus *Chrionema* comprises six species, *S. squamentum* (minimum depth 256 m) from the Florida straits and Caribbean area of the W Atlantic, *C. chlorotaenia* (minimum depth 350 m) in the Indo-West Pacific and four species from the Pacific Ocean.

4.3.23.36 Family Uranoscopidae (183) (Stargazers)

The Uranoscopidae are well-known shallow-water bottom-living fishes of average maximum length 35 cm with upward looking eyes, venomous spines behind the operculae, that adopt a sit-and-wait foraging strategy, attracting prey fish with a lure that extends from the floor of the mouth. There are 52 species in eight genera. A few species, especially in the genus *Kathetostoma,* range down to upper slope depths such as *K. cubana* (Spiny stargazer, 600 m_{fbd}) usually found at depths greater than 200 m in the Central Western Atlantic.

Suborder Pholidichthyoidei

No deep-water species.

Suborder Blenniodei

No deep-water species.

Suborder Icosteoidei
4.3.23.37 Family Icosteidae (184) (Ragfishes)

This suborder comprises one family, one genus and one species, *Icosteus aenigmaticus* (Ragfish, 1420 m_{fbd}), a very unusual fish that grows to over 2 m in length and is found exclusively in the North Pacific Ocean (66°N–32°N) from Japan to Alaska and Southern California (Fig. 4.32f). The skeleton is largely cartilaginous and the muscles rather flabby. *I. aenigmaticus* are very conspicuous, but rare, with one specimen per year on average recorded by ichthyologists in the NE Pacific Ocean. They occur over a very wide depth range from the epipelagic to the bathypelagic. Feeding habits are very generalised and include jellyfish, fishes and squids, both benthic and pelagic. They appear to be associated with cold water, and Allen (2001) suggests that spawning occurs at depths greater than 220 m and that canyons and deep channels on bathyal slopes may be important areas of habitat for this species.

Suborder Callionymoidei

This suborder comprises two families: the mainly shallow-water Callionymidae and the deeper-living Draconettidae.

4.3.23.38 Family Callionymidae (185) (Dragonets)

These are elongated demersal fishes with large fins, often colourful showing sexual dimorphism with a ventral small mouth and gill opening reduced to a small opening on the upper side of the head. They thrive on sandy and muddy bottoms feeding on benthic infauna. They are distributed in all the oceans in warm and temperate seas mainly in the Indo-West Pacific. There are 185 species of which 25 are regarded as bathydemersal living just beyond the shelf edge or on the summits of seamounts and ridges down to a maximum depth of 850 m. There are 17 genera. A few of the ca. 100 species of *Callionymus* are considered to be deep, including *C. africanus* (African deepwater dragonet, 220 m_{fbd}) and *C. carebares* (Indian deep-water dragonet, 330 m_{fbd}) from the Indian Ocean. The genus *Foetorepus* tends to occur deeper including *F. agassizii* (Spotfin dragonet, 700 m_{fbd}) from the W Atlantic. The monotypic *Paracallionymus costatus* (Ladder dragonet, 457m_{fbd}) is distributed on slopes off South Africa where it is preyed on by *Genypterus capensis* (Kingklip). *Protogrammus sousai* (320 m_{fbd}) is described from the south Meteor Bank in the E Atlantic. The genus *Synchiropus* has 10 deep-water species from islands and seamounts of the West Pacific and *S. monacanthus* (Deep-water dragonet, 428 m_{fbd}) from the SE Atlantic and W Indian Ocean.

4.3.23.39 Family Draconettidae (186) (Slope dragonets)

The Draconettidae comprises 14 species that occur at an average depth range of 300–430 m on the upper slopes in all the major oceans in tropical and subtropical areas, between 35°N and 35°S. They are extremely rare with most species known only from a few specimens (Fricke, 1992). Compared with the Callionymidae, they have a relatively broad gill opening but live deeper on soft bottoms of continental margins, submarine ridges and seamounts. There is often sexual dimorphism as in the Callionymidae. Two genera are recognised: *Centrodraco* has two species in the Atlantic Ocean, *C. acanthopoma* (594 m_{fbd}) in the North and *C. oregonus* (411 m_{fbd}) in the SW Atlantic off Brazil, two species in the Indian

Ocean *C. insolitus* (350 m_{fbd}) in the East off NW Australia and *C. lineatus* (415 m_{fbd}) in the west, and nine species in the Pacific Ocean including *C. atrifilum* (640 m_{fbd}), the deepest recorded species (Fig. 4.32g) from Eastern Australia (Fricke, 2010). The monotypic *Draconetta xenica* (376 m_{fbd}) with characteristic long and slender spines on the first dorsal fin is found throughout the Indo-West Pacific region from East Africa to the Hawaiian Islands.

Suborder Gobioidei

With over 1100 species this is one of major Perciform groups and includes eight marine families. They are overwhelmingly shallow-water coastal fishes with only a few species considered as deep-sea.

4.3.23.40 Family Gobiidae (187) (Gobies)

Comprising 1000 species, average maximum length 6 cm, these are familiar small demersal fishes associated with coastal sublittoral habitats. Misleadingly the genus *Bathygobius* only reaches an average maximum depth of less than 10 m. A few species from other genera are found in the tropics beyond shelf depths including *Cabillus macrophthalmus* (Bigeye cabillus, 400 m_{fbd}) from the W Pacific, *Priolepis goldshmidtae* (400 m_{fbd}) and *Obliquogobius turkayi* (496 m_{fbd}) from the Red Sea, *Robinsichthys arrowsmithensis* (600 m_{fbd}) from the W Atlantic and *Thorogobius rofeni* (650 m_{fbd}) from the E Atlantic.

Suborder Acanthuroidei

No deep-water species.

Suborder Scombrolabracoidei
4.3.23.41 Family Scombrolabracidae (188) (Longfin escolars)

This family and order comprise one species (Fig. 4.32g) *Scombrolabrax heterolepis* (Longfin escolar, black mackerel, 900 m_{fbd}) first described from a specimen caught in the commercial fishery for black scabbardfish (*Aphanopus carbo*) at 800–900 m depth off Madeira in the NE Atlantic Ocean (Roule, 1922).

(g)

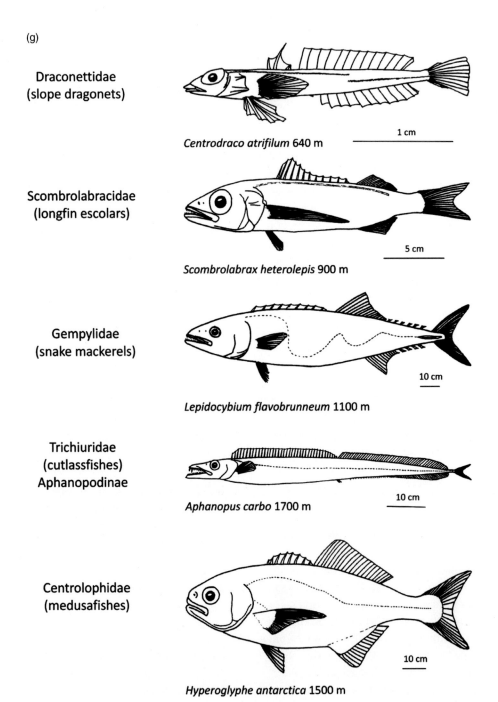

Draconettidae
(slope dragonets)

Centrodraco atrifilum 640 m

1 cm

Scombrolabracidae
(longfin escolars)

Scombrolabrax heterolepis 900 m

5 cm

Gempylidae
(snake mackerels)

Lepidocybium flavobrunneum 1100 m

10 cm

Trichiuridae
(cutlassfishes)
Aphanopodinae

Aphanopus carbo 1700 m

10 cm

Centrolophidae
(medusafishes)

Hyperoglyphe antarctica 1500 m

10 cm

Figure 4.32g ORDER PERCIFORMES. *Centrodraco atrifilum*, W Pacific, Australia. *Scombrolabrax heterolepis* (longfin escolar) circumglobal tropical and subtropical except E Pacific and SE Atlantic. *Lepidocybium flavobrunneum* (escolar) circumglobal 50°N–56°S. *Aphanopus carbo* (black scabbardfish) North Atlantic 71°N–22°N. *Hyperoglyphe antarctica* (bluenose warehou) Southern Hemisphere 28°S–55°S, Australia, New Zealand, Argentina and South Africa.

Though the species occurs in tropical latitudes in all the major oceans, it is nonetheless quite rare, and little is known of its biology. Larvae are found in the upper 200 m of the water column and appear to be absent from the eastern sides of the Atlantic and Pacific Oceans (Potthoff et al., 1980). Adults grow to 30 cm maximum length, are dark brown in colour and are associated with bathyal slopes of continents, islands and seamounts. *S. heterolepis* appear in small quantities as bycatch in the Hawaii-based deep-sea longline fishery for tuna but are not normally landed (National Marine Fisheries Service, 2013). They are also found in the stomachs of tunas and billfishes.

Suborder Scombroidei

The Scombroidei comprises 160 species including some of the fastest swimming oceanic fishes such as the tunas, swordfish and marlins. There are six families. The Sphyraenidae (barracudas) are entirely shallow-water inshore and neritic species. Some of the pelagic oceanic species have been recorded as diving to great depths, for example in the Scombridae (mackerels and tunas) *Thunnus maccoyii* (Southern bluefin tuna, 2743 m_{fbd}), Xiphiidae *Xiphias gladius* (Swordfish, 800 m_{fbd}) and Istiophoridae (bill fishes). *Tetrapturus angustirostris* (Shortbill spearfish, 1830 m_{fbd}) are the deepest recorded species in their respective families. However, because they spend most time near the surface they cannot be considered deep-sea species, although they may be capable of exploiting mesopelagic and bathypelagic prey. Two families contribute significantly to the deep-sea ichthyofauna and fishery resources: Gempylidae and Trichiuridae.

4.3.23.42 Family Gempylidae (189) (Snake mackerels)

The Gempylidae are benthopelagic or mesopelagic, with brownish coloration and body shape, which varies from fusiform mackerel-like fast-swimming forms to elongated slow-swimming forms. They can be considered to be collectively intermediate between the epipelagic scombridae and the benthopelagic Trichiuridae (Nakamura and Parin, 1993). They occur

in all the major oceans over an average depth range of 185–760 m. There are 24 species in 15 genera. The monotypic *Diplospinus multistriatus* (Striped escolar, 1000 m_{fbd}) has an elongated body form, maximum length 33 cm and occurs in the mesopelagic of central water masses of the Atlantic, Indian and Pacific Oceans, where it migrates up to 100–200 m depth during the night, feeding on crustacea and small fishes. *Lepidocybium flavobrunneum* (Escolar, 1100 m_{fbd}) is fusiform in shape and circumglobal in its distribution (50°N–56°S), growing to 2 m length and (Fig. 4.32g) is taken as a bycatch in fisheries off continental slopes. However, the oily flesh has purgative properties and is not particularly desirable as food fish. *Nealotus tripes* (Black snake mackerel, 600 m_{fbd}) has a more elongated body shape, reaching a maximum length of 25 cm and occurs worldwide (47°N–35°S) in the mesopelagic. The genus *Neoepinnula* comprises two fusiform sackfish species: *N. americana* (American sackfish, 457 m_{fbd}), and *N. orientalis* (sackfish, 570 m_{fbd}), which are benthopelagic on upper slopes. *Nesiarchus nasutus* (Black gemfish, 1200 m_{fbd}) has an elongated body shape up to 130 cm total length and occurs almost worldwide in tropical to subtropical regions in the bathypelagic to mesopelagic. The genus *Paradiplospinus* comprises two Southern Hemisphere species: *P. antarcticus* (2830 m_{fbd}), which is pelagic and circumglobal in the Southern Ocean (31°S–79°S) feeding on krill, squid and myctophids, and *P. gracilis* (626 m_{fbd}), which occurs on the continental slopes of SW Africa. They have an elongated body form and grow to over 50 cm length. *Promethichthys prometheus* (Roudi escolar, 800 m_{fbd}) is benthopelagic on slopes in tropical and warm temperate waters of all the major oceans and grows to over 1 m maximum length.

The genus *Rexea* comprises seven benthopelagic species, three of which occur across the Indo-West Pacific, *R. bengalensis* (Bengal escolar, 820 m_{fbd}), *R. nakamurai* (Nakamura's escolar, 370 m_{fbd}) and *R. prometheoides* (Royal escolar, 540 m_{fbd}) and four species that are endemic to the Pacific Ocean. The largest species in the genus, *R. solandri* (Silver gemfish, 800 m_{fbd}) grows up to 1.1 m long and is the target of major fisheries off Australia and New Zealand. *Rexichthys johnpaxtoni* (Paxton's escolar,

470 m$_{fbd}$) is known from NE Australia and New Caledonia. *Ruvettus pretiosus* (Oilfish, 800 m$_{fbd}$) is a circumglobal oceanic species, including the Mediterranean Sea, that is associated with upper slope depths and grows to 3 m maximum length. Like *Lepidocybium*, the oily flesh is not particularly palatable but is taken in fisheries. *Thyrsitoides marleyi* (Black snoek, 400 m$_{fbd}$) from the Indo-West Pacific, *Thyrsitops lepidopoides* (White snake mackerel, 350 m$_{fbd}$) from the SW Atlantic and SE Pacific and *Tongaichthys robustus* (Tonga escolar, 312 m$_{fbd}$) from the SW Pacific are all slope-associated mesobenthopelagic species. *T. marleyi* grows to 2 m length. The genera, *Epinnula*, *Gempylus* and *Thyrsites* are confined predominantly to shelf depths. Most species of Gempylidae are taken in fisheries to some extent.

4.3.23.43 Family Trichiuridae (190) (Cutlassfishes, Hairtails)

The Trichiuridae all have an elongated laterally flattened body form with several species growing to over 2 m in length, with characteristic protruding lower jaw and fang-like teeth. The deepest living species are black or dark-skinned, whereas shallower species are light-coloured (Nakamura and Parin, 1993). They are demersal species associated with slope depths down to over 1000 m. Following Nelson (2006) here we consider three subfamilies.

Subfamily Aphanopodinae

With an average depth range of 330–1030 m this is the deepest-living subfamily of Trichiuridae with 18 species in two genera. The genus *Aphanopus* comprises seven species distributed in all the major oceans. There are two species endemic to the Atlantic Ocean, *A. carbo* (Black scabbardfish, 1700 m$_{fbd}$) in the North (71°N–22°N (Fig. 4.32g) and *A. intermedius* (Intermediate scabbardfish, 1350 m$_{fbd}$) further south (50°N–35°S). They are benthopelagic on slopes where they prey on fishes, cephalopods and crustacea. *A. carbo* are captured by longline fisheries off Portugal, Madeira and the Azores and appear as a bycatch in trawl fisheries west of the British Isles and

in the NW Atlantic. Catches in the south probably include *A. intermedius*. *A. mikhailini* (Mikhailin's scabbardfish, 2000 m$_{fbd}$) occurs in the South Atlantic and Indian Oceans and in the SW Pacific off New Zealand and Australia. *A. microphthalmus* (Smalleye scabbardfish, 1022 m$_{fbd}$) is found in the tropical W Indian Ocean and the W Pacific. *A. beckeri* (843 m$_{fbd}$) has been described from West Australia and *A. arigato* (Pacific black scabbardfish, 1350 m$_{fbd}$) from the NW Pacific and *A. capricornis* (1024 m$_{fbd}$) from the S Pacific. Only the North Atlantic species are exploited to a significant degree. The genus *Benthodesmus* comprises 11 species of elongated ribbon-like fishes that grow to an average maximum length of 80 cm and is distributed in all tropical and temperate oceans on bathyal slopes. Two species are circumglobal: *B. tenuis* (Slender frostfish, 850 m$_{fbd}$) at low latitudes (45°N–36°S) and *B. elongatus* (Elongate frostfish, 950 m$_{fbd}$) in the Southern Hemisphere (9°S–44°S). There is a cluster of four species endemic to the Indo-Australian Archipelago as well as *B. tuckeri* (Tucker's frostfish, 790 m$_{fbd}$) extending from this area into the Indian Ocean and *B. vityazi* (Vityaz' frostfish, 900 m$_{fbd}$) extending across the tropical Indian and Pacific Oceans. *B. oligoradiatus* (Sparse-rayed frostfish, 1000 m$_{fbd}$) is endemic to the N Indian Ocean, *B. pacificus* (North-Pacific frostfish, 1000 m$_{fbd}$) to N Pacific (50°N–21°N) and *B. simonyi* (Simony's frostfish, 900 m$_{fbd}$) to the North Atlantic (72°N–16°N). The genus *Benthodesmus* is of no interest to fisheries.

Subfamily Lepidopodinae

The fishes in this subfamily are silvery in colour, have a small or absent caudal fin, and live somewhat shallower that the Aphanopodinae. Only the deepest-living species are considered here. The monotypic *Assurger anzac* (Razorback scabbardfish, 400 m$_{fbd}$) is a large fish up to 2.5 m long, living circumglobally in the tropics (37°N–37°S), and is benthopelagic feeding on fishes and squid between 150 and 400 m depth. The genus *Evoxymetopon* comprises three species found on seamounts and upper slopes. The genus *Lepidopus* comprises six species distributed through all the oceans. *L. caudatus* (Silver scabbardfish, 620

m_{fbd}) grows to over 2 m length and is highly valued in fisheries throughout its range in the Eastern Atlantic, Western Mediterranean, Southern Indian Ocean and off Australia and New Zealand.

Subfamily Trichiurinae (Hairtails)

The Trichiurinae are the shallowest-dwelling of the three Trichiurid subfamilies. They are characterised by the absence of caudal and pelvic fins and their long tapering hair tail. The genus *Trichiurus* is distributed circumglobally with nine species. *T. lepturus* (Largehead hairtail, 620 m_{fbd}) is commercially important worldwide and is commonly caught inshore as well as in deep-water trawling offshore. It grows to over 2 m total length, and the flesh quality is considered excellent.

Suborder Stromateoidei

The Stromateoidei are generally pelagic fishes with laterally flattened elongated oval body form that grow to a mean maximum length of 50 cm. There are over 70 species in six families that range from the surface down to over 2000 m and are distributed worldwide in warm to temperate seas. Here four families that have deep-water representatives are considered.

4.3.23.44 Family Centrolophidae (191) (Medusafishes)

Also known as ruffs, barrelfishes and blackfishes, these are epipelagic, mesopelagic and demersal fishes that range in maximum length from 14 to 140 cm. They have a single long dorsal fin with spiny rays at the anterior end. The juveniles are often associated with floating objects including jellyfish. The adults feed on crustacea, salps and small fishes. They occur in all the tropical and temperate seas. There are 30 species of which 10 can be considered deep-sea.

The monotypic *Centrolophus niger* (Rudderfish, 1050 m_{fbd}) usually occurs at 300–700 m depth, is dark-coloured, grows to 1.5 m length and has an antitropical distribution in the temperate North Atlantic and Mediterranean in the north, and around

much of the Southern Hemisphere including off New Zealand and South Africa. The genus *Hyperoglyphe* comprises six species, which occur in deep water over continental slopes and are commercially exploited to varying degrees. The most important species is *H. antarctica* (Bluenose warehou, 1500 m_{fbd}), which is circumglobal in the Southern Hemisphere (28°S–55°S) usually at depths of 260–490 m. It is also known as the Blue eye trevalla in Australia, Blue Nose in New Zealand, Antarctic butterfish, Deep-sea trevally or Bluenose sea bass (Fig. 4.32g). Off New Zealand it is taken as bycatch in fisheries for *Polyprion* by longline, and Alfonsino (*Beryx splendens*) by trawl (Horn, 2003) as well as directed fisheries. There are four species of *Hyperoglyphe* in the Atlantic Ocean, two in the North and two in the South and *H. japonica* (Pacific barrelfish, 400 m_{fbd}) in the NW Pacific. *Icichthys australis* (Southern driftfish, 2141 m_{fbd}) is circumglobal in the Southern Hemisphere (34°S–61°S). The genus *Psenopsis* comprises four species from the West Pacific, one Indian Ocean species and *P. obscura* (Obscure ruff, 800 m_{fbd}), which is found just beyond the shelf edge across the Indo-West Pacific. *Schedophilus* comprises eight species including *S. huttoni* (New Zealand ruffe, 1000 m_{fbd}), which is demersal on the continental slopes from 800 to 1000 m depth in the S Pacific and the W Indian Ocean off South Africa. *Seriolella* comprises six species distributed around the Southern Hemisphere and *S. caerulea* (White warehou, 800 m_{fbd}), which occurs over much of this range (27°S–56°S) and is listed as a deep-sea commercial species (Tandstad et al., 2011). The monotypic *Tubbia tasmanica* (Tasmanian ruffe, 850 m_{fbd}) is also circumglobal in temperate waters of the Southern Hemisphere.

4.3.23.45 Family Nomeidae (192) (Driftfishes)

The Nomeidae are pelagic fishes that occur worldwide in tropical and subtropical areas. They have two dorsal fins, the first with long spines. Of the 16 species a few occur in deep water. *Cubiceps pauciradiatus* (Bigeye cigarfish, 1000 m_{fbd}) is circumtropical (40°N–40°S) and benthopelagic above the continental slope down to 1000 m depth (Fig. 4.32h). It ascends to near

(h)

Nomeidae
(driftfishes)

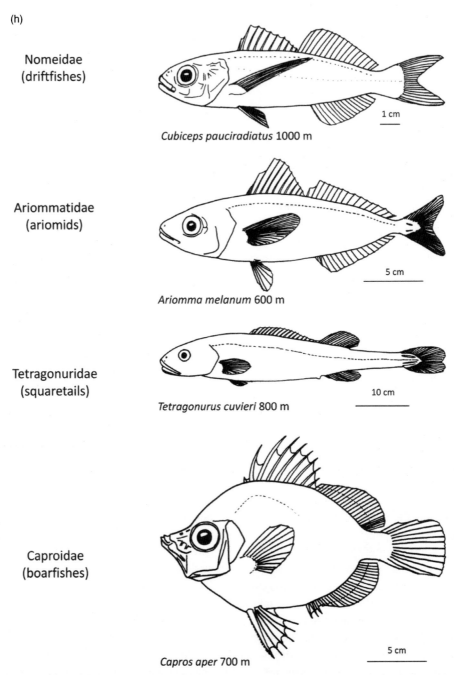

Cubiceps pauciradiatus 1000 m

1 cm

Ariommatidae
(ariomids)

Ariomma melanum 600 m

5 cm

Tetragonuridae
(squaretails)

Tetragonurus cuvieri 800 m

10 cm

Caproidae
(boarfishes)

Capros aper 700 m

5 cm

Figure 4.32h ORDER PERCIFORMES. *Cubiceps pauciradiatus* (bigeye cigarfish) circumglobal 40°N–40°S. *Ariomma melanum* (brown driftfish) N Atlantic 42°N–8°N. *Tetragonurus cuvieri* (smalleye squaretail) probably circumglobal including the Mediterranean. *Capros aper* (boarfish) NE Atlantic and Mediterranean 62°N–10°N.

the surface at night and feeds mainly on salps. In the genus *Psenes* two species are circumglobal: *P. cyanophrys* (Freckled driftfish, 550 m$_{fbd}$) at 49°N–43°S, *P. pellucidus* (Bluefin driftfish, 1000 m$_{fbd}$) in the Northern Hemisphere (42°N–24°N) associated with jellyfish and floating weed.

4.3.23.46 Family Ariommatidae (193) (Ariomids)

This family comprises eight species in one genus of mainly demersal fishes, average maximum length 35 cm, found along tropical and subtropical coastlines. Three species occur in the deep sea: *Ariomma luridum* (Ariommid, 350 m minimum depth), in the Pacific Ocean, *A. parini* (Parin's ariomma, 205 minimum depth) in the Indian Ocean and *A. melanum* (Brown driftfish, 600 m$_{fbd}$) from the Atlantic Ocean where it is usually found at 180–550 m depth on soft substrates feeding on small crustacea (Fig. 4.32h). There is a minor fishery interest in *A. melanum*.

4.3.23.47 Family Tetragonuridae (194) (Squaretails)

The Tetragonidae are the most elongated of the Stromateoidei. There are three species, *Tetragonurus cuvieri* (Smalleye squaretail, 800 m$_{fbd}$) in the Mediterranean, Indian, Pacific and Atlantic Oceans (Fig. 4.32h), *T. atlanticus* (Bigeye squaretail, 800 m$_{fbd}$) in the Atlantic, Indian and Pacific Oceans and *T. pacificus* (Pacific squaretail) in the Pacific and Indian Oceans. They are pelagic oceanic fishes strongly associated with gelatinous pelagic tunicates colonies. Young fishes are observed to live inside salps, eating the guts and the gill bars. The fish stomach contents typically include planktonic items that would have been captured by the tunicates. The elongated body form and ability to retract the dorsal and pelvic fins into grooves enable *Tetragonus* to move forward and backward in and out of the body cavities of the salps. Long sharp mandibular teeth and a muscular pharyngeal sac with tooth-like papillae that acts like a gizzard facilitate consumption of gelatinous tunicates. At 40–60 mm body length *Tetragonus* resorb the swim bladder and become mesopelagic growing up to 50–70 cm maximum length. Janssen and Harbison (1981) suggest that adult *Tetragonus* live in the cloaca of large pyrosomes or in the body cavities of larger deep-living salps. *Tetragonus* should not be considered to be free-living fishes swimming in the vast void of the deep ocean but rather they are adapted to exploit the three-dimensional network of tubular substrate provided by aggregations of colonial tunicates. This structure in the deep ocean is poorly understood because it is destroyed by nets and most types of sampling equipment. Risso (1810) reported from personal experience that the flesh is very dangerous to eat, leading to severe abdominal pains and nausea and suggested the toxin is derived from the consumption of jellyfish. This and other early reports that *Tetragonus* prey on jellyfish and ctenophores have proved to be erroneous, as they are clearly tunicate specialists (Janssen and Harbison, 1981) but because Risso was a pharmacist by profession his descriptions of the toxicity symptoms should be respected.

Suborder Caproidei

This suborder comprises one family of 18 species of laterally flattened, rhomboid-shaped fishes similar to Dories (Zeiformes), in particular the Oreos such as *Neocyttus helgae* (false boarfish) with which they were formerly classified (Nelson, 2006). They are small fishes less than 30 cm long, and their sharp spines tend to damage other species when caught as bycatch in trawl fisheries.

4.3.23.48 Family Caproidae (195) (Boarfishes)

The Caproidae are circumglobal in their distribution at tropical to temperate latitudes except for the NE Pacific. They often occur in abundance in shoals on continental slopes, seamounts, ridges and around islands at an average depth range of 155–430 m. Although not found particularly deep, they never occur at the surface so are considered to be deep-sea fishes and are listed as a deep-water fishery resource by FAO (Tanstad et al., 2011). There are two genera. *Antigonia capros* (Deepbody boarfish, 900 m$_{fbd}$) is almost worldwide in subtropical and tropical waters (44°N–24°S) typically at 100–300 m feeding benthically on small molluscs and crustacea. Four species of *Antigonia* are endemic to different parts of the Pacific Ocean, one species *A. combatia* (Shortspine boarfish, 585 m$_{fbd}$) in the Western Atlantic, three species occur across the Indo-Pacific and seven species have been described from the Western Indian Ocean (Parin and Borodulina, 2006). The monotypic *Capros aper* (boarfish, 700 m$_{fbd}$) is confined to NE Atlantic (62°N–10°N) and the Mediterranean Sea, forming schools and feeding on benthic fauna including crustacea, worms and molluscs (Fig. 4.32h). It is recorded at 288–700 m depth in bottom trawl surveys on slopes of the Eastern Ionian Sea (Mytilineou et al., 2005). Since 2001 a fishery for *Capros aper* has developed to the SW of Ireland with a total allowable catch of 82000 tonnes in 2012 and 2013 caught by pelagic trawl for fish meal (Strange, 2016). There are also minor fisheries for *Antigonia capros* and other species of boarfish around the world.

4.3.24 Order Pleuronectiformes (Flatfishes)

Flatfishes of the order Pleuronectiformes are amongst the most distinctive of fishes with the body flattened, both eyes on one side enabling them to lie with the blind side on the seafloor (Fig. 4.33). In this way they are sheltered from currents and protected from predators with the aid of camouflage or varying degrees of burial within the substrate. There are over 700 species varying in size from *Hippoglossus hippoglossus* (Atlantic halibut, 2000 m$_{fbd}$) maximum reported length 4.7 m, to the postage-stamp-sized *Rhinosolea microlepidota* from Japanese waters that grows to 2.4 cm. Most species have the eyes on the right side (dextral) and those with eyes on the left side are known as sinistral. Pleuronectiforms are well adapted for demersal life on soft sediment of shelf seas with relatively strong tidal currents. Their depth distribution extends onto the slopes beyond the continental shelf but they are absent from the abyssal depths >3000 m. The widely cited report of a flatfish seen by the crew of the Bathyscaphe *Trieste* at the bottom of the Marianas Trench at over 10,000 m depth is dismissed as erroneous (Jamieson and Yancey, 2012). Many species of Pleuronectiformes are fished commercially by trawl, traps and longlines. There are 12 families, and here we follow the sequence given by Van Der Laan et al. (2014) omitting those with no deep-sea representatives.

4.3.24.1 Family Citharidae (196) (Large-Scale flounders)

This is a small family with six species in four genera. Three species in the sinistral genus *Citharoides* occur down to upper slope depths of 290 m in the tropical Indo-West Pacific, and the rare monotypic dextral *Lepidoblepharon ophthalmolepis* (Scale-eyed flounder, 428 m$_{fbd}$) is reported from depths greater than 300 m in the W. Pacific.

4.3.24.2 Family Scophthalmidae (197) (Turbots)

This sinistral family with nine species is confined to the North Atlantic, Baltic, Mediterranean and Black Seas. One genus, *Lepidorhombus*, comprises two species found from shelf depths to the deep sea, *L. boscii* (Four-spot megrim, 800 m$_{fbd}$) and *L. whiffiagonis* (Megrim, 700 m$_{fbd}$; Fig. 4.33a). The latter occurs widely in the NE Atlantic from Iceland to off West Sahara, and in the Eastern Mediterranean its depth range is 288–700 m (Mytilineou et al., 2005). It is exploited commercially.

4.3.24.3 Family Paralichthyidae (198) (Sand flounders)

The Paralichthyidae comprises over 100 sinistral species found in all three major oceans. Three species from the Western Atlantic *Ancylopsetta antillarum* (458 m$_{fbd}$) off the Bahamas and the more widespread *Citharichthys cornutus* (Horned whiff, 400 m$_{fbd}$) and *C. dinoceros* (Spined whiff, 2000 m$_{fbd}$) are considered bathydemersal. A few other species such as *Pseudorhombus micrognathus* (392 m$_{fbd}$) from the Indo-West Pacific also occur at upper slope depths.

4.3.24.4 Family Bothidae (199) (Lefteye flounders)

This family comprises over 160 species from the temperate and tropical Atlantic, Indian and Pacific Oceans. A few of the 35 *Arnoglossus* species occur at upper slope depths, such as *A. rueppelii* (Rüppell's scaldback, 897 m) in the Mediterranean Sea and *A. boops* (732 m$_{fbd}$) in waters off New Zealand. The genus *Chascanopsetta* comprises eight specialised deep-water species recorded at 150–650 m depth

(a)

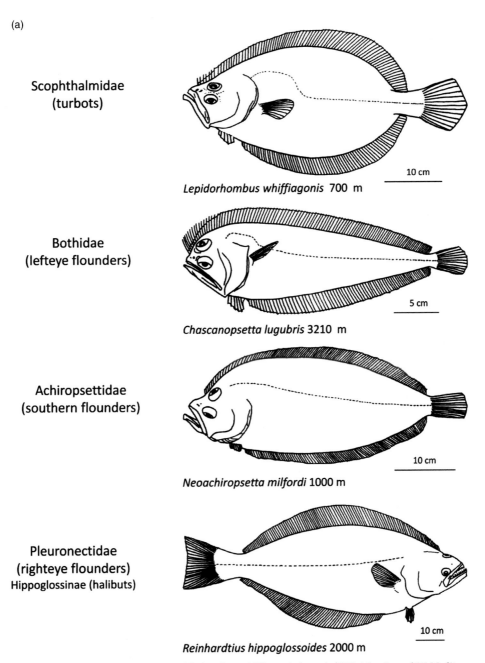

Scophthalmidae
(turbots)

Lepidorhombus whiffiagonis 700 m

10 cm

Bothidae
(lefteye flounders)

Chascanopsetta lugubris 3210 m

5 cm

Achiropsettidae
(southern flounders)

Neoachiropsetta milfordi 1000 m

10 cm

Pleuronectidae
(righteye flounders)
Hippoglossinae (halibuts)

Reinhardtius hippoglossoides 2000 m

10 cm

Figure 4.33a ORDER PLEURONECTIFORMES. *Lepidorhombus whiffiagonis* (megrim) NE Atlantic and W Mediterranean 69°N–26°N. *Chascanopsetta lugubris* (Pelican flounder) Atlantic, Indo-W Pacific. *Neoachiropsetta milfordi* (finless flounder) sub-Antarctic circumglobal New Zealand, Chile, Argentina, South Africa, Crozet Islands. *Reinhardtius hippoglossoides* (Greenland halibut) N Pacific, Arctic and N Atlantic 79°N–33°N.

(Amaoka and Parin, 1990) with a flexible strongly compressed body and extremely large mouth (Amaoka and Yamamoto, 1984; Fig. 4.33a).

C. lugubris (Pelican flounder, 3210 m$_{fbd}$) occurs in the tropical and subtropical Atlantic, Indian and West Pacific Oceans, two species are endemic to the Indian

Ocean, one to the Indo-West Pacific and four to the Pacific Ocean. Two monotypic species from the W Pacific are *Japonolaeops dentatus* (500 m_{fbd}) and *Kamoharaia megastoma* (Wide-mouthed flounder, 800 m_{fbd}). The latter has a more extreme wide-opening mouth than any of the *Chascanopsetta* species. About half of the species of the genus *Laeops* in the Indo-West Pacific region occur at depths greater than 250 m, the deepest being *L. clarus* (Clear fin-base flounder, 1025 m_{fbd}) from the Philippines. The genus *Monolene* comprises deep-water species with a mean minimum depth of occurrence of 173 m, including *M. maculipinna* (Pacific deepwater flounder, 384 m_{fbd}) from the E Pacific and *M. sessilicauda* (Deepwater flounder, 550 m_{fbd}) from the W Atlantic, both of which are fished commercially. *Parabothus* includes three deep-water species from the tropical W Pacific and Indo-Australia archipelago with a mean depth range 250–440 m, *P. coarctatus* (580 m_{fbd}), *P. filipes* (330 m_{fbd}) and *P. kiensis* (400 m_{fbd}). The genus *Tosarhombus* comprises six species from the Indo-West Pacific distributed from the outer shelf down to 500 m depth.

4.3.24.5 Family Achiropsettidae (200) (Southern flounders)
This family comprises four monotypic genera of sinistral flounders endemic to Antarctic and sub-Antarctic waters characterised by rudimentary or absent pectoral fins. Three are found deep, *Achiropsetta tricholepis* (Prickly flounder, 1186 m_{fbd}), *Mancopsetta maculata* (Antarctic armless flounder, 1115 m_{fbd}) and *Neoachiropsetta milfordi* (Finless flounder, 1000 m_{fbd}; Fig. 4.33a).

4.3.24.6 Family Pleuronectidae (201) (Righteye flounders)
The family Pleuronectidae comprises approximately 100 species of, dextral flatfishes found in the Arctic, Atlantic, Indian and Pacific Oceans from inshore down to 2000 m depth.

Subfamily Lyopsetinae (Slender soles)
This family comprises one monotypic genus occurring in the NE Pacific; *Lyopsetta exilis* (Slender sole,

800 m_{fbd}) is a small flatfish <35 cm that is of minor commercial interest in deep-water fisheries.

Subfamily Hippoglossinae (Halibuts)
This family includes the genera *Atheresthes*, *Clidoderma*, *Hippoglossoides*, *Reinhardtius* and *Verasper* (Nelson, 2006), which are all from the Northern Hemisphere. They are relatively large fish with a wide mouth gape and are amongst the deepest-living Pleuronectiforms with their range extending down the slopes to a mean depth of 1400 m. Most are commercially exploited in fisheries. *Atheresthes* comprises two species endemic to the North Pacific, *A. evermanni* (Kamchatka flounder, 1200 m_{fbd}) and *A. stomias* (Arrow-tooth flounder, 950 m_{fbd}). The monotypic *Clidoderma asperrimum* (Roughscale sole, 1900 m_{fbd}) is also from the North Pacific and is usually caught at 400–600 m depth. *Hippoglossoides* comprises six species, four from the North Pacific and two from the North Atlantic including *H. hippoglossus* (Atlantic halibut, 2000 m_{fbd}) and *H. stenolepis* (Pacific halibut, 1200 m_{fbd}). The monotypic *Reinhardtius hippoglossoides* (Greenland halibut, 2000 m_{fbd}) is distributed across high latitudes (79°N–33°N) in the Pacific, Arctic and Atlantic Oceans (Fig. 4.33a). It is most commonly caught at 500–1000 m depth, often pelagically, and forms the basis for a major deepwater fishery with global production averaging over 100,000 tonnes per annum since the 1970s (FAO, 2015). The genus *Verasper* comprises two relatively shallow-water species from the NW Pacific.

Within the tribe Microstomini (small-mouthed flounders) there are some deep-water representatives such as *Embassichthys bathybius* (Deep-sea sole, 1800 m_{fbd}), several species of *Glyptocephalus* (1600 m_{fbd}) and *Microstomus pacificus* (Dover sole, 1370 m_{fbd}) all in the Northern Hemisphere but because they are often caught at shelf depths they are not necessarily considered deep-sea species.

Subfamily Poecilopsettinae (Bigeye flounders)
This subfamily (Nelson, 1994) comprises the genera *Marleyella*, *Nematops* and *Poecilopsetta*

(Kawai et al., 2010). They are predominantly small (average length 14 cm) flatfish of the upper slope (average minimum depth 205 m) and maximum depth 550 m in tropical latitudes. *Marleyella* are two relatively shallow-water species from the W. Indian Ocean, and *Nematops* comprises four deeper-living species from the Indo-Pacific. The genus *Poecilopsetta* (Fig. 4.33b) comprises two species from the Western Atlantic including *P. beanii* (Deep-water dab, 1636 m_{fbd}), six species endemic to the Indian Ocean, eight species endemic to the Pacific Ocean and two Indo-West Pacific species.

4.3.24.7 Family Samaridae (202) (Crested flounders)

This family comprises 27 species of small dextral flatfish from the tropical and subtropical Indo-West Pacific. Average maximum length is 10 cm. Although noted by Nelson (2006) as primarily deep-water species, some occur mainly in shallow water, and the overall average depth range is 150–210 m. They are known as crested flounders because of the very elongated anterior dorsal fin rays in the genus *Samaris*. There are three genera: *Plagiopsetta* comprising three species from the tropical W Pacific at depths 65–350 m; the shallow-water genus *Samaris*; and *Samariscus* (Fig. 4.33b) comprising 19 species, including the deeper-living *S. longimanus* (Longfinned flounder, 333 m_{fbd}) and *S. multiradiatus* (430 m_{fbd}).

4.3.24.8 Family Soleidae (203) (Soles)

This family comprises about 150 dextral marine species in 27 genera of which three contain deep-water representatives, *Aseraggodes*, *Austroglossus* and *Bathysolea*. Two species of *Aseraggodes* are recorded from depths greater than 200 m off Indonesia. *Austroglossus microlepis* (West coast sole, 400 m_{fbd}) occurs on slopes of SW Africa from Namibia to the Cape and is taken in fisheries. The genus *Bathysolea* comprises four species (Desoutter and Chapleau, 1997): *B. profundicola* (Deep-water sole, 1350 m_{fbd}) in the E Atlantic from Ireland to Angola including the W Mediterranean usually at depths of 200–600 m (Fig. 4.33b), *B. polli* (420 m_{fbd}) and

B. lactea (620 m_{fbd}) off West Africa and *B. lagarderae* (310 m_{fbd}) from the W Indian Ocean. *B. profundicola* is taken in fisheries as a bycatch in the Bay of Biscay (Blanchard, 2001) and adjacent areas.

4.3.24.9 Family Cynoglossidae (204) (Tonguefishes)

This family comprises 130 species of sinistral flatfishes found mainly in tropical and subtropical seas. There are three genera, two of which are found in the deep sea; a few species of *Cynoglossus* and *Symphurus* with a high proportion of deep-water species including some that are endemic to hydrothermal vents of the West Pacific. The genus *Cynoglossus* occurs across the Indo-W Pacific; *C. acutirostris* (Sharpnose tonguesole, 1425 m_{fbd}) occurs in the Red Sea and the Gulf of Aden, living on sandy bottoms feeding on benthic invertebrates, *C. carpenteri* (Hooked tonguesole, 420 m_{fbd}) is more widespread around the Indian subcontinent and *C. suyeni* (421 m_{fbd}) is found in the east around the Philippines and the West Pacific. The genus *Symphurus* is characterised by the continuous marginal fin around the body of the fish comprising the dorsal, caudal and anal fins. There are 75 species, a third of which occur at upper slope depths in tropical and subtropical seas with a circumglobal distribution. Six deep-sea species are endemic to the W Atlantic, *S. billykrietei* (Kriete's tonguefish, 650 m_{fbd}) in the North, *S. marginatus* (Margined tonguefish, 750 m_{fbd}), *S. nebulosus* (Freckled tonguefish, 810 m_{fbd}), *S. piger* (Deep-water tonguefish, 549 m_{fbd}) and *S. stigmosus* (Blotchfin tonguefish, 373 m_{fbd}) at lower latitudes and *S. ginsburgi* (Ginsburg's tonguefish, 300 m_{fbd}) in the SW off Brazil and Uruguay. Two species are endemic to the E Atlantic: *S. vanmelleae* (Vanmelle's tonguefish, 945 m_{fbd}; 15°N–12°S) and *S. variegatus* (800 m_{fbd}) off South Africa. Three species are endemic to the Indian Ocean: *S. trifasciatus* (Threeband tonguesole, 737 m_{fbd}) from the Gulf of Oman to the Bay of Bengal, *S. maldivensis* (293 m_{fbd}) in the West and *S. woodmasoni* (1046 m_{fbd}) in the East. There are five deep-sea species endemic to the W Pacific: *S. hondoensis* (815 m_{fbd}) off S Japan, *S. schultzi* (494 m_{fbd}) in the S China Sea, *S. bathyspilus*

(b)

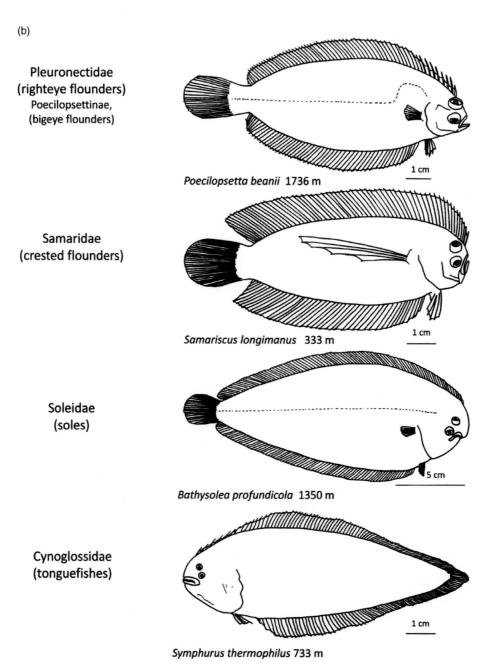

Pleuronectidae
(righteye flounders)
Poecilopsettinae,
(bigeye flounders)

Poecilopsetta beanii 1736 m

1 cm

Samaridae
(crested flounders)

Samariscus longimanus 333 m

1 cm

Soleidae
(soles)

Bathysolea profundicola 1350 m

5 cm

Cynoglossidae
(tonguefishes)

Symphurus thermophilus 733 m

1 cm

Figure 4.33b ORDER PLEURONECTIFORMES. *Poecilopsetta beanii* (deep-water dab) W Atlantic and Gulf of Mexico 45°N–8°S. *Samariscus longimanus* (longfinned flounder) Indo W Pacific. *Bathysolea profundicola* (deep-water sole) E Atlantic 55°N–17°S. *Symphurus thermophilus* (Western Pacific tonguefish) endemic to volcanic hydrothermal vent sites of the W Pacific margin 27°N–36°S.

(500 m_{fbd}), *S. regani* (1473 m_{fbd}) and *S. undatus* (357 m_{fbd}), which is also found off Hawaii. Two rare deep-sea species have been recorded from the tropical East Pacific, *S. diabolicus* (Devil's tonguefish, 501 m_{fbd}) and *S. microlepis* (Smallfin tonguefish, 531 m_{fbd}). Four species are distributed across the Indo-West Pacific,

S. strictus (Blackbelly tonguesole, 730 m_{fbd}) from Mozambique to Japan and Hawaii, *S. septemstriatus* (Sevenband tonguesole, 479 m_{fbd}) from Sri Lanka to the Philippines, *S. gilesii* (1080 m_{fbd}) and *S. australis*.

There are three species of *Symphurus* (Fig. 4.33b) endemic to hydrothermal vent fields associated with the back arc underwater volcanoes of the W Pacific (Tunnicliffe et al., 2010). *S. thermophilus* (Western Pacific tonguefish, 733 m_{fbd}) occurs in the Northern Hemisphere at vent sites (23°–15°N) on the northern Mariana Arc, Bonin Arc and the Okininawa Trough between 215 and 733 m depth (Tunnicliffe et al., 2013). *S. maculopinnis* (561 m_{fbd}) has been described from the Tonga arc in the Southern Hemisphere (25 °S) and a third undescribed species is reported to range from 25°S to 36°S on the Tonga and Kermadec ridges (Munroe et al., 2011). *S. thermophilus* is seen concentrated in large numbers on consolidated sulphur crusts and swimming near pools of molten sulphur. They feed on carrion falls of pelagic fishes that have succumbed to the effects of vulcanism in the overlying water column and also on a wide variety of invertebrates. The distinctive nature of these West Pacific volcanic sites is that they are on summits of seamounts and therefore are not particularly deep; *Symphurus* spp. have been observed as shallow as 83 m and are absent from adjacent deeper vent sites at 900–1000 m depth (Tunnicliffe et al., 2010). Munroe (1998) draws attention to the small size of many *Symphurus* species that has enabled them to exploit a wide range of habitats, including the shelf edge and upper slopes. The vent- associated species, which are less than 10 cm long with females gravid at 4.4 cm, fits this concept of a diminutive versatile group that has successfully colonised a peculiar new habitat.

4.3.25 Order Tetraodontiformes[1]

This order comprises about 400 species in 10 families inhabiting a wide range of habitats (Fig. 4.34) and with a vast range in body size; some species mature at 2 cm length, whereas the sunfishes (*Mola*) attain over 3 m in length (Matsuura, 2015). A common feature is a small mouth with few teeth or beak-like tooth plates. Over half of the species are shallow-reef associated, and most of the deep-water representatives belong to one family, the Triacanthodidae.

4.3.25.1 Family Triacanthodidae (205) (Spike fishes)

The Triacanthodidae are small fishes, mean maximum length 12 cm, found over an average depth range of 240–540 m on the upper slopes in tropical regions of all the major oceans. In contrast to their shallower-living relatives (Family Triacanthidae), which are all similar to one another, the Triancanthodidae are remarkably diverse in body form, ranging from

---

and Ophidiiformes, whereas the Tetraodontiformes were listed as the most advanced of the Acanthopterygii. The Tetraodontiformes encompasses a diversity of spiny-rayed fishes including spikefishes, triggerfishes, filefishes, boxfishes, puffers and ocean sunfishes. The Lophiiformes includes goosefishes, frogfishes, sea toads, batfishes and the deep-sea anglerfishes. Molecular genetic studies have shown these groups to be closely related to one another (Miya et al., 2010), and Near et al. (2013) place them together in clade XIV at the top of the evolutionary tree of spiny-rayed fishes. The Caproidei (boarfishes) and other advanced Perciformes also appear in this clade. This rather unexpected molecular evidence has prompted a re-evaluation of morphological evidence. Chanet et al. (2013) point out important similarities in the soft anatomy of the Tetraodontiformes and Lophiiformes. They are all rather large-headed shortened fishes conforming to varying degrees to the description of the Ocean sunfish as 'an enormous swimming head with a short trunk'. They have other common features, such as small dorsal gill opening, rounded kidneys (superficially similar to mammalian kidneys) anteriorly placed in the body cavity, compact thyroid gland, abbreviated spinal cord, asymmetric liver and the presence of supramedullary giant neurons in the anterior spinal cord. Pending a more comprehensive revision of the taxonomy of higher teleosts, here we place the Tetraodontiformes and Lophiiformes adjacent to one another.

---

[1] The Tetraodontiformes and Lophiiformes had until recently been considered as distinct, distantly related orders of teleosts. Nelson (2006) placed the Lophiiformes in the Paracanthopterygii, together with the Gadiformes

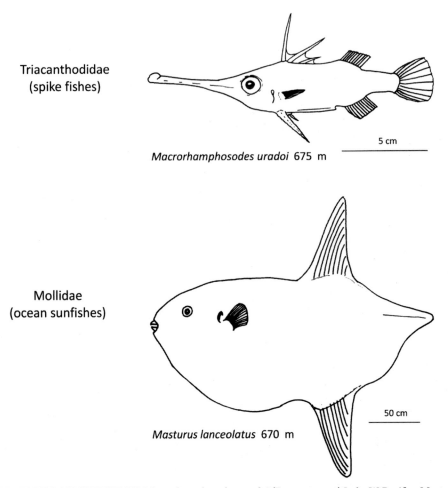

Triacanthodidae
(spike fishes)

5 cm

Macrorhamphosodes uradoi 675 m

Mollidae
(ocean sunfishes)

50 cm

Masturus lanceolatus 670 m

Figure 4.34 ORDER TETRAODONTIFORMES *Macrorhamphosodes uradoi* (Trumpetsnout) Indo-W Pacific. *Masturus lanceolatus* (sharptail mola) circumglobal 37°N–35°S.

relatively normal-looking species through intermediates to weirdly specialized forms with extremely long tubular snouts (Tyler, 1968). There are 23 species in 11 genera. The monotypic *Atrophacanthus japonicus* (2000 m_{fbd}) is found across the Indo-West Pacific from East Africa to Japan. The genus *Bathyphylax* comprises three species: *B. bombifrons* (615 m_{fbd}) from the Indo-West Pacific, *B. omen* (291 m_{fbd}) from the Indian Ocean and *B. pruvosti* (500 m_{fbd}) from the Central Pacific. The genus *Halimochirurgus* (610 m_{fbd}) comprises two long-snouted species found across the Indo-West Pacific. *Hollardia goslinei* (Hawaiian spikefish, 366 m_{fbd}) is found in the Pacific Ocean, and two other

species in this genus occur in the W Atlantic, *H. hollardi* (Reticulate spikefish, 915 m_{fbd}) and *H. meadi* (Spotted spikefish, 450 m_{fbd}). The monotypic, *Johnsonina eriomma* (549 m_{fbd}) is also endemic to the W Atlantic. There are two species of trumpetsnout spike fish in the genus *Macrorhamphosodes* that occur across the Indo-West Pacific (Fig. 4.34). *M. uradoi* (675 m_{fbd}) feeds almost exclusively on fish scales, which it removes from living fishes by stalking from behind and snatching a scale from the tail region with its asymmetric jaws. The direction and degree of jaw asymmetry varies between individuals, those with jaws that curve to the left are presumed to attack from the right and vice versa. It is suggested that this variation makes it difficult for prey species to learn or evolve appropriate escape responses. In the NW Pacific *M. uradoi* around the shelf edge have been found to feed predominantly on *Glossanodon semifasciatus* (Argentinidae), *Chlorophthalmus* (Chlorophthalmidae) and *Ventrifossa garmani* (Macrouridae), whereas offshore on the Kyushu–Palau Ridge stomachs mainly contained scales of *Emmelichthys struhsakeri* (Emmelichthyidae) at 400 m depth and *Scombrops boops* (Scombropidae) at 600 m (Nakae and Sasaki, 2002). The monotypic *Mephisto fraserbrunneri* (291 m_{fbd}) is endemic to the Bay of Bengal, and two species of *Parahollardia* (396 m_{fbd}) are from the W. Atlantic. The genus *Paratriacanthodes* comprises three Indo-West Pacific species, the deepest and most widespread being *P. retrospinis* (Sawspine spikefish, 920 m_{fbd}) found from Mozambique to Japan and New Caledonia. *Triacanthodes* comprises four Indo-West Pacific species found down to depths of 500 m. The monotypic *Tydemania navigatoris* (Fleshy-lipped spikefish, 500 m_{fbd}) found in the Indo-West Pacific from E Africa to Japan is reported to feed entirely on fish scales in a similar manner to *Macrorhamphosodes* (Mok, 1978). In general the Triacanthodidae are benthic fishes that, apart from the scale feeders, feed on benthic invertebrates. They are of no commercial importance.

4.3.25.2 Family Tetraodontidae (206) (Puffers)

The pufferfishes in this family usually occur in shallow warm tropical seas, but some species have been recorded at depths over 300 m (Matsuura, 2015), including the circumglobal *Lagocephalus lagocephalus* (Oceanic puffer, 476 m_{fbd}) and Indo-West Pacific *Tylerius spinosissimus* (Spiny blaasop, 435 m_{fbd}). The latter may be a truly deep-water species because it has never been recorded at depths less than 250 m.

4.3.25.3 Family Molidae (207) (Ocean sunfishes)

This family comprises four species of large pelagic oceanic fishes, three of which grow to over 3 m length

and have been recorded at depth: *Mola mola* (Ocean sunfish, 480 m_{fbd}), *M. ramsayi* (Southern sunfish, 300 m_{fbd}) and *Masturus lanceolatus* (Sharptail mola, 670 m_{fbd}; Fig. 4.34). Like many large oceanic fishes, *M. mola* shows diel vertical migrations, generally deeper during the day. The two *Mola* species cannot be regarded as deep-sea fishes because they are also often sighted at the surface during the day (Pope et al., 2010). However, Harbison and Janssen (1987) report that a *M. lanceolatus* collided with the *Johnson Sea Link II* submersible at over 670 m depth off the Bahamas, and they concluded it was feeding on the large gelatinous zooplankton, medusa, cnetophores and siphonophores that were abundant at this depth. In view of the rarity of sightings and captures at the surface, they suggest that *M. lanceolatus* should be considered as a permanent member of the mesopelagic community. Its distribution is circumglobal in tropical to subtropical waters (37°N–35°S).

4.3.26 Order Lophiiformes

The Lophiiformes comprises about 350 marine species in 18 families of which 67 per cent of species and 13 families occur predominantly in the deep sea, including bathydemersal slope-dwelling species and a major diversification into the bathypelagic environment (Fig. 4.35). Here we have accepted molecular genetic evidence (Miya et al., 2010) in placing the Lophiiformes at the top of the evolutionary tree of fishes. However, there is a major incongruence between the most recent morphological studies (Pietsch and Orr, 2007) and molecular data regarding the sequence of families within the Lophiiformes. For the purposes of this text we generally retain the order of families as listed in most widely accepted taxonomic authorities, Nelson (2006) and Van Der Laan et al. (2014). In contrast to the Tetraodontiformes, the mouth in Lophiiformes is large, and the first ray of the dorsal fin, if present, is not a rigid spine but typically forms the classic whip-like lure, comprising the illicium and terminal esca, used by angler-fishes to attract their prey. There are seven demersal families of which three have deep-water representatives: Lophiidae, Ogcocephalidae and Chaunacidae. The superfamily Ceratioidea comprises 11 families of almost entirely bathypelagic species.

4.3.26.1 Family Lophiidae (208) (Goosefishes)

The Lophiidae is considered to be the primitive sister group of the rest of the Lophiiformes comprising 28 species in four genera. They are demersal fishes that occur over an average depth range of 240–748 m and are often caught in trawls at upper slope depth. They feed using a sit-and-wait strategy, pouncing with the aid of the pectoral and pelvic fins as well as tail movements to grab often large prey in the enormous mouth (Laurensen et al., 2004). The genus *Lophiodes* is circumglobal at tropical and subtropical latitudes with one species in the E Atlantic, three species in the W Atlantic, four in the E Pacific, five in the W Pacific, one species endemic to the Indian Ocean and three Indo-West Pacific species. They are relatively small fishes, average maximum length 28 cm, and of no interest to fisheries. The deepest-living species is *L. infrabrunneus* (Shortspine goosefish, 1560 m_{fbd}) is found throughout the Indo-West Pacific region on the slopes at depths greater than 500 m. The monotypic *Lophiomus setigerus* (Blackmouth angler, 800 m_{fbd}) occurs throughout the Indo-West Pacific and is often found at shelf depths, where it is commercially exploited. The genus *Lophius* comprises six species endemic to the Atlantic Ocean and *Lophius litulon* (Yellow goosefish, 560 m_{fbd}) from the NW Pacific. These are relatively large fishes, average maximum length 1.1 m, that are all taken in fisheries. The largest individuals live deepest and often form a valuable component of deep-water trawl catches although smaller-size classes are found at shelf depths. *L. budegassa* (Blackbellied angler, 1013 m_{fbd}) and *L. piscatorius* (Angler, 1000 m_{fbd}) from the NE Atlantic and Mediterranean are amongst the most commercially important and deepest-living species. The genus *Sladenia* comprises three rare deep-sea tropical species confined to depths greater than 750 m: *S. shaefersi* (1200 m_{fbd}) from the W Atlantic and *S. remiger* (Celebes monkfish, 1540 m_{fbd}) and *S. zhui* (979 m_{fbd}) from the

W Pacific (Fig. 4.35a). They are of no commercial significance. The genus *Lophius*, and probably the other Lophiidae, spawn large gelatinous egg masses that float near the sea surface and the larvae are planktonic.

4.3.26.2 Family Chaunacidae (209) (Sea toads, Coffinfishes)

The Chaunacidae are demersal angler fishes that occur in all oceans except the polar seas, predominantly in the Indo-West Pacific, at depths ranging from 50 to 2000 m mainly at upper slope depths. They are generally less than 30 cm long, often brightly coloured red or pink, with a very loose flaccid skin and small spine-like scales (Caruso, 1989; Ho and Last, 2013). In contrast to the flattened head of the Lophiidae, the head is rounded with a conspicuous pattern of sensory canals. The illicium or angling apparatus is very short (Fig. 4.35a). There are 23 species in two genera. The genus *Chaunacops*, also previously known as *Bathychaunax* (Caruso et al., 2006), comprises the four rare deep-living species: *C. coloratus* (depth range 1250–3297 m) from the E Indian Ocean and E Pacific, *C. roseus* (depth range 1023–2200 m_{fbd}) from the W Atlantic, *C. melanostomus* (Tadpole coffinfish, depth range 1320–1760 m_{fbd}) from the E Indian Ocean and *C. spinosus* (eastern tadpole coffinfish, 1056 m_{fbd}) from the SW Pacific. *C. coloratus* has been observed by ROV at 2313–2585 m depth on the slopes of seamounts in the E Pacific off California walking on rocky substrates using its paired fins and swimming up to over 4 m off the seafloor when startled by the submersible (Lundsted et al., 2012). Colour varies from blue in individuals <14 cm long to pink or deep red in larger individuals. *C. melanostomus* have been observed by an oil industry ROV at 1714 m depth off W Australia on a sandy substrate and appeared uniformly black in colour (Ho and McGrouther, 2015). The genus *Chaunax* comprises 19 species, including three endemic to the Atlantic Ocean: *C. pictus* (Pink frogmouth, 978 m_{fbd}), which is also found in the Mediterranean, *C. stigmaeus* (Redeye gaper, 730 m_{fbd}) and *C. suttkusi* (1060 m_{fbd}), both from the W Atlantic. Three species are endemic to the Indian Ocean, there

(a)

Lophiidae
(goosefishes)

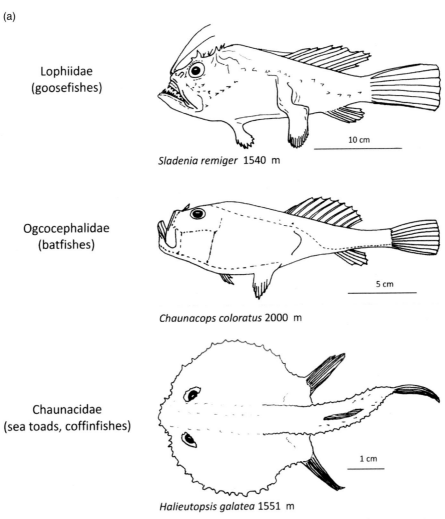

Sladenia remiger 1540 m

Ogcocephalidae
(batfishes)

Chaunacops coloratus 2000 m

Chaunacidae
(sea toads, coffinfishes)

Halieutopsis galatea 1551 m

Figure 4.35a ORDER LOPHIIFORMES. *Sladenia remiger* (Celebes monkfish) W Pacific tropical. *Chaunacops coloratus*, Indo-Pacific. *Halieutopsis galatea*, W Indian Ocean.

are two widespread Indo-West Pacific species, *C. penicillatus* (Pencil coffinfish, 658 m$_{fbd}$) and *C. russatus* (1200 m$_{fbd}$), and 10 species are endemic to the W Pacific, including recent discoveries off New Zealand (Ho et al., 2013). Only one species is known from the E Pacific, *C. latipunctatus* (800 m$_{fbd}$) from the Galapagos Islands region. The average depth range is 325–790 m. Molecular genetic studies place the Chaunacidae in the same clade as the Ceratioidei, most closely related to the major bathypelagic radiation of Lophiiform fishes (Miya 2010; Lundsted et al., 2012).

4.3.26.3 Family Ogcocephalidae (210) (Batfishes)

The batfishes are small, average maximum length 14 cm, demersal fishes found worldwide in all tropical and subtropical seas except the Mediterranean. They have a large head, which is greatly flattened into a distinctive disc or triangular shape with large protruding pectoral fins and the tail behind. The paired fins are highly mobile, enabling the fish to walk on the seafloor feeding on small invertebrates and fishes. The skin is rough with tubercles and spines giving a hairy appearance in some species. The illicium is very short and retracts into a cavity at the front of the head above the mouth (Ho et al., 2013). They range from inshore to 4000 m depth; the average depth range is 480 to 800 m, and a few species are bathypelagic. There are 77 species in 10 genera. *Coelophrys* has a less-flattened rounded body than is typical for the family, possibly reflecting a benthopelagic habit. This genus is confined to the Indo-West Pacific with *C. micropa* (1400 m_{fbd}) found from South Africa to the Philippines and six other species endemic to the West Pacific from Japan to Indonesia. The genus *Dibranchus* was originally so named because it has two functional gills on each side, only the second and third gill arches carry filaments, a feature that has since been found in six other genera (Bradbury, 1999). There are 14 species of *Dibranchus* distributed mainly around the American continents, three species in the W Atlantic, eight species in the E Pacific, two species in the Indian Ocean and one in the NW Pacific generally at slope depths ranging down to over 2000 m depth. The genus *Halicmetus* comprises three upper slope species: *H. ruber* (549 m_{fbd}) occurs from the Arabian Sea to Japan, and two species are endemic to the West Pacific. *Halieutopsis* is the deepest-living genus ranging between 391 and 4000 m depth. There are nine species, three endemic to the Indian Ocean, five endemic to the W. Pacific Ocean and one Indo-W Pacific species. They are small rare fish, average maximum length 6.6 cm (Fig. 4.35a). Some of the deeper-living species have poorly calcified bones (Bradbury, 1988). *H. margaretae* (1185 m_{fbd}) from the North Pacific is considered to be benthopelagic. The genus *Malthopsis* comprises 16 species distributed over an average depth range of 215–550 m. They are characterised by a markedly triangular shape and multispined body scales known as 'bucklers' (Ho, 2013). Maximum length is less than 15 cm. Their geographic distribution is centred on the Indo-West Pacific: *M. gigas* (Giant triangular batfish, 540 m_{fbd}) occurs from Madagascar to Japan and New Zealand, *M. mitrigera* (650 m_{fbd}) is also widespread throughout the area, four species are endemic to the Indian Ocean, five species to the

W Pacific and *M. jordani* (520 m_{fbd}) to Hawaii. An interesting outlier is *M. gnoma* (475 m_{fbd}) found in the Caribbean area of the W Atlantic. In the genus *Solocisquama*, *S. stellulata* (550 m_{fbd}) is most widespread, ranging from South Africa to Hawaii, and two other species are endemic to the East Pacific. The genera *Halieutichthys*, *Ogcocephalus* and *Zalieutes* are confined to shelf depths. The genus *Halieutaea* is also generally quite shallow, average maximum depth of 290 m, but includes *H. retifera* (386 m_{fbd}) reported as bathypelagic around Hawaii.

Superfamily Ceratioidea (Deep-sea anglerfishes)

The Ceratioids are extraordinarily specialised in their adaptation to the deep-sea pelagic environment (Pietsch, 2009). The walking ability, which is widespread in the demersal lophiform families, is lost, pelvic fins are absent (lost at metamorphosis), the body is usually scaleless, the skeleton reduced and there is highly marked sexual dimorphism with females up to more than 10 times longer than the males. Only the females have an illicium, which is usually equipped with a light organ to attract prey to the large well-developed jaws. All families have bioluminescent organs except Caulophrynidae and Neoceratiidae. The females typically adopt a 'float-and-wait' foraging strategy, suspended neutrally buoyant in the meso- to bathypelagic at an average depth range of 400–1650 m. The males of most species have well-developed eyes and huge nostrils and swim actively in pursuit of food and mates. Average female length is 16 cm, and the much smaller dwarf males attach themselves to the body of the female to varying degrees. In some species the males continue to move independently and only attach temporarily to the female during spawning. In other cases the males can be facultative or obligate external parasites (Miya et al., 2010); in the latter attachment becomes permanent with fusion of epidermal and dermal tissues so that the male is sustained by nutrition from the female circulatory system. Spawning occurs at depth with the female producing a Lophioid gelatinous mass of eggs, which is fertilised by the attached males. The eggs float towards the surface, where they develop in the upper 200 m of the water column. At metamorphosis the juveniles descend rapidly to 2000–2500 m depth, and the females gradually ascend to their adult depth range taking several years to reach maturity. Drifting on the deep-sea circulation with limited swimming capability Ceratoidae occur in all the oceans where there is sufficient depth except for the Mediterranean Sea (Pietsch, 2009). They never occur in large numbers; in the survey of deep-sea pelagic fishes of the Sargasso Sea ceratoids comprised less than 0.5 per cent of the catch (Sutton et al., 2010). There are about 170 species in 12 families. A major difficulty is matching the males to the correct species of female, and in many cases the males remain unknown. Owing to their distinctive morphology and biology they have received much attention from different authors who were necessarily constrained by the small number of specimens available for examination. Most information to date has been collated in the large volume by Pietsch (2009).

4.3.26.4 Family Caulophrynidae (211) (Fanfins)

The adult females of this family are distinguished from other Ceratoids by having extremely long dorsal and anal fin rays, greater than 60 per cent of the standard length (Fig. 4.35b). The illicium varies from short (16 per cent of standard length) to very long (270 per cent), and the esca is simple. The males attach to the belly of the female and are anchored by means of tissue outgrowths from the upper and lower jaws. Apart from the loss of the large olfactory organs there is no degeneration of the body of the male, the eyes remain intact, and it is interpreted as facultative parasite. This family comprises five species in two genera.

Caulophryne has two circumglobal species *C. jordani* (Fanfin angler, 1510 m_{fbd}) and *C. pelagica* (2500 m_{fbd}), *C. polynema* (1250 m_{fbd}) from the E Atlantic and *C. bacescui* from the E Pacific. The monotypic *Robia legula* (1500 m_{fbd}), noted for its extremely long illicium, is recorded from the W Central Pacific.

4.3.26.5 Family Neoceratiidae (212) (Toothed seadevil)

There is only one species in this family, *Neoceratias spinifer* (Spiny seadevil, 1200 m_{fbd}), which occurs worldwide between 40°N and 15°S. There is no illicium in the adult females, a feature unique to this family. The body is elongated, and there are two or three irregular rows of long hinged teeth on the outer edges of the jaws (Fig. 4.35b). The males are obligate parasites, but the heart and gills show no sign of degeneration so it is assumed that respiration remains functional and the males depend on the female only for nutrition.

4.3.26.6 Family Melanocetidae (213) (Black Seadevils)

The sole genus, *Melanocetus* is distributed worldwide between the Arctic and Antarctic polar fronts. Two species are circumglobal, *M. johnsonii* (Humpback anglerfish, 4500 m_{fbd}) usually found at depths of 500 to 1500 m and the deeper-living *M. murrayi* (Murray's abyssal anglerfish, 6000 m_{fbd}?), caught predominantly at 1000–2500 m but probably occurring throughout much of the abyss. A cited capture at 6370 m is based on the maximum depth reached by the net (Pietsch and Van Duzer, 1980) so does not provide direct evidence that the fish was caught at hadal depths. *M. eustalus* (1675 m_{fbd}) and *M. niger* are from the E Pacific, and *M. rossi* (420 m_{fbd}) is known only from the Ross Sea. *M. murrayi* is thought to be the most phylogenetically derived member of the family, with reduced skeletal ossification reflecting the poor food supply in its abyssal habitat. Adult female *Melanocetus* are black or dark-coloured globular fishes with a short illicium (23–60 per cent of body length) with a simple escal bulb (Fig. 4.35b). Males continue to feed and grow after metamorphosis and retain large well-developed eyes and nostrils. They are presumed to attach to females temporarily for spawning.

4.3.26.7 Family Himantolophidae (214) (Footballfishes)

This family comprises 21 species in one genus *Himantolophus* (Plate 8b) with the adult females

(b)

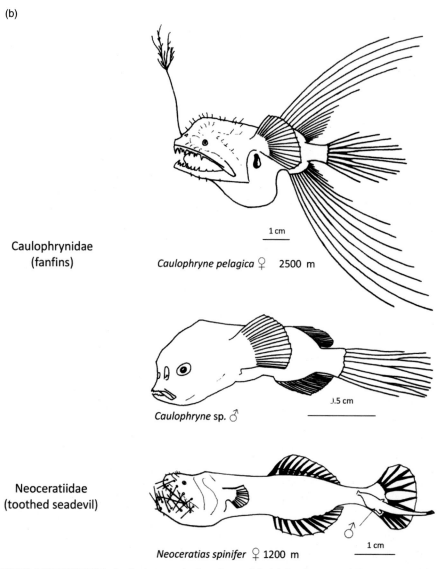

Caulophrynidae
(fanfins)

Caulophryne pelagica ♀   2500 m

1 cm

Caulophryne sp. ♂

Neoceratiidae
(toothed seadevil)

Neoceratias spinifer ♀ 1200 m

Figure 4.35b ORDER LOPHIIFORMES. *Caulophryne pelagica*, circumglobal (after Regan and Trewavas, 1932). *Neoceratias spinifer* (Spiny seadevil) W Pacific (after Bertelsen, 1951). (Permissions, Zoological Museum Copenhagen).

characterised by globose shape, short blunt snout and chin, which are padded with thick skin. The illicium is stout and thick and the esca has well-developed light-guiding appendages. Five species groups are recognised within the genus based on adult females (Pietsch, 2009). In the *groenlandicus* group, most samples of *H. groenlandicus* (Atlantic footballfish, 830 m minimum depth) are from the Atlantic Ocean, but there is evidence of occurrence in the Pacific and Indian Oceans (Fig. 4.35b). Two other species in the group are from the Atlantic and two from the Pacific. The *cornifer* group comprises six species: *H. cornifer* (1900 m_{fbd}) occurs in all three major oceans, three species are described from the E Atlantic, one from the SW Pacific and one from the E Pacific. The *albinares* group comprises four species from the Atlantic Ocean including *H. albinares* (845 m minimum) and one species from the NW Pacific. The *nigricornis* group has just two tropical to subtropical species: *H. nigricornis* from the Pacific and *H. melanolophus* (550 m$_{fbd}$) from the W Atlantic Ocean. The *appelii* group comprises two species that are probably circumglobal in the Southern Hemisphere (Pietsch and Kenaley, 2011). The adult males of *Himatolophus* are the largest known in the ceratioids reaching a maximum length of 39 mm with large olfactory organs but relatively small eyes. They are not parasitic, only attaching to females temporarily during spawning. However, evidence of feeding in male adults remains elusive, and it is possible that their large size provides sufficient reserves to survive a long period of fasting before mating.

4.3.26.8 Family Diceratiidae (215) (Double anglers)

The globose adult females of this family are superficially similar to the Himantolophidae but distinguished from those and all the other Ceratioids by the presence of a second light-bearing dorsal fin spine directly behind the base of the illicium (Pietsch et al., 2004). This second spine is most conspicuous in young specimens and becomes enclosed in a cavity in mature females. There are two genera each with three species, which are distributed circumglobally in tropical and subtropical latitudes but absent from the East Pacific. They are associated with high productivity areas adjacent to continents. Analysis of stomach contents of larger individuals has revealed remains of fish, coelenterates and crustacea as well as benthic prey items such as sea urchins, polychaetes and gastropods suggesting a more demersal lifestyle than most Ceratioids. *Bufoceratias shaoi* (1200 m$_{fbd}$) is found across the Indo-West Pacific, *B. thele* (1500 m$_{fbd}$) is confined to the W Pacific and *B. wedli* (1750 m$_{fbd}$) occurs in both E and W Atlantic. The genus *Diceratias* has a similar distribution with *D. bispinosus* (Two-rod anglerfish, 1400 m$_{fbd}$) in the Indo-Pacific, *D. trilobus* (1211 m$_{fbd}$) in the NW Pacific and *D. pileatus* (1430 m$_{fbd}$) in the Atlantic (Fig. 4.35c). There is no evidence of sexual parasitism; the males appear to be free-living with relatively large eyes.

4.3.26.9 Family Oneirodidae (216) (Dreamers)

This is the most speciose of the Ceratioid families with 65 species in 16 genera, which are so diverse that it is impossible to provide a description of a typical specimen. The only consistent diagnostic feature of the family is that, on the top of the skull on each side, a process on the anterior edge of the pterotic bone overlaps the sphenotic bone that lies in front (Pietsch and Orr, 2007). In all other respects the morphology of adult females is highly variable, and this is reflected in the evocative common names coined by scientists for the genera, such as smoothhead dreamers, spike head dreamers, spiny dreamers and fattail dreamers. Males are free-living and not parasitic except for the finding of attached males in two genera, *Bertella* and *Leptocanthichthys*, presumed to represent facultative sexual parasitism.

The genera are here described in the sequence given by Pietsch (2009), and morphology refers to adult females unless otherwise stated. *Lophodolos* (pugnose dreamers) has an elongated body, short snout and large mouth (Fig. 4.35c). There are two species, both distributed circumglobally in the tropics but *L. acanthognathus* (Whalehead dreamer, 1050 m$_{fbd}$) has a broader latitudinal range, 65°N to 40°S, in the Atlantic Ocean. *Pentherichthys* (Thickjaw dreamer) has an elongated body, thick lower jaw, short illicium and esca with a trailing posterior appendage. There is one species, *P. venustus* (2500 m$_{fbd}$), found in the Atlantic, Indian and Pacific Oceans. *Chaenophryne* has a globose body shape reminiscent of *Himantolophus* with thick skin so that specimens are recovered relatively intact. Three species are circumglobal, *C. draco* (Smooth dreamer, 1750 m$_{fbd}$), *C. longiceps* (Can-opener smoothdreamer, 1000 m$_{fbd}$) and *C. ramifera* (1000 m$_{fbd}$), and two species are described from the Pacific Ocean only. *Spiniphryne* (spiny dreamers) is distinguished from other Oneirodids by spiny skin and an elongated laterally flattened body shape. There are two rare species: *S. duhameli* (2500 m$_{fbd}$) from the Central Pacific and *S. gladisfenae* (Prickly dreamer, 1955 m$_{fbd}$) from the tropical Atlantic Ocean. The monotypic *Dermatias platynogaster* (fattail dreamer, 1342 m$_{fbd}$) has a rounded body, very deep caudal peduncle and short

(c)

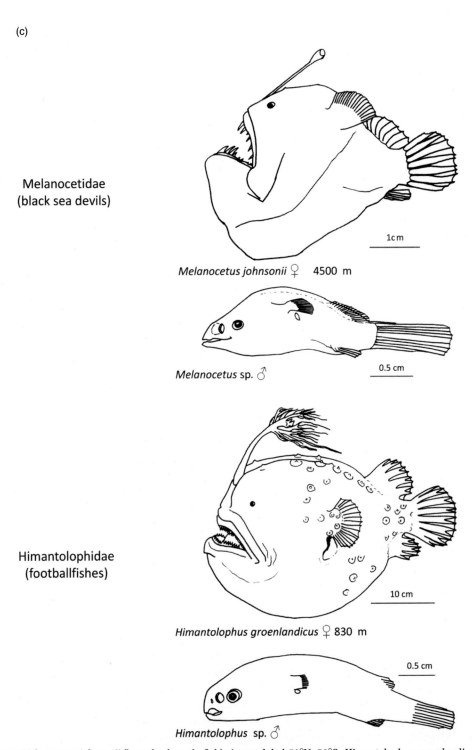

Melanocetidae
(black sea devils)

Melanocetus johnsonii ♀   4500 m

1 cm

Melanocetus sp. ♂

0.5 cm

Himantolophidae
(footballfishes)

Himantolophus groenlandicus ♀ 830 m

10 cm

Himantolophus sp. ♂

0.5 cm

Figure 4.35c *Melanocetus johnsonii* (humpback anglerfish) circumglobal 50°N–50°S. *Himantolophus groenlandicus* (Atlantic footballfish) circumglobal tropics and subtropics (after Susan Laurie Bourque © Canadian Museum of Nature).

blunt snout (Fig. 4.35c). It is known only from a few specimens taken in the W. Pacific. *Oneirodes* (Common dreamers) comprises 36 species, making it the most species-rich ceratioid genus. Most of the species have a short deep body that is almost spherical, and the illicium is usually short (Fig. 4.35d). The genus *Oneirodes* has a worldwide distribution between 66 °N and 60°S, three species are circumglobal: *O. eschrichtii* (Bulbous dreamer, 1455 m_{fbd}) from 65°N to 35°S, *O. kreffti* (Krefft's dreamer, 2000 m_{fbd}) in the Southern Hemisphere 18°S–45°S and *O. notius* (2000 m_{fbd}) in circum-Antarctic waters 35°S–60°S. The other species in the genus are all regional species endemic to particular water masses or biogeographic provinces. 13 *Oneirodes* species are endemic to different parts of Pacific Ocean including *O. bulbosus* (Bulb-fish, 2000 m_{fbd}) and *O. thompsoni* (2014 m_{fbd}) from the North Pacific. Seven species are endemic to the Atlantic Ocean, one to the Indian Ocean, seven species have an Indo-Pacific distribution and four species are Atlanto-Pacific. The monotypic *Tyrannophryne pugnax* (tyrannical toad, 2100 m_{fbd}) has an extremely large oblique mouth with the jaws extending to behind the pectoral fins (Fig. 4.35d) and is known from few tiny (<50 mm long) specimens widely distributed across the tropical and South Pacific Ocean. The genus *Phyllorhinichthys* (Leafysnout dreamers) comprises two species with variable flaps on the snout known from very few specimens captured in all three major oceans. The monotypic *Microlophichthys microlophus* (Short-rod anglerfish, 2200 m_{fbd}) with an elongated body, large mouth extending behind the eye and a short illicium is found in tropical and subtropical parts of all oceans. Also monotypic, *Danaphryne nigrifilis* (Dana dreamer, 2000 m_{fbd}), with an unusually large esca and small mouth, occurs in tropical to temperate latitudes of all oceans. The genus *Dolopichthys* (longsnout dreamers) comprises seven species with an elongated body and snout. Relatively deep-living, three species are circumglobal in tropical and temperate latitudes, *D. jubatus* (1353 m_{fbd}), *D. longicornis* (2200 m_{fbd}) and *D. pullatus* (2000 m_{fbd}). All other known species are from the Atlantic Ocean including *D. allector* (2000 m_{fbd}) also recorded from the SE Pacific. The following genera are all monotypic:

Bertella idiomorpha (spikehead dreamer, 3475 m_{fbd}) is confined to the N Pacific, *Ctenochirichthys longimanus* (combfin dreamer) to the E Pacific, *Leptacanthichthys gracilispinis* (Lightline dreamer, 2000 m_{fbd}) occurs in the N Atlantic and Pacific, *Puck pinnata* (Mischievous dreamer, 4073 m_{fbd}) in the Pacific and *Chirophryne xenolophus* (longhand dreamer) in the W. Pacific.

4.3.26.10 Family Thaumatichthyidae (217) (Wolftrap anglers)

This family is very distinctive and quite unlike any other ceratioids. The body is elongated, and the upper jaws are very long, extending forward far beyond the lower jaws. The left and right upper jaw bones do not come together but are parallel to one another, connected by an elastic membrane. The esca bears teeth. There is no evidence of sexual parasitism. There are eight species in two genera. *Lasiognathus* comprises five pelagic species known from a few widely scattered specimens collected in the Atlantic and Pacific Oceans at latitudes of 45°N–35°S down to an average maximum depth of 1800 m. The head and body are narrow and laterally compressed. The enlarged upper jaw is equipped with numerous long teeth (Fig. 4.35d). The function of two or three teeth on the esca at the end of the whip-like illicium remains enigmatic; Pietsch (2009) suggests they may serve to warn predators rather than to snag or injure prey. The genus *Thaumatichthys* comprises three species of very unusual benthic ceratioids. The illicium is very short, and the esca with one tooth is reversed so that it hangs inside the front of the mouth from the membrane connecting the two widely separated upper jaws. Unlike *Lasiognathus*, *Thaumatichthys* is rather dorso-ventrally flattened. Most specimens have been captured by bottom trawl at slope depths of 1300–2200 m at tropical latitudes. *Thaumatichthys axeli* (3570 m_{fbd}) is known from the E Pacific (Fig. 4.35d), *T. binghami* (3200 m_{fbd}) from the W Atlantic and *T. pagidostomus* (1440 m_{fbd}) from the Atlantic and Indian Oceans. One specimen of *T. binghami*, 29 cm long was captured with abyssal benthic holothurians (*Benthodytes typica*) in its

(d)

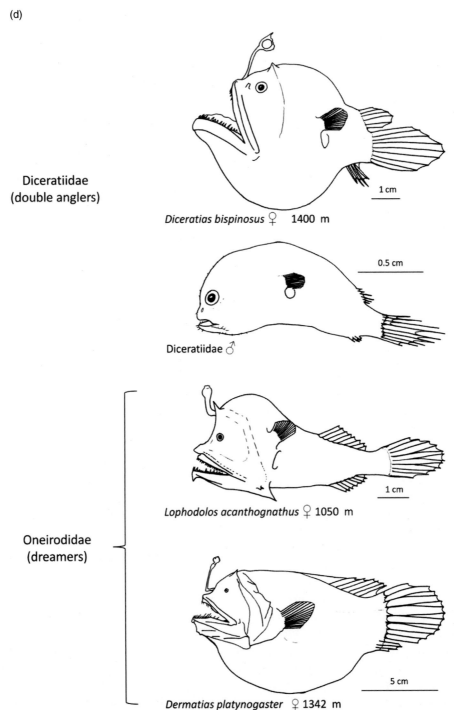

Diceratiidae
(double anglers)

Diceratias bispinosus ♀   1400 m

Diceratiidae ♂

Oneirodidae
(dreamers)

Lophodolos acanthognathus ♀ 1050 m

Dermatias platynogaster ♀ 1342 m

Figure 4.35d ORDER LOPHIIFORMES. *Diceratias bispinosus* (two-rod anglerfish) Indo-Pacific. *Lophodolos acanthognathus* (whalehead dreamer) circumglobal tropical to temperate seas (after Susan Laurie Bourque © Canadian Museum of Nature). *Dermatias platynogaster*, W Pacific.

stomach. Whilst indicative of benthic feeding it is unlikely that holothurians would have been attracted to the luminescent esca. No other data exist on food consumed.

4.3.26.11 Family Centrophrynidae (218) (Deep-sea anglers, Prickly seadevils)

This family comprises one species *Centrophryne spinulosa* (Horned lantern fish, 2325 m$_{fbd}$), which occurs in all three major oceans, 40°N–43°S. The adult females are elongated, laterally compressed fishes with a mouth that opens almost horizontally, standard length up to 23 cm (Fig. 4.35d). The males are free-living with no evidence of sexual parasitism.

4.3.26.12 Family Ceratiidae (219) (Seadevils)

The adult females of this family are laterally compressed elongated fishes somewhat similar to *Centrophryne,* but they uniquely have two or three fleshy bioluminescent caruncles located on the midline of the trunk just in front of the first dorsal fin. The mouth opening is almost vertical. The metamorphosed free-living males have large eyes but small olfactory organs. The males are obligate sexual parasites, attaching to the female apparently as soon as possible; a 10 mm male has been recorded attached to a 16 mm female *Cryptopsaras* (Pietsch, 2009). The genus *Ceratias* has three circumglobal species, *C. holboelli* (Kroyer's deep-sea anglerfish, 4400 m$_{fbd}$) females can grow to 1.2 m long in tropical to temperate waters usually at 400–2000 m depth, *C. tentaculatus* (Southern seadevil, 2900 m$_{fbd}$) in the Southern Hemisphere (26°S–69°S) and *C. uranoscopus* (Stargazing seadevil, 2091 m$_{fbd}$) in tropical waters (Fig. 4.35e) The monotypic *Cryptopsaras couesii* (Triplewart seadevil, 3085 m) is also circumglobal in tropical to subtropical latitudes usually at 500–1250 m depth. The bioluminescent caruncles decrease in size during growth in *Cryptopsaras* so they are inconspicuous in the mature individuals, and escal bulb tissue almost encloses the short illicium. In contrast to the rarity of many ceratioids *Cryptopsaras couesii* is the most commonly caught deep-sea angler species with over 1000 adult females catalogued in museums around the world.

4.3.26.13 Family Gigantactinidae (220) (Whipnose anglers)

Of all the ceratioids, the adult females of the Gigantactinidae have the most streamlined archetypical fish shape with a small head, elongated laterally flattened body tapering to a slim caudal penducle. They reach a maximum size of 40 cm. The illicium on the tip of the snout is very long, between and one and five times the body length. The mouth is large, opening horizontally. The metamorphosed males have unusually small eyes but large olfactory lobes and are free living with no evidence of sexual parasitism. The males continue to grow after metamorphosis indicating they feed successfully. There are 23 species in two genera. The genus *Gigantactis* comprises 20 species: *G. vanhoeffeni* (5300 m$_{fbd}$) and *G. perlatus* (2000 m$_{fbd}$) are circumglobal in tropical and temperate regions of all the oceans and *G. meadi* is circumglobal between 30°S and 55°S. Four species are recorded as endemic to the Pacific Ocean, eight species to the Atlantic, one species has an Indo-Pacific distribution and four species are Atlanto-Pacific in their distribution (Fig. 4.35e). Moore (2002) describes remarkable video observations from the ROV *Jason* of *G. vanhoeffeni* or *G. perlatus*, at 5000 m depth in the North Pacific at the site of the H2O deep-sea observatory. The fish was drifting upside-down motionless in an almost horizontal posture above the seafloor with the illicium curving forward and downward so that the esca was a few centimetres above the bottom. When disturbed, it swam vigorously with body undulations but continuing in the inverted posture. *Gigantactis* has multiple long teeth on the lower jaw outside the mouth as an aid in grasping prey. In the benthic feeding *Thaumatichthys* similar teeth are on the upper jaw. It suggested that the arrangement in *Gigantactis* may facilitate upside-down foraging on the seafloor. The genus *Rhynchactis* comprises three rare species, distinguished from *Gigantactis* by reduction of the jaws, lack of teeth and absence of a bulbous esca on the illicium. They are distributed

(e)

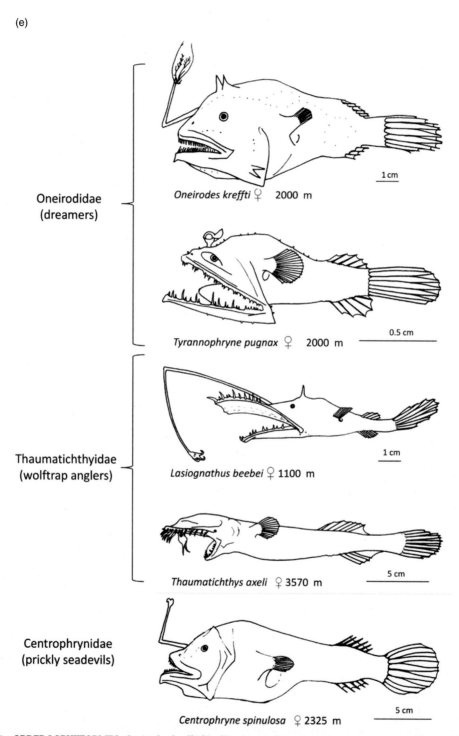

Oneirodidae
(dreamers)

Oneirodes kreffti ♀ 2000 m

1 cm

Tyrannophryne pugnax ♀ 2000 m

0.5 cm

Thaumatichthyidae
(wolftrap anglers)

Lasiognathus beebei ♀ 1100 m

1 cm

Thaumatichthys axeli ♀ 3570 m

5 cm

Centrophrynidae
(prickly seadevils)

Centrophryne spinulosa ♀ 2325 m

5 cm

Figure 4.35e ORDER LOPHIIFORMES. *Oneirodes kreffti* (Krefft's dreamer) Southern Hemisphere circumglobal 18°S–45°S. *Tyrannophryne pugnax* Pacific (after Regan and Trewavas, 1932, Permission, ZMUC). *Lasiognathus beebei*, N Atlantic and E Pacific (after Susan Laurie Bourque © Canadian Museum of Nature, Coad, 1995). *Thaumatichthys axeli*, E central Pacific (after Bertelsen and Struhsaker, 1977, Permission, ZMUC). *Centrophryne spinulosa* (horned lantern fish) circumglobal 40°N–43°S.

circumglobally between 32°N and 12°S (Bertelsen and Pietsch, 1998), *R. leptonema* (1500 m$_{fbd}$) and *R. macrothrix* (2000 m$_{fbd}$) are probably cosmopolitan and *R. microthrix* (2250 m$_{fbd}$) is known only from the Western Indian Ocean.

4.3.26.14 Family Linophrynidae (221) (Leftvent seadevils)

The common name of this family derives from one of the key distinguishing features common to all members: the displacement of the anal opening to the left of centreline of the body. Any functional significance of this peculiarity is obscure. In other respects the Linophrynidae is the family in which all the definitive features of the deep-sea anglers are expressed in the most extreme way. Some species have the largest mouth and longest teeth (relative to body size) of any vertebrate. Sexual dimorphism is extreme, some of the males may be the world's smallest vertebrates and all are obligate parasites. In addition to the bioluminescent esca, the genus *Linophryne* has elaborate luminescent chin barbels, often longer than the body length. There are 27 species in five genera in this family occurring in all the oceans between 65°N and 45°S. The genus *Acentrophryne* comprises two species from the tropical E Pacific with no chin barbel and a short illicium less than 10 per cent of the body length. The two species appear to differ in their ecology; *A. longidens* (910 m$_{fbd}$) are caught far above the seabed, whereas *A. dolichonema* appear in bottom trawls between 201 and 1105 m depth (Pietsch and Simazaki, 2005). Three genera are monotypic. *Borophryne apogon* (Netdevil, 1750 m$_{fbd}$) is from the E Pacific. Adult females usually have a single male attached midventrally anterior to the anus, but the largest known female specimen, 10.1 cm standard length, has two males attached 16 and 22 mm long (Pietsch 2009). *Haplophryne mollis* (Soft leafvent angler, 2250 m$_{fbd}$) is unique amongst the ceratioids for its lack of pigmentation. In life it may be largely transparent, but preserved specimens appear translucent white. *H. mollis* is found in all the major oceans from 54°N in the Atlantic to 45°S in the SW

Pacific around Australia and New Zealand. The males are obligate parasites becoming so fused to the female that a connection is established between the blood-vascular systems. Up to three males per female have been observed (Fig. 4.35f). *Photocorynus spiniceps* (1420 m$_{fbd}$) is lightly pigmented and has conspicuous spines on the top of the head and a short illicium. It is restricted to a narrow latitudinal range 35°N to 13°S around the world at depths greater than 1000 m. *P. spiniceps* is a small fish, adult female maximum size is 6.9 cm and the length range of the parasitic males 6.2–7.3 mm makes them arguably the world's smallest known sexually mature vertebrates compared with the 7.7 mm long tropical frog, *Paedophryne amanuensis* (Rittmeyer et al., 2012).

Linophryne, with 22 species is the most diverse genus in the family. They are characterised by the presence of a long bioluminescent barbel which often branches to form an elaborate structure as long as the body. The illicium is not excessively long, between 20 and 40 per cent of standard length. The caudal peduncle is short and the mouth large, conforming to the 'swimming head' appearance of the orders Tetraodontiformes and Lophiiformes. The length of the jaws can be up to 45 per cent of the standard length (SL) of the fish, and teeth of the lower jaw up to 24 per cent of SL. The largest known female is 27.5 cm long and the largest free-living male 21 mm. Attached parasitic males grow to 30 mm, indicating significant growth based on nutrition derived from the female. The genus *Linophryne* occurs worldwide between 40°N and 40°S, possibly associated with continental margins and islands slopes. *L. densiramus* (Thickbranch angler, 1400 m$_{fbd}$) is the most widespread species probably occurring in all three major oceans. *L. indica* (Headlight angler, 1900 m$_{fbd}$) is the best-known species, caught most often and occurring across the Indo-Pacific region. Other species are more localised or represented by very few samples; seven species including *L. algibarbata* (2200 m$_{fbd}$) are recorded only from the Atlantic Ocean (Fig. 4.35f), four from the Pacific and one from the Indian Ocean. Seven species occur in the Atlantic and Pacific, and *L. bicornis* is recorded from the Indian and Atlantic Oceans.

(f)

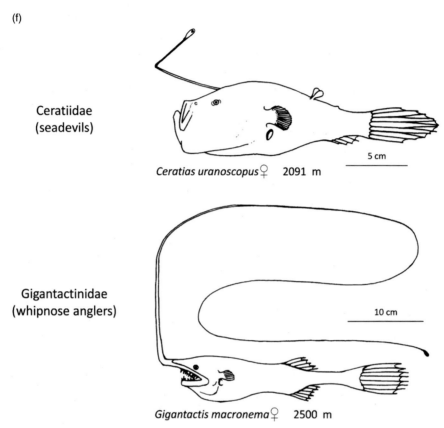

Ceratiidae
(seadevils)

Ceratias uranoscopus♀ 2091 m

Gigantactinidae
(whipnose anglers)

Gigantactis macronema♀ 2500 m

Figure 4.35f ORDER LOPHIIFORMES. *Ceratias uranoscopus* (stargazing seadevil) circumglobal tropical (after Regan and Trewavas, 1932, Permission, Zoological Museum Copenhagen). *Gigantactis macronema,* Atlantic and Pacific.

Order Mugiliformes

No deep-water representatives.

4.4 Class Sarcopterygii

Following Nelson (2006) this class includes lungfishes, coelacanths and all the higher vertebrates including amphibia, reptiles, birds and mammals. Here we consider only one order, the Coelacanthiformes.

4.4.1 Order Coelacanthiformes

There is good fossil record of the Coelacanthiformes from the Upper Devonian to the Cretaceous with nine families now represented by just one extant family, the Latimeridae (Fig. 4.36). Although much interest has surrounded the evolutionary proximity of the coelacanth to the first terrestrial vertebrates, it has been established that it is the freshwater lungfishes and not the coelacanths that are the closest living relatives of the tetrapods (Amemiya et al., 2013).

4.4.1.1 Family Latimeridae (222) (Gombessas, Coelacanths)

The finding of a living Coelacanth off the coast of South Africa in 1938, a representative of a group of fishes previously thought to have been extinct since the end of the Cretaceous 70 Ma, is a well-known story of zoological discovery (Smith, 1939). With their lobed fins and other morphological features they are

(g)

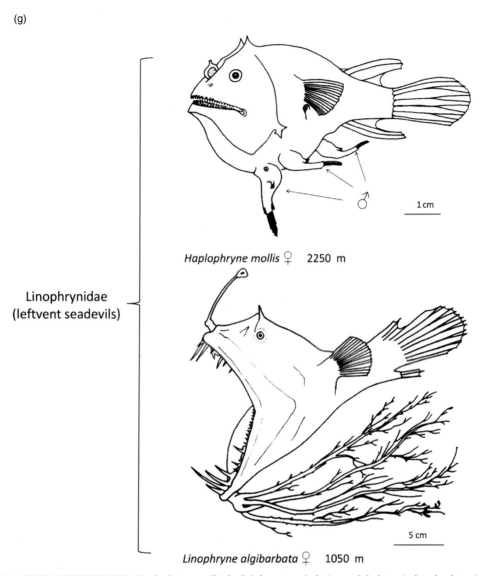

Linophrynidae
(leftvent seadevils)

Haplophryne mollis ♀   2250 m

Linophryne algibarbata ♀   1050 m

Figure 4.35g ORDER LOPHIIFORMES. *Haplophryne mollis* (soft leftvent angler) circumglobal tropical and subtropical (after Regan and Trewavas, 1932). *Linophryne algibarbata*, N Atlantic (after Susan Laurie Bourque © Canadian Museum of Nature, Coad, 1995). Images of free-living males in the previous panels after Bertelsen (1951, 1982). (Permissions, Zoological Museum Copenhagen).

unique amongst marine fishes and have been the subject of much research collated in special volumes on biology, evolution (Musick et al., 1991) and the genome (Amemiya et al., 2014). It is now thought that there are two living species in one genus: *Latimeria chalumnae* (African coelacanth, 700 m_{fbd}) from the Western Indian Ocean off the Comoro Islands, South Africa, Madagascar and Mozambique (Fig. 4.36) and

a more recently discovered second species, *L. menadoensis* (Sulawesi coelacanth, 200 m_{fbd}), found on the North coast of the Indonesian island of Sulawesi (Erdmann, 1999; Fricke et al., 2000). It is likely that coelacanths survived the end of Cretaceous K/Pg mass extinction by retreat into deep-water refuges (Guinot et al., 2013), and the depth range of *L. chalumnae* falls within the definition of a deep-sea

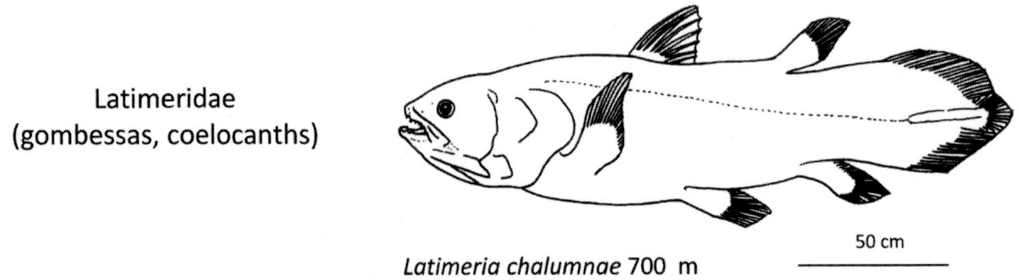

Latimeridae
(gombessas, coelocanths)

Latimeria chalumnae 700 m

50 cm

Figure 4.36 ORDER COELACANTHIFORMES. *Latimeria chalumnae* (Coelacanth) W Indian Ocean.

species. Tracking studies show that *L. chalumae* off the island of Grand Comore spend the daytime in volcanic caves at around 200 m depth and forage to greater depths down to 550 m during the night (Hissman and Fricke, 2000). They feed on upper slope demersal fishes such as deep-sea cardinalfishes (Epigonidae), *Diaphus metopoclampus* (Myctophidae, lanternfish), *Polymixia nobilis* (Polymixiidae, beardfishes), *Symphysanadon* (slope fishes), *Ilyophis brunneus* (Synaphobranchidae) and *Beryx decadactylus* (Alfonsino, Berycidae), often drifting in headstand position just above the seafloor possibly using a rostral electroreceptor organ on the head to detect prey movement (Uyeno and Tsutsumi, 1991). The diel migration pattern, shallow during the day, deep during the night is the reverse of the typical movement pattern of mesopelagic fauna but is possibly defined by the depths at which cave refuges are available. Downward movement at sunset enables interception of upward migrating prey species. Coelacanths are accidentally captured in the fishery for oilfish (*Ruvettus pretiosus*, Gempylidae) that are found at depths down to 800 m and in mixed species trawl fisheries. Scuba divers encountered several *L. chalumnae* at 104–108 m depth underneath overhangs in the Jesser Canyon, Sodwana Bay, South Africa; this was well beyond normal depth limit of 30 m for recreational diving, and indeed one of authors of the report died following the diving operations (Venter et al., 2000). Most observations of living coelacanths in their natural environment have been made at greater depths using manned research submersibles with the crew at normal atmospheric pressure (Fricke et al., 1991). Although the depths at which coelacanths live are not great, attempts to keep them alive in aquaria or in shallow water have proved futile. Coelacanths are large (up to 2 m long), slow-growing, nonmigratory, ovoviviparous fishes with low fecundity (typically 20 pups per female every three years). The population size in the Comoro Islands has been estimated as less than 300 individuals and declining owing to an increase in bycatch in fisheries (Hissman et al., 1998). Fricke (1997) reviewed the eco-ethical challenges posed by such a potentially vulnerable population. To some extent those concerns have been ameliorated by subsequent discoveries of *L. chalumnae* at more widespread locations and *L. menadoensis* at several locations; nevertheless, conservation of coelacanths in the face of increasing fishing pressure is a continuing urgent issue (Robinson and Amemiya, 2014).

5

The Deep-Sea Fish Faunas

The purpose of this chapter is to describe the fishes that occur in the various ecological settings of the deep sea. In some cases these might be referred to as the 'community' of fishes in a given habitat; however, it is rare that fishes alone constitute the community. They co-occur with crustacea, molluscs and a variety of other invertebrates.

5.1 Mesopelagic and Bathypelagic Fish Fauna

The waters beneath the photic zone at depths greater than 200 m represent over 60 per cent of the total world ocean volume and are the largest living space on the planet. The fauna is dominated by relatively small fishes, a 'Lilliputian' fauna from the families Gonostomatidae (Bristlemouths), Phosichthyidae (Light fishes), Sternoptychidae (hatchet-fishes), Stomiidae (barbeled dragonfishes) and Myctophidae (lantern fishes; Sutton et al., 2010), which are mostly less than 10 cm long. At nighttime large numbers of these ascend to the surface layers to feed, the largely piscivorous Stomiidae pursuing the zooplanktivorous species. In a study by Roe and Badcock, (1984), the number of fish at 100 m depth in the NE Atlantic (44°N) increased more than tenfold at night mainly due to high abundances of the myctophid *Benthosema glaciale,* whereas the abundance at 450 m depth correspondingly decreased. This major upward shift is detectable by echo-sounders in the form of scattering layers that move upward at dusk and down again at dawn (Sutton, 2013). However, not all fishes participate in this vertical migration. At 30°N in the NE Atlantic Badcock and Merrett (1976) found that Gonostomatidae represented over 60 per cent of the

mesopelagic catch of which *Cyclothone braueri* was 62 per cent, *C. microdon* 26 per cent and *C. pseuodopallida* 2 per cent. These three species occupy very broad depth ranges down to over 2000 m (Fig. 5.1), but their biomass is concentrated in relatively narrow species-specific depth bands; *C. braueri* (400–600 m), *C. microdon* (500–900 m) and *C. pseuodopallida* (500–800 m) with little change between day and night. It is the myctophids that mostly show the classic vertical migration with ascent during the night, though in several species part of the population does not migrate (*Hygophum hygomi*, *H. reinhardti*, *Benthosema suborbital*, *Diogenichthys atlanticus* and *Notolychnus valdiviae*; Fig. 5.1). There is also variation between life history stages. Badcock and Merrett (1976) conclude that although individual *Chauliodus danae* do not show diel migration, there is a tendency for more of the population to move up in the water column at night, presumably in pursuit of prey. In general it is the shallower-living component of the mesopelagic fish fauna that shows the clearest diel vertical migrations.

In the Sargasso Sea, Sutton et al. (2010) found that 58 per cent of the fish biomass in the water column was in the mesopelagic zone 0–1000 m, 38 per cent at 1000–2000 m, 0.9 per cent at 2000–3000 m, 8 per cent at 3000–4000 m and 0.2 per cent at 4000–5000 m. 27 species were found to occur both above and below the 1000 m depth horizon, indicating that there is considerable interchange between the mesopelagic and bathypelagic. 17 species were classified as spanners, species that have very broad depth ranges above and below the 1000 m stratum, including the *Cyclothone* species (as in Fig. 5.1) Stomiidae, Melamphaidae and the genera *Serrivomer, Gigantura* and *Melanonus*. 10 species were classified as

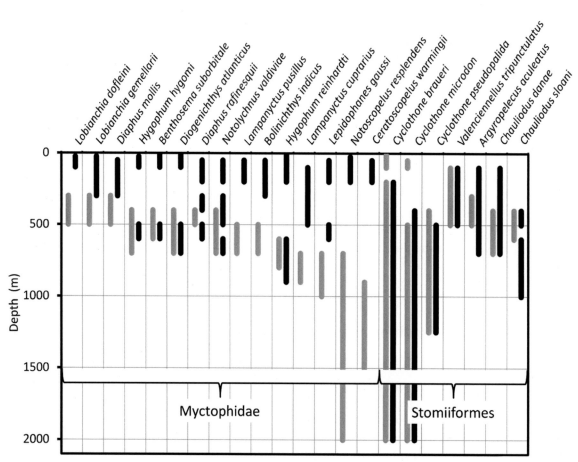

Figure 5.1 Pelagic depth distribution of Myctophidae, Gonostomidae, Sternoptychidae and Stomiidae in the NE Atlantic 30°N, 23°W from net tows during 31 March–8 April, 1972, showing the patterns of diel vertical migration. Grey bars – daytime distribution (1 h after sunrise to 1 h before sunset) Black bars – nighttime distribution (1 h after sunset to 1 h before sunrise). After Badcock and Merrett (1976).

vacillators that occur around 1000 m depth and do not migrate (*Sternoptyx, Scopeloberyx robustus*) or migratory species that descend to the bathypelagic at >1000 m depth during daylight (*Ceratoscopelus, Hygophum, Lampanyctus* and *Melamphaes*). Species found exclusively at depths greater than 1000 m are termed 'holobathypelagic' and include the adult ceratioid anglerfishes (11 families), whalefishes (Cetomimidae), gulper eels (Eurypharyngidae), swallower eels (Saccopharyngidae), tubeshoulders (Platytroctidae) and some species of Chiasmodontidae (swallowers). The bathypelagic fauna remains the least well-known of all fish faunas owing to the

chronic lack of sampling effort (Webb et al., 2010). The secondary peak in biomass at 3000–4000 m in the Sargasso Sea suggests that this zone may be more important than has hitherto been supposed.

Sea floor topography has an important effect on pelagic species. Over the Mid-Atlantic Ridge in the North Atlantic (Sutton et al., 2008) found a biomass maximum at 1500–2300 m at depths corresponding to the flanks of the ridge explained by the larger average body size of fishes in this layer comprising Gonostomatidae (28 per cent of the biomass); Microstomatidae, Melamphaidae and Serrivomeridae (15–16.5 per cent each); Stomiidae (6.3 per cent); and

Platytroctidae, Alepocephalidae and Myctophidae (3 per cent each). Close to the seafloor, in the benthic boundary layer, the biomass was 40 per cent higher than in the surface layers. In the benthic boundary layer, *Cyclothone* species were absent, and the main species comprising the biomass were *Eurypharynx pelecanoides* (Eurypharyngidae), *Bathylagus euryops* (Bathylagidae), *Sigmops bathyphilum* (Gonostomatidae), *Scopeloberyx robustus* (Melamphaidae), *Serrivomer beanii* (Serrivomeridae), *Borostomias antarcticus*, *Chiasmodon niger* and *Chauliodus sloani* (Stomiidae). Seamounts and mid-ocean ridges generally have enhanced abundance and biomass resulting from a number of mechanisms but with reduced mesopelagic biomass observed directly over some shallow-summit seamounts where enhanced predation may be occurring (Sutton, 2013).

The depth distribution of deep pelagic fishes is therefore defined by changes in adaptations from upper mesolepelagic types with silvery reflective bodies and large ventral light organs that ascend to the surface at night (e.g. myctophids) to lower mesopelagic types with reduced silveriness and no tendency to migrate (e.g. *Cyclothone*) to the least-active bathypelagic dark- bodied fishes with variable distribution of light organs that stalk their prey (e.g. Ceratioid anglerfishes and Eurypharyngidae). The horizontal geographic distribution of the mesopelagic species can generally be understood in terms of the biogeographic zones of the oceans as expounded by Longhurst (1998; see Chapter 1). Species of myctophids can be recognised as falling into the following categories: cold or warm water species, Northern or Southern Hemisphere species, polar or tropical species, oceanic, pseudo-oceanic or slope-associated species, oligotrophic gyre species, eutrophic upwelling species, low oxygen-tolerant species, and so on (e.g. Hulley and Lutjeharms, 1995; Krefft, 1976). Myctophids are typically dominant in the more productive regions of the world, whereas in the low latitude gyres of the North Atlantic (Fig. 5.1) and the North Pacific, *Cyclothone* sp. are dominant. This leads to the concept that mesopelagic fish faunas can be characterised according to biogeographic province. Some species are ubiquitous and span

across many regions, but the assemblage in each province is a combination of endemic species (if any) and species from adjacent regions, resulting in more or less unique assemblages in each province. In some parts of the world such as in the Southern Ocean, the North Atlantic, California Current and Eastern Australia, there are sufficient data to statistically discriminate different ecoregions on the basis of myctophid zoogeography (Flynn and Marshall, 2013). Using these and other data Sutton et al. (2017) have constructed a global biogeographic classification of the mesopelagic comprising 33 ecoregions (Fig. 5.2). These broadly correspond to Longhurst provinces related to patterns of productivity. The mesopelagic fish faunas from representative studies within each ecoregion are listed in Table 5.1. In the text that follows, numbers in square brackets refer to the ecoregion numbers in Table 5.1 and Figure 5.2.

5.1.1 Arctic Ocean

The Central Arctic [1] ocean basins are generally depauperate in deep pelagic fishes but the myctophid *Benthosema glaciale* has been reported from ice-free areas (Catul et al., 2013), and *Boregadus saida* (Arctic cod) can be deep-pelagic in its habits (Mueter et al., 2013).

5.1.2 Pacific Ocean

The Pacific Sub-Arctic [2] is dominated by an abundance of the myctophids, *Stenobrachius leucopsarus, S. nannochir* and high biomass of deep-sea smelts (Bathylagidae) in some areas (Sinclair and Stabeno, 2002). Beamish et al. (1999) estimated that *Stenobrachius leucopsarus* in this region could have a biomass of 21 million tonnes, providing vast feeding opportunities for a variety of fishes, birds and mammals. The California Current [3] region is a transition zone between sub-Arctic species to the north and Eastern tropical species to the south. In the Southern California current system the mesopelagic fish abundance is dominated by *Cyclothone* (64 per cent), but the biomass is dominated by Myctophidae (45 per cent) followed by *Cyclothone* spp. (35 per cent)

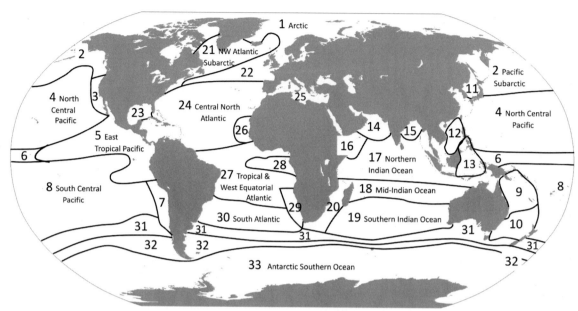

Figure 5.2 The mesopelagic ecoregions of the world's oceans. (After Sutton et al., 2017).
Key to small ecoregions not labelled: 3 – California Current, 6 – Equatorial Pacific, 7 – Peru Upwelling / Humboldt Current,
9 - Coral Sea, 10 – Tasman Sea, 11 – Sea of Japan, 12 – South China Sea, 13 – Indo-Pacific Pocket Basins, 14 – Arabian Sea,
15 – Bay of Bengal, 16 – Somali Current, 20 – Agulhas current, 22 – North Atlantic Drift, 23 – Gulf of Mexico,
25 – Mediterranean Sea, 26 – Mauritania/Cape Verde, 28 – Guinea Basin and East Equatorial Atlantic, 29 – Benguela
Upwelling, 31 – Circumglobal Subtropical Front, 32 – Sub-Antarctic. Depths < 200 m are not shown, boundaries are
approximate.

and Sternoptychthidae (18 per cent; Davison et al., 2015). The vertically migrating planktivores include the native species *Leuroglossus stilbius*, *Bathylagoides wesethi*, as well as the warm-water *Diogenichthys atlanticus* and *Triphoturus mexicanus*, and the cold-water *Diaphus theta* and *Stenobrachius leucopsarus*. The nonmigratory planktivores comprise *Melamphaes* spp., *Poromitra* spp., *Cyclothone* spp., and Sternoptychidae, the cold-water *Bathylagus pacificus*, *Microstoma* and *Protomyctophum crockeri* and warm-water *Scopelogadus bispinosus*. The piscivores include the cold-water *Chauliodus macouni* and warm-water *Idiacanthus antrostomus* and *Stomias atriventer* (Koslow et al., 2014). Samples in the Gulf of California are dominated by the Myctophids *Triphoturus mexicanus* and *Diaphus pacificus*, *Vinciguerria lucetia* (Phosichthyidae), and *Cyclothone* spp. (Robison, 1972). In the low productivity area of the North Central Pacific [4] gyre the mesopelagic fish assemblage is dominated by several species of

Cyclothone followed by hatchet-fishes, myctophids and Melamphaidae (Table 5.1). Throughout this and other warm-water areas of the Pacific there are higher trophic level carnivores, which though rare are ubiquitous and play an important role in the food chain, such as *Chauliodus sloani*, *Echiostoma barbatum*, *Leptostomias haplocaulus*, *Odontostomops normalops*, *Bathyleptus lisae*, *Sternonosudis macrura*, and *Avocettina infans* (Barnett, 1984). Around seamounts of the Emperor Seamount chain, the lightfish *Maurolicus imperatorius* is extremely dominant representing over 99 per cent of catches (Sassa et al., 2002). Tropical East Pacific [5] upwelling areas off Central and South America are dominated by the myctophid *Diogenichthys laternatus* (Evseenko and Shtaut, 2005). The Central Equatorial Pacific [6] area has relatively high productivity compared with the gyre provinces to the north and south, the important species being *Cyclothone acclinidens*, *C. signata* and *Sternoptyx obscura*, which are rare in

Table 5.1 List of Mesopelagic Ecoregions (after Sutton et al., 2017) as in Figure 5.1. Mesopelagic Fish Species Are Listed from Representative Studies in Each Ecoregion

* Denotes that the species are ranked in descending order of importance.

Ecoregion			Fish species		
	Name	Description	Myctophids	Other species	References
1	Arctic	Cold, low salinity water. Circulation restricted by land masses. Depauperate fauna	*Benthosema glaciale* on fringes	*Boreogadus saida*	Catul et al. (2013), Mueter et al. (2013)
Pacific Ocean					
2	Pacific Sub-Arctic	Productive waters, permanent halocline in offshore regions. Very few species in common with Pacific Central gyres. Low diversity.	*Stenobrachius leucopsarus, Stenobrachius nannochir, Diaphus theta, Protomyctophum thompsoni, Ceratoscopelus townsendi, Electrona risso, Lampanyctus jordani, Lampanyctus regalis, Lampanyctus ritteri, Notoscopelus japonicus, Symbolophorus californiensis, Tarletonbeania crenularis, Tarletonbeania taylori*	*Leuroglossus schmidti, Bathylagus pacificus, Pseudobathylagus milleri*	Sinclair & Stabeno (2002)
3	California Current	Productive system driven by seasonal coastal upwelling. Distinctive fauna, some endemism.	*Ceratoscopelus townsendi, Diogenichthys atlanticus, Hygophum reinhardtii, Myctophum nitidulum, Notolychnus valdiviae, Notoscopelus resplendens, Triphoturus mexicanus, Diaphus theta, Nannobrachium spp., Stenobrachius leucopsarus,*	*Leuroglossus stilbius, Bathylagoides wesethi, Bathylagus pacificus, Lipolagus ochotensis, Microstomamicrostoma, Cylothone, Chauliodus macouni, Idiacanthus antrostomus, Stomias atriventer*	Koslow et al. (2014), Robison (1972)

Table 5.1 (*cont.*)

	Ecoregion Name	Description	Fish species		References
			Myctophids	Other species	
			Symbolophorus californiensis, Tarletonbeania crenularis, Protomyctophum crockeri		
4	North Central Pacific	Northern Pacific gyre system. Oligotrophic; fauna distinct from the Equatorial Pacific.	*Ceratoscopelus warming, Triphotururus nigrescens, Lampanyctus sp., Notolychnus valdiviae, Benthosema suborbital, Bolinichthys longipes, Lampanyctus steinbecki, Diaphus mollis, Lobianchia gemellari	*Cyclothone alba, Cyclothone pallida, Cyclothone pseudopallida, Sternoptyx diaphana, Sternoptyx pseudobscura, Cyclothone atraria, Vinciguerria nimbaria, (Sternoptyx sp.), Scopeloberyx, opisthopterus, Scopeloberyx robustus, Melamphaes simus, Vinciguerria poweriae	Barnett (1984)
5	East Tropical Pacific	Extensive stratification, shallow thermocline and very low oxygen, Eutrophic; endemic species adapted to low oxygen	*Diogenichthys laternatus, Lampanyctus parvicauda, Gonichthys tenuiculus,	*Vinciguerria lucetia, Bathylagus nigrigenys, Scopelarchoides nicholsi, Bathophilus filifer	Evseenko and Shtaut (2005)
6	Equatorial Pacific	Complex of both eastward and westward currents; zones of convergence and divergence, semi-permanent upwelling. Mesopelagic fauna differs markedly from gyres to north and south	*Lampanyctus sp., Notolychnus valdiviae, Lampanyctus hubbsi, Diaphus splendidus, Bolinichthys photothorax, Diaphus similis, Diaphus sp., Ceratoseopelus warming, Lampanyctus steinbecki, Hygophum proximum, Triphoturus nigrescens,	Cyclothone acclinidens, Cyclothone signata, Viniguerria nimbaria, Sternoptyx obscura, Gonostoma elongatum, Scopeloberyx opisthopterus, Scopelogadus mizolepis	Barnett (1984)

	Region	Description	Species		Reference
			Symbolophorus evermanni, Idiolychnus urolampus		
7	Peru Upwelling /Humboldt Current	Northward flow of surface waters of sub-Antarctic origin; strong upwelling. Highly productive zone. Sharp faunal transition at western boundary.	*Diogenichthys atlanticus, Triphoturus mexicanus*	*Cyclothone acclinidens, *Vinciguerria lucetia, Scopelogadus mizolepis, Sternoptyx diaphana, Bathylagus nigrigenys, Melamphaes acanthomus*	Sielfeld et al. (1995)
8	South Central Pacific	Gyral analog to that of North Pacific. Oligotrophic; fauna distinct from N. Pacific gyre with lower biomass	*Notolychnus valdiviae, *Ceratoscopelus warmingi, Lampanyctus steinbecki, Diogenichthys atlanticus, Lampanyctus sp., Scopelopsis multipunctatus, Lampadena urophaos, Bolinichthys photothorax.*	*Cyclothone alba, Cyclothone *pseudopallida, Cyclothone pallida, Cyclothone microdon, Sternoptyx diaphana, Scopelosaurus smithi, Argyropelecus hemigymnus, Argyropeleceus lychnus, Vinciguerria nimbaria, Vinciguerria poweriae, Valenciennellus tripunctulatus, Sternoptyx pseudobscura , Sternoptyx sp.*	Barnett (1984)
9	Coral Sea	Core associated with North Queensland Current and the Coral Sea gyre. Distinct mesopelagic fish fauna, notably lantern fishes	*Diaphus aliciae, Diaphus jenseni, Diaphus malayanus, Diaphus regani, Diaphus signatus, Diogenichthys panurgus, Lampanyctus vadulus, Myctophum aurolaternatum, Myctophum lychnobium, Nannobrachium nigrum.*		Flynn and Marshall, (2013)
10	Tasman Sea	Semi-enclosed circulation bounded by the Tasman Front, Fish fauna includes temperate	*Bolinichthys nikolayi, Centrobranchus nigroocellatus,*		Flynn and Marshall, (2013)

Table 5.1 (cont.)

#	Ecoregion Name	Description	Fish species		References
			Myctophids	Other species	
		species, with greater diversity than Coral Sea ecoregion.	*Diaphus anderseni, Diaphus bertelseni, Diaphus brachycephalus, Diaphus fragilis, Diaphus lucidus, Diaphus mollis, Diaphus parri, Diaphus perspicillatus, Diogenichthys atlanticus, Hygophum reinhardtii, Lobianchia gemellarii, Myctophum asperum, Myctophum nitidulum, Myctophum selenops, Triphoturus nigrescens* (Subtropical). *Diaphus effulgens, Diaphus kapalae, Diaphus meadi, Hygophum hygomii, Lampadena notialis, Lampanyctus pusillus, Myctophum phengodes, Notoscopelus caudispinosus, Notoscopelus resplendens, Scopelopsis multipunctatus* (Temperate).		
11	Sea of Japan	Atidal; very cold deep waters (0.5°C). Depauperate mesopelagic fauna.			
12	South China Sea	Deep, warm marginal sea, exchanges with Kuroshio.	*Benthosema fibulatum, Benthosema pterotum, Bolinichthys longipes,*	*Gonostoma elongatum, Argyripnus ephippiatus,*	Bourret (1985), Wang and Chen (2001)

#	Ecoregion	Description	Species	Species	Reference
		Very speciose fish fauna, with species not found in the east.	*Diaphus coeruleus, Diaphus diademophilus, Diaphus garmani, Diaphus lucidus, Diaphus regani, Diaphus rivatoni, Diaphus whitleyi. Diaphus garmani, Diaphus sagamiensis, Diaphus suborbitalis, Diaphus watasei, Hygophum proximum, Lampadena luminosa, Lampanyctus nobilis, Myctophum brachygnathum, Mytophum nitidulum, Myctophum obtusirostrae, Neoscopelus microchir*	*Polyipnus spinifer, Polymetme elongate, Idiacanthus fasciola*	
13	Indo-Pacific Pocket Basins	Very deep (> 4000 m) basins of the Indo-Malayan Archipelago, Sulu, Celebes, and Banda Seas. Complex oceanography due to topography. Unique Indo-Pacific fauna. Celebes Sea and Sulu Sea may constitute distinct ecoregions	*Benthasema fibulatum, Bolinichthys pyrsobolus, Bolinichthys sp., Ceratoscopelus warming, Diaphus fragilis, Diaphus garmani, Diaphus lutkeni, Diaphus parri, Diaphus rafnesquei, Diaphus sagamiensis, Diaphus suborbitalis, Lampadena sp., Lampadena luminosa, Lampanyctus alatus, Lampanyctus ater, Lampanyctus idostigma, Lampanyctus nobilis, Lampanyctus simulator, Myctophum asperum, Myctophum aurolatematum,*	*Diremoides parini, Melamphaes sp., Scopeloberyx sp., Scopelogadus sp., Neoscopelus, Lestidiops similus?, Lestiolepis intermedia, Notolepis sp., Sudis sp., Howella sp., Cubiceps pauciradiatus, Bathylagus sp., Gonostomatidae, Polyipnus indicus, Polyipnus lateratus, Chauliodus sloani*	Dolar et al. (2003), data from Sulu sea dolphin stomach contents

Table 5.1 (cont.)

Ecoregion Name	Description	Fish species Myctophids	Other species	References
		Myctophum brachygnathum, Myctophum nitidulum, Myctophum phengoides, Notoscopelus resplendens, Symbolophorus evermanni, Symbolophorus boops		
Indian Ocean				
14 Arabian Sea	Highly seasonal: upwelling during the SW Monsoon, convective overturn during the NE Monsoon. OMZ region. Very productive; biodiversity intermediate, but abundance, biomass and dominance of few species is extremely high.	*Benthosema pterotum, Benthosema fibulatum, Bolinichthys longipes, Ceratoscopelus warmingii, Diaphus arabicus, Diaphus fragilis, Diaphus aliciae, Diaphus lucidus, Diaphus thiollierei, Diogenichthys panurgus, Hygophum proximum, Lampanyctus turneri, Myctophum spinosum, Symbolophorus evermanni*	*Vinciguerria nimbaria, Vinciguerria leucetia, Stomias boa, Astronesthes sp., Chauliodus pammelas, Chauliodus sloani, Stomias affinis, Stomias nebulosus*	Gjøsaeter (1984), Kinzer et al. (1993), Butler et al. (2001), Karuppasamy et al. (2010)
15 Bay of Bengal	Smallest of four major OMZ regions, separated from Arabian Sea. Mesopelagic fauna adapted to low oxygen concentrations	*Benthosema pterotum*	*Chauliodus pammelas, Astronesthes lamellosus*	Raman and James, (1990)
16 Somali Current	Seasonal monsoon conditions, with strong upwelling.	*Benthosema fibulatum, Diaphus watasei*	*Argyropelecus affinis, Argyropelecus hemigymnus, Argyropelecus sladeni, Polyipnus omphus, Sternoptyx obscura*	Sutton et al. (2017), Wessel and Johnson (1995)

17	Northern Indian Ocean	Gyre system affected by complex seafloor topography, Fish fauna distinct from Southern Indian Ocean	*Lampanyctus turneri, Diogenichthys panurgus, Diaphus signatus, Diaphus lucidus, Ceratoscopelus warmingii, Diaphus aliciae, Benthosema fibulatum	Karuppasamy et al. (2010), southern stations
18	Mid-Indian Ocean	A broad frontal region with characteristic temperature–salinity-oxygen (TSO) signature, and a diverse mesopelagic fish fauna.	Benthosema fibulatum, Hygophum hygomi, Myctophum spinosum, Myctophum aurolaternatum, Myctophum asperum, Symbolophorus evermanni, Diaphus garmani, Diaphus nielseni, Diaphus watasei, Diaphus perspicillatus, Diaphus suborbital, Diaphus thiollierei	Gjøsæter (1979)
19	Southern Indian Ocean	A broad band of subtropical surface waters north of the Subtropical Front. Well oxygenated by oligtrophic.	Myctophum nitidulum, Myctophum phengodes, Myctophum asperum, Symbolophorus evermanni, Gonichthys barnesi, Centrobranchus nigroocelatus	Olivar et al. (2016)
20	Agulhas current	Largest western boundary current in the world ocean, drives upwelling to the south. Enhanced productivity; fauna reflects combination of sub-Antarctic and tropical spp.	Diaphus parri, Diaphus richardsoni, Diogenichthys panurgus, Lampanyctus turneri, Myctophum spinosum, Triphoturus nigrescens, Diaphus diadematus, Symbolophorus evermanni, Benthosema fibulatum, Hygophum hygomii	Vinciguerria attenuata, Vinciguerria nimbaria, Cyclothone alba, Cyclothone braueri, Cyclothone microdon, Cyclothone pallida, Diplophos taenia, Gonostoma atlanticum, Acanthochaenus luetkenii, Argyropelecus aculeatus, Astronesthes trifibulatus,	Froese and Pauly (2016)

Table 5.1 (cont.)

Ecoregion Name	Description	Fish species		References
		Myctophids	Other species	
			Melanostomias paucilaternatus, Stomias nebulosus	
Atlantic Ocean				
21 NW Atlantic Sub-Arctic	The only major ocean region with deep-water connection with the Arctic. High seasonal productivity; distinct cold-water assemblages.	*Benthosema glaciale, Protomyctophum arcticum, Myctophum punctatum, Notoscopelus kroeyeri, Lampanyctus macdonaldi, Lampanyctus crocodilus, Lampanyctus intricarius, Lampadena speculigera	*Nansenia groenlandica, Chauliodus sloani, Gonostomatidae, Bathylagus euryops, Nemichthys scolopaceus, Schedophilus medusophagus, Borostomias antarcticus, Normichthys operosus, Arctozenus rissoi, Serrivomer beani, Paralepis coregonoides, Stomias boa ferox, Xenodermichthys copei, Argyropelecus hemigymnus, Maurolicus muelleri, Holtbyrnia anomala, Linophryne lucifer, Holtbyrnia macrops, Scopelogadus beanii	Sutton and Sigurðsson (2008)
22 North Atlantic Drift	Eddy field region, expanding eastwards as the tail of the Gulf Stream. Transition ecotone with admixture of boreal and subtropical species	*Benthosema glaciale, Protomyctophum arcticum, Myctophum punctatum	*Cyclothone microdon, Maurolicus muelleri, Chauliodus sloani, Serrivomer beanii, Bathylagus euryops, Scopelogadus beanii, Borostomias antarcticus	Sutton et al. (2008)
23 Gulf of Mexico	Enclosed, deep sea; flow dominated by the Loop Current. Extremely diverse ecotonal fauna of tropical,	*Notolychnus valdiviae, Lepidophanes guentheri, Benthosema suborbitale, Hygophum benoiti,	*Cyclothone pallida, Cyclothone pseudopallida, Cyclothone braueri, Cyclothone alba,	Ross et al. (2010)

		subtropical and temperate taxa.	Diaphus dumerilii, Myctophum affine, Ceratoscopelus warmingii, Lampanyctus alatus, Hygophum taaningi, Diaphus mollis, Diaphus splendidus, Diogenichthys atlanticus, Diaphus perspicillatus	Valenciennellus tripunctulatus, Vinciguerria poweriae, Cyclothone acclinidens, Pollichthys mauli, Gonostoma elongatum, Chauliodus sloani, Vinciguerria nimbaria, Argyropelecus aculeatus, Argyropelecus hemigymnus, Bregmaceros atlanticus, Vinciguerria attenuata, Melamphaes simus, Synagrops spinosus, Bathophilus pawneei
24	Central North Atlantic	Broad area of warm and consistent TSO conditions, including the Sargasso Sea. Faunal composition distinct from the North Atlantic Drift and Equatorial Atlantic	*Benthosema suborbital, Hygophum hygomi, Diogenichthys atlanticus, Ceratoscopelus warmingii, Notoscopelus resplendens, Notolychnus valdiviae, Lobianchia dofleini, Hygophum reinhardti, Bolinichthys indicus, Lampanyctus cuprarius, Diaphus rafinesquei, Diaphus mollis	*Cyclothone braueri, Cyclothone microdon, Valenciennellus tripunctulatus, Argyropelecus hemigymnus, Cyclothone pseudopallida, Argyropelecus aculeatus, Gonostoma elongatum, Chauliodus danae, Cyclothone pallida, Bonapartia pedaliota, Sternoptyx diaphana, Vinciguerria poweriae, Idiacanthus fasciola, Vinciguerria attenuata	Badcock and Merrett, (1976)
25	Mediterranean Sea	Landlocked, with single shallow strait; deep waters uniformly warm and highly saline.	*Ceratoscopelus maderensis, Benthosema glaciale, Notoscopelus elongatus, Hygophum benoiti, Lampanyctus crocodilus,	*Cyclothone braueri, Maurolicus muelleri, Argyropelecus hemigymnus, Cyclothone pygmaea, Vinciguerria attenuata,	Olivar et al. (2012)

Table 5.1 (cont.)

Ecoregion Name	Description	Fish species		References
		Myctophids	Other species	
		Lobianchia dofleini, Myctophum punctatum, Lampanyctus pusillus, Diaphus holti, Symbolophorus veranyi, Hygophum hygomii, Notoscopelus bolini, Electrona risso	*Stomias boa, Lestidiops jayakari, Notolepis risso, Bathophilus nigerrimus, Chauliodus sloani, Evermannella balbo, Ichthyococcus ovatus*	
26 Mauritania/ Cape Verde	Upwelling region with discrete faunal communities, including endemics and relict populations of 'cool water' taxa.	**Benthosema glaciale, Lampanyctus alatus, Lampanyctus pusillus, Hygophum benoiti, Notolychnus valdiviae, Lobianchia dofleini, Diaphus rafinesquii, Myctophum punctatum, Diaphus holti, Ceratoscopelus maderensis, Benthosema suborbitale, Bolinichthys indicus, Lampanyctus crocodilus*		Badcock (1981)
27 Tropical & West Equatorial Atlantic	Oligotrophic waters of the South Atlantic subtropical gyre but easterly winds cause localised divergence and upwelling.	**Ceratoscopelus warmingii, Bolinichthys photothorax, Lepidophanes gaussi, Lepidophanes guentheri, Nannobrachium cuprarium, Lampanyctus photonotus*	**Vinciguerria nimbaria, Cyclothone sp., Sternoptyx diaphana, Argyropelecus hemigymnus*	Kobyliansky et al. (2010)
28 Guinea Basin and East	Trade wind reversal causes shoaling of pycnocline (Atlantic monsoon analog).	**Diogenichthys atlanticus, Diaphus problematicus, Hygophum macrochir,*	**Vinciguerria nimbaria*	Kobyliansky et al. (2010)

	Region	Description	Species	Species	Reference
	Equatorial Atlantic	Marked OMZ at about 400 m. High diversity of mesopelagic fish except in OMZ areas.	Benthosema suborbitale, Diaphus vanhoeffeni, Ceratoscopelus warmingii, Lampanyctus alatus, Lepidophanes guentheri, Diaphus mollis, Notoscopelus resplendens, Lampanyctus photonotus, Diaphus brachycephalus,		
29	Benguela Upwelling	Strong upwelling system with highly productive pelagic fisheries and a severe seasonal OMZ.	*Lampanyctodes hectoris, Diaphus garmani, Diaphus taaningi, Lampadena pontifer, Diaphus dumerilii	*Maurolicus muelleri	Hulley and Prosch (1987)
30	South Atlantic	Gyral system of complex circulation affected by local topography (e.g., Walvis Ridge and Rio Grande Rise). Mostly oligotrophic except for borders.	Lobianchia dofleini, Diogenichthys atlanticus, Diaphus meadi, Lampanyctus pussilis, Lampanyctus procerus, Nannobrachium atrum	Argyropelecus hemigymnus, Valenciennellus tripunctulatus, Vinciguerria poweriae, Vinciguerria attenuata, Cyclothone alba, Cyclothone braueri, Cyclothone pallida	Kobyliansky et al. (2010)
Antarctic Ocean					
31	Circumglobal Subtropical Front	Broad convergence band of subtropical and sub-Antarctic waters. Frontal zone highly variable. Highly productive with distinct deep-pelagic assemblages.	*Gymnoscopelus nicholsi, Lampichthys procerus, Lampadena notialis, Protomyctophum tenisoni, Cerastocopelus warmingii, Scopelopsis multipunctatus, Diaphus hudsoni, Electrona subaspera, Gymnoscopelus piabilis, Symbolophorus barnardi, Taaningichthys bathyphilus	*Bathylagus antarcticus, Sternoptyx diaphana, Photichthys argenteus, Argyropelecus aculeatus, Argyropelecus gigas, Sternoptyx diaphana, Stomias gracilis, Stomias boa boa	Figueroa et al. (1998)

Table 5.1 (cont.)

Ecoregion Name	Description	Fish species			References
		Myctophids	Other species		
32 Sub-Antarctic	The zone between the Antarctic Polar and Subtropical Fronts. Cold, low salinity waters. High primary productivity. deep-pelagic fauna either confined to, or centred within the sub–Antarctic Front; high number of endemic species	*Electrona carlsbergi, Protomyctophum bolini, Gymnoscopelus braueri, Gymnoscopelus nicholsi, Gymnoscopelus fraseri, Electrona antarctica, Protomyctophum choriodon, Krefftichthys anderssoni, Lampanyctus achirus, Protomyctophum parallelum, Protomyctophum gemmatum, Gymnoscopelus bolini, Protomyctophum tenisoni, Electrona subaspera	Borostomias antarcticus, Bathylagus tenuis, Stomias gracilis, Nansenia antarctica, Benthalbella elongata, Bathylagus antarcticus, Benthalbella macropinna, Bathylagus gracilis, Poromitra crassiceps, Melanostigma gelatinosum, Argyropelecus hemigymnus, Oneirodes notius, Stomias boa		Collins et al. (2008)
33 Antarctic/ Southern Ocean	Characterised by cold, dense bottom water. The Polar Front is a strong barrier. Highly productive seasonally.	[1] Electrona antarctica, Gymnoscopelus opisthopterus, Gymnoscopelus braueri, Gymnoscopelus nicholsi, Krefftichthys anderssoni, Protomyctophum bolini, Protomyctophum tenisoni			Duhamel et al. (2014)

[1] 'Antarctic' and 'Broadly Antarctic' pattern species after Duhamel et al. (2014).

the oligotrophic areas (Barnett, 1984). In the highly productive Humboldt current area [7] off Peru and Chile, Sielfeld et al. (1995) also found *Cyclothone acclinidens* to be the dominant mesopelagic fish species representing 62 per cent of the catch by abundance and 13 per cent of biomass. The myctophids *Diogenichthys atlanticus* and *Triphoturus mexicanus* represent 29 per cent of abundance and 22 per cent of biomass. In the South Central Pacific [8] the dominant species are the same as in the North Pacific gyre (*Cyclothone alba, Cyclothone pseudopallida* and *Cyclothone pallida*), and although there are strong similarities between the two areas, Barnett (1984) points out that amongst the rarer species some only occur in the North and vice versa. The Coral Sea [9] in the Western Pacific has a gyre circulation separating it from the neighbouring Pacific equatorial waters and Australian waters. It has a unique myctophid fauna comprising tropical species. The Tasman Sea [10] lying between Australia and New Zealand and defined by major fronts, has a mixed myctophid fauna comprising subtropical and temperate species (Flynn and Marshall, 2013). The Sea of Japan [11] has temperatures close to 0°C at depths below 400 m, the mesopelagic is sparsely populated by a few plankton species (Ikeda and Hirakawa, 1998) and there are no mesopelagic fishes. The South China Sea [12] has an average depth of 1212 m, with a maximum depth of over 5000 m and is connected to the West Pacific by the 2000 m deep Bashi Channel (Liu, 2013). Inflow of warm waters of the Kuroshio Current result in a species-rich mesopelagic fish fauna with strong affinities to the central and Eastern tropical Pacific. Yang and Huang (1986) recorded 62 species of myctophid, Wang and Chen (2001) found 36 species to the south and west of Taiwan and Bourret (1985) provides information on Gonostomatidae and Sternoptychidae. The deep basins of the Sulu, Celebes and Banda Seas of Indo-Malayan Archipelago are termed collectively by Sutton et al. (2017) as the Indo-Pacific Pocket Basins [13]. They are to a large extent closed off from one another as well as from the surrounding seas. They lie within the East Indies centre of speciation (Briggs, 2003), one of the centres considered as important for the origin of deep-sea species (Zezina, 1997). The fauna includes the Neoscopelid *Solivomer arenidens* that has no light organs and is endemic to the Sulu Sea. The different basins should probably be considered as separate biogeographic provinces, for example the myctophid *Diaphus impostor* occurs in the Banda and Celebes Seas but not in the Sulu Sea (Nafpaktitis et al., 1995). Dalzell and Pauly (1989) predicted a high biomass of mesopelagic fishes in the Banda Sea as a potential resource, and Dolar et al. (2003) found myctophids to be the most important component of the diet of spinner dolphins (*Stenella longirostris*) in the Sulu Sea. Dolphin stomach content analysis provides information on the fish fauna of the Sulu Sea, which has not otherwise been well documented.

5.1.3 Indian Ocean

The Arabian Sea [14] is characterised by extremely high biomass of the myctophid *Benthosema pterotum*, which has been proposed as a candidate for commercial exploitation (Gjøsaeter, 1984). *B. pterotum* is the only species that occurs in the Gulf of Oman, whereas in the wider Arabian Sea the dominant myctophid is *Diaphus arabicus*. They remain within the oxygen-minimum zone during the day and ascend into oxygen-rich surface waters to feed during the night (Kinzer et al., 1993; Karuppasamy et al., 2010). Predatory stomiids including the endemic *Chauliodus pammelas* also follow the vertical migratory cycle of myctophids between the OMZ and surface layers (Butler et al., 2001). The Bay of Bengal [15] also has an oxygen-minimum layer, and the mesopelagic fish fauna is similar to that of the Arabian Sea with *Benthosema pterotum* the dominant myctophid (Raman and James, 1990). The Somali Current [16] is within the monsoon area, but there is nevertheless a faunal contrast with the Arabian Sea, the dominant myctophid being *Benthosema fibulatum* and *Diaphus watasei*, a benthic feeder on slopes. There is a distinctive assemblage of hatchetfishes (Wessel and Johnson, 1995). The Northern Indian Ocean [17] is an oligotrophic gyre system with low primary productivity except in localized areas of upwelling. The mesopelagic fish fauna differs from that in the Southern Indian Ocean. The Mid-Indian Ocean [18] comprises a broad equatorial frontal region with relatively high productivity, absence of monsoon influence and a diverse mesopelagic fish fauna (Gjøsæter, 1979). The Southern Indian Ocean [19] has well-oxygenated waters but low mesopelagic fish diversity and biomass relative to the rest of the Indian Ocean. Survey data are provided by Olivar et al. (2016) from surface assemblages at night. The Agulhas Current [20] is the western boundary current of the southwest Indian creating regions of high productivity but also carrying tropical fauna southward resulting in a rich fauna with warm and cold water elements. It contains a characteristic Algulhas Current myctophid assemblage (Hulley and Lutjeharms, 1995).

5.1.4 Atlantic Ocean

In the North Atlantic Sub-Arctic [21] north of the subpolar front, mesopelagic fish samples from the Irminger Sea and SW of Iceland are dominated by the myctophid *Benthosema glaciale* (92 per cent of the catch) and *Nansenia groenlandica* 3 per cent (Microstomatidae, pencil smelts; Sutton and Sigurðsson, 2008). In the East, high abundances of sub-Arctic species *Maurolicus mülleri, Benthosema glaciale,* and *Arctozenus risso* can extend to Arctic latitudes off Norway (Siegelman-Charbit and Planque, 2016). In the North Atlantic Drift [22] there is an eddy zone with water moving eastward south of the subpolar front, *Benthosema glaciale* becomes less dominant than in the north comprising 51 per cent of abundance and 29 per cent of biomass with *Cyclothone microdon* (24 per cent abundance) the second most important species (Sutton et al., 2008). The Gulf of Mexico [23] is relatively isolated from the main Atlantic Ocean Basin and has a distinctive fauna including *Synagrops spinosus* (Acropomatidae), which is endemic to the slopes of the Western Atlantic. Species dominance is less strong than at high latitudes; the top ranked species is the Myctophid, *Notolychnus valdiviae,* which comprises

only 15 per cent of abundance followed by four species of *Cyclothone* each at 10–12 per cent (Ross et al., 2010).

The Central North Atlantic [24] is a broad region of rather stable subtropical waters between the North Atlantic Drift to the north and equatorial waters to the south. Studies both in the Sargasso Sea in the West (Sutton et al., 2010) and at 30°N, 23°W in the East (Badcock and Merrett, 1976) show *Cyclothone braueri* to be the dominant species representing numerically 48 per cent and 37 per cent of the catch, respectively. The top ranked myctophids are *Lepidophanes guentheri* (2.5 per cent) in the West and *Benthosema suborbitale* (2.5 per cent) in the east.

The Mediterranean Sea [25] is relatively depauperate in mesopelagic species with 17 species of Myctophidae compared with 59 in the NE Atlantic (Papaconstantinou et al., 1997). In the Western Mediterranean in the region of the Balearic Islands mesopelagic samples are dominated in abundance by *Cyclothone braueri* comprising 78–91 per cent of the catch in small nets (Olivar et al., 2012) and *Cyclothone pygmaea,* which is endemic to the Mediterranean. In the northern basin of the Aegean Sea in the Eastern Mediterranean samples are dominated by myctophids *Hygophum benoii, Myctophum punctatum* and *Benthosema glaciale*; mesopelagic diversity in Greek waters is not markedly lower than in the Western Mediterranean (Papaconstantinou et al., 1997).

The Mauritania to Cape Verde [26] area off West Africa is influenced by the southward Canary current with a transition from a northern to a southern fauna. However, there are also changes in faunal composition with distance offshore from an inshore slope fauna, through the upwelling area to an offshore oceanic fauna. It is possible to identify several discrete mesopelagic assemblages. The myctophid *Benthosema glaciale* dominates in abundance, and Badcock (1981) suggests a relationship with Mediterranean subpopulations of this species that have diverged from the subpolar populations in the North Atlantic.

The Tropical and West Equatorial Atlantic [27] region is mostly oligotrophic and over the Mid-Atlantic Ridge mesopelagic fish assemblages are of low abundance with catches dominated by the

Phosichthyid *Vinciguerria nimbaria,* which comprises over 90 per cent by abundance in shallow hauls (Kobyliansky et al., 2010). The Guinea Basin and East Equatorial Atlantic [28] is more productive with higher mesopelagic diversity, *Vinciguerria nimbaria* amounting to 25 per cent and myctophids better represented. The Benguela Upwelling [29] is one of the most productive regions of the world, and vast quantities of the pseudo-oceanic lantern fish, *Lampanyctodes hectoris,* were targeted by commercial fisheries (see Chapter 6). The silvery light fish (*Maurolicus muelleri*: Sternoptychidae) also occurs in large quantities (Hulley and Prosch, 1987). Small amounts of *Diaphus garmani* from the Agulhas current with *Diaphus taaningi, Lampadena pontifex* and *Diaphus dumerilii* from the north also occur. The other deep pelagic ichthyofauna is relatively poorly known.

5.1.5 The Antarctic Ocean

The Subtropical Front region [31] is a circumglobal zone of convergence between subtropical and sub-Antarctic waters. In the SW Atlantic where the zone impinges on the continental slopes of Argentina, Figueroa et al. (1998) found seven sub-Antarctic species of myctophid, four sub-Antarctic/subtropical species, four tropical/subtropical species and one widespread species, *Taaningichthys bathyphilus.* Similar mixing of species across the frontal region occurs all around the Southern Ocean (Barange et al., 1998). The Sub-Antarctic region lies between the Sub-tropical Front and the Antarctic Polar Front and hence incorporates the Sub-Antarctic Front, an area of very high productivity. Duhamel et al. (2014) list 18 species of myctophid that are considered to be transitional zone species and thirteen Sub-Antarctic pattern species. The Sub-Antarctic is generally regarded as a krill-based ecosystem, but it is increasingly becoming apparent that myctophids also comprise a substantial biomass (25–50 per cent) that can act as an alternative ecosystem pathway supporting predator populations in years of low krill abundance. In the Northern Scotia Sea, Collins et al. (2008) found myctophids numerically comprised 76 per cent of the catch with

predatory stomiids present at greater depths. One specimen of the ceratioid anglerfish *Oneirodes notius* was also captured. In the true Antarctic [33] south of the Polar Front the oceanic *Electrona antarctica* and bathypelagic *Gymnoscopelus opisthopterus* are myctophids adapted to the high latitude conditions, and several other species can occur in this region (Duhamel et al., 2014).

5.2 Bathyal Demersal Fishes (200–3000 m Depth)

The bathyal demersal fish fauna represents some of the highest diversity of deep-sea fishes and the most significant accessible commercial deep-sea fish resources. The habitat is distributed in narrow strips along the ocean margins, mid-ocean ridges and around oceanic islands and seamounts (Figs. 1.10 and 1.11). The zone down to 2250 m depth is characterised as the slope, comprising upper slope (200–750 m), mid-slope (750–1500 m) and lower slope (1500–2250 m). Greater depths with gentler gradients are defined as the rise with the range 2250–3000 m, known as the upper rise (Merrett and Haedrich, 1997). This terminology, however, varies between studies (Fig. 1.3).

In the Porcupine Seabight region of the NE Atlantic Priede et al. (2010) found 99 species in 39 families (Table 5.2), including representatives of the Class Myxini (1 species), Class Chondrichthyes (22 species) and Class Actinopterygii (76 species). Each species has a characteristic depth range, and there is a transition from upper slope species to deeper-living species on the upper rise and beyond 3000 m depth into the abyss (Fig. 5.3). The maximum in species richness is at mid to lower slope depths with 48 species present at 1600 m depth (Fig. 5.3b).

In the North Atlantic Ocean biomass decreases with depth from a maximum around 1000 m depth (Bergstad et al., 2012; Fig. 5.4.). There is very wide variation with apparently substantially lower biomass on the American slope in the Mid-Atlantic Bight and Georges Bank than on European slopes, although there is evidence that off Newfoundland biomass is

Table 5.2 **Bathyal Demersal Fishes. List of Species Occurring in the Porcupine Seabight. Data from Priede et al. (2010). Depth Rank – Ranked in Terms of Minimum Depth of Occurrence Cross Reference to Figure 5.1.**
MAR are species listed as also occurring in the Mid-Atlantic Ridge by Bergstad et al. (2012).

| Order | Family | Species | Depths | | | |
			Rank	Min (m)	Max (m)	MAR
Myxiniformes	Myxinidae	*Myxine ios*	59	1110	1540	
Chimaeriformes	Rhinochimaeridae	*Harriotta raleighana*	82	1527	1900	
		Rhinochimaera atlantica	69	1360	1533	
	Chimaeridae	*Chimaera monstrosa*	34	695	1541	
		Hydrolagus affinis	85	1720	2397	1
		Hydrolagus mirabilis	35	695	2058	
Hexanchiformes	Hexanchidae	*Hexanchus griseus*	44	763	960	
Squaliformes	Dalatiidae	*Dalatias licha*	29	545	1360	
	Centrophoridae	*Centrophorus squamosus*	52	986	986	
		Deania calcea	40	740	740	
	Somniosidae	*Centroscymnus coelolepis*	58	1027	1720	1
		Scymnodon ringens	36	706	982	
	Etmopteridae	*Etmopterus princeps*	83	1533	1720	1
		Etmopterus spinax	15	407	1790	
Carcharhiniformes	Scyliorhinidae	*Apristurus laurussonii*	43	750	1448	
		Galeus melastomus	25	506	848	
		Galeus murinus	57	1024	1024	
Rajiformes	Rajidae	*Bathyraja richardsoni*	91	1987	2530	1
		Dipturus nidarosiensis	66	1263	1312	
		Leucoraja circularis	32	695	695	
		Neoraja caerulea	72	1390	1390	
		Rajella bigelowi	53	1000	4118	
		Rajella fyllae	46	808	1430	
Notacanthiformes	Halosauridae	*Halosauropsis macrochir*	75	1440	3485	1
		Halosaurus johnsonianus	71	1379	1448	
	Notacanthidae	*Notacanthus bonaparte*	18	470	2504	1
		Notacanthus chemnitzii	31	685	2500	1
		Polyacanthonotus challengeri	95	2410	2040	1
		Polyacanthonotus rissoanus	41	740	2500	1
Anguilliformes	Synaphobranchidae	*Histiobranchus bathybius*	87	1790	4842	1
		Ilyophis arx	86	1789	1900	
		Ilyophis blachei	67	1284	1789	
		Ilyophis brunneus	84	1533	2875	1
		Synaphobranchus kaupii	16	407	2500	1
	Nettastomatidae	*Venefica proboscidea*	81	1527	1527	

Table 5.2 (*cont.*)

Order	Family	Species	Rank	Min (m)	Max (m)	MAR
Osmeriformes	Alepocephalidae	*Alepocephalus agassizii*	70	1368	2567	1
		Alepocephalus australis	78	1500	2572	1
		Alepocephalus bairdii	38	706	2500	
		Alepocephalus productus	77	1462	2737	1
		Alepocephalus rostratus	39	706	2504	
		Bathylaco nigricans	94	2292	2292	
		Bathytroctes microlepis	89	1845	4840	1
		Bellocia koefoedi	98	2500	4840	
		Bellocia michaelsarsi	96	2410	4222	
		Conocara macropterum	73	1430	2670	1
		Conocara murrayi	74	1440	2875	1
		Narcetes stomias	90	1867	2737	1
		Rouleina attrita	68	1312	1927	1
	Argentinidae	*Argentina silus*	8	240	330	
		Argentina sphyraena	1	265	695	
Aulopiformes	Ipnopidae	*Bathypterois dubius*	55	1016	2434	1
	Bathysauridae	*Bathysaurus ferox*	80	1519	3639	1
		Bathysaurus mollis	99	2645	4245	1
Zeiformes	Oreosomatidae	*Neocyttus helgae*	76	1462	1462	1
Gadiformes	Macrouridae	*Coelorinchus caelorhincus*	14	407	1312	
		Coelorinchus labiatus	19	472	1900	1
		Coryphaenoides armatus	93	2016	4865	1
		Coryphaenoides brevibarbis	88	1845	4292	1
		Coryphaenoides carapinus	79	1500	3485	1
		Coryphaenoides guentheri	62	1200	2875	1
		Coryphaenoides leptolepis	92	1993	4865	1
		Coryphaenoides mediterraneus	42	743	2645	1
		Coryphaenoides rupestris	37	706	1932	1
		Malacocephalus laevis	2	240	685	
		Nezumia aequalis	20	472	2058	
		Paracetonurus flagellicauda	97	2486	3089	
	Trachyrinchidae	*Trachyrincus murrayi*	64	1205	1600	1
		Trachyrincus scabrus	26	506	1541	
	Moridae	*Antimora rostrata*	48	853	2970	1
		Guttigadus latifrons	51	985	1867	
		Halargyreus johnsonii	30	545	1900	1
		Lepidion eques	27	506	2420	1
		Mora moro	23	500	1312	
	Merlucciidae	*Merluccius merluccius*	6	247	785	
	Gadidae	*Gaidropsarus argentatus*	11	1024	1024	

Table 5.2 (*cont.*)

Order	Family	Species	Rank	Depths Min (m)	Max (m)	MAR
		Gaidropsarus macrophthalmus	56	265	2078	
		Molva dypterygia	21	490	1462	
		Molva macrophthalma	17	470	695	
		Phycis blennoides	10	265	1144	
Ophidiiformes	Carapidae	*Echiodon drummondii*	13	407	545	
	Ophidiidae	*Spectrunculus grandis*	49	853	4298	1
	Bythitidae	*Cataetyx alleni*	63	1205	1205	
		Cataetyx laticeps	60	1144	2500	1
Beryciformes	Trachichthyidae	*Hoplostethus atlanticus*	50	960	1677	1
		Hoplostethus mediterraneus	28	527	1430	
	Berycidae	*Beryx decadactylus*	24	506	506	
Scorpaeniformes	Sebastidae	*Helicolenus dactylopterus*	7	247	853	
		Trachyscorpia cristulata echinata	45	763	1144	
	Psychrolutidae	*Cottunculus thomsonii*	54	1016	1541	
	Liparidae	*Paraliparis hystrix*	33	695	695	
Perciformes	Epigonidae	*Epigonus telescopus*	22	490	2567	
	Zoarcidae	*Lycodes terraenovae*	65	1217	2462	
		Pachycara crassiceps	61	1144	2504	
	Trichiuridae	*Aphanopus carbo*	47	853	1993	1
Pleuronectiformes	Scophthalmidae	*Lepidorhombus boscii*	5	247	785	
		Lepidorhombus whiffiagonis	4	247	545	
	Pleuronectidae	*Glyptocephalus cynoglossus*	12	293	506	
	Soleidae	*Microchirus variegatus*	3	247	506	
Lophiiformes	Lophiidae	*Lophius piscatorius*	9	265	808	
		Total number	99			40

closer to European values. The Mid-Atlantic Ridge has generally higher biomass than comparable depths on the continental slopes. On the European slopes to the west of Scotland there is evidence of depletion by fishing activity with a 60 per cent lower biomass at 1000 m depth in surveys in the 2000s compared with early surveys in the 1970s. Biomass beyond 3000 m depth is uniformly low.

The Gadiformes and Ophidiformes form a dominant component of the deep demersal fauna in the North Atlantic comprising 34 per cent of species at slope depths and 41 per cent on the rise (Haedrich and Merrett, 1988) with gadidae decreasing in importance with depth whilst Macrourids and Ophidiids become more important on the lower slope and rise (Merrett and Haedrich, 1997). The Chondrichthyes are significant at slope depths in the Porcupine Seabight (Table 5.2), Chimaeriformes at average depth range 1200–1900 m, sharks at 800 –1200 m and rays at 1200–1900 m. The Halosaurs are found across the lower slope and rise and Notacanthidae across slope depths. A succession of Synaphobranchid eel species

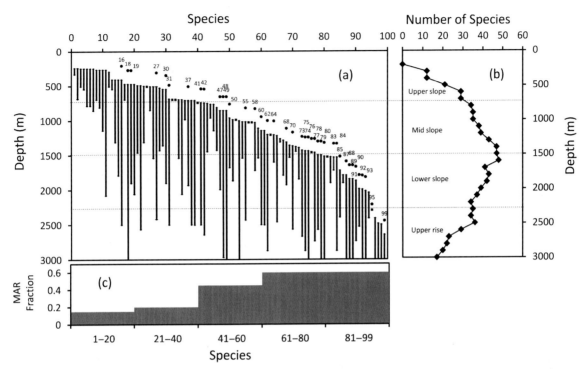

Figure 5.3 Bathyal demersal fish distribution on the slopes of the Porcupine Seabight, NE Atlantic Ocean.

(a) Depth ranges of 99 recorded bathyal species after Priede et al. (2010). Each vertical bar shows the depth range of an individual species with depth rank reference numbers listed in Table 5.2. Points and reference numbers indicate that the species also occur on the Mid-Atlantic Ridge according to Bergstad et al. (2008).

(b) The total number of fish species as a function of depth. Horizontal dashed lines indicate slope and rise depth zones according to Haedrich and Merrett (1988).

(c) The fraction of species that also occur on the Mid-Atlantic Ridge.

is found from the upper slope to the abyss. The Alepocephalids form an important part of the species richness at mid-slope to abyssal depths and are occasionally abundant. The Berciformes are represented by two species of roughy (*Hoplostethus*) and the Alfonsino (*Beryx*) at slope depths. The Scorpaeniformes comprise *Helicolenus dactylopterus* (Blackbelly rosefish) at upper slope depths, a second Sebastid, a Psycholutrid and a liparid all at slope depths. In the Perciformes the cardinal fish (*Epigonus*) is found mainly on the upper slope, the black scabbardfish (*Aphanopus*) on the mid-slope and two species of eelpout (Zoarcidae) on the lower slope. Flatfishes, Pleuronectiformes are only found on the upper slope.

Similar species combinations, dominated by the same orders and families, occur around the North Atlantic basin. However, the species change with latitude or longitude along a depth stratum, leading Haedrich and Merrett (1990) to conclude that there is 'little support for continuity in horizontal zones' and suggesting there is little evidence for faunal zonation or communities in deep-sea demersal fishes. Koslow (1993) challenged this view and was able to identify at least five and possibly nine species clusters around the North Atlantic. In one scheme he recognised seven groupings; 1 – Eastern Temperate shallow (<1500 m,), 2 – Eastern Temperate-Boreal deep (>1500 m), 3 – Western Temperate Deep, 4 – Eastern Tropical to Subtropical Shallow, 5 – Eastern Tropical to Sub-tropical Deep; 6 – Western Tropical; 7 – Western Temperate-Boreal Shallow. These were related to latitude, longitude and differences in prevailing water masses.

With the addition of the Mid-Atlantic Ridge and seamounts, Bergstad et al. (2012) recognise ten geographic subareas around the North Atlantic Basin (Fig. 5.5.). Species richness is consistently higher in the Western Atlantic than in the Eastern Atlantic, both at the family and species level. The Mid-Atlantic Ridge has fewer families and species than the continental margin and shows the greatest similarity to the Eastern North Atlantic slopes. Distinctive patterns of differentiation between depth zones (Bergstad et al., 2008) were detected in each of the major areas sampled. The upper slope has the highest proportion of unique species (26 per cent) decreasing to 6 per cent on the upper rise. Comparing species found in the Porcupine Seabight (Priede et al., 2010) and the Mid-Atlantic Ridge (Bergstad et al., 2008), the highest percentage of common species is found amongst the deeper-living species. There are two reasons for this: firstly the area of the Mid-Atlantic Ridge shallower than 1000 m is very small so that there is only a small amount of habitat available for upper slope species; secondly there is a general trend that deeper-living species are more widespread resulting in less differentiation between areas.

In the Norwegian Basin, to the north of the ridge between Scotland and Iceland, there is a distinctly different fish fauna. Fish abundance decreases abruptly at 720 m depth, beneath which the water temperature falls to less than 0°C. In this cold deep zone fish biomass is very low, with only seven species living beyond 1500 m depth dominated by the Zoarcid *Lycodes frigidus* (glacial eelpout, 3000 m_{fbd}) and the Liparid *Paraliparis bathybius* (Black seasnail, 4009 m_{fbd}). In the region of the steep temperature gradient there is a high biomass comprising large boreo-arctic species: *Raja hyperborea* (Arctic skate, 2500 m_{fbd}), *Macrourus berglax* (Roughhead grenadier), *Reinhardtius hippoglossoides* (Greenland halibut) and the redfishes *Sebastes mentella* and *Sebastes marinus*. The shallower warmer water harbours species known also from the North Atlantic (Bergstad et al., 1999).

The Mediterranean Sea is presumed to have been colonised by Atlantic species passing through the

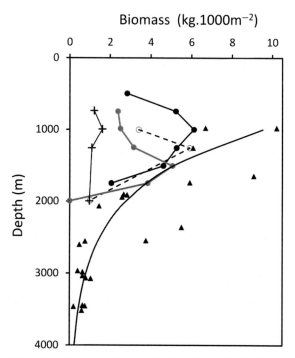

Figure 5.4 Biomass of demersal bathyal fishes in the North Atlantic Ocean. Circles and solid lines – NE Atlantic, Black – Scotland Slope 1970s, Grey –Scotland Slope 2000s, Open circles and dashed line – Bay of Biscay, Ireland and Scotland Slopes. Black Triangles – Mid-Atlantic Ridge 2004–2009, an exponential curve is fitted to the data $Y = 32.6e^{(-0.00123X)}$, $R^2= 0.75$. Crosses – NW Atlantic, Georges Bank and Mid Atlantic Bight. The NE and NW Atlantic points are means for each depth stratum. The Mid-Atlantic Ridge points are from single trawl tows. (After Bergstad et al., 2012).

Straits of Gibraltar during the last 5.33 Ma as the sea refilled following the dry period of the Messinian Salinity Crisis (see Chapter 1). Reflecting the oligotrophic conditions in the Mediterranean Sea the deep-sea ichthyofauna has a very low biomass declining from 0.22 kg. 1000 m^{-2} in the Western Basin to 0.034 kg.1000 m^{-2} in the East at mid-slope depths, 800–1300 m (D'Onghia et al., 2004), an order of magnitude lower than the values reported for the North Atlantic in Figure. 5.4. In a survey of the main deep basins in the Mediterranean Sea D'Onghia et al. (2004) documented 44 demersal species living deeper than 600 m, representing 63 per cent of the families listed in Table 5.2 for the Eastern North Atlantic. An additional seven families of mainly warmer water

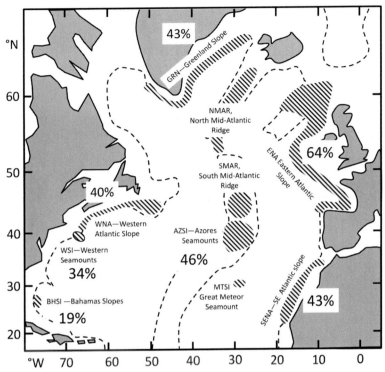

Figure 5.5 The North Atlantic Ocean, showing the bathyal demersal ichthyofaunal subareas according to Bergstad (2012). Sorensen's index of fish species overlap between the Mid-Atlantic Ridge, (NMAR and SMAR) and each subarea is indicated as a percentage.

representatives such as the Oxynotidae (rough sharks), Chlorophthalmidae (green eyes), and Cynoglossidae (tongue fishes) brings the family diversity to over 80 per cent of that commonly observed in the Atlantic Ocean. The Macrourids are represented by seven species including *Coelorinchus mediterraneus*, which is endemic to the waters around Italy. Despite recent colonisation and the unusual warm oligotrophic conditions the Mediterranean Sea therefore harbours a surprisingly typical deep-sea demersal fish fauna. Canyons in the western basin are dominated by *Alepocephalus rostratus*, *Trachyrincus scrabrus* and *Mora moro* (Stefansecu, 1994). On the Greek slopes of the Eastern Ionian Sea Mytilineou et al. (2005) recorded 83 demersal species living between 300 and 1200 m depth. *Argentina sphyraena* and *Chlorophthalmus agassizi* were dominant on the upper slope at 300–500 m, *Chlorophthalmus agassizi* and *Phycis blennoides* at 500–700 m and *Galeus melastomus* and *Nezumia sclerorhynchus* at 700 to 1200 m. A remarkable feature is the 22 species of

Chondrichthyes including 11 species of skates and rays (Batoidea) as well as a relatively high abundance of fishes on the upper slopes attributed to absence of any commercial fishing activity at depths greater than 500 m in Greek waters.

On the northern slopes of the Gulf of Mexico, Powell et al. (2003) found 106 species living between 200 to 3000 m depth, which they divided into three assemblages: the upper slope (315–785 m) dominated by *Coelorinchus carribaeus* (Blackfin grenadier, 700 m_{fbd}) and *Steindachneria argentea* (Luminous hake, 500 m_{fbd}), the mid-slope (686–1369 m) dominated by the macrourids *Nezumia cyrano* (915 m_{fbd}) and *Coryphaenoides zaniophorus* (Thickbeard grenadier, 2375 m_{fbd}) and the deep zone (1533–3075 m) dominated by ophidiids *Dicrolene kanazawai* (2342 m_{fbd}) and *Acanthonus armatus* (Bony-eared assfish, 4415 m_{fbd}). The peak of richness at 48 species is on the upper slope.

In the Indian Ocean Williams et al. (1996) surveyed the slopes of Western Australia between 200 and

1500 m depth from tropical latitudes in the North (20°S) to the temperate waters in the South (35°). Although the region was found to be highly speciose with 388 species from 108 families, nevertheless biomass was very low in global terms, i.e., 0.35–0.57 kg.1000m^{-2} compared with values of 1.7–4.82 kg. 1000m^{-2} off Eastern Australia and the North Atlantic values shown in Figure 5.4. This low biomass is attributed to the influence of the warm Leeuwin current from the North, the absence of coastal upwelling and consequent low productivity of overlying waters off Western Australia (Williams et al., 2001). This results in a fauna made up of many relatively small, dispersed benthic fishes with no aggregations of large commercial species. The Macrourids, for example, are mostly represented by small-mouthed *Coelorinchus* species, *Cetonurus globiceps*, *Bathygadus spongiceps* and the genus *Gadomus* rather than the more robust *Coryphaenoides* species such as those found off Eastern Australia and New Zealand. There is evidence of a latitudinal transition from a northern fauna to a distinctive southern fauna on the mid-slope. The latter has affinities with the biogeographic province that extends from the Great Australian Bight to Tasmania, New South Wales and the Chatham Rise off New Zealand. By depth the most distinct delineations occurred between the shelf break community down to 250–350 m, the upper slope to 700–800 m and the mid-slope at greater depths. At shelf break and upper slope depths the catches were numerically dominated by Acropomatidae and Chlorophthalmidae. At greater depths the dominant families were Macrouridae, Bathygadidae, Synaphobranchidae, Alepocephalidae and Oreosomatidae. Overall, the Macrouridae was the most speciose family with 50 species, followed by the Squalidae (22 species), Alepocephalidae (17), Ophidiidae (17), Moridae (13), Triglidae (13), Scyliorhinidae (10) and Scorpaenidae (10).

The region of SE Australia from the Eastern Australian Bight, around Tasmania and New South Wales to 35°S has been surveyed at mid-slope depths from 800 to 1200 m, where a fishery had developed for orange roughy (*Hoplostethus atlanticus*). Koslow et al. (1994) found 111 species of demersal fishes from 39 families. Of these, seven families were dominant, comprising 96 per cent of the catch by weight, Trachichthyidae (23 per cent, one species, *Hoplostethus atlanticus*), Squalidae (22 per cent, 13 species) Oreosomatidae (20 per cent, five species), Macrouridae (13 per cent, 26 species), Synaphobranchidae (8 per cent, 1 species, *Diastobranchus capensis*), Alepocephalidae (5 per cent, 5 species) and Moridae (4 per cent, 9 species). Significant differences were detected in species composition between the east and west of Tasmania and between 100 m depth strata. On a broader scale there was moderate species overlap with the Great Australian Bight to the west and waters off New Zealand on Challenger Plateau and Chatham Rise (Sorensen's coefficient $S = 38$–40 per cent) suggesting a continuous mid-slope community extending over 6500 km from SW Australia to east of New Zealand. Virtually no overlap was found with the North Pacific, but there was significant overlap with the temperate North Atlantic ($S = 9$–15 per cent); 14 of the species from the mid-slopes of SE Australia also occur in the Porcupine Seabight in the NE Atlantic, including *Halosauropsis macrochir*, *Antimora rostrata*, *Mora moro* and *Hoplostethus atlanticus* (Table 5.2; Priede et al., 2010). Koslow et al. (1994) suggest that this is linked to the pattern of global deep-water circulation (Fig. 1.18) whereby components of deep water from the North Atlantic circulate around the Southern Hemisphere creating bipolar distributions of some temperate deep-water fishes.

Francis et al. (2002) have analysed the demersal fish assemblages of the New Zealand EEZ between 34.0° and 53.99°S and identified two deep-water strata: the upper slope assemblage with 29 species distributed from 50–1099 m depth and a mid-slope assemblage with 42 species from 600–1499 m. This analysis excluded those rare species that were found in less than 1 per cent of the samples and species with identification issues (*Beryx splendens*, *Dipturus nasutus*, *Dipturus innominatus*, *Hydrolagus novaezealandiae*, *Kathetostoma giganteum*). Therefore, although this cannot be considered as an exhaustive list of the New Zealand deep-sea ichthyofauna, with 17 species the

Table 5.3 **The Most Frequently Occurring Deep Demersal Fish Species in the New Zealand EEZ % = Percentage of Samples in which the Species Occurs. (Data from Francis et al., 2002) Upper Slope = Species Ranging 50–1099 m Depth. Mid-Slope = Species Ranging 600–1499 m Depth**

Family	Species	New Zealand Name	%
Upper slope			
Merlucciidae	*Macruronus novaezelandiae*	Hoki	63
Macrouridae	*Lepidorhynchus denticulatus*	Javelinfish	44
Ophidiidae	*Genypterus blacodes*	Ling	43
Chimaeridae	*Hydrolagus bemisi*	Pale ghost shark	41
Sebastidae	*Helicolenus percoides*	Sea perch	33
Merlucciidae	*Merluccius australis*	Hake	32
Rhinochimaeridae	*Harriotta raleighana*	Longnose spookfish	29
Mid-Slope			
Trachichthyidae	*Hoplostethus atlanticus*	Orange roughy	69
Etmopteridae	*Etmopterus baxteri*	Baxter's dogfish	58
Moridae	*Halargyreus johnsonii*	Johnson's cod	55
Macrouridae	*Coryphaenoides subserrulatus*	Four rayed rattail	52
Centrophoridae	*Deania calcea*	Shovelnose dogfish	51
Macrouridae	*Coryphaenoides serrulatus*	Serrulate rattail	49
Oreosomatidae	*Pseudocyttus maculatus*	Smooth oreo	49
Moridae	*Mora moro*	Ribaldo Mora	45
Synaphobranchidae	*Diastobranchus capensis*	Basketwork eel	43
Somniosidae	*Centroscymnus crepidater*	Longnose velvet	43
Macrouridae	*Caelorinchus innotabilis*	Notable rattail	43
Trachyrinchidae	*Trachyrincus aphyodes*	Unicorn rattail	35*
Trachyrinchidae	*Trachyrincus longirostris*	White rattail	
Oreosomatidae	*Neocyttus rhomboidalis*	Spiky oreo	29
Macrouridae	*Macrourus carinatus*	Ridge scaled	29
Somniosidae	*Centroscymnus owstoni*	Owston's dogfish	28
Rhinochimaeridae	*Rhinochimaera pacifica*	Pacific spookfish	28
Alepocephalidae	*Alepocephalus antipodianus*	Small scaled brown slickhead	27
Alepocephalidae	*Alepocephalus australis*	Big scaled brown slickhead	26

Macrouridae is clearly the most speciose family. The analysis gives a very useful ranking of species occurrence from thousands of trawl tows from the 1960s onward; the most frequently occurring species (found in > 25 per cent of tows) are listed in Table 5.3. The two commercially important species, Hoki and orange roughy, are at the top of the list in their respective depth ranges. The family and species composition show strong affinities to the SE Australian fauna, and some species are common to the North Atlantic.

Merrett and Haedrich (1997) point out that almost 40 per cent of the deep demersal genera in the Atlantic Ocean also occur in the Indian Ocean; there is 25 per cent overlap with the South Australia region but only 14 per cent overlap with the North Pacific. In this, and the analysis by Koslow et al. (1994), the North Pacific Ocean stands out as an area with an unusual deep-demersal fish fauna. Of 94 species listed by Pearcy et al. (1982) only seven, mostly those with a depth range extending into abyssal depths, also occur in the Atlantic Ocean. The faunal composition is greatly influenced by a remarkable diversification in two families of Scorpaeniformes. On the mid- to upper slope there are 12 species of rockfishes (Sebastidae) and 25 species of snail fishes (Liparidae; Table 5.4). Within the Perciformes the Zoarcidae is represented by 14 species. Unusually for a deep demersal fish fauna, the macrourids represent less than 7.5 per cent of species and the entire gadiformes less than 10 per cent of the species richness. There is only one family of flatfishes, the Pleuronectidae with six species.

In terms of the modern deep-water circulation, the North Pacific is remote from the main centres of deep-water formation in the Atlantic sector of the globe. From the late Cretaceous (*ca.* 80 Ma) when the Atlantic Ocean opened up, one consequence was the convergence of North America and Asia closing off the connection between the Pacific and the Arctic Oceans. At 3.5 Ma the Bering Straits opened for a short period, which resulted in the Great Trans-Arctic Biotic interchange, largely with Pacific species invading the Atlantic rather than vice versa. Thus the Gadidae, which originate in the Atlantic, make little contribution to the North Pacific. Briggs (2003) points out that the North Pacific Ocean is a major area of evolutionary origin of marine species including, amongst fishes, the Liparidae, Zoarcidae and the genus Sebastes, which have dispersed into cold waters around the world. It is probable that the proximity of NE Pacific sampling sites to this 'evolutionary engine' explains the unusual composition of the deep-sea ichthyofauna. Taxa that originate elsewhere such as the Gadidae (Atlantic) and Macrouridae (Indo West-Pacific) have limited opportunities to enter suitable niches.

The Antarctic Ocean to the south of the Polar Front is relatively isolated from the rest of the world's oceans and is well known for its endemic fish fauna of the suborder Notothenioidei (see Chapter 4). However, the deep-sea demersal fish fauna is primarily made up of families known in other parts of the world (Kock, 1992; Table 5.5). Amongst these the snailfishes (Liparidae)

are by far the most speciose family with *ca.* 150 species in the Antarctic region, most of which live at depths greater than 800 m. Having invaded from the North Pacific they then further diversified in the Antarctic region. The Zoarcids followed a similar invasion route from the Northern Hemisphere along the East Pacific and there are *ca.* 28 species in the Antarctic (Matschiner et al., 2015). Macrourids and Morids are also well represented with four genera in each family. The rays (Rajidae) may be a relic of an old chondrichthyan fauna of the Antarctica, occurring in deep-water catches, although they are not considered to be primarily deep-sea fishes. The Notothenioidei are primarily coastal fishes, but four of eight known families have species that occur in the deep sea (Table 5.5; Chapter 4), and several species including the tooth fishes *Dissostichus* sp. are important resources in deep-water fisheries of the Southern Ocean.

In view of the difficulties of sampling slopes around the world, Clark et al. (2010) have suggested that seamounts might be used as a proxy to examine whether deep-sea demersal fish assemblages are globally homogenous. Seamounts are generally populated by a subset of species already known from the nearest adjacent continental slopes and other bathyal habitat in the region (Christiansen et al., 2015; Tracey et al., 2004). Clark et al. (2010) examined five seamount groups for which there were comparable replicate data for a full range of species, the Azores area of the Mid-Atlantic Ridge, Chilean seamounts in the SE Pacific and three areas in the SW Pacific: New Caledonia, New Zealand and SE Australia. The geographic regions nearest to one another, New Zealand and SE Australia show the greatest similarity, whereas the others are clearly different (Fig. 5.6). *Hoplostethus atlanticus* (orange roughy) occurs widely but in low catch rates in the Azores and is absent from New Caledonia. Conversely *Pseudopentaceros richardsoni* (pelagic armorhead) is only found in New Caledonia and *Aphanopus carbo* (black scabbard fish) only in the North Atlantic. *Neoscopelus microchir* is an unusual example of a large benthic feeding Myctophiform (Chapter 4) that is characteristic of the New Caledonia seamounts. It is evident that there is a distinctive deep demersal fish

Table 5.4 **List of Deep Demersal Fishes Occurring at Bathyal Depths in the NE Pacific Ocean on the Slopes Off Oregon (After Pearcy et al., 1982). Depth Ranges are from FishBase (Froese and Pauly, 2016).**

Order	Family	Species	Depths Min (m)	Max (m)
Myxiniformes	Myxinidae	*Eptatretus deani*	103	2743
Chimaeriformes	Chimaeridae	*Hydrolagus colliei*	0	913
Hexanchiformes	Hexanchidae	*Hexanchus griseus*	1	2500
Squaliformes	Squalidae	*Squalus acanthias*	0	1463
	Somniosidae	*Somniosus pacificus*	0	2000
Carcharhiniformes	Scyliorhinidae	*Apristurus brunneus*	32	1306
	Triakidae	*Galeorhinus zyopterus*	0	1100
Rajiformes	Rajidae	*Bathyraja abyssicola*	362	2910
		Bathyraja inornata	800	2938
		Bathyraja interrupta	23	1500
		Bathyraja trachura	213	2550
		Raja binoculata	3	800
		Raja inornata	13	1600
		Raja rhina	9	1069
Notacanthiformes	Notacanthidae	*Notacanthus chemnitzii*	125	1000
		Polyacanthonotus challengeri	777	4560
Osmeriformes	Alepocephalidae	*Alepocephalus tenebrosus*	46	5500
		Conocara salmoneum	2400	4500
		Leptochilichthys agassizii	2000	3100
Aulopiformes	Bathysauridae	*Bathysaurus mollis*	1550	4903
Gadiformes	Macrouridae	*Albatrossia pectoralis*	140	3500
		Coryphaenoides acrolepis	300	3700
		Coryphaenoides armatus	282	5180
		Coryphaenoides cinereus	150	3500
		Coryphaenoides filifer	1285	2904
		Coryphaenoides leptolepis	610	4000
		Nezumia stelgidolepis	277	909
	Moridae	*Antimora microlepis*	175	3048
	Merlucciidae	*Merluccius productus*	0	1000
Ophidiiformes	Ophidiidae	*Dicrolene filamentosa*	935	1867
		Spectrunculus grandis	800	4300
	Bythitidae	*Cataetyx rubrirostris*	300	1000
	Aphyonidae	*Sciadonus pedicellaris*	1847	4880
Scorpaeniformes	Sebastidae	*Sebastes aleutianus*	25	900
		Sebastes alutus	0	825
		Sebastes aurora	124	769

Table 5.4 (*cont.*)

Order	Family	Species	Min (m)	Max (m)
			Depths	
Scorpaeniformes (continued)		*Sebastes babcocki*	49	625
		Sebastes crameri	25	600
		Sebastes diploproa	0	800
		Sebastes flavidus	0	800
		Sebastes helvomaculatus	25	549
		Sebastes ruberrimus	15	549
		Sebastes saxicola	9	547
		Sebastolobus alascanus	17	1600
		Sebastolobus altivelus	201	1757
	Anoplopomatidae	*Anoplopoma fimbria*	0	2740
	Cottidae	*Icelinus filamentosus*	18	800
	Agonidae	*Bathyagonus nigripinnis*	18	1290
	Psychrolutidae	*Psychrolutes phrictus*	500	2800
	Liparidae	*Acantholiparis caecus*	1300	2122
		Acantholiparis opercularis	227	3610
		Careproctus cypselurus	35	1775
		Careproctus filamentosus	2265	2940
		Careproctus gilberti	73	886
		Careproctus gilberti	1900	3334
		Careproctus melanurus	89	2286
		Careproctus microstomus	2721	3585
		Careproctus microstomus	1900	2818
		Careproctus ovigerus	1920	2910
		Elassodiscus caudatus	1920	2910
		Genioliparis ferox	2884	2884
		Osteodiscus caudatus	1900	3000
		Paraliparis cephalus	294	1800
		Paraliparis dactylosus	541	1000
		Paraliparis deani	18	1008
		Paraliparis latifrons	2030	3279
		Paraliparis megalopus	2830	3585
		Paraliparis mento	800	1253
		Paraliparis paucidens	1536	2275
		Paraliparis pectoralis	681	1636
		Paraliparis rosaceus	1050	3358
		Paraliparis ulochir	700	1900
		Rhinoliparis attenuatus	350	2189
		Rhinoliparis barbulifer	28	2189
Perciformes	Zoarcidae	*Bothrocara brunneum*	129	2570
		Bothrocara molle	60	2688
		Bothrocara pusillum	221	2189

Table 5.4 (*cont.*)

Order	Family	Species	Depths	
			Min (m)	Max (m)
Perciformes (continued)		*Lycenchelys camchatica*	200	2100
		Lycenchelys crotalinus	200	2816
		Lycenchelys jordani	200	2816
		Lycodapus camchatica	933	2225
		Lycodapus fierasfer	50	2212
		Lycodapus pachysoma	1160	2600
		Lycodes cortezianus	73	620
		Lycodes diapterus	146	844
		Lycodes pachysoma	25	973
		Pachycara bulbiceps	2400	4780
		Taranetzella lyoderma	500	3000
Pleuronectiformes	Pleuronectidae	*Atheresthes stomias*	18	950
		Embassichthys bathybius	41	1800
		Glyptocephalus zachirus	0	900
		Lyopsetta exilis	25	800
		Microstomus pacificus	10	1370
		Parophrys vetulus	0	550

fauna around seamounts in different parts of the world. Clark et al. (2010) found that environmental covariates (salinity, temperature, depth, nutrients, etc.) could explain only 30 per cent of the variance in assemblage composition. Clearly other factors such as the oceanographic setting, size and shape of the seamount and the regional biogeographic pattern of colonisation by fishes are also important. Rowden et al. (2010) point out that fewer than 300 seamounts of the thousands estimated to occur worldwide have been biologically sampled. Reliable fish faunal data are available for far fewer seamounts.

An important factor contributing to environmental heterogeneity on deep-sea slopes is the presence of biogenic structures such as cold-water corals (Buhl-Mortensen et al., 2010). Ross and Quatrini (2007) have shown that of 99 species observed at 365–910 m depth on the Atlantic continental slopes of the SE United States, 92 are associated to varying degrees with coral reef habitats. Prime reef was dominated by the Morid *Laemonema melanurum*, the macrourid *Nezumia sclerorhynchus*, Alfonsino *Beryx*

decadactylus and Sebastid *Helicolenus dactylopterus*, whereas the intervening off-reef areas were dominated by the ray *Fenestraja plutonia*, Morid *Laemonema barbatulum*, hagfish *Myxine glutinosa* and greeneye *Chlorophthalmus agassizi*. Further north in the N. Atlantic off Newfoundland, Baker et al. (2012a) found a positive association between cup corals and the grenadier *C. rupestris*, whereas *Anarhichas minor*, *Neocyttus helgae*, *Hoplostethus atlanticus* and *Lepidion eques* were also associated with outcrops or complex habitats (Baker et al., 2012b) These and studies elsewhere (Costello, 2005) are drawing attention to the importance of habitat structure in understanding the distribution of fishes on deep-sea slopes.

5.3 Abyssal Demersal Species (3000–6000 m Depth)

Although the abyssal zone covers more than 50 per cent of the area of the planet (see Chapter 1), it

Table 5.5 **List of Demersal Fish Families in the Antarctic Deep Sea. After Kock (1992).**

Deep-Sea Families		
Order	Family	Rank
Notacanthiformes	Halosauridae	
	Notacanthidae	
Anguilliformes	Synaphobranchidae	
Alepocephaliformes	Alepocephalidae	4
Zeiformes	Oreosomatidae	
Gadiformes	Macrouridae	3
	Moridae	5
Ophidiiformes	Carapidae	
	Ophidiidae	
Scorpaeniformes	Psychrolutidae	
	Bathylutichthyidae	
	Liparidae	1
Perciformes	Zoarcidae	2
Antarctic Coastal families with deep representatives		
Order	Family	
Perciformes	Artedidraconidae	
	Bathydraconidae	
	Channichthyidae	
	Nototheniidae	

has been relatively poorly surveyed because of the length of time required to trawl at great depths as well as inherent technical difficulties. Depths greater than 3000 m are occasionally sampled at the deep end of fisheries surveys of national EEZs, but few studies target locations in the vast areas between 4000 to 6000 m in the open ocean. Haedrich and Merrett (1988) found approximately 70 species of fish living at depths greater than 3000 m around the North Atlantic Basin (Table 5.6). This list is certainly not complete (Merrett, 1994; Priede et al., 2010) but gives a useful baseline overview of the taxonomic composition, depth and geographic distribution fishes in the abyss. Apart from one remarkable observation of a sea lamprey (*Petromyzon marinus*) at over 4000 m depth off New England (Haedrich, 1977), the Class Myxini, mainly hagfishes, is absent from these depths. The chondrichthyes are represented by a few rays (Rajidae) and catches of the longnose chimaera

(*Harriotta raleighana*) at the foot of the slopes off NW Africa. Sharks are absent. Twenty species of teleosts appear to be endemic to abyssal depths but of these, six are reported from depths less than 3000 m in other studies leaving only fourteen truly endemic abyssal species: two Ipnopidae, two Macrouridae, one Zoarcidae, three Ophidiidae, four Aphyonidae and two Liparidae.

The abyssal endemic species are not dominant either in numbers or biomass. The important species caught in greatest numbers are all types in which the depth range extends from the bathyal slopes down into the abyss. The Notacanthiformes are found around most of the North Atlantic Basin and on the Mid-Atlantic Ridge down to 3500 m depth but *Polyacanthonotus challengeri* extends down to over 4000 m depth mainly in the East Atlantic Basin in the Porcupine Seabight. Amongst the Anguilliformes the deep-water arrowtooth eel, *Histiobranchus bathybius*,

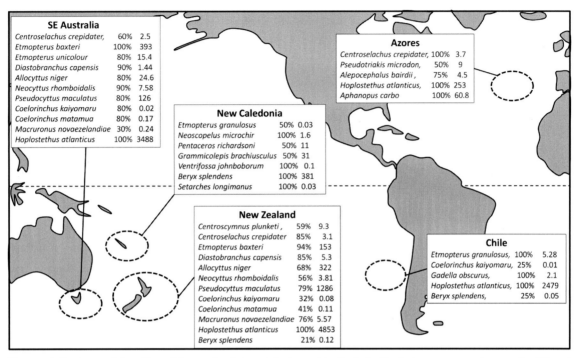

Figure 5.6 Seamount Fishes. Lists of the species that contribute most to dissimilarities between seamount groups in different regions of the world. Percentage presence and average catch weight (kg) each is given for each species in each region. Data from Clark et al. (2010).

is the most important abyssal species, comprising up to 15 per cent of the catch in samples at over 4000 m depth in the Eastern Basin. The Alepocephalids are distributed around the ocean margins and the Mid-Atlantic Ridge with *Bellocia koefoedi* caught in abundance off the slopes of NW Africa and also found on the Porcupine Abyssal Plain. *Conocara salmoneum* and *Rinoctes nasutus* occur down to maximum abyssal depths. Six species of Ipnopidae (deep-sea tripod fishes) are found in the abyssal North Atlantic with *Bathypterois grallator* most abundant on the ocean margins and ridges, whereas *B. longipes* can comprise a large proportion of the catch on abyssal plains at depths > 4000 m. Both known species of Bathysauridae are present: *Bathysaurus ferox* and the deeper-living *B. mollis*.

Abyssal samples are overwhelmingly dominated in terms of numbers and biomass by the Macrouridae. The most abundant is *Coryphaenoides armatus*, which occurs from mid-bathyal depths down to the deepest

parts of the abyssal plains throughout the Atlantic Ocean. The Morid, *Antimora rostrata*, is confined to slopes of ocean margins, the Mid-Ocean Ridge, islands and seamounts. The Ophidiiformes, both the Ophidiidae (cusk eels) and the viviparous Aphyonidae (blind cusk-eels), make the greatest contribution to abyssal fish species richness. They are mostly relatively rare species except for occasional high catches of large specimens of *Spectrunculus grandis*. The pricklefish (*Acanthochaenus luetkenii*) is recorded from mid latitudes off the slopes of New England and Morocco. The Liparidae (snailfishes) and Zoarcids (eelpouts) account for a third of the abyssal endemic species, though with relatively small numbers of individuals. Observations of dolphin carcasses deployed at abyssal depths in the North Atlantic show Macrouridae to be the first fishes to arrive, followed by Ophidiids, and after all the soft flesh has been removed from the carcase, Zoarcids take up residence in the skeleton (Jones et al., 1998).

Table 5.6 Abyssal Demersal Fishes of the North Atlantic Ocean. After Haedrich and Merrett (1988) with Some Modifications. Depths Indicate the Shallow Boundary of Each 750 m Depth Stratum. 1 = Presence of the Species Within the Stratum. E = Abyssal Endemic, E? = Abyssal Endemic in the Atlantic but Shallower Occurrences on Record, NE = Not Abyssal Endemic Maximum Abundance - Highest Percentage of Catch of Dominant Species. – = Few or Rare

Order	Family	Genus	Species	Depth strata (m) Bathyal 750	1500	2250	Abyssal 3000	3750	4250	Abyssal Endemic	Maximum Abundance
Petromyzoniformes	Petromyzontidae	*Petromyzon*	*marinus*	1					1	NE	–
Chimaeriformes	Rhinochimaeridae	*Harriotta*	*raleighana*		1		1			NE	0–5%
Rajiformes	Rajidae	*Bathyraja*	*spinicauda*			1	1			NE	–
		Raja	*bigelowi*		1			1		NE	–
		Raja	*hyperborea*		1				1	NE	–
Notacanthiformes	Halosauridae	*Halosauropsis*	*macrochir*	1	1	1	1			NE	15–30%
	Notacanthidae	*Notacanthus*	*chemnitzi*	1		1	1			NE	0–5%
		Polyacanthonotus	*challengeri*	1		1	1			NE	0–5%
Anguilliformes	Synaphobranchidae	*Haptenchelys*	*texis*			1	1			NE	–
		Histiobranchus	*bathybius*			1	1	1	1	NE	5–15%
		Ilyophis	*arx*					1		E?	–
		Ilyophis	*brunneus*	1	1	1	1			NE	0–5%
	Nettastomatidae	*Venefica*	*proboscidea*		1	1	1	1		NE	0–5%
Alepocephaliformes	Alepocephalidae	*Alepocephalus*	*agassizi*	1		1	1			NE	–
		Alepocephalus	*australis*	1				1		NE	–
		Bathytroctes	*microlepis*		1	1	1			NE	0–5%
		Bellocia	*koefoedi*		1	1		1		NE	>30%
		Bellocia	*michaelsarsi*		1	1	1	1		NE	0–5%
		Conocara	*salmoneum*				1	1		E?	–
		Narcetes	*stomias*		1	1	1	1		NE	5–15%
		Rinoctes	*nasutus*			1	1	1	1	NE	–
Aulopiformes	Ipnopidae	*Bathymicrops*	*regis*				1		1	E	–
		Bathypterois	*grallator*		1	1	1			NE	0–5%
		Bathypterois	*longipes*			1	1	1	1	NE	15–30%
		Bathytyphlops	*sewelli*			1	1	1		NE	–

Order	Family	Genus	species							Status	%
		Discoverichthys	praecox						1	E	–
		Ipnops	murrayi		1		1		1	NE	–
	Bathysauridae	Bathysaurus	ferox	1	1	1			1	NE	0–5%
		Bathysaurus	mollis	1	1	1	1	1	1	NE	0–5%
Gadiformes	Macrouridae	Coryphaenoides	armatus	1	1	1		1	1	NE	>30%
		Coryphaenoides	brevibarbis	1	1	1	1	1	1	NE	5–15%
		Coryphaenoides	leptolepis	1	1	1	1	1	1	NE	15–30%
		Coryphaenoides	mediterraneus	1		1	1		1	NE	0–15%
		Coryphaenoides	profundicola					1	1	E	–
		Echinomacrurus	mollis						1	E	–
		Lionurus	carapinus	1	1	1	1		1	NE	>30%
		Paracetonurus	flagellicauda	1	1		1		1	NE	–
	Moridae	Antimora	rostrata	1	1	1	1	1	1	NE	5–15%
Ophidiiformes	Ophidiidae	Abyssobrotula	galatheae						1	E	–
		Acanthonus	armatus	1	1					NE	–
		Barathrites	iris						1	E	–
		Bassozetus	compressus	1	1			1	1	E?	–
		Bassozetus	normalis	1	1				1	NE	–
		Bassozetus	oncerocephalus			1			1	E	–
		Bassozetus	taenia	1	1			1	1	NE	–
		Bathyonus	laticeps	1	1			1	1	NE	–
		Bathyonus	pectoralis	1	1			1	1	NE	–
		Holcomycteronus	squamosus	1	1				1	NE	–
		Penopus	macdonaldi	1	1	1				NE	–
		Penopus	microphthalmus	1	1	1				NE	–
		Porogadus	catena	1	1			1	1	NE	–
		Porogadus	miles	1	1	1			1	NE	–
		Porogadus	subarmatus	1	1	1				NE	–
		Spectrunculus	grandis	1	1	1		1	1	NE	0–5%
	Aphyonidae	Aphyonus	brevidorsalis						1	E	–
		Aphyonus	rassi	1	1	1	1		1	NE	–
		Barathronus	bicolor	1	1	1	1	1	1	NE	–
		Barathronus	multidens			1				E	–

Table 5.6 (*cont.*)

Order	Family	Genus	Species	Depth strata (m)						Abyssal Endemic	Maximum Abundance
				Bathyal				Abyssal			
				750	1500	2250	3000	3750	4250		
		Barathronus	*parfaiti*		1	1	1		1	NE	–
		Barathronus	*unicolor*				1	1		E	–
		Nybelinella	*erikssoni*			1			1	NE	–
		Sciadonus	*galatheae*					1	1	E?	–
		Sciadonus	*jonassoni*						1	E	–
Stephanoberyciformes	Stephanoberycidae	*Acanthochaenus*	*luetkenii*		1	1	1			NE	15–30%
Scorpaeniformes	Liparidae	*Careproctus*	*aciculipunctatus*					1		E	–
		Careproctus	*merretti*					1		E	–
		Paraliparis	*bathybius*					1		E?	–
Perciformes	Zoarcidae	*Lycenchelys*	*alba*					1		E	–
		Lycenchelys	*crassiceps*		1			1		NE	–
		Pachycara	*obesa*				1	1		E?	–
Total number of species within depth stratum							44	35	33		

In an AUV imaging survey of the Porcupine Abyssal Plain at 4800 m depth Milligan et al. (2016) estimated that Macrouridae numerically comprise over 90 per cent of the fish assemblage followed by the eel *Histiobranchus bathybius* (3.5 per cent), *Bathysaurus mollis* (1.5 per cent), Alepocephalids (1 per cent, half of which were *Conocara salmoneum*) and Zoarcids (0.5 per cent). The fish density was estimated at 723 individuals km^{-2} compared with previous estimates of 152 km^{-2} from trawl catches and 188 km^{-2} using baited cameras (Priede and Merrett, 1996).

Merrett and Haedrich (1997) point out that the abundance and composition of the abyssal fish fauna is almost certainly influenced by the supply of organic matter from the surface. The Porcupine Abyssal Plain (48°50'N) lies under the NADR (North Atlantic Drift Province), an area of high seasonal surface productivity (Longhurst, 1998). On the Maderia Abyssal Plain (31°N) at 4900 m depth under the oligotrophic North Atlantic Sub-tropical Gyre (NAST), Armstrong et al. (1992) estimated the abundance of macrourids to be 5 per cent of standing stock on the Porcupine Abyssal Plain. Other species present on the Madeira Abyssal Plain were *Histiobranchus bathybius* (Synaphobranchidae), and Ophidiidae, *Spectrunculus grandis* and *Barathrites* sp., in correspondingly low numbers. On the nearby Seine Abyssal Plain (33°46'N), Christiansen et al. (2015) also found the most abundant family to be the Macrouridae followed by Ipnopidae and Alepocephalidae (Table 5.7). In the Eastern Basin of the Atlantic Ocean Merrett (1987) proposed a faunal divide at a latitude of 34° and 41°N between a northern fish assemblage (41°–49°N) and a southern assemblage with fewer macrourids (20°–22°N; Merrett and Haedrich, 1997).

Further evidence of the link with surface productivity is provided by Cousins and Priede (2012) on the Crozet Plateau in the Southern Indian Ocean. This is a High Nutrient Low Chlorophyll (HNLC) region with low surface productivity but to the east of the Crozet Islands natural iron enrichment from the land results in a carbon flux 2.5 times greater to the deep than in the surrounding ocean. In this high productivity area the fish abundance and biomass

were, respectively, 1.5 and 1.9 times higher than in the reference HNLC sampling site at the same depth (Table 5.7). The predominance of macrourids (60–70 per cent of the catch) is again evident. In the NE Pacific Ocean on the Tufts Abyssal Plain, Pearcy et al. (1982) observed the same predominance of macrourids with *Coryphaenoides armatus* amounting to 77–100 per cent of the catch in the deepest trawls. The rest of the catch comprised mainly Ophidiiformes and Liparids. In the West Pacific a similar pattern has been observed by Linley et al. (2016) with the macrourids, *C. armatus*, *C. yaquinae* as well as some Ophidiids dominant at abyssal depths in the Kermadec trench region north of New Zealand (34°S). In this area, liparids were confined to adjacent hadal trench depths. In the more tropical abyssal areas of the South Fiji Basin (25°S) and around the New Hebrides Trench (21°S) marcourids are absent. Instead there is a simplified ophidiid-dominated community, similar to that observed by Anderson et al. (1985) in the Caribbean Sea (Table 5.7). Ophidiid-dominated communities have also been observed in the Sargasso Sea (Fleury and Drazen, 2013) and the Arabian Sea (Janßen et al., 2000).

The Mediterranean Sea fish fauna is relatively impoverished with only three abyssal demersal species: the macrourid *Coryphaenoides mediterraneus* comprising 88 per cent by number followed by *Bathypterois mediterraneus* (Ipnopidae, 8 per cent) and *Cataetyx laticeps* (Bythitidae, Viviparous brotulas, 4 per cent; D'Onghia et al., 2004). In the oligotrophic Eastern Mediterranean, *Coryphaenoides mediterraneus* are stunted in their growth, typically 15–25 cm long compared with up to over 70 cm length for the same species in the North Atlantic. *C. mediterraneus* are found at the deepest point in the Mediterranean Sea, the 5111 m Calypso Deep of the Hellenic Trench.

It has been speculated that the presence of abyssal hills in areas previously assumed to be featureless flat plains may enhance environmental heterogeneity and biodiversity. In their AUV study of the Porcupine Abyssal Plain, Milligan et al. (2016) found no difference in abundance or species composition of fishes occurring on elevated terrain compared with

Table 5.7 Abyssal demersal fish fauna composition at different sampling locations.

X = present, XX = relatively abundant, XXX = dominant abundant species. (X) = Family present not identified to genus or species. For the Indian Ocean, Crozet plateau abundance is given in number of individuals km⁻². High = area of high surface production. Low = low surface production.

Order	Family	Genus	Species	Atlantic	Indian		Pacific		
				Seine Abyssal Plain[1]	Crozet Plateau[2] High	Low	Tufts Abyssal Plain[3]	Kermadec Trench margin[4]	Fiji and New Hebrides[4]
Depth (m)				4416–4419	4182–4195	4258–4270	3000–5180	3000–6000	3000–6000
Notacanthiformes	Halosauridae	Aldrovandia	affinis						X
Anguilliformes	Synaphobranchidae	Histiobranchus	bathybius	X	70	22		X	
		Ilyophis	robinsae					X	X
Alepocephaliformes	Alepocephalidae	Conocara	salmoneum	X			X		
		Rinoctes	nasutus	X					
Aulopiformes	Ipnopidae	Bathymicrops	regis	XX					
		Bathypterois	oddi		278				
		Bathypterois	longipes	X		203			
	Bathysauridae	Bathysaurus	mollis				X		
Gadiformes	Macrouridae	Coryphaenoides	armatus	XX	200	113	XXX	XXX	
		Coryphaenoides	ferrieri					X	
		Coryphaenoides	filicauda		296	295			
		Coryphaenoides	lecointei		447	147		X	
		Coryphaenoides	leptolepis				X	X	
		Coryphaenoides	profundicolus	X					
		Coryphaenoides	yaquinae				X	XX	
		Echinomacrurus	mollis	X					
Ophidiiformes	Ophidiidae	Apagesoma	australis		2	11			
		Bassozetus	sp.					X	XX
		Barathrites	iris					X	X
		Bathyonus	caudalis					X	

Order	Family	Genus	species					
		Bathyonus	*pectoralis*					2
		Holcomycteronus	*brucei*				33	2
		Holcomycteronus	*profundissimus*		X			
		Holcomycteronus	*squamosus*			X		
		Spectrunculus	*grandis*		X	XX		
	Aphyonidae	*Barathronus*	*pacificus*		X			
		Sciadonus	*pedicellaris*		X			
Scorpaeniformes	Liparidae	*Careproctus*	sp.					5
		Careproctus	*crozetensis*					10
		Careproctus	*discoveryae*					5
		Careproctus	*microstomus*		X			
		Paraliparis	*latifrons*		X			
		Paraliparis	*megalopus*		X			
		Paraliparis	*mento*		X			
		Paraliparis	*wolffi*				23	
Perciformes	Zoarcidae	*Lycenchelys*	*antarctica*				11	4
		Pachycara	*cousinsi*	(X)				5
		Pachycara	*priedei*	(X)			23	10
			Total (n.km^{-2})				881	1336

[1] Christiansen et al. (2015),
[2] Cousins and Priede (2012),
[3] Pearcy et al. (1982),
[4] Linley et al. (2016).

the flat plain. Macrourids in particular appear to move freely and randomly across the topography with no evidence of aggregations related to these putative landmarks in the deep sea.

The Polar Abyssal fish fauna is little known. In the Arctic, Zoarcids including *Lycodes frigidus* (Glacial eelpout, 3000 m$_{fbd}$) and Liparids including *Paraliparis bathybius* (Black seasnail, 4009 m$_{fbd}$) have been recorded at abyssal depths (Stein et al., 2005). In the Antarctic, Zoarcids, Liparids (*Notoliparis kurchatovi*, 5474 m$_{fbd}$) and Ophidids (*Holcomycteronus brucei*, 4575 m$_{fbd}$) are reported from the greatest depths (Kock, 1992).

In summary, therefore, the global abyssal demersal fish fauna is made up of a small number of families, Macrouridae, Ophidiidae, Ipnopidae, Synaphobranchidae, Liparidae and Zoarcidae and is dominated by a few species of the genus *Coryphaenoides* that account for more than half of numbers and biomass. Biomass and abundance are lower in oligotrophic areas with macrourids absent from some tropical oligotrophic regions where they are replaced mainly by Ophidiids. In Polar regions macrourids are also absent, the sparse fauna mainly comprising Zoarcids and Liparids.

5.4 Fishes of the Hadal Zone (> 6000 m Depth)

The hadal zone is made up of an archipelago of trenches, troughs and basins that extend to depths greater than 6000 m (Table 1.1) with the deepest and the largest trenches located in the West Pacific Rim, the Japan-Kuril-Kamachatka, Kermadec, Tonga, Mariana and Philippine Trenches extending to over 10 km depth (see Chapter 1). Because fishes cannot survive at depths greater than 8400 m (Yancey et al., 2014) they are confined to the margins of such trenches: the dominant mobile fauna of the deep axial basins are scavenging amphipods (Blankenship and Levin, 2007). At least 18 species of fishes have been observed or captured from hadal trenches (Jamieson et al., 2015; Linley et al., 2016, Table 5.8). These can be divided into Abyssal-hadal transition species,

which are also found on the abyssal plains and the Hadal endemic species.

The only Anguilliform recorded at hadal depths is the Synaphobranchid, *Ilyophis robinsae*, an abyssal species endemic to the central West Pacific found in the New Hebrides Trench at 2087–5344 m and the Kermadec Trench area at 997–6068 m (Linley et al., 2016). The macrourid, *Coryphaenoides yaquinae*, is the dominant large scavenging fish species over most of the deeper parts (> 4700 m) of the abyssal Pacific Ocean and is a frequent visitor to the great hadal trenches of the West Pacific. Cuskeels of the family Ophidiidae, subfamily Neobythitinae are also often seen around the trench margins. *Abyssobrotula galatheae* is known as the world's deepest fish based on a single specimen trawled from the Puerto Rico Trench (Nielsen, 1977), and it has also been recorded from the Izu-Bonin Trench (Machida, 1989). It generally occurs at much shallower depths, has never been observed in situ at hadal depths and the hadal specimens were all obtained with nonclosing nets (Jamieson, 2015) so that they could be pelagic captures well above the nominal bottom depth of the net tow. *Barathrites iris* is a widely occurring abyssal cuskeel, and whereas a very similar fish has been observed at 6116 m in the Kermadec Trench, it could not be definitively identified to species. Very high abundances of a *Barathrites* sp. have been seen at baited cameras at 6173 m in the Peru–Chile Trench, although very few were seen nearby at 5329 m (Jamieson, 2015). Similarly the genus *Bassozetus* has often been seen in camera images in the trenches of the SW Pacific and Rass (1955) reported *Bassozetus zenkevitchi* from the Kurile-Kamchatka Trench. *Holcomycetronus profundissumus* and *Leucicorus atlanticus* are relatively rare lower abyssal to upper hadal species. As the pearl fish *Echiodon neotes*, recorded from the Kermadec Trench (Markle and Olney, 1990) was captured by a nonclosing net, its depth record of 8300 m is considered doubtful (Jamieson 2015). Linley et al. (2016) for the first time recorded the presence of eelpouts (Zoarcidae) at hadal depths down to 6162 m in the New Hebrides and the Mariana Trench and suggested the presence of more than one species.

Table 5.8 **Fishes of the Hadal Zone**

AH - abyssal-hadal, *HE* - hadal endemic

Order	Family	Genus Species	Endemic	Depth (m) Min	Max	Trenches	References
Anguilliformes	Synaphobranchidae	*Ilyophis robinsae*	AH	2087	6068	Kermadec	Linley et al. (2016)
Gadiformes	Macrouridae	*Coryphaenoides yaquinae*	AH	3400	7012	Japan, Mariana, Kermadec	Linley et al. (2016)
Ophidiiformes	Ophidiidae Neobythitinae	*Abyssobrotula galatheae*	AH	3110	8370	Izu-Bonin, Puerto Rico	Jamieson (2015)
		Barathrites sp.	AH		6173	Kermadec, Peru–Chile	Jamieson (2015)
		Bassozetus zenkevitchi	AH		6930	Kurile-Kamchatka	Rass (1955)
		Bassozetus sp.	AH		6898	New Hebrides, Kermadec, Mariana	Linley et al. (2016)
		Holcomycetronus profundissumus	AH	5600	7160	Java	Roule (1913)
	Carapidae	*Leucicorus atlanticus*	AH	4580	6800	Cayman	Jamieson (2015)
		Echiodon neotes	HE	8200	8300	Kermadec	Markle & Olney (1990)
Scorpaeniformes	Liparidae	*Notoliparis antonbruuni*	HE	6150	6150	Peru–Chile	Stein (2005)
		Notoliparis kermadecensis	HE	6474	7669	Kermadec	Linley et al. (2016)
		Pseudoliparis amblystomopsis	HE	7210	7450	Japan-Kuril-Kamachatka	Linley et al. (2016)
		Pseudoliparis belyaevi	HE	6945	7703	Japan,	Linley et al. (2016)
		Mariana snailfish	HE	6198	8078	Mariana	Linley et al. (2016)
		Ethereal snailfish	HE	8007	8145	Mariana	Linley et al. (2016)
		Peru–Chile snailfish	HE		7049	Peru–Chile	Jamieson (2012)
		Puerto Rico snailfish	HE		7300	Puerto Rico	Pérès (1965)
Perciformes	Zoarcidae	Unknown eel-pouts	AH		6162	New Hebrides, Mariana	Linley et al. (2016)

The only fishes that are endemic to hadal depths are snailfishes (Liparidae). Linley et al. (2016) propose that there are probably eight species at depths between 6150 and 8145 m (Table 5.8). Each species is endemic to a given contiguous trench system, thus *Pseudoliparis amblystomopsis* is found throughout the Japan-Kuril-Kamachatka Trench system in the NW Pacific, *Notoliparis kermadecensis* in the Kermadec Trench in the SW Pacific and *Notoliparis antonbruuni* in the Peru–Chile Trench of the SE Pacific. There can more than one species in a given trench, e.g. *Pseudoliparis belyaevi* coexisting with *P. amblystomopsis* in the Japan Trench, two putative new species in the Mariana Trench and a second species in the Peru–Chile Trench together with *Notoliparis antonbruuni*. Here we presume that the snailfishes observed by Pérès (1965) through the ports of a bathyscaphe, identified tentatively as belonging to the genus *Careproctus* is a distinctive Atlantic hadal species. When snailfishes are present, they can appear in large aggregations of up to 30 individuals attracted to bait within view of a camera at around 7500 m depth. The smaller individuals, 10–15 cm long, are found at greatest depths (8000 m) and the largest ones, 20–30 cm long, at the shallowest (6500 m), suggesting ontogenic ascent of the walls of the trench. Although attracted to bait, hadal liparids prefer to suction-feed on amphipods and other crustacea attracted to the bait; so-called necrophagivory. The reproductive strategy of liparids producing small batches of large eggs that hatch as benthic juveniles (Stein, 1980) lends itself to a self-sustaining life cycle contained within a single trench system. It is tempting to regard the aggregations of hadal snailfishes seen by benthic cameras as family groups comprising parents and offspring reared with some degree of parental care. It is not known how the eggs are brooded.

The hadal snailfishes are clearly benthic fishes, and Jamieson (2015) suggests they would be unable to swim across the trench in the pelagic open water above trench axis whilst remaining above the 8400 m lethal limiting depth. In large trenches the deeper water may act as a partial barrier restricting movement of individuals and gene flow between the two sides. Whether there are any fishes in the pelagic waters of the hadal trenches is not known. It is conceivable that benthopelagic abyssal-hadal transition species such as *Coryphaenoides yaquinae* and the Ophidiidae could move freely across the trench from rim to rim.

A large proportion of the food supply to hadal trenches is derived from the ocean surface as part of the continuous rain of sinking particulate organic matter. There is a tendency for gradual concentration of such material in the axis of the trench either through gradual downslope transport by currents or major sediment slides during slope failures (Jamieson et al., 2010). Ichino et al. (2015) modelled this funnelling effect and predicted high spatial variability in benthic biomass with the greatest biomass on the flat plains of the trench axis. Such downslope transport is likely to bypass the fishes confined to depths less than 8400 m, but high concentrations of organic matter have been reported from the Atacama segment of the Peru–Chile Trench at 7800 m depth (Danovaro et al., 2003) and observed by Pérès (1965) at 7300 m, where he saw the liparids in the Puerto Rico Trench. Also Fujikara et al. (1999) found chemosynthetic communities at a depth of 7326 m in the Japan Trench. Such locations within their depth range can provide hotspots of feeding opportunities for trench-endemic liparids. Proximity to land masses also means that trenches are susceptible to terrestrial inputs providing further enrichment despite their remoteness in terms of vertical distance from the surface. This, however, raises concern regarding pollution and adverse anthropogenic inputs (Jamieson et al., 2010).

5.5 Fishes of Hydrothermal Vents and Cold Seeps

Hydrothermal vents and cold seeps provide oases of enriched food supply in an otherwise impoverished deep-sea environment. It is therefore not surprising that shortly after their discovery in 1977 that the presence of 50–75 opportunistic deep-sea benthopelagic fish species was catalogued by Cohen and Haedrich (1983) around the Galapagos Riftt hydrothermal vent sites in the E Pacific. These included rays (Rajidae), slickheads (Alepocephalidae), tripod fishes (Ipnopidae), Halosaurs (Halosauridae), deep-sea cods (Moridae), grenadiers (Macrouridae), cuskeels (Ophidiidae), blind cuskeels (Aphyonidae) and eelpouts (Zoaricidae). The most abundant species were *Coryphaenoides bulbiceps*, *Bassozetus*, Zoarcids, *Antimora*, and *Spectrunculus*, common representatives of the deep benthopelagic community anywhere. Biscoito et al. (2002) point out that Atlantic vent sites tend to have more species possibly because of their shallower depth compared with the Pacific Ocean and identify about 90 species worldwide living in the vicinity of vents and seeps, the top five families in rank of importance being Macrouridae, Ophidiidae, Squalidae, Moridae and Synaphobranchidae. Whether such peripheral species make significant use of the vent site is difficult to ascertain, but they provide a pathway for the potential export of energy to the wider deep sea.

Over 20 fish species have been recognised as living inside either a hydrothermal vent field or cold seep and can be regarded as endemic to varying degrees (Table 5.9). The hagfishes (Myxinidae) are represented by a single specimen of a very slender species 314 mm long, *Eptatretus strickrotti*, from the East Pacific Rise, although there are reports of video observations of hagfish at other vent sites (Møller and Jones, 2007). At least four species of Synaphobranchid eel have been recorded living in vent sites or seeps. *Ilyophis saldanhai* and *Thermobiotes mytilogeiton* (Geistdoerfer, 1991) are probably vent-endemic but *I. saldanhai* has a remarkable distribution found at vents both in the Mid-Atlantic Ridge and the East Pacific Rise (Cause et al., 2005), suggesting it may be not be an obligate vent-endemic species. Staudigel et al. (2006) found large swarms of *Dysommina rugosa* at 710 m depth in the volcanic cone of the Vailulu'u seamount off Samoa; however, it is clearly not vent-endemic because the species was known from the West Atlantic and across the Pacific Ocean long before hydrothermal vents were discovered. As eels have planktonic eggs and leptocephalus larvae that are readily dispersed, it is unlikely that species endemic to particular vent sites can evolve. The Ophidiiformes *Ventichthys biospeedoi* (presumed to be oviparous) and *Thermichthys hollisi*, which is viviparous, have been observed at multiple sites in the East Pacific Rise, swimming in shimmering vent fluids and are presumed to be vent-endemic (Nielsen and Cohen, 2005; Nielsen et al., 2006). The liparid *Careproctus hyaleius*, which probably produces benthic eggs, has a lifecycle well suited to the establishment of endemic populations in vent areas (Geistdoerfer, 1994). The Zoarcidae are the most characteristic of vent species with 11 species more or less endemic to hydrothermal vents or cold seeps, with new discoveries as exploration continues (Anderson et al. 2016). Small flatfishes or tonguefishes of the genus *Symphurus* are characteristic of relatively shallow hydrothermal vent fields associated with underwater volcanoes of the West Pacific (Tunnicliffe et al., 2010).

Symphurus and *Dysommina rugosa* have both been observed to feed on the carcasses of pelagic fishes and other organisms that succumb to the low pH and adverse conditions of the hydrothermal plume. This is in addition to the nutritional benefit that may accrue from feeding on the vent organisms that have derived their growth from chemosynthesis. *Symphurus* avoid bacterial mats, and it is unlikely that any fishes feed directly on bacteria. Sancho et al. (2005) show that *Thermarces cerberus* swimming amongst vestimentiferan tube worms (*Rifta* spp.) and vent mussels (*Bathymodiolus thermophilus*) feed preferentially on gastropod molluscs (mainly *Lepetodrilus elevatus*) and amphipod crustaceans (mainly *Ventiella sulfuris*), selecting the largest prey. They would also take the amphipod *Halice hesmonectes* when present in large swarms and the snail *Cyathermia naticoides*. These prey items ultimately derive their nutrition from chemo-autotrophic bacteria that are not capable of producing polyunsaturated fatty acids (PUFAs), which are essential for fish life. Pond (2008) shows that although the prey do contain sufficient PUFA to meet the requirements of *Thermarces cerberus*, the concentrations of PUFA in the tissues of *T. cerberus* are low compared with most marine fishes, indicating the deficiency of the hydrothermal vent diet. The ovaries of *T. cerberus* contained higher

Table 5.9 **Fishes of Hydrothermal vents and cold seeps**

E - vent endemic, NE - nonendemic

ORDER Family	Genus Species	Endemic	Depth (m)	Location comments	References
MYXINIFORMES					
Myxinidae	Eptatretus strickrotti	E	2211	East Pacific Rise	Møller and Jones (2007)
ANGUILLIFORMES					
Synaphobranchidae	Dysommina rugosa	NE	260–775	W Atlantic, E Central Pacific to Hawaii. Nafanua hydrothermal vent, Samoa W Pacific	Staudigel et al. (2006)
	Ilyophis saldanhai	E	2838–3100	Mid-Atlantic Ridge, Broken Spur, SE Pacific Rise Grommit vent site	Karmovskaya and Parin (1999), Cause et al. (2005)
	Thermobiotes mytilogeiton	E	1750	W Pacific, Vai Lili vent, Valu Fa Rise	Geistdoerfer (1991)
	Haptenchelys texis	NE	2121–4086	W Atlantic: Bahamas, E Atlantic: NW Africa.	Robins & Martin (1976),
OPHIDIIFORMES					
Ophidiidae	Ventichthys biospeedoi	E	2586	East Pacific Rise	Nielsen et al. (2006)
Bythitidae	Thermichthys hollisi		2500	East Pacific Rise	Nielsen and Cohen (2005)
SCORPAENIFORMES					
Liparidae	Careproctus hyaleius	E	2630	East Pacific Rise	Geistdoerfer (1994)
PERCIFORMES					
Zoarcidae	Pachycara caribbaeum	E	1049–2300	W Atlantic. Hydrothermal vent and methane seeps Mid-Cayman Spreading Centre of the Caribbean Sea.	Anderson et al. (2016)
	Pachycara gymninium	NE	1829–3225	NE Pacific: Juan de Fuca Ridge	Anderson & Peden (1988)
	Pachycara rimae	E	2500	SE Pacific Galapagos Rise	Anderson (1989)
	Pachycara saldanhai	E	2280	Mid Atlantic Ridge Rainbow.	Biscoito & Almeida (2004)
	Pachycara sulaki	NE	2000–3510	W Atlantic, Florida escarpment cold seeps	Anderson (1989)
	Pachycara thermophilum	E	3020–3480	Mid-Atlantic Ridge hydrothermal vents, Snake pit, Logatchev, Broken Spur	Geistdoerfer (1994)

Pyrolycus manusanus	E	1623–1931	Manus Basin, Bismarck Sea	Machida & Hashimoto (2002)
Pyrolycus moelleri	E	1336	W Pacific, Brothers seamount Kermadec Ridge	Anderson (2006)
Thermarces andersoni	E	2620	East Pacific Rise	Rosenblatt and Cohen (1986)
Thermaces cerberus	E	2630	East Pacific Rise Lucky Strike	Rosenblatt and Cohen (1986)
Thermarces pelophilum	E	1947	NW Atlantic: cold seeps of the Barbados accretionary complex	Geistdoerfer (1999)
PLEURONECTIFORMES				
Cynoglossidae				
Symphurus maculopinnis	E	561	SW Pacific Tonga Arc	Munroe et al. (2011)
Symphurus thermophilus	E	215–733	NW Pacific, Mariana Arc, Bonin Arc and the Okininawa Trough	Munroe and Hashimoto (2008)
Symphurus species A	E	83–562	SW Pacific Tonga & Kermadec Ridges	Tunnicliffe et al. (2010)

proportions of PUFA than the muscle or liver tissue, showing that it is preferentially passed on from mother to offspring because it is essential for early development.

Although hydrothermal vent fishes are likely to experience higher temperatures than the surrounding ambient conditions and are often described as swimming in the shimmering water of distinct refractive index, they do not directly contact the very high temperatures (350°C) encountered in fluid plumes. Fishes tend to swim in waters of no more than 5–10°C. For respiration, the main challenges are the effects of hypoxia, low pH, high CO and high H_2S levels that can reduce or block aerobic metabolism. Weber et al. (2003) show that the haemoglobin of *T. ceberus* is relatively temperature insensitive and has a very high oxygen affinity compared with that of shallow-water eelpouts and other deep-sea fishes. Because zoarcids have no swim bladder, the haemoglobin shows no root effect as found in deep-sea fishes with swim bladders such as *Antimora* and *Coryphaenoides*. Dahlhoff et al. (1990) also found that the Lactate dehydrogenase (LDH) in the muscles of the zoarcid *Thermarces andersoni* is insensitive to temperature change from 5–20°C, but the Bythitid *Thermichthys hollisi* is sensitive to high temperatures, suggesting *T. andersoni* is the more truly vent-endemic fish with appropriate metabolic adaptations.

It is likely that most vent fishes can survive away from vent systems, but there is clear evidence that several species have physiological and behavioural adaptations that enable them to exploit this patchy environment more effectively than irregular visitors. In this sense they can be regarded as vent-endemic species.

5.6 The Circumglobal Deep-Sea Fish Fauna

Briggs (1960) first published a checklist circumtropical fishes and found only 107 species, a miniscule proportion of the global fish fauna that is truly circumglobal. Gaither et al. (2016) have updated this to 284 circumglobal species. This analysis excludes Southern Ocean and Antarctic circumglobal species, which, while appearing impressive in a Mercator map projection, in reality simply circulate in a single ocean system. Truly global fish species that occur in all three major ocean basins at low latitudes are rare, and deep-sea species account for over 80 per cent of the species listed by Gaither et al. (2016). These include representatives from 55 families: Hexanchidae (3), Dalatiidae (1), Squalidae(2), Odontaspididae (1), Mitsukurinidae (1), Cetorhinidae (1), Halosauridae (2), Notacanthidae (1), Synaphobranchidae (1), Derichthyidae (1), Bathylagidae (2), Nemichthyidae (4), Platytroctidae (3), Alepocephalidae (8), Microstomatidae (1), Opisthoproctidae (3), Gonostomatidae (12), Sternoptychidae (11), Phosichthyidae (4), Ipnopidae (1), Giganturidae (1), Scopelarchidae (3), Omosudidae (1), Paralepididae (2), Neoscopelidae (2), Myctophidae (47), Lampridae (1), Trachipteridae (1), Regalicidae (1), Stylephoridae (1), Macrouridae (5), Moridae(2), Melanonidae (1), Ophidiidae (3), Melamphaidae (7), Anoplogastridae (1), Diretmidae (1), Trachichthyidae (1), Rondeletiidae (1), Barbourisiidae (1), Cetomimidae (2), Centriscidae (2), Setarchidae (2), Chiasmodontidae (6), Gempylidae (6), Trichiuridae (2), Nomeidae (5), Tetragonuridae (1), Tetraodontidae (2), Molidae (1), Caulophrynidae (1), and the Ceratioid anglerfishes Melanocetidae (1), Oneirodidae (5), Gigantactinidae (2) and Linophrynidae (1). About 25 per cent of deep-sea families have circumglobal species, and it is amongst the pelagic families, Myctophidae, Gonostomatidae and Sternoptychidae that there is the greatest number species, presumably because of the lack of barriers to gene flow. There is a possibility that some putative circumglobal species are in fact species complexes. Compared with other habitats, deep-sea regions are relatively well interconnected, particularly in the pelagic and abyssal realms, giving rise to a disproportionally high number of widespread species, termed by Gaither et al. (2016): 'Fishes that rule the world'!

6 Deep-Sea Fisheries and Conservation

6.1 Global Trends in Deep Fishing

Small-scale fisheries have probably exploited deep-sea fishes for centuries. In those places where the seafloor slopes down steeply from the coast, deep-sea fishing can be carried out from small boats. Such traditional artisanal fisheries in the N Atlantic have been documented off Madeira (Lowe, 1843–1860), the mainland of Portugal (Wright, 1870) and fishing through ice in Arctic Greenland for *Reinhardtius hippoglossoides* (Greenland halibut; Rink, 1852). Lines were made of natural fibres, hemp (*Cannabis sativa*) grown in Europe, manila hemp (*Musa textilis*) from SE Asia, sisal (*Agave sisalana*) native to Mexico, olona (*Touchardia latifolia*) endemic to Hawaii and whalebone baleen (*Eubalaena glacialis*) off Greenland. Polynesian fishermen of the South Pacific traditionally used wooden hooks up to 23 cm long for catching *Ruvettus pretiosus* (oilfish; Merrett and Haedrich, 1997). Foster (1973) studied such incurved hooks and found that for certain species of deep-sea fish they were more effective than standard modern hooks. A traditional deep-bottom fishery from small boats is widespread around Pacific Islands fishing down to 300–400 m depth using baited hooks targeting ruby snapper, *Etelis carbunculus*, deep-water Epinephelinae, Gempylidae and oilfish *Ruvettus pretiosus* aided by modern equipment and materials (Preston et al., 1999). Fishermen of Madeira continued to use traditional sisal lines until the 1980s, when they switched to synthetic monofilament lines. This change increased the catch per hook, enabled deployment of more hooks and exploitation of new areas (Martins and Fereira, 1995). This kind of change has been replicated around the world with improvements in the propulsion of boats, equipment as well as materials.

The major growth in deep-sea fishing came with massive growth of the fishing fleet of the Soviet Union from 1946 onward so that by 1967 it had the world's largest national fishing fleet with 11 per cent of the world's total landings, 5.8 Mt (Sealy, 1974). The fleet included large freezer trawlers, as well as mother ships with flotillas of catching vessels, tankers, fish carriers and supply ships. This enabled fishing to be conducted on a global basis, and from 1965 a research and development programme developed new techniques for fishing at depths greater than 500 m. From 1969 the USSR was responsible for almost the entire world catch of macrourids in the N Atlantic, reaching 82,600 t in 1971. The commercial operations were supported by exploratory fishing and research cruises in the Atlantic (Troyanovsky and Lisovsky 1995), the Pacific Ocean (Tuponogov et al., 2008) and elsewhere. By the 1970s the USSR faced competition from other fishing nations; although the Japanese fleet was smaller in terms of total fishing vessel tonnage, its catches were greater. Sealy (1974) reported that although the USSR began developing trawling capability down to 2000 m, the planned growth in fish production did not materialise. Rewards from fishing deeper waters were meagre, and coastal states were imposing 200-mile exclusive fishing zones that gradually excluded the USSR and other distant-water fleets from continental slope areas around the world. Chile, Ecuador and Peru declared 200-mile zones in 1952, followed by most other South American countries by 1970 and most of the world by 1978. These unilateral declarations were confirmed by the United Nations Convention on the Law of the Sea (UN, 1982) giving coastal states the rights to a 200-mile Exclusive Economic Zone (EEZ). Coastal states such as New Zealand then began exploiting deep-water fisheries that had been discovered and developed initially by the USSR and Japan (NZ, 2009).

6.2 Types of Fishing

Fishing both for shallow- and deep-water species uses a variety of methods that depend on the behaviour of the species concerned. As some of these methods have already been discussed in Chapter 1, key points only are dealt with here.

6.2.1 Baited Traps and Lines

Attracting fish to bait is one of the most ancient methods used by humans and continues to be widely used with longliners capable of deploying thousands of hooks at a time either on the seafloor or in the water column. The method is very effective for most sharks, rays and many gadiform fishes. However, species that naturally feed on pelagic live prey and/or gelatinous zooplankton, or specialise in certain types of benthic invertebrates, are not attracted to carrion. Priede et al. (2010) found that only 20 per cent of deep-sea fish species captured by bottom trawl on the slopes of the Porcupine Seabight in the NE Atlantic are attracted to baits. On the Mid-Atlantic Ridge baited longlines caught only 50 per cent of the species captured by trawl (Bergstad et al., 2008). Several commercially important deep-sea species including *Coryphaenoides rupestris* (Roundnose grenadier) and *Hoplostethus atlanticus* (orange roughy) are never attracted to baits, an important consideration in discussion of possible phasing out of deep-sea trawling.

6.2.2 Static Nets

Bottom set trammel nets, gill nets or entangling nets have been utilised in the NE Atlantic Ocean for the capture of hake (*Merluccius merluccius*) at 100–600 m depth, angler- or monkfish (*Lophius piscatorius*, *Lophius budegassa*) at 100–800 m depth, and deep-water sharks (*Centrophorus squamosus*, *Centroscymnus coelolepis*) at 800–1600 m depth. Each vessel can deploy several fleets of nets with a total of up to 100 km of net per trip. Over 60 vessels were involved in the hake fishery. This method is highly effective and to some extent size-selective depending on the mesh size; nevertheless a bycatch of up to 15 nontarget species has been recorded. A major concern is that gear that is lost or abandoned through accident or adverse weather will continue 'ghost fishing'. Regulations have been issued in an attempt to ensure that vessels remain in attendance near the gear throughout its soak time of 50–120 h, but it is tempting for skippers to leave nets in place while landing previous catches in port (CEC, 2006). Deep-sea gillnetting is widely perceived as a potentially destructive fishing method, and measures are being taken in a number of jurisdictions to restrict or prohibit the practice.

6.2.3 Pelagic Trawling

With modern scanning sonars and large pelagic trawl nets equipped with sensors, fishing vessels are able to target and enclose schools of fishes to catch large quantities of a single species in a relatively short time. This is the method used for northern and southern blue whiting (*Micromesistius poutassou*, *M. australis*) and Greater argentine (*Argentina silus*), also known as the Greater silver smelt). The fish in the school swim close together and are herded by the net, the outer panels of which may have very large mesh sizes to reduce drag. Despite the fact that the mesh size may be several metres, much larger than the fish, visual and vibration stimuli cause the fish to concentrate in the centre of the net, enabling capture in the smaller meshes of the cod end (Wardle, 1983). This kind of gear is capable of highly selective removal of an almost predictable quantity of fish from the sea with no collateral damage from seafloor contact or bycatch. However, the fishing power of such vessels is so great that they should be considered as similar to their equivalent on land, combine harvesters, which are kept in storage for most of the year and only used during a relatively short harvesting season. Regulation of their deployment needs to be draconian, although it is potentially a benign form of fishing with no effects on the environment other than the removal of target species.

Deeper-living pelagic species such as lantern fishes (myctophidae) have no schooling habit, tending to be dispersed in the sound-scattering layer until disturbed (Barham, 1966). They do not have the sensory reflexes that could fool them into being herded into the mouth of a trawl cod end. Conversely Kaartvedt et al. (2012) show that mycotophids, *Benthosema glaciale* (Glacier lantern fish) have extremely effective escape reflexes that enable them to avoid pelagic nets and potentially predatory fishes. Fishing for species of the deep-scattering layer therefore requires towing at depths with the highest possible densities for long periods of time to accumulate sufficient catch and is quite distinct from the corralling technique used for schooling pelagic fishes.

6.2.4 Bottom Trawling

This is conducted mainly using the classic otter trawl towed along the sea bed. For nonaggregating bottom-living species such as rays, flatfish and some gadiforms, the technique entails choosing a suitable trawl track on the seafloor that avoids obstructions, possibly following a depth contour at an appropriate depth. The net is then towed for the requisite time so that sufficient area is covered by the swath to realise a worthwhile catch. If a virgin seafloor is being fished, this is obviously extremely damaging as it removes everything standing above the seafloor. In heavily fished shelf areas, because the trawl tracks become smoothed and established as preferred fishing areas, fish are caught as they cross the trawling 'highway' and indeed may be attracted to the disturbances created by previous fishing activity (Kaiser and Spencer, 1994). In some shelf sea areas it has been suggested that the seafloor has become a modified environment, analogous to pastures on land, so that continued trawling is necessary to maintain productivity (van Denderen et al., 2013). In the deep sea, bottom trawling causes irreversible damage to the ecosystem, which probably does not have the resilience of a shallow-water system. High levels of discarding of bycatch are also a concern.

Instead of towing for tens of kilometres to accumulate a catch, for some species bottom trawling can be targeted at aggregations of fish or 'marks' visible on the sonar. This is appropriate for benthopelagic species that may congregate around topographic features such as the summit of seamounts, canyons or ridges. Species including orange roughy (*Hoplostethus atlanticus*), pelagic armorheads (*Pentaceros* spp.) and Alfonsino (*Beryx* spp.) behave in this way, and a skilled fishing skipper may be able to capture fish with the trawl making no contact at all with the bottom. However, because fishing activities are carried out repeatedly at favoured fish aggregation sites, localised damage to the seafloor is inevitable though not as widespread as with traditional bottom trawling. The latest semipelagic trawls allow the mouth of the trawl to touch the seafloor while the otter boards remain in the water column.

6.3 Global Trends in Fish Catch

The total world marine fish catch in 1950 was ca. 16.7 Mt (million tonnes), increasing rapidly to 60 Mt by 1970 when there was a hiatus caused by collapses in some major stocks such as herring (*Clupea harengus*) in the North Sea and the Peruvian anchovy (*Engraulis ringens*) in the SE Pacific (Fig. 6.1a). From 1975 onward, growth in world catches resumed partly through exploitation of new resources including the deep sea (Fig. 6.1b), reaching a peak of 87.7 Mt in 1996. Since then production has fluctuated though with a general downward trend to about 80 Mt (Ye and Cochrane, 2011). Global deep-sea fisheries first reached 1 Mt around 1978, increasing to a peak of about 3.7 Mt in 2003, when it was followed by a decline. Deep-sea fisheries therefore increased from less than 1 per cent of world catch in the 1950s to an average of about 3.5 per cent in the first decade of the twenty-first century.

Morato et al. (2006) showed that from 1950 onward there was a gradual increase in the mean depth of fishing by the global fleet, creeping deeper at a rate of about 1 m per decade. For bottom-living fishes only,

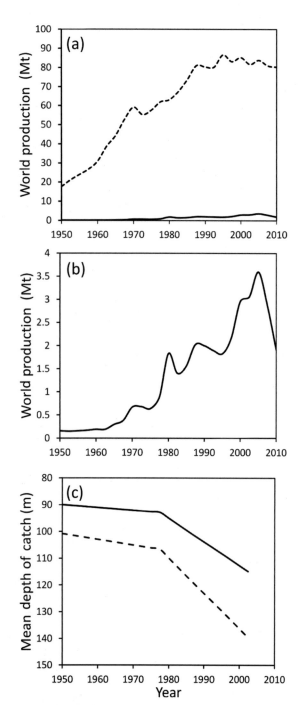

the rate was faster, over 2 m per decade. From 1978 onward the shift to the deep sea accelerated to 9 m and 13 m per decade, respectively (Fig. 6.1c). Because deep-sea fishing only accounts for a low percentage of the global catch, it is remarkable that these trends can be discerned in the global dataset. For high seas demersal fishes beyond the EEZs, the mean depth of fishing since 1950 has increased by 22 m per decade. Deep-sea fisheries now reach down to a depth of about 2000 m (Tandstad et al., 2011).

Global marine fisheries exploit about 3000 different species of fish (FishBase, 2016), and comparing their depth and length distribution (Fig. 6.2) with the overall distribution of fishes, it is evident that fishing is targeting large body size and shallow depth. Few species with a body length less than 10 cm are fished commercially, and the average length of exploited species is 70.5 cm (SD ± 86.9) compared with 35.2 cm (SD ± 54.3) for the global marine fish fauna. Amongst exploited deep-sea fishes the largest include *Somniosus microcephalus* (Greenland shark, maximum length 7.3 m), *Hexanchus griseus* (Bluntnose sixgill shark, 4.82 m), *Ruvettus pretiosus* (oilfish, 3 m), *Dissostichus eleginoides* (Patagonian toothfish, 2.15 m) and *Genypterus blacodes* (pink cusk-eel, 2 m). Fishing extends to species with maximum depths greater than 3000 m. The deepest exploited species are *Spectrunculus grandis* (Pudgy cusk eel, 4300 m_{fbd}) globally distributed but occasionally caught; *Gymnoscopelus bolini* (Grand lanternfish, 4200 m_{fbd}) minor catches around the Antarctic; *Dissostichus eleginoides* (Patagonian toothfish, 3850 m_{fbd}) highly commercial around the Antarctic; *Isistius brasiliensis* (cookie cutter shark, 3700 m_{fbd}) minor fisheries circumglobal; *Centroscymnus coelolepis* (Portuguese dogfish, 3700 m_{fbd}) minor circumglobal and two grenadier species from the North Pacific *Albatrossia pectoralis* (giant grenadier, 3500 m $_{fbd}$) and *Coryphaenoides cinereus*

Figure 6.1 Trends in the world catch of marine fishes as recorded by FAO.

(a) Dashed line: total marine species captured (after Ye and Cochrane, 2011). Solid line: total deep-sea species (after Tandstad et al., 2011).

(b) Total deep-sea species captured, same data is in solid line above in (a).

(c) Mean depth of capture (after Morato et al., 2006). Solid line: all species. Dashed line: demersal species only.

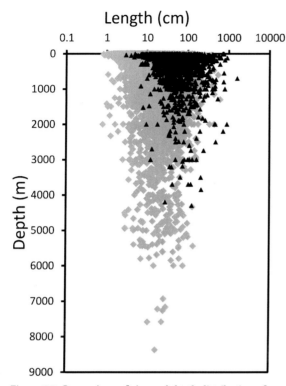

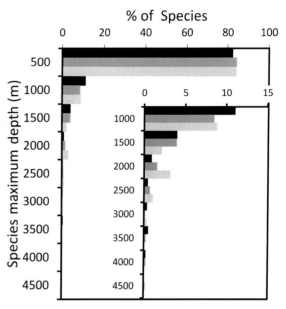

Figure 6.2 Comparison of size and depth distribution of exploited and unexploited fish species. Each point represents a species maximum recorded depth and maximum length from FishBase (Froese and Pauly, 2016). Grey symbols: unexploited species (n = 11753), Black symbols; exploited species (n = 2915). Note that fisheries exploit predominantly shallower and larger species.

Figure 6.3 The depth distribution of exploited fish species by maximum depth of occurrence.
Black: Commercial or highly commercial species (n = 1329). Dark grey: minor commercial species (n = 1268). Pale grey: subsistence fisheries only (n = 281).

(Popeye grenadier, 3500 m$_{fbd}$). Fisheries for these and other species are carried out predominantly at much shallower depths, but Figure 6.2 indicates that the effects of deep-sea fisheries can now reach far down the slopes to the margins of the abyssal plains.

Exploited fish species can be conveniently divided into three categories: commercial species for which there may be a directed fishery, minor commercial species caught as a bycatch or in small-scale fisheries and subsistence species exploited only by artisanal fishermen. Over 80 per cent of exploited species in all three categories have maximum depths < 500 m (Fig. 6.3). A total of 474 exploited species have maximum depths > 500 m, comprising 233 commercial species, 195 minor commercial and

44 subsistence species. Subsistence fishermen also compete for and also utilise commercial species. It is interesting to note that the depth trends of all three groups are similar, with subsistence fishermen exploiting species living as far down as 3000 m depth. Exploitation of deep-sea fishes therefore is not solely the monopoly of large- scale commercial operations, although they land the greatest tonnages.

6.4 Composition of the World Deep-Sea Fish Catch

There are probably over 300 species of exploited deep-sea fish. Table 6.1 lists all the marine deep-water species in FishBase (Froese and Pauly, 2016) for which there are records of exploitation plus some species regarded as comprising part of the global deep-sea catch recorded by FAO (Tandstad et al., 2011). There are 276 individually identified species as well as some mixed-species categories such as Ratfishes, Boarfishes

Table 6.1 **Deep-Sea exploited fishes**

ORDER / Family	Scientific Name	Common Name	Depth (m) min	Depth (m) max	AREA	Impor- tance	FAO catch (tonnes)
MYXINIFORMES							
Myxinidae	Eptatretus atami	Brown hagfish	300	536	NW Pacific	***	
	Eptatretus cirrhatus	Broadgilled hagfish	0	1,100	W Pacific	**	
	Eptatretus stouti	Pacific hagfish	16	966	NE Pacific	**	
	Myxine glutinosa	Atlantic hagfish	30	1,200	N Atlantic	**	
	Myxinidae	Hagfishes NEI				**	2,335
CHIMAERIFORMES							
Callorhinchidae	Callorhinchus capensis	Cape elephantfish	10	374	SE Atlantic	**	679
	Callorhinchus milii	Ghost shark	0	227	SW Pacific	***	1,426
	Callorhinchidae	Elephantfishes NEI					
	Chimaera monstrosa	Rabbit fish	40	1,400	E Atlantic	**	284
	Chimaera phantasma	Silver chimaera	90	540	W Pacific	***	1
	Hydrolagus novaezealandiae	Dark ghost shark	25	950	SW Pacific	**	2,083
	Hydrolagus ogilbyi	Hydrolagus	120	350	Australia	**	
	Hydrolagus spp.	Ratfishes NEI					1,308
	Chimaeriformes	Chimaeras, etc. NEI					122
SQUATINIFORMES							
Squatinidae	Squatina dumeril	Atlantic angel shark	128	1,375	W Atlantic	**	
HEXANCHIFORMES							
Chlamydoselachidae	Chlamydoselachus anguineus	Frilled shark	0	1,570	Atlantic, Indian, Pacific	**	2
Hexanchidae	Heptranchias perlo	Sharpnose sevengill shark	0	1,000	Circumglobal	**	
	Hexanchus griseus	Bluntnose sixgill shark	1	2,500	Circumglobal	***	22
	Hexanchus nakamurai	Bigeyed sixgill shark	0	600	Atlantic, Indian, Pacific	*	
SQUALIFORMES							
Echinorhinidae	Echinorhinus brucus	Bramble shark	10	900	Atlantic, Indian, Pacific	**	1
Dalatiidae	Dalatias licha	Kitefin shark	37	1,800	Atlantic, Indian, Pacific	***	628
	Isistius brasiliensis	Cookie cutter shark	0	3,700	Circumglobal	**	
Oxynotidae	Oxynotus centrina	Angular roughshark	40	777	E Atlantic, Mediterranean	**	62
	Oxynotus paradoxus	Sailfin roughshark	265	720	E Atlantic	**	1

Family	Species	Common name			Distribution		
Centrophoridae	*Centrophorus granulosus*	Gulper shark	50	1,440	Atlantic, Indian, Pacific	**	268
	Centrophorus lusitanicus	Lowfin gulper shark	300	1,400	E Atlantic, Indian, W Pacific	**	348
	Centrophorus moluccensis	Smallfin gulper shark	125	823	Indo-W Pacific	**	996
	Centrophorus squamosus	Leafscale gulper shark	145	2,400	E Atlantic, Indian, W Pacific	**	
	Centrophorus uyato	Little gulper shark	50	1,400	Atlantic, Indian, W Pacific	**	224
	Deania calcea	Birdbeak dogfish	60	1,490	E Atlantic, Indian, Pacific	**	
	Deania hystricosa	Rough longnose dogfish	600	1,000	NE Atlantic, NW Pacific	**	
	Deania quadrispinosa	Longsnout dogfish	150	1,360	SE Atlantic, S Indo-W Pacifc	**	
Squalidae	*Squalus japonicus*	Japanese spurdog	150	300	W Pacific	**	
	Squalus megalops	Shortnose spurdog	30	750	E Indian	***	
Somniosidae	*Centroscymnus coelolepis*	Portuguese dogfish	150	3,700	Atlantic, Indo-W Pacific	**	1,346
	Centroscymnus crepidater	Longnose velvet dogfish	230	1,500	E Atlantic, Indian, Pacific	**	162
	Centroscymnus owstonii	Roughskin dogfish	100	1,500	Atlantic, Indian, Pacific	**	
	Scymnodon ringens	Knifetooth dogfish	200	1,600	E Atlantic, SW Pacific	**	
	Somniosus microcephalus	Greenland shark	0	2,200	Arctic, N Atlantic	**	41
	Somniosus rostratus	Little sleeper shark	200	1,330	NE Atlantic, Mediterranean	**	1
Etmopteridae	*Centroscyllium fabricii*	Black dogfish	180	1,600	Atlantic	*	32
	Etmopterus hillianus	Caribbean lanternshark	180	717	W Atlantic	*	
	Etmopterus polli	African lanternshark	300	1,000	E Atlantic	**	
	Etmopterus spinax	Velvet belly	200	2,490	E Atlantic, Mediterranean	**	14
	Etmopterus NEI						67

CARCHARHINIFORMES

Family	Species	Common name			Distribution	
Triakidae	*Iago garricki*	Longnose houndshark	250	475	W Pacific	**
	Iago omanensis	Bigeye houndshark	110	2,195	W Indian	*
Pseudotriakidae	*Pseudotriakis microdon*	False catshark	173	1,890	Atlantic, Indian, Pacific	*
Scyliorhinidae	*Apristurus herklotsi*	Longfin catshark	533	864	W Pacific	**
	Apristurus japonicus	Japanese catshark			NW Pacific	**
	Apristurus kampae	Longnose catshark		1,888	NE Pacific	*
	Apristurus longicephalus	Longhead catshark	680	950	Indo-W Pacific	**
	Apristurus parvipinnis	Smallfin catshark	600	1,380	W Atlantic	*
	Figaro boardmani	Australian sawtail catshark	128	823	Indo-W Pacific	**
	Galeus eastmani	Gecko catshark			W Pacific	***
	Galeus polli	African sawtail catshark	200	720	E Atlantic	**

Table 6.1 (cont.)

ORDER Family	Scientific Name	Common Name	Depth (m) min	max	AREA	Importance	FAO catch (tonnes)
MYLIOBATIFORMES							
Plesiobatidae	*Plesiobatis daviesi*	Deep-water stingray	44	780	Indo-Pacific	**	
Urolophidae	*Urolophus expansus*	Wide stingaree	140	420	Australia	*	
RAJIFORMES							
Rajidae	*Amblyraja georgiana*	Antarctic starry skate	20	350	Southern Ocean	*	
	Amblyraja taaf	Whiteleg skate	300	500	S Indian	*	
	Bathyraja meridionalis	Dark-belly skate	65	2,240	Southern Ocean	**	0
	Bathyraja spp. NEI	Bathyraja rays					4
	Dipturus doutrei	Violet skate	163	1,200	E Atlantic	***	
	Dipturus oxyrinchus	Longnosed skate	15	900	NE Atlantic, Mediterranean	**	
	Leucoraja fullonica	Shagreen ray	30	550	NE Atlantic, Mediterranean	**	
	Raja maderensis	Madeiran ray		500	E Atlantic	***	
	Raja rhina	Longnose skate	9	1,069	E Pacific	**	1
ALBULIFORMES							
Albulidae	*Pterothrissus belloci*	Longfin bonefish	20	500	E Atlantic	**	
OSMERIFORMES							
Alepocephalidae	*Alepocephalus bairdii*	Baird's slickhead	365	1,700	N Atlantic	**	2,680
ARGENTIFORMES							
Argentinidae	*Argentina silus*	Greater argentine	140	1,440	N Atlantic	***	3,865
	Argentina sphyraena	Argentine	50	700	NE Atlantic, Mediterranean	**	2,756
	Argentina spp.						30,476
	Glossanodon semifasciatus	Deep-sea smelt	70	1,017	NW Pacific	***	4,370
STOMIIFORMES							
Sternoptychidae	*Maurolicus muelleri*	Silvery lightfish	271	1,524	Atlantic, Pacific	**	14,840

Order / Family	Species	Common name			Distribution		
AULOPIFORMES							
Chlorophthalmidae							
	Chlorophthalmus agassizi	Shortnose greeneye	50	1,000	Circumglobal	**	
	Chlorophthalmus albatrossis	-	300	350	Indo-W Pacific	***	
	Chlorophthalmus borealis	-	244	262	W Pacific	***	
	Chlorophthalmus corniger	Spinyjaw greeneye	265	458	Indo-W Pacific	*	
	Chlorophthalmus punctatus	Spotted greeneye			SE Atlantic, SW Indian	*	
	Chlorophthalmidae	Greeneyes					
MYCTOPHIFORMES							
Myctophidae							
	Diaphus coeruleus	Blue lantern fish	457	549	Indo-W Pacific	**	
	Gymnoscopelus braueri	Brauer's lanternfish		2,700	Southern Ocean	**	
	Gymnoscopelus fraseri	Fraser's lanternfish	50	250	Southern Ocean	**	
	Gymnoscopelus nicholsi	Nichol's lanternfish	300		Southern Ocean	**	0
	Lampanyctodes hectoris	Hector's lanternfish	0		SE Atlantic, SW SE Pacific	**	7,288
	Benthosema pterotum	Skinnycheek lanternfish	10	300	Indo-W Pacific	**	7,216
	Myctophidae	Lanternfishes NEI					958
LAMPRIFORMES							
Lampridae	*Lampris guttatus*	Opah	100	500	Circumglobal	**	891
Trachipteridae	*Desmodema polystictum*	Polka-dot ribbonfish	0	500	Circumglobal	***	
	Trachipterus arcticus	Dealfish	300	600	N Atlantic		330
	Trachipterus trachypterus	Mediterranean dealfish	100	600	Atlantic, Mediterranean, Pacific	**	48
	Trachipturus spp.	Dealfishes					23
POLYMIXIIFORMES							
Polymixiidae	*Polymixia japonica*	Silver eye	160	628	N Pacific	***	
	Polymixia lowei	Beardfish	50	600	W Atlantic	**	
	Polymixia nobilis	Stout beardfish	100	770	Atlantic	***	1
ZEIFORMES							
Cyttidae	*Cyttus traversi*	King dory	200	978	SE Atlantic, S Indo-W Pacifc	***	299
Oreosomatidae	*Allocyttus niger*	Black oreo	560	1,300	SW Pacific	***	5,311
	Allocyttus verrucosus	Warty dory	0	1,800	Circumglobal	**	210
	Neocyttus rhomboidalis	Spiky oreo	200	1,240	S Atlantic, Indian, Pacific	**	113
	Pseudocyttus maculatus	Smooth oreo dory	400	1,500	S Atlantic, Indian, Pacific	***	10,254
	Oreosomatidae	Oreo dories NEI					
Parazenidae	*Cytopsis rosea*	Rosy dory	150	730	Atlantic, Indo-W Pacific	***	1,336

Table 6.1 (*cont.*)

ORDER Family	Scientific Name	Common Name	Depth (m) min	Depth (m) max	AREA	Impor-tance	FAO catch (tonnes)
GADIFORMES							
Macrouridae	*Albatrossia pectoralis*	Giant grenadier	140	3,500	N Pacific	***	
	Coelorinchus anatirostris	Duckbill grenadier	300	550	NW Pacific	**	
	Coelorinchus caribbaeus	Blackfin grenadier	200	700	W Atlantic	**	
	Coelorinchus chilensis	Chilean grenadier	260	1,480	SW Atlantic, SE Pacific	**	74
	Coelorinchus fasciatus	Banded whiptail	73	1,086	S Atlantic, Indian, Pacific	**	
	Coelorinchus innotabilis	Notable whiptail	554	1,463	SW Pacific	**	
	Coelorinchus japonicus	Japanese grenadier	300	1,000	NW Pacific	**	
	Coelorinchus kamoharai	Kamohara grenadier	220	400	NW Pacific	**	
	Coelorinchus kishinouyei	Mugura grenadier	200	600	W Pacific	**	
	Coelorinchus matamua	Mahia whiptail	450	1,000	SE Atlantic, SW Pacific	**	
	Coelorinchus occa	Swordsnout grenadier	400	2,220	Atlantic	*	
	Coelorinchus oliverianus	Hawknose grenadier	85	1,245	New Zealand	***	
	Coelorinchus parallelus	Spiny grenadier	630	990	Indo-W Pacific	**	
	Coelorinchus productus	Unicorn grenadier	271	600	W Pacific	**	
	Coryphaenoides acrolepis	Pacific grenadier	300	3,700	N Pacific	***	
	Coryphaenoides anguliceps	Loosescale grenadier	720	2,400	E Pacific	*	
	Coryphaenoides ariommus	Humboldt grenadier	768	1,860	E Pacific	*	
	Coryphaenoides cinereus	Popeye grenadier	150	3,500	N Pacific	**	
	Coryphaenoides delsolari	Trident grenadier	300	1,650	E Pacific	**	
	Coryphaenoides longifilis	Longfin grenadier	550	3,000	N Pacific	**	
	Coryphaenoides marginatus	Amami grenadier	250	790	NW Pacific	**	
	Coryphaenoides mexicanus	Mexican grenadier	110	1,600	W Atlantic	**	
	Coryphaenoides nasutus	Largenose grenadier		1,537	NW Pacific	**	
	Coryphaenoides rupestris	Roundnose grenadier	180	2,600	N Atlantic	***	14,889
	Coryphaenoides serrulatus	Serrulate whiptail	540	2,070	Indo-SW Pacific	**	
	Coryphaenoides subserrulatus	Longrayed whiptail	900	1,180	S Atlantic, Indian, Pacific	**	
	Cynomacrurus piriei	Dogtooth grenadier	500	3,800	Southern Ocean	**	0

Species	Common name			Region		
Lepidorhynchus denticulatus	Thorntooth grenadier	180	1,000	SW Pacific	**	
Lucigadus nigromaculatus	Blackspotted grenadier	200	1,463	S Indo-W Pacific	**	
Macrourus berglax	Roughhead grenadier	100	1,000	N Atlantic	***	1,622
Macrourus carinatus	Ridge scaled rattail	200	1,200	S Atlantic, Indian, Pacific	***	5,293
Macrourus holotrachys	Bigeye grenadier	300	1,400	SW Atlantic	**	12
Macrourus whitsoni	Whitson's grenadier	400	3,185	Southern Ocean	***	102
Macrourus spp.	Grenadiers NEI					2,152
Nezumia aequalis	Common Atlantic grenadier	200	2,320	Atlantic 70°N – 12°S	***	25
Nezumia atlantica	Western Atlantic grenadier	366	1,097	W Atlantic	**	
Nezumia brevibarbata	Shortbeard grenadier	549	1,737	SE Atlantic	*	
Nezumia condylura	Japanese pugnose grenadier	200	720	NW Pacific	**	
Nezumia convergens	Peruvian grenadier	600	1,870	E Pacific	*	
Nezumia duodecim	Twelve-rayed grenadier	329	1,261	E Atlantic	**	
Nezumia latirostrata	Broadsnout grenadier	589	1,400	E Pacific	*	
Nezumia liolepis	Smooth grenadier	768	1,660	E Pacific	*	
Nezumia loricata	Parrot grenadier	600	1,500	E Indian	**	
Nezumia orbitalis	Spectacled grenadier	523	800	E Pacific	***	
Nezumia proxima	Short-tail grenadier	355	910	W Pacific	***	
Nezumia pudens	Atacama grenadier	580	1,238	SE Pacific	**	
Nezumia pulchella	Thumb grenadier	250	960	SE Pacific	**	
Nezumia sclerorhynchus	Roughtip grenadier	130	3,200	Atlantic	***	
Nezumia stelgidolepis	California grenadier	277	909	E Pacific	**	
Bathygadus macrops	Bullseye grenadier	200	777	Atlantic	*	
Malacocephalus laevis	Softhead grenadier	200	1,000	Atlantic, Indian, Pacific	***	
Malacocephalus occidentalis	Western softhead grenadier	140	1,945	Atlantic	***	
Sphagemacrurus grenadae	Pugnose grenadier	1,000	1,500	W Atlantic	*	
Trachonurus sulcatus	Bristly grenadier	700	1,500	N Atlantic	**	
Ventrifossa divergens	Plainfin grenadier	183	772	Indo-W Pacific	***	
Ventrifossa garmani	Sagami grenadier	200	720	NW Pacific	***	
Ventrifossa macropogon	Longbeard grenadier	439	1,000	W Atlantic	*	
Ventrifossa mucocephalus	Slimehead grenadier	450	732	W Atlantic	*	
Ventrifossa nigrodorsalis	Spinaker grenadier	270	700	W Pacific	**	
Ventrifossa petersonii	Peterson's grenadier	296	1,019	Indian	**	
Macrouridae	Grenadiers, rattails NEI					24,313

Table 6.1 (cont.)

ORDER Family	Scientific Name	Common Name	Depth (m) min	max	AREA	Impor- tance	FAO catch (tonnes)
Macruronidae	*Macruronus magellanicus*	Patagonian grenadier	30	500	SW Atlantic, SE Pacific	****	193,586
	Macruronus novaezelandiae	Blue grenadier	0	1,000	SW Pacific	****	136,268
Trachyrinchidae	*Trachyrincus scabrus*	Roughsnout grenadier	395	1,700	E Atlantic, Mediterranean	**	372
	Trachyrincus villegai	Grey grenadier	250	980	SE Pacific	**	
Moridae	*Antimora microlepis*	Finescale mora	175	3,048	N Pacific	*	
	Antimora rostrata	Blue antimora	1,371	75	Circumglobal except N Pacific	**	133
	Laemonema longipes	Longfin codling	80	1,830	N Pacific	***	
	Lepidion lepidion	Mediterranean codling	500	2,307	Mediterranean	**	
	Mora moro	Common mora	450	2,500	Atlantic, Indian, Pacific	**	1,260
	Physiculus japonicus	Japanese codling	139	1,007	NW Pacific	*	
	Physiculus roseus	Rosy cod	300	549	Indo-Pacific	**	
Merlucciidae	*Merluccius albidus*	Offshore silver hake	80	1,170	W Atlantic	**	16
	Merluccius capensis	Shallow-water Cape hake	50	1,000	SE Atlantic	**	2,369
	Merluccius gayi	South Pacific hake	50	500	SE Pacific	****	89,195
	Merluccius capensis/paradoxus	Cape hakes	50	1,000	SE Atlantic, SW Indian	****	285,655
	Merluccius polli	Benguela hake	50	910	E Atlantic	**	3,664
Gadidae	*Arctogadus glacialis*	Arctic cod	0	1,000	Arctic, NE Atlantic	**	
	Micromesistius australis	Southern blue whiting	50	900	SW Atlantic, SE SW Pacific	****	95,054
	Micromesistius poutassou	Blue whiting	150	3,000	N Atlantic	****	1,357,076
	Brosme brosme	Tusk	18	1,000	N Atlantic	****	25,450
	Gaidropsarus argentatus	Arctic rockling	150	2,260	N Atlantic	***	
	Molva dypterygia	Blue ling	150	1,000	N Atlantic, W Mediterranean	***	9,755
	Molva molva	Ling	100	1,000	N Atlantic, W Mediterranean	****	38,371
	Phycis blennoides	Greater forkbeard	10	1,047	NE Atlantic, Mediterranean	***	3,390

Taxon	Common name	Min	Max	Distribution		Catch
OPHIDIIFORMES						
Ophidiidae						
Genypterus blacodes	Pink cusk-eel	22	1,000	SW Atlantic, SE, SW Pacific	****	42,310
Genypterus capensis	Kingklip	50	500	SE Atlantic	***	8,477
Genypterus chilensis	Red cusk-eel			SE Pacific	***	534
Genypterus maculatus	Black cusk-eel			SE Pacific	***	608
Genypterus spp.	Cusk-eels NEI					556
Homostolus acer	Filament cusk	300	1,000	W Pacific	**	
Hoplobrotula armata	Armoured cusk	200	350	W Pacific	***	
Spectrunculus grandis	Pudgy cuskeel	800	4,300	Atlantic, Pacific	**	
Ophidiidae	Cusk-eels, brotulas NEI					879
STEPHANOBERYCIFORMES						
Melamphaidae						
Sio nordenskjoldii	Nordenskjold's bigscale	200	3,000	S Atlantic, Indian, Pacific	**	
BERYCIFORMES						
Trachichthyidae						
Gephyroberyx darwinii	Darwin's slimehead	9	1,210	Atlantic, Indo-W Pacific	***	
Hoplostethus atlanticus	Orange roughy	180	1,809	Atlantic, Indian, Pacific	****	17,002
Hoplostethus cadenati	Black slimehead	70	1,000	E Atlantic, NW Indian	***	
Trachichthyidae	Slimeheads NEI					
Berycidae						
Beryx decadactylus	Alfonsino	110	1,000	Atlantic, Indian, Pacific	***	2,194
Beryx splendens	Splendid alfonsino	25	1,300	Circumglobal	***	326
Beryx spp.	Alfonsinos NEI					7,462
Centroberyx affinis	Redfish	10	450	W Pacific	***	
Ostichthys kaianus	Kai soldierfish	310	640	Indo-Pacific	***	
Holocentridae						1,106
CETOMIMIFORMES						
Barbourisiidae						
Barbourisia rufa		25	600	Circumglobal	**	210
SYNGNATHIFORMES						
Centriscidae						
Macroramphosus scolopax	Longspine snipefish	25	600	Circumglobal	***	74

Table 6.1 (cont.)

ORDER Family	Scientific Name	Common Name	Depth (m) min	max	AREA	Importance	FAO catch (tonnes)
SCORPAENIFORMES							
Sebastidae	*Helicolenus avius*	-	450	600	NW Pacific	***	
	Helicolenus dactylopterus	Blackbelly rosefish, Bluemouth	50	1,100	Atlantic, Mediterranean	***	6,048
	Helicolenus lengerichi	-			SE Pacific	**	
	Sebastes aleutianus	Rougheye rockfish	25	900	N Pacific	***	
	Sebastes alutus	Pacific ocean perch	0	825	N Pacific	****	34,000
	Sebastes aurora	Aurora rockfish	124	769	NE Pacific	***	
	Sebastes diploproa	Splitnose rockfish	0	800	NE Pacific	**	14
	Sebastes flammeus	-	300	500	NW Pacific	***	
	Sebastes iracundus	-	300	1,400	NW Pacific	***	
	Sebastes matsubarai	-	310	383	NW Pacific	***	
	Sebastes mentella	Beaked redfish	300	1,441	N Atlantic	***	67,992
	Sebastes proriger	Redstripe rockfish	12	425	N Pacific	***	
	Sebastes reedi	Yellowmouth rockfish	137	375	NE Pacific	***	
	Sebastolobus macrochir	Broadbanded thornyhead	257	1,537	N Pacific	***	
	Trachyscorpia cristulata echinata	Spiny scorpionfish	200	2,500	NE Atlantic, Mediterranean	***	26
Setarchidae	*Ectreposebastes imus*	Midwater scorpionfish	150	2,000	Circumglobal	**	
Scorpaenidae	*Pontinus kuhlii*	Offshore rockfish	100	600	NE Atlantic, Mediterranean	***	67
	Scorpaena scrofa	Red scorpionfish	20	500	E Atlantic, Mediterranean	***	43
Triglidae	*Pterygotrigla picta*	Spotted gurnard	160	500	S Pacific	*	193
	Trigla lyra	Piper gurnard	100	700	NE Atlantic, Mediterranean	**	
Anoplopomatidae	*Anoplopoma fimbria*	Sablefish	175	2,740	N Pacific	****	24,342
	Erilepis zonifer	Skilfish	0	680	N Pacific	***	165
Hoplichthyidae	*Hoplichthys haswelli*	Armoured flathead	140	700	SW Pacific	**	
Cottidae	*Hemilepidotus jordani*	Yellow Irish lord	257	604	N Pacific	***	
PERCIFORMES							
Acropomatidae	*Acropoma japonicum*	Glowbelly	100	500	Indo-W Pacific	***	
	Synagrops adeni	Aden splitfin	60	600	Indian	***	
Serranidae	*Hyporthodus mystacinus*	Misty grouper	30	400	W Atlantic	***	
	Hyporthodus octofasciatus	Eightbar grouper	150	300	Indo-W Pacific	**	

Family	Species	Common name			Distribution		
Polyprionidae	*Polyprion americanus*	Wreckfish	40	600	Atlantic, Mediterranean, SW Pacific	**	1,257
	Polyprion oxygeneios	Hapuku wreckfish	50	854	Circumglobal 28°S - 50°S	***	1,667
	Stereolepis doederleini	-	400	600	NW Pacific	***	
Serranidae	*Pronotogrammus eos*	Bigeye bass	155	325	W Pacific, W Atlantic	**	
Epigonidae	*Epigonus telescopus*	Black cardinal fish	75	1,200	Atlantic, Indian, SW Pacific	***	1,781
Malacanthidae	*Lopholatilus chamaeleonticeps*	Great Northern tilefish	80	540	W Atlantic	***	1,255
Bramidae	*Brama brama*	Atlantic pomfret	0	1,000	Atlantic, Indian, S Pacific	***	
Emmelichthyidae	*Emmelichthys nitidus nitidus*	Cape bonnetmouth	86	500	S Indo-W Pacific	**	3,694
	Plagiogeneion rubiginosum	Rubyfish	50	600	SE Atlantic, Indian, SW Pacific	***	514
	Emmelichthyidae	Bonnetmouths, rubyfishes NEI					9
Lutjanidae	*Etelis oculatus*	Queen snapper	100	450	W Atlantic	***	
	Paracaesio stonei	Cocoa snapper	200	320	W Pacific	*	
Nemipteridae	*Parascolopsis rufomaculatus*	Red-spot dwarf monocle bream	210	320	NW Australia	*	
Sparidae	*Pagellus bogaraveo*	Red Sea Bream	?	700	NE Atlantic, Mediterranean	**	
Pentacerotidae	*Pentaceros richardsoni*	Pelagic armourhead	0	1,000	SE Atlantic,Indian, S Pacific	***	198
	Pentaceros wheeleri	Slender armorhead	146	800	N Pacific	*	
Zoarcidae	*Lycodes diapterus*	Black eelpout	146	844	N Pacific	*	
	Lycodes pacificus	Blackbelly eelpout	9	399	NE Pacific	*	
	Lycodes vahlii	Vahl's eelpout	65	1,200	Arctic, NW Atlantic	*	
Nototheniidae	*Dissostichus eleginoides*	Patagonian toothfish	50	3,850	SW Atlantic, Southern Ocean	***	24,350
	Dissostichus mawsoni	Antarctic toothfish	0	2,200	Southern Ocean	***	3,709
Channichthyidae	Channichthyidae	Crocodile icefishes NEI	90	600	Southern Ocean	**	8
Uranoscopidae	*Kathetostoma cubana*	Spiny stargazer	200	600	W Atlantic	*	
Gempylidae	*Lepidocybium flavobrunneum*	Escolar	200	1,100	Circumglobal	**	230
	Nealotus tripes	Black snake mackerel	914	1,646	Circumglobal	*	
	Rexea prometheoides	Royal escolar	135	540	Indo-W Pacific	***	
	Ruvettus pretiosus	Oilfish	100	800	Circumglobal	**	17,117
	Thyrsitops lepidopoides	White snake mackerel	30	350	SW Atlantic, SW Pacific	**	
Trichiuridae	*Aphanopus carbo*	Black scabbardfish	200	1,700	N Atlantic	***	10,012
	Lepidopus caudatus	Silver scabbardfish	42	620	NE SE Atlantic, Med., SE SW Pacific	****	11,551
Centrolophidae	*Hyperoglyphe antarctica*	Bluenose warehou	40	1,500	SE SW Atlantic SW Pacific	***	2,450
	Psenopsis cyanea	Indian ruff	179	300	W Indian	**	
	Seriolella caerulea	White warehou	1	800	SE SW Pacific, SW Atlantic	***	2,263

Table 6.1 (*cont.*)

ORDER Family	Scientific Name	Common Name	Depth (m) min	max	AREA	Importance	FAO catch (tonnes)
Nomeidae	*Psenes pellucidus*	Bluefin driftfish		1,000	Atlantic, Indian, Pacific	***	
Ariommatidae	*Ariomma melanum*	Brown driftfish	180	600	Atlantic	**	
Caproidae	*Antigonia capros*	Deepbody boarfish	50	900	Circumglobal	**	
	Antigonia combatia	Shortspine boarfish	115	585	W Atlantic	**	
	Capros aper	Boarfish	40	700	NE Atlantic, Mediterranean	**	38,009
	Caproidae	Boarfishes NEI					172
PLEURONECTIFORMES							
Citharidae	*Citharoides macrolepis*	Twospot largescale flounder	182	200	W Indian	***	
Scophthalmidae	*Lepidorhombus whiffiagonis*	Megrim	100	700	NE Atlantic, Mediterranean	****	10,756
Bothidae	*Chascanopsetta lugubris*	Pelican flounder	60	3,210	Atlantic, Indian, Pacific	***	
	Japonolaeops dentatus	-	300	500	W Pacific	***	
	Monolene maculipinna	Pacific deepwater flounder	205	384	E Pacific	***	
	Monolene microstoma	Smallmouth moonflounder	25	460	E Atlantic	***	
	Monolene sessilicauda	Deepwater flounder	150	550	W Atlantic	**	
	Parabothus coarctatus	-	253	580	Pacific	***	
	Parabothus kiensis	-	200	400	Indo-W Pacific	***	
Pleuronectidae	*Lyopsetta exilis*	Slender sole	25	800	E Pacific	**	
	Reinhardtius hippoglossoides	Greenland halibut	1	2,000	Arctic, N Atlantic, N Pacific	****	110,897
Soleidae	*Austroglossus microlepis*	West coast sole	100	400	SE Atlantic	**	4,306
	Bathysolea profundicola	Deepwater sole	200	1,350	E Atlantic	***	
LOPHIIFORMES							
Lophiidae	*Lophius budegassa*	Blackbellied angler	300	1,013	NE Atlantic, Mediterranean	***	33
	Lophius gastrophysus	Blackfin goosefish	40	700	W Atlantic	**	2,580
	Lophius litulon	Yellow goosefish	25	560	NW Pacific	***	
	Lophius piscatorius	Angler	20	1,000	NE Atlantic, Mediterranean	****	29,926
	Lophius vaillanti	Shortspine African angler	200	800	E Atlantic	***	98
	Lophius vomerinus	Devil anglerfish	150	400	SE Atlantic	***	15,102
						Total	2,904,384

Key: NEI - not elsewhere included, Importance: **** - highly commercial, *** - commercial, ** - minor commercial, * - subsistence, FAO catch: Average catch for 10 years 2003–2012

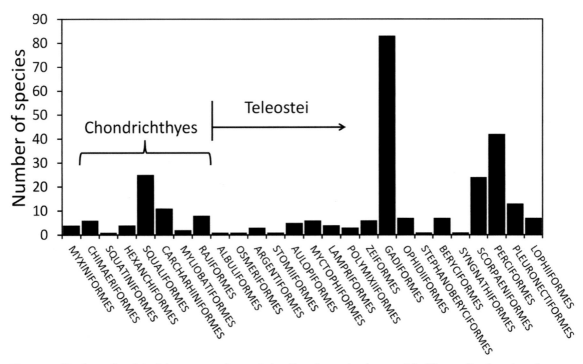

Figure 6.4 Numbers of exploited deep-sea species per Order. Note the predominance of Gadiformes. By class, Agnatha: 4 species, Chondrichthyes: 57 species, Teleostei: 215 species.

and Grenadiers, which are not elsewhere included (NEI). Their importance to fisheries is listed as highly commercial, commercial, minor commercial or subsistence according to FishBase. The exploited species fall into 26 orders, and the preponderance of 83 species of Gadiformes stands out, accounting for 30 per cent of all deep-sea fishery species (Fig. 6.4). Furthermore, one family, the Macrouridae dominates within the Gadiformes with 59 listed species approximately equal to the entire Chondrichthyes (chimaeras, sharks and rays), which comprises 57 species.

FAO (2014) records the world catches of 125 deep-sea species, and the average reported landings of those species during the decade 2003 to 2012 are listed in Table 6.1. Over this period, the average annual total reported deep-sea catch was 2,904,499 tonnes, of which Gadiformes comprised 78.8 per cent of landings (Fig. 6.5a). A single species, Blue whiting (*Micromesistius poutassou*) from the North Atlantic accounted for 47 per cent of the world deep-sea catch (Table 6.2). Together with the much

smaller stock of southern blue whiting, *Micromesistius australis*, the two blue whiting species amount to 50 per cent of the world deep-sea catch. These together with the Cape hakes, *Merluccius capensis* and *M. paradoxus* from the SW Atlantic, the Macruronids *Macruronus novaezelandiae* (hoki) off New Zealand and *M. magellanicus* (Patagonian grenadier) off South America and *Reinhardtius hippoglossoides* (Greenland halibut) from the Arctic, North Pacific and North Atlantic Oceans produce 75 per cent of the world's deep-sea fish catch.

The evaluation of the importance of deep-sea fishing therefore depends greatly on these few species. Hakes (Merlucciidae) and cods (Gadidae) were fished long before the development of the modern deep-sea fishing industry and arguably should not be included in deep-sea fishing statistics. If these two families are removed, the total remaining annual average deep-sea catch decreases to less than 1 Mt (Fig. 6.5b). The Gadiformes remain dominant with 38 per cent of

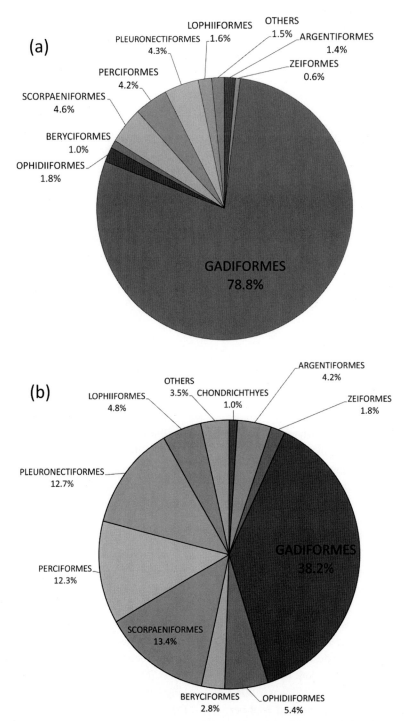

Figure 6.5 Proportion per order of the average world annual catch world catch of deep-sea fishes during 2003–2012. Data from FAO.

(a) All deep-sea species listed in Table 6.1 as a percentage of the total average annual catch 2,904,499 tonnes.

(b) All deep-sea species listed in Table 6.1 excluding the families Merlucciidae (hakes) and Gadidae (cods) as percentage of the average annual catch 994,507 tonnes.

Table 6.2 **Top 50 Deep-Sea Exploited Fish Species (Mean annual catch 2003–2012)**

Rank	Order	Family	Scientific Name	Common Name	Tonnes	%
1	Gadiformes	Gadidae	*Micromesistius poutassou*	Blue whiting	1,357,076	46.73
2	Gadiformes	Merlucciidae	*Merluccius capensis/ paradoxus*	Cape hakes	285,655	9.84
3	Gadiformes	Macruronidae	*Macruronus magellanicus*	Patagonian grenadier	193,586	6.67
4	Gadiformes	Macruronidae	*Macruronus novaezelandiae*	Blue grenadier, hoki, blue hake	136,268	4.69
5	Pleuronectiformes	Pleuronectidae	*Reinhardtius hippoglossoides*	Greenland halibut	110,897	3.82
6	Gadiformes	Gadidae	*Micromesistius australis*	Southern blue whiting	95,054	3.27
7	Gadiformes	Merlucciidae	*Merluccius gayi*	South Pacific hake	89,195	3.07
8	Scorpaeniformes	Sebastidae	*Sebastes mentella*	Beaked redfish	67,992	2.34
9	Ophidiiformes	Ophidiidae	*Genypterus blacodes*	Pink cusk-eel	42,310	1.46
10	Gadiformes	Gadidae	*Molva molva*	Ling	38,371	1.32
11	Perciformes	Caproidae	*Capros aper*	Boarfish	38,009	1.31
12	Scorpaeniformes	Sebastidae	*Sebastes alutus*	Pacific ocean perch	34,000	1.17
13	Argentiformes	Argentinidae	*Argentina* spp.	Argentine, silver smelts	30,476	1.05
14	Lophiiformes	Lophiidae	*Lophius piscatorius*	Angler	29,926	1.03
15	Gadiformes	Gadidae	*Brosme brosme*	Tusk	25,450	0.88
16	Perciformes	Nototheniidae	*Dissostichus eleginoides*	Patagonian toothfish	24,350	0.84
17	Scorpaeniformes	Anoplopomatidae	*Anoplopoma fimbria*	Sablefish	24,342	0.84
18	Gadiformes	Macrouridae	Macrouridae	Grenadiers, rattails NEI	24,313	0.84
19	Perciformes	Gempylidae	*Ruvettus pretiosus*	Oilfish	17,117	0.59

Table 6.2 (*cont.*)

Rank	Order	Family	Scientific Name	Common Name	Tonnes	%
20	Beryciformes	Trachichthyidae	*Hoplostethus atlanticus*	Orange roughy	17,002	0.59
21	Lophiiformes	Lophiidae	*Lophius vomerinus*	Devil anglerfish	15,102	0.52
22	Gadiformes	Macrouridae	*Coryphaenoides rupestris*	Roundnose grenadier	14,889	0.51
23	Stomiiformes	Sternoptychidae	*Maurolicus muelleri*	Silvery lightfish	14,840	0.51
24	Perciformes	Trichiuridae	*Lepidopus caudatus*	Silver scabbardfish	11,551	0.40
25	Pleuronectiformes	Scophthalmidae	*Lepidorhombus whiffiagonis*	Megrim	10,756	0.37
26	Zeiformes	Oreosomatidae	*Pseudocyttus maculatus*	Smooth oreo dory	10,254	0.35
27	Perciformes	Trichiuridae	*Aphanopus carbo*	Black scabbardfish	10,012	0.34
28	Gadiformes	Gadidae	*Molva dypterygia*	Blue ling	9,755	0.34
29	Ophidiiformes	Ophidiidae	*Genypterus capensis*	Kingklip	8,477	0.29
30	Beryciformes	Berycidae	*Beryx* spp.	Alfonsinos NEI	7,462	0.26
31	Myctophiformes	Myctophidae	*Lampanyctodes hectoris*	Hector's lantern fish	7,288	0.25
32	Myctophiformes	Myctophidae	*Benthosema pterotum*	Skinnycheek lantern fish	7,216	0.25
33	Scorpaeniformes	Sebastidae	*Helicolenus dactylopterus*	Blackbelly rosefish, Bluemouth	6,048	0.21
34	Zeiformes	Oreosomatidae	*Allocyttus niger*	Black oreo	5,311	0.18
35	Gadiformes	Macrouridae	*Macrourus carinatus*	Ridge scaled rattail	5,293	0.18
36	Argentiformes	Argentinidae	*Glossanodon semifasciatus*	Deep-sea smelt	4,370	0.15
37	Pleuronectiformes	Soleidae	*Austroglossus microlepis*	West coast sole	4,306	0.15

Table 6.2 (*cont.*)

Rank	Order	Family	Scientific Name	Common Name	Tonnes	%
38	Argentiformes	Argentinidae	*Argentina silus*	Greater argentine, Greater silver smelt	3,865	0.13
39	Perciformes	Nototheniidae	*Dissostichus mawsoni*	Antarctic toothfish	3,709	0.13
40	Perciformes	Emmelichthyidae	*Emmelichthys nitidus nitidus*	Cape bonnetmouth	3,694	0.13
41	Gadiformes	Merlucciidae	*Merluccius polli*	Benguela hake	3,664	0.13
42	Gadiformes	Gadidae	*Phycis blennoides*	Greater forkbeard	3,390	0.12
43	Argentiformes	Argentinidae	*Argentina sphyraena*	Argentine, lesser silver smelt	2,756	0.09
44	Osmeriformes	Alepocephalidae	*Alepocephalus bairdii*	Baird's slickhead	2,680	0.09
45	Lophiiformes	Lophiidae	*Lophius gastrophysus*	Blackfin goosefish	2,580	0.09
46	Perciformes	Centrolophidae	*Hyperoglyphe antarctica*	Bluenose warehou	2,450	0.08
47	Gadiformes	Merlucciidae	*Merluccius capensis*	Shallow-water Cape hake	2,369	0.08
48	Perciformes	Centrolophidae	*Seriolella caerulea*	White warehou	2,263	0.08
49	Myxiniformes	Myxinidae	Myxinidae	Hagfishes NEI	2,335	0.08
50	Beryciformes	Berycidae	*Beryx decadactylus*	Alfonsino	2,194	0.08
				Other species	32,120	1.11
				Total world annual catch	2,904,384	tonnes

the total catch, but the contribution from Scorpaeniformes, Pleuronectiformes and Perciformes becomes evident.

Even this reduced pool of species includes Lophiiformes and Pleuronectiformes, which are often caught at shelf depths. Whilst sharks were probably the first and most successfully exploited deep-sea species, the contribution of Chondrichthyes to the global deep-sea catch since 2002 has become very small; less than 0.5 per cent of the total. Chondrichthyes have been severely depleted and in most jurisdictions, targeting of sharks and rays is now prohibited, and only small bycatches are reported (though landings may be unreported to circumvent such regulations).

For the purposes of this chapter we accept a rather inclusive definition of deep-sea fisheries encompassing species that although caught at depths greater than 200 m, also occur at shelf depths and might not be regarded biologically as truly deep-sea fishes. In New Zealand waters at depths between 200 and 800 m are considered to be 'middle depths', with deep water defined as depths greater than 800 m (Colman, 1995).

6.4.1 The Top Twenty Deep-Sea Species

Over time, there have been significant changes in the composition of the global deep-sea catch, including the exploitation of species that had previously been almost unknown. The time course of these changes is illustrated in Figure 6.6 showing trends for the top 20 taxa from Table 6.2 from 1950 to 2014 in chronological order of onset of increase in their exploitation.

In the 1950s fisheries began moving deeper beyond the shelf break in search of unexploited stocks of familiar species such as ling (*Molva molva*) and tusk (*Brosme brosme*) in the North Atlantic (for which there are continuous catch records in some areas from 1904 onward). Reported landings for these species grew in parallel with the general increase in the global fishing industry (Fig. 6.6a). Both species are now managed in European waters under advice from the ICES Working Group on Biology and Assessment of Deep-Sea Fisheries Resources (ICES, 2015) with total allowable catches in the twenty-first century set somewhat below the peak landings reached in the 1970s and 1980s. The Pacific ocean perch (*Sebastes alutus*) is the first example of a major 'boom and bust' deep-sea fishery. In the North Pacific a fishery was developed by the USSR and Japan reaching a peak of 480,000 tonnes in 1965 taken largely on the high seas outside the then narrow management jurisdiction of coastal states. Catches decreased to a minimum of 8,578 tonnes in 1985 and are now stable at just over 30,000 tonnes with management largely under the control of USA and Canada within their respective EEZs. The Greater Argentine or Greater Silver Smelt (*Argentina silus*) had generally been harvested as a bycatch in mixed industrial trawl fisheries in the NE Atlantic but from the 1970s a directed fishery was

developed off Norway (Johannessen and Monstad, 2003) and has since expanded across the NE Atlantic to Faroe, Iceland and NW Scotland. It also occurs as a bycatch in fisheries for blue whiting and deep-sea shrimps. Catches have increased as trawl fisheries have moved deeper from shelf depths to 300–600 m, the preferred depth range for this species (Fig. 6.6d). The Argentine or Lesser Silver smelt (*Argentina sphyraena*) occurs further south and in the Mediterranean and is often combined with *A. silus* in landing statistics (ICES, 2015). The Faroese semipelagic trawl fishery for *Argentina silus* is certified as sustainable by the Marine Stewardship Council with a catch of 15,000 to 20,000 tonnes per annum (MSC, 2016). The fluctuating ascending trend in Figure 6.6d represents shifts from stock to stock and general global increase in exploitation.

The Cape Hakes are taken in a mixed fishery for two species (the shallow-water Cape hake (*Merluccius capensis*) and the deep-water Cape hake (*Merluccius paradoxus*)), which began off South Africa in the 1920s and grew to 50,000 tonnes by 1950 (Payne and Punt, 1995). During the 1960s foreign distant-water vessels from, Japan, Spain, the Soviet Union, Poland, Bulgaria, Romania and Eastern Europe began fishing off South Africa and adjacent waters, leading to a vast increase in catches that peaked at 1,122,000 tonnes in 1971 of which South African vessels took 27 per cent. In 1977 South Africa declared a 200-mile exclusive fishing zone, removed most of the foreign fishing effort and together with Namibia, the fishery has been gradually brought under control with a catch of 200,000–250,000 tonnes per annum (Fig. 6.6e). The South African trawl fishery for Cape Hakes is certified as sustainable (MSC 2016) at 134,000 tonnes per annum. Greenland halibut (*Reinhardtius hippoglossoides*) has a long history of exploitation by the Inuit people of the Arctic. Exports of small quantities of salted fish from Canada began from 1857 onward, but the flesh is too oily for successful long-term preservation in this way (Bowering and Brodie, 1995). Processing of fish for frozen fillet, new fishing techniques including synthetic fibre gill nets

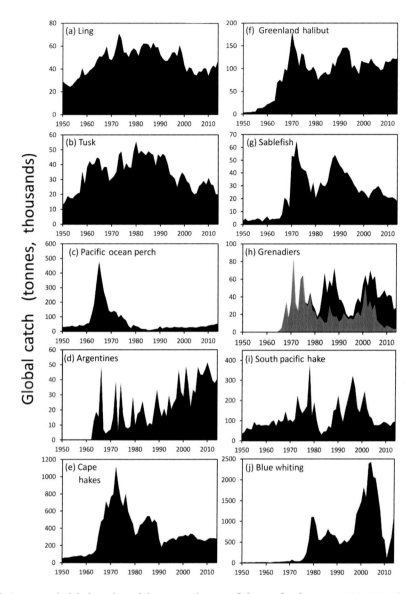

Figure 6.6 Trends in reported global catches of the top 20 deep-sea fish taxa for the years 1950–2014. Arranged in chronological order of the onset of major exploitation. Catch in thousands of tonnes, please note the different scales for each species. Data from FAO.

(a) Ling, *Molva molva*. (b) Tusk, *Brosme brosme*. (c) Pacific ocean perch, *Sebastes alutus*. (d) Argentines, including Greater argentine or Greater silver smelt, *Argentina silus*, Argentine or lesser silver smelt, *Argentina sphyraena* and Argentines NEI. (e) Cape hakes, deep-water Cape hake *Merluccius paradoxus* and Shallow-water Cape hake *Merluccius capensis*. (f) Greenland halibut, *Reinhardtius hippoglossoides*. (g) Sablefish, *Anoplopoma fimbria*. (h) Grenadiers, grey: Roundnose grenadier, *Coryphaenoides rupestris*. Black: All other species of grenadier, Macrouridae NEI plus Chilean grenadier, *Coelorinchus chilensis*. Dogtooth grenadier, *Cynomacrurus piriei*. Roughhead grenadier, *Macrourus*. Ridge scaled rattail, *Macrourus carinatus*. Whitson's grenadier, *Macrourus whitsoni*. Common Atlantic grenadier, *Nezumia aequalis*. (i) South Pacific hake, *Merluccius gay*. (j) Blue whiting, *Micromesistius poutassou*. (k) Orange roughy, *Hoplostethus atlanticus*. (l) Southern blue whiting, *Micromesistius australis*. (m) Pink cusk-eel, *Genypterus blacodes*. (n) Angler, *Lophius piscatorius*. (o) Blue grenadier, Hoki, *Macruronus novaezelandiae*. (p) Patagonian grenadier, *Macruronus magellanicus*. (q) Patagonian toothfish, *Dissostichus eleginoides*. (r) Beaked redfish, *Sebastes mentella* (s) Oilfish, *Ruvettus pretiosus* (t) Boarfish, *Capros aper*.

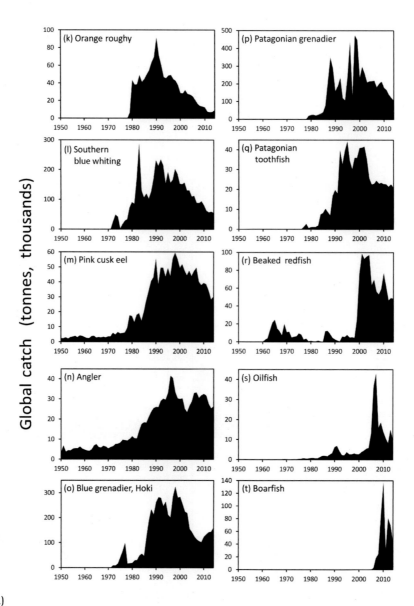

Figure 6.6 (*cont.*)

and offshore catches by trawlers from the USSR, Poland and East Germany led to a great increase in catches during the 1960s, reaching a peak global catch of 181,000 tonnes in 1971 (Fig. 6.6f), approximately half of which was taken in the Northeast Arctic Area (Barents Sea). The North East Arctic stock declined greatly, with the 1989–1994 year classes appearing to be absent from stock surveys with the result that in 1992 targeted fishing

was entirely prohibited, and it was estimated that 12–15 years would be required for the stock to recover before targeted fishing could recommence (Høines and Gundersen, 2008). The global catch has been maintained at around 100,000 tonnes per annum as fisheries have expanded geographically to new areas including the Okhotsk Sea, the Bering Sea, and the Pacific coast of North America as far south as Mexico.

The Sablefish or Blackcod (*Anoplopoma fimbria*) has a wide distribution in the North Pacific from Japan to the Bering Sea and Baja California, but abundance is centred on Northern British Columbia and the Gulf of Alaska. There is a long history of exploitation with recorded landings from 1913 onward but in common with many stocks this fishery experienced a massive increase in exploitation by international fleets in the 1970s with a peak of 65,000 tonnes in 1972 (Fig. 6.6g), also followed by fishery restrictions as the coastal states asserted 200-mile exclusive fishery zones. The fishery in US waters off Alaska was certified as sustainable from 2006 with an annual catch of 18,100 tonnes (MSC, 2006) representing a large proportion of the global catch.

Macrourids or grenadiers had not been considered for exploitation until surveys by the USSR in the early 1960s discovered concentrations of Roundnose grenadier (*Coryphaenoides rupestris*) on slopes of the North Atlantic and the Mid-Atlantic Ridge often associated with Greenland halibut (Atkinson, 1995). The first reports of landings came in 1967, and only four years later in 1971 the peak catch of 84,000 tonnes was recorded (Fig. 6.6h). In the NE Atlantic west of the British Isles a fishery has been conducted by vessels from France and to a lesser extent Spain and is managed under a precautionary approach to allow the stocks to recover with a global reported catch of only 3843 tonnes in 2014. There is a probably distinct stock of *C. rupestris* in the deep waters of the Skaggerak channel between Norway and Denmark; long-term studies over almost three decades show that throughout that time there has been only one pronounced recruitment event, in 1992 (Bergstad et al., 2013), and the directed fishery has been closed since 2006. Roundnose grenadiers are long-lived, reaching maturity at 8–12 years and have a potential life span of over 70 years. With major recruitment events possibly so rare, it is not surprising that in the NW Atlantic the species is considered endangered (Devine et al., 2006). There is some evidence that stocks of macrouridae in the NE Atlantic have stabilised since the introduction of management measures, albeit at a much lower level than the virgin

biomass (Neat and Burns, 2010). The FAO statistics show that catches of other species of macrourids from the N Atlantic, the Southern Ocean, the South Atlantic and elsewhere have increased as the Roundnose grenadier has declined.

South Pacific hake (*Merluccius gayi*) is exploited in the SW Pacific off the coasts of Ecuador, Peru and Chile with stocks in the latter two EEZs recognised as subspecies. The Chilean fishery began in 1940, reaching a peak of 130,000 t in 1968, which was followed by a decline (Aguayo-Hernández, 1995). At this time distant-water vessels from Poland, Cuba, Spain, Soviet Union and Japan together with Peruvian fishing vessels began fishing for *M. gayi* on a large scale leading to a peak catch of 382,439 t in 1978. The stock biomass declined by about 75 per cent, and in 1983 the reported landings were 31,160 tonnes, 10 per cent of the previous peak. Since that time the catch has grown but with major fluctuations associated with El Niño events that have a great influence in the region (Espino et al., 1995).

The blue whiting (*Micromesistius poutassou*) is a highly migratory pelagic species in the North Atlantic. It reaches sexual maturation at 2–7 years of age and typical average age in the population is 5 years. The adults undertake extensive feeding migrations, but the fishery is concentrated on large schools that congregate each spring for spawning at the edge of the continental shelf west of the British Isles, around the Faroes and at offshore banks at depths of 300–600 m. The onset of large-scale exploitation of blue whiting from the late 1970s was enabled by the development of new pelagic fishing technology. The fishery is now dominated by large pelagic trawlers that either process and freeze the catch on board or hold the fish in tanks of chilled seawater from which they can be rapidly pumped direct into shore-based processing plants when the vessel arrives in port. The remarkable catching power of such vessels is shown by the report that in 2013 over 95 per cent of the 51,600 tonne total catch by the Netherlands was taken in 18 fishing trips, an average of 2,867 tonnes per trip (ICES, 2014). These fishing vessels are able to catch their quota rapidly and then switch to other pelagic species such as mackerel (*Scomber scombrus*), Horse mackerel (*Trachurus trachurus*) and herring (*Clupea harengus*) in order to realise a return on the high capital investment. The blue whiting fishery attained a peak catch of 2,428,955 tonnes in 2004 after a period of apparent increase in the spawning stock biomass. Fishing mortality at this time was 0.5–0.75 in some year classes, indicative of excess fishing effort. A period of historically low recruitment followed, and the catches declined to a low of 108,077 tonnes with an estimated spawning stock biomass of 2.9 million tonnes in 2010. A period of fishing restraint appears to have been successful in quickly restoring the stock so that the spawning stock biomass had grown to 5.5 million tonnes with a reported catch of 1,160,872 tonnes in 2014 (ICES, 2014). Optimism regarding successful management and the future of this stock is reflected in the leading fishery stakeholders entering into assessment for sustainability status (MSC, 2016). As a relatively fast-growing species with year-to-year differences in recruitment, blue whiting are likely to display volatility in spawning stock biomass and hence catches.

Orange roughy (*Hoplostethus atlanticus*) fisheries began with some exploratory catches of a then almost unknown species by Soviet vessels in the 1970s around New Zealand (Koslow, 2007). After New Zealand enforced a 200-mile exclusive fishing zone in 1977 there was rapid expansion in fishing aided by demand stimulated after the collapse of major fisheries in the Northern Hemisphere and joint ventures with displaced fishery interests searching for new fishing opportunities. The fishery grew from zero in 1978 to 49,000 tonnes in 1983. This was achieved by progressive exploitation of new grounds beginning with the Chatham Rise in 1978, the Challenger Plateau in 1982 and expanding to eight different fishing areas by 1986, making it New Zealand's most valuable fish species (Clark, 1995). The fishery peaked in 1989 at 55,000 tonnes and has since declined as stock biomass was depleted to less than 20 per cent of prefishing levels. Off Southern Australia the fishery started in 1985 and peaked at ca. 40,000 tonnes in 1990 (Clark, 2001). Catches off Chile, Namibia, South Africa, the

Southern Indian Ocean and the NE Atlantic Ocean added to the global catch, but none reached the quantities found off Southern Australia and New Zealand. In the NE Atlantic stocks are small, and no targeted fishing has been allowed since 2010. In the NE Atlantic it has been concluded that 'Due to its very low productivity, orange roughy can only sustain very low rates of exploitation. Currently, it is not possible to manage a sustainable fishery for this species' (ICES, 2015). Clark (2001) points out that for smaller stocks it is not possible to gain enough timely information for precautionary management during the development phase of the fishery but that for larger stocks management might be possible. All orange roughy stocks have passed the 'fish down' phase during which the stock biomass has been reduced to a fraction of the virgin. Many fisheries are closed, and evidence of continuing recruitment has remained sparse (Francis and Clark, 2005). Kloser et al. (2015) have found signs of recovery of stocks in Eastern Australian waters around Tasmania following closure of the fishery in 2004. Stock biomass has increased, and there is evidence of recruitment of young fish. However, in view of slow maturation of orange roughy (ca. 26 years), this increase is still largely based on progeny from spawning events before the start of commercial fishing. Evidence of a large aggregation of smaller and younger fish is cited by Kloser et al. (2015) as evidence of the recovery in New Zealand waters also. Fishing in New Zealand waters continues with low levels of Total Allowable Catch, which the industry argues will probably prove to be sustainable (Clement et al., 2008; MSC, 2016).

The fishery for the southern blue whiting (*Micromesistius australis*) developed slightly later than that for the northern blue whiting. The fishery off New Zealand occurs mainly south of South Island on the Beaufort Platform and Campbell Island focussing on spawning aggregations during August and September. Separate fisheries occur off Chile, Argentina and around the South Atlantic Islands. The New Zealand Fishery is certified as sustainable (MSC, 2016) with total allowable catch set on the basis of annual acoustic surveys of spawning stock abundance. Recruitment is variable from year to year, and management needs to be responsive to interannual changes. Recent average landings (< 60,000 t) are well below the peak of 287,470 t recorded in 1983.

The pink cusk-eel (*Genypterus blacodes*) occurs in fisheries off Southern Australia, New Zealand, Chile and Argentina. Various common or marketing names are used, including pink ling or banded ling off Australia, in New Zealand ling, the Chilean kingklip, and kingklip off Argentina, as well as various names in Spanish and other languages. There is a long history of landings of *G. blacodes* as a bycatch in fisheries for hake off South America. From the late 1970s a more targeted fishery developed, and by 1992, when Chilean stocks were deemed to be fully exploited, TACs were introduced (Baker et al., 2014). In New Zealand waters the fishery began in the mid-1970s with Japanese and Korean longliners. During the 1980s the fishery diversified with trawlers as well as longliners and increased New Zealand ownership of vessels. Landings peaked in 2000, and since then there has been a declining trend with sustainability certification at a TAC of around 20,000 t (MSC, 2016). There is an extensive fishery in the SW Atlantic along the edge of the Patagonian shelf and around islands.

The angler- or monkfish (*Lophius piscatorius*) is a species for which there is a long history of exploitation at shelf depths in the NE Atlantic from Iceland and Norway to the Mediterranean Sea. They are caught mainly in mixed species trawl fisheries but also by gill nets set beyond the shelf edge. From the early 1980s there was a considerable increase in landings (Laurenson, 2008) and expansion of the fishery to greater depths, peaking at 41,509 t in 1996. The recommended TAC is currently set at less than half of the peak landings. Despite modest tonnages this is a high value species, and prices have increased as supply has declined.

Small quantities of blue grenadier or hoki (*Macruronus novaezelandiae*) ca. 100 t per annum have been recorded as being taken off New Zealand from 1955 onward, but significant commercial fisheries were developed by Japanese trawlers around 1975 reaching 98,967 t in 1977 before the assertion of the New Zealand 200-mile exclusive fishing zone. The fishery was then developed by New Zealand from 1986 onward under a quota management system reaching 324,819 t in 1998. In response to a hiatus in recruitment the TAC was reduced to 180,000 t in 2003–2004 and 90,000 t in 2007–2008. Presumably as a result of these measures there appears to have been an improvement in stock status permitting an increase in TAC to 160,000 from 2014–2015 (McKenzie, 2016). The fishery is conducted by bottom and pelagic trawls at depths of 200–700 m depth and is regarded as well managed, having been certified as sustainable since 2001 (MSC, 2016). The species has high fecundity and a fast growth rate, becoming sexually mature at 4–5 years, which sustains high productivity but with considerable interannual variation in recruitment. There is also a small fishery off southern Australia, with landings of 3,000 to 4,000 t per annum in recent years.

The fishery for *Macruronus magellanicus*, Patagonian grenadier, longtail hake or Patagonian whiphake off Chile and Argentina, originally developed as part of a mixed demersal trawl fishery directed primarily at hakes (*Merluccius gayi* and *M. australis*), and pink cusk-eel (*Genypterus blacodes*). As these other species became overfished more *Macruronus magellanicus* were taken from the 1980s onward. After New Zealand had opened up international markets for hoki the species began to be targeted from 1995–1998 and was marketed as Chilean hoki (Tascheri et al., 2010). The fishery for Argentine hoki developed over the same time scale and from 2012 has been certified as sustainable with a typical catch of around 100,000 t per annum using demersal and semipelagic trawl gears (MSC, 2016). Management of the Chilean fishery has proved more difficult owing to bycatch and unknown levels of discarding in the mixed demersal fishery. The TAC in 2013 had been decreased to 40,000 t, and the status of the stock does not currently warrant certification of sustainability.

Patagonian toothfish (*Dissostichus eleginoides*) were originally caught as a bycatch in trawl fisheries in the SW Atlantic, but during the 1980s a targeted fishery developed using new deep-water long-lining techniques first of all off Chile, then the Patagonian shelf, South Georgia and eventually spreading across the entire range of the species around the Antarctic. The species is now fished using both trawl and longline methods (Collins et al., 2010) and is marketed as Chilean seabass in North America. According to FAO catches, it peaked in 1995 at 44,219 t and since 2005 has remained just above 20,000 t (Fig. 6.6q). Because of the fragmented and remote nature of the fishing grounds, illegal, unreported and unregulated (IUU) catches are a particular problem for this species. Lack (2008) estimates that an average of 17 per cent of world trade in *Dissostichus eleginoides* is represented by IUU, which would clearly compromise management measures, although attempts are made to adjust for the effect of IUU. Despite these uncertainties five fisheries for Patagonian toothfish have been certified as sustainable: Kerguelen and Crozet Islands (5100 t), Macquarie Island (290 t), Falkland Islands (1,085 t), South Georgia (3,500 t), Heard Island and McDonald Islands (2,500; MSC, 2016). However, the total tonnage in these TACs is approximately half of the total global landings and cannot be taken to imply that all fisheries for the species are sustainably managed.

Although the Beaked Redfish or pelagic redfish (*Sebastes mentella*) is found across the North Atlantic from Long Island in the west to the Barents Sea in the east, the major fishery has developed in the area around Iceland, Southern Greenland, the Irminger Sea and the Reykjanes Ridge segment of the Mid-Atlantic Ridge. Following exploratory fishing by the Soviet Union in 1981 a commercial fishery began the following year and expanded further with the development of deep-water pelagic fishing techniques in the 1990s. There is a discrepancy between catch data reported to FAO (Fig. 6.6r), which show a peak catch of 98,662 t in 2001 and higher values reported by ICES with a peak catch of 180,000 t in 1996 and 129,000 t in 2001 (Sigurðsson et al., 2006). Some workers recognise different stocks, demersal shelf and slope stocks, which may also include *Sebastes marinus* (golden redfish) and two pelagic stocks, a shallow epipelagic or oceanic redfish at depths < 500 m and the deep pelagic redfish found at

depths > 500 m, which are solely *Sebastes mentella* (Thompson, 2003). The catch for the deep pelagic stock is estimated by ICES to have peaked at 104,000 t in 2003. Thompson (2003) discussed problems raised by the movement of some of the stock westward into NW Atlantic management jurisdictions. Fifteen nations are involved in the fishery with Germany, Iceland, Russia, Norway, the Faroe Islands and Greenland accounting for about 80 per cent of the catch. There is a fundamental disagreement between participants as to whether *Sebastes mentella* should be managed as a single stock with a common overall TAC or whether the different putative stocks can be managed separately. In the absence of agreement some self-allocated unilateral TACs have been set (ICES NWWG Report, 2015). It is agreed that *S. mentella* is a slow-growing, late-maturing deep-sea species and hence vulnerable to overexploitation. It is a matter of concern that during 2013–2104 there was no evidence of recruitment and that in several areas the stock appears to be at a historical low. The ICES working group also reports problems of IUU fishing activity and misreporting of catches. The ICES stock advice for 2016 is that the deep pelagic fishery catch should be less than 10,000 t (ICES Advice, 2015), a 90 per cent decrease from the heyday of the fishery during 1995–2005. There is no management plan for the fishery, and this appears to be a classic example of failure owing to conflicts of interests in exploiting a stock that straddles multiple jurisdictions.

There is a long history of the capture of oilfish (*Ruvettus pretiosus*) by subsistence and artisanal fishermen in tropical and subtropical regions. The maximum size of over 2 m also makes them attractive to recreational game fishermen. Such traditional landings, however, are not reflected in official statistics. The FAO data (Fig. 6.6s) show small-scale catches of oilfish by Taiwan of *ca.* 100 t per annum from 1958 onward, increasing through the 1980s and 1990s before surging to a peak of 42,935 t in 2007, 98 per cent of which is logged by Taiwan. In addition to targeted fisheries, oilfish appear as a bycatch in line fisheries for tuna and swordfish (Collette et al., 2015). The highly fatty flesh with purgative properties makes marketing problematic; several countries ban imports or give health advice against excessive consumption. From time to time scandals erupt regarding the mislabelling of *Ruvettus pretiosus* under names such as codfish or butterfish. Oilfish are also processed into fish meal, fish cakes and other manufactured products.

Since 2001 a fishery for boarfish (*Capros aper*) has grown rapidly in the NE Atlantic off the South West of Ireland, reaching a peak of 137,676 t in 2010. This is a highly profitable industrial fishery for fish meal (Strange, 2016), which uses pelagic trawlers that are also engaged in fishing for mackerel (*Scomber scombrus*), horse mackerel (*Trachurus trachurus*), herring (*Clupea harengus*) and blue whiting (*Micromesistius poutassou*). Two-thirds of the catch in 2010 was taken by Ireland and most of the rest by Denmark. The concentrations of boarfish were discovered in a well-known fishery area and appear to be the result of exceptional recruitment events possibly attributable to changes in oceanographic conditions. There is evidence of previous episodic high abundances of boarfish from 1840 onward with reports of the nuisance this has caused to regular trawl fisheries (ICES, 2014). In 2011 a TAC of 33,000 t was introduced, increasing to 82,000 t in 2012, to 133,957 t in 2014 before being reduced to 53,292 t in 2015 with a further reduction proposed for 2016. There is some debate as to whether *Capros aper* is primarily a shelf-dwelling species that congregates at the shelf edge for spawning or whether the large abundances are representative of deep-water stocks swept up onto the shelf. It is not clear whether the new stock of boarfish will become a permanent sustainable feature of fisheries in the NE Atlantic or whether this is part of a pattern of episodic outbursts of this species in different parts of its range.

6.4.2 Other Species

6.4.2.1 Myxiniformes

Eptatretus atami (Brown hagfish) is fished commercially in traps off Japan and Korea. In recent years the fishery has expanded geographically to exploitation of *Eptatretus stouti* off the West Coasts of Canada and the USA, *Eptatretus cirrhatus* off New Zealand and *Myxine glutinosa* in the NW Atlantic reaching over 2000 t per annum off Atlantic Canada

by 2010 (Ellis et al., 2015). In fishing trials off Newfoundland no hagfish were caught in traps shallower than 384 m, and the largest and hence most valuable *Myxine glutinosa* were caught at 591–644 m depth (Grant, 2006). This is likely to remain a small-scale fishery serving a specialist market for 'eel-skin' leather and Korean cuisine.

6.4.2.2 Chimaeriformes

Chimaeras are susceptible to capture by longlines, set nets and bottom trawls. Their large size, average maximum length about 1 m, and large liver make them potentially desirable for fisheries, but they generally do not occur in very large numbers and are usually caught as part of a mixed fishery. The Callorhincidae should not strictly be listed as deep-sea fishes because they are mainly captured in small quantities by bottom-set gill nets at depths less than 150 m: *Callorhinchus capensis* (Cape elephant fish) off South Africa and Namibia and C. *milli* (Ghost shark) off South Australia and New Zealand. The Chimaeridae are deeper-living. *Chimaera monstrosa* (rabbit fish) is taken together with *Hydrolagus* sp, as a bycatch in fisheries in the NE Atlantic and is largely discarded. Estimated landings have fluctuated up to just over 1000 t per annum since 2000 (ICES, 2015). In the Pacific Ocean *Chimaera phantasma* (silver chimaera) is widely taken in fisheries from Japan, China, Taiwan, Korea, the Philippines and as a bycatch in trawl fisheries off Western Australia. *Hydrolagus novaezealandiae* (dark ghost shark) is captured as a bycatch in the targeted fisheries for hoki (*Macruronus novaezelandiae*), silver warehou (*Seriolella punctata*), barracouta or snoek (*Thyrsites atun*, Gempylidae) and squid off New Zealand. Since 2000, the catch in New Zealand waters has been around 2000 t per annum.

6.4.2.3 Squatiniformes

The Atlantic Angel Shark (*Squatina dumeril*) is of minor fishery interest in the Western Atlantic from the coast of Massachusetts to Venezuela. In US waters it appears as a bycatch in trawl fisheries off the Gulf of Mexico and Atlantic coast but from

1993 commercial landings were prohibited. Females mature at 85 cm, and the reproductive cycle may take three to four years including two years' vitellogenesis, one year gestation and one year resting after giving birth to an average of 7 pups per cycle (Baremore, 2010). Small quantities, < 200 t of angel sharks are landed from the Mediterranean and off Australia.

6.4.2.4 Hexanchiformes

The listed Hexachiform sharks are large cosmopolitan species occurring in all the main oceans. The frilled shark (*Chlamydoselachus anguineus*) is rather rare, grows up to 2 m long and appears as an incidental catch, which can be utilised as fishmeal or as food fish. *Heptranchias perlo* (sharpnose sevengill shark) and *Hexanchus nakamurai* (bigeyed sixgill shark) similarly occur as bycatch or are taken in subsistence fisheries. *Hexanchus griseus* (bluntnose sixgill shark) is the largest (up to 4.8 long) and the most abundant species of the group with very large litter sizes from 22 to 108 pups per female. It is listed in Annex 1 of UNCLOS as a 'Highly Migratory Oceanic Shark' and is taken by longlines, gillnets, traps, pelagic and bottom trawls for use as fresh, frozen, dried or salted for human consumption, fishmeal and oil (Maguire et al., 2006). The reported global catch has increased from 5 t in 2000 to 64 t in 2013 (FAO, 2016). Attempts to establish sustainable fisheries in the NE Pacific have not been successful, and the species is listed as 'near threatened' by the International Union for the Conservation of Nature (IUCN; Cook and Compagno, 2005). In the N Atlantic around the Azores the average catch of *Hexanchus griseus* is less than 1 t in a mixed deep-water fishery that lands a total of 2000–2500 t per annum (ICES WGDEEP). Kabasakal (2006) describes occasional landings of *Hexanchus griseus* from waters around Turkey, which attract much attention when the fish is displayed in the marketplace but which fail to find a buyer and are discarded within a couple of days due to their rapid putrefaction. However in the shark market in Cochin, India, *Hexanchus griseus*, although rare, are sold for the large liver and meat quantity (Akhilesh et al.,

2011). In Indonesia the meat commands a higher price than that of other deep-water chondrichthyans (White et al., 2006).

6.4.2.5 Squaliformes

None of the 25 listed squaliform shark species supports a major targeted species-specific commercial fishery; most are caught as a bycatch in bottom trawl, set net or line fisheries or in fisheries directed to deep-sea sharks collectively. For a group of species from the NE Atlantic *Centrophorus squamosus*, leafscale gulper shark, *Centroscymnus coelolepis*, Portuguese dogfish, *Deania calcea*, birdbeak dogfish, *Centroscymnus crepidater*, longnose velvet dogfish and *Centroscyllium fabricii*, black dogfish, Kjerstad et al. (2003) estimated that as a percentage of body weight these fishes can produce 24 per cent of skinless back meat, 17 per cent of fillet and 18–26 per cent of oil. Oil from the liver has squalene content of 39–83 per cent and is the most valuable product. On this basis about 1000 sharks of typical body weight 5 kg are required to produce 1 t of oil. Sharks also yield 1.4 per cent of dried cartilage by weight, but fins were considered of poor quality for the market, although tails could be sold.

The two most important species in the NE Atlantic are *Centrophorus squamosus* and *Centroscymnus coelolepis*. The total catch of these species combined reached 10000 t in 2001, and since 2010 the TAC has been reduced to zero following concerns regarding the sustainability of the stocks (ICES, 2015). A targeted fishery for *Dalatias licha* (kitefin shark) off the Azores yielded up to 900 t per annum until 1998 when the fishery was closed and a few hundreds of tonnes were caught elsewhere until the zero TAC regime was introduced. The other deep-water species in the NE Atlantic, including gulper shark (*Centrophorus granulosus*), birdbeak dogfish (*Deania calcea*, black dogfish *Centroscyllium fabricii*, longnose velvet dogfish (*Centroscymnus crepidater*), velvet belly *Etmopterus spinax*, angular rough shark *Oxynotus centrina*, lowfin gulper shark (*Centrophorus lusitanicus*) and knifetooth dogfish (*Scymnodon ringens*) yielded a total declared catch of 3236 t in

1996 and declined thereafter until the zero TAC was introduced.

In the Indian Ocean Akhilesh et al. (2011) noted an increase in deep-sea shark species landed in Cochin, India, during 2008–2009 with *Centrophorus* spp. comprising 7.3 per cent of the total shark catch and *Hexanchus griseus* and *Deania profundorum* (arrowhead dogfish) appearing for the first time. Targeted shark fishing is carried out using longlines, gillnets and hooks and lines down to 1000 m depth. The deep-sea shark fishery is driven by high market demand for squalene-rich oil, particularly from *Centrophorus* but also from *Echinorhinus brucus* (Bramble shark) and *Hexanchus griseus*. The meat is also utilised. As in the Atlantic Ocean, it is noted that the price of deep-sea shark fins is lower than for fins of pelagic sharks. During the 1980s a fishery for *Centrophorus* developed in the Maldives resulting in oil exports increasing rapidly from 10,000 litres in 1980 to 90,000 litres in 1982 followed by a decline and collapse of the fishery by 1995 and very small amounts of oil exported from 2000 onward (Kyne and Simpfendorfer, 2007).

Squalus megalops (Shortnose spurdog) is one of the most abundant chondrichthyes captured in trawl fisheries off SW Australia, mostly taken as a bycatch amounting to ca. 600 t per annum; however, studies show that, given its low productivity, increased fishing mortality could quickly put the species at risk (Braccini et al., 2006). Artisanal fishermen from Eastern Indonesia fish for deep-sea sharks on longlines set down to 600 m depth. The catch is dominated by unidentified *Squalus* spp. and possibly seven different species of *Centrophorus*. The catch is processed for the liver oil, and ca. 48000 kg per annum is exported. The meat is salted and dried. The fins again are considered of lower value than those of shallow-water species. The yolked ova of *Centrophorus* are considered a particular delicacy in Indonesia (White et al., 2006).

In New Zealand waters trawls targeted at orange roughy (*Hoplostethus atlanticus*) and smooth oreos (*Pseudocyttus maculatus*) on the Chatham Rise captured large numbers of *Deania calcea* (birdbeak dogfish), but few of the larger squalene oil-rich

squalid species (*Centrophorus squamosus*, *Scymnodon plunketi*, and *Dalatias licha*) giving little incentive for the development of a directed fishery (Wetherbee, 2000).

6.4.2.6 Carcharhiniformes

The Carcharhiniformes are primarily surface-dwelling sharks, but the deep-water representatives of the order in the families Triakidae (houndsharks) and Pseudotriakidae (false catsharks) and Scyliorhinidae (catsharks) do appear in fishery catches. None of these species supports a targeted fishery, being exploited only to a minor degree or taken in subsistence fisheries. The family Triakidae (hound sharks) are widely exploited in coastal and surface waters, but the deep-sea representatives are of minor fishery interest, *Iago garricki* (longnose houndshark) is probably taken as bycatch in the Northwestern Western Australia slope fishery at depths > 200 m. *Iago omanensis* (bigeye houndshark) occurs rarely in deep-sea shrimp trawls off India (Akhilesh et al., 2011) and is taken in artisanal gill net and line fisheries in the Red Sea (Baranes and McCormack, 2009). *Pseudotriakis microdon* (false catshark) is a worldwide species taken as a sporadic bycatch in fisheries for halibut in the North Atlantic Ocean, orange roughy in the Southern Indian Ocean and in fishery surveys in the North Pacific Ocean. In the family Scylliiorhnidae the deep-water genus *Apristurus* comprises over 30 species, often with very localised distributions; for example, the Japanese catshark (*Apristurus japonicus*) is found only along a 500 km stretch of deep-sea slope off SE Japan and is taken as bycatch in bottom trawl and gill net fisheries. Wetherbee (2000) found five putative new species of *Apristurus* in trawls around the Chatham Rise off New Zealand. In general *Apristurus* spp. are poorly documented in fishery returns, and there is great uncertainty regarding their status, although they are not targeted. *Figaro boardmani* is endemic to the southern Australian Ocean and is only of minor importance as bycatch in some fisheries. *Galeus eastmani* (Gecko catshark) is a bycatch of commercial deep-water trawl fisheries in the northwest Pacific and remains common off Japan, possibly owing to its extended bathymetric range beyond fishery depths. *Galeus polli* (African sawtail catshark) is taken in deep-water fisheries off Namibia but is not very susceptible to trawl capture.

6.4.2.7 Myliobatiformes

Plesiobatis daviesi (Deep-water stingray) is taken in small quantities in deep-water fisheries off South Africa, Indonesia, Taiwan, off Western Australia and Queensland and is not targeted. *Urolophus expansus* (wide stingaree) occurs in the Western Deep-water trawl fishery off Australia.

6.4.2.8 Rajiformes

The Rajiformes are a very specious group with 125 deep-sea species, most of which are probably susceptible to fisheries to some degree. There is great sensitivity regarding the exploitation of skates and rays since *Dipturus batis* (blue skate, 1000 m_{fbd}) or common skate was fished almost to extinction on shelf seas around Europe (Brander, 1981), and Casey and Myers (1998) suggested a similar fate for the barndoor skate, *Dipturus laevis* in the NW Atlantic. Rajiformes are morphologically very conservative, and there is considerable difficulty in identifying species at sea in a mixed fishery, so they are often grouped together as skates and rays. Iglesias et al. (2010) pointed out that *Dipturus batis* was often landed as *Dipturus oxyrinchus* or under other names either deliberately or inadvertently, masking the true decline in the species. Owing to these difficulties there are few well-documented species examples of fisheries for rajiformes.

Amblyraja georgiana (Antarctic starry skate), *A. taaf* (whiteleg skate) and *Bathyraja meridionalis* (dark-belly skate) all occur as bycatch in Southern ocean fisheries for toothfish (*Dissostichus* spp.). The aim of management in recent years has been to avoid catches of these species, and the Commission for the Conservation of Antarctic Marine Living Resources (CCAMLR) recommends that any live skate should be cut from the line and released during fishing operations. Off South Georgia the bycatch of skates

and rays peaked at 90 t in 1994 and had declined to no more than 2 or 3 t per annum by 2012-2103. A bycatch limit for skates and rays in the CCAMLR area was set at 130 t per annum for 2012 (Rogers et al., 2015). Outside the CCAMLR area around the Falkland/Malvinas islands there is evidence that *Bathyraja meridionalis* is increasingly retained for landing by some vessels.

Dipturus doutrei (javelin skate, violet skate) appears in bycatch in fisheries for hakes (*Merluccius* spp.) in the SW Atlantic off Southern Africa. In the NE Atlantic and Mediterranean *Dipturus oxyrinchus* (longnosed skate) is severely depleted at the shallow end of its depth range, and it is increasingly caught at greater depths, predominantly at 200-500 m in the Mediterranean Sea. It is a large skate, up to 1.5 m long and presumed to be slow growing and unable to survive high levels of exploitation. In the Mediterranean it is caught either by longline or as a bycatch in the deep trawl fishery for prawns, *Nephrops* and *Aristeus* spp. In the NE Atlantic the TAC for skates and rays has been progressively decreased with a view to avoiding possible species extinctions. Deep-water species fall within this TAC. *Leucoraja fullonica* (Shagreen ray) is taken throughout its range in the NE Atlantic and Mediterranean as a bycatch in deep-water trawl, longline and gillnet fisheries. *Raja maderensis* (Madeiran ray) occurs in E Atlantic around the islands of Madeira and the Azores, where it is taken in longline fisheries. There is concern about an apparent decline in the stock and its vulnerability considering the restricted range. In the NE Pacific Ocean *Raja rhina* (longnose skate) appears as a bycatch in deep-water commercial longline and trawl fisheries from the Gulf of Alaska to Southern California. For skate (all species) off California the recent trend of catches has been for a decrease from over 600 t in 2000. Off Canada, 300-400 t per annum are caught in trawl fisheries and a similar amount as bycatch in the halibut longline fishery. Since 2007 no directed fisheries for skate are allowed in Canadian waters but a directed fishery has developed in the Gulf of Alaska. Generally for deep-sea skates the trend is that fishing mortality as bycatch in fisheries for other species exceeds safe limits and where there is a management regime in place, directed fisheries are being prohibited and TACs decreased.

6.4.2.9 Albuliformes

Pterothrissus belloci (longfin bonefish) is caught in fisheries off Angola in the E Atlantic. Trials on processing the catch found the flesh to be very bony, quick to spoil but high in protein content (Batista et al., 2001).

6.4.2.10 Osmeriformes

A fishery for *Alepocephalus bairdii* (Baird's slickhead) developed in the NE Atlantic from around 1990. The catches peaked at 13 500 t in 2002 but thereafter have decreased to generally less than 500 t from 2011 onward. The largest proportion of the catch was on the Mid-Atlantic ridge together with grenadiers and redfish (*Sebastes mentella*). Allain et al. (2003) reported that most *A. bairdii* were caught at 1000–1250 m and were discarded by French vessels targeting roundnose grenadier (*Coryphaenoides rupestris*).

6.4.2.11 Argentiniformes

The fisheries for Argentines or silver smelts in the NE Atlantic have been considered earlier.

In the North Pacific *Glossanodon semifasciatus* (deep-sea smelt, Nigisu) has been an important component of fisheries in Japanese waters (Nishimura, 1966) comprised of distinct stocks with a catch of about 1000 tonnes in 2004 in the Pacific Ocean and 3000 tonnes in the Sea of Japan (Nashida et al., 2007) with evidence of decline.

6.4.2.12 Stomiiformes

Maurolicus muelleri (silvery lightfish)has been reported to occur worldwide with high abundances off Japan (Ikeda, 1996) around seamounts in the Pacific Ocean (Boehlert et al. 1994) off South Africa (Armstrong and Prosch, 1991) and on ridges and slopes in the North Atlantic. The genus is now regarded as comprising 15 allopatric species with *Maurolicus muelleri* confined to the NE Atlantic (Parin and Kobyliansky, 1996). The USSR and Russian Federation recorded catches of 300–9000 t between 1980 and 1988. From 2008 Iceland and the Faroe Islands began fishing trials for *Maurolicus muelleri* in the North Atlantic landing 46,500 t in 2009, but there is little evidence of development of a sustained fishery.

6.4.2.13 Aulopiformes

The greeneyes (*Chlorophthalmus* spp.) are probably widely exploited in small-scale fisheries around the world, but catches are not well reported. The shortnose green eye, *Chlorophthalmus agassizi*, is abundant in the central basin of the Mediterranean, occurs in deep water trawl catches in Greek waters (Anastasopoulou et al., 2006) but is usually discarded around the Mediterranean. *C. borealis* (big-eyed greeneye, mehikari) is exploited in deep-water fisheries off Japan and Taiwan, and a market is being developed. *C. albatrossis* (aome-eso) was formerly discarded from Japanese offshore bottom trawl fisheries but has become an important commercial species with over 500 t landed into Fukushima province in 2003 (Hirakawa et al., 2007). During 1979 to 1983 the USSR reporting catches of up to 2023 tpa of greeneyes.

6.4.2.14 Myctophiformes

The mesopelagic lantern fishes of the order Myctophiformes occur in enormous abundance in the world's oceans and might be regarded as the major hitherto underexploited fishery resource. Gjøsæter and Kawaguchi (1980) estimated that the global biomass of mesopelagic fishes to be approximately 1 Gt; a large proportion of which are myctophiformes. Kaartvedt et al. (2012) revised that estimate upward to *ca.* 10^{10} t. If only 1 per cent of that could be harvested, this would be equal to the current global fish catch of *ca.* 100 Mt. Against this background, the current total global reported harvest of lantern fishes of less than

20,000 t is very disappointing. Gjøsæter and Kawaguchi (1980) identified the NW Indian Ocean as the region with the greatest potential resource. High abundances of Hector's lantern fish (*Lampanyctodes hectoris*) were also reported in the SE Atlantic. A fishery for Hector's lantern fish began off South Africa in 1970, peaked at 42,400 t in 1973, continued to 1999 with varying quantities and little has been landed since then. The high oil content creates difficulties in processing of the catch and has restricted market demand.

At least 17 nations have undertaken experimental fisheries for lantern fishes from the 1970s onward, and the South African fishery for Hector's lantern fish has been most successful. The USSR reported a peak catch of 30,000 t in 1988. Iran accounts for the entire reported catch of skinnycheek lanternfish (*Benthosema pterotum*) in the NW Indian Ocean from around 2000 tpa in the 1990 reaching a peak of 14,000 in 2010. There is a fundamental problem that the efficiency of capture of myctophids by nets is extremely low, of the order of 1–2 per cent (Gjøsæter and Kawaguchi, 1980). Commercial fisheries can probably only be developed through the use of a completely new technology, possibly an artificial intelligence robotic vehicle that can target and capture individual myctophids targeting areas of highest abundance. However, it is probably better to harvest and efficiently manage the stocks of predatory species that feed on myctophids rather than to fish further down the food chain, which has other inherent dangers. Pauly et al. (1998) assert that owing to their small size and extreme dilution in the oceans (average biomass $<1\text{g m}^{-3}$) lantern fish are likely to continue to remain latent rather than exploitable resources.

6.4.2.15 Lampriformes

The Lampriformes are pelagic, tend to be solitary and do not occur in high abundances. However, they are large, and the Opah or moonfish (*Lampris guttatus*) has become popular in restaurants and fish markets of Hawaii as well as around the Pacific region. Since 1995, an upward trend in landings has been reported from French Polynesia, New Zealand and the USA,

with the USA taking about 80 per cent of the global reported catch of 1300 t in 2014. They are generally captured incidentally in fisheries for tunas and billfish as a high-value bycatch. Although small-scale landings of dealfish and ribbonfishes are reported (family Trachipteridae), peaking at 1940 t of dealfish (*Trachipterus arcticus*) in the North Atlantic taken by the Faroe Islands in 2009, there is no evidence of market demand for these as food fish.

6.4.2.16 Polymixiiformes

The beardfishes of the genus *Polymixia* occur around oceanic islands and seamounts in tropical and subtropical regions and are widely exploited in small-scale fisheries. The silver eye (*Polymixia japonica*) is listed by Chapman et al. (2008) as a species potentially caught by Pacific Ocean deep-bottom snapper fishermen. The only official landings statistics (< 1 t) are from Spain and Portugal from fisheries around the Canary Islands, Madeira and the Azores in the NE Atlantic where the Stout beardfish (*Polymixia nobilisis*) appears as a bycatch in fisheries for Alfonsinos.

6.4.2.17 Zeiformes

Large-scale exploitation of dories and oreos began with the Soviet Union in 1977 followed by New Zealand fishing in its newly established 200-mile fishery zone. In 1981 the total catch peaked at 47,000 t. The total catch then declined to 23,000 t in 2000 and to 10,500 t by 2015. Small quantities of Warty dory (*Allocyttus verrucosus*) are taken off Japan. The catch of spiky oreo (*Neocyttus rhomboidalis*) off Australia peaked at 620 t in 1998, followed by the smooth dory (*Pseudocyttus maculatus*) off New Zealand at 14,192 t in 2004 and black oreo (*Allocyttus niger*) off New Zealand at 6136 t in 2006. The latter two species have been managed off New Zealand under a quota management system both as a targeted fishery and as a bycatch with orange roughy. These are valuable species, but catches have been close to zero since 2012.

6.4.2.18 Gadiformes

As the dominance of gadiformes in deep-sea fisheries has already been discussed, here we mention the lesser taxa not previously dealt with. The Moridae or deep-sea cod family seems an obvious candidate for exploitation because they form a conspicuous component of the deep demersal fish fauna, with large body size and occur in reasonable abundance (Merrett and Haedrich, 1997). In the original description of the species Risso (1810) states that the flesh of the Common mora (*Mora moro*) is tender white with a good taste but with a strong odour. They are probably captured quite widely in quantities that do not appear in fishery statistics. According to FAO, landings of blue antimora (*Antimora rostrata*) have been reported from 18 different nations since 1995 with just over 200 t landed in 2008 followed by a gradual decline. The common mora (*Mora moro*) is exploited in two distinct fisheries, in the North Atlantic including the Mid-Atlantic Ridge, with peak landings of about 500 t in 2003 (ICES, 2015) and a larger fishery off New Zealand where it is known as the Ribaldo. In the mid-1970s Japanese and Korean vessels captured up to 4,920 t per annum of Ribaldo but under the New Zealand quota management system the TAC since 2006 has been 1,683 t and recorded catches somewhat lower. There is no TAC set for *Mora moro* in the North Atlantic.

6.4.2.19 Ophidiiformes

The fisheries for the pink cusk-eel (*Genypterus blacodes*) has already been described (Fig. 6.6m). There is a long history of exploitation of kingklip (*Genypterus capensis*) in the SE Atlantic by South African vessels, averaging ca. 1,400 t per annum from 1950 to 1965, after which fishing effort began to increase as foreign vessels appeared, culminating in a reported catch of 21,200 t in 1973, over half of which was taken by Spain. Vessels from 19 nations participated at various times in the SW Atlantic fishery. Since 2000 annual landings have been less than 10,000 t, mostly taken by vessels from Namibia and South Africa. The red-cusk eel (*Genypterus chilensis*) is exploited apparently exclusively by Chilean vessels with a decline in catches from an average of 2,800 t throughout the 1950s to ca. 500 t in recent years. The fishery for black-cusk eel (*Genypterus maculatus*) was developed by Chile from 1980 onward, reaching a peak of ca. 7,000 t in 1990, though recent landings are below 300 t. Other species of Ophidiiformes are caught in small quantities, Filament cusk (*Homostolus acer*) and Armoured cusk (*Hoplobrotula armata*) are reported from fish markets in Japan (Nielsen et al., 1999), and the pudgy cusk-eel (*Spectrunculus grandis*) is captured in deep-water trawls in the Eastern Atlantic Ocean (Knudsen et al., 2015).

6.4.2.20 Stephanoberyciformes

The Melamphaidae are generally regarded as of no commercial importance (Anderson, 1986). The Southern Ocean Nordenskjold's bigscale (*Sio nordenskjoldii*) is recorded as a rare bycatch in fisheries for orange roughy off New Zealand and mesopelagic net tows in the Scotia Sea (Collins et al. 2012). Although widely cited as a minor commercial fishery species, records of exploitation are elusive but the species may have been considered as a potential deep pelagic fishery resource together with lantern fishes.

6.4.2.21 Beryciformes

The orange roughy and alfonsinos are commercially the most important species in this order and have been dealt with earlier. Two of the relatives of the orange roughy in the family Trachichthyidae, Darwin's slimehead (*Gephyroberyx darwinii*) and black slimehead (*Hoplostethus cadenati*), are reported from deep-water trawl fisheries in the Atlantic Ocean off West Africa and are utilised to a small extent (Iwamoto, 2015).

Centroberyx affinis, known as redfish in Australia and red snapper in New Zealand, probably has a long history of exploitation but appears in FAO statistics only from 1995 onward. Catches in New Zealand waters peaked at ca. 200 t in 2002, whereas off South Australia catches reached ca. 2,000 t in 1998, since when they have declined to ca. 300 t and are now considered to be overfished off New South Wales. The Holocentridae, Kai soldierfish (*Ostichthys kaianus*, 640 m$_{fbd}$) and Japanese soldierfish (*O. japonicus* 350 m$_{fbd}$) are captured in the Pacific Ocean artisanal deep-bottom snapper fishery (Chapman et al. 2008).

6.4.2.22 Cetomimiformes

Although it occurs circumglobally the velvet whalefish or red whalefish, *Barbourisia rufa* is found in greatest abundance in the Western Pacific off Japan and New Zealand. In 2010, 210 t were reported as captured by the Cook Islands declining to an average of 45 t per annum since that time (FAO, 2016).

6.4.2.23 Syngnathiformes

The longspine snipefish, *Macroramphosus scolopax*, is caught in pelagic fisheries at upper slope depths with reported catches of up to 32,500 t per annum during 1976–1986 (average 18,000 t) mainly off Portugal. Although the fishery has since declined to almost zero (Marques et al., 2005), a small fishery has developed in the Southern Hemisphere off New Zealand and the Cook Islands landing no more than 200 t per annum.

6.4.2.24 Scorpaeniformes

Species of the Scorpaeniformes are widely exploited, and the major deep-sea species, Pacific ocean perch, *Sebastes alutus*, beaked redfish *Sebastes mentella* and sablefish, *Anoplopoma fimbria*, have already been discussed. About 20 other species (Table 6.1) can be regarded as deep-sea exploited species often in mixed species fisheries such as the ocean perch fishery for *Helicolenus* spp. off Australia (Rowling et al., 2010). Blackbelly rosefish or bluemouth, *Helicolenus dactylopterus*, is exploited throughout its range in the North and South Atlantic and the Mediterranean. There was a long-established fishery off South Africa that averaged *ca.* 750 t per annum from 1950–1975. Argentina began a similar size fishery from 1965 onward. From 1976 to 1986 there was a boom in blackbelly rosefish fishery with over 25,000 t reported in 1976 mostly by the USSR, reaching a peak of 37,000 t in 1984 involving seven nations. However, since 1987 the average global catch has been less than 10,000 t. By virtue of its widespread fragmented distribution the species is not considered as threatened (Nunoo et al., 2015). Other species of Sebastidae do not necessarily appear in statistics, and the recent appearance of small quantities of splitnose rockfish (*Sebastes diploproa*) on the US West Coast from 2010 and spiny scorpion fish (*Trachyscorpia cristulata echinata*) from 2004 in the NE Atlantic probably reflects improved reporting. The offshore rockfish (*Pontinus kuhlii*) is reported by Portugal from 2000 onward, and red scorpion fish (*Scorpaena scrofa*) from 1999 by Portugal, Spain, Croatia and the United Kingdom. Catches of spotted gurnard (*Pterygotrigla picta*) are reported from New Zealand from 1996. The skilfish (*Erilepis zonifer*) is represented by one year of reported of catches in 2009 by Russia in the N Pacific.

6.4.2.25 Perciformes

The 40 or so exploited deep-sea Perciform species represent a minute fraction (0.5 per cent) of the total number of marine species in this order. This reflects the fact that most Perciform species live in shallow seas. The major exploited species, Patagonian toothfish (*Dissostichus eleginoides*), oilfish (*Ruvettus pretiosus*) and boarfish (*Capros aper*) have been discussed earlier. The next most important species are the Trichiuridae, black scabbardfish (*Aphanopus carbo*) and silver scabbardfish (*Lepidopus caudatus*). The black scabbardfish has been captured by artisanal longline fishermen off the Island of Madeira in the NE Atlantic since at least about 1800 (Machete et al., 2011). This fishery expanded to offshore mainland Portugal from 1983 (Figuerdo et al., 2003) and more recently around the Azores. From 1986 onward black scabbardfish began to be landed in the developing deep-water trawl fishery for roundnose grenadier (*Coryphaenoides rupestris*) and blue ling (*Molva dypterygia*) in the NE Atlantic to the west of the British Isles and around Iceland. Total landings peaked at over 20,000 t in 2011 (FAO, 2016). Spawning occurs in the south of the species distribution around Madeira and the Azores, and it is likely that trawl fisheries in the north are exploiting the juvenile and prespawning components of a single stock during a migratory cycle, which covers a large part of the NE Atlantic (Farias et al., 2013). The silver scabbardfish or frostfish is exploited throughout its range in the Northern and Southern Hemispheres. A small-scale Portuguese fishery landed around 3,000 t per annum throughout the 1950s, which gradually grew and together with landings mainly by Poland peaked at 21,000 t in 1977. The fishery off South Africa and New Zealand developed from 1980 onward, reaching a second peak of 27,000 t combining North Atlantic and Southern Hemisphere landings (FAO, 2016).

None of the other Perciform species supports a major fishery. Wreckfish (*Polyprion americanus*) are typically taken in small-scale fisheries or as bycatch throughout their range, with Spain and Portugal recording the largest catches. The Hapuki wreckfish (*Polyprion oxygeneios*) is mainly taken by New Zealand with a consistent average catch of about 1500 tpa since 1950. The black cardinal fish (*Epigonus telescopus*) is mainly exploited off New Zealand; the fishery peaked at over 4000 t during 1994–2000 (FAO, 2016), but has since declined, and the stock is regarded as depleted. Great Northern tilefish (*Lopholatilus chamaeleonticeps*) is a highly desirable fish in US waters of the W Atlantic. Since 1950 the average total US reported catch has been *ca.* 1,400 tpa, but Fisher et al. (2014) show that the catches have historically been highly variable related to environmental oscillations and that the stocks are now depleted. The Cape Bonnetmouth (*Emmelichthys nitidus nitidus*) is exploited in two main fisheries, off New Zealand, where it is known as Red Pearl Fish at *ca.* 2,500 tpa and off South Africa *ca.* 400 tpa. The main fishery for rubyfish (*Plagiogeneion rubiginosum*) is off New Zealand at ca. 500 tpa.

The Lutjanidae (snappers) contains over 100 species in tropical and subtropical regions of the world.

Gomez et al. (2014) identified the five genera of the subfamily Etelinae: *Aphareus, Aprion, Etelis, Pristipomoides* and *Randallichthys* as 'deep-sea snappers'. They are exploited by artisanal fishermen in the Pacific region and provide an important source of protein for human populations. These fisheries tend not to appear in official statistics and the typical depth range (100–400 m) is shallower than the usual definition of deep-sea fisheries.

The North Pacific armorhead or slender armorhead (*Pentaceros wheeleri*) has a remarkable exploitation history. Trawlers from the Soviet Union and Japan began targeting this species on the Emperor seamounts in international waters NW of Hawaii in 1968. The fishery only lasted for 10 years, peaking at 177,300 t, plunging down to 900 t by 1978 as the entire stock had been exhausted through a period of intense fishing with no chance of recovery (Koslow, 2007). For a short time this was the largest deep-water fish resource in the world. The Southern Hemisphere Pelagic armorhead (*Pentaceros richardsoni*) was never as productive, and the peak reported catch was 1,400 t in 1996.

The bluenose warehou (*Hyperoglyphe antarctica*) and white warehou (*Seriolella caerulea*) are mainly exploited off New Zealand with peak catches of 3200 t and 3300 t in 2004 and 2007, respectively.

6.4.2.26 Pleuronectiformes

The flatfishes of the order Pleuronectiformes are widely exploited at shelf depths, and those species whose range extends into or is predominantly in deep water are pursued by deep-sea fisheries. To some extent the choice of species listed in Table 6.1 is arbitrary excluding, for example, the Atlantic halibut (*Hippoglossus hippoglossus*) that is recorded down to 2000 m depth. Catches of Megrim (*Lepidorhombus whiffiagonis*) in the NE Atlantic and Mediterranean increased steadily from 3700 t in 1950 to 18,000 t in 1995 since when there has been a decline to 15,000 t in 2014 with evidence of decline in spawning stock biomass. The West Coast sole (*Austroglossus microlepis*) was exploited by South Africa with an average catch of 1,400 t during the 1950s reaching a peak of over 11,000 t in 2007 when landings from Angola and Namibia were included.

6.4.2.27 Lophiiformes

The various species of the genus *Lophius* are all exploited in respective regions. The blackbellied angler (*Lophius budegassa*) is often caught with *L. piscatorius*. The blackfin goosefish (*L. gastrophysus*) is mainly caught in SW Atlantic off Brazil. The shortspine African angler (*L. vaillanti*) is caught in small quantities off West Africa and may be simply a southern stock of *L. piscatorius*. The fishery for devil anglerfish (*L. vomerinus*) in the SE Atlantic was developed from 1967 initially by the USSR and then by South Africa, Spain and Namibia, reaching a peak of 24,500 t in 1998, since when there has been a decline to less than 10,000 tpa in recent years.

None of the rarer deeper-living angler fishes such as the pelagic Ceratioids are of any commercial interest.

6.5 Overview of the Fishing Trends

The global growth of deep-sea fish production over the past 50 years has been the result of exploitation of a succession of different species and stocks, each of which peaked for a few years before being replaced by another species. Some stocks such as the orange roughy (*Hoplostethus atlanticus*) in the North Atlantic and slender armorhead (*Pentaceros wheeleri*) of the North Pacific seamounts were fished to economic extinction, whereas others have been stabilised at a much lower level of production. The massive contribution of the blue whiting accounting for over half the total deep-sea fish catch in recent years masks the underlying trend that the global deep-sea fishery production from all other deep-sea species has plateaued at about 1 Mt, and there is little prospect of that quantity increasing. Deep water fishing has the potential for providing important protein supplies in developing countries such as in Pacific Polynesia

(Gomez, 2015) and Southern Africa (Payne and Punt, 1995) but it is the developed countries (including US, Japan and Europe) that consume over 73 per cent of total available seafood exports (Tacon and Metian, 2015).

Merrett and Haedrich (1997) pointed out that deep-sea fisheries are inherently less productive than those in shelf seas and compare the deep-sea fishing industry to a roving predator in a sparse environment. The strategy for such a predator is to search, exploit, deplete, move on and search again. This is exemplified by the trends in Figure 6.6, as resource after resource has been depleted. Merrett and Haedrich (1997) commented on the development of the Patagonian toothfish fisheries occurring as they wrote their book. Two decades later, it is remarkable that new resources continue to be found including the boarfish in a previously well-explored area of the North East Atlantic. Consumers in developed countries increasingly demand that the fish they buy is derived from sustainable sources. The response of the fishing industry has been to seek certification of sustainability of particular stocks or fisheries from organisations such as the Marine Stewardship Council (MSC, 2016). Some deep-sea stocks such as the Faroese fishery for Greater silver smelt, Cape hakes off South Africa, sablefish off Alaska, Argentine hoki, some fisheries for Patagonian toothfish and New Zealand fisheries for southern blue whiting, pink cusk-eel and hoki have been certified as sustainable. These generally produce much smaller tonnages than in the past and represent a small fraction of the total deep-sea catch. Good practice by a certain group of fishers and managers is recognised, and it is presumed that the stocks concerned can keep producing more or less constant amounts of fish.

6.6 The Impacts of Deep-Sea Fishing

It is axiomatic that fishing removes target species. In shelf-seas fisheries it is assumed that having reduced the population size, density-dependent effects result in enhanced growth and recruitment tending to replace the lost biomass. Continuous harvesting thus stimulates production and a maximum sustainable yield is typically achieved when the stock biomass is reduced to *ca.* 30 per cent of the unexploited stock size (Beverton and Holt, 1957). With annual spawning and recruitment, fast growth and early maturation these assumptions work reasonably well. However in deep-sea species with slow growth rate, long lifespan, episodic spawning and sporadic recruitment the stimulatory effect of fishing activity is absent. Fishing simply removes biomass from the stock. Article 61 of the United Nations Law of the Sea requires that harvested living resources within exclusive economic zones are not endangered through exploitation and should be maintained at levels that can produce a maximum sustainable yield (UN, 1982). For many deep-sea species it is doubtful if a sustainable yield in this sense is possible. In Canadian waters of the NW Atlantic, Devine et al. (2006) showed that over a 17-year period five species caught in the deep bottom trawl fisheries either targeted roundnose grenadier (*Coryphaenoides rupestris*) and roughhead grenadier (*Macrourus berglax*) or as bycatch; blue antimora (*Antimora rostrata*), spiny eel (*Notacanthus chemintzi*) and spinytail skate (*Bathyraja spinicauda*) had been fished down to a level defined as critically endangered by the IUCN. The two targeted species were reduced by 99.6 per cent and 93.6 per cent, respectively, clearly in violation of UNCLOS. Gordon (2003) presented evidence that in the NE Atlantic most deep demersal fish stocks in the Rockall Trough area had been fished to below safe biological limits. In the Porcupine seabight area orange roughy (*Hoplostethus atlanticus*) was reduced by 94 per cent and roundnose grenadier (*Coryphaenoides rupestris*) by 41 per cent after the onset of the commercial fishery for these species (Priede et al., 2011). It appears that most small stocks of orange roughy around the world were fished to economic extinction before effective management could be implemented (Norse et al., 2012). The largest orange roughy stock in New Zealand waters, the Chatham Rise stock, was reduced to 20 per cent of virgin biomass, and there are few signs of biological compensation such as increased growth rates and improved recruitment (Clark, 2001).

Deep-sea demersal fisheries typically occur at depths where fish biodiversity is high (Priede et al., 2010) and inevitably catch numerous nontarget species. In the study by Devine et al. (2006) in the NW Atlantic three of the species identified as endangered were not the targets of any fishery. Many deep-sea fishes are fragile and are unlikely to survive passing through the meshes of a trawl or if discarded back into the sea. In a rare study spanning the time from before a fishery began, data on abundance of demersal fishes were collected using research trawls at depths of 800 to 4800 m during the years 1977–1989 in the Porcupine Seabight area of the NW Atlantic (Merrett et al., 1991a, 1991b). A commercial fishery trawl targeting roundnose grenadier, orange roughy and black scabbardfish developed from 1988 onward at depths down to 1500 m. A second phase of scientific sampling was then undertaken during 1997–2002 when the commercial fishery had been well established for about 10 years (Bailey et al., 2009). The effect of the commercial fishery has been to reduce by 53 per cent the total abundance of demersal fishes at 1,000 m depth (Fig. 6.7). Significant decreases were detected in nine nontarget species, the abyssal halosaur (*Halosauropsis macrochir*), smallmouth spiny eel, (*Polyacanthonotus rissoanus*), Kaup's arrowtooth eel (*Synaphobranchus kaupii*), spearsnouted grenadier (*Coelorinchus labiatus*), Günther's grenadier (*Coryphaenoides guentheri*), Mediterranean grenadier (*C. mediterraneus*), common Atlantic grenadier (*Nezumia aequalis*), roughnose grenadier (*Trachyrincus murrayi*) and blue antimora (*Antimora rostrata*; Bailey et al., 2009). The fishery impinged on the depth ranges of 74 nontarget fish species, and its effect was highly significant down to depth of 2520 m and depletion extended down to 3086 m. Fish are mobile, moving up and down the slope during the course of their lifecycle, hence although the commercial fishery was confined to depths < 1500 m, its effects after 10 years were evident down to twice that depth. The area directly targeted by the fishery was 52,000 km^2, but the effect was transmitted by fish movements to 2.74 times that area, i.e., 142,000 km^2. The fish biomass decreased by 57 per cent over the entire area (Godbold et al., 2013).

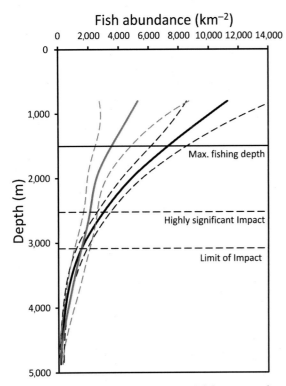

Figure 6.7 The effect of a deep-sea trawl fishery on total abundance of demersal fishes (all species) in the Porcupine Seabight area of the NE Atlantic Ocean. (Data from Bailey et al., 2009).
Black line – Abundance of fish from surveys during 1977–1989 before commercial fishing began.
Grey line – Abundance of fish from surveys during 1997–2002 after the onset of commercial fishing.
Dashed curves – upper and lower 95 per cent confidence limits on the fitted curves. Horizontal lines indicate, 1500 m the maximum depth of commercial fishing activity, 2520 m the depth at which the 95 per cent confidence intervals intersect, 3086 m the maximum limit of the depletion effect.

Deep-water trawling is therefore indiscriminate, exerting mortality on 20 times more species than are actively harvested and impacting a much larger area than is actually targeted by the fishing vessels.

Where trawls contact the seafloor there is an obvious destructive effect on conspicuous biological structures that are a source of heterogeneity and biodiversity in the deep ocean margins. These include corals, sponges, stalked crinoids (sea lilies), worm tubes and other organisms that vary in size from small solitary individuals (< 25 cm high), medium-sized (25–300 cm) to large reef-forming corals and sponges (> 3 m; Buhl-Mortensen, 2010). These three-dimensional structures provide refuges and habitats for a wide variety of life including fishes (Ross and Quatrini, 2007; Milligan et al. 2016). Together with hydrothermal vents and cold seeps they are recognised as vulnerable marine ecosystems (VMEs) that should be protected from damage by fishing under the 2006 United Nations General Assembly resolution 61/105 (Auster et al., 2011). Growth rates of these habitat-forming fauna are slow with estimated ages ranging from tens to thousands of years. It is hardly surprising that on previously trawled seamounts off New Zealand and Australia on which fishing is now prohibited there has been no sign of recovery after 5–10 years, and full recovery could occur on time scales of centuries or longer (Clark et al., 2015).

In addition to removing biogenic structures, persistent trawling on the upper slopes has the effect of smoothing the topography by removing protuberances and suspending sediment in the water column, which is exported down-slope by gravity flow, eventually settling down to fill any depressions. This smoothing effect has become apparent in changes in sea-bed mapping in the vicinity of canyons in the Western Mediterranean; the amount of sediment transported is equal to that produced by natural geological processes. Trawling is therefore accelerating the rate of erosion of the continental slope (Puig et al., 2012). The seafloor sediments in the trawled area, owing to the constant winnowing effect of trawling, are increased in density and impoverished in organic carbon (Martin et al., 2014). The trawling removes 60–100 per cent of the daily input of organic carbon leading to 80 per cent reduction in small meiofauna living in the sediments, 50 per cent reduction in biodiversity and 25 per cent reduction in species richness of nematodes (Pesceddu et al., 2014). Thus even in open sediment-covered areas where there are no visually obvious VMEs, the environment is profoundly impoverished impairing deep-sea biodiversity and ecosystem functioning.

6.7 Conclusions

There is growing consensus that the damage caused by deep-sea bottom trawling outweighs the economic and social benefits of the industry (Norse et al., 2012). Whilst deep-sea fishing is stagnating at less than 3 Mt production per year, global aquaculture continues to expand, producing 66.6 Mt in 2012, of which 24.7 Mt was marine species, more than 30 and 10 times the global deep-sea fish production, respectively. Furthermore total aquatic food production, marine and freshwater, is very small compared with the total for terrestrial agriculture, 3,982 Mt in 2011 (Tacon and Metian, 2015) so that deep-sea fishing contributes less than 0.1 per cent to global food security. Other marine industries such as deep-sea mining, gas and oil extraction are expected to produce environmental impact assessments before any activity can commence. Such a procedure for fisheries is provided for in official guidelines (FAO, 2009) but is rarely implemented (Auster et al., 2011). Following environmental assessments and public consultation, in 1995 the Shell Oil Company was prevented from disposing of the disused Brent Spar oil storage buoy in the deep waters of the NE Atlantic (Owen and Rice, 1999). This would have directly impacted *ca.* 0.005 km^2 of seafloor and arguably might have enhanced habitat heterogeneity because the deep-water coral *Lophelia pertusa* was growing on it. At the same time the deep-sea fishing industry has been allowed to alter the environment over thousands of km^2 with minimal public scrutiny. Benn et al. (2010) showed that in the NW Atlantic at depths greater than 200 m, bottom trawling has an impact at least an order of magnitude greater in area than all other human activities combined.

The General Fisheries Commission for the Mediterranean (GFCM) in 2005 adopted a total ban on trawling in the Mediterranean Sea at depths greater than 1000 m (Garcia et al., 2014). Clarke et al. (2015) propose that in view of the diminishing value of catch and the increase in bycatch with depth, bottom trawling should be restricted to depths less than 600 m in European waters. Such a limit, however, would prohibit most of the fisheries in New Zealand waters; depth limits should be tailored to realities in different regions. Norse et al. (2012), arguing that deep-water fisheries in general are not particularly profitable and are sustained by subsidies capable of generating a much greater return if invested in rebuilding stocks in highly productive coastal areas, propose the ending of all deep-sea fishing on the high seas outside EEZs. In these high seas areas, fisheries are managed by Regional Fishery Management Organisations (RFMOs) that have the power to manage fisheries according to ecosystem approaches. In the North Atlantic the RFMOs have declared large areas of the Mid-Atlantic Ridge as no-fishing zones (O'Leary et al., 2012); nevertheless large areas of the globe, the Arctic, Central and SW Atlantic Oceans are not yet covered by RFMOs.

Throughout the twentieth century fishery management can be perceived as a failure with catches only sustained by improvements in technology and continual increases in effort at the expense of a decline in the underlying resources. Thurstan et al. (2010) show that catch per unit effort in British demersal fisheries has declined by 94 per cent from 1889 to 2007. Ling declined by 95.7 per cent and the continuing catches shown in Fig. 6.6a reflect a shift of the fishery to the deep-sea and a vast increase in effort. In 1889 a fleet of mainly sailing vessels landed twice the catch attained by the present-day high-powered fleet. Thus the pattern of successive depletion of resources in deep water from the 1950s onward can be regarded as simply a continuation of what has happened previously in shelf seas (Koslow et al., 2000). It is argued by Fernandes and Cook (2013) that since 2002 there has been a revolution in European fishery management with strict controls on effort leading to increases in stock biomasses. Various other stocks around the world are beginning to be regarded as sustainable (MSC, 2016). These reversals, however, come from extremely low baselines relative to the virgin biomasses before industrialised fishing began. Such optimism is based on the lack of a collective memory of how abundant fish stocks were in the past: the so-called shifting baseline syndrome (Pauly, 1995).

Epilogue

Two centuries have passed since Risso (1810, 1826) first began the process of systematically collecting and describing deep-sea fishes. The circumnavigation of the globe by the HMS *Challenger* (1873–1876) revealed the existence of deep-sea fishes in all of the world's major oceans, and the voyage of *Valdivia* (1898–1899) found fishes throughout the water column from the surface to the abyss, the largest living space on the planet. New species continue to be described, and Yancey et al. (2014) finally showed that there is a maximum depth limit, 8400 m, beyond which fishes cannot live. From best-selling books describing early voyages of discovery (Wyville Thomson, 1873) to twenty-first-century electronic media, there has been continuing widespread public fascination with the natural history of the creatures of the deep. Children's books and magazines feature pictures of weird deep-sea fishes that rival dinosaurs in exciting the imagination of the young. The discovery of life in the deep has produced some of the most iconic images of nature on this planet.

In parallel with the sense of wonder and curiosity regarding deep-sea fishes, from the beginning there has been an interest in their potential for exploitation. Antoine Risso described the flavour and palatability of the new species he was discovering whilst also expressing concern about the depleted state of fisheries. He argued for efforts to determine the cause of the decline in fish stocks and the means of restoration of their former glory (Risso, 1826). He called for regulations to restrict the use of destructive nets and the resources necessary to enforce those regulations. During the 1910 North Atlantic Deep-Sea expedition on board the RV *Michael Sars*, Murray and Hjort (1912) made some large catches of deep-sea fishes and speculated that improved technology might *render fishing profitable even down to 500 fathoms (914m) or more*. Such improvements in technology did not materialise until the 1960s, when industrial-scale fishing of the deep sea began and catches rapidly grew up to 3 million tonnes per annum by the year 2000.

Rowe et al. (1986) cast doubt on the sustainability of deep-sea fishing. They calculated that the area of deep-sea floor required to sustain an idealised catch of one fish per day, equivalent to 200 kg per year that could support a family or small community, is approximately 100 km^2. This showed that there is insufficient food energy in the deep sea for a realistic fishery. How is it that deep-sea fishing has continued to grow apparently in violation of the laws of thermodynamics? First, commercial deep-sea fishing is shallower than the 4000 m depth used in these calculations, but even at mid-slope depths where the subsistence area would be 10 km^2, economic viability seems untenable. The calculation assumes an equilibrium fishery at a fixed location, but the deep-sea fishing industry has depended on taking high initial yields from a succession of virgin stocks and moving on once a particular stock had been exhausted. Species that aggregate on seamounts and other seabed features also import production from a much wider area than the footprint of the fishery, further increasing the potential yield. In Chapter 6, the rise and fall of the top 20 deep-sea exploited species today is described. However there are other species that were greatly exploited in the past that have dropped out of the top 20, such as the pelagic armorheads (*Pentaceros* spp.), which averaged over 150,000 tonnes per annum for a short period of time but now produce only a few hundred tonnes. Over 60 species of deep-sea Chondrichthyes, chimaeras, sharks and rays, which were formerly exploited, in some cases for centuries, are now protected and can no longer be fished legally in most jurisdictions. The growth of deep-sea fishing has been at the expense of the devastation of at least one species per year since the 1960s. Antoinne Risso's original aim was to catalogue all of the natural resources of his native region; it is possibly significant that he had the foresight to direct his attention also to citrus fruits (Risso and Poiteau, 1818–1822), the global production of which now far exceeds that of deep-sea fisheries.

Deep-sea fishing is in inevitable decline as opportunities to exploit new areas, species and stocks diminish. Efforts at creating sustainable fishing must ultimately result in lower catches than the peaks achieved in the past. Depths greater than 1500 m are generally beyond the scope of direct fishing impact and may provide a refuge from exploitation for some deeper-living species. Major long-term threats to deep-sea fishes, as with all life on the planet, derive from trends of global climate change. Although deep-sea fishes are generally cold-water species, warming of the oceans itself may not be a direct threat. Much of the deep-sea fish fauna originated during the early Cretaceous when the deep sea was warm (Priede and Froese, 2013), and the Mediterranean Sea, which is warm down to a depth of over 5000 m, is populated by deep-sea fishes. Substantial changes may be expected in ocean ecosystems over the next 100 years driven by an increase in dissolved carbon dioxide (CO_2) and consequent ocean acidification resulting from burning of fossil fuels (IGBP, IOC, SCOR, 2013). Although the effects on deep-sea fishes are likely to be indirect through loss of coral habitats and changes in prey availability, larval stages of deep-sea fishes in the surface layers of the ocean may be directly affected by acidity (Wittmann and Pörtner, 2013). Breakdown of the thermohaline ocean circulation, predicted by several global climate change models (Hansen et al., 2016), could have a direct effect on life in the deep sea through lack of replenishment of dissolved oxygen below the thermocline resulting in a global anoxic event such as has occurred seven times during the last 500 million years of earth history (Takashima et al., 2006). This would theoretically lead to the extinction of all fish life in the deep sea. Rahmstorf (2006) rates breakdown of the global thermohaline circulation as one of a number of high-impact but low-probability risks associated with global warming. Wilson (1999) argues that, although widespread, global anoxia has never been complete during past events, and deep-sea life has survived through such apparent catastrophes. It seems probable that deep-sea fishes will outlive the perturbations of the planet's ecosystem perpetrated by the human species.

List of Deep-Sea Fish Families and Their Taxonomic Classification

The family reference numbers are as used in the text.

	Ref. No.
Class Myxine	
ORDER MYXINIFORMES	
Family Myxinidae	1
ORDER PETROMYZONTIFORMES	
Family Petromyzontidae	2
Class Chondrichthyes	
Subclass Holocephali	
ORDER CHIMAERIFORMES	
Family Rhinochimaeridae	3
Family Chimaeridae	4
Subclass Elasmobranchii	
Subdivision Selachii	
SUPERORDER SQUALOMORPHI	
ORDER HEXANCHIFORMES	
Family Chlamydoselachidae	5
Family Hexanchidae	6
ORDER PRISTIOPHORIFORMES	
Family Pristiophoridae	7
ORDER SQUATINIFORMES.	
Family Squatinidae	8
ORDER SQUALIFORMES	
Family Echinorhinidae	9
Family Dalatiidae	10
Family Oxynotidae	11
Family Centrophoridae	12
Family Squalidae	13
Family Somniosidae	14
Family Etmopteridae	15
SUPERORDER GALEOMORPHI	
ORDER LAMNIFORMES	
Family Odontaspididae	16

(cont.)

	Ref. No.
Family Mitsukurinidae	17
Family Megachasmidae	18
Family Alopiidae	19
Family Cetorhinidae	20
Family Lamnidae	21
ORDER CARCHARHINIFORMES	
Family Scyliorhinidae	22
Family Proscylliidae	23
Family Pseudotriakidae	24
Family Triakidae	25
Subdivision Batoidea	
ORDER TORPEDIFORMES	
Family Narkidae	26
Family Narcinidae	27
Family Torpedinidae	28
ORDER PRISTIFORMES	
ORDER MYLIOBATIFORMES	
Family Hexatrygonidae	29
Family Plesiobatidae	30
Family Urolophidae	31
Family Myliobatidae	32
ORDER RAJIFORMES	
Family Rajidae	33
Class Actinopterygii	
Subdivision Elopomorpha	
ORDER ELOPIFORMES	
ORDER ALBULIFORMES	
Family Albulidae	34
ORDER NOTACANTHIFORMES	
Family Halosauridae	35
Family Notacanthidae	36
ORDER ANGUILLIFORMES	
Anguilliformes I (demersal)	
Family Synaphobranchidae	37
Family Myrocongridae	38
Family Muraenidae	39
Family Chlopsidae	40
Family Derichthyidae	41
Family Colocongridae	42
Family Congridae	43
Family Nettastomatidae	44

(cont.)

	Ref. No.
Family Ophichthidae	45
Family Muraenesocidae	46
Anguilliformes II (bathypelagic)	
Family Saccopharyngidae	47
Family Eurypharyngidae	48
Family Cyematidae	49
Family Monognathidae	50
Family Nemichthyidae	51
Family Serrivomeridae	52
Family Anguillidae	53
Subdivision Otocephala	
ORDER ALEPOCEPHALIFORMES	
Family Platytroctidae	54
Family Bathylaconidae	55
Family Alepocephalidae	56
Subdivision Euteleostei	
ORDER ARGENTINIFORMES	
Family Argentinidae	57
Family Microstomatidae	58
Family Opisthoproctidae	59
Family Bathylagidae	60
ORDER STOMIIFORMES	
Family Gonostomatidae	61
Family Sternoptychidae	62
Subfamily Maurolicinae	
Subfamily Sternoptychinae	
Family Phosichthyidae	63
Family Stomiidae	64
Subfamily Astronesthinae	
Subfamily Stomiinae	
Subfamily Melanostomiinae	
Subfamily Idiacanthinae	
Subfamily Malacosteinae	
ORDER ATELEOPODIFORMES	
Family Ateleopodidae	65
ORDER AULOPIFORMES	
SUBORDER SYNODONTOIDEI	
SUBORDER CHLOROPHTHALMOIDEI	
Family Chlorophthalmidae	66
Family Bathysauropsidae	67
Family Bathysauroididae	68

(*cont.*)

	Ref. No.
Family Notosudidae	69
Family Ipnopidae	70
SUBORDER GIGANTUROIDEI	
Family Bathysauridae	71
Family Giganturidae	72
SUBORDER ALEPISAUROIDEI	
Family Scopelarchidae	73
Family Evermannellidae	74
Family Alepisauridae	75
Family Paralepididae	76
ORDER MYCTOPHIFORMES	
Family Neoscopelidae	77
Family Myctophidae	78
Tribe Notolychnini	
Subfamily Lampanytinae	
Tribe Diaphini	
Tribe Gymnoscopelini	
Tribe Lampanyctini	
Subfamily Myctophinae	
Tribe Electronini	
Tribe Myctophini	
Tribe Gonichthyini	
ORDER LAMPRIFORMES	
SUBORDER BATHYSOMI	
Family Veliferidae	79
Family Lampridae	80
SUBORDER TAENOSOMI	
Family Lophotodidae	81
Family Radiicephalidae	82
Family Trachipteridae	83
Family Regalicidae	84
ORDER POLYMIXIIFORMES	
Family Polymixiidae	85
ORDER ZEIFORMES	
Family Cyttidae	86
Family Oreosomatidae	87
Family Parazenidae	88
Family Zenionidae	89
Family Grammicolepididae	90
Family Zeidae	91
ORDER STYLEPHORIFORMES	
Family Stylephoridae	92

(cont.)

	Ref. No.
ORDER GADIFORMES	
SUBORDER MURAENOLEPIDOIDEI	
Family Muraenolepididae	93
SUBORDER MACROUROIDEI	
Family Macrouridae	94
Subfamily Macrourinae	
Subfamily Bathygadinae	
Family Macruronidae	95
Family Steindachneriidae	
SUBORDER GADOIDEI	
Family Trachyrinchidae	97
Sub-family Trachyrincinae	
Sub-family Macrouroidinae	
Family Moridae	98
Family Merlucciidae	99
Family Melanonidae	100
Family Euclichthyidae	101
Family Gadidae	102
Subfamily Gadinae	
Subfamily Lotinae	
Subfamily Gaidropsarinae	
Subfamily Phycinae	
Family Bregmacerotidae	103
ORDER OPHIDIIFORMES	
SUBORDER OPHIDIOIDEI	
Family Carapidae	104
Subfamily Pyramodontinae	
Subfamily Carapinae	
Family Ophidiidae	105
Subfamily Brotulinae	
Subfamily Brotulotaeniinae	
Subfamily Ophidiinae	
Subfamily Neobythitinae	
SUBORDER BYTHITOIDEI	
Family Bythitidae	106
Subfamily Bythitinae	
Subfamily Brosmophycinae	
Family Aphyonidae	107
Family Parabrotulidae	108
ORDER BATRACHOIDIFORMES	
Family Batrachoididae	109

(*cont.*)

	Ref. No.
ORDER GOBIESOCIFORMES	
Family Gobiesocidae	110
ORDER STEPHANOBERYCIFORMES	
Family Stephanoberycidae	111
Family Gibberichthyidae	112
Family Melamphaidae	113
Family Hispidoberycidae	114
ORDER BERYCIFORMES	
SUBORDER TRACHICHTHYOIDEI	
Family Anoplogastridae	115
Family Diretmidae	116
Family Anomalopidae	117
Family Trachichthyidae	118
SUBORDER BERYCOIDEI	
Family Berycidae	119
SUBORDER HOLOCENTROIDEI	
Family Holocentridae	120
ORDER CETOMIMIFORMES	
Family Rondeletiidae	121
Family Barbourisiidae	122
Family Cetomimidae	123
ORDER SYNGNATHIFORMES	
Family Centriscidae	124
Family Syngnathidae	125
ORDER SCORPAENIFORMES	
SUBORDER DACTYLOPTEROIDEI	
Family Dactylopteridae	126
SUBORDER SCORPAENOIDEI	
Family Sebastidae	127
Family Setarchidae	128
Family Scorpaenidae	129
Family Tetrarogidae	130
Family Plectrogeniidae	131
Family Congiopodidae	132
SUBORDER PLATYCEPHALOIDEI	
Family Triglidae	133
Family Peristediidae	134
Family Bembridae	135
Family Parabembridae	136
Family Hoplichthyidae	137
Family Anoplopomatidae	138

(*cont.*)

	Ref. No.
SUBORDER HEXAGRAMMOIDEI	
SUBORDER COTTOIDEI	
Family Ereuniidae	139
Family Cottidae	140
Family Hemitripteridae	141
Family Agonidae	142
Family Psychrolutidae	143
Subfamily Cottunculinae	
Subfamily Psychrolutinae	
Family Bathylutichthyidae	144
Family Cyclopteridae	145
Family Liparidae	146
ORDER PERCIFORMES	
SUBORDER PERCOIDEI	
Family Howellidae	147
Family Acropomatidae	148
Family Symphysanodontidae	149
Family Polyprionidae	150
Family Serranidae	151
Family Ostracoberycidae	152
Family Callanthiidae	153
Family Grammatidae	154
Family Opistognathidae	155
Family Epigonidae	156
Family Malacanthidae	157
Family Scombropidae	158
Family Carangidae	159
Family Bramidae	160
Family Caristiidae	161
Subfamily Caristiinae	
Subfamily Paracaristiinae	
Family Emmelichthyidae	162
Family Lutjanidae	163
Family Nemipteridae	164
Family Sparidae	165
Family Sciaenidae	166
Family Mullidae	167
Family Bathyclupeidae	168
Family Chaetodontidae	169
Family Pentacerotidae	170

(cont.)

	Ref. No.
Subfamily Pentacerotinae	
Subfamily Histiopterinae	
Family Cepolidae	171
SUBORDER ELASSOMATOIDEI	
SUBORDER ZOARCOIDEI	
Family Bathymasteridae	172
Family Zoarcidae	173
Subfamily Zoarcinae	
Subfamily Gymnelinae	
Subfamily Lycodinae	
Subfamily Lycozoarcinae	
Subfamily Neozoarcinae	
Family Stichaeidae	174
Family Anarhichadidae	175
SUBORDER NOTOTHENIOIDEI	
Family Nototheniidae	176
Family Artedidraconidae	177
Family Bathydraconidae	178
Family Channichthyidae	179
SUBORDER TRACHINOIDEI	
Family Chiasmodontidae	180
Family Champsodontidae	181
Family Percophidae	182
Subfamily Brembropinae	
Family Uranoscopidae	183
SUBORDER ICOSTEOIDEI	
Family Icosteidae	184
SUBORDER CALLIONYMOIDEI	
Family Callionymidae	185
Family Draconettidae	186
SUBORDER GOBIOIDEI	
Family Gobiidae	187
SUBORDER SCOMBROLABRACOIDEI	
Family Scombrolabracidae	188
SUBORDER SCOMBROIDEI	
Family Gempylidae	189
Family Trichiuridae	190
Subfamily Aphanopodinae	
Subfamily Lepidopodinae	
Subfamily Trichiurinae	
SUBORDER STROMATEOIDEI	
Family Centrolophidae	191

(*cont.*)

REFERENCES

Aarestrup K., Økland F., Hansen M.M., Righton D., Gargan P., Castonguay M., Bernatchez L., Howey P., Sparholt H., Pedersen M. I., & McKinley, R.S. (2009). Oceanic spawning migration of the European eel (*Anguilla anguilla*). *Science* **325**:1660. doi: 10.1126/science.1178120

Agnew D.J., Heaps L., Jones C., Watson A. Berkieta K., & Pearce J. (1999). Depth distribution and spawning pattern of *Dissostichus eleginoides* at South Georgia. *CCAMLR Science*, **6**, 19–36.

Aguayo-Hernández, M. (1995). Biology and fisheries of Chilean hakes (*M. gayi* and *M. australis*). In J. Alheit & T.J. Pitcher (eds.) *Hake: Biology, Fisheries and Markets*. London: Chapman & Hall, pp. 305–337.

Ainsley S.M., Ebert D.A., & Cailliet G.M. (2011). Age, growth, and maturity of the whitebrow skate, *Bathyraja minispinosa*, from the eastern Bering Sea. *ICES Journal of Marine Science* **68**: 1426–1434. doi:10.1093/icesjms/fsr072

Akhilesh K.V., Ganga U., Pillai N.G.K., Vivekanandan E., Bineesh K.K. Shanis C.P.R. & Hashim M. (2011). Deep-sea fishing for chondrichthyan resources and sustainability concerns – a case study from the southwest Indian Coast. *Indian Journal of Geo-Marine Sciences* **40**: 347–355.

Alcock A.W. (1899). *A Descriptive Catalogue of the Indian Deep-Sea Fishes in the Indian Museum, Being a Revised Account of the Deep-Sea Fishes Collected by the Royal Indian Marine Survey Ship Investigator*. Calcutta: Trustees of the Indian Museum.

Aldrovandi U. (1613). *De piscibus libri V, et de cetis liber unus*. Bononia: Bellgambam.

Alexander R.M. (1966). Physical aspects of swimbladder function. *Biological Reviews* **41**: 141–176.

Alexander R.M. (1975). *The Chordates*. Cambridge: Cambridge University Press.

Alexander R.M. (1990). Size, speed and buoyancy adaptations in aquatic animals. *American Zoologist* **30**: 189–196.

Alexander R.M. (1999). *Energy for Animal Life*. Oxford: Oxford University Press.

Alheit, J. & Pitcher T.J. (1995). *Hake: Biology, Fisheries and Markets*. London: Chapman & Hall.

Ali H.A., Mok H.-K. & Fine M.L. (2016). Development and sexual dimorphism of the sonic system in deep sea neobythitine fishes: The upper continental slope. *Deep-Sea Research I* **115**: 293–308.

Allain V., Biseau A. & Kergoat B. (2003). Preliminary estimates of French deepwater fishery discards in the Northeast Atlantic Ocean. *Fisheries Research*, **60**, 185–92.

Allen G. H. (2001). The Ragfish, *Icosteus aenigmaticus* Lockington, 1880: A synthesis of historical and recent records from the North Pacific Ocean and the Bering Sea. *Marine Fisheries Review* **63**(4): 1–31.

Amaoka K. (2005). Ateleopodiformes: Ateleopodidae: Jellynoses. In: Richards W.J. (ed.) *Early Stages of Atlantic Fishes: An Identification Guide for the Western Central North Atlantic*, pp. 295–300. Volume 1 Boca Ratan, FL: CRC Press, 2672 pages.

Amaoka K. & Parin N.V. (1990). A new flounder, *Chascanopsetta megagnatha*, from the Sala-y-Gomez Submarine Ridge, Eastern Pacific Ocean (Teleostei: Pleuronectiformes: Bothidae). *Copeia* **1990** (3): 717–722.

Amaoka K. & Yamamoto E. (1984). Review of the genus *Chascanopsetta*, with the description of a new species. Bulletin of the Faculty of Fisheries. *Hokkaido University* **35**(4): 201–224

Ambrose D.A. (1996a). Alepocephalidae: Slickheads. In: Moser H.G. (ed.) *The Early Stages of Fishes in the California Current Region. California Cooperative Oceanic Fisheries Investigations*, pp. 224–233. Atlas No 33. La Jolla, CA: South West Fisheries Centre.

Ambrose D.A. (1996b). Anotopteridae: Daggertooth. In: Moser H.G. (ed.) *The Early Stages of Fishes in the California Current Region. California Cooperative Oceanic Fisheries Investigations*, pp. 369–371. Atlas No 33. La Jolla, CA: South West Fisheries Centre.

Ambrose D.A. (1996c). Bythitidae: Brotulas. In: Moser H.G. (ed.) *The Early Stages of Fishes in the California Current Region. California Cooperative Oceanic Fisheries Investigations*, pp. 538–545. Atlas No. 33. La Jolla, CA: South West Fisheries Centre.

Ambrose D.A. (1996d). Carapidae: Pearlfishes. In: Moser H.G. (ed.) *Early Stages of Fishes in the California Current Region. California Cooperative Oceanic Fisheries Investigations*, pp. 532–537. Atlas No 33. La Jolla, CA: South West Fisheries Centre.

Ambrose D.A. (1996e). Macrouridae: Grenadiers. In: Moser H.G. (ed.) *Early Stages of Fishes in the California Current Region. California Cooperative Oceanic Fisheries Investigations*, pp. 483–499. Atlas No 33. La Jolla, CA: South West Fisheries Centre.

Amemiya C.T., Alföldi J., Lee A.P., Fan S., Philippe H. MacCallum I., Braasch I., Manousaki T., Schneider I., et al. (2013). The African coelacanth genome provides insights into tetrapod evolution. *Nature* 496(7445): 311–316.

Amemiya C.T., Dorrington R., & Meyer A. (2014). The coelacanth and its genome (Introduction to Special Issue: Genome of the African Coelocanth). *Journal of Experimental Zoology Part B Molecular and Developmental Evolution* 322B: 317–321.

Anastasopoulou A. & Kapiris K. (2008). Feeding ecology of the shortnose greeneye *Chlorophthalmus agassizi* Bonaparte, 1840 (Pisces: Chlorophthalmidae) in the eastern Ionian Sea (eastern Mediterranean). *Journal of Applied Ichthyology* 24: 170–179

Anastasopoulou A., Yiannopoulos C., Megalofonou P., & Papaconstantinou C. (2006). Distribution and population structure of the *Chlorophthalmus agassizi* (Bonaparte, 1840) on an unexploited fishing ground in the Greek Ionian Sea. *Journal of Applied Ichthyology* 22: 521–529.

Anderson M.E. (1986). Family No 95: Parabrotulidae. In: Smith M.M. & Heemstra P.C. (eds.) Smith's Sea Fishes. Berlin: Springer Verlag.

Anderson M.E. (1989). Review of the eelpout genus *Pachycara* Zugmayer, 1911 (Teleostei: Zoarcidae), with description of six new species. *Proceedings of the California Academy of Sciences* 46: 221–242.

Anderson M.E. (2006). Studies on the Zoarcidae (Teleostei: Perciformes) of the Southern Hemisphere. XI. A new species of Pyrolycus from the Kermadec Ridge. *Journal of the Royal Society of New Zealand* 36(2): 63–68.

Anderson M.E., Crabtree R.E., Carter H.J., Sulak K.J., & Richardson M.D. (1985). Distribution of demersal fishes of the Caribbean Sea found below 2,000 meters. *Bulletin of Marine Science* 37: 794–807.

Anderson M.E. & Federov V.V. (2004). Family Zoarcidae, Swainson 1839 eelpouts. *California Academy of Sciences, Annotated Checklists of Fishes* 34: 1–58.

Anderson M.E. & Peden A.E. (1988). The eelpout genus *Pachycara* (Teleostei: Zoarcidae) in the northeastern Pacific Ocean, with descriptions of two new species. *Proceedings of the California Academy of Sciences* 46: 83–94.

Anderson M.E., Somerville R., & Copley J.T. (2016). A new species of *Pachycara* Zugmayer, 1911 (Teleostei: Zoarcidae) from deep-sea chemosynthetic environments in the Caribbean Sea. *Zootaxa* **4066** (1): 071–077.

Anderson T.R. & Rice T. (2006). Deserts on the sea floor: Edward Forbes and his azoic hypothesis for a lifeless deep ocean. *Endeavour* 30: 131–137.

Anderson W.D. & Springer V.G. (2005). Review of the perciform fish genus *Symphysanodon* Bleeker (Symphysanodontidae), with description of three new species, *S. mona*, *S. parini*, and *S. rhax*. *Zootaxa* 996: 1–44.

Andriashev A.P. (1953). *Ancient Deep-Water and Secondary Deep-Water Fishes and Their Importance in Zoogeographical Analysis. Notes of Special Problems in Ichthyology.* Moscow, Leningrad: Akademie. Nauk. SSSR. Ikhiol.Kom. [Translation by A. R. Gosline, Ichthyological Laboratory, Bureau of Commercial Fisheries. Fish & Wildlife Service Translation Series 6. Washington DC: US National Museum].

Andriashev A.P. (1955). On a new Liparid fish from a depth over 7000 meters. *Trudy Instituta Okeanologii, Akademiya Nauk, SSSR* (Proceedings of the Institute of Oceanology, Academy of Sciences, USSR) 12: 340–344 (in Russian).

Andriashev, A.P., (1998). A review of recent studies of Southern Ocean Liparidae (Teleostei: Scorpaeniformes). *Cybium* 22(3): 255–266.

Anon. (2014). ftp.fao.org/FI/DOCUMENT/vme/VME_NPFC_workshop_11-13March2014/Day2/Day 2 StateFisheryResources-EmperorSeamounts.pdf accessed 15 October 2015.

Armstrong J.D., Bagley P.M., & Priede I.G. (1992). Photographic and acoustic tracking observations of the behaviour of the grenadier, *Coryphaenoides (Nematonurus) armatus*,the eel, *Synaphobranchus bathybius*, and other abyssal demersal fish in the

North Atlantic Ocean. *Marine Biology* **112**: 535–544.

Armstrong J.D., Johnstone A.D.F., & Lucas M.C. (1992). Retention of intragastric transmitters after voluntary ingestion by captive cod, *Gadus morhua* L. *Journal of Fish Biology* **40**: 135–137

Armstrong M.J. & Prosch R.M. (1991). Abundance and distribution of the mesopelagic fish *Maurolicus muelleri* in the southern Benguela system, *South African Journal of Marine Science*, **10**(1): 13–28, doi: 10.2989/02577619109504615

Artedi P. (1738). *Ichthyologia, Sive, Opera Omnia de Piscibus Scilicet Bibliotheca Ichthyologica, Philosophia Ichthyologica, Genera Piscium, Synonymia Specierum, Descriptiones Specierum: Omnia in hoc Genere Perfectiora, Quam Antea Ulla / Posthuma Vindicavit, Recognovit, Coaptavit & Edidit Carolus Linnaeus*. Leiden: Wishoff.

Aschliman N.C., Nishida M., Miya M., Inoue J.G., Rosana K.M., & Naylor G.J.P. (2012). Body plan convergence in the evolution of skates and rays (Chondrichthyes: Batoidea). *Molecular Phylogenetics and Evolution* **63** (2012) 28–42.

Ast J.C. & Dunlap P.V. (2005). Phylogenetic resolution and habitat specificity of members of the *Photobacterium phosphoreum* species group. *Environmental Microbiology* **7**(10): 1641–1654.

Atkinson D.B. (1995). The biology and fishery of Roundnose grenadier (*Coryphaenoides rupestris*, Gunnerus, 1765) in the North West Atlantic. In: Hopper A.G. (ed.) *Deep-water Fisheries of the North Atlantic Slope*. Dordrecht, Netherlands: Kluwer.

Auster P. J., Gjerde K., Heupel E., Watling L., Grehan A., & Rogers A.D. (2011). Definition and detection of vulnerable marine ecosystems on the high seas: problems with the 'move-on' rule. *ICES Journal of Marine Science*, **68**: 254–264.

Backus R.H. (1966). The 'Pinger' as an aid in deep trawling. *Journal du Conseil Permanent International pour l'Exploration de la Mer* **30**(2): 270–277.

Backus R.H., Craddock J.E., Haedrich R.L., Shores D.L., Teal J.M., Wing A.S., Mead G.W., & Clarke W.D. (1968). *Ceratoscopelus maderensis*: Peculiar sound-scattering layer identified with this Myctophid fish. *Science* **160**: 991–993.

Badcock J., (1981). The significance of meristic variation in *Benthosema glaciale* (Pisces, Myctophoidei) and of the species distribution off northwest Africa. *Deep-Sea Research*, **28**A: 1477–1491.

Badcock J. & Merrett N.R. (1976). Midwater fishes in the eastern North Atlantic–I. Vertical distribution and associated biology in 30°N, 23°W, with developmental notes on certain myctophids. *Progress in Oceanography* **7**: 3–58.

Badcock, J.R. & Merrett N.R. (1972). On *Argyripnus atlanticus*, Maul 1952 (Pisces, Stomiatoidei), with a description of post larval forms. *Journal of Fish Biology*, **4**: 277 287.

Badenhorst A. (1988). Aspects of the South African longline fishery for kingklip *Genypterus capensis* and the Cape hakes *Merluccius capensis* and *M. paradoxus*. *South African Journal of Marine Science* **6**: 33–42. doi: 10.2989/025776188784480708

Bagley P.M., Smith A. & Priede I.G. (1994). Tracking movements of deep demersal fishes, *Coryphaenoides (Nematonurus) armatus*, *Antimora rostrata* and *Centroscymnus coelolepis* in the Porcupine Seabight, N.E. Atlantic Ocean. *Journal of the Marine Biological Association of the UK* **74**: 473–480.

Bailey D.M., Bagley, P.M., Jamieson, A.J., Collins A.M. & Priede, I.G. (2003). *In situ* investigation of burst swimming and muscle performance in the deep sea fish *Antimora rostrata* (Günther, 1878). *Journal of Experimental Marine Biology & Ecology* **285-6**: 295–311.

Bailey D.M., Collins M.A., Gordon J.D.M., Zuur A.F. & Priede I.G. (2009). Long-term changes in deep-water fish populations in the North East Atlantic: deeper-reaching effect of fisheries? *Proceedings of the Royal Society of London B.* **275**: 1965–1969.

Bailey D.M., Genard B., Collins M.A., Rees F., Unsworth S.K., Battle E.J.V., Bagley P.M., Jamieson A.J., & Priede I.G. (2005). High swimming and metabolic activity in the deep-sea eel *Synaphobranchus kaupii* revealed by integrated in situ and in vitro measurements. *Physiological & Biochemical Zoology* **78**: 335–346.

Bailey D.M. Jamieson A.J., Bagley P.M., Collins M.A., & Priede I.G. (2002). Measurement of in situ oxygen consumption of deep-sea fish using an autonomous lander vehicle. *Deep Sea Research Part I: Oceanographic Research* **49**(8): 1519–1529.

Bailey D.M., King N.J., & Priede I.G. (2007). Cameras and carcasses: Historical and current methods for using artificial food falls to study deep-water animals.

Marine Ecology Progress Series, **350**: 179–191, doi: 10.3354/meps07187

Bailey D.M., Ruhl H.A., & Smith K.L. Jr. (2006). Long-term change in benthopelagic fish abundance in the abyssal Northeast Pacific Ocean. *Ecology*. **87**(3): 549–555.

Bailey D.M., Wagner H.-J., Jamieson A.J., Ross M.F., & Priede I.G. (2007). A taste of the deep-sea: The roles of gustatory and tactile searching behaviour in the grenadier fish *Coryphaenoides armatus*. *Deep-sea Research I* **54**; 99–108.

Baird R.C. & Hopkins T.L. (1981). Trophodynamics of the Fish *Valenciennellus tripunctulatus*. II. Selectivity, Grazing Rates and Resource Utilization. *Marine Ecology Progress Series* **5**: 11–19.

Baird R.C. & Jumper G.Y. (1993). Olfactory organs in the deep sea hatchetfish *Sternoptyx diaphana* (Stomiiformes, Sternoptychidae) *Bulletin of Marine Science* **53**(3): 1163–1167.

Baird R.C., Jumper G.Y., & Gallaher E.E. (1990). Sexual dimorphism and demography in two species of oceanic midwater fishes (Stomiiformes: Sternoptychidae) from the eastern Gulf of Mexico. *Bulletin of Marine Science* **47**: 561–566.

Baird R.C., Wilson D.F., & Milliken D.M. (1973). Observations on *Bregmaceros nectabanus* Whitley in the anoxic, sulfurous water of the Cariaco Trench. *Deep-Sea Research* **20**: 503–504.

Baker K.D., Haedrich R.L., Snelgrove P.V.R., Wareham V.E., Edinger E.N., & Gilkinson K.D. (2012b). Small-scale patterns of deep-sea fish distributions and assemblages of the Grand Banks, Newfoundland continental slope. *Deep-Sea Research I* **65**: 171–188.

Baker K.D., Haedrich R.L., Fifield D.A., & Gilkinson K.D. (2012). Grenadier abundance examined at varying spatial scales in deep waters off Newfoundland, Canada, with special focus on the influence of corals. *Journal of Ichthyology* **52**: 678–689.

Baker K.D., Haedrich R.L., Snelgrove P.V.R., Wareham V.E., Edinger E.N., & Gilkinson K.D. (2012a). Small-scale patterns of deep-sea fish distributions and assemblages of the Grand Banks, Newfoundland continental slope. *Deep-Sea Research II* **65**: 171–188.

Baker L.L., Wiff R., Quiroz J.C., Flores A., Céspedes R., Barrientos M.A., Ojeda V., & Gatica C. (2014). Reproductive ecology of the female pink cusk-eel (*Genypterus blacodes*): evaluating differences between fishery management zones in the Chilean austral zone. *Environmental Biology of Fishes* **97**: 1083–1093. doi: 10.1007/s10641-013-0199-2

Baldwin C.C. & Johnson G.D. (1996). Interrelationships of Aulopiformes. In: Stiassny, M.L.J., Parenti, L.R., & Johnson G.D. (eds.) *Interrelationships of Fishes*, pp. 355–404. San Diego, CA: Academic Press.

Ballara S.L. (2014). Fishery characterisation and standardised CPUE analyses for lookdown dory, *Cyttus traversi* (Hutton, 1872) (Zeidae), 1989–90 to 2011–12. New Zealand Fisheries Assessment Report 2014/62.

Balushkin A.V. (1996). A new genus and species of liparid fish *Palmoliparis beckeri* from the northern Kurile Islands (Scorpaeniformes, Liparidae) with consideration of phylogeny of the family. *Journal of Ichthyology*. **36**: 281–287.

Balushkin A.V. (2012). *Volodichthys* gen. nov. New species of the primitive snailfish (Liparidae: Scorpaeniformes) of the Southern Hemishpere. Description of vew species *V. solovjevae* sp. nov. (Cooperation Sea, the Antarctic). *Journal of Ichthyology* **52**(1): 1–10.

Balushkin A.V. & Prirodina V.P. (2010). A new species of Muraenolepididae (Gadiformes) *Muraenolepis evseenkoi* sp. nova from continental seas of Antarctica. *Journal of Ichthyology* **50**(7): 495–502.

Balushkin A.V. & Voskoboinikova O.S. (2008). Revision of the genus *Genioliparis* Andriashev et Neelov (Liparidae, Scorpaeniformes) with description of a new species *G. kafanovi* sp. n. from the Ross Sea (Antractica). *Journal of Ichthyology* **48**(3): 201–208.

Bambach RK (2006). Phanerozoic biodiversity mass extinctions. *Annual Review of Earth and Planetary Sciences* **34**: 127–155.

Bambach R.K., Knoll A.H., & Wang S.C. (2004). Origination, extinction, and mass depletions of marine diversity. *Paleobiology* **30**(4): 522–542.

Baranes A. & McCormack C. (2009). Iago omanensis. *The IUCN Red List of Threatened Species 2009*: e.T161501A5437914. http://dx.doi.org/10.2305/IUCN.UK.2009-2.RLTS.T161501A5437914.en. Downloaded on 21 March 2017.

Barange, M., Pakhomov, E.A., Perissinotto, R., Froneman, P.W., Verheye, H.M., Taunton-Clark, J., & Lucas, M.I. (1998). Pelagic community structure of the subtropical convergence region south of Africa and in the mid-Atlantic ocean. *Deep-Sea Research I*, 45: 1663–1687.

Bardack D. (1991). First fossil hagfish (Myxinoidea): a record from the Pennsylvanian of Illinois. *Science* 254: 701–703.

Baremore I.E. (2010). Reproductive aspects of the Atlantic angel shark *Squatina dumeril*. *Journal of Fish Biology* 76: 1682–95. doi: 10.1111/j.1095-8649.2010.02608.x.

Barham A.G., Ayer, N.J., & Boyce R.E. (1967). Macrobenthos of the San Diego Trough: Photographic census and observations from bathyscaphe, *Trieste. Deep-Sea Research* 14: 773–784.

Barham, E.G. (1966). Deep scattering layer migration and composition: *Observations from a Diving Saucer Science*, 151: 1399–1403.

Barnett, M.A. (1984). Mesopelagic fish zoogeography in the central tropical and subtropical Pacific Ocean: Species composition and structure at representative locations in three ecosystems *Marine Biology* 82: 199–208. doi: 10.1007/BF00394103

Barreiros, J.P., Machado L.F., Vieira R.P., & Porteiro F.M. (2011). Occurrence of *Grammicolepis brachiusculus* Poey, 1873 (Pisces: Grammicolepididae) in the Azores Archipelago. *Life and Marine Sciences* 28: 83–88.

Bartlett J.M.S. & Stirling D. (2003). A short history of the polymerase chain reaction. *PCR Protocols* 3–6. doi:10.1385/1-59259-384-4:3.

Bassot, J.-M. (1966). On the comparative morphology of some luminous organs. In: Johnson F.H. &.Haneda Y. (eds.) *Bioluminescence in Progress*, pp. 557–610. Princeton NJ: Princeton University Press.

Batista I., Pires C., Bandarra N.M., & Goncalves A. (2001). Chemical characterization and preparation of salted minces from bigeye grunt and longfin bonefish. *Journal of Food Biochemistry* 25: 527–540.

Beamish R.J., Leask K.D., Ivanov O.A., Balanov A.A., Orlov A.M., & Sinclair B. (1999). The ecology, distribution, and abundance of midwater fishes of the Subarctic Pacific gyres. *Progress in Oceanography* 43: 399–442.

Beaulieu S.E. (2002). Accumulation and fate of phytodetritus on the sea floor. *Oceanography and Marine Biology Annual Review.* 40: 171–232.

Becker J.J., Sandwell D.T., Smith W.H.F., Braud J., Binder B., Depner J., Fabre D., Factor J., Ingalls S., Kim S.-H., Ladner R., Marks K., Nelson S., Pharaoh A., Trimmer R., von Rosenberg J., Wallace G., & Weatherall P., (2009). Global bathymetry and elevation data at 30 arc-seconds resolution: SRTM30 PLUS. *Marine Geodesy* 32: 355–371.

Becker T.W & Faccenna C. (2009). A review of the role of subduction dynamics for regional and global plate motions. In: Lallemand S. & Funiciello F. (eds.) *Subduction Zone Geodynamics*, Berlin Heidelberg: Springer-Verlag.

Beebe W. (1932). Nineteen new species and four postlarval deep-sea fish. *Zoologica* 13(4): 47–107.

Beebe W. (1934). *Half Mile Down*. New York: Harcourt, Brace & Co.

Béguer-Pon M., Castonguay M., Shan S., Benchetrit J,. & Dodson J.J. (2015). Direct observations of American eels migrating across the continental shelf to the Sargasso Sea. *Nature Communications* 6:8705 doi: 10.1038/ncomms9705

Ben-Avraham Z., Woodside J., Lodolo E., Gardosh M., Grasso M., Camerlenghi A., & Vai G.B. (2006). Eastern Mediterranean basin systems. *Geological Society, London, Memoirs* 32: 263–276. doi: 10.1144/GSL.MEM.2006.032.01.15

Benfield M.C., Caruso J.H., & Sulak K.J. (2009). *In situ* video observations of two manefishes (Perciformes: Caristiidae) in the mesopelagic zone of the Northern Gulf of Mexico. *Copeia* 2009: 637–641.

Benn A.R., Weaver P.P., Billet D.S.M., van den Hove S., Murdock A.P., Doneghan G.B., & Le Bas T. (2010). Human activities on the deep seafloor in the North East Atlantic: An assessment of spatial extent. *PLoS ONE* 5(9): e12730. doi:10.1371/journal.pone.0012730

Benton M. (2005). *Vertebrate palaeontology*, 3rd edition. Oxford: Blackwell.

Berelson W. (2002). Particle settling rates increase with depth in the ocean. *Deep-Sea Research II* 49: 237–251.

Berenbrink M., Koldkjær P., Kepp O., & Cossins A.R. (2005). Evolution of oxygen secretion in fishes and the emergence of a complex physiological system. *Science* 307: 1752–1757.

Bergstad O.A. (1995). Age determination of deep-water fishes: experiences, status and challenges for the future. In: Hopper A.G., ed. *Deep-water Fisheries of the North Atlantic Oceanic Slope*, pp. 267–283. NATO ASI Series, Series E: Applied Sciences, Vol. 296. London: Kluwer Academic Publishers.

Bergstad O.A., Bjelland O., & Gordon J.D.M. (1999). Fish communities on the slope of the eastern Norwegian Sea. *Sarsia* **84**: 67–78.

Bergstad O.A., Clark L., Hansen H.Ø., & Cousins N. (2012). Distribution, population biology, and trophic ecology of the deepwater demersal fish *Halosauropsis macrochir* (Pisces: Halosauridae) on the Mid-Atlantic Ridge. *PLoS ONE* **7**(2): e31493. doi:10.1371/journal.pone.0031493

Bergstad O.A., Hansen H.Ø., & Jørgensen T. (2013). Intermittent recruitment and exploitation pulse underlying temporal variability in a demersal deepwater fish population. *ICES Journal of Marine Science*. doi: 10.1093/icesjms/fst202

Bergstad O.A., Menezes G., & Høines Å.S. (2008). Demersal fish on a mid-ocean ridge: Distribution patterns and structuring factors. *Deep-Sea Research II* **55**: 185–202.

Bergstad O.A., Menezes G.M.M., Høines Å., Gordon J.D.M., & Galbraith J.K. (2012). Patterns of distribution of deepwater demersal fishes of the North Atlantic mid-ocean ridge, continental slopes, islands and seamounts. *Deep-Sea Research I* **61**: 74–83.

Bernardino A.F., Levin L.A., Thurber A.R., & Smith C.R. (2012). Comparative composition, diversity and trophic ecology of sediment macrofauna at vents, seeps and organic falls. *PLoS ONE* **7**(4): e33515. doi:10.1371/journal.one.0033515

Bertelsen E. (1951). The ceratioid fishes. Ontogeny, taxonomy, distribution and biology. *Dana Reports* **39**: 1–276.

Bertelsen E. (1958) A new type of light organ in the deepsea fish *Opisthoproctus. Nature* **181**: 862–863. doi:10.1038/181862b0

Bertelsen E. (1982). First records of metamorphosed males of the families Diceratiidae and Centrophrynidae (Pisces: Ceratioidei). *Steenstrupia* **8**(16): 309–315.

Bertelsen E. & Pietsch T.W. (1998). Revision of the deepsea anglerfish genus Rhynchactis Regan (Lophiiformes: Gigantactinidae), with descriptions of two new species. *Copeia* **1998**(3): 583–590.

Bertelsen E. & Struhsaker P.J. (1977). The ceratioid fishes of the genus *Thaumatichthys*: Osteology, relationships and biology. *Galathea Reports* **14**: 7–40.

Best A.C.G. & Bone Q. (1976). On the integument and photophores of the Alepocephalid fishes *Xenodermichthys* and *Photostylus. Journal of the Marine Biological Association of the U.K.* **56**: 227–236.

Best P.B. & Photopoulou T. (2016). Identifying the 'demon whale-biter': Patterns of scarring on large whales attributed to a cookie-cutter shark Isistius sp. *PLoS ONE* **11**(4): e0152643. doi:10.1371/journal.pone.0152643

Betancur-R R., Broughton R.E., Wiley E.O., Carpenter K., López J.A., Li C., Holcroft N.I., Arcila D., Sanciangco M., Cureton II J.C., Zhang F., Buser T., Campbell M.A., Ballesteros J.A., Roa-Varon A., Willis S., Borden W.C., Rowley T., Reneau P.C., Hough D.J., Lu G., Grande T., Arratia G. & Ortí G. (2013). The tree of life and a new classification of bony fishes. *PLOS Currents Tree of Life.* 2013 Apr 18. edition 1. doi: 10.1371/currents.tol.53ba26640df0ccaee75bb165c8c26288.

Beverton R.J.H. & Holt S.J. (1957). On the dynamics of exploited fish populations. *Fishery Investigation Series II*, Vol. **XIX**.

Bianchi G., Carpenter K.E., Roux J.-P., Molloy F.J., Boyer D., & Boyer H.J. (1999). FAO species identification field guide for fishery purposes. The living marine resources of Namibia. Rome: FAO. http://www.fao.org/docrep/009/x3478e/x3478e00.HTM

Bilecenoglu M., Kaya M., & Irmak E. (2006). First records of the slender snipe eel, *Nemichthys scolopaceus* (Nemichthyidae), and the robust cusk-eel, *Benthocometes robustus* (Ophidiidae), from the Aegean Sea. *Acta Ichthyologica et Piscatoria* **36** (1): 85–88.

Billett D.S.M., Bett B.J., Jacobs C.L., Rouse I.P. & Wigham B.D. (2006). Mass deposition of jellyfish in the deep Arabian Sea. *Limnology & Oceanography* **51**: 2077–2083.

Billett D.S.M., Lampitt R.S., Rice A.L., & Mantoura R.F.C. (1983). Seasonal sedimentation of phytoplankton to the deep-sea benthos. *Nature* **302**: 520–522 doi:10.1038/302520a0

Biscoito M. & Almeida A.J. (2004). New Species of *Pachycara* Zugmayer (Pisces: Zoarcidae) from the Rainbow Hydrothermal Vent Field (Mid-Atlantic Ridge). *Copeia*, **2004** (3): 562–568.

Biscoito M., Segonzac M., Almeida A.J., Desbruyères D., Geistdoerfer P., Turnipseed M., & Van Dover C. (2002). Fishes from the hydrothermal vents and cold seeps – an update. *Cahiers de Biologie Marine* **43**: 359–362.

Blakey R. (2011). Library of Paleogeography. Colorado Plateau Geosystems, Flagstaff, AZ: Author. http://cpgeosystems.com/paleomaps.html Accessed 20 June 2014.

Blanchard F. (2001). The effect of fishing on demersal fish community dynamics: an hypothesis. *ICES Journal of Marine Science* **58**: 711–718.

Blankenship L. E. & Levin L. A. (2007). Extreme food webs: foraging strategies and diets of scavenging amphipods from the ocean's deepest 5 kilometers. *Limnology and Oceanography* **52**, 1685–1697.

Blaxter J.H.S. (1979). The effect of hydrostatic pressure on fishes. In: Ali M.A. (ed.) *Environmental Physiology of Fishes*. New York: Plenum Press.

Blaxter J.H.S., Wardle C.S. & Roberts B.L. (1971). Aspects of the circulatory physiology and muscle systems of deep-sea fish. *Journal of the Marine Biological Association of the UK*, **51**: 991–1006.

Block B.A., Booth D.T., & Carey F.G. (1992). Direct measurement of swimming speeds and depth of blue marlin. *Journal of Experimental Biology* **166**: 267–284.

Block B.A., Dewar H., Blackwell S.B., Williams T.D. et al. (2001). Migratory movements, depth preferences, and thermal biology of Atlantic bluefin tuna. *Science* **293**: 1310–1314.

Boehlert G.W., Wilson C.D., & Mizuno K. (1994). Populations of the Sternotychid fish *Maurolicus muelleri* on seamounts in the Central North Pacific. *Pacific Science.* **48**: 57–69.

Boetius A. & Wenzhöfer F. (2013). Seafloor oxygen consumption fuelled by methane from cold seeps. *Nature Geoscience* **6**: 725–734.

Bone Q. (1972). Buoyancy and hydrodynamic functions of integument in the castor oil fish, *Ruvettus pretiosus* (Pisces: Gempylidae). *Copeia* **1972**: 78–87.

Bone, Q. (1973). A note on the buoyancy of some lantern-fishes. (Myctophoidei). *Journal of the Marine Biological Association. U.K.* 53:619–633

Bone Q. & Moore R.H. (2008). *Biology of Fishes*, 3rd edition. Abingdon. England: Taylor & Francis.

Bone Q. & Roberts.BL. (1969). The density of elasmobranchs. *Journal of the Marine Biological Association U K.* **49**: 913–937. doi: 10.1017/s0025315400038017

Bourret P. (1985). Poissons Téléostéens: Gonostomatidae, Sternoptychidae, et Myctophidae (MUSORSTOM II).

Mémoires du Muséum National d'Histoire Naturelle., séries. A, *Zoologie. Paris.* **133**: 55–82.

Bowering W.R. & Brodie W.B. (1995). Greenland halibut (*Reinhardtius hippoglossoides*). A review of the dynamics of its distribution and fisheries off Eastern Canada and Greenland. In: Hopper A.G. (ed.) *Deep-Water Fisheries of the North Atlantic Slope*, pp. 113–160. Dordrecht, Netherlands: Kluwer.

Braccini, J.M. (2008). Feeding ecology of two high-order predators from south-eastern Australia: The coastal broadnose and the deepwater sharpnose sevengill sharks. *Marine Ecology Progress Series* **371**: 273–284. doi:10.3354/meps07684.

Braccini, J.M., Gillanders, B.M. & Walker, T.I. (2006). Hierarchical approach to the assessment of fishing effects on non-target chondrichthyans: case study of *Squalus megalops* in southeastern Australia. *Canadian Journal of Fisheries and Aquatic Sciences* **63** (11): 2456–2466 doi: 10.1139/f06-141

Bradbury, M.G. (1988). Rare fishes of the deep-sea genus *Halieutopsis*: a review with descriptions of four new species (Lophiiformes: Ogcocephalidae). *Fieldiana* **44**: 1–22

Bradbury, M.G. (1999). A review of the fish genus *Dibranchus* with descriptions of new species and a new genus, *Solocisquama* (Lophiiformes, Ogcocephalidae). *Proceedings of the California Academy of Sciences* **51**(5): 259–310.

Brander K. (1981). Disappearance of common skate *Raia batis* from Irish Sea. *Nature* **290**: 48–49.

Brauer, A. (1906). *Die Tiefsee-Fische. I. Systematischer Teil. Wissenschaftliche Ergebnisse der deutschen Tiefsee-Expedition 'Valdivia', 1898–99* (editor C. Chun) Volume. 15(1): 1–432, Gustav Fischer, Jena.

Brauer, A. (1908). *Die Tiefsee-Fische. II. Anatomischer Teil. Wissenschaftliche Ergebnisse der deutsch fsee-Expedition 'Valdivia', 1898–99* (editor C. Chun) Volume. 15(2): 1–266, Gustav Fischer, Jena.

Bray D.J. (2011). Toothed Whiptail, *Lepidorhynchus denticulatus*, in Fishes of Australia, accessed 27 Jul 2015, www.fishesofaustralia.net.au/home/species/4355

Briggs J.C (2003). Marine centres of origin as evolutionary engines. *Journal of Biogeography* **30**: 1–18.

Bray R.A. (2004). The bathymetric distribution of the digenean parasites of deep-sea fishes. *Folia Parasitologica* **51**: 268–274.

Bray D.J. & Gomon M.F. (2011). Atlantic Sabretooth, *Coccorella atlantica*, in Fishes of Australia, accessed 22 May 2015, www.fishesofaustralia.net.au/home/species/2796

Briggs, J.C. (1960). Fishes of worldwide (circumtropical) distribution. *Copeia*, 171–180.

Britton J.C. & Morton B. (1994). Marine carrion and scavengers. *Oceanography and Marine Biology: An Annual Review* 32: 369–434.

Brodeur R.D., Seki M.P., Pakhomov E.A., & Suntsov A.V. (2005). Micronekton – What are they and why are they important? *PICES Press* 13(1): 7–11. Online publication accessed 18 July 2016 www.pices.int/publications/pices_press

Broecker, W.S. (1991). The great ocean conveyor. *Oceanography* 4: 79–89.

Brown, S.K., Mahon R., Zwanenburg K.C.T, Buja K.R., Claflin L.W., O'Boyle R.N., Atkinson B., Sinclair M., Howell G., & Monaco M.E. (1996). *East Coast of North America Groundfish: Initial Explorations of Biogeography and Species Assemblages*. Silver Spring, MD: National Oceanic and Atmospheric Administration, and Dartmouth, NS: Department of Fisheries and Oceans.

Bruland, K.W. & Silver M. W. (1981). Sinking rates of fecal pellets from gelatinous zooplankton (salps, pteropods, doliolids). *Marine Biology* 63: 295–300, doi: 10.1007/BF00395999

Brulla L., Nizet E., & Verney E.B. (1953). Blood perfusion of the kidney of *Lophius Piscatorius* L. *Journal of the Marine Biological Association of the UK* 32: 329–336.

Bruun A.F. (1951). The Philippine Trench and its bottom fauna. *Nature* 168: 692–693 doi: 10.1038/168692b0

Buckland S.T., Anderson D.R., Burnham K.P., Laake J.L., Borchers D.L., & Thomas L. (2001). *Introduction to Distance Sampling: Estimating Abundance of Biological Populations*. Oxford: Oxford University Press.

Buesseler K.O. Lamborg C.H., Boyd P.W., Lam P.J., Trull T.W., Bidigare R.R., Bishop J.K.B., Casciotti K.L., Dehairs F., Elskens M., Honda M., Karl D.M., Siegel D.A., Silver M.W., Steinberg D.K., Valdes J., Van Mooy B., & Wilson S. (2007). Revisiting carbon flux through the ocean's twilight zone. *Science* 316: 567–570.

Buhl-Mortensen L., Vanreusel A., Gooday A.J., Levin L.A., Priede I.G., Buhl-Mortensen P., Gheerardyn H., King N.J., & Raes M. (2010). Biological structures as a source of habitat heterogeneity and biodiversity on the deep ocean margins. *Marine Ecology* 31: 21–50.

Bullough L.W., Turrell W.R., Buchan P. & Priede I.G. (1998). Commercial deep water trawling at sub-zero temperatures – observations from the Faroe-Shetland channel. *Fisheries Research* 39: 33–41.

Burne, R.H. (1926). A contribution to the anatomy of the ductless glands and lymphatic system of the angler fish (*Lophius piscatorius*). *Philosophical Transactions of the Royal Society of London. B* 215: 1–57.

Busby M.S. & Orr J.W. (1999). A pelagic basslet *Howella sherborni* (Family Acropomatidae) off the Aleutian Islands. *Alaska Fishery Research Bulletin* 6(1): 49–53.

Bussing W.A. & López-S. M.I. (1977). *Guentherus altivela* Osorio, the first ateleopodid fish reported from the eastern Pacific Ocean. *Revista de Biologia Tropical* 25(2): 179–190.

Butler J.L. & Ahlstrom E.H. (1976). Review of the deep-sea fish genus *Scopelengys* (Neoscopelidae) with a description of a new species, *Scopelengys clarkei*, from the Central Pacific. *Fishery Bulletin* 74: 142–150.

Butler M., Bollens S.M., Burkhalter B., Madin L.P., & Horgan E. (2001). Mesopelagic fishes of the Arabian Sea: Distribution, abundance and diet of *Chauliodus pammelas*, *Chauliodus sloani*, *Stomias affinis*, and *Stomias nebulosus*. *Deep-Sea Research II* 48: 1369–1383.

Byrkjedal I., Poulsen J.Y. & Galbraith J. (2011). *Leptoderma macrophthalmum* n.sp., a new species of smooth-head (Otocephala: Alepocephalidae) from the Mid Atlantic Ridge. *Zootaxa* 2876: 49–56.

Caira J.N., Benz G.W., Borucinska J., & Kohler N.E. (1997). Pugnose eels, *Simenchelys parasiticus* (Synaphobranchidae) from the heart of a shortfin mako, *Isurus oxyrinchus* (Lamnidae). *Environmental Biology of Fishes* 49: 139–144.

Calistry. (2017). Van Der Waals equation calculator. http://calistry.org/calculate/vanDerWaalsCalcula tor. On line resource accessed 9 March 2017.

Cavin L. & Martin M. (1995). Les Actinoptérygiens et la limite Crétacé-Tertiaire. *Geobios* 28 (supplement 2): 183–188.

Campbell M.A., Chen W.-J., & López A. (2013). Are flatfishes (Pleuronectiformes) monophyletic? *Molecular Phylogenetics and Evolution* 69 (2013): 664–673.

Campbell R.A., Haedrich R.L,. & Munroe T.A. (1980) Parasitism and ecological relationships among deep-sea benthic fishes. *Marine Biology* **57**: 301–313.

Campillo A. (1992). Les Pecheries Françaises de Mediterranee, Synthese des Connaissances. Réf. RI DRV 92-019 RH/Sete. Accessed online 2 March 2015: http://archimer.ifremer.fr/doc/1992/rapport-1125.pdf

Canals, M., Puig P., Durrieu de Madron X., Heussner S., Palanques A., & Fabrés J. (2006). Flushing submarine canyons. *Nature* **444**: 354–357.

Carrasson M. & Matallanas J. (1998). Feeding habits of *Alepocephalus rostratus* (Pisces: Alepocephalidae) in the Western Mediterranean Sea. *Journal of the Marine Biological Association of the U.K.* **78**: 1295–1306.

Carrete Vega G. & Wiens J.J. (2012). Why are there so few fish in the sea? *Proceedings of the Royal Society of London. B* **279**: 2323–2329. doi:10.1098/rspb.2012.0075

Carter H.J. & Musick J.A. (1985). Sexual dimorphism in the deep-sea fish *Barathrodemus manatinus* (Ophidiidae). *Copeia* **1985**: 69–73.

Caruso J.H. (1989). Systematics and distribution of Atlantic chaunacid anglerfishes (Pisces: Lophiiformes). *Copeia* **1989**(1): 153–165.

Caruso J.H., Ho H.-C., & Pietsch T.W. (2006). *Chaunacops* Garman, 1899, a senior objective synonym of *Bathychaunax* Caruso, 1989 (Lophiiformes: Chaunacoidei: Chaunacidae). *Copeia* (1): 120–121.

Casey J.M. & Myers R.A. (1998). Near extinction of a large, widely distributed fish. *Science* **31**: 690–692. doi: 10.1126/ science.281.5377.690.

Castle P.H.J. (1959). A large leptocephalid (Teleostei, Apodes) from off South Westland, New Zealand. *Transactions of the Royal Society of New Zealand* **87** (pts 1–2): 179–184.

Castle P.H.J. (1991). First Indo-Pacific Record of the Eel Family Myrocongridae, with the Description of a New Species of *Myroconger. Copeia* **1991**: 148–150.

Castonguay L. & McCleave J.D. (1987a). Distribution of leptocephali of the oceanic species *Derichthys serpentinus and Nessorhamphus ingolfianus* (Family Derichthyidae) in the western Sargasso Sea in relation to physical oceanography. *Bulletin of Marine Science* **41**: 807–821.

Castonguay M. & McCleave J.D. (1987b) Vertical distributions, diel and ontogenetic vertical migrations and net avoidance of leptocephali of *Anguilla* and other common species in the Sargasso Sea. *Journal of Plankton Research* **9**: 195–214.

Catul V., Gauns M., & Karuppasamy P.K. (2011). A review on mesopelagic fishes belonging to family Myctophidae. *Reviews in Fish Biology and Fisheries* **21**: 339–354.

Causse R., Biscoito M., & Briand P (2005). First record of the deep-sea eel *Ilyophis saldanhai* (Synaphobranchidae, Anguilliformes) from the Pacific Ocean. *Cybium* **29**: 413–416.

Cavin L. (2001). Effects of the Cretaceous–Tertiary boundary event on bony fishes. In: Buffetaut E. & Koeberl C. (eds.) *Geological and Biological Effects of Impact Events*, pp. 141–158. Berlin: Springer.

CCAMLR. (2015). Toothfish fisheries. Commission for the Conservation of Antarctic Marine Living Resources. www.ccamlr.org/en/fisheries/toothfish-fisheries Accessed 10 December 2015.

CEC. (2006). Deep-sea gillnet fisheries. Report of the Scientific, Technical and Economic Committee for Fisheries. Accessed 24 February 2016. https://stecf .jrc.ec.europa.eu/documents/43805/122924/06–11_ Adhoc+06-01+Deep-Sea+Gillnet+Fisheries.pdf

Cember R.P. (1988). On the sources, formation, and circulation of Red Sea deep water. *Journal of Geophysical Research* **93** (C7): 8175–8191.

Chanet B., Guintard C., Betti E., Gallut C., Dettaï A., & Lecointre G. (2013). Evidence for a close phylogenetic relationship between the teleost orders Tetraodontiformes and Lophiiformes based on an analysis of soft anatomy. *Cybium* **37**(3): 179–198.

Chapman L., Desurmont A., Choi Y., Boblin P., Sokimi W. & Beverly S. (2008). *Fish species identification manual for deep-bottom snapper fishermen*. Secretariat of the Pacific Community, Noumea, New Caledonia.

Charter S.R. (1996). Nemichthyidae: Snipe Eels. In: Moser H.G. (ed.) *Early Stages of Fishes in the California Current Region. California Cooperative Oceanic Fisheries Investigations*, pp. 122–129. Atlas No 33. La Jolla, CA: South West Fisheries Centre.

Chave, E.H. & Mundy B.C. (1994). Deep-sea benthic fish of the Hawaiian Archipelago, Cross Seamount, and Johnston Atoll. *Pacific Science* **48**: 367–409.

Chavez F.P. & Messié M. (2009). A comparison of Eastern Boundary Upwelling Ecosystems. *Progress in Oceanography* **83**: 80–96.

Chembian A.J. (2007). New record of *Rhinochimaera atlantica* (Chimaeriformes: Rhinochimaeridae) spawning ground in the Gulf of Mannar along the south-east coast of India. *Indian Journal of Fisheries* 54(4): 345–350.

Chernova N.V. & Stein D.L. (2002). Ten new species of *Psednos* (Pisces, Scorpaeniformes: Liparidae) from the Pacific and North Atlantic Oceans. *Copeia* 2002: 755–778.

Chernova N.V. & Stein D.L. (2004). A remarkable new species of *Psednos* (Teleostei: Liparidae) from the western North Atlantic Ocean. *Fishery Bulletin* 102: 245–250.

Childress, J.J., Barnes, A.T., Quetin, L.B., & Robison, B.H. (1978). Thermally protecting cod ends for the recovery of living deep-sea animals. *Deep-Sea Research* 25: 419–422.

Childress J.J. & Somero G. (1979). Depth-related enzymic activities in muscle, brain and heart of deep-living pelagic marine teleosts. *Marine Biology* 52: 273–283.

Childress J.J., Taylor S.M., Cailliet G.M. & Price M.H. (1980). Patterns of growth, energy utilization and reproduction in some meso- and bathypelagic fishes off Southern California. *Marine Biology* 61: 27–40.

Childress J.J. & Thuesen E.V. (1995). Metabolic potentials of deep-sea fishes: A comparative approach. In: Hochachka P.W. & Mommsen T.P. (eds.) *Biochemistry and Molecular Biology of Fishes*, pp. 175–196, Vol. 5. Amsterdam: Elsevier.

Christiansen B., Vieira R.P., Christiansen S., Denda A., Oliveira F., & Gonçalves J.M.S. (2015). The fish fauna of Ampère Seamount (NE Atlantic) and the adjacent abyssal plain. *Helgoland Marine Research* 69: 13–23. doi:10.1007/s10152-014-0413-4

Claes J.M., Ho H.C., & Mallefet J. (2012). The control of luminescence from pygmy shark (*Squaliolus aliae*) photophores. *Journal of Experimental Biology* 215: 1691–1699

Claes J.M., Nilsson D.-E., Straube N., Collin S.P., & Mallefet J. (2014). Iso-luminance counterillumination drove bioluminescent shark radiation. *Scientific Reports* 4: 4328. doi:10.1038/srep04328 (2014).

Claes J.M., Partridge J.C., Hart N.S., Garza-Gisholt E., Ho H.-C., et al. (2014) Photon Hunting in the Twilight Zone: Visual Features of Mesopelagic Bioluminescent Sharks. *PLoS ONE* 9(8): e104213. doi:10.1371/journal.pone.0104213

Clark M. (1995). Experience with management of Orange Roughy (*Hoplostethus atlanticus*) in New Zealand waters, and the effects of commercial fishing on stocks over the period 1980–1993. In: Hopper A.G. (ed.) *Deep-Water Fisheries of the North Atlantic Slope*, pp. 151–266. Dordrecht, Netherlands: Kluwer.

Clark M. (2001). Are deep water fisheries sustainable? the example of orange roughy (*Hoplostethus atlanticus*) in New Zealand. *Fisheries Research* 51: 123–135.

Clark M.R., Althaus F., Schlacher T.A., Williams A., Bowden D.A., & Rowden A.A. (2015). The impacts of deep-sea fisheries on benthic communities: a review. *ICES Journal of Marine Science.* doi: 10.1093/icesjms/fsv123.

Clark M.R., Althaus F., Williams A., Niklitschek E., Menezes G.M., Hareide N.-R., Sutton P., & O'Donnell C. (2010). Are deep-sea demersal fish assemblages globally homogenous? Insights from seamounts. *Marine Ecology* 31(Suppl. 1): 39–51.

Clarke J., Milligan T.J., Bailey D.M., & Neat F.C. (2015). A scientific basis for regulating deep-sea fishing by depth. *Current Biology* 25: 2425–2429.

Clarke M.R. (1969). A new midwater trawl for sampling discrete depth horizons. *Journal of the Marine Biological Association UK* 49: 945–960.

Clarke M.R. (2003). Searching for deep sea squids. *Berliner paläobiologische Abhandlungen* 3: 49–59.

Clarke M.W., Kelly C.J., Connolly P.L., & Molloy J.P. (2003). A life history approach to the assessment and management of deepwater fisheries in the North east Atlantic. *Journal of Northwest Atlantic Fisheries Science* 31: 401–411.

Clarke T.A. (1982). Feeding habits of stomiatoid fishes from Hawaiian waters. *Fishery Bulletin* 80: 287–304.

Clausen D.M. (2008). The giant grenadier in Alaska. *American Fisheries Society Symposium* 63: 413–450.

Clement G., Wells R., & Gallagher C.M. (2008). Industry management within the New Zealand quota management system: The Orange Roughy Management Company. In: Townsend R.E., Shotton R., & Uchida H. (eds.) *Case Studies in Fisheries Self-governance*, pp. 277–290. FAO Fisheries Technical Paper 504. Rome: FAO.

Coad, B.W. (1995). *Encyclopedia of Canadian fishes.* Canadian Museum of Nature and Canadian Sportfishing Productions Inc. Singapore.

Coghlan A. (2008). Deep-sea fishing. In: Lück E. (ed.) *The Encyclopedia of Tourism and Recreation in Marine Environments*, pp. 132–133. Wallingford, England: CABI.

Cohen D.H. & Haedrich R.L. (1983). The fish fauna of the Galapagos thermal vent region. *Deep-Sea Research* 30A: 371–379.

Cohen D.M. (1977). Swimming performance of the gadoid fish *Antimora rostrata* at 2400 meters. *Deep-Sea Research* 24: 275–277.

Cohen D.M. & Rohr B.A. (1993). Description of a giant Circumglobal Lamprogrammus species (Pisces: Ophidiidae). *Copeia* 1993(2): 470–475.

Cohen D.M., Rosenblatt R.H., & Moser H.G. (1990). Biology and description of a bythitid fish from deep-sea thermal vents in the tropical eastern Pacific. *Deep-Sea Research* 37: 267–283.

Collett R. (1880). Fishes. *The Norwegian North-Atlantic Expedition 1876–1878*. 3(1): 1–166.

Collett, R. (1896). *Poissons provenant des campagnes du Yacht "L'Hirondelle" (1885–1888). Résultats des campagnes scientifiques accomplis sur son yacht par Albert I, Prince Souverain de Monaco. Résultats des campagnes scientifiques du Prince de Monaco*. Monaco: Imprimerie de Monaco.

Collette, B.B., Curtis, M., Smith-Vaniz, W.F., Pina Amargos, F., Williams, J.T., & Grijalba Bendeck, L. (2015). Ruvettus pretiosus. *The IUCN Red List of Threatened Species 2015*: e.T190432A16644022. http://dx.doi.org/10.2305/IUCN.UK.2015-4.RLTS.T190432A16644022.en. Downloaded on 09 April 2016.

Collin, S.P., Hoskins, R. V., & Partridge, J. C. (1998). Seven retinal specialisations in the tubular eye of the deepsea pearleye, *Scopelarchus michaelsarsi*: a case study in visual optimisation. *Brain Behaviour and Evolution*, 51: 291–314.

Collin S.P. & Partridge J.C. (1996). Retinal specialisations in the eyes of deep-sea teleosts. *Journal of Fish Biology* 49A: 157–174.

Collins M.A., Bailey D.M., Ruxton G.D., & Priede I.G. (2005). Trends in body size across an environmental gradient: a differential response in scavenging and non-scavenging demersal deep-sea fish. *Proceedings of the Royal Society B* 272: 2051–2057.

Collins M.A., Brickle P., Brown J. & Belchier M. (2010). The Patagonian Toothfish: biology, ecology and fishery. *Advances in Marine Biology* 58: 227–300.

Collins, M.A., Priede, I.G., & Bagley, P.M. (1999). *In situ* comparison of activity of two deep-sea scavenging fish occupying different depths zones. *Proceedings of the Royal Society, Series B* 266:2011–2016.

Collins M.A., Stowasser G., Fielding S., Shreeve R., Xavier J.C., Venables H.J., Enderlein P., Cherel Y., & Van de Putte A. (2012). Latitudinal and bathymetric patterns in the distribution and abundance of mesopelagic fish in the Scotia Sea. *Deep-Sea Research II* 59–60: 189–198.

Collins M.A., Xavier J.C., Johnston N.M., North A.W., Enderlein P., Tarling G.A., Waluda C.M., Hawker E.J., & Cunningham N.J. (2008). Patterns in the distribution of myctophid fish in the northern Scotia Sea ecosystem. *Polar Biology* 31: 837–851. DOI 10.1007/s00300-008-0423-2

Colman J.A. (1995). Biology and fisheries of New Zealand hake (*M.australis*). In: Alheit J. & Pitcher T.J. (eds.) *Hake: Biology, Fisheries and Markets*, pp. 365–388. London: Chapman & Hall.

Company J.B., Puig P., Sardà F., Palanques A., Latasa M., et al. (2008). Climate influence on deep sea populations. *PLoS ONE* 3(1): e1431. doi:10.1371/journal.pone.0001431

Connelly, D.P. et al. (2012). Hydrothermal vent fields and chemosynthetic biota on the world's deepest seafloor spreading centre. *Nature Communications* 3, Article number: 620. doi:10.1038/ncomms1636

Cook S.F. & Compagno L.J.V. (2005). Hexanchus griseus. *The IUCN Red List of Threatened Species 2005*: e.T10030A3155348. http://dx.doi.org/10.2305/IUCN.UK.2005.RLTS.T10030A3155348.en. Downloaded on 25 February 2016.

Corbella C. (2013). *Taxonomy and phylogeny of Lyconidae (Teleostei:Zoarcidae) from the Southern Ocean and Magellan Province*. Doctoral Thesis. Univeritat Autonoma de Barcelona.

Corliss J.B. & Ballard R.D., (1977). Oases of life in the cold abyss. *National Geographic Magazine* 152: 441–453.

Correia A.T., Faria R., Alexandrino P., Antunes C., Isidro E.J., & Coimbra J. (2006). Evidence for genetic differentiation in the European conger eel *Conger conger* based on mitochondrial DNA analysis. *Fisheries Science* 72: 20–27.

Correia A.T., Ramos A.A., Barros F., Silva G., Hamer P., Morais P., Cunha R.L., & Castilho R. (2012). Population structure and connectivity of the European conger eel (*Conger conger*) across the north-eastern Atlantic and western Mediterranean: Integrating molecular and otolith elemental approaches. *Marine Biology* **159**:1509–1525. doi 10.1007/s00227-012-1936-3

Cossins A.R. & MacDonald A.G. (1986). Homeoviscous adaptation under pressure: III The fatty acid composition of liver mitochondrial phospholipids of deep-sea fish. Biochim. *Biophys Acta* **860**: 325–335.

Costello M.J., Cheung A. & De Hauwere N. (2010). Surface area and seabed area, volume, depth slope and topographic variation for the World's Sea, Oceans and countries. *Environmental Science & Technology* **44**: 8821–8828.

Costello M.J., May R.M., & Stork N.E. (2013). Can we name Earth's species before they go extinct? *Science* **339**: 413–416. doi: 10.1126/science.1230318

Costello M.J., McCrea M., Freiwald A., Lundalv T., Jonsson L., Bett B.J., van Weering T.C.E., de Haas H., Roberts J.M. & Allen D. (2005). Role of cold-water *Lophelia pertusa* coral reefs as fish habitat in the NE Atlantic. In: Freiwald A. & Roberts J.M. (eds.) *Cold-Water Corals and Ecosystems*. Berlin: Springer-Verlag.

Cousins N.J., Linley T., Jamieson A.J., Bagley P.M., Blades H., Box T., Chambers R., Ford A., Shields M.A., & Priede I.G. (2013). Bathyal demersal fishes of Charlie-Gibbs fracture zone region (49°–54°N) of the Mid Atlantic Ridge: II. Baited camera lander observations. *Deep-Sea Research II* **98**: 397–406. http://dx.doi.org/10.1016/j.dsr2.2013.08.002

Cousins N.J. & Priede I.G. (2012). Abyssal demersal fish fauna composition in two contrasting productivity regions of the Crozet Plateau, Southern Indian Ocean. *Deep-Sea Research I* **64**: 71–77. doi:10.1016/j.dsr.2012.02.003

Cousins N.J., Shields M.A., Crockard D. & Priede I.G. (2013). Bathyal demersal fishes of Charlie Gibbs fracture zone region (49°–54°N) of the Mid-Atlantic Ridge: I. Results from trawl surveys. *Deep-Sea Research II* **98**: 388–396. http://dx.doi.org/10.1016/j.dsr2.2013.08.012

Cowie G.L. & Levin L.A. (2009). Benthic biological and biogeochemical patterns and processes across an oxygen minimum zone (Pakistan margin, NE Arabian Sea). *Deep-Sea Research II* **56**: 261–270.

Cox G.K., Sandblom E., Richards J.G. & Farrell A.P. (2011). Anoxic survival of the Pacific hagfish

(*Eptatretus stoutii*). *Journal of Comparative Physiology B* **181**, 361–371. 10.1007/s00360-010-0532-4

Crabtree R.E. (1995). Chemical composition and energy content of deep-sea demersal fishes from tropical and temperate regions of the western North Atlantic. *Bulletin of Marine Science* **56** (2): 434–449.

Crabtree R.E. & Sulak K.J. (1986). A contribution to the life history and distribution of Atlantic species of the deep-sea fish genus *Conocara* (Alepocephalidae). *Deep-Sea Research Part A, Oceanographic Research Papers*. **33**: 1183–1201.

Crabtree R.E., Sulak K.J., & Musick J.A. (1985). Biology and distribution of the species of *Polyacanthonotus* (Pisces: Notacanthiformes) in the western North Atlantic. *Bulletin of Marine Science* **36**:235–248.

Craig,J., Jamieson, A.J., Hutson R., Zuur A.F., & Priede I.G. (2010). Factors influencing the abundance of deep pelagic bioluminescent zooplankton in the Mediterranean Sea. *Deep-Sea Research I* **57**: 1474–1484. doi:10.1016/j.dsr.2010.08.005

Craig J., Priede I.G., Aguzzi J., Company J.B. & Jamieson A.J. (2015). Abundant bioluminescent sources of low-light intensity in the deep Mediterranean Sea and North Atlantic Ocean. *Marine Biology* **162**:1637–1649.

Craig J., Youngbluth M., Jamieson A.J., & Priede I.G. (2015). Near seafloor bioluminescence, macrozooplankton and macroparticles at the Mid-Atlantic Ridge. *Deep-Sea Research I* **98**: 62–75.

Crane J.M. (1968). Bioluminescence in the batfish *Dibranchus atlanticus. Copeia* **1968**: 410–411.

CSIRO Marine & Atmospheric Research. (2011). Deepsea Flathead, Hoplichthys haswelli. In: *Fishes of Australia*, accessed 16 October 2015, http://www.fishesofaustralia.net.au/home/species/3374

Currie R.I. (1983). Research ship design & logistics. In: MacDonald A.G. & Priede I.G. (eds.) *Experimental Design at Sea*, pp. 387–402. London: Academic Press.

Cuvier G., Pietsch T. (ed.). (1995). *Historical Portrait of the Progress of Ichthyology, from Its Origins to Our Own Time*. Baltimore, MD: Johns Hopkins.

Cuvier G. & Vallenciennes M. (1828). *Histoire Naturelle des Poissons*. Paris: Tome Premier. Levrault.

D'Onghia G., Indennidate A., Giove A., Savini A., Capezzuto F., Sion L., Vertino A., & Maiorano P. (2011). Distribution and behaviour of deep-sea benthopelagic fauna observed using towed cameras in the Santa Maria di Leuca cold-water coral province. *Marine Ecology Progress Series* **443**: 95–110.

D'Ongia G., Maiorano P., & Sion L. (2008) A review on the reproduction of Grenadiers in the Mediterranean with new data on the gonad Maturity and fecundity. *American Fisheries Society Symposium* **63**: 169–184.

D'Onghia G., Politou C.-Y., Bozzano A., Lloris D., Rotllant G., Sion L., & Mastrototaro F. (2004). Deepwater fish assemblages in the Mediterranean Sea. *Scientia Marina* **68** (Suppl. 3): 87–99.

D'Onghia G., Sion L., Maiorano P., Mytilineou C., Dalessandro S., Carlucci R., & Desantis S. (2006). Population biology and life strategies of *Chlorophthalmus agassizii* Bonaparte, 1840 (Pisces: Osteichthyes) in the Mediterranean Sea. *Marine Biology* **149**: 435–446. doi:10.1007/s00227-005-0231-y

Dagit, D.D., Hareide, N., & Clò, S. (2007). *Chimaera monstrosa. The IUCN Red List of Threatened Species.* Version 2014.3. <www.iucnredlist.org>. Downloaded on 31 January 2015.

Dahlhoff E., Schneidemann S., & Somero G.N. (1990). Pressure-temperature interactions on M4-Lactate Dehydrogenases from hydrothermal vent fishes: Evidence for adaptation to elevated temperatures by the Zoarcid *Thermarces andersoni*, but not by the Bythitid, *Bythites hollisi. Biological Bulletin* **179**: 134–139.

Dalzell P. & Pauly D. (1989) Assessment of the fish resources of Southeast Asia, with emphasis on the Banda and Arafura seas. *Netherlands Journal of Sea Research* **24**: 641–650.

Danovaro R., Company J.B., Corinaldesi C., D'Onghia G., Galil B.,et al. (2010). Deep-sea biodiversity in the Mediterranean Sea: The known, the unknown, and the unknowable. *PLoS ONE* **5**(8): e11832. doi:10.1371/journal.pone.0011832

Danovaro R., Della Croce N., Dell'Anno A., & Pusceddu A. (2003). A depocenter of organic matter at 7800m depth in the SE Pacific Ocean. *Deep-Sea Research I* **50**: 1411–1420.

Darwin C. (1859). *On the Origin of Species by Means of Natural Selection.* London: Murray.

Das, M.K. & Nelson J.S. (1996). Revision of the percophid genus *Bembrops* (Actinopterygii: Perciformes). *Bulletin of Marine Science* **59**(1):9–44.

Davenport J. & Kjørsvik E. (1986). Buoyancy in the Lumpsucker *Cyclopterus lumpus. Journal of the Marine Biological Association of the UK* **66**: 159–174.

Davesne D., Friedman M., Barriel V., Lecointre G., Janvier P., Gallut C., & Otero O. (2014). Early fossils illuminate character evolution and interrelationships of Lampridiformes (Teleostei, Acanthomorpha). *Zoological Journal of the Linnean Society* **2014** (172): 475–498.

Davies A.J., Hosein S., & Merrett N.R. (2012). Haematozoans from deep water fishes trawled off the Cape Verde Islands and over the Porcupine Seabight, with a revision of species within the genus *Desseria* (Adeleorina: Haemogregarinidae). *Folia Parasitologica* **59** (1): 1–11.

Davies A.J. & Merrett N.R. (1998). Presumptive viral erythrocytic necrosis (VEN) in the benthopelagic fish *Coryphaenoides (Nematonurus) armatus* from the abyss west of Portugal. *Journal of the Marine Biological Association of the UK* **78**: 1031–1034.

Davies I.E. & Barham, E.G. (1969). The Tucker opening-closing micronekton net and its performance in a study of the deep scattering layer. *Marine Biology* **2**: 127–131.

Davies R., Cartwright J., Pike J., & Line C. (2001). Early Oligocene initiation of North Atlantic Deep Water formation. *Nature* **410**: 917–920.

Davis, M.P. (2015). Evolutionary relationships of the deep-sea pearleyes (Aulopiformes: Scopelarchidae) and a new genus of pearleye from Antarctic waters. *Copeia*, **2015**(1): 64–71.

Davis M.P. & Fielitz C. (2010). Estimating divergence times of lizardfishes and their allies (Euteleostei: Aulopiformes) and the timing of deep-sea adaptations. *Molecular Phylogenetics and Evolution* **57**: 1194–1208.

Davis M.P., Holcroft N.I., Wiley E.O., Sparks J.S., & Smith W.L. (2014). Species-specific bioluminescence facilitates speciation in the deep sea. *Marine Biology* **161**:1139–1148.

Davison P.C., Checkley Jr D.M., Koslow J.A., & Barlow J. (2013). Carbon export mediated by mesopelagic fishes in the northeast Pacific Ocean. *Progress in Oceanography* **116**: 14–30.

Davison P.C., Lara-Lopez A., & Koslow J.A. (2015). Mesopelagic fish biomass in the southern California current ecosystem. *Deep-Sea Research II* 112:129–142.

De Leo F.C., Smith C.R., Rowden A.A., Bowden D.A., & Clark M.R. (2010). Submarine canyons: hotspots of benthic biomass and productivity in the deep sea. *Proceedings of the Royal Society of London Series B.* 277:2783–2792. doi:10.1098/rspb.2010.0462

Deng X., Wagner H.-J. ,& Popper A.N. (2011). The inner ear and its coupling to the swim bladder in the deep-sea fish *Antimora rostrata* (Teleostei: Moridae). *Deep-Sea Research I* 58: 27–37. doi:10.1016/j. dsr.2010.11.001

Denton E.J. (1970). On the organization of reflecting surfaces in some marine animals. *Philosophical Transactions of the Royal Society of London* 258B: 285–313.

Denton E.J., Gilpin-Brown, J.B., & Wright P.G. (1972). The angular distribution of the light produced by some mesopelagic fish in relation to their camouflage. *Proceedings of the Royal Society B* 182:145–158. doi:10.1098/rspb.1972.0071

Denton E.J. & Marshall N.B. (1958). The buoyancy of bathypelagic fishes without a gas-filled swimbladder. *Journal of the Marine Biological Association of the UK.* 69: 409–435.

Denton J.E. & Yousef M.K. (1976). Body composition and organ weights of rainbow trout, *Salmo gairdneri*. *Journal of Fish Biology* 8: 489–499.

Desoutter, M. & Chapleau F. (1997). Taxonomic status of *Bathysolea profundicola* and *B. polli* (Soleidae; Pleuronectiformes) with notes on the genus. *Ichthyological Research* 44(4): 399–412.

DeVaney S.C., Hartel K.E., & Themelis D.E. (2009). The first records of *Neocyema* (Teleostei: Saccopharyngiformes) in the Western North Atlantic with comments on its relationship to *Leptocephalus holti* Schmidt 1909. *Northeastern Naturalist* 16(3): 409–414.

Devine J.A., Baker K.D., & Haedrich R.L. (2006). Fisheries: Deep-sea fishes qualify as endangered. *Nature* 439: 29.

Diaz H.F. & Bradley R.S. (eds.). (2004). *The Hadley Circulation: Present, Past and Future.* Dordrecht: Kluwer Academic Publishers.

Doel, R.E., Levin, T.J., & Marker, M.K. (2006). Extending modern cartography to the ocean depths: military patronage, Cold War priorities, and the Heezen-Tharp mapping project, 1952–1959. *Journal of Historical Geography* 32, 605–626.

Dolar M.L., Walker W.A., Kooyman G.L., & Perrin W.F. (2003). Comparative feeding ecology of Spinner dolphins (*Stenella longirostris*) and Fraser's dolphins (*Lagenodelphis hosei*) in the Sulu Sea. *Marine Mammal Science* 19: 1–19.

Dolgov A.V., Drevetnyak V., Sokolov K.M., Grekov A.A., & Shestopal I.P. (2008). Biology and fisheries of roughhead grenadier in the Barents Sea. *American Fisheries Society Symposium* 63: 343–363.

Douglas R.H., Partridge J.C. & Marshall, N. J. (1998). The eyes of deep-sea fish I: Lens pigmentation, tapeta and visual pigments. *Progress in Retinal and Eye Research.* 17: 597–636.

Drazen J.C. (2002). Energy budgets and feeding rates of *Coryphaenoides acrolepis* and *C. armatus*. *Marine Biology* 140: 677–686.

Drazen J.C. (2007). Depth related trends in proximate composition of demersal fishes in the eastern North Pacific. *Deep-Sea Research I* 54: 203–219.

Drazen J.C., Bailey D.M., Ruhl H.A., & Smith, Jr. K.L. (2012), The role of carrion supply in the abundance of deep water fish off California. *PLoS ONE* 7(11): e49332. doi:10.1371/journal.pone.0049332

Drazen J.C., Bird L.E., & Barry J.P. (2007). Development of a hyperbaric trap-respirometer for the capture and maintenance of live deep-sea organisms. *Limnology & Oceanography: Methods* 3: 488–498.

Drazen J.C., Buckley, T.W., & Hoff, G.R. (2001). The feeding habits of slope dwelling macrourid fishes in the eastern North Pacific. *Deep-Sea Research I* 48: 909–995.

Drazen J.C., Dugan B., & Friedman J.R. (2013). Red muscle proportions and enzyme activities in deep-sea demersal fishes *Journal of Fish Biology* 83: 1592–1612. doi:10.1111/jfb.12268

Drazen J.C., Friedman J.R., Condon N.E., Aus E.J., Gerringer M.E., Keller A.A., & Clarke M.E. (2015). Enzyme activities of demersal fishes from the shelf to the abyssal plain. *Deep-Sea Research I* 100: 117–126.

Drazen J.C. & Haedrich R.L. (2012). A continuum of life histories in deep-sea demersal fishes. *Deep-Sea Research I* 61: 34–42.

Drazen J.C., Popp, B., & Choy, C. (2008). Bypassing the abyssal benthic food web: Macrourid diet in the eastern North Pacific inferred from stomach content and stable isotopes analyses. *Limnology and Oceanography* 53: 2644–2654.

Drazen J.C. & Robison B.H. (2004). Direct observations of the association between a deep-sea fish and a giant scyphomedusa. *Marine and Freshwater Behaviour and Physiology* **37**(3): 209–214, doi: 10.1080/10236240400006190

Drazen J.C. & Seibel B.A. (2007). Depth-related trends in metabolism of benthic and benthopelagic deep-sea fishes. *Limnology & Oceanography* **52**(5): 2306–2316.

Drazen J.C. & Sutton T.T. (2016). Dining in the deep: The feeding ecology of deep-sea fishes. *Annual Reviews in Marine Science* **9**: 337–366.

Drazen J.C. & Yeh J. (2012). Respiration of four species of deep-sea demersal fishes measured *in situ* in the eastern North Pacific. *Deep-Sea Research I* **60**:1–6.

Ducklow H.W., Steinberg D.K., & Buessler K.O. (2001). Upper ocean carbon export and the biological pump. *Oceanography* **14**: 50–58.

Duffy C.A.J. (1997). Further records of the goblin shark, *Mitsukurina owstoni* (Lamniformes: Mitsukurinidae), from New Zealand. *New Zealand Journal of Zoology* **24**: 2,167–171. doi: 10.1080/03014223.1997.9518111

Duhamel, G. (1995). Révision des genres *Centriscops* et *Notopogon*, Macroramphosidae des zones subtropicale et tempérée de l'hémisphére sud. *Cybium* **19**(3): 261–303.

Duhamel G., Hulley P.-A. Causse R., Koubbi P., Vacchi M., Pruvost P., et al. (2014). Chapter 7, Biogeographic patterns of fish. In: De Broyer C., Koubbi P., Griffiths H.J., Raymond B., Udekem d'Acoz C. d', et al. (eds.) *Biogeographic Atlas of the Southern Ocean*, pp. 328–498. Cambridge: Scientific Committee on Antarctic Research.

Duhamel G. & King N. (2007). Deep-sea snailfish (Scorpaeniformes: Liparidae) of genera *Careproctus* and *Paraliparis* from the Crozet Basin (Southern Ocean). *Cybium* **31**(3): 379–387.

Dulčić J. (2002). First record of scalloped ribbon fish, *Zu cristatus* (Pisces: Trachipteridae), eggs in the Adriatic Sea. *Journal of Plankton Research*. **24**: 1245–1246.

Duméril A.M.C. (1806). *Zoologie analytique, ou méthode naturelle de classification des animaux, rendue plus facile à l'aide de tableaux synoptiques*. Paris: Allais.

Dunlap P.V., Takami M., Wakatsuki S., Hendry T.A., Sezaki K., & Fukui A. (2014). Inception of bioluminescent symbiosis in early developmental stages of the deep-sea fish, *Coelorinchus kishinouyei* (Gadiformes: Macrouridae). *Ichthyological Research* **61**: 59–67. doi: 10.1007/s10228-013-0374-7

Eastman J.T., DeVries A.L., Coalson R.E., Nordquist R.E., & Boyd R.B. (1979). Renal conservation of antifreeze peptide in Antarctic eelpout, *Rhigophila dearborni*. *Nature* **282**: 217–221.

Eastman T., Hikida R.S., & Devries A.L. (1994). Buoyancy studies and microscopy of skin and subdermal extracellular matrix of the Antarctic snailfish, *Paraliparis devriesi*. *Journal of Morphology* **220**: 85–101.

Ebert D.A. (2013). *Deep-sea Cartilaginous Fishes of the Indian Ocean*. Volume 1. *Sharks. FAO Species Catalogue for Fishery Purposes*. No. 8, Vol. 1. Rome: FAO.

Eddy F.B. & Handy R.D. (2012). *Ecological and Environmental Physiology of Fishes*. Oxford: Oxford University Press

Ellerby D.J. (2010). How efficient is a fish? *Journal of Experimental Biology* **213**: 3765–3767. doi: 10.1242/jeb.034520

Ellis J.E., Rowe S., & Lotze H.K. (2015). Expansion of hagfish fisheries in Atlantic Canada and worldwide. *Fisheries Research* **161**: 24–33.

Enns T., Scholander P.F., & Bradstreet E.D. (1965). Effect of hydrostatic pressure on gases dissolved in water. *Journal of Physical Chemistry* **69**: 389–391.

Erdmann M.V. (1999). An account of the first living coelacanth known to scientists from Indonesian waters. *Environmental Biology of Fishes* **54**: 439–443.

Erlandson J,M,, Moss M.L., & Des Lauriers M. (2008). Life on the edge: early maritime cultures of the Pacific Coast of North America. *Quaternary Science Reviews* **27**: 2232–2245.

Eschmeyer, W. N. (ed.) (2014). *Catalog of fishes: genera, species, references*. (http://research.calacademy.org/research/ichthyology/catalog/fishcatmain.asp). Electronic version accessed 5 May 2014.

Espino M., Castillo R., & Fernández F (1995). Biology and fisheries of the Peruvian Hake. In: Alheit J. & Pitcher T.J. (eds.) *Hake: Biology, Fisheries and Markets*, pp. 339–363. London: Chapman & Hall.

Etnoyer P.J., Wood J., & Shirley T.C. (2010). How large is the seamount biome? *Oceanography* **23**:206–209.

Evseenko S.A. & Bol'shakova Y.Y. (2014). First finding of a juvenile of the streamer fish *Agrostichthys parkeri*

(Regalecidae) near the Walvis Ridge (Southern Atlantic). *Journal of Ichthyology* **54**: 608–610.

Evseenko, S.A. & Shtaut, M.I. (2005). On the species composition and distribution of ichthyoplankton and micronekton in the Costa Rica Dome and adjacent areas of the tropical Eastern Pacific. *Journal of Ichthyology* **45**: 513–525.

Eyring, C.F., Christensen R.J., & Raitt R.W. (1948). Reverberation in the sea. *Journal of the Acoustical Society of America* **20**:462–475.

FAO. (2009). *Report of the Technical Consultation on International Guidelines for the Management of Deep-sea Fisheries in the High Seas, Rome. 4–8 February and 25–29 August 2008, FAO Fisheries and Aquaculture Report, 881.*

FAO. (2014). Deep-sea fisheries. www.fao.org/fishery/topic/4440/en accessed 18 June 2014.

FAO. (2015). Species Fact Sheets Reinhardtius hippoglossoides (Walbaum, 1792). www.fao.org/fishery/species/2544/en, Accessed 21 December 2015.

FAO. (2016). Species Fact Sheets *Hexanchus griseus* (Bonnaterre, 1788) www.fao.org/fishery/species/2003/en accessed 25 February 2016

Farias I., Morales-Nin B., Lorance P., & Figueiredo I. (2013). Black scabbardfish, *Aphanopus carbo*, in the northeast Atlantic: distribution and hypothetical migratory cycle. *Aquatic Living Resources* **26**: 333–342.

Farr H.K. (1980). Multibeam bathymetric sonar: Sea beam and hydro chart. *Marine Geodesy* **4**: 77–93.

Fasham M.J.R. (ed.). (2003). *Ocean Biogeochemistry, The Role of the Ocean Carbon Cycle in Global Change.* Berlin: Springer.

Feagans-Bartow, J.N. & Sutton T.T. (2014). Ecology of the oceanic rim: pelagic eels as key ecosystem components. *Marine Ecology Progress Series* **502**: 257–266.

Feldman G.C., Kuring N.A., Ng C., Esaias W.E., Mcclain C.R., Elrod J.A., Maynard N., Endres D., Evands R., Brown J., Walsh S., Carle M. & Podesta G. (1989). Ocean Color: Availability of the Global Data Set. *Eos* **70**: 634–641.

Fernandes P.G. & Cook R.M. (2013). Reversal of fish stock decline in the Northeast Atlantic. *Current Biology* **23**: 1432–1437. http://dx.doi.org/10.1016/j.cub.2013.06.016

Fernandez-Arcaya U., Drazen J.C., Murua H., Ramirez-Llodra E., Bahamon N., Recasens L., Rotllant G., &

Company J.B. (2016). Bathymetric gradients of fecundity and egg size in fishes: a Mediterranean case study. *Deep Sea Research I* **116**:106–117. doi.org/10.1016/j.dsr.2016.08.005

Fernholm B. & Holmberg K. (1975). The eyes in three genera of hagfish (*Eptatretus, Paramyxine* and *Myxine*) – a case of degenerative evolution. *Vision Research* **15**: 253–259.

Figueiredo I., Bordalo-Machado P., Reis S., Sena-Carvalho D., Blasdale T., Newton A., & Gordo L. S. (2003). Observations on the reproductive cycle of the black scabbardfish (*Aphanopus carbo* Lowe, 1839) in the NE Atlantic. *ICES Journal of Marine Science* **60**: 774–779.

Figueroa D.E., Díaz de Astarloa J.M., & Martos P. (1998). Mesopelagic fish distribution in the southwest Atlantic in relation to water masses. *Deep-Sea Research I* **45**: 317–332.

Fischer W. & Bianchi G. (1984). FAO Species identification sheets for fishery purposes, Western Indian Ocean, Fishing Area 51. Rome: Food and Agriculture Organization of the United Nations.

Fisher J.A.D., Frank K.T., Petrie B. & Leggett W.C. (2014) Life on the edge: environmental determinants of tilefish (*Lopholatilus chamaeleonticeps*) abundance since its virtual extinction in 1882. *ICES Journal of Marine Science.* doi: 10.1093/icesjms/fsu053.

Fleury A.G. & Drazen J.C. (2013). Abyssal scavenging communities attracted to *Sargassum* and fish in the Sargasso Sea. *Deep Sea Research I* **72**: 141–147. doi:10.1016/j.dsr.2012.11.004

Flynn A.J. & Marshall N.J. (2013). Lanternfish (Myctophidae) zoogeography off Eastern Australia: a comparison with physicochemical biogeography. *PLoS ONE* **8**(120): e80950. doi:10.1371/journal.pone.0080950

Fock H., Uiblein F., Köster F., & von Westernhagen H. (2002). Biodiversity and species-environment relationships of the demersal fish assemblage at the Great Meteor Seamount (subtropical NE Atlantic), sampled by different trawls. *Marine Biology* **141**:185–199. doi: 10.1007/s00227-002-0804-y

Follesa M.C., Porcu C., Cabiddu S., Davini M.A., Sabatini A., & Cau A. (2007). First observations on the reproduction of *Alepocephalus rostratus* Risso, 1820 (Osteichthyes, Alepocephalidae) from the Sardinian Channel (Central-Western Mediterranean). *Marine Ecology* **28** (Suppl. 1), 75–81.

Foote K.G. (1983). Linearity of fisheries acoustics, with addition theorems. *Journal of the Acoustical Society of America* **73**: 1932–1940.

Forbes, E. (1844). Report on the Mollusca and Radiata of the Aegean Sea, and on their distribution, considered as bearing on geology. *Report of the British Association for the Advancement of Science.* **13**: 129–193

Forbes E. & Godwin-Austen R. (1859). *The Natural History of the European Seas.* London: John Van Voorst.

Forman J.S. & Dunn M.R. (2010). The influence of ontogeny and environment on the diet of lookdown dory, *Cyttus traversi. New Zealand Journal of Marine and Freshwater Research* **44**:4, 329–342. doi: 10.1080/00288330.2010.523080

Fornshell J.A. & Tesei A. (2013). The development of SONAR as a tool in marine biological research in the twentieth century. *International Journal of Oceanography* (2013), Article ID 678621. http://dx.doi .org/10.1155/2013/678621

Forster G.R. (1964). Line-fishing on the continental slope. *Journal of the Marine Biological Association of the UK* **44**, 277–284.

Forster G.R. (1968). Line-fishing on the continental slope II. *Journal of the Marine Biological Association of the UK* **48**, 479–483.

Forster, G.R. (1973). Line fishing on the continental slope: the selective effect of different hook patterns. *Journal of the Marine Biological Association of the UK* **53**, 749–751.

Fossen I. & Bergstad O.A. (2006). Distribution and biology of blue hake, *Antimora rostrata* (Pisces: Moridae), along the mid-Atlantic Ridge and off Greenland. *Fisheries Research* **82**: 19–29.

Fossen I., Cotton C.F., Bergstad O.A., & Dyb J.E. (2008). Species composition and distribution patterns of fishes captured by longlines on the Mid-Atlantic Ridge. *Deep-Sea Research II* **55**: 203–217.

Francis M.P. & Duffy C. (2002). Distribution, seasonal abundance and bycatch of basking sharks (*Cetorhinus maximus)* in New Zealand, with observations on their winter habitat. *Marine Biology* **140** (4):831–842.

Francis M.P., Hurst R.J., McArdle B.H., Bagley N.W., & Anderson O.F. (2002). New Zealand demersal fish assemblages. *Environmental Biology of Fishes* **65**: 215–234.

Francis R.I.C.C. & Clark M.R. (2005). Sustainability issues for orange roughy fisheries. *Bulletin of Marine Science* **76**(2): 337–351.

Franco M.A.L., Braga A.C., Nunan G.W.A., & Costa P.A.S. (2009). Fishes of the family Ipnopidae (Teleostei: Aulopiformes) collected on the Brazilian continental slope between 11° and 23°S. *Journal of Fish Biology* **75**: 797–815.

Franz V. (1907). Bau des Eulenauges und theorie des Teleskopauges. *Biologisches Zentralblatt* **27**: 271–278, 341–350.

Fraser P.J., Cruickshank S.F., & Shelmerdine R.L. (2003). Hydrostatic pressure effects on vestibular hair cell afferents in fish and crustacea. *Journal of Vestibular Research* **13**: 235–242.

Fricke H. (1997). Living coelacanths: values, eco-ethics and human responsibility. *Marine Ecology Progress Series.* **161**:1–15.

Fricke H., Hissmann K., Schauer J., Erdmann M., Moosa M.K., & Plante R. (2000). Biogeography of Indonesian coelacanths. *Nature* **403**: 38.

Fricke H.W., Schauer J., Hissmann K., Kasang L., & Plante R. (1991). Coelacanth *Latimeria chalumnae* aggregates in caves: observations in their resting habitat and social behaviour. *Environmental Biology of Fishes* **30**: 281–285.

Fricke R. (1992). Revision of the family Draconettidae (Teleostei), with descriptions of two new species and a new subspecies. *Journal of Natural History* **26**(1):165–195.

Fricke R. (2010). *Centrodraco atrifilum*, a new deepwater dragonet species from eastern Australia (Teleostei: Draconettidae). *Stuttgarter Beiträge zur Naturkunde A, Neue Serie* **3**: 341–346.

Friedman J.F., Condon N.E., & Drazen J.C. (2012). Gill surface area and metabolic enzyme activities of demersal fishes associated with the oxygen minimum zone off California. *Limnology & Oceanography* **57**(6): 1701–1710.

Friedman M. & Sallan L.C. (2012). Five hundred million years of extinction and recovery: a phanerozoic survey of Large-scale diversity patterns in fishes. *Palaeontology* **55**: 707–742.

Froese R. (2014) The making of FishBase. http://www .fishbase.de/manual/English/fishbasethe_making_ of_fishbase.htm (accessed 5 May 2014)

Froese R. & Pauly D. (eds.). (2016). FishBase. World Wide Web electronic publication. www.fishbase.org, version (01/2016).

Froese R. & Sampang A. (2004). Taxonomy and biology of seamount fishes. In: Morato T. & Pauly D. (eds.) *Seamounts: Biodiversity and Fisheries*, pp. 25–31. Fisheries Centre Research Reports. Vancouver: University of British Columbia 12(5).

Fudge D. S., Levy N., Chiu S., & Gosline J. M. (2005). Composition, morphology and mechanics of hagfish slime. *Journal of Experimental Biology* **208**: 4613–4625.

Fujikura K., Kojima S., Tamaki K, Maki Y., Hunt J., & Okutan T. (1999). The deepest chemosynthesis-based community yet discovered from the hadal zone, 7326 m deep, in the Japan Trench. *Marine Ecology Progress Series* **190**: 17–26.

Fukui A. & Kitagawa Y. (2006). *Dolichopteryx rostrata*, a new species of spookfish (Argentinoidea: Opisthoproctidae) from the eastern North Atlantic Ocean. *Ichthyological Research* **53**: 7–12.

Gage J.D. & Tyler P. (1991). *Deep-Sea Biology: A Natural History of Organisms at the Deep-Sea Floor*. Cambridge: Cambridge University Press.

Gaither M.R., Bowen B.W., Rocha L.A., & Briggs J.C. (2016). Fishes that rule the world: circumtropical distributions revisited. *Fish and Fisheries* **17**: 664–679.

Gallo N.D., Cameron J., Hardy K., Fryer P., Bartlett D.H., & Levin L.A. (2015). Submersible- and lander-observed community patterns in the Mariana and New Britain Trenches: Influence of productivity and depth on benthic community structure epibenthic and scavenging communities. *Deep-Sea Research I* **99**: 119–133.

Garcia D.M., Rice J., & Charles A. (eds.). (2014). *Governance of Marine Fisheries and Biodiversity Conservation: Interaction and Co-evolution*, Oxford: Wiley-Blackwell.

Garcıa V.B., Lucifora L.O. & Myers R.A. (2008). The importance of habitat and life history to extinction risk in sharks, skates, rays and chimaeras. *Proceedings of the Royal. Society B* **275**: 83–89.

Garcia-Castellanos D., Estrada F., Jiménez-Munt I., Gorini C., Fernàndez M., Vergés J., & De Vicente R. (2009). Catastrophic flood of the Mediterranean after the Messinian salinity crisis. *Nature* **462**: 778–782.

García-Mederos A.M., Tuset V.M., Santana J.I., & González J.A. (2010). Reproduction, growth and feeding habits of stout beardfish *Polymixia nobilis*

(Polymixiidae) off the Canary Islands (NE Atlantic). *Journal of Applied Ichthyology* **26**: 872–880.

Garman S.H. (1899). The fishes. Reports on an exploration off the west coast of Mexico, Central America, South America, and off the Galapagos Islands, in charge of Alexander Agassiz, by the U.S. Fish Commission steamer 'Albatross', during 1891, Lieut. Commander Z. L. Tanner, U.S.N. commanding. Pt. 26. *Memoirs of the Museum of Comparative Zoology* **24**: 1–421.

Gartner J.V., Crabtree R.E., & Sulak K.J. (1997). Feeding at depth. In: Randall D.J., & Farrell A.P. (eds.) *Deep-Sea Fishes*, pp. 115–93. San Diego: Academic Press.

Gaskett A.C., Bulman C., He X., & Goldsworthy S.D. (2001). Diet composition and guild structure of mesopelagic and bathypelagic fishes near Macquarie Island, Australia. *New Zealand Journal of Marine and Freshwater Research* **35**(3): 469–476. doi: 10.1080/00288330.2001.9517016

Gebbie G. & Huybers P. (2012). The mean age of ocean waters inferred from radiocarbon observations: sensitivity to surface sources and accounting for mixing histories. *Journal of Physical Oceanography* **42**: 291–305. doi: http://dx.doi.org/10.1175/JPO-D-11-043.1

GEBCO. (2014). General bathymetric chart of the oceans. Accessed 3 June 2014. www.gebco.net/data_and_products/gridded_bathymetry_data/

Gebruk A.V., Chevaldonné P., Shank T., Lutz R.A., & Vrijenhoek R.C. (2000). Deep-sea hydrothermal vent communities of the Logatchev area (14°45'N Mid-Atlantic Ridge): diverse biotopes and high biomass. *Journal of the Marine Biological Association of the U.K.* **80**: 383–393.

Geistdoerfer P. (1991). Ichtyofaune associée à l'hydrothermalisme océanique et description de *Thermobiotes mytiligeiton*, nouveau genre et nouvelle espèce de Synaphobranchidae (Pisces, anguilliformes) de l'océan Pacifique. Comptes rendus de l'Académie des sciences. *Série 3, Sciences de la vie* **312**(3): 91–97.

Geistdoerfer P. (1994a) *Careproctus hyaleius*, nouvelle espèce de poisson cyclopteridae (Liparinae) de l'écosystème hydrothermal de la dorsale de l'Océan Pacifique oriental (13°N). *Cybium* **18**(3): 325–333.

Geistdoerfer P. (1994b) *Pachycara thermophilum*, une nouvelle espèce de poisson Zoarcidae des sites hydrothermaux de la dorsale Médio-Atlantique. *Cybium* **18**(2): 109–115.

Geistdoerfer P. (1999). *Thermarces pelophilum*, espece nouvelle de Zoarcidae associee a l'emission de fluides froids au niveau du prisme d'accretion de la barbade, Ocean Atlantique Nord-Ouest. *Cybium* **23** (1): 5–11.

Gessner C. (1555). *Historiae animalium* Vol 3 Bk 4. *De piscium et aquatilium animantium natura.* Zurich: Christoph Froschoverus.

Gibbs A.G. (1997). Biochemistry at depth. In Randall D.J. & Farrell A.P. (eds.) *Deep-Sea Fishes.* San Diego: Academic Press.

Gibbs A. & Somero G. N. (1990). Na$^+$-K$^+$-adenosine tri-phosphatase activities in gills of marine teleost fishes: changes with depth, size and locomotory activity level. *Marine Biology* **106**: 315–321.

Gibbs R.H., Clarke T.A., & Gomon J.R. (1983). *Taxonomy and Distribution of the Stomioid Fish Genus* Eustomias *(Melanostomiidae), I: Subgenus* Nominosto-mias. Washington, DC: Smithsonian Contributions to Zoology, no. 380.

Gilbert C.H. (1905). The deep-sea fishes. Pt 2. In: Jordan D.S. & Everyman B.W. (eds.) *The Aquatic Resources of the Hawaiian Islands*, pp. 577–713. Bulletin of the U.S. Fisheries Commission. 23(2)

Gilbert C.H. & Hubbs C.L. (1928). The macrourid fishes of the Philippine Islands and the East Indies. *U.S. National Museum Bulletin 100*, 1(pt. 7): 369–588.

Gilchrist, J. D. F. (1922). Deep-sea fishes procured by the S.S. "Pickle" (Part I). *Report of the Fisheries and Marine Biological Survey, Union of South Africa* (Rep. 2, art. 3): 41–79.

Gillibrand E.J.V., Bagley P.M., Jamieson A., Herring P.J., Partridge J.C., Collins M.A., Milne R., & Priede I.G. (2007). Deep sea benthic bioluminescence at artifi-cial food falls, 1000 to 4800m depth, in the Porcu-pine Seabight and Abyssal Plain. *North East Atlantic Ocean. Marine Biology* **150**: 1053–1060. doi :10.1007/s00227-006-0407-0.

Gillibrand E.J.V., Jamieson A.J., Bagley P.M., Zuur A.F., & Priede I.G. (2007). Seasonal development of a deep pelagic bioluminescent layer in the temperate Northeast Atlantic Ocean. *Marine Ecology Progress Series* **341**: 37–44

Gilmore R.G., (1997). *Lipogramma robinsi*, a new basslet from the tropical western Atlantic, with descriptive and distributional notes on *L. flavescens* and *L. anabantoides* (Perciformes: Grammatidae). *Bulletin of Marine Science* **60**(3): 782–788.

Gjøsæter J. (1979). Mesopelagic fish. In: Saetre R. & Silva R.P. (eds.) *The Marine Fish Resources of Mozam-bique*, pp. 101–114. Bergen: Institute of marine Research.

Gjøsaeter, J. (1984). Mesopelagic fish, a large potential resource in the Arabian Sea. *Deep-Sea Research.* Part A **31**: 1019–1035. On line publication www.fao.org/WAIRDOCS/FNS/FN130E/ch11.htm accessed 2 October 2016.

Gjøsaeter J. & Kawaguchi K. (1980). A review of the world resources of Mesopelagic Fish, Food and Agriculture Organization of the United Nations. FAO Fisheries Technical Paper 193.

Glover A.G., Higgs N., & Horton T. (2014). World register of deep-sea species. Accessed at www .marinespecies.org/deepsea on 2014-05-0

Godbold J.A., Bailey D.M. Collins M.A., Gordon J.D.M., Spallek W.A., & Priede I.G. (2013). Putative fishery-induced changes in biomass and population size structures of demersal deep-sea fishes in ICES Sub-area VII, North East Atlantic Ocean. *Biogeosciences* **10**: 529–539. doi:10.5194/bg-10-529-2013, 2013

Gomez C., Williams A.J., Nicol S.J., Mellin C., Loeun K.L., & Bradshaw C.J.A. (2015). Species distribution models of tropical deep-sea snappers. *PLoS ONE* **10** (6): e0127395. doi:10.1371/journal.pone.0127395

González-Costas F. & Murua H. (2008). An analytical assessment of the Roughhead grenadier stock in NAFO subareas 2 and 3. *American Fisheries Society Symposium* **63**: 319–342.

Goode G.B. & Bean T.H. (1895). *Oceanic Ichthyology, a treatise on the Deep-Sea and Pelagic Fishes of the world based chiefly upon collections made by the steamers Blake, Albatross and Fish Hawk in the Northwestern Atlantic Ocean.* Smithsonian Institu-tion, Special Bulletin 2. Washington, DC: Govern-ment Printing Office.

Gordeeva N.V. (2013). Genetic divergence in the tribe Electronini (Myctophidae). *Journal of Ichthyology* **53**: 575–584.

Godø O.R., Huse I., & Michalsen K. (1997). Bait defence behaviour of wolffish and its impact on long-line catch rates. *ICES Journal of Marine Science* **54**: 273–275.

Gordon J.D.M. (2003). The Rockall Trough, Northeast Atlantic: the cradle of deep-sea biological oceanography that is now being subject to unsustainable fishing activity. *Journal of Northwest Atlantic Fisheries Science* 31: 57–83.

Gore M.A., Rowat D., Hall J., Gell F.R., & Ormond R.F. (2008). Transatlantic migration and deep mid-ocean diving by basking shark. *Biology Letters* 4: 395–398.

Govoni J.J., Olney J.E., Markle D.F., & Curtsinger W.R. (1984). Observations on structure and evaluation of possible functions of the vexillum in larval Carapidae (Ophidiiformes). *Bulletin of Marine Science* 34: 60–70.

Graae M. J. F. (1967). *Lestidium bigelowi*, a new species of paralepidid fish with photophores. *Breviora* 277:1–10.

Grant S.M. (2006). An exploratory fishing survey and biological resource assessment of Atlantic hagfish (*Myxine glutinosa*) occurring on the southwest slope of the Newfoundland Grand Bank. *Journal of Northwest Atlantic Fishery Science.* 36: 91–110.

Grassle, J.F. (2000). The Ocean Biogeographic Information System (OBIS): an on-line, worldwide atlas for accessing, modeling and mapping marine biological data in a multidimensional geographic context. *Oceanography* 13: 5–7.

Grassle J.F., Sanders H.L., Hessler R.R., Rowe G.T., & MacLennan T. (1975). Pattern and zonation: a study of the bathyal megafauna using the research submersible Alvin. *Deep-Sea Research* 22: 643–659.

Gray E.W. (1789). Observations on the class of animals called, by Linnaeus, Amphibia; particularly on the means of distinguishing those serpents which are venomous, from those which are not so. *Philosophical Transactions of the Royal Society of London* 79: 21–36.

Greer-Walker M. & Pull G. A. (1975). A survey of red and white muscle in marine fish. *Journal of Fish Biology* 7: 295–300.

Greer-Walker M., Santer R.M., Benjamin M., & Norman D. (1985). Heart structure of some deep-sea fish (Teleostei: Macrouridae). *Journal of Zoology London* (A) 205: 75–89.

Gregory S., Brown J., & Belchier M. (2014). Ecology and distribution of the grey notothen, *Lepidonotothen squamifrons*, around South Georgia and Shag Rocks, Southern Ocean. *Antarctic Science* 26: 239–249. doi:10.1017/S0954102013000667

Greven H., Walker Y., & Zanger K. (2009). On the structure of teeth in the viperfish *Chauliodus sloani*

Bloch & Schieder, 1801 (Stomiidae). *Bulletin of Fish Biology* 11: 87–98.

Grogan D. & Lund R. (2004). The origin and relationships of early Chondrichthyes. In: Carrier J.C., Musick J.A., & Heithanus M.R. (eds.) *Biology of Sharks and Their Relatives*, pp. 3–31. London: CRC Press.

Guinot G, Adnet S., Cavin L., & Cappetta H. (2013). Cretaceous stem chondrichthyans survived the end-Permian mass extinction. *Nature Communications* 4: 2669. doi: 10.1038/ncomms3669

Günther A.C.L.G. (1880). *An introduction to the study of fishes*. Edinburgh: Black.

Günther A.C.L.G. (1887). Report on the Deep-Sea Fishes collected by the H.M.S. Challenger during the years, 1873–1876. *Report on the Scientific Results of the Voyage of the HMS Challenger during the years 1873–1876* 22.

Haedrich R. L. (1964). Food habits and young stages of North Atlantic *Alepisaurus. Breviora* 201: 1–15.

Haedrich R.L. (1965). Identification of a deep-sea mooring-cable biter. *Deep-Sea Research* 12: 773–776.

Haedrich R.L. (1977). A sea lamprey from the deep ocean. *Copeia* 1977: 767–768.

Haedrich R.L. & Craddock J.E. (1968). Distribution and biology of the opisthoproctid fish *Winteria telescopa Brauer* 1901. *Breviora* 294: 1–11.

Haedrich R.L. & Merrett N.R. (1988). Summary atlas of deep-living demersal fishes in the North Atlantic Basin. *Journal of Natural History* 22(5): 1325–1362.

Haedrich R.L. & Merrett N.R. (1990). Little evidence for faunal zonation or communities in the deep sea demersal fish faunas. *Progress in Oceanography* 24: 239–250.

Haffner R.E. (1952). Zoogeography of the Bathypelagic Fish, *Chauliodus. Systematic Zoology* 1: 112–133.

Halliday R.G., Themelis D.E., & Hickey W.M. (2012). Demersal fishes caught with bottom gillnets and baited gears at 500–2 800 m on the Continental Slope off Nova Scotia, Canada. *Journal of Northwest Atlantic Fisheries Science* 44: 31–40.

Hamon N., Sepulchre O., Lefebvre V. & Ramstein G. (2013). The role of eastern Tethys seaway closure in the Middle Miocene Climatic Transition (ca. 14 Ma). *Climate of the Past* 9: 2687–2702. www.clim-past .net/9/2687/2013/ doi:10.5194/cp-9-2687-2013

Hansen C. A. & Sidell B. D. (1983). Atlantic hagfish cardiac-muscle metabolic basis of tolerance to anoxia. *American Journal of Physiology* **244**: R356–R362.

Hansen J., Sato M., Hearty P., Ruedy R., Kelley M., Masson-Delmotte V., Russell G., Tselioudis G., Cao J., Rignot E., Velicogna I., Tormey B., Donovan B., Kandiano E., von Schuckmann K., Kharecha P., Legrande A.N., Bauer M., & Lo K.-W. (2016). Ice melt, sea level rise and superstorms: evidence from paleoclimate data, climate modeling, and modern observations that 2°C global warming could be dangerous. *Atmospheric Chemistry and Physics* **16**: 3761–3812.

Harbison G.R. & Janssen J. (1987). Encounters with a Swordfish (*Xiphias gladius*) and Sharptail Mola (*Masturus lanceolatus*) at Depths Greater Than 600 Meters. *Copeia* **1987** (2): 511–513.

Harbison G.R., Smith K.L., & Backus R.H. (1973). *Stygiomedusa fabulosa* from the North Atlantic, its taxonomy, with a note on its natural history. *Journal of the Marine Biological Association of the UK* **53**: 615–617.

Hare J.A. & Marancik (2005). Psychrolutidae: tadpole or fathead sculpins. In: Richards W.J. (ed,) *Early Stages of Atlantic Fishes: An Identification Guide for the Western Central North Atlantic*, pp. 1191–1192. Boca Raton, FL: CRC Press.

Harold A.S (1994). A taxonomic revision of the Sternoptychid genus *Polyipnus* (Teleostei: Stomiiformes) with an analysis of phylogenetic relationships. *Bulletin of Marine Science* **54**(2): 428–534.

Harold A.S., Hartel K.E., Craddock J.E., & Moore J.A. (2002). Hatchetfishes and relatives. Family Sternoptychidae. In: Collette B.B. & Klein-MacPhee G. (eds.) *Bigelow and Schroeder's Fishes of the Gulf of Maine*, pp. 184–190. Washington, DC: Smithsonian Institution Press.

Harold A.S. & Lancaster K. (2003). A new species of hatchetfish genus *Argyripnus* (Stomiiformes: Sternoptychidae) from the Indo-Pacific. *Proceedings of the Biological Society of Washington* **116**: 883–891.

Harper A. A., Macdonald A. G., Wardle C. S. & Pennec J.-P. (1987). The pressure tolerance of deep-sea fish axons: results of Challenger cruise 6B/85. *Comparative Biochemistry and Physiology* **88**A: 647–53.

Harrison C.M.H. (1966). On the first halosaur leptocephalus: from Madeira. *Bulletin of the British Museum (Natural History) Zoology* **14**: 445–486.

Harrison C.M.H. & Palmer G. (1968). On the neotype of *Radiicephalus elongatus* Osorio with remarks on its biology. *Bulletin of the British Museum (Natural History) Zoology* **16**(5): 185–208.

Hartman O. & Emery K. O. (1956). Bathypelagic coelenterates. *Limnology and Oceanography* **1**: 304–312.

Hawkins A.D. & Amorim M.C.P. (2000). Spawning sounds of the male haddock, *Melanogrammus aeglefinus*. *Environmental Biology of Fishes* **59**: 29–41.

Hayes A.J. & Sim A. J. W. (2011). Ratfish (*Chimaera*) spine injuries in fishermen. *Scottish Medical Journal* **56**: 161–163. doi: 10.1258/smj.2011.011115

Hayes D.B., Ferreri C.P., & Taylor W.W. (2013). Active Fish Capture Methods. In Zale A.V., Parrish D.L., & Sutton T.M. (eds.) *Fisheries Techniques*, 3rd edition, Chapter 7. Bethesda, MD: American Fisheries Society.

Haygood M.G. (1993). Light organ symbioses in fishes, *Critical Reviews in Microbiology* **19**(4): 191–216. doi: 10.3109/10408419309113529

Hebert P.D.N., Cywinska A., Ball S.L., & deWaard J.R. (2003), Biological identifications through DNA barcodes. *Proceedings of the Royal Society London B* **270**: 313–321. doi:10.1098/rspb.2002.2218

Hector, J. (1875). Descriptions of five new species of fishes obtained in the New-Zealand seas by H. M. S -"Challenger" Expedition, July 1874. *Annals and Magazine of Natural History* (Series 4) **15**(85) (art. 11): 78–82.

Heezen B.C. (1969). The world rift system: an introduction to the symposium. *Technophysics* **8**: 269–279.

Heezen B.C. & Hollister C.D. (1971). *Face of the deep*. New York: Oxford University Press.

Heger A., King N., Wigham B.D., Jamieson A.J., Bagley P.M., Allan L., Pfannkuche O., & Priede I.G. (2007). Benthic Bioluminescence in the bathyal North East Atlantic: luminescent responses of *Vargula norvegica* (Ostracoda: Myodocopida) to predation by the deep water eel (*Synaphobranchus kaupii*). *Marine Biology* **151** (4): 1471–1478. doi:10.1007/s00227-006-0587-7

Heimberg A.M., Cowper-Sallari R., S´emon M., Donoghue P.C.J., & Peterson K.J. (2010). MicroRNAs reveal the interrelationships of hagfish, lampreys, and gnathostomes and the nature of the ancestral vertebrate. *Proceedings of the National Academy of Sciences of the USA* **107**: 19379–19383. doi: 10.1073/pnas.1010350107

Heinke F. (1913). Untersuchungen über die Scholle-Generalbericht I Schollen-fischerei und Schonmassregeln. Vorlaeufige kurze Euebersicht über die wichtigsten Ergebnisse de Berichts. *Rapports et Procès-Verbaux des Réunions du Conseil Permanent pour L'exploration de la Mer* **180**: 1–70.

Helfman G.S., Collette B.B., Facey D.E., & Bowen B.W. (2009). *The Diversity of fishes, Biology Evolution and Ecology*, 2nd edition. Oxford: Wiley-Blackwell.

Henriques C. (2004). *In situ lander observations of deep-sea fishes in the Eastern Atlantic Ocean*. PhD thesis University of Aberdeen.

Henry L.-A., Stehmann M.F.W., De Clippele L., Findlay H.S., Golding N., & Roberts J.M. (2016). Seamount egg-laying grounds of the deep-water skate Bathyraja richardsoni. *Journal of Fish Biology*. doi:10.1111/jfb.13041

Herring P. (2002). *The Biology of the Deep Ocean*. Oxford: Oxford University Press.

Herring P.J. (1971). Bioluminescence in an evermannellid fish. *Journal of Zoology (London)* **181**: 297–307.

Herring P.J. (1987). Systematic distribution of bioluminescence in living organisms. *Journal of Bioluminescence and Chemiluminescence* **1**: 147–163.

Herring P.J. (1992). Bioluminescence of the oceanic apogonid fishes *Howella brodiei* and *Florenciella lugubris*. *Journal of the Marine Biological Association of the UK* **72**: 139–148. doi:10.1017/S0025315400048840

Herring P.J. (2000). Species abundance, sexual encounter and bioluminescent signalling in the deep sea. *Philosophical Transactions: Biological Sciences* **355**: 1273–1276.

Herring P.J. & Cope C. (2005). Red bioluminescence in fishes: on the suborbital photophores of *Malacosteus*, *Pachystomias* and *Aristostomias*. *Marine Biology* **148**: 383–394. doi: 10.1007/s00227-005-0085-3

Herring P.J., Gaten E., & Shelton P.M.J. (1999). Are vent shrimps blinded by science? *Nature* **398**: 116.

Hirakawa N., Suzuki N., Narimatsu Y., Saruwatari T., & Ohno A. (2007). The spawning and settlement season of *Chlorophthalmus albatrossis* along the Pacific Coast of Japan. *The Raffles Bulletin of Zoology* **14**: 167–170.

Hissmann K., Fricke H., & Schauer J. (1998). Patterns of time and space utilization in coelacanths (*Latimeria chalumnae*) determined by ultrasonic telemetry. *Marine Biology* **136**: 943–952.

Hissmann K., Fricke H., & Schauer J. (2008). Population monitoring of the Coelacanth (*Latimeria chalumnae*). *Conservation Biology* **12**(4): 759–765.

Ho H.-C. (2013). Two new species of the batfish genus *Malthopsis* (Lophiiformes: Ogcocephalidae) from the Western Indian Ocean. *Zootaxa* **3716**: 289–300.

Ho H.-C. & Last P.R. (2013). Two new species of the coffinfish genus *Chaunax* (Lophiiformes: Chaunacidae) from the Indian Ocean. *Zootaxa* **3710**(5): 436–448.

Ho H.-C. & McGrouther M. (2015). A new anglerfish from eastern Australia and New Caledonia (Lophiiformes: Chaunacidae: *Chaunacops*), with new data and submersible observation of *Chaunacops melanostomus*. *Journal of Fish Biology* **86**(3): 940–951.

Ho H.-C., Roberts C.D., & Shao K.-T. (2013). Revision of batfishes (Lophiiformes: Ogcocephalidae) of New Zealand and adjacent waters, with description of two new species of the genus *Malthopsis*. *Zootaxa* **3626**(1): 188–200.

Ho H.-C., Roberts C. D., & Stewart A. L. (2013). A review of the anglerfish genus *Chaunax* (Lophiiformes: Chaunacidae) from New Zealand and adjacent waters, with descriptions of four new species. *Zootaxa* **3620**(1): 89–111.

Hochachka P.W., Storey K.B., & Baldwin J. (1975). Gill citrate synthase from an Abyssal fish. *Comparative Biochemistry & Physiology* **52B**: 43–49.

Høines Å. S. & Gundersen A. C. (2008). Rebuilding the stock of Northeast Arctic Greenland halibut (*Reinhardtius hippoglossoides*). *Journal of Northwest Atlantic Fisheries Science* **41**: 107–117. doi:10.2960/J.v41.m618

Holdsworth E.W.H. (1874). *Deep-Sea Fishing and Fishing Boats*. London: Stanford.

Honjo S., Manganini S.J., Krishfield R.A., & Francois R. (2008). Particulate organic carbon fluxes to the ocean interior and factors controlling the biological pump: a synthesis of global sediment trap programs since 1983. *Progress in Oceanography* **76**: 217–285.

Hopkins T.L. & Baird R.C. (1973). Diet of the Hatchetfish *Sternoptyx diaphana*. *Marine Biology* **21**: 34–46.

Hopkins T.L. & Baird R.C. (1981). Trophodynamics of the Fish *Valenciennellus tripunctulatus*. I. Vertical Distribution, Diet and Feeding Chronology. *Marine Ecology Progress Series* **5**: 1–10.

Horn P.L. (2003). Stock structure of bluenose (*Hypero-glyphe antarctica*) off the north-east coast of New Zealand based on the results of a detachable hook tagging programme. *New Zealand Journal of Marine and Freshwater Research* **37**(3): 623–631. doi:10.1080/00288330.2003.9517193

Horn P.L., Dunn M.R., & Forman J. (2013). The diet and trophic niche of orange perch, *Lepidoperca aurantia* (Serranidae: Anthiinae) on Chatham Rise, New Zealand. *Journal of Ichthyology* **53**: 310–316.

Hort J. (1911). The '*Michael Sars*' North Atlantic Deep-Sea Expedition, 1910. *The Geographical Journal* **37**: 349–377.

Hovis W.A., Clark D.K., Anderson F., Austin R.W., Wilson W.H., Baker E.T., Ball D., Gordon H.R., Mueller J.L., El-Sayed S.Z., Sturm B., Wrigley R.C., & Yentsch C.S. (1980). Nimbus-7 Coastal Zone Color Scanner: system description and initial imagery. *Science* **210**: 60–63.

Howes G.J. (1991). Biogeography of gadoid fishes. *Journal of Biogeography* **18**: 595–622.

Hsü K.J., Ryan W.B.F. & Cita M.B. (1973). Late Miocene desiccation of the Mediterranean. *Nature* **242**: 240–244. doi:10.1038/242240a0

Hubbs C. L. (1935). Half mile down by William Beebe Review. *Copeia* **1935** (2): 105.

Hubbs C. L. & Iwamoto T. (1977). A new genus (*Mesobius*), and three new bathypelagic species of Macrouridae (Pisces, Gadiformes) from the Pacific Ocean. *Proceedings of the California Academy of Sciences* (Series 4), **41** (7): 233–251.

Hughes G.M. (1966). The dimensions of fish gills in relation to their function. *Journal of Experimental Biology* **45**: 177–195.

Hughes G.M. (1979). Morphometry of fish gas exchange organs in relations to their respiratory function. In: Ali M.A. (ed.) *Environmental Physiology of Fishes*, pp. 33–46. New York: Plenum Press.

Hughes G.M. & Iwai T. (1978). A morphometric study of the gills in some Pacific deep-sea fishes. *Journal of Zoology* **184**: 155–170, doi:10.1111/ j.1469–.7998.1978.tb03272.x

Hughes S.E. (1981). Initial U.S. Exploration of Nine Gulf of Alaska Seamounts and Their Associated Fish and Shellfish Resources. *Marine Fisheries Review*. **42**: 26–33.

Hulley P.A. & Lutjeharms J.R.E. (1995). The south-western limit for the warm-water, mesopelagic ichthyofauna of the Indo-West Pacific: lanternfish (Myctophidae) as a case study. *South African Journal of Marine Science* **15**(1): 185–205. doi: 10.2989/ 02577619509504843

Hulley P.A. & Prosch R.M. (1987). Mesopelagic fish derivatives in the southern Benguela upwelling region *South African Journal of Marine Science* **5** (1): 597–611, doi: 10.2989/025776187784522289

Humphreys R.L. Jr., Winans G.A., & Tagami D.T. (1989). Synonymy and life history of the North Pacific pelagic armorhead, *Pseudopentaceros wheeleri* Hardy (Pisces: Pentacerotidae). *Copeia* **1989**(1): 142–153.

ICES. (2012). *Report of the Working Group on the Biology and Assessment of Deep-Sea Fisheries Resources (WGDEEP)*, 28 March–5 April, Copenhagen, Denmark: ICES, CM 2012/ACOM:17.

ICES. (2013). *Manual for the Midwater Ring Net sampling during IBTS Q1: The International Bottom Trawl Survey Working Group*. Copenhagen. Denmark: International Council for the Exploration of the Sea.

ICES. (2014). *International Council for Exploration of the Sea. Report of the Report of the Working Group on Widely Distributed Stocks*. (WGWIDE), 26 August–1 September 2014, ICES Headquarters. Copenhagen, Denmark: ICES CM 2014/ACOM:15.

ICES. (2015). *Report of the Working Group on Biology and Assessment of Deep-Sea Fisheries Resources (WGDEEP), 20–17 March 2015*. Copenhagen, Denmark: ICES, CM 2015/ACOM:17.

ICES Advice. (2015). ICES Advice on fishing opportunities, catch, and effort Iceland Sea and Oceanic Northeast Atlantic Ecoregions. 2.3.4a Beaked redfish (Sebastes mentella) in Subareas V, XII, and XIV (Iceland and Faroes grounds, north of Azores, east of Greenland) and NAFO Subareas 1+2 (deep pelagic stock > 500 m). Accessed online 8 April 2016. www.ices.dk/sites/pub/Publication%20Reports/ Advice/2015/2015/smn-dp.pdf

ICES.NWWG. (2015). *Report of the North-Western Working Group (NWWG), 28 April–5 May, ICES HQ*. Copenhagen Denmark: ICES, CM 2015/ ACOM:07.

Ichino M.C., Clark M.R., Drazen J.C., Jamieson A., Jones D.O.B., Martin A.P., Rowden A.A., Shank T.M., Yancey P.H., & Ruhl H.A. (2015). The distribution of benthic biomass in hadal trenches: a modelling approach to investigate the effect of vertical and lateral organic matter transport to the seafloor. *Deep-Sea Research I* **100**: 21–33.

IGBP, IOC, SCOR. (2013). *Ocean Acidification Summary for Policymakers – Third Symposium on the Ocean in a High-CO2 World*. Stockholm, Sweden: International Geosphere-Biosphere Programme.

Iglésias S.P., Dettai A., & Ozouf-Costaz C. (2012). *Barbapellis pterygalces*, new genus and new species of a singular eelpout (Zoarcidae: Teleostei) from the Antarctic deep waters. *Polar Biology* **35**(2): 215–220.

Iglésias S.P., Toulhout L. & Sellos D.P. (2010). Taxonomic confusion and market mislabelling of threatened skates: important consequences for their conservation status. *Aquatic Conservation* **20**: 319–333.

Ikeda T. (1996). Metabolism, body composition, and energy budget of the mesopelagic fish *Maurolicus muelleri* in the Sea of Japan. *Fishery Bulletin* **94**: 49–58.

Ikeda T. & Hirakawa K. (1998). Metabolism and body composition of zooplankton in the cold mesopelagic zone of the southern Japan Sea. *Plankton Biology and Ecology* **45**(1): 31–44.

Inoue J. G., Miya M., Lam K., Tay B. H., Danks J. A., Bell J., Walker T. I., & Venkatesh B. (2010). Evolutionary origin and phylogeny of the modern holocephalans (Chondrichthyes: Chimaeriformes): a mitogenomic perspective. *Molecular Biology and Evolution* **27**: 2576–2586.

Inoue J.G., Miya M., Miller M.J., Sado T., Hanel R., Hatooka K., Aoyama J., Minegishi Y., Nishida M., & Tsukamoto K. (2010). Deep-ocean origin of the freshwater eels. *Biology Letters* **6**: 363–366. doi:10.1098/rsbl.2009.0989

International Commission on Zoological Nomenclature. (1999). *International Code of Zoological Nomenclature*, 4th edition. Singapore: International Commission on Zoological Nomenclature. Online version: http://iczn.org/iczn Accessed 24 June 2014.

Irigoien X., Klevjer A., Røstad A., Martinez U., Boyra G., Acuña J.L., Bode A., Echevarria F., Gonzalez-Gordillo J.I., Hernandez-Leon S., Agusti S., Aksnes D.L., Duarte C.M., & Kaartvedt S. (2014). Large mesopelagic fishes biomass and trophic efficiency in the open ocean. *Nature Communications* **5**: 3271. doi: 10.1038/ncomms4271

Isaacs J.D. & Schick G.B. (1960). Deep-sea free instrument vehicle. *Deep-Sea Research* **7**: 61–67.

Isaacs J.D. & Schwartzlose R.A. (1975). Active animals of the deep-sea floor. *Scientific American* **233**(4): 84–91. doi: 10.1038/scientificamerican1075-84

Ivanov V.V., Shapiro G.I., Huthnance J.M., Aleynik D.L., & Golovin P.N. (2004). Cascades of dense water around the world ocean. *Progress in Oceanography* **60**: 47–98.

Iwamoto T. (1975). The abyssal fish *Antimora rostrata* (Günther). *Comparative Biochemistry and Physiology Part B: Comparative Biochemistry* **52**: 7–11.

Iwamoto T. (2008a). A brief taxonomic history of grenadiers. *American Fisheries Society Symposium* **63**: 3–13.

Iwamoto T. (2008b). A note on the correct spelling of *Coelorinchus* (*Caelorinchus*) and C.*coelorhincus* (*caelorhincus*). *American Fisheries Society Symposium* **63**: xi.

Iwamoto T. (2015). Hoplostethus cadenati. *The IUCN Red List of Threatened Species 2015*: e. T21110117A21907969. http://dx.doi.org/10.2305/IUCN.UK.2015-4.RLTS.T21110117A21907969.en. Downloaded on 21 March 2017.

Iwamoto T. & Arai T. (1987). A new grenadier *Malacocephalus okamurai* (Pisces: Gadiformes: Macrouridae) from the Western Atlantic. *Copeia* **1987** (1): 204–208.

Iwamoto T. & Orlov A. (2006). *Paracetonurus flagellicauda* (Keofoed, 1927) (Macrouridae, Gadiformes, Teleostei), new records from the Mid-Atlantic Ridge and Madagascar Plateau. *Proceedings of the California Academy of Sciences 4th Series* **57**(11): 379–386.

Iwamoto T. & Orlov A. (2008). First Atlantic Record of *Asthenomacrurus victoris* Sazonov & Shcherbachev (Macrouridae, Gadiformes, Teleostei). *Proceedings of the California Academy of Sciences*, Series 4, **59**(4): 125–131.

Iwamoto T. & Ungaro N. (2002). A new grenadier (Gadiformes, Macrouridae) from the Mediterranean. *Cybium* **26**(1): 27–32.

Iwamoto T. & Sazonov Y.I. (1994). Revision of the genus *Kumba* (Pisces, Gadiformes, Macrouridae), with the description of three new species. *Proceedings of the California Academy of Sciences* **48**(11): 221–237.

Jackson T.L. (2002). Preliminary guide to the identification of the early life history stages of giganturid fishes of the Western Central North Atlantic. *NOAA Technical Memorandum* NMFS-SEFSC-484.

Jacobs D.K. & Lindberg D.R. (1998). Oxygen and evolutionary patterns in the sea: Onshore/offshore trends and recent recruitment of deep-sea faunas. *Proceedings of the National Academy of Sciences USA* **95**: 9396–9401.

James P.S.B.R. (1981). Exploited and potential capture fishery resources in the inshore waters of India. *Central Marine Fisheries Research Institute Bulletin* **30**A: 72–82. http://eprints.cmfri.org.in/726/1/James_psbr_72.pdf accessed on line 27 August 2015.

Jamieson A.J. (2015). *The Hadal Zone: Life in the Deepest Oceans.* Cambridge: Cambridge University Press.

Jamieson A.J., Fujii T., Mayor D.J., Solan M., & Priede I.G. (2010). Hadal trenches: the ecology of the deepest places on Earth. *Trends in Ecology & Evolution* **25**: 190–197.

Jamieson A.J., Fujii T., Solan M., Matsumoto A.K., Bagley P.M., & Priede I.G. (2009). Liparid and macrourid fishes of the hadal zone: *in situ* observations of activity and feeding behaviour. *Proceedings of the Royal Society of London B.* **276**: 1037–1045. doi:10.1098/rspb.2008.1670

Jamieson A.J., Godø O.R., Bagley P.M., Partridge J.C., & Priede I.G. (2006). Illumination of trawl gear by mechanically stimulated bioluminescence. *Fisheries Research* **81**: 276–282.

Jamieson A.J., Lacey N.C., Lörz A.-N., Rowden A.A., & Piertney S.B. (2013). The supergiant amphipod *Alicella gigantea* (Crustacea: Alicellidae) from hadal depths in the Kermadec Trench, SW Pacific Ocean. *Deep-Sea Research II* **92**: 107–113.

Jamieson A.J., Priede I.G.,& Craig J. (2012). Distinguishing between the abyssal macrourids *Coryphaenoides yaquinae* and *C. armatus* from *in situ* photography. *Deep Sea Research I* **64**: 78–85.

Jamieson A.J., Boorman B., & Jones, D.O.B. (2013). Deepsea sampling. In: Eleftheriou A. (ed.) *Methods for the Study of Marine Benthos*, pp. 285–347, 4th edition. Oxford: Wiley-Blackwell.

Jamieson A.J. & Yancey P.H. (2012). On the validity of the trieste flatfish: dispelling the myth. *Biological Bulletin* **222**: 171–175.

Janßen F., Treude T., & Witte U. (2000). Scavenger assemblages under differing trophic conditions: a case study in the deep Arabian Sea. *Deep Sea Research II* **47**: 2999–3026.

Janssen J. (2004). Lateral line sensory ecology. In Von der Emde G., Mogdans J., & Kapoor B.G. (eds.) *The Senses of Fish; Adaptations for the Reception of Natural Stimuli*, pp. 231–264. Berlin: Springer.

Janssen J., Gibbs R.H., Jr., & Pugh P.R. (1989). Association of *Caristius* Sp. (Pisces: Caristiidae) with a Siphonophore, *Bathyphysa conifer. Copeia* **1989**: 198–201.

Janssen J. & Harbison G.R. (1981). Fish in salps: the association of squaretails (*Tetragonurus* spp.) with pelagic tunicates. *Journal of the Marine Biological Association of the UK* **61**(4): 917–927.

Janvier P. (2010). MicroRNAs revive old views about jawless vertebrate divergence and evolution. *Proceedings of the National Academy of Sciences of the USA* **107**: 19137–19138.

Jeffreys R.M., Lavaleye M.S.S., Bergman M.J.N., Duineveld G.C.A., Witbaard R., & Linley T. (2010). Deepsea macrourid fishes scavenge on plant material: Evidence from *in situ* observations. *Deep-Sea Research I* **57**: 621–627.

Jennings S., Reynolds J.D., Mills S.C. (1998). Life history correlates of responses to fisheries exploitation. *Proceedings of the Royal Society of London B* **265**: 333–339.

Johannessen A. & Monstad T. (2003). Distribution, growth and exploitation of Greater Silver Smelt (*Argentina silus*, (Ascanius, 1775) in Norwegian waters. *Journal of North Atlantic Fisheries Science* **31**: 319–332.

Johnson G.D., Paxton J.R., Sutton T.T., Satoh T.P., Sado T., Nishida M., & Miya M. (2009). Deep-sea mystery solved: astonishing larval transformations and extreme sexual dimorphism unite three fish families. *Biology Letters* **5**: 235–239.

Johnson J.Y. (1862). Descriptions of some new genera and species of fish obtained at Madiera. *Proceedings of the Zoological Society of London* **1862**: 167–180.

Johnson J.Y. (1863). Descriptions of five new species of fishes obtained at Madeira. *Proceedings of the Zoological Society of London* **1863**(33): 36–46.

Johnson M.W. (1948). Sound as a tool in marine ecology, from data on biological noises and the deep scattering layer. *Journal of Marine Research* **7**(3): 443–458.

Johnson R.K. & Bertelsen E. (1991). The fishes of the family Giganturidae: systematics, development, distribution and aspects of biology. *Dana Reports* **91**: 1–45.

Johnson R.K. (1974). Five new species and a new genus of alepisauroid fishes of the Scopelarchidae (Pisces: Myctophiformes). *Copeia* **1974**(2): 449–457.

Johnston I.A. & Herring P.J. (1985). The transformation of muscle into bioluminescent tissue in the fish *Benthabella infans* Zugmayer. *Proceedings of the Royal Society of London B* **255**: 213–218.

Jones D.O.B., Bett B.J., Wynn R.B., & Masson D.G. (2009). The use of towed camera platforms in deep-water science. *Underwater Technology* **28**(2): 41–50.

Jones E.C. (1971). *Isistius brasiliensis*, a squaloid shark, the probable cause of crater wounds on fishes and cetaceans. *Fisheries Bulletin* **69**(4): 791–798.

Jones E.G. (1999). *'Burial at sea'. Consumption and dispersal of large fish and cetacean food-falls by deep-sea scavengers in the Northeast Atlantic Ocean and the Eastern Mediterranean Sea*. PhD thesis University of Aberdeen.

Jones E.G., Collins M.A., Bagley P.M., Addison S., & Priede I.G. (1998). The fate of cetacean carcasses in the deep sea: observations on consumption rates and succession of scavenging species in the abyssal north-east Atlantic Ocean. *Proceedings of the Royal Society of London. B* **265**: 1119–1127.

Jones E.G., Tselepides A., Bagley P.M., Collins M.A., & Priede I.G. (2003). Bathymetric distribution of some benthic and benthopelagic species attracted to baited cameras and traps in the Eastern Mediterranean. *Marine Ecology Progress Series* **251**: 75–86.

Jones M.R.L. & Breen B.B. (2013) Food and feeding relationships of three sympatric slickhead species (Pisces: Alepocephalidae) from northeastern Chatham Rise, New Zealand. *Deep-Sea Research I* **79**: 1–9.

Jordan D.S. (1917–1920). *The genera of fishes from Linnaeus to Cuvier, 1758–1833, seventy five years with the accepted type of each*. A contribution to the stability of scientific nomenclature. Stanford, CA: Stanford University.

Jordan D.S. (1919). *The Genera of Fishes, Part III from Guenther to Gill, 1859–1880, Twenty Two Years, with the Accepted Type of Each*. Stanford, CA: Stanford University.

Jordan D.S. & Starks E.C. (1904). List of fishes dredged by the steamer Albatross off the coast of Japan in the summer of 1900, with descriptions of new species and a review of the Japanese Macrouridae. *Bulletin of the U.S. Fish Commission* **22**[1902]: 577–630, Pls. 1–8.

Jørgensen J.M., Lomholt J.P., Weber R.E., & Malte H. (eds.). (1998). *The Biology of Hagfishes*. London: Chapman & Hall

Jumper G. Y. & Baird R.C. (1991). Location by olfaction: a model and application to the mating problem in the deep sea hatchetfish Argyropelecus hemigyrnnus. *The American Naturalist* **138**: 1431–1458.

Kaartvedt S., Staby A., & Aksnes D.L. (2012). Efficient trawl avoidance by mesopelagic fishes causes large underestimation of their biomass. *Marine Ecology Progress Series* **456**: 1–6. doi: 10.3354/meps09785

Kabasakal H. (2006). Distribution and biology of the bluntnose sixgill shark *Hexanchus griseus* (Bonnaterre, 1788) (Chondrichthyes, Hexanchidae), from Turkish waters. *Annales Series Historia Naturalis* **16**(1): 29–36.

Kaiser M.J. & Spencer B.E. (1994) Fish scavenging in recently trawled areas. *Marine Ecology Progress Series*. **112**: 41–49.

Kanwisher, J. & Ebeling, A. (1957). Composition of the swimbladder gas in bathypelagic fishes. *Deep-Sea Research* **4**(3): 211–217.

Karmovskaya E.S. & Merrett N.R. (1998). Taxonomy of the deep-sea eel genus, *Histiobranchus* (Synaphobranchidae, Anguilliformes), with notes on the ecology of *H. bathybius* in the eastern North Atlantic. *Journal of Fish Biology* **53**: 1015–1037.

Karmovskaya E.S. & Parin N.V. (1999). [ref. 24869] A new species of the genus *Ilyophis* (Synaphobranchidae, Anguilliformes) from the hydrothermal area Broken Spur (Mid-Atlantic Submarine Ridge). *Voprosy Ikhtiologii* **39**(3): 316–325.

Karuppasamy P.K., Muraleedharan K.R., Dineshkumar P.K., & Nair M. (2010). Distribution of mesopelagic micronekton in the Arabian Sea during the winter monsoon. *Indian Journal of Marine Sciences* **39**: 227–237.

Katsanevakis S., Maravelias, C. D., & Kell, L. T. (2010). Landings profiles and potential métiers in Greek set longliners. *ICES Journal of Marine Science*, **67**: 646–656.

Kaufmann R.S., Wakefield W.W., & Genin A. (1989). Distribution of epibenthic megafauna and lebensspuren on two central North Pacific seamounts. *Deep-Sea Research* **36**: 1863–1896.

Kawaguchi K. & Butler J.L. (1984). Fishes of the genus *Nansenia* (Microsomatidae) with descriptions of seven new species. *Los Angeles County Museum Contributions in Science* **352**: 1–22.

Kawai T. (2013). Revision of the peristediid genus *Satyrichthys* (Actinopterygii: Teleostei) with the description of a new species, *S. milleri* sp. nov. *Zootaxa* **3635**(4): 419–438.

Kawai T., Amaoka K., & Serét (2010). A new righteye flounder, *Poecilopsetta multiradiata* (Teleostei: Pleuronectiformes: Poecilopsettidae), from New Zealand and New Caledonia (South-West Pacific). *Ichthylogical Research* **57**: 193–198.

Keeling R.F., Körtzinger A., & Gruber N. (2010). Ocean deoxygenation in a warming world. *Annual Review of Marine Science* **2**: 199–229.

Kemp K.M., Jamieson A.J., Bagley P.M., McGrath H., Bailey D.M., Collins M.A., & Priede I.G. (2006). Consumption of large bathyal food fall, a six month study in the NE Atlantic. *Marine Ecology-Progress Series* **310**: 65–76.

Kenaley C.P. (2007). Revision of the stoplight loosejaw genus *Malacosteus* (Teleostei: Stomiidae: Malacosteinae), with description of a new species from the temperate Southern Hemisphere and Indian Ocean. *Copeia* **2007**(4): 886–900.

Kenaley C.P. (2009). Revision of Indo-Pacific species of the loosejaw dragonfish genus *Photostomias* (Teleostei: Stomiidae: Malacosteinae). *Copeia* **2009**(1): 175–189.

Kenaley C.P., DeVaney S.C., & Fjeran T.T. (2013). The complex evolutionary history of seeing red: molecular phylogeny and the evolution of an adaptive visual system in deep-sea dragonfishes (Stomiiformes: Stomiidae). *Evolution* **68**–4: 996–1013.

Kim S.-Y. (2012). Phylogenetic systematics of the family Pentacerotidae (Actinopterygii: Order Perciformes). *Zootaxa* **3366**: 1–111.

Kimura S., Kohno Y., Tsukamoto Y., & Okiyama M. (1990). Record of the Parabrotulid fish *Parabrotula plagiophthalma* from Japan. *Japanese Journal of Ichthyology* **37**: 318–320.

King N.J., Bagley P.M., & Priede I.G. (2006). Depth zonation and latitudinal distribution of deep sea scavenging demersal fishes of the Mid-Atlantic Ridge, 42°–53°N. *Marine Ecology Progress Series* **319**: 263–274.

King N.J., Bailey D.M., & Priede I.G. (2007). Introduction; role of scavengers in marine ecosystems. *Marine*

Ecology Progress Series **350**: 175–178. doi: 10.3354/meps07186

King N.J. & Priede I.G. (2008). *Coryphaenoides armatus*, the abyssal grenadier: distribution, abundance, and ecology as determined by baited landers. *American Fisheries Society Symposium* **63**: 139–161.

Kinzer J., Böttger-Schnack R., & Schulz K. (1993). Aspects of horizontal distribution and diet of myctophid fish in the Arabian Sea with reference to the deep water oxygen deficiency. *Deep-Sea Research II* **40**: 783–800.

Kious J.W. & Tilling R. (1996). *This Dynamic Earth: The Story of Plate Tectonics*. Online booklet. Washington, DC: U.S. Geological Survey. http://pubs.usgs .gov/publications/text/dynamic.html Accessed 23 September 2014.

Kjerstad M., Fossen I., & Willemsen H.M. (2003). Utilization of deep-sea sharks at Hatton Bank in the N Atlantic. *Journal of NorthWest Atlantic Fishery Science* **31**: 333–338.

Klepadlo C. (2011). Three new species of the genus *Photonectes* (Teleostei: Stomiiformes: Stomiidae: Melanostomiinae) from the Pacific Ocean. *Copeia* **2011**(2): 201–210.

Klimpel S., Busch M.W., Kellermanns E., Kleinertz S., & Palm H.W. (2009). *Metazoan Deep-Sea Fish Parasites*. Solingen,Germany: Verlag Natur & Wissenschaft.

Kloser R.J., Sutton C., Krusic-Golub K., & Ryan T.E. (2015). Indicators of recovery for orange roughy (*Hoplostethus atlanticus*) in eastern Australian waters fished from 1987. *Fisheries Research* **167** (2015): 225–235.

Klug S. & Kriwet J. (2010). Timing of deep-sea adaptation in dogfish sharks: insights from a supertree of extinct and extant taxa. *Zoologica Scripta* **39**: 331–342.

Knudsen S.W., Møller P.R., & Gravlund P. (2007). Phylogeny of the snailfishes (Teleostei: Liparidae) based on molecular and morphological data. *Molecular Phylogenetics and Evolution* **44**(2007): 649–666.

Knudsen S., Nielsen J., & Uiblein F. (2015). Spectrunculus grandis. *The IUCN Red List of Threatened Species 2015*: e.T18139045A60799962. http://dx.doi.org/ 10.2305/IUCN.UK.2015-4.RLTS.T18139045A60 799962.en. Downloaded on 21 March 2017.

Kobylianskii S.G. (2006). *Bathylagus niger* sp. nova (Bathylagidae, Salmoniformes) a new species of

Bathylagus from the subpolar waters of the Southern Ocean. *Journal of Ichthyology* **46**(6): 413–417.

Kobyliansky S.G., Orlov A.M., & Gordeeva N.V. (2010). Composition of deep-sea pelagic ichthyocenes of the Southern Atlantic, from waters of the area of the Mid-Atlantic and Walvis Ridges. *Journal of Ichthyology* **50**(10): 932–949.

Kock K.-H. (1992). *Antarctic fish and fisheries.* Cambridge: Cambridge University Press.

Kooistra W.H.C.F. & Medlin,L.K. (1996). Evolution of the diatoms (Bacillariophyta). IV A reconstruction of their age from small subunit rRNA coding regions and the fossil record. *Molecular Phylogenetics and Evolution (Academic Press)* **6**(3): 391–407. doi:10.1006/mpev.1996.0088

Koslow A.J., Davison P., Lara-Lopez A., & Ohman M.D. (2014). Epipelagic and mesopelagic fishes in the southern California Current System: ecological interactions and oceanographic influences on their abundance. *Journal of Marine Systems* **138**: 20–28.

Koslow J.A. (1993) Community structure in North Atlantic deep-sea fishes. *Progress in Oceanography* **31**: 321–338.

Koslow J.A. (2007). *The Silent Deep; the Discovery, Ecology and Conservation of the Deep-Sea.* Chicago: University of Chicago Press.

Koslow J.A., Boehlert G.W., Gordon J.D.M., Haedrich R.L., Lorance P., & Parin N. (2000). Continental slope and deep-sea fisheries: implications for a fragile ecosystem. *ICES Journal of Marine Science* **57**: 548–557.

Koslow J.A., Bulman C.M., & Lyle J.M. (1994). The mid-slope demersal fish community off southeastern Australia. *Deep-Sea Research I* **41**: 113–141.

Koslow J.A., Kloser R.J., & Williams A. (1997). Pelagic biomass and community structure over the mid-continental slope off southeastern Australia based upon acoustic and midwater trawl sampling. *Marine Ecology Progress Series* **146**: 21–35.

Kotlyar A.N. (2008). Revision of the genus *Poromitra* (Melamphaeidae): Part 2. New species of the group *P. crassiceps. Journal of Ichthyology* **48**(8): 581–592.

Kotlyar A.N. (2009). Revision of the genus *Poromitra* (Melamphaidae): Part 4. Species of *P. cristiceps* group: *P. atlantica, P. oscitans,* and *P. agofonovae* Kotlyar, species nova. *Journal of Ichthyology* **49**(8): 563–574.

Kotrschal K., Van Staaden M.J., & Huber R. (1998). Fish brains: evolution and environmental relationships. *Reviews in Fish Biology and Fisheries* **8**: 373–408.

Krefft G. (1976). Distribution patterns of oceanic fishes in the Atlantic Ocean. *Revue Des Travaux De l'Institut Des Pêches Maritimes* **40**(3/4): 439–460.

Krefft G. (1990). Chimaeridae. In: Quero J.C., Hureau J.C., Karrer C., Post A., & Saldanha L. (eds.) *Check-List of the Fishes of the Eastern Tropical Atlantic (CLOFETA),* pp. 111–113. Lisbon: JNICT; Paris: SEI; and Paris: UNESCO. Vol. 1.

Kriwet J. (2003). Lancetfish teeth (Neoteostei, Alepisauroidei) from the early cretaceous of Alcaine, Spain. *Lethaia* **36**: 323–332.

Kriwet J. & Benton M.J. (2004). Neoselachian (Chondrichthyes, Elasmobranchii) diversity across the Cretaceous–Tertiary boundary. *Palaeogeography, Palaeoclimatology, Palaeoecology* **214**: 181–194.

Kriwet J., Kiessling W., & Klug S. (2009). Diversification trajectories and evolutionary life-history traits in early sharks and batoids. *Proceedings of the Royal Society of London B* **276**: 945–951. doi:10.1098/rspb.2008.1441

Krueger W.H. & Gibbs R.H. (1966). Growth changes and sexual dimorphism in the stomiatoid fish *Echiostoma barbatum. Copeia* **1966**(1): 43–49.

Kubota T., Shiobara Y., & Kubodera T. (1991). Food habits of the frilled shark *Chlamydoselachus anguineus* collected from Suruga bay, central Japan. *Nippon Suisan Gakkaishi* **57**(1): 15–20. doi:10.2331/suisan.57.15

Kukuev E.I., Parin N.V., & Trunov I.A. (2013). Materials for the revision of the family Caristiidae (Perciformes): 3. Manefishes (genus *Caristius*) from moderate warm waters of the Pacific and Atlantic oceans with a description of three new species from the southeast Atlantic (*C. barsukovi* sp. n., *C. litvinovi* sp. n., *C. walvisensis* sp. n.). *Journal of Ichthyology* **53**(8): 541–561.

Kulczykowska E, Popek W., & Kapoor B.G. (eds.). (2010). *Biological Clock in Fish.* Boca Raton, FL: CRC Press.

Kulka D., Hood C., & Huntington J. (2007). *Recovery strategy for northern wolffish (Anarhichas denticulatus) and spotted wolffish (Anarhichas minor), and management plan for Atlantic wolffish (Anarhichas lupus) in Canada.* Ottawa, Canada: Department of Fisheries and Oceans.

Kunzig R. (2003) Deep-sea biology: living with the endless frontier. *Science* 302: 991.

Kunzmann A. & Zimmerman C. (1992). *Aethotaxis mitopteryx*, a high-Antarctic fish with benthopelagic mode of life. *Marine Ecology Progress Series* 88: 33–40.

Kuraku S. & Kuratani S. (2006). Timescale for cyclostome evolution inferred with a phylogenetic diagnosis of hagfish and lamprey cDNA sequences. *Zoological Science* 23: 1053–1064.

Kusukawa S. (2000). The *Historia piscium* (1986). *Notes and Records of the Royal Society of London* 54(2): 179–197.

Kyne P.M. & Simpfendorfer C.A. (2007). *A collation and summarization of available data on deepwater chondrichthyans: biodiversity, life history and fisheries*. Burnaby, BC: IUCN SSC Shark Specialist Group for the Marine Conservation Biology Institute. At: www.flmnh.ufl.edu/fish/organizations/SSG/SSG.htm

Lack, M. (2008). *Continuing CCAMLR's Fight against IUU Fishing for Toothfish*. Sydney: WWF Australia and TRAFFIC International.

La Mesa M., Eastman J.T., & Licandro P. (2007). Feeding habits of *Bathydraco marri* (Pisces, Notothenioidei, Bathydraconidae) from the Ross Sea, Antarctica. *Polar Biology* 30: 541–547. doi: 10.1007/s00300-006-0211-9

Lampitt R.S., Bett B.J., Kiriakoulakis K., Popova E.E., Ragueneau O., Vangriesheim A., & Wolff G.A. (2001). Material supply to the abyssal seafloor in the Northeast Atlantic. *Progress in Oceanography* 50: 27–63.

Lampitt R.S. & Burnham M.P. (1983). A free fall time lapse camera and current meter system "Bathysnap" with notes on the foraging behaviour of a bathyal decapod shrimp. *Deep-Sea Research* 30: 1009–1017.

Lampitt R.S., Merrett N.R., & Thurston M.H. (1983). Inter-relations of necrophagous amphipods, a fish predator, and tidal currents in the deep sea. *Marine Biology* 74: 73–78.

Land M.F. & Osorio D.C. (2011). Marine optics: dark disguise. *Current Biology* 21, R918–R92. doi: 10.1016/j.cub.2011.10.009

Laptikhovsky V. (2010). Migrations and structure of the species range in ridge-scaled rattail *Macrourus carinatus* (Southwest Atlantic) and their application to fisheries management. ICES Journal of Marine Science doi:10.1093/icesjms/fsq081

Laptikhovsky V., Arkhipin A., & Brickle P. (2008). Biology and distribution of grenadiers of the family Macrouridae around the Falkland Islands. *American Fisheries Society Symposium* 63: 261–284.

Lauerman L.M.L. & Kaufmann R.S. (1998). Deep-sea epibenthic echinoderms and a temporally varying food supply: results from a one year time series in the N.E. Pacific. *Deep-Sea Research II* 45(1998): 817–842.

Laurenson C. H., Dobby H., McLay H. A. & Leslie B. (2008). Biological features of the *Lophius piscatorius* catch in Scottish waters. *ICES Journal of Marine Science* 65: 1281–1290.

Laurenson C.H., Hudson I.R., Jones D.O.B., & Priede I.G. (2004). Deep water observations of anglerfish (*Lophius piscatorius* L.) in the North-eastern Atlantic Ocean by means of remotely operated vehicle. *Journal of Fish Biology* 65: 947–960. doi: 10.1111/j.0022-1112.2004.00496.x

Lawver L.A. & Gahagan L.M. (2003). Evolution of Cenozoic seaways in the circum-Antarctic region. *Palaeogeography, Palaeoclimatology, Palaeoecology* 198: 11–37.

Laxson C.J., Condon N.E., Drazen J.C., & Yancey P.H. (2011). Decreasing urea: trimethylamine N-oxide ratios with depth in chondrichthyes: a physiological depth limit? *Physiological and Biochemical Zoology* 84: 494–505.

Le Pichon X. (1968). Sea-floor spreading and continental drift. *Journal of Geophysical Research* 73(12): 3661–3697.

Lecointre G., Gallut C., Bonillo C., Couloux A., Ozouf-Costaz C., & Dettaï A. (2011). The Antarctic fish genus *Artedidraco* is paraphyletic (Teleostei, Notothenioidei, Artedidraconidae). *Polar Biology* 34(8): 1135–1145.

Levin L.A. (2000). Oxygen minimum zone benthos: adaptation and community response to hypoxia. *Oceanography and Marine Biology: An Annual Review* 2003(41): 1–45.

Licht M., Schmuecker K., Huelsken T., Hanel R., Bartsch P., & Paeckert M. (2012). Contribution to the molecular phylogenetic analysis of extant holocephalan fishes (Holocephali, Chimaeriformes). *Organisms Diversity and Evolution* 12: 421–432. doi: 10.1007/s13127-011-0071-1

Lim J., Fudge D.S., Levy N., & Gosline J.M. (2006). Hagfish slime ecomechanics: testing the gill-clogging hypothesis. *Journal of Experimental Biology* **209**: 702–710.

Lindsay D.J., Hunt J.C., & Hayashi K. (2001). Associations in the midwater zone: the penaeid shrimp *Funchalia sagamiensis* Fujino 1975 and pelagic tunicates (Order: Pyrosomatida). *Marine and Fresh-Water Behaviour and Physiology* **2001**: 157–170.

Linley T., Gerringer M.E., Yancey P.H., Drazen J.C., Weinstock C.L., & Jamieson A.J. (2016). Fishes of the hadal zone including new species, in situ observations and depth records of Liparidae. *Deep-sea Research I* **114**: 99–110.

Linley T.D., Alt, C.H.S., Jones D.O.B., & Priede I.G. (2013). Bathyal demersal fishes of Charlie Gibbs Fracture Zone region (49°-54°N) of the Mid-Atlantic Ridge: III. Results from remotely operated vehicle (ROV) video transects. *Deep-Sea Research II* **98**: 407–411. http://dx.doi.org/10.1016/j.dsr2.2013.08.013

Linley T.D., Stewart A.L., McMillan P.J., Clark M.R., Gerringer M.E., Drazen J.C., Fujii T., & Jamieson A.J. (2016). Bait attending fishes of the abyssal zone and hadal boundary: community structure, functional groups and species distribution in the Kermadec, New Hebrides and Mariana trenches. *Deep-Sea Research I* **121**:38–53.

Linnaeus C. (1735). *Systemae Naturae*. Leiden: Haak.

Linnaeus C. (1758). *Systemae Naturae*, 10th edition. Stockholm: Salvius.

Lins L.S.F., Ho S.Y.W., Wilson G.D.F., & Lo N. (2012). Evidence for Permo-Triassic colonization of the deep sea by isopods. *Biology Letters*. doi:10.1098/rsbl.2012.0774

Liu J.Y. (2013). Status of marine biodiversity of the China Seas. *PLoS ONE* **8**(1): e50719. http://doi.org/10.1371/journal.pone.0050719

Livermore R., Nankivell A., Eagles G., & Morris P. (2005) Paleogene opening of Drake Passage. *Earth and Planetary Science Letters* **236**(2005): 459–470.

Lloris D., Matallanas J., & Oliver P. (2005). *Hakes of the World (Family Merlucciidae): An Annotated and Illustrated Catalogue of Hake Species Known to Date*. Rome: Food and Agriculture Organization of the United Nations. Available for download at www.fao.org

Lloris D., Stefanescu C., & Rucabado J. (1994). New data on the distribution and biology of *Rhynchogadus*

hepaticus and *Eretmophorus kleinenbergi* (Osteichthyes: Moridae). *Cybium* **18**: 129–134.

Locket N.A. (1977). Adaptations to the deep-sea environment. In: Crescitelli F. (ed.) *The Visual System in Vertebrates*, pp. 67–192. Berlin: Springer-Verlag.

Løkkeborg S. (2005). *Impacts of Trawling and Scallop Dredging on Benthic Habitats and Communities*. FAO Fisheries Technical Paper. No. 472. Rome, FAO.

Lombarte A. & Cruz A. (2007). Otolith size trends in marine fish communities from different depth strata. *Journal of Fish Biology* **71**: 53–76.

Longhurst A. (1998). *Ecological Geography of the Sea*. New York: Academic Press.

Lopez J.A., Westneat M.W., & Hanel R. (2007). The phylogenetic affinities of the mysterious anguilliform genera *Coloconger* and *Thalassenchelys* as supported by mtDNA sequences. *Copeia* **2007**(4): 959–966.

Lorance P., Large P.A., Bergstad O.A., & Gordon J.D.M. (2008). Grenadiers of the NE Atlantic-distribution, biology, fisheries and the impacts, and developments in stock assessment and management. *American Fisheries Society Symposium* **63**: 365–397.

Love R.H., Fisher R.A., Wilson M.A., & Nero R.W. (2004). Unusual swimbladder behavior of fish in the Cariaco Trench. *Deep-Sea Research I* **51**: 1–16.

Lowe R. T. (1833). Description of a new genus of acanthopterygian fishes, *Alepisaurus ferox*. *Proceedings of the Zoological Society of London* **1**: 104.

Lowe R.T. (1843-1860). *A History of the Fishes of Madeira*, London: Van Voorst.

Ludwig H. & Macdonald A.G. (2005). The significance of the activity of dissolved oxygen, and other gases, enhanced by high hydrostatic pressure. *Comparative Biochemistry and Physiology* **140**A: 387–395.

Lundsten L., Johnson S.B., Cailliet G.M., DeVogelaere A.P,. & Clague D.A. (2012). Morphological, molecular, and *in situ* behavioural observations of the rare deep-sea anglerfish *Chaunacops coloratus* (Garman, 1899), order Lophiiformes,in the eastern North Pacific. *Deep-Sea Research I* **68**: 46–53.

Lütken C. F. (1892). Spolia Atlantica. Scopelini Musei zoologici Universitatis Hauniensis. Bidrag til Kundskab om det aabne Havs Laxesild eller Scopeliner. Med et tillaeg om en anden pelagisk fiskeslaegt. *Kongelige Danske Videnskabernes Selskab Series* **6**(3): 221–297.

Machete M., Morato T., & Menezes G. (2011). Experimental fisheries for black scabbardfish (*Aphanopus carbo*) in the Azores, Northeast Atlantic. *ICES Journal of Marine Science* **68**: 302–308.

Machida Y (1989). Record of *Abyssobrotula galatheae* (Ophidiidae: Ophidiiformes) from the Izu-Bonin Trench, Japan. *Bulletin of Marine Science and Fisheries* Kochi University, Japan **11**: 23–25.

Machida Y. & Hashimoto J. (2002). *Pyrolycus manusanus*, a new genus and species of deep-sea eelpout from a hydrothermal vent field in the Manus Basin, Papua New Guinea (Zoarcidae, Lycodinae). *Ichthyological Research* **49**:1–6.

Machida Y. & Shiogaki M. (1988). *Leptochilichthys microlepis*, a new species of the family Leptochilichthyidae, Salmoniformes, from Aomori, northern Japan. *Japanese Journal of Ichthyology* **35**: 1–6.

Macpherson E. (1985). Daily ration and feeding periodicity of some fishes off the coast of Namibia. *Marine Ecology Progress Series* **26**: 253–60.

Macpherson E. (1989). Influence of geographical distribution, body size and diet on population density of benthic fishes off Namibia (South West Africa). *Marine Ecology Progress Series* **50**: 295–299.

Maguire J.-J., Sissenwine M., Csirke J., Grainger R., & Garcia S. (2006). *The State of World Highly Migratory, Straddling and Other High Seas Fishery Resources and Associated Species*. FAO Fisheries Technical Paper. No. 495. Rome: FAO.

Mann D.A. & Jarvis S.M. (2004). Potential sound production by a deep-sea fish. *Journal of the Acoustical Society of America* **115**: 2331–2333.

Mantyla A.W. & Reid, J.L. (1983). Abyssal characteristics of the world ocean waters. *Deep-Sea Research* **30**A: 805–833.

Marine Stewardship Council. (2015). New Zealand EEZ ling trawl and longline fishery. www.msc.org/track-a-fishery/fisheries-in-the-program/certified/pacific/new-zealand-eez-ling-trawl-and-longline-fishery accessed 15 September 2015.

Markle D.F. & Merrett N.R. (1980). The abyssal alepocephalid, *Rinoctes nasutus* (Pisces: Salmoniformes), a redescription and an evaluation of its systematic position. *Journal of Zoology, London* **190**: 225–239.

Markle D.F. & Olney J.E.(1990). Systematics of the pearlfishes (Pisces: Carapidae). *Bulletin of Marine Science* **47**: 269–410.

Markle D.F & Sazonov Y.I. (1996). Review of the rare deep-sea genus, *Aulastomatomorpha* (Teleostei: Salmoniformes), with a discussion of relationships. *Copeia* **1996**(2): 497–500.

Marques V., Chaves C., Morai A., Cardador F., & Stratoudakis Y. (2005). Distribution and abundance of snipefish (*Macroramphosus* spp.) off Portugal (1998–2003). *Scientia Marina* **69**: 563–576.

Marshall A., Bennett M.B., Kodja G., Hinojosa-Alvarez S., Galvan-Magana F., et al. (2011). Manta birostris. *IUCN 2011. IUCN Red List of Threatened Species*. Version 2011.2. Available: <www.iucnredlist.org>. Accessed: 17 February 2015. www.iucnredlist.org/apps/redlist/details/198921/0.

Marshall N.B. (1953). Egg size in Arctic, Antarctic and deep-sea fishes. *Evolution* **7**: 328–341.

Marshall N.B. (1954). *Aspects of Biology*. London: Hutchinson.

Marshall N.B. (1960). Swimbladder structure of deep-sea fishes in relation to their systematics and biology. *Discovery Reports* **31**: 1–122.

Marshall N.B. (1965). Systematic and biological studies of the Macrourid fishes (Anacanthini-Teleostii) *Deep-Sea Research* **12**: 299–322.

Marshall N.B. (1966a). *Bathyprion danae* a new genus and species of alepocephaliform fishes. *Dana-Reports* **68**: 1–9.

Marshall N.B. (1966b). *The Life of Fishes*. London: Weidenfeld and Nicolson.

Marshall N. B. (1967). The olfactory organs of bathypelagic fishes. In: *Aspects of Marine Zoology*, pp. 57–70. London: Academic Press.

Marshall N.B. (1972). Swimbladder organisation and deep ranges of deep-sea teleosts. *Symposia of the Society of Experimental Biology* **26**: 261–272.

Marshall N.J. (1996). The lateral line systems of three deep-sea fish. *Journal of Fish Biology* **49** (A): 239–258.

Martín J., Puig P., Masqué P., Palanques A., & Sánchez-Gómez A. (2014). Impact of bottom trawling on deep-sea sediment properties along the flanks of a submarine canyon. *PLoS ONE* **9**(8): e104536. doi:10.1371/journal.pone.0104536

Martini F.H. & Beulig A. (2013). Morphometrics and gonadal development of the hagfish *Eptatretus cirrhatus* in New Zealand. *PLoS ONE* **8**(11): e78740. doi:10.1371/journal.pone.0078740

Martins R. & Fereira C. (1995). Line fishing for black scabbardfish (*Aphanopus carbo* Lowe, 1839) and other deep water species in the eastern mid-Atlantic to the north of Madeira. In: Hooper A.G. (ed.) *Deep-Water Fisheries of the North Atlantic Oceanic Slope*, pp. 323–335. Dordrecht, Netherlands: Kluwer Academic Publishers.

Matallanas J. (2000). On Mediterranean and some northeastern Atlantic Liparidae (Pisces: Scorpaeniformes) with the restoration of *Eutelichthys*. *Journal of the Marine Biological Association of the UK* **80**: 935–939.

Matschiner M., Colombo M., Damerau M., Ceballos S., Hanel R., & Salzburger W. (2015). The adaptive radiation of Notothenioid fishes in the waters of Antarctica. In Riesch R. et al. (eds.) *Extremophile Fishes*, pp. 35–57. Heidelberg, Germany: Springer International. doi:10.1007/78-3-319-13362-1_3

Matsui T. & Rosenblatt R.H. (1987). Review of the deep-sea fish family Platytroctidae (Pisces: Salmoniformes). *Bulletin of the Scripps Institution of Oceanography* **26**: 1–158. http://escholarship.org/uc/item/35v4k0ks

Matsunaga T. & Rahman A. (1998). What brought the adaptive immune system to vertebrates? The jaw hypothesis and the seahorse. *Immunology Reviews* **166**: 177–186.

Matsuura K. (2015). Taxonomy and systematics of tetraodontiform fishes: a review focusing primarily on progress in the period from 1980 to 2014. *Ichthyological Research* **62**(1): 72–113.

Mauchline J. & Gordon J.D.M. (1984a). Feeding and bathymetric distribution of the gadoid and morid fish of the Rockall Trough. *Journal of the Marine Biological Association of the UK* **64**: 657–665.

Mauchline J. & Gordon J.D.M. (1984b). Occurrence and feeding of berycomorphid and percomorphid teleost fish in the Rockall Trough. *Journal du Conseil - Conseil international pour l'exploration de la mer* **41**: 239–247.

Mauchline J. & Gordon J.D.M. (1991). Oceanic pelagic prey of benthopelagic fish in the benthic boundary layer of a marginal oceanic region. *Marine Ecology Progress Series* **74**: 109–115.

Maury M.F. (1855). *The Physical Geography of the Sea*. New York: Harper and Bros.

McClain C.R. & Hardy S.M. (2010). The dynamics of biogeographic ranges in the deep sea. *Proceedings of the Royal Society of London B* **277**: 3533–3546. doi:10.1098/rspb.2010.1057

McClain C.R., Rex M.A., & Jabbour R. (2005). Deconstructing bathymetric patterns of body size in deep-sea gastropods. *Marine Ecology Progress Series* **297**, 181–187.

McCleave J.D. & Miller M.J. (1994). Spawning of *Conger oceanicus* and *Conger triporiceps* (Congridae) in the Sargasso Sea and subsequent distribution of leptocephali. *Environmental Biology of Fishes* **39**: 339–355.

McDowall R.M. & Stewart A.L. (1999). *Further specimens of Agrostichthys parkeri (Teleostei: Regalecidae), with natural history notes.* Proceedings of the 5th Indo-Pacific Fish Conference, Noumea, 1997, pp. 165–174.

McEachran J.D. & Dunn K.A. (1998). Phylogenetic Analysis of Skates, a Morphologically Conservative Clade of Elasmobranchs (Chondrichthyes: Rajidae). *Copeia* **1998** (2): 271–290.

McIntyre F.D., Collie N., Stewart M., Scala L., & Fernandes P.G. (2013). A visual survey technique for deep water fish: Estimating anglerfish abundance in closed areas. *Journal of Fish Biology* **83**: 739–753.

McKenzie A. (2016). *Assessment of hoki (*Macruronus novaezelandiae*) in 2015.* New Zealand Fisheries Assessment Report 2016/01. Wellington: Ministry for Primary Industries.

McMillan P., Iwamoto T., Stewart A., & Smith P.J. (2012). A new species of grenadier, genus *Macrourus* (Teleostei, Gadiformes, Macrouridae) from the southern hemisphere and a revision of the genus. *Zootaxa* **3165**: 1–24.

McMillan P.J., Griggs L.H., Francis M.P., Marriott P.J, Paul L.J., Mackay E., Wood B.A., Sui H., & Wei F. (2011). *New Zealand Fishes. Volume 3: A Field Guide to Common Species Caught by Surface Fishing.* New Zealand Aquatic Environment and Biodiversity Report No. 69. Wellington: Ministry of Fisheries.

McMillan P.J. & Paulin C.D. (1993). Descriptions of nine new species of rattails of the Genus *Caelorinchus* (Pisces, Macrouridae) from New Zealand. *Copeia* **1993**: 819–840.

Mead G.W., Bertelsen E., & Cohen D.M. (1964). Reproduction among deep-sea fishes. *Deep-Sea Research* **11**: 569–596.

Meléndez R.C. & Markle D.F. (1997). Phylogeny and zoogeography of *Laemonema* and *Guttigadus* (Pisces: Gadiformes: Moridae). *Bulletin of Marine Science* **61**(3): 593–670.

Melo M.R.S. (2009a). Revision of the genus *Chiasmodon* (Acanthomorpha: Chiasmodontidae), with description of two new species. *Copeia* **2009**(3): 583–608.

Melo M.R.S. (2009b). *Taxonomic and phylogenetic revision of the family Chiasmodontidae (Perciformes: Acanthomorpha)*. PhD thesis Auburn University, Alabama, USA.

Merrett N.R. (1981). First record of a Notacanth leptocephalus; a benthopelagic capture in the slope waters off Ireland. *Journal of Fish Biology* **18**: 53–57.

Merrett N.R. (1987). A zone of faunal change in assemblages of abyssal demersal fish in the eastern North Atlantic: a response to seasonality in production? *Biological Oceanography* **5**(2): 137–151.

Merrett N.R. (1994). Reproduction in the North Atlantic oceanic ichthyofauna and the relationship between fecundity and species' sizes. *Environmental Biology of Fishes* **41**: 207–245.

Merrett N.R. & Barnes S.H. (1996). Preliminary survey of egg envelope morphology in the Macrouridae and the possible implications of its ornamentation. *Journal of Fish Biology* **48**: 101–119.

Merrett N.R., Gordon J.D.M., Stehman M., & Haedrich R.L. (1991a). Deep demeral fish assemblage structure in the Porcupine Seabight (Eastern North Atlantic): slope sampling by three different trawls compared. *Journal of the Marine Biological Association of the UK* **71**: 329–358.

Merrett N.R. & Haedrich R.L. (1997). *Deep-Sea Demersal Fish & Fisheries*. London: Chapman & Hall.

Merrett N.R., Haedrich R.L., Gordon J.D.M., & Stehman M. (1991b). Deep demeral fish assemblage structure in the Porcupine Seabight (Eastern North Atlantic): results of single warp trawling at lower slope to abyssal soundings. *Journal of the Marine Biological Association of the UK* **71**: 359–373.

Merrett N.R. & Marshall N.B. (1981). Observations on the ecology of deep-sea bottom-living fishes collected off northwest Africa (08°–27°N). *Progress in Oceanography* **9**: 185–244.

MerrettN.R. & Moore J.A. (2005). A new genus and species of deep demersal fish (Teleostei: Stephanobercidae) from the tropical eastern North Atlantic. *Journal of Fish Biology* **67**: 1699–1710.

Merrett N.R. & Nielsen J.G. (1987). A new genus and species of the family Ipnopidae (Pisces, Teleostei) from the eastern North Atlantic, with notes on its ecology. *Journal of Fish Biology* **31**: 451–464.

Merrett N.R., Sazonov Y.I., & Shcherbachev Y.N. (1983). A new genus and species of rattail fish (Macrouridae) from the eastern North Atlantic and the eastern Indian Ocean, with notes on its ecology. *Journal of Fish Biology* **22**: 549–561.

Mienert J., Berndt C., Laberg J.S., & Vorren T.O. (2003). Slope instability of continental margins. In: Wefer G., Billett D., Hebbeln D., Jørgensen B.B., Schlüter M., & van Weering T. eds.) *Ocean Margin Systems*, pp. 179–193. Berlin: Springer-Verlag.

Miller M.J., Chikaraishi Y., Ogawa N.O., Yamada Y., Tsukamoto K., & Ohkouchi N. (2013). A low trophic position of Japanese eel larvae indicates feeding on marine snow. *Biology Letters* **9**: 20120826. http://dx.doi.org/10.1098/rsbl.2012.0826

Miller M.J. & McCleave J.D. (2007). Species assemblages of leptocephali in the southwestern Sargasso Sea. *Marine Ecology Progress Series* **344**: 197–212. doi: 10.3354/meps06923

Miller R.R. (1947). A new genus and species of deep-sea fish of the family Myctophidae from the Philippine Islands. *Proceedings of the US National Museum* **97**: 81–90.

Milligan R.J., Morris K.J., Bett B.J., Durden J.M., Jones D.O.B., Robert K., Ruhl H.A., & Bailey D.M. (2016). High resolution study of the spatial distributions of abyssal fishes by autonomous underwater vehicle. *Scientific Reports* **6**, Article number: 26095. doi:10.1038/srep26095

Milliken D.M. & Houde E.D. (1984). A new species of Bregmacerotidae (pisces), *Bregmaceros cantori*, from the Western Atlantic Ocean. *Bulletin of Marine Science* **35**: 11–19.

Mills E.L. (1983). Problems of deep-sea biology: An historical perspective. In: Rowe G.T. (ed.) *The Sea*, pp. 1–79. Volume 8: Deep-Sea Biology. Cambridge, MA: Harvard University Press.

Mitson R.B. & Knudsen H.P. (2003). Causes and effects of underwater noise on fish abundance estimation. *Aquatic Living Resources* **16**: 255–263.

Miya M. (1994). *Cyclothone kobayashii*, a new gonostomatid fish (Teleostei: Stomiiformes) from the Southern Ocean, with notes on its ecology. *Copeia* **1994**(1): 191–204.

Miya M. & Markle D.F. (1993). *Bajacalifornia aequatoris*, new species of Alepocephalid fish (Pisces: Salmoniformes) from the central equatorial Pacific. *Copeia* **1993**(3): 743–747.

Miya M. & Nielsen J.G. (1991). A new species of the deep-sea fish genus *Parabrotula* (Parabrotulidae) from Sagami Bay with notes on its biology. *Japanese Journal of Ichthyology* **38**: 1–5.

Miya M., Pietsch T.W., Orr J.W., Arnold R.J., Satoh T.P., Shedlock A.M., Ho H.-C., Shimazaki M., Yabe M., & Nishida M. (2010). Evolutionary history of anglerfishes (Teleostei: Lophiiformes): a mitogenomic perspective. *BMC Evolutionary Biology 2010* **10**: 58. www.biomedcentral.com/1471-2148/10/58

Mizusawa N., Takami M., & Fukui A. (2015). Redescription of the spookfish *Dolichopteryx anascopa* Brauer 1901 (Argentinoidei: Opisthoproctidae). *Ichthyological Research* **62**: 236–239.

Mok H.-K. (1978). Scale-feeding in *Tydemania navigatoris* (Pisces: Triacanthodidae). *Copeia* **1978**(2): 338–340.

Møller P.R. & Jones W.J. (2007). *Eptatretus strickrotti* n. sp. (Myxinidae): first hagfish captured from a hydrothermal vent. *Biological Bulletin* **212**: 55–66.

Møller P.R. & Stewart A.L. (2006). Two new species of eelpouts (Teleostei, Zoarcidae) of the genus *Seleniolycus* from the Ross Dependency, Antarctica. *Zootaxa* **1376**: 53–56.

Montgomery J. & Pankhurst N. (1997). Sensory physiology. In: Randall D.J. & Farrell A.P. (eds.) *Deep-Sea Fishes*, pp. 325–249. San Diego: Academic Press.

Moore J.A. (2002). Upside-down swimming behavior in a whipnose anglerfish (Teleostei: Ceratioidei: Gigantactinidae). *Copeia* **2002**: 1144–1146.

Mora C., Tittensor D. P., & Myers R. A. (2008). The completeness of taxonomic inventories for describing the global diversity and distribution of marine fishes. *Proceedings of the Royal Society B* **275**: 149–155.

Morato T., Machete M., Kitchingham A., Tempura F., Lai S., Menezes G., Pitcher T.J., & Santos R.S. (2008). Abundance and distribution of seamounts in the Azores. *Marine Ecology Progress Series* **357**: 17–21.

Morato T., Watson R., Pitcher T.J., & Pauly D. (2006). Fishing down the deep. *Fish and Fisheries* **7**: 24–34.

Morita T. (1999). Molecular phylogenetic relationships of the deep-sea fish genus *Coryphaenoides* (Gadiformes: Macrouridae) based on mitochondrial DNA. *Molecular Phylogenetics and Evolution* **13**: 447–454.

Morita T. (2008). Comparative sequence analysis of myosin heavy chain proteins from congeneric shallow- and deep-living rattail fish (genus *Coryphaenoides*). *Journal of Experimental Biology* **211**: 1362–1367.

Morris K.J., Bett B.J., Durden J.M., Huvenne V.A.I., Milligan R., Jones D.O.B., McPhail S., Bailey D.M., Robert K., Bailey D.M., & Ruhl H.A. (2014). A new method for ecological surveying of the abyss using autonomous underwater vehicle photography. *Limnology and Oceanography Methods* **12**: 795–809.

Moser H.G. (1996a). Anoplopomatidae: sablefish and skilfish. In: Moser H.G. (ed.) *Early Stages of Fishes in the California Current Region. California Cooperative Oceanic Fisheries Investigations*, pp. 807–809. Atlas No 33. La Jolla, CA: South West Fisheries Centre.

Moser H.G. (1996b). Caristiidae: manefishes or veilfins. In: Moser H.G. (ed.) *Early Stages of Fishes in the California Current Region. California Cooperative Oceanic Fisheries Investigations*, pp. 973–975. Atlas No 33. La Jolla, CA: South West Fisheries Centre.

Moser H.G. (1996c). Chauliodontidae: viperfishes. In: Moser H.G. (ed.) *Early Stages of Fishes in the California Current Region. California Cooperative Oceanic Fisheries Investigations*, pp. 297–299. Atlas No 33. La Jolla, CA: South West Fisheries Centre.

Moser H.G. (ed.). (1996d). *Early Stages of Fishes in the California Current Region. California Cooperative Oceanic Fisheries Investigations*. Atlas No 33. La Jolla, CA: South West Fisheries Centre

Moser H.G. (1996e). Idiacanthidae: blackdragons. In: Moser H.G. (ed.) *Early Stages of Fishes in the California Current Region. California Cooperative Oceanic Fisheries Investigations*, pp. 325–327. Atlas No 33. La Jolla, CA: South West Fisheries Centre.

Moser H.G. (1996f). Malacosteidae: loosejaws. In: Moser H.G. (ed.) *Early Stages of Fishes in the California Current Region. California Cooperative Oceanic Fisheries Investigations*, pp. 321–323. Atlas No 33. La Jolla, CA: South West Fisheries Centre.

Moser H.G. (1996g). Myctophiformes: lanternfishes. In: Moser H.G. (ed.) *Early Stages of Fishes in the California Current Region. California Cooperative Oceanic Fisheries Investigations*, pp. 383–385. Atlas No 33La Jolla, CA: South West Fisheries Centre.

La Jolla, CA: South West Fisheries Centre.

Moser H.G. & Ahlstrom D.E.H. (1996). Bathylagidae: Blacksmelts and Smoothtongues. In: Moser H.G. (ed.) *Early Stages of Fishes in the California Current*

Region. California Cooperative Oceanic Fisheries Investigations, pp. 188–207. Atlas No 33. La Jolla, CA: South West Fisheries Centre.

Moser H.G. & Ahlstrom E.H. (1974). Role of larval stages in systematic investigations of marine teleosts: the myctophidae, a case study. *Fishery Bulletin* 72: 391–413.

Moser H.G., Ahlstrom E.H., & Paxton J.R. (1984). Myctophidae: development. In: Moser H.G., Richards W.J., Cohen D.M., Fahay M.P., Kendall M.P., & Richardson S.L. (eds.) *Ontogeny and Systematics of Fishes*, pp. 218–239. Special Publication. Lawrence, KS: American Society of Ichthyologists and Herpetologists.

Moser H.G. & Charter S.R. (1996). Notacanthiformes. In: Moser H.G. (ed.) *Early Stages of Fishes in the California Current Region. California Cooperative Oceanic Fisheries Investigations*, pp. 82–85. Atlas No 33. La Jolla, CA: South West Fisheries Centre, La Jolla.

MSC. (2016). Marine Stewardship Council, Certified sustainable seafood. Online publication accessed 1 April 2016: www.msc.org/

Mueter F.J., Reist J.D., Majewski A.R., Sawatzky C.D., Christiansen J.S., Hedges K.J., Coad B.W., Karamushko O.V., Lauth R.R., Lynghammar A., MacPhee S.A., & Mecklenburg C.W. (2013). *Marine fishes of the Arctic*. Online publication www.arctic.noaa.gov/report13/marine_fish.html accessed 29 September 2016.

Müller R. D., Sdrolias M., Gaina C., & Roest W. R. (2008). Age, spreading rates, and spreading asymmetry of the world's ocean crust. *Geochemistry, Geophysics, Geosystems* 9, Q04006. doi:10.1029/2007GC001743.

Munk O., (1959). The eyes of *Ipnops murrayi* Günther, 1878. *Galathea Reports* 3: 79–87.

Munk O. (1965). Ocular degeneration in deep-sea fishes. *Galathea Reports* 8: 21–31.

Munk O. (1977). The visual cells and retinal tapetum of the foveate deep-sea fish *Scopelosaurus lepidus* (Teleostei). *Zoomorphologie* 87: 21–49.

Munk O. & Frederiksen R.D. (1974). On the function of aphakic apertures in teleosts. *Videnskabelige Meddelelser Dansk Naturhistorisk Forening* 137: 65–94.

Munk O. & Jørgensen J.M. (2010). Putatively luminous tissue in the abdominal pouch of a male dalatiine shark, *Euprotomicroides zantedeschia* Hulley &

Penrith, 1966. *Acta Zoologica* 69: 247–251. doi: 10.1111/j.1463-6395.1988.tb00921.

Munroe T.A. (1998). Systematics and ecology of tonguefishes of the genus *Symphurus* (Cynoglossidae: Pleuronectiformes) from the western Atlantic Ocean. *Fishery Bulletin* 96(1): 1–182.

Munroe T.A. & Hashimoto J. (2008). A new Western Pacific tonguefish (Pleuronectiformes: Cynoglossidae): The first Pleuronectiform discovered at active hydrothermal vents. *Zootaxa* 1839: 43–59.

Munroe T.A., Tyler J., & Tunnicliffe V. (2011). Description and biological observations on a new species of deep water symphurine tonguefish (Pleuronectiformes: Cynoglossidae: *Symphurus*) collected at Volcano-19, Tonga Arc, West Pacific Ocean. *Zootaxa* 3061: 53–66.

Murray J. & Hjort J. (1912). *The Depths of the Ocean*. London: Macmillan.

Murray J.W., Jannasch H.W., Honjo S., Anderson R.F., Reeburgh W.S., Top Z., Friederich G.E., Codispoti L.A., & Izdar E. (1989). Unexpected changes in the oxic/anoxic interface in the Black Sea. *Nature* 338: 411–413. doi:10.1038/338411a0

Musick J.A., Bruton M.N., & Balon E. K. (eds.). (1991). The biology of *Latimeria chalumnae* and evolution of coelocanths. *Environmental Biology of Fishes* 32(1–4): 1–445.

Musick J. A. & Cotton C. F. (2015). Bathymetric limits of chondrichthyans in the deep sea: A re-evaluation. *Deep-Sea Research II* 115: 73–80.

Mytilineou C., Politou C.-Y., Papaconstantinou C., Kavadas S., D'Onghia G., & Sion L. (2005). Deepwater fish fauna in the Eastern Ionian Sea. *Belgian Journal of Zoology* 135: 229–233.

Nafpaktitis B.G., Robertson D.A., & Paxton J.A. (1995). Four new species of *Diaphus* (Myctophidae) from the Indo-Pacific. *New Zealand Journal of Marine and Freshwater Research* 29: 335–344.

Nakae M. & Sasaki K. (2002). A scale-eating triacanthodid, *Macrorhamphosodes uradoi*: prey fishes and mouth 'handedness' (Tetraodontiformes, Triacanthoidei). *Ichthyological Research* 49: 7–14.

Nakamura I., Meyer C.G. & Sato K. (2015). Unexpected positive buoyancy in deep sea sharks, *Hexanchus griseus*, and an *Echinorhinus cookei*. *PLoS ONE* 10(6): e0127667. doi:10.1371/journal.pone.0127667

Nakamura I. & Parin N.V. (1993). FAO Species Catalogue. Vol. 15. Snake mackerels and cutlassfishes of the world (families Gempylidae and Trichiuridae). An annotated and illustrated catalogue of the snake mackerels, snoeks, escolars, gemfishes, sackfishes, domine, oilfish, cutlassfishes,. scabbardfishes, hairtails, and frostfishes known to date. *FAO Fisheries Synopsis* **125**(15): 136.

Nakanishi M. & Hashimoto J. (2011). A precise bathymetric map of the world's deepest seafloor, Challenger Deep in the Mariana Trench. *Marine Geophysical Researches* **32**(4): 455–463. doi:10.1007/s11001-011-9134-0

Nashida K., Sakaji H., & Honda H. (2007). Spawning seasons of adult and growth of 0-year-old Deepsea smelt *Glossanodon semifasciatus* in Tosa Bay, Pacific coast of Shikoku. *Bulletin of the Japanese Society of Fisheries and Oceanography* **71**(4): 270–278.

National Geographic. (1967). *Indian Ocean Floor Map.* October 1967. Washington DC.

National Geographic. (1968). *Atlantic Ocean Floor Map.* June 1968. Washington DC.

National Geographic. (1969). *Pacific Ocean Floor Map.* October 1969. Washington DC.

National Marine Fisheries Service. (2013). U.S. National Bycatch Report First edition Update 1. In: Benaka L.R., Rilling C., Seney E.E., & Winarsoo H. (eds.). Washington, DC: U.S. Department of Commerce. www.st.nmfs.noaa.gov/observer-home/first-edition-update-1 accessed 16 December 2015.

Near T.J, Dornburg A., Eytan R.I., Keck B.P., Smith W.L., Kuhn K.L., Moore J.A. Price A.A., Burbrink F.T, Friedman M., & Wainwright P.C. (2013). Phylogeny and tempo of diversification in the superradiation of spiny-rayed fishes. *Proceedings of the National Academy of Sciences of the USA* **1101**: 2738–12743. doi:10.1073/pnas.1304661110

Near T. J., Eytan R. I., Dornburg A., Kuhn K. L., Moore J. A., Davis M. P., Wainwright P. C., Friedman M., & Smith, W. L. (2012). Resolution of ray-finned fish phylogeny and timing of diversification. *Proceedings of the National Academy of Sciences of the United States of America* **109**: 13698–13703. doi: 10.1073/pnas.1206625109

Near T.J., Jones C.D., & Eastman J.T. (2009). Geographic intraspecific variation in buoyancy within Antarctic notothenioid fishes. *Antarctic Science* **21**(2): 123–129.

Neat F. & Burns F. (2010). Stable abundance, but changing size structure in grenadier fishes (Macrouridae) over a decade (1998–2008) in which deepwater fisheries became regulated. *Deep Sea Research Part I Oceanographic Research Papers* **57**: 434–440. doi: 10.1016/j.dsr.2009.12.003

Neat F.C. & Campbell N. (2013). Proliferation of elongate fishes in the deep sea. *Journal of Fish Biology* **83**: 1576–1591.

Neira F.J. & Lyle J.M. (2011). DEPM-based spawning biomass of *Emmelichthys nitidus* (Emmelichthyidae) to underpin a developing mid-water trawl fishery in south-eastern Australia. *Fisheries Research* **110**: 236–243.

Nelson J. S. (1994). *Fishes of the World*, 3rd edition. New York: Wiley.

Nelson J.S. (2006). *Fishes of the World*, 4th edition. Hoboken, NJ: Wiley.

Nemeth D. (1994). Systematics and distribution of fishes of the family Champsodontidae (Teleostei: Perciformes), with descriptions of three new species. *Copeia* **1994**(2): 347–371.

New Zealand. (2011). *Southern Blue Whiting, Fisheries Chapter Plan.* Online publication. www.mpi.govt.nz. Accessed 18 August 2015.

Niass F. & Ozawa T. (2000). Morphological differences between North Pacific and Atlantic specimens of *Lampadena anomala* (Family Myctophidae). *Ichthyological Research* **47**: 299–302.

Nicol J.A.C. (1958). Observations on luminescence in pelagic animals. *Journal of the Marine Biological Association of the UK* **37**(3): 705–752.

Nicol J.A.C. (1960). Studies on luminescence. On the subocular light-organs of stomiatoid fishes. *Journal of the Marine Biological Association of the UK* **39**: 529–548.

Nielsen J., Badcock J., & Merrett N.R. (1990). New data elucidating the taxonomy and ecology of the Parabrotulidae (Pisces: Zoarcoidei). *Journal of Fish Biology* **37**: 437–448.

Nielsen J., Hedeholm R.B., Heinemeier J., Bushnell P.G., Christiansen J.S., Olsen J., Ramsey C.W., Brill R.W., Simon M., Steffensen K.F., & Steffensen J.F. (2016). Eye lens radiocarbon reveals centuries of longevity in the Greenland shark (*Somniosus microcephalus*). *Science* **353**: 702–704. doi: 10.1126/science.aaf1703

Nielsen J.G. (1964). Fishes from depths exceeding 6000 meters. *Galathea Report* **7**: 113–124.

Nielsen J. G. (1966). Synopsis of the Ipnopidae (Pisces, Iniomi) with description of two new abyssal species. *Galathea Report* **8**: 49–75.

Nielsen J. G. (1977). The deepest living fish *Abyssobrotula galathea*. A new genus and species of oviparous ophidiids (Pisces, Brotulidae). *Galathea Rep.* **14**: 41–48.

Nielsen J.G. & Cohen D.M. (2005). *Thermichthys* (Bythitidae), a replacement name for preoccupied *Gerhardia* Nielsen & Cohen, 2002 and a second specimen of *Thermichthys hollisi* from the southeast Pacific. *Cybium* **29**(4): 395–398.

Nielsen J.G., Cohen D.M., Markle D.F. & Robins C.R. (1999). *FAO species catalogue*. Volume 18. *Ophidiiform fishes of the world (Order Ophidiiformes). An annotated and illustrated catalogue of pearlfishes, cusk-eels, brotulas and other ophidiiform fishes known to date.* FAO Fisheries Synopsis. No. 125, Vol. 18. Rome: FAO.

Nielsen J.G. & Hartel K.E. (1996). *Monognathus berteli* sp. nov. from the Indian Ocean (Pisces, Monognathidae). *Ichthyological Research* **43**(2): 113–115.

Nielsen J.G. & Larsen V. (1968). Synopsis of the Bathylaconidae (Pisces, Isospondyli) with a new eastern Pacific species. *Galathea Reports* **9**: 221–238.

Nielsen J.G. & Machida Y. (2006). *Neobythitoides serratus*, a new bathyal genus and species from the East China Sea (Teleostei: Ophidiidae). *Zootaxa* **1227**: 63–68.

Nielsen J.G. & Merrett N.R. (2000). Revision of the cosmopolitan deep-sea Genus *Bassozetus* (Pisces: Ophiliidae) with two new species. *Galathea Report* **18**: 7–56.

Nielsen J.G., Mincarone M.M., & Di Dario F. (2015). A new deep-sea species of *Barathronus* Goode & Bean from Brazil, with notes on *Barathronus bicolor* Goode & Bean (Ophidiiformes: Aphyonidae). *Neotropical. Ichthyology* **13**: 53–60.

Nielsen J.G. & Møller P.R. (2011). Revision of the bathyal cusk-eels of the genus *Bassogigas* (Ophidiidae) with description of a new species from off Guam, west Pacific Ocean. *Journal of Fish Biology* **78** : 783–795.

Nielsen J.G., Møller P.R., & Segonzac M. (2006). *Ventichthys biospeedoi* n. gen. et sp. (Teleostei, Ophidiidae) from a hydrothermal vent in the South East Pacific. *Zootaxa* **1247**: 13–24.

Nielsen J.G., Ross S.W. & Cohen D.M. (2009). Atlantic occurrence of the genus *Bellottia* (Teleostei, Bythitidae) with two new species from the Western North Atlantic. *Zootaxa* **2018**: 45–57.

Nikaido M., Sasaki T., Emerson J.J., Aibara M., Mzighani S.I., Budeba Y.L., Ngatunga B.P., Iwata M., Abe Y., Li W.H., & Okada N. (2011). Genetically distinct coelacanth population off the northern Tanzanian coast. *Proceedings of the National Academy of Sciences of the USA* **108**: 18009–18013. doi: 10.1073/pnas.1115675108

Nishimura S. (1966). Early life history of the deep-sea smelt, *Glossanodon semifasciatus* (Kishinouye) (Teleostei: Clupeida) PART I. *Publications of the Seto Marine Biological Laboratory* **XIII**(5): 349–360.

Norse E.A., Brooke S., Cheung W.W.L., Clark M.R., Ekelandd I., Froese R., Gjerde K.M., Haedrich R.L., Heppell S.S., Morato T., Morgan L.E., Pauly D., Sumaila R., & Watson R. (2012). Sustainability of deep-sea fisheries. *Marine Policy* **36**: 307–320.

Nunoo F., Bannermann P., Russell B., & Poss S. (2015). Helicolenus dactylopterus. *The IUCN Red List of Threatened Species 2015*: e.T195093A15592445. http://dx.doi.org/10.2305/IUCN.UK.2015-4.RLTS .T195093A15592445.en. Downloaded on 21 March 2017.

Nybelin O. (1957). Deep-sea bottom fishes. *Report of the Swedish Deep Sea Expedition 2. Zoology* **20**: 247–345.

NZ. (2009). History of fishing in New Zealand: growth and the EEZ. NZ Fisheries Info site: http://fs.fish .govt.nz/Page.aspx?pk=51&tk=166 accessed 6 May 2016.

O'Connor S. & Veth P. (2005). Early Holocene shell fish hooks from Lene Hara Cave, East Timor establish complex fishing technology was in use in Island South East Asia five thousand years before Austronesian settlement. *Antiquity* **79**: 249–256.

Okamura O. (1970). Studies on the macrourid fishes of Japan. Morphology, ecology and phylogeny. *Reports of the USA Marine Biological Station Kochi University.* **17**: 1–179.

Okiyama M., Tominaga Y., & Ida H. (2007). A megapterygium larva of *Discoverichthys praecox* (Aulopiformes: Ipnopidae) from the tropical western Pacific. *Ichthyological Research* **54**: 262–267. doi: 10.1007/s10228-007-0399-x

O'Leary B.C., Brown R.L., Johnson D.E., vonNordheim H., Ardron J., Packeiser T., & Roberts C.M. (2012). The first network of marine protected areas (MPAs) in the high seas: The process, the challenges and where next. *Marine Policy* **36**: 598–605.

Olivar M.P., Bernal A., Molí B., Peña M., Balbín R., Castellón A., Miquel J., & Massuti E. (2012). Vertical distribution, diversity and assemblages of mesopelagic fishes in the western Mediterranean. *Deep-Sea Research I* 62: 53–69.

Olivar M.P., González-Gordillo J.I., Salat J., Chust G., Cózar A., Hernández-León A., Fernández de Puelles M.L., & Irigoien X. (2016). The contribution of migratory mesopelagic fishes to neuston fish assemblages across the Atlantic, Indian and Pacific Oceans. *Marine and Freshwater Research* 67: 1114–1127.

Olney J.E. (2005). Stylephoridae: Tube eyes. In: Richards W.J. (ed.) *Early Stages of Atlantic Fishes: An Identification Guide for the West Central North Atlantic.* Volume 1, pp.1013–1014. Boca Raton, FL: CRC Press.

Olney J.E., Johnson G.D., & Baldwin C.C. (1993). Phylogeny of lampridiform fishes. *Bulletin of Marine Science* 52: 137–169.

Olson K. R. (1996). Secondary circulation in fish: Anatomical organization and physiological significance. *Journal of Experimental Zoology* 275: 172–185.

Oppo D.W. & Curry W.B. (2012). Deep Atlantic circulation during the last glacial maximum and deglaciation. *Nature Education Knowledge* 3(10): 1. www.nature.com/scitable/knowledge/library/deep-atlantic-circulation-during-the-last-glacial-25858002

Orlov A.M. & Tokranov A.M. (2011). Some rare and insufficiently studied snailfish (Liparidae, Scorpaeniformes, Pisces) in the PacificWaters off the Northern Kuril Islands and Southeastern Kamchatka, Russia. *International Scholarly Research Network ISRN Zoology* 2011, Article ID 341640, 12 pages doi:10.5402/2011/341640

Orr, J. W. & Busby M.S. (2001). *Prognatholiparis ptychomandibularis*, a new genus and species of the fish family Liparidae (Teleostei: Scorpaeniformes) from the Aleutian Islands, Alaska. *Proceedings of the Biological Society of Washington* 114(1): 51–57.

Overdick A.A., Busby M.S., & Blood D.M. (2014). Descriptions of eggs of snailfishes (family Liparidae) from the Bering Sea and eastern North Pacific Ocean. *Ichthyological Research* 61(2): 131–141. doi: 10.1007/s10228-013-0384-5

Owen P. & Rice T. (1999). *Decommissioning of Brent Spar.* London: Spon Press.

Ozaka C., Yamamoto N., & Somiya H. (2009). The Aglomerular Kidney of the Deep-sea Fish, *Ateleopus japonicus* (Ateleopodiformes: Ateleopodidae): Evidence of wider occurrence of the aglomerular condition in Teleostei. *Copeia* 2009: 609–617.

Padilla A., Zeller D., & Pauly D. (2015). The fish and fisheries of Bouvet Island. In: Palomares M.L.D. & Pauly D. (eds.) *Marine Fisheries Catches of Sub-Antarctic Islands, 1950–2010*, pp. 21–30. Fisheries Centre Research Report 23(1). Vancouver, BC: Fisheries Centre, University of British Columbia.

Papaconstantinou C., Anastasopoulou K., & Caragitsou E. (1997). Comments on the mesopelagic fauna of the North Aegean Sea. *Cybium* 21: 281–288.

Parin N.V. (2004). A new mesopelagic fish *Ioichthys kashkini* Parin, gen.et sp, nova (Opisthoproctidae) from the North Western part of the Indian Ocean. *Journal of Ichthyology* 44(7): 485–488.

Parin N.V., Belyanina T.N., & Evseenko S.A. (2009) Materials to the revision of the genus *Dolichopteryx* and closely related taxa (*Ioichthys, Bathylychnops*) with the separation of a new genus Dolichopteroides and Description of three new species (Fam. Opisthoproctidae). *Journal of Ichthyology* 49(10): 839–851.

Parin N.V. & Borodulina O.D. (2006). Antigonias (*Antigonia*, Caproidae) of the Western Indian Ocean: 2. Species with eight spiny rays in the dorsal fin. *Journal of Ichthyology* 46 3): 203–211.

Parin N.V. & Kobyliansky S.G. (1996). Diagnoses and distribution of fifteen species recognized in genus *Maurolicus* Cocco (Sternoptychidae, Stomiiformes) with a key to their identification. *Cybium* 20(2): 185–195.

Parmentier E. & Vandewalle P. (2003). Morphological adaptations of Pearlfish (Carapidae) to their various habitats. In: Val A. L. & Kapoor B. G. (eds.) *Fish Adaptations*, pp. 261–276. Oxford: Science Publisher.

Parr A.E. (1948). The classification of the fishes of the genera *Bathylaco* and *Macromastax*, possible intermediates between the Isospondyli and Iniomi. *Copeia* 1948: 48–54.

Parrish R.H. (1972). Symbiosis in blacktail snailfish, *Careproctus melanurus*, and box crab,

Lopholithodes foraminatus. California Fish and Game 58: 239–240.

Partridge J.C., Douglas R.H., Marshall N.J., Chung W.-S., Jordan T.M., & Wagner H.-J. (2014). Reflecting optics in the diverticular eye of a deep-sea barreleye fish (*Rhynchohyalus natalensis*). *Proceedings of the Royal Society. B* 281: 2013–3223. http://dx.doi.org/10.1098/rspb.2013.3223

Passow U. & Carlson C.A. (2012). The biological pump in a high CO_2 world. *Marine Ecology Progress Series* 470: 249–271.

Paulin C.D. & Moreland J.M. (1979). *Congiopodus coriaceus*, a new species of pigfish, and a redescription of *C. leucopaecilus*(Richardson), from New Zealand. (Pisces:Congiopodidae). *New Zealand Journal of Zoology* 6: 601–608.

Pauly D. (1995). Anecdotes and the shifting baseline syndrome of fisheries. *Trends in Ecology and Evolution* 10: 430.

Pauly D., Christensen V., Dalsgaard J., Froese R., & Torres F. Jr. (1998). Fishing down marine food webs. *Science* 279: 860–863. doi: 10.1126/science.279.5352.860

Pavlov D.S., Parin N.V., & Balushkin A.V. (2009). In memory of Anatole Petrovich Andriashev (August 19, 1910 to January 4, 2009). *Journal of Ichthyology* 49: 547–562. doi: 10.1134/S003294520907008X

Paxton J.R. (1972). Osteology and relationships of the lanternfishes (family Myctophidae). *Bulletin of the Natural History Museum of Los Angeles City* 13: 1–81.

Paxton J.R. (1989). Synopsis of the Whalefishes (Family Cetomimidae) with descriptions of four new genera. *Records of the Australian Museum* 41: 135–206. ISSN 0067 1975

Payne A.I.L. & Punt A.E. (1995). Biology and fisheries of South African Cape Hakes (*M. capensis* and *M. paradoxus*). In: Alheit J. & Pitcher T.J. (eds.) *Hake, Biology, Fisheries and Markets*, pp. 15–47. London: Chapman & Hall.

Pearcy W.G. & Ambler J.W. (1974). Food habits of deep-sea macrourid fishes off the Oregon coast. *Deep-Sea Research* 21: 745–759.

Pearcy W.G., & Hubbard L. (1964). A modification of the Isaacs–Kidd midwater trawl for sampling at different depth intervals. *Deep-Sea Research* 11: 263–264.

Pearcy W.G., Meyer S.L., & Munk O. (1965). A 'four-eyed' fish from the deep-sea: *Bathylychnops exilis* Cohen, 1958. *Nature* 207: 1260–1262.

Pearcy W.G., Stein D.L., & Carney R.S. (1982). The deep-sea benthic fish fauna of the Northeastern Pacific Ocean on Cascadia and Tufts Abyssal Plains and adjoining continental slopes. *Biological Oceanography* 1(4): 375–428.

Pelster B. (1997). Buoyancy at depth. In: Randall D.J. & Farrell A.P. (eds.) *Deep-Sea Fishes*, pp. 195–237. Academic Press, San Diego, California.

Pepperell J. (2010). *Fishes of the Open Ocean: A Natural History and Illustrated Guide.* Chicago, IL: University of Chicago.

Pequeño G. (2008). Grenadier fishes from Chilean waters: some aspects in relation to fisheries. *American Fisheries Society Symposium* 63: 41–48.

Pérès J.M. (1965). Aperçu sur les résultats de deux plongées effectuées dans le ravin de Puerto-Rico par le bathyscaphe *Archimède. Deep-Sea Research* 12: 883–891.

Petrov A.F. (2011). New data on the diet of deepsea icefish *Chionobathyscus dewitti* (Channichthyidae) in the Ross Sea in 2010. *Journal of Ichthyology* 51(8): 692–694.

Pettersson O. & Drechsel C.F. (1913). Memorandum on investigations in the Atlantic Ocean and programme for same. *Rapports Et Procès – Verbaux Des Réunions Du Conseil Permanent International Pour L'exploration De La Mer* 16: 1–21, 2 plates.

Pham C.K., Canha A., Diogo H., Pereira J.G., Prieto R., & Morato T. (2013). Total marine fishery catch for the Azores (1950–2010). *ICES Journal of Marine Science*, doi.10.1093/icesjms/fst024

Pham C.K., Diogo H., Menezes G., Porteiro F., Braga-Henriques A., Vandeperre F., & Morato T. (2014). Deep-water longline fishing has reduced impact on Vulnerable Marine Ecosystems. *Scientific Reports* 4, 4837. doi:10.1038/srep04837 (2014)

Phleger C.E. & Grigor M.R. (1990). Role of wax esters in determining buoyancy in *Hoplostethus atlanticus* (Beryciformes: Trachichthyidae). *Marine Biology* 105: 229–233.

Pianka E.R. (1970). On r and K selection. *American Naturalist* 104(940): 592–597. doi:10.1086/282697

Piccard J. & Dietz R.S. (1961). *Seven Miles Down.* London: Longmans.

Pietsch T.W. (1978). The feeding mechanism of *Stylephorus chordatus* (Teleostei: Lampridiformes): functional and ecological implications. *Copeia* 1978 (2): 255–262.

Pietsch T.W. (2009). *Oceanic Anglerfishes, Extraordinary Diversity in the Deep Sea*. Berkeley: University of California Press.

Pietsch T.W., Ho H.-C., & Chen H.-M. (2004). Revision of the deep-sea anglerfish genus *Bufoceratias* Whitley (Lophiiformes: Ceratioidei: Diceratiidae), with description of a new species from the Indo-West Pacific Ocean. *Copeia* **2004** (1): 98–107.

Pietsch T.W. & Kenaley C.P. (2011). A new species of deep-sea ceratioid anglerfish, genus *Himantolophus* (Lophiiformes: Himantolophidae), from southern waters of all three major oceans of the world. *Copeia* **2011**(4): 490–496.

Pietsch T.W. & Orr J.W. (2007). Phylogenetic relationships of deep-sea anglerfishes of the suborder Ceratioidei (Teleostei: Lophiiformes) based on morphology. *Copeia*, **2007**(1): 1–34.

Pietsch T.W. & Shimazaki M. (2005). Revision of the deep-sea anglerfish genus *Acentrophryne* Regan (Lophiiformes: Ceratioidei: Linophrynidae), with the description of a new species from off Peru. *Copeia* **2005**(2): 246–251.

Pietsch T.W. & Van Duzer J.P. (1980). Systematics and distribution of ceratioid anglerfishes of the family *Melanocetidae* with the description of a new species from the Eastern North Pacific Ocean. *Fishery Bulletin* **78**: 59–87.

Pinto G. (2007) Global distribution of passive margins. Online Map www.unalmed.edu.co/rrodriguez/Passive%20Margin/Passive%20margin.htm accessed 23 September 2014.

Pirrera L., Bottari T., Busalacchi B., Giordano D., Modica L., Perdichizzi A., Perdichizzi F., Profeta A., & Rinelli P. (2009). Distribution and population structure of the fish *Helicolenus dactylopterus dactylopterus* (Delaroche, 1809) in the Central Mediterranean (Southern Tyrrhenian Sea). *Marine Ecology* **30** (Suppl.1): 161–174.

Pollard D. (2014). Microichthys coccoi. *The IUCN Red List of Threatened Species* 2014: e.T194859A49087606. http://dx.doi.org/10.2305/IUCN.UK.2014-3.RLTS.T194859A49087606.en. Downloaded on 24 November 2015.

Poltev Yu. N. (2013). Carcinophyly of fish of the genus *Careproctus* (Scorpaeniformes: Liparidae) in waters of Southeastern Sakhalin (Sea of Okhotsk). *Journal of Ichthyology* **53**: 416–424.

Pond D.W., Fallick A.E., Stevens C.J., Morrison D.J. & Dixon D.R. (2008). Vertebrate nutrition in a deep-sea hydrothermal vent ecosystem: Fatty acid and stable isotope evidence. *Deep-Sea Research I* **55**:1718–1726.

Pope E.C., Hays G.C., Thys T.M., Doyle T.K., Sims D.W., Queiroz N., Hobson V.J., Kubicek L., & Houghton J.D.R. (2010). The biology and ecology of the ocean sunfish *Mola mola*: a review of current knowledge and future research perspectives *Reviews in Fish Biology and Fisheries* **20**(4): 471–487.

Popper A. N. & Fay R. R. (1993). Sound detection and processing by fish: critical review and major research questions. *Brain, Behaviour and Evolution* **41**: 14–38.

Porcu C., Follesa M.C., Gastoni A., Mulas A., Pedoni C., & Cau A. (2013). The reproductive cycle of a deep-sea eel, *Nettastoma melanurum* (Nettastomatidae: Anguilliformes) from the south-eastern Sardinian Sea (central-western Mediterranean). *Journal of the Marine Biological Association of the UK* **93**(4): 1105–1115. doi:10.1017/S0025315412001452

Post A. & Quèro J.-C. (1991.) Distribution et taxinomie de *Howella* (Perciformes, Perchthyidae) de L'Atlantique. *Cybium* **15**(2): 111–128.

Potthoff T., Richards W.J., & Ueyanagi S. (1980). Development of *Scombrolabrax heterolepis* (Pisces: Scombrolabracidae) and comments on familial relationships. *Bulletin Marine Science* **30**(2): 329–357.

Poulsen J.Y., Byrkjedal I., Willassen E., Rees D., Takeshima H., Satoh T.P., Shinohara G., Nishida M., & Miya M. (2013). Mitogenomic sequences and evidence from unique gene rearrangements corroborate evolutionary relationships of myctophiformes (Neoteleostei). *BMC Evolutionary Biology 2013* **13**: 111. doi:10.1186/1471-2148-13-111

Powell M.L., Kavanaugh S.I., & Sower S.A. (2005). Current knowledge of hagfish reproduction: implications for fisheries management. *Integrative and Comparative Biology* **45**: 158–165.

Powell S.M., Haedrich R.L., & McEachran J.D. (2003). The deep-sea demersal fish fauna of the Northern Gulf of Mexico. *Journal of NorthWest Atlantic Fisheries Science* **31**: 19–33.

Preston G.L., Mead P.D., Chapman L.B., & Taumaia P. (1999). *Deep-bottom Fishing Techniques for the Pacific Islands: A Manual for Fishermen*.

Secretariat of the Pacific Community. Noumea, New Caledonia: Secretariat of the Pacific Community.

Priede I.G. (1983). Use of satellites in marine biology. In: MacDonald A.G. & Priede I.G. (1983). *Experimental Biology at Sea*, pp. 3–50. New York: Academic Press.

Priede I.G. (1984). A basking shark (*Cetorhinus maximus*) tracked by satellite together with simultaneous remote sensing. *Fisheries Research* 2: 201–216.

Priede I.G. (1985). Metabolic scope in fishes. In: Tytler P. & Calow P. (eds). *Fish Energetics: New Perspectives* , pp. 33–64. London: Croom Helm.

Priede I.G. (1992). Wildlife telemetry: an introduction. In: Priede I.G. & Swift S.M. (eds.) *Wildlife Telemetry: Remote Monitoring and Tracking of Animals*, pp. 3–25. Chichester: Ellis Horwood.

Priede I.G. (2013). Biogeography of the Oceans: a Review of Development of Knowledge of Currents, Fronts and Regional Boundaries from Sailing Ships in the Sixteenth Century to Satellite Remote Sensing. *Pure and Applied Geophysics* 171(6): 1013–1027. doi: 10.1007/s00024-013-0708-4

Priede I.G. (2014). Biogeography of the oceans: a review of development of knowledge of currents, fronts and regional boundaries from sailing ships in the sixteenth century to satellite remote sensing. *Pure and Applied Geophysics* 171: 1013–1027. DOI 10.1007/s00024-013-0708-4

Priede I.G. & Bagley P.M. (2000). *In situ* studies on deep-sea demersal fishes using autonomous unmanned lander platforms. *Oceanography & Marine Biology, Annual Review* 38: 357–392.

Priede I.G., Bagley P., Armstrong J.D., Smith K.L., & Merrett N.R. (1991). Direct Measurement of Active dispersal of Food-falls by abyssal demersal fishes. *Nature* 351: 647–649.

Priede I.G., Bagley P.M., Smith A., Creasey S., & Merrett N.R. (1994). Scavenging deep demersal fishes of the Porcupine Seabight (NE Atlantic Ocean); observations by baited camera, trap and trawl. *Journal of the Marine Biological Association of the UK* 74: 481–498.

Priede I.G., Bergstad O.A., Miller P.I., Vecchione M., Gebruk A., Falkenhaug T., Billett D.S.M., Craig J., Dale A.C., Shields M.A., Tilstone G.H., Sutton T.T., Gooday A.J., Inall M.E., Jones D.O.B., Martinez-Vicente V., Menezes G.M., Niedzielski T., Sigurðsson Þ., Rothe N., Rogacheva A., Alt C.H.S., Brand T.,

Abell R., Brierley A.S., Cousins N.J., Crockard D., Hoelzel A.R., Høines Å., Letessier T.B., Read J.F., Shimmield T., Cox M.J., Galbraith J.K., Gordon J.D.M., Horton T., Neat F., & Lorance P., (2013). Does presence of a Mid Ocean Ridge enhance biomass and biodiversity? *PLoS ONE* 8(5): e61550. doi:10.1371/journal.pone.0061550

Priede I.G., Billett D.S.M., Brierley, A.S., Hoelzel A.R., Inall M., Miller P.I., Cousins N.J., Shields M.A., & Fujii T. (2013). The ecosystem of the Mid-Atlantic ridge at the sub-polar front and Charlie Gibbs Fracture Zone; ECO-MAR project strategy and description of the sampling programme 2007–2010. *Deep-Sea Research II* 98: 220–230. http://dx.doi .org/10.1016/j.dsr2.2013.06.012

Priede I.G. & Froese R. (2013). Colonisation of the deep-sea by fishes. *Journal of Fish Biology* 83: 1528–1550. doi:10.1111/jfb.12265

Priede I.G., Froese R., Bailey D.M., Bergstad O.A., Collins M.A., Dyb J.E., Henriques C., Jones E.G. & King N. (2006). The absence of sharks from abyssal regions of the world's oceans. *Proceedings of the Royal Society B*. 273: 1435–1441. doi:10.1098/ rspb.2005.3461

Priede I.G., Godbold J.A., King N.J., Collins M.A., Bailey D.M., & Gordon J.D.M. (2010). Deep-sea demersal fish species richness in the Porcupine Seabight, NE Atlantic Ocean: global and regional patterns. *Marine Ecology* 31: 247–260. 10.1111/j.1439-0485.2009.00330.x

Priede I.G., Godbold J.A., Niedzielski T., Collins M.A., Bailey D.M., Gordon J.D.M., & Zuur A.F. (2011). A review of the spatial extent of fishery effects and species vulnerability of the deep-sea demersal fish assemblage of the Porcupine Seabight, Northeast Atlantic Ocean (ICES Subarea VII). *ICES Journal of Marine Science* 68: 281–289. doi:10.1093/icesjms/ fsq045

Priede I.G. & Holliday F.G.T. (1980). The use of tilting tunnel respirometer to investigate some aspects of metabolism and swimming activity of the plaice (*Pleuronectes platessa* L.). *Journal of Experimental Biology* 85: 295–309.

Priede I.G. & Merrett N.R. (1996). Estimation of abundance of abyssal demersal fishes; a comparison of data from trawls and baited cameras. *Journal of Fish Biology* 49(Suppl. A): 207–216. doi: 10.1111/ j.1095-8649.1996.tb06077.x

Priede I.G. & Smith K.L. Jr. (1986). Behaviour of the abyssal grenadier, *Coryphaenoides yaquinae*, monitored using ingestible acoustic transmitters in the Pacific Ocean. *Journal of Fish Biology* 29(Suppl. A): 199–206. doi: 10.1111/j.1095-8649.1986. tb05011.x

Priede I.G., Smith K.L. Jr. & Armstrong J.D. (1990). Foraging behaviour of abyssal grenadier fish: inferences from acoustic tagging and tracking in the North Pacific Ocean. *Deep-sea Research* 37: 81–101.

Priede I.G., Williams L.M., Wagner H.-J., Thom A, Brierley I., Collins M.A., Collin S.P., Merrett N.R., & Yau C. (1999). Implication of the visual system in regulation of activity cycles in the absence of solar light: 2-[125I] iodomelatonin binding sites and melatonin receptor gene expression in the brains of demersal deep-sea gadiform fishes. *Proceedings of the Royal Society. Series B* 266: 2295–2302.

Prokofiev A.M. (2007). Osteology and some other morphological characters of *Howella sherborni*, with a discussion of the systematic position of the genus (Perciformes, Percoidei). *Journal of Ichthyology* 47 (6): 413–426.

Prokofiev A.M. (2014). Taxonomy and Distribution of Deepsea Herring (Bathyclupeidae) in Oceans. *Journal of Ichthyology* 54(8): 493–500.

Puig P., Canals M., Company J.B., Martín J., Amblàs D., Lastras G., Palanques A., & Calafat A.M. (2012). Ploughing the deep sea floor. *Nature* 489(7415): 286–289.

Pusceddu A., Bianchellia S., Martín J., Puig P., Palanques A., Masqué P., & Danovaro R. (2014). Chronic and intensive bottom trawling impairs deep-sea biodiversity and ecosystem functioning. *Proceedings of the National Academy of Sciences* 111: 8861–8866.

Pusineri C., Magnin V., Meynier L., Spitz J., Hassani S., & Ridoux V. (2007). Food and feeding ecology of the common dolphin (*Delphinus delphis*) in the oceanic northeast atlantic and comparison with its diet in neritic areas. *Marine Mammal Science* 23: 30–47.

Pusineri C., Vasseur Y., Hassani S., Meynier L., Spitz J., & Ridoux V. (2005). Food and feeding ecology of juvenile albacore, *Thunnus alalunga*, off the Bay of Biscay: a case study. *ICES Journal of Marine Science* 62: 116–122. doi:10.1016/j.icesjms.2004.09.004

Quattrini A.M. & Demopoulos A.W.J. (2016). Ectoparasitism on deep-sea fishes in the western North Atlantic: In situ observations from ROV surveys. *International Journal for Parasitology: Parasites and Wildlife* 5: 217–228.

Quéro J.-C., Du Buit M.-H., & Vayne J.-J. (1998). Les observations de poissons tropicaux et le réchauffement des eaux dans l'Atlantique Européen. *Oceanologica Acta* 21(2): 345–351.

Quéro J.-C. & Ozouf-Costaz C. (1991). *Ostracoberyx paxtoni*, nouvelle espèce des côtes est de l'Australie. Remarques sur les modifications morphologiques des *Ostracoberyx* au cors de leur croissance (Perciformes, Ostracoberycidae). *Cybium* 15(1): 43–54.

Rabindranath A., Daase M., Falk-Petersen S., Wold A., Wallace M.I., Berge J. & Brierley A.S. (2011). Seasonal and diel vertical migration of zooplankton in the High Arctic during the autumn midnight sun of 2008. *Marine Biodiversity* 41: 365–382.

Radchenko O.A. (2015). The system of the suborder Zoarcoidei (Pisces, Perciformes) as inferred from molecular genetic data. *Russian Journal of Genetics* 51(11): 1096–1112.

Radchenko O.A., Chereshnev I.A., & Petrovskaya A.V. (2010). Relationships and position of the genus *Neozoarces* of the subfamily Neozoarcinae in the system of the suborder Zoarcoidei (Pisces, Perciformes) by molecular-genetic data. *Journal of Ichthyology* 50(3): 246–251.

Radchenko V.I. (2007). Mesopelagic fish community supplies 'biological pump'. *The Raffles Bulletin of Zoology: Supplement* 14: 265–271.

Rahmstorf S. (2006). Thermohaline ocean circulation. In: Elias S. A. (ed.) *Encyclopedia of Quaternary Sciences*. Amsterdam: Elsevier.

Raman M. & James P.S.B.R. (1990). Distribution and abundance of Lanternfishes of the family Myctophidae in the EEZ of India. In: *Proceedings of the first workshop on scientific results of FORV Sagar Sampada, 5–7 June 1989, Kochi*. http://eprints .cmfri.org.in/5189/

Ramsing N. & Gundersen J. (2013). Seawater and gases: tabulated physical parameters of interest to people working with microsensors in marine systems. www.unisense.com

Rass T. S. (1955). Deep-sea fishes of the Kurile-Kamchatka trench. Trudy Instituta Okeanologii, Akademiya Nauk SSSR (Proceedings of the Institute of Oceanology, Academy of Sciences, USSR) (12: 328–339. [In Russian]

Rass T.S., Grigorasch V.A. & Spanovskaya V.D. (1982). Deep-sea bottom fishes caught on the 14th cruise of the *R/V Akademik Kurchatov*. *Proceedings of the Institute of Oceanology, Academy of Sciences, USSR* 100: 337–347 (1975). Translation Series No. 29 Virginia Institute of Marine Science, Gloucester Point, Virginia 23062, USA

Rees J.-F., De Wergifosse B., Noiset O., Dubuisson M., Janssens B., & Thompson E.M. (1998). The origins of marine bioluminescence: turning oxygen defence mechanisms into deep-sea communication tools. *Journal of Experimental Biology* 201: 1211–1221.

Regan, C.T. (1925). Dwarfed males parasitic on the females in oceanic angler-fishes (Pediculati, Ceratioidea). *Proceedings of the Royal Society of London B* 97: 386–400.

Regan C.T. & Trewavas E. (1932). Deep-sea angler fishes (Ceratioidea). *The Carlsberg Foundation's Oceanographical Expedition Round the World 1928–30 and Previous 'Dana' Expeditions under the Leadership of the Late Professor Johannes Schmidt*. Dana Report. 2:1–113, Pls. 1–10.

Reichart G. J., Lourens L. J., & Zachariasse W. J. (1998). Temporal variability in the northern Arabian Sea Oxygen Minimum Zone (OMZ) during the last 225,000 years. *Paleoceanography* 13: 607–621.

Reif W.-E. (1985). Functions of scales and photophores in mesopelagic luminescent sharks. *Acta Zoologica (Stockh.)* 66: 111–118.

Rex M. & Etter R. (2010). *Deep-Sea Biodiversity: Pattern and Scale*. Cambridge MA: Harvard University Press.

Rex M. A., Etter R. J., Morris J. S., Crouse J., McClain C. R., Johnson N. A., Stuart C. T., Deming J. W., Thies R. & Avery R. (2006). Global bathymetric patterns of standing stock and body size in the deep-sea benthos. *Marine Ecology Progress Series* 317: 1–8.

Ribas D., Muñoz M., Casadevall M., &, Gil de Sola L. (2006). How does the northern Mediterranean population of *Helicolenus dactylopterus dactylopterus* resist fishing pressure? *Fisheries Research* 79 (2006): 285–293.

Rice A.L. (1986). *British Oceanographic Vessels 1800–1950*. London: The Ray Society.

Rice A.L., Aldred R.G., Darlington E., & Wild R.A. (1982). The quantitative estimation of the deep-sea megabenthos: a new approach to an old problem. *Oceanologica Acta* 5: 63–72.

Rigby S. & Milsom C.V. (2000). Origins, evolution, and diversification of zooplankton. *Annual Review of Ecology, Evolution, and Systematics* 31: 293–313.

Rink H. (1852). De danske Handelsdistricketer I Nordgrønland, deres geografiske Beskanffendhed og productive. Forste Deel. In : *Grønland geografisk og statistisk beskrevet*. 1 (1857)

Risso A. (1810). *Ichthyologie de Nice, ou histoire naturelle des poisons du Departement des Alpes Maritimes*. Paris: Schoell.

Risso A. (1820). Mémoire sur quelques poissons observes dans la mer de Nice. *Journal De Physique, De Chimie, D'histoire Naturelle et Des Arts* 91: 241–255.

Risso A. (1826). *Histoire naturelle des principales productions de l'Europe Méridionale et particulièrement de celles des environs de Nice et des Alpes Maritimes*; volume 3 Paris: Levrault.

Risso A. & Poiteau A. (1818–1822). *Histoire naturelle des Orangiers*. Paris: Audot.

Rittmeyer, E.N., Allen A., Gründler, M.C, Thompson D.K., & Austin C.C. (2012). Ecological guild evolution and the discovery of the world's smallest vertebrate. *PLoS ONE (Public Library of Science)* 7(1): e29797. doi:10.1371/journal.pone.0029797. PMC 3256195. PMID 22253785. Retrieved 11 January 2012.

Roa-Varón A. & Ortí G. (2009). Phylogenetic relationships among families of Gadiformes (Teleostei, Paracanthopterygii) based on nuclear and mitochondrial data. *Molecular Phylogenetics and Evolution* 52: 688–704.

Robalo J.I., Sousa-Santos C., Cabral H., Castilho R., & Almada V.C. (2009). Genetic evidence fails to discriminate between *Macroramphosus gracilis* Lowe 1839 and *Macroramphosus scolopax* Linnaeus 1758 in Portuguese waters. *Marine Biology* 156: 1733–1737. doi: 10.1007/s00227-009-1197-y

Robbins E.I., Porter K.G., & Haberyan K.A. (1985). Pellet microfossils: possible evidence for metazoan life in early Proterozoic time. *Proceedings of the National Academy of Sciences USA* 82: 5809–5813.

Roberts C.D. & Gomon M.F. (2012). A review of giant roughies of the genus *Hoplostethus* (Beryciformes, Trachichthyidae), with descriptions of two new Australasian species. *Memoirs of Museum Victoria* 69: 341–354.

Robertson D.A. (1977). Planktonic eggs of the lanternfish, *Lampanyctodes hectoris* (family Myctophidae). *Deep-Sea Research* 24: 849–852.

Robertson D.A. (1981). Possible functions of surface structure and size in some planktonic eggs of marine fishes. *New Zealand Journal of Marine and Freshwater Research* 15: 147–153.

Robins C.H. & Martin D.M. (1976). *Haptenchelys texi.* In: Robins C.H. & Robins C.R. (eds.) *New Genera and Species of Dysommine and Synaphobranchine Eels (Synaphobranchidae) with an Analysis of the Dysomminae,* pp. 267–274. *Proceedings of the Academy of Natural Sciences of Philadelphia,* 127: 249–280.

Robinson A.R., Leslie W.G., Theocharis A., & Lascaratos A. (2001). *Mediterranean Sea Circulation. Encyclopedia of Ocean Sciences.* San Diego, CA: Academic Press, 1689–1706.

Robinson M. & Amemiya C.T. (2014). Coelacanths. *Current Biology* 24(2): R62–R63.

Robison B.H. (1972). Distribution of mid water fishes in the Gulf of Califonia. *Copeia* 1972(3): 448–461.

Robison B.H. (1984). Herbivory by the myctophid fish *Ceratoscopelus warmingii. Marine Biology* 84: 119–123.

Robison B.H. (2000). The coevolutionof undersea vehicles and deep-sea research. *Marine Technology Society Journal* 33: 65–73.

Robison B.H. (2004). Deep pelagic biology. *Journal of Experimental Marine Biology and Ecology* 300: 253–272.

Robison B.H. & Reisenbichler K.R. (2008). *Macropinna microstoma* and the paradox of its tubular eyes. *Copeia* 2008(4): 780–784.

Rodríguez-Cabello C., González-Pola C., & Sánchez F. (2016). Migration and diving behaviour of *Centrophorus squamosus* in the NE Atlantic. *Combining Electronic Tagging and Argo Hydrography to Infer Deep Ocean Trajectories. Deep-Sea Research I.* 115: 48–62.

Rodríguez-Cabello C. & Sánchez F. (2014). Is *Centrophorus squamosus* a highly migratory deep-water shark? *Deep-Sea Research I* 92: 1–10.

Roe H.S.J. (1988). Midwater biomass profiles over the Madeira Abyssal Plain and the contribution of copepods. *Hydrobiologia* 167/168: 169–181.

Roe H. S. J. & Badcock J. (1984). The diel migrations and distributions within a mesopelagic community in the North East Atlantic. 5. Vertical migrations and feeding of fish. *Progress in Oceanography* 13: 389–424.

Roe H. S. J. & Shale D. M. (1979). A new multiple rectangular midwater trawl (RMT l+8M) and some modifications to the Institute of Oceanographic Sciences' RMT 1+8. *Marine Biology* 50: 283–288.

Rogers A.D. (2000). The role of the oceanic oxygen minima in generating biodiversity in the deep sea. *Deep-Sea Research. II* 47: 119–148. doi:10.1016/S0967-0645(99)00107-1

Rogers A.D. (1994). The biology of seamounts. *Advances in Marine Biology* 30: 305–350.

Rogers A.D., Yesson C., & Gravestock P. (2015). A biophysical and economic profile of South Georgia and the South Sandwich Islands as Potential Large Scale Antarctic Protected areas. *Advances in Marine Biology* 70: 1–286. doi: 10.1016/bs.amb.2015.06.001.

Romer A.S. (1970). *The Vertebrate Body,* 4th edition. Philadelphia: W.B. Saunders.

Rondelet. (1554). *Libri de piscibus marinus, in quibus verae piscium effigies expressae sunt.* Lyons: Matthias Bonhomme.

Rosenblatt R.H. & Butler J.L. (1977). The ribbonfish genus *Desmodema,* with the description of a new species (Pisces, Trachipteridae) *Fishery Bulletin* 5: 843–855.

Rosenblatt R.H. & Cohen D.M. (1986). Fishes living in deep sea thermal vents in the tropical eastern Pacific, with descriptions of a new genus and two new species of eelpouts (Zoarcidae). *Transactions of the San Diego Society of Natural History* 21: 71–79.

Ross L.G. & Gordon J.D.M. (1978). Guanine and permeability in swimbladders of slope-dwelling fish. In: McLusky D.S. et al. (eds.) *Physiology and Behaviour of Marine Organisms: Proceedings of the 12th European Symposium on Marine Biology Stirling, Scotland, September 1977,* pp. 113–121. Oxford: Pergamon Press.

Ross S.W. & Quattrini A.M. (2007). The fish fauna associated with deep coral banks off the southeastern United States. *Deep-Sea Research I* 54: 975–1007.

Ross S.W., Quattrini A.M., Roa-Varón A.Y., & McClain J.P. (2010). Species composition and distributions of mesopelagic fishes over the slope of the north-central Gulf of Mexico. *Deep-Sea Research II* 57: 1926–1956.

Roule L. (1913). Notice préliminaire sur *Grimaldichthys profundissimus* nov. gen. nov.sp. Possion abyssal recueilli à 6036 mètre de profondeur dans l'Ocean Atlantique par S.A.S le Prince de Monaco. *Bulletin de l'Institut Oceanographique (Monaco)* 261: 1–8.

Roule M. L. (1922). Description de *Scombrolabrax heterolepis* nov. gen. nov. sp., poisson abyssal noveau de I'lle Madère. *Bulletin de L' Institut Océanographique, Monaco* **408**: 1–8.

Rountree R.A., Juanes F., Goudey C.A., & Ekstrom K.E. (2012). Is biological sound production important in the deep sea? In: Popper, A.N. & Hawkins A. (eds.) *The Effects of Noise on Aquatic Life*, pp. 181–183. New York: Springer Science & Business Media.

Rowat D. & Gore M. (2007). Regional scale horizontal and local scale vertical movements of whale sharks in the Indian Ocean off Seychelles. *Fisheries Research* **84**:32–40

Rowden A.A., Dower J.F., Schlacher T.A., Consalvey M., & Clark M.R. (2010). Paradigms in seamount ecology: fact, fiction and future. *Marine Ecology* **31**: 226–241.

Rowe G.T. (1983). Biomass and production of deep-sea macrobenthos. In: Rowe G.T. (ed.) *Deep-Sea Biology*, pp. 97–121. New York: Wiley-interscience.

Rowe G.T. & Menzies R. J. (1967). Use of sonic techniques and tension recordings as improvements in abyssal trawling. *Deep-sea Research* **14**: 271–274.

Rowe G.T., Merrett N., Shepherd J., Needler G., Hargrave B., & Marietta M. (1986). Estimates of direct transport of radioactive waste in the deep sea with species reference to organic carbon budgets. *Oceanologica Acta* **9**: 199–208.

Rowling, K. Hegarty A., & Ives M. (eds.). (2010). Ocean Perch (*Helicolenus* spp.). In: *Status of Fisheries Resources in New South Wales, 2008/09*, pp. 217–220. Cronulla, NSW: Industry and Investment NSW.

Ruxton G. D. & Houston D. C. (2004). Energetic feasibility of an obligate marine scavenger. *Marine Ecology Progress Series* **266**: 59–63.

Ryther J.H. (1956). Photosynthesis in the ocean as a function of light intensity. *Limnology and Oceanography* **1**(1): 61–70.

Sabatés A., Bozzano A., & Vallvey I. (2003). Feeding pattern and the visual light environment in myctophid fish larvae. *Journal of Fish Biology* **63**: 1476–1490.

Sallan L.C. & Coates M.I. (2010). End-Devonian extinction and a bottleneck in the early evolution of modern jawed vertebrates. *PNAS* **107**: 10131–10135.

Sameoto D., Wiebe P., Runge J., Postel L., Dunn J., Miller C., & Coombs S. (2000). Collecting zooplankton In: Harris R. et al. (eds.) *ICES Zooplankton Methodology Manual*, pp. 55–81. San Diego, CA: Academic Press.

Sancho G., Fisher C.R., Mills S., Micheli F., Johnson G.A., Lenihan H.S., Peterson C.H., & Mullineaux L.S. (2005). Selective predation by the zoarcid fish *Thermarces cerberus* at hydrothermal vents. *Deep-Sea Research I* **52**: 837–844.

Sassa C., Kawaguchi K., Kinoshita T., & Watanabe C. (2002). Assemblages of vertical migratory mesopelagic fish in the transitional region of the western North Pacific. *Fisheries Oceanography* **11**(4): 193–204.

Sato K., Stewart A.L., & Nakaya K. (2013). *Apristurus garricki* sp. nov., a new deep-water catshark from the northern New Zealand waters (Carcharhiniformes: Scyliorhinidae). *Marine Biology Research* **9**: 758–767. http://dx.doi.org/10.1080/17451000.2013.765586

Sazonov Y.I. (1996). Morphology and significance of the luminous organs in alepocephaloid fishes. In: Uiblein F., Ott J., & Stachowitsch M. (eds.) *Deep-Sea and Extreme Shallow-Water Habitats: Affinities and Adaptations.* Biosystematics and Ecology Series **11**:151–163. Wien: Österreichische Akademie der Wissenschaften.

Schauer J., Hissmann K., & Fricke H. (1997). A method for the deployment of externally attached, sonic fish tags from a manned submersible and their effects on coelacanths. *Marine Biology* **128**: 359–362.

Schlee S. (1973). *A History of Oceanography; The Edge of an Unfamiliar World.* London: Robert Hale.

Schmidt-Nielsen K. (1987). Per Fredrik Thorkelsson Scholander, November 29, 1905–June 13, 1980. *Biographical Memoir.* Washington, DC: National Academy of Sciences.

Schmitter-Soto J.J. (2008). The Oarfish, *Regalecus glesne* (Teleostei: Regalecidae), in the Western Caribbean. *Caribbean Journal of Science* **44**: 125–128.

Schmittner A., Sarnthein M.,.Kinkel H., Bartoli G., Bickert T. et al. (2004). Global impact of the Panamanian Seaway closure. *Eos* **85**(49): 526–528.

Schopf T.J.M. (1980). *Paleoceanography.* Cambridge, MA: Harvard University Press.

Schott F.A. & McCreary J.P. (2001). The monsoon circulation of the Indian Ocean. *Progress in Oceanography* **51**: 1–123.

Schulte P., Alegret L., Arenillas I., Arz J.A., Barton P.J. et al. (2010). The Chicxulub asteroid impact and mass extinction at the Cretaceous-Paleogene boundary. *Science* **327**: 1214–1218. doi: 10.1126/science.1177265

Scotese, C.R. (2002). PALEOMAP website. www.scotese.com, accessed 20 June 2014.

Scotese C.R., Bambach R.K., Barton C., Van Der Voo R., & Ziegler A.M. (1979). Paleozoic base maps. *The Journal of Geology* **87**: 217–277.

Sealy T.S. (1974). Soviet Fisheries: A review. Paper 1075. *Marine Fisheries Review* **36**(8): 5–22.

Searle R. (2013). *Mid-Ocean Ridges*. Cambridge: Cambridge University Press.

Sébert P. (1997). Pressure effects on shallow-water fishes. In: Randall D.J. & Farrell A.P. (eds.) *Deep-Sea Fishes*, pp. 279–323. San Diego, CA: Academic Press.

Secombes C.J. & Ellis A.E. (2012). The immunology of teleosts. In: Roberts R.J. (ed.) *Fish Pathology*, 4th edition, pp. 144–166. London: Blackwell.

Seibel B.A. & Drazen J.C. (2007). The rate of metabolism in marine animals: environmental constraints, ecological demands and energetic opportunities. *Philosophical Transactions of the Royal Society B* **362**: 2061–2078. doi:10.1098/rstb.2007.2101

Seki M.P. & Somerton D.A. (1994). Feeding ecology and daily ration of the pelagic armorhead, *Pseudopentaceros wheeleri*, at Southeast Hancock Seamount. *Environmental Biology of Fishes* **39**: 73–84.

Shandikov G.A., Eakin R.R., & Usachev S. (2013). *Pogonophryne tronio*, a new species of Antarctic short-barbeled plunderfish (Perciformes: Notothenioidei: Artedidraconidae) from the deep Ross Sea with new data on *Pogonophryne brevibarbata*. *Polar Biology* **36**: 273–289. doi: 10.1007/s00300-012-1258-4

Shao K.-T., Iwamoto T., Ho H.-C., Cheng T.-Y., & Chen C.-Y. (2008). Species composition and distribution pattern of grenadiers (Family Bathygadidae, Macrouridae and Macrourididae) fromTaiwan. *American Fisheries Society Symposium* **63**: 17–29.

Sherrard K.M. (2000). Cuttlebone morphology limits habitat depth in eleven species of Sepia (Cephalopoda: Sepiidae) *Biological Bulletin* **198**: 404–414.

Shibanov V.N. & Vinnichenko V.I. (2008). Russian investigations and the Fishery of Roundnose Grenadier in the North Atlantic. *American Fisheries Society Symposium* **63**: 399–412.

Shimokawa T., Amaoka K., Kajiwara Y., & Suyama S. (1995). Occurrence of *Thalasenchelys coheni* (Anguilliformes; Chlopsidae) in the West Pacific Ocean. *Japanese Journal of Ichthyologgy* **42**: 89–92.

Shinohara G. & Imamura H. (2007). Revisiting recent phylogenetic studies of 'Scorpaeniformes'. *Ichthyological Research* **54**: 92–99. doi: 10.1007/s10228-006-0379-6

Shreeve R.S., Collins M.A., Tarling G.A., Main C.E., Ward P., & Johnston N.M. (2009). Feeding ecology of myctophid fishes in the northern Scotia Sea. *Marine Ecology Progress Series* **386**: 221–236.

Sidell B.D. & O'Brien K.M. (2006). When bad things happen to good fish: the loss of hemoglobin and myoglobin expression in Antarctic icefishes. *Journal of Experimental Biology* **209**: 1791–1802.

Siegelman-Charbit L. & Planque B. (2016). Abundant mesopelagic fauna at oceanic high latitudes. *Marine Ecology Progress Series* **546**: 277–282. doi: 10.3354/meps11661

Sielfeld W.K., Vargas M.F., & Fuenzalida R.F. (1995). Peces mesopelágicos frente a la costa norte de Chile (18°25'-21°47'S). *Investigaciones Marinas, Valparaiso* **23**: 83–97.

Sigsbee C.D. (1880). *Deep-Sea Sounding and Dredging: A Description and Discussion of Methods and Appliances Used on Board the Coast and Geodetic Survey Steamer 'Blake'*. Washington, DC: Government Printing Office.

Sigurðsson T., Kristinsson K., Rätz H-J., Nedreaas K.H., Melnikov S.P., & Reinert J. (2006). The fishery for pelagic redfish (*Sebastes mentella*) in the Irminger Sea and adjacent waters. *ICES Journal of Marine Science* **63**(4): 725–736. doi: 10.1016/j.icesjms.2005.12.010

Sigurðsson T., Thorsteinsson V. & Gustafsson L. (2006). In situ tagging of deep-sea redfish: application of an underwater, fish-tagging system. *ICES Journal of Marine Science* **63**: 523–531.

Silverberg N., Edenborn H.M., Ouellet G. & Beland P. (1987). Direct evidence of a mesopelagic fish, *Melanostigma atlanticum* (Zoarcidae) spawning within bottom sediments. *Environmental Biology of Fishes* **20**: 195–202.

Sims D.W., Southall E.J., Richardson A.J., Reid P.C., & Metcalfe J.D. (2003). Seasonal movements and behaviour of basking sharks from archival tagging:

no evidence of winter hibernation. *Marine Ecology Progress Series* **248**: 187–196.

Sinclair E.H. & Stabeno P.J. (2002). Mesopelagic nekton and associated physics of the southeastern Bering Sea. *Deep-Sea Research II* **49**: 6127–6145.

Skov P.V. & Steffensen J.F. (2003). The blood volumes of the primary and secondary circulatory system in the Atlantic cod *Gadus morhua* L., using plasma bound Evans Blue and compartmental analysis. *Journal of Experimental Biology* **206**: 591–599.

Smith C.R. & Baco A.R. (2003). Ecology of whale falls at the deep-sea floor. *Oceanography and Marine Biology: An Annual Review* **2003**(41): 311–354.

Smith C.R., Glover A.G., Treude T., Higgs N.D., & Amon D.J. (2015). Whale-fall ecosystems: recent insights into ecology, paleoecology, and evolution. *Annual Review of Marine Science* **7**: 571–596. doi: 10.1146/annurev-marine-010213-135144

Smith D.G. (2002). Saccopharyngidae, Swallower eels. In: Carpenter K.E. (ed.) *FAO Species Identification Guide for Fishery the Living Marine Resources of the Western Central Atlantic*, Volume 2 *Bony Fishes Part 1 (Acipenseridae to Grammatidae)*, pp. 758–759. Rome: FAO. www.fao.org/3/a-y4161e/y4161e11.pdf accessed 3 March 2015.

Smith J.L.B. (1939). A living fish of Mesozoic type. *Nature* **143**: 455–456. doi:10.1038/143455a0

Smith K.L., Jr. (1978). Metabolism of the abyssopelagic rattail *Coryphaenoides armatus* measured in situ. *Nature* **274**: 362–364.

Smith K.L., Jr. (1982). Zooplankton of a bathyal benthic boundary layer: in situ rates of oxygen consumption and ammonium excretion. *Limnology and Oceanography* **27**(3): 461–471.

Smith K.L., Jr. (1987). Food energy supply and demand: A discrepancy between particulate organic carbon flux and sediment community oxygen consumption in the deep ocean. *Limnology and Oceanography* **32**: 201–220.

Smith K.L., Jr. & Hessler R. R. (1974). Respiration of benthopelagic fishes: *In situ* measurements at 1230 metres. *Science* **184**: 72–73.

Smith K.L., Jr., Kaufmann R.S., Baldwin R.J., & Carlucci A.F. (2001) Pelagic-benthic coupling in the abyssal eastern North Pacific: An 8-year time-series study of food supply and demand. *Limnology & Oceanography* **46**:543–556.

Smith K.L., Jr., Kaufmann R.S., & Wakefield W.W. (1993). Mobile megafaunal activity monitored with a time-lapse camera in the abyssal North Pacific. *Deep-Sea Research I* **40**: 2307–2324.

Smith K.L., Jr., Ruhl H.A., Bett B.J., Billett D.S.M., Lampitt R.S., & Kaufmann R.S. (2009). Climate, carbon cycling, and deep-ocean ecosystems. *Proceedings of the National Academy of Sciences USA* **106**: 19211–19218.

Smith K.L. Jr., Sherman A.D., Huffard C.L., McGill P.R., Henthorn R., Von Thun S., Ruhl H.A., Kahru M., & Ohman M.D. (2014). Large salp bloom export from the upper ocean and benthic community response in the abyssal northeast Pacific: Day to week resolution. *Limnology & Oceanography* **59**: 745–757.

Smith W.H.F. & Sandwell D.T. (1997). Global seafloor topography from satellite altimetry and ship depth soundings. *Science* **277**: 1956–1962.

Smith W.L. & Wheeler W.C. (2006). Venom evolution widespread in fishes: a phylogenetic road map for the bioprospecting of piscine venoms. *Journal of Heredity* **97**: 206–217.

Somero G.N. (1992). Adaptations to high hydrostatic pressure. *Annual Reviews of Physiology* **54**: 557–577.

Somero G.N. & Siebenaller J.F. (1979). Inefficient lactate dehydrogenases of deep-sea fishes. *Nature* **282**: 100–102.

Somiya H. (1977). Bacterial bioluminescence in chlorophthalmid deep-sea fish: a possible interrelationship between the light organ and the eyes. *Experientia* **33**: 906–909.

Somiya H., Yamakawa T., & Okiyama M. (1996). Okiyama *Bathysauropis gigas*, a deep-sea aulopiform fish with a peculiar iris process and a pure-cone retina. *Journal of Fish Biology* **49**(Supplement A): 175–181.

Soofiani N.M. & Hawkins A.D. (1982). Energetics costs at different levels of feeding in juvenile cod, Gadus morhua L. *Journal of Fish Biology* **21**: 577–592.

Sorensen P.W. & Wisenden B.D. (eds.). (2015). *Fish Pheromones and Related Cues.* Chichester: Wiley Blackwell.

Sorenson L., Santini F., & Alfaro M.E. (2014). The effect of habitat on modern shark diversification. *Journal of Evolutionary Biology* **27**: 1536–1548.

Springer S. & Burgess G.H. (1985). Two new dwarf dog-sharks (*Etmopterus*, Squalidae), found off the

Caribbean Coast of Colombia. *Copeia* **1985**(3): 584–591.

Staiger J. C. (1972). *Bassogigas profundissimus* (Pisces; Brotulidae) from the Puerto Rico Trench. *Bulletin of Marine Science* **22**(l): 26–33.

Star B. & Jentoft S. (2012). Why does the immune system of Atlantic cod lack MHC II? *Bioessays* **34**: 648–651.

Star B., Nederbragt A.J., Jentoft S., Grimholt U., Malmstrøm M. et al. (2011). The genome sequence of Atlantic cod reveals a unique immune system. *Nature* **477**: 207–210. doi:10.1038/nature10342

Starr R.M. & Green K. (2010). *Fishes of Las Gemelas Seamounts and Isla Del Coco: Preliminary Findings of September 2009 Submersible Surveys*. San Diego, CA: Extension Publications University of California. https://escholarship.org/uc/item/9w01v9h0 Accessed 25 April 2015.

Staudigel H., Hart S.R., Pile A., Bailey B.E., Baker E.T., Brooke S., Connelly D.P., Haucke L., German C.R., Hudson I., Jones D., Koppers A.A.P., Konter J., Lee R., Pietsch T.W., Tebo B.M., Templeton A.S., Zierenberg R., & Young C.M. (2006). Vailulu'u Seamount, Samoa: life and death on an active submarine volcano. *Proceedings of the National Academy of Sciences of the USA* **103**: 6448–6453.

Stefanescu C., Morales-Nin B., & Perri F. (1994). Fish assemblages on the slope in the Catalan Sea (Western Mediterranean): influence of a submarine canyon. *Journal of the Marine Biological Association of the UK* **74**: 499–512.

Stehmann M. (1990). Rajidae. In: Quéro J. C. et al. (eds.) *Check-list of the Fishes of the Eastern Tropical Atlantic*, vol. 1, pp. 29–50. Lisbon, Portugal: Junta Nacional de Investigacao Cientifica e Tecnológica.

Stehmann M. & Pompert J. (2009). Bathyraja meridionalis. *The IUCN Red List of Threatened Species 2009*: e.T161619A5466024. http://dx.doi.org/10.2305/IUCN.UK.2009-2.RLTS.T161619A5466024.en. Downloaded on 22 February 2016.

Stehmann M.F.W. & Merrett N.R. (2001). First records of advanced embryos and egg capsules of *Bathyraja* skates from the deep north-eastern Atlantic. *Journal of Fish Biology* **59**: 338–349.

Stein D.L. (1980). Aspects of reproduction of liparid fishes from the continental slope and abyssal plain off Oregon, with notes on growth. *Copeia* **1980**: 687–699.

Stein D.L. (1985). Towing large nets by single warp at abyssal depths: methods and biological results. *Deep-Sea Research* **32**: 183–200.

Stein D.L. (2005). Descriptions of four new species, redescription of *Paraliparis membranaceus*, and additional data on species of the fish family Liparidae (Pisces, Scorpaeniformes) from the west coast of South America and the Indian Ocean. *Zootaxa* **1019**: 1–25.

Stein D.L. & Chernova N.V. (2009). Anatole Petrovich Andriashev (1910–2009) *Copeia* **2009**(3): 628–634. 2009.doi:10.1643/OT-09-099

Stein D.L., Drazen J.C., Schlining K.L., Barry J.P., & Kuhnz L. (2006). Snailfishes of the central California coast: video, photographic and morphological observations. *Journal of Fish Biology* **69**: 970–98.

Stein D.L., Felley J.D., & Vecchione M. (2005). ROV observations of benthic fishes in the Northwind and Canada Basins. *Arctic Ocean Polar Biology* **28**: 232–237. doi:10.1007/s00300-004-0696-z

Stein D.L. & Pearcy W.G. (1982). Aspects of reproduction, early life history, and biology of macrourid fishes off Oregon, U.S.A. *Deep-Sea Research*. **29**: 1313–1329.

Stenseth N.C., Mysterud A., Ottersen G., Hurrell J.W., Chan K.-S., & Lima M. (2002). Ecological effects of climate fluctuations. *Science* **297**: 1292–1296. doi: 10.1126/science.1071281

Stergiou K.I., Machias A., Somarakis S., & Kapantagakis A. (2003). Can we define target species in Mediterranean trawl fisheries? *Fisheries Research* **59**: 431–435.

Stevenson D.E. & Kenaley C.P. (2011). Revision of the manefish genus *Paracaristius* (Teleostei: Percomorpha: Caristiidae), with descriptions of a new genus and three new species. *Copeia* **2011**(3): 385–399.

Stevenson D.E. & Kenaley C.P. (2013). Revision of the manefish genera *Caristius* and *Platyberyx* (Teleostei: Percomorpha: Caristiidae), with description of five new species. *Copeia* **2013**(3): 415–434.

Stiassny M.L.J. (1996). Basal Ctenosquamate relationships and the interrelationships of the myctophiform (Scopelomorph) fishes. In: Stiassny M.L.J., Parenti L.R., & Johnson G.D. (eds.) *Interrelationships of Fishes*, pp. 405–426. San Diego, CA: Academic Press.

Stockton W.L. & DeLaca T.E. (1982). Food falls in the deep sea: occurrence, quality and significance. *Deep-Sea Research* **29**: 157–169.

Stommel H. & Arons A.B. (1960). On the abyssal circulation of the world ocean- II. An idealized model of the circulation pattern and amplitude in oceanic basins. *Deep Sea Research* **6**: 217–218.

Stoner A.W., Ryer C.H., Parker S.J., Auster P.J., & Wakefield W.W. (2008). Evaluating the role of fish behavior in surveys conducted with underwater vehicles. *Canadian Journal of Fisheries and Aquatic Sciences* **65**: 1230–1243.

Stowasser G., Pond D.W., & Collins M.A. (2009). Using fatty acid analysis to elucidate the feeding habits of Southern Ocean mesopelagic fish. *Marine Biology* **156**: 2289–2302. doi: 10.1007/s00227-009-1256-4

Stramma L., Schmidtko A., Levin L.A., & Johnson G.C. (2010). Ocean oxygen minima expansions and their biological impacts. *Deep Sea Research Part I: Oceanographic Research Papers* **57**(4): 587–595.

Strange K. (2016). Building a knowledge base for management of a new fishery: boarfish (*Capros aper*) in the Northeast Atlantic. *Fisheries Research* **174**: 94–102.

Sturges W. (2005). Deep-Water Exchange between the Atlantic, Caribbean and Gulf of Mexico. In: *Circulation in the Gulf of Mexico: Observations and Models*, pp. 263–278. Geophysical Monograph Series 161. Washington, DC: American Geophysical Union. 10.1029/161GM019

Suarez S.S. (1975). The reproductive biology of *Ogilbia cayorum*, a viviparous brotulid fish. *Bulletin of Marine Science* **25**: 143–173.

Sudarto Lalu X.C., Kosen J.D., Tjakrawidjaja A.H., Kusumah R.V., Sadhotomo B., Kadarusman, P.L., Slembrouck J., & Paradis E. (2010). Mitochondrial genomic divergence in coelacanths (*Latimeria*): slow rate of evolution or recent speciation? *Marine Biology* **157**: 2253–2262. doi: 10.1007/s00227-010-1492-7

Sulak K.J. (1975). *Talismania mekistonema*, a new Atlantic species of the family Alepocephalidae (Pisces: Salmoniformes). *Bulletin of Marine Science* **25**: 88–93.

Sulak K.J., Wenner C.A., Sedberry G.R., & Van Guelpen L. (1985). Life history and systematics of deep-sea lizardfishes, genus *Bathysaurus* (Synodontidae). *Canadian Journal of Zoology* **63**: 623–642.

Summers A.P., Hartel, K.E., & Koob T.J. (1999). Agassiz, garman, albatross, and the collection of deep-sea fishes. *Marine Fisheries Review* **61**: 58–68.

Suntsov A.V., Widder E.A., & Sutton T.T. (2008). Bioluminescence in larval fishes. In: Finn R.N. & Kapoor B.G. (eds.) *Fish Larval Physiology*, pp. 51–88. Bergen, Norway: University of Bergen Press.

Sutton T.T. (2005). Trophic ecology of the deep-sea fish *Malacosteus niger* (Pisces: Stomiidae): an enigmatic feeding ecology to facilitate a unique visual system? *Deep-Sea Research I* **52**: 2065–2076.

Sutton T.T. (2013). Vertical ecology of the pelagic ocean: classical patterns and new perspectives. *Journal of Fish Biology* **83**: 1508–1527.

Sutton T.T., Clark M.R., Dunn D.C., Halpin P.N., Rogers A.D., Guinotte J., Bograd S.J., Angel M.V., Perez J.A.A., Wishner K., Haedrich R.L., Lindsay D.J., Drazen J.C., Vereshchakam A., Piatkowski U., Morato T., Błachowiak-Samołyk K., Robison B.H., Gjerder K.M., Pierrot-Bults A., Bernalt P., Reygondeau G., & Heino M. (2017). A global biogeographic classification of the mesopelagic zone. Deep-Sea Research I. doi.org/10.1016/j.dsr.2017.05.006

Sutton T.T. & Hartel K.E. (2004). New species of *Eustomias* (Teleostei: Stomiidae) from the Western North Atlantic, with a review of the subgenus *Neostomias*. *Copeia* **2004**(1): 116–121.

Sutton T.T. & Hopkins T.L. (1996a). Species composition, abundance, and vertical distribution of the stomiid (Pisces: Stomiiformes) fish assemblage of the Gulf of Mexico. *Bulletin of Marine Science* **59**: 530–42.

Sutton T.T. & Hopkins T.L. (1996b). Trophic ecology of the stomiid (Pisces: Stomiidae) fish assemblage of the eastern Gulf of Mexico: strategies, selectivity and impact of a top mesopelagic predator group. *Marine Biology* **127**: 179–192.

Sutton T.T., Porteiro F.M., Heino M., Byrkjedal I., Langhelle G., Anderson C.I.H., Horne J., Søiland H., Falkenhaug T., Godø O.R., & Bergstad O.A. (2008). Vertical structure, biomass and topographic association of deep-pelagic fishes in relation to a mid-ocean ridge system. *Deep-Sea Research II* **55**: 161–184.

Sutton T.T. & Sigurðsson T. (2008). Vertical and horizontal distribution of mesopelagic fishes along a transect across the northern Mid-Atlantic Ridge. *International Council for Exploration of the Sea CM 2008/C* **16**: 1–12.

Sutton T.T., Wiebe P.H., Madin L.P., & Bucklin A. (2010). Diversity and community structure of pelagic fishes to 5000 m depth in the Sargasso Sea. *Deep-Sea Research II* **57**: 2220–2233.

Suziumov E.M. (1970–1979). Vessels, Scientific Research. *The Great Soviet Encyclopedia*, 3rd edition (1970–1979). New York: Macmillan.

Svendsen F.M. & Byrkjedal I. (2013). Morphological and molecular variation in *Synaphobranchus* eels (Anguilliformes: Synaphobranchidae) of the Mid-Atlantic Ridge in relation to species diagnostics. *Marine Biodiversity* 43: 407–420. doi: 10.1007/s12526-013-0168-1

Sweetman A.K., Smith C.R., Dale T., & Jones D.O.B. (2014). Rapid scavenging of jellyfish carcasses reveals the importance of gelatinous material to deep-sea food webs. *Proceedings of the Royal Society B* 281: 20142210. http://dx.doi.org/10.1098/rspb.2014.2210

Swezey R. R. & Somero, G. N. (1982). Polymerization thermodynamics and structural stabilities of skeletal muscle actins from vertebrates adapted to different temperatures and pressures. *Biochemistry* 21: 4496–4503.

Swinney, G.N. (1994). Comments on the Atlantic species of the genus *Evermannella* (Scopelomorpha, Aulopiformes, Evermannellidae) with a re-evaluation of the status of *Evermanella melanoderma*. *Journal of Fish Biology* 44(5): 809–819.

Tacon A.G.J & Metian M. (2015). Feed matters: satisfying the feed demand of aquaculture. *Reviews in Fisheries Science & Aquaculture* 23(1): 1–10. doi:10.1080/23308249.2014.987209

Takashima R., Nishi H., Huber B.T., & Leckie M. (2006). Greenhouse world and the Mesozoic ocean. *Oceanography* 19: 82–92.

Talley L.D., Lobanov V., Ponomarev V., Salyuk A., Tishchenko P., & Zhabin I. (2003). Deep convection and brine rejection in the Japan Sea. *Geophysical Research Letters* 30(4): 1159.

Tamburini C., Canals M., Durrieu de Madron X., Houpert L., Lefèvre D., et al. (2013). Deep-sea bioluminescence blooms after dense water formation at the ocean surface. *PLoS ONE* 8(7): e67523. doi:10.1371/journal.pone.0067523

Tandstad M., Shotton R., Sanders J., & Carocci F. (2011). Deep-sea fisheries. In: *Review of the State of World Marine Fishery Resources*, pp. 265–277. FAO Fisheries and Aquaculture Technical Paper No. 569. Rome: FAO.

Tascheri R., Saavedra-Nievas J.C., & Roa-Uretad R. (2010). Statistical models to standardize catch rates in the multi-species trawl fishery for Patagonian grenadier (*Macruronus magellanicus*) off Southern Chile. *Fisheries Research* 105: 200–214.

Taylor J.R. & Ferrari R. (2011). Ocean fronts trigger high latitude phytoplankton blooms. *Geophysical Research Letters* 38, L23601.

Taylor L.R., Compagno L.J.V., & Struhsaker P.J. (1983). Megamouth – a new species, genus, and family of lamnoid shark (*Megachasma pelagios*, family Megachasmidae) from the Hawaiian Islands. *Proceedings of the California Academy of Sciences* 43: 87–110.

Tchernavin V.V. (1953). Summary of the Feeding Mechanisms of a Deep Sea Fish, *Chauliodus sloani* Schneider. London: British Museum (Natural History).

Thacker C.E. & Roje D.M. (2009). Phylogeny of cardinalfishes (Teleostei: Gobiiformes: Apogonidae) and the evolution of visceral bioluminescence. *Molecular Phylogenetics and Evolution* 52: 735–745.

The International Hydrographic Organisation. (2003). *The History of GEBCO 1903–2003*. Principality of Monaco: IHO.

Themelis D.E. & Halliday R.G. (2012). Species composition and relative abundance of the mesopelagic fish fauna in the slope sea off Nova Scotia. *Northeastern Naturalist* 19(2): 177–200.

Thiel H. (1975). The size structure of the deep-sea benthos. *Internationale Revue der gesamten Hydrobiologie und Hydrographie* 60: 575–606.

Thompson E.M. & Rees J.-F. (1995). Origins of luciferins: ecology of bioluminescence in marine fishes. *Biochemistry and Molecular Biology of Fishes* 4: 435–466. doi:10.1016/S1873-0140(06)80021-4

Thomsen L. & McCave I.N. (2000). Aggregation processes in the benthic boundary layer at the Celtic Sea continental margin. *Deep-Sea Research I* 47: 1389–1404.

Thomson A. (2003). The management of redfish (*Sebastes mentella*) in the north Atlantic Ocean - a stock in movement. In: *Papers Presented at the Norway-FAO Expert Consultation on the Management of Shared Fish Stocks*, pp. 192–199. Bergen, Norway, 7–10 October 2002. FAO Fisheries Report. No. 695, Suppl. Rome: FAO.

Thurstan R.H., Brockington S., & Roberts C.M. (2010). The effects of 118 years of industrial fishing on UK bottom trawl fisheries. *Nature Communications* 1: 15 doi: 10.1038/ ncomms1013

Tighe K.A. & McCosker J.E. (2003). Two new species of the genus *Chlopsis* (Teleostei: Anguilliformes: Chlopsidae) from the southwestern Pacific. *Zootaxa* **236**: 1–8.

Torres J. J., Belman B.W., & Childress J.J. (1979). Oxygen consumption rates of midwater fishes as a function of depth of occurrence. *Deep-Sea Research.* **26**: 185–197.

Tota B. (1978). Functional cardiac morphology and biochemistry in Atlantic Bluefin Tuna. In: Sharp G.D. & Dizon A.E. (eds.) *The Physiological Ecology of Tunas*, pp. 89–112. New York: Academic Press.

Tozer H. & Dagit D.D. (2004). Husbandry of spotted ratfish, *Hydrolagus colliei*. In: Smith M., Warmolts D., Thoney D., & Hueter R. (eds.) *The Elasmobranch Husbandry Manual: Captive Care of Sharks, Rays and Their Relatives*, pp. 487–491. Columbus, OH: Ohio Biological Survey.

Tracey D.M., Bull B., Clark M.R., & MacKay K. (2004). Fish species composition on seamounts and adjacent slope in New Zealand waters. *New Zealand Journal of Marine and Freshwater Research* **38**(1): 163–182.

Trayanovsky F.M. & Lisovsky S.F. (1995). Russian (USSR) fisheries research in deep waters (below 500 m) in the North Atlantic. In: Hopper A.G. (ed.) *Deep-water Fisheries of the North Atlantic Slope*, pp. 357–365. Dordrecht, Netherlands: Kluwer.

Treberg J.R., and Speers-Roesch B. (2016). Does the physiology of chondrichthyan fishes constrain their distribution in the deep sea? *Journal of Experiment Biology* **219**: 615–625.

Tsukamoto Y. (2002). Leptocephalus larvae of *Ptcrothrissus gissu* collected from the Kuroshio–Oyashio transition region of the western North Pacific, with comments on its metamorphosis. *Ichthyological Research* **49**: 267–269.

Tucker G. H. (1951). Relation of fishes and other organisms to the scattering of underwater sound. *Journal of Marine Research* **10**: 215–238.

Tunnicliffe V., Koo B.F., Tyle J. & So S. (2010). Flatfish at seamount hydrothermal vents show strong genetic divergence between volcanic arcs. *Marine Ecology* **31**: 158–167.

Tunnicliffe V., Tyler J., & Dower J.F. (2013). Population ecology of the tongue fish *Symphurus thermophilus*

(Pisces; Pleuronectiformes; Cynoglossidae) at sulphur-rich hydrothermal vents on volcanoes of the northern Mariana Arc. *Deep-Sea Research II* **92**: 172–182.

Tuponogov V.N., Orlov A.M., & Kodolov L.S. (2008). The most abundant grenadiers of the Russian Far East EEZ: distribution and basic biological patterns. *American Fisheries Society Symposium* **63**: 285–316.

Turner J.R., White E.M., Collins M.A., Partridge J.C., & Douglas R.H. (2009). Vision in lanternfish (myctophidae): adaptations for viewing bioluminescence in the deep-sea. *Deep-Sea Research-I* **56**: 1003–1017.

Tyco Electronics Corporation. (2012). Rochester Brand Engineered Cable Solutions for Harsh Environments.

Tyler J.C. (1968). A monograph of Triacanthoidea. *Academy of Natural Sciences Philadelphia Monographs* **16**: 1–364.

Tytler P. & Blaxter J. H. S. (1973). Adaptation by cod and saithe to pressure changes. *Netherlands Journal of Sea Research* **7**: 31–45.

Uiblein F., Lorance P., & Latrouite D. (2003). Behaviour and habitat utilisation of seven fish species on the Bay of Biscay continental slope, NE Atlantic. *Marine Ecology Progress Series* **257**: 223–232.

Uiblein F., Nielsen J.G. & Møller P.R. (2008). Systematics of the ophidiid genus Spectrunculus (Teleostei: Ophidiiformes) with resurrection *of S. crassus*. *Copeia* **2008**(5): 542–551.

UN. (1982). United Nations Convention on the Law of the Sea. www.un.org/Depts/los/convention_agreements/texts/unclos/unclos_e.pdf Accessed 20 June 2014

UNESCO. (2009). Global Open Oceans and Deep Seabed (GOODS) – Biogeographic Classification. Paris: UNESCO-IOC. (IOC Technical Series, 84) 96pp.

Uyeno T. & Tsutsumi T. (1991). Stomach contents of *Latimeria chalumnae* and further notes on its feeding habits. *Environmental Biology of Fishes* **32**: 275–279.

Vaillant L. (1888). *Expéditions scientifiques du Travailleur et du Talisman pendant les annees 1880, 1881, 1882, 1883*. Paris: Masson.

van Denderen P.D., van Kooten T., & Rijnsdorp A.D. (2013). When does fishing lead to more fish? Community consequences of bottom trawl fisheries in demersal food webs. *Proceedings of the Royal*

Society. B **280**: 20131883. http://dx.doi.org/ 10.1098/rspb.2013.1883

Van der Grient J.M.A. & Rogers A.D. (2015). Body size versus depth: regional and taxonomical variation in deep-sea meio- and macrofaunal organisms. *Advances in Marine Biology* **71**: 71–108. http://dx .doi.org/10.1016/bs.amb.2015.07.002

Van Der Laan R., Eschmeyer W.M., & Fricke R. (2014). Family-group names of Recent fishes. *Zootaxa* **3882**(2): 001–230.

Van Dover C.L. (2000). *The Ecology of Deep-Sea Hydrothermal Vents*. Princeton, NJ: Princeton University Press.

van Ginneken V., Antonissen E., Müller U.K., Booms R., Eding E., Verreth J., & van den Thillart G. (2005). Eel migration to the Sargasso: remarkably high swimming efficiency and low energy costs. *Journal of Experimental Biology* **208**: 1329–1335.

van Haren H. (2004). Spatial variability of deep-ocean motions above an abyssal plain. *Journal of Geophysical Research* **109**: C12014. doi:10.1029/ 2004JC002558.

Vardaro M.F., Parmley D., & Smith K.L., Jr. (2007). A study of possible 'reef effects' caused by a long-term time-lapse camera in the deep North Pacific. *Deep-Sea Research. I* **54**: 1231–1240.

Velasco E.M., González F., Amez M., Méndez E., & Punzón A. (2010). First occurrence of the deepwater scorpionfish *Setarches guentheri* (Scorpaeniformes: Setarchidae) in Cantabrian waters: a northernmost occurrence in the eastern Atlantic. *Marine Biodiversity Records* **3**: e82. doi:10.1017/S1755267210000722.

Venter P., Timm P., Gunn G., le Roux E., Serfontein C., Smith P., Smith E., Bensch M., Harding D., & Heemstra P. (2000) Discovery of a viable population of coelacanths (*Latimeria chalumnae* Smith, 1939) at Sodwana Bay, South Africa. *South African Journal of Science* **96**(11/12): 567–568.

Videler J.J. (1993). *Fish Swimming*. London: Chapman and Hall.

Vitale S., Ragonese S., Cannizzaro L., Fiorentino F., & Mazzola S. (2014). Evidence of trawling impact on *Hoplostethus mediterraneus* in the central–eastern Mediterranean Sea. *Journal of the Marine Biological Association of the UK* **94**(3): 631–640.

VLIZ. (2009). Longhurst Biogeographical Provinces. Available online at www.vliz.be/vmdcdata/vlimar/ downloads.php. Consulted on 8 July 2014.

Wagner H.-J. (2001a). Brain areas in abyssal demersal fishes. *Brain Behavior and Evolution* **57**: 301–316.

Wagner H.-J. (2001b). Sensory brain areas in mesopelagic fishes. *Brain Behavior and Evolution* **57**: 117–133.

Wagner H.-J. (2002). Sensory brain areas in three families of deep-sea fish (slickheads, eels, and grenadiers): comparison of mesopelagic and demersal species. *Marine Biology* **141**: 807–817.

Wagner H.-J. (2003). Volumetric analysis of brain areas indicates a shift in sensory orientation during development in the deep-sea grenadier *Coryphaenoides armatus*. *Marine Biology* **142**: 791–797.

Wagner H.-J., Douglas R.H., Frank T.M., & Roberts N.W. (2009). A novel vertebrate eye using both refractive and reflective optics. *Current Biology* **19**: 108–114.

Wagner H.-J., Fröhlich E., Negishi K., & Collin S.P. (1998). The eyes of deep-sea fish II. Functional morphology of the retina. *Progress in Retinal and Eye Research* **17**: 637–685.

Wagner H.-J., Kemp K., Mattheus U., & Priede I.G. (2007). Rhythms at the bottom of the deep sea: cyclic current flow changes and melatonin patterns in two species of demersal fish. *Deep-Sea Research I* **54**: 1944–1956.

Wahlberg M., Westerberg H., Aarestrup K., Feunteun E., Gargan P., & Righton D. (2014). Evidence of marine mammal predation of the European eel (*Anguilla anguilla* L.) on its marine migration. *Deep-Sea Research I* **86**: 32–38.

Wakai N., Takemura K., Morita T., & Kitao A. (2014). Mechanism of deep-sea fish a-actin pressure tolerance investigated by molecular dynamics simulations. *PLoS ONE* **9**(1): e85852. doi:10.1371/journal. pone.0085852

Wall C.C., Rountree R.A., Pomerleau C., & Juanes F. (2014). An exploration for deep-sea fish sounds off Vancouver Island from the NEPTUNE Canada ocean observing system. *Deep-Sea Research I* **83**: 57–64.

Wang J.T.-M. & Chen C.-T. (2001). A review of lanternfishes (Families: Myctophidae and Neoscopelidae) and the distributions around Taiwan and the Tungsha Islands with notes on seventeen new records. *Zoological Studies* **40**: 103–126.

Wardle C.S. (1975). Limit of fish swimming speed. *Nature (Lond.)* **225**: 725–727. doi:10.1038/255725a0

Wardle C.S. (1983). Fish reactions to towed fishing gears. In Priede I.G. & MacDonald A.G. (eds.) *Experimental*

Biology at Sea, pp. 168–195. London: Academic Press.

Wardle C.S., Tetteh-Lartey N., Macdonald A.G., Harper A.A., & Penneq J.-P. (1987). The effect of pressure on the lateral swimming muscle of the European eel *Anguilla anguilla* and the deep sea eel *Histiobranchus bathybius*; results of Challenger Cruise 6B/85. *Comparative Biochemistry & Physiology* 88A: 595–598.

Watanabe H., Fujikura K., Kojima S., Miyazaki J. I., & Fujiwara Y. (2010). Japan: vents and seeps in close proximity. In: Kiel S. (ed.) *The Vent and Seep Biota: Aspects from Microbes to Ecosystems*, pp. 379–401. Netherlands: Springer. (doi:10.1007/978-90-481-9572-5_12)

Watson W. (1996). Sternoptychidae: Hatchetfishes. In: Moser H.G. (ed.) *Early Stages of Fishes in the California Current Region. California Cooperative Oceanic Fisheries Investigations*, pp. 268–283. Atlas No 33. La Jolla, CA: South West Fisheries Centre.

Watson W. & Sandknop E.M. (1996). Notosudidae: Paperbones. In: Moser H.G. (ed.) *Early Stages of Fishes in the California Current Region. California Cooperative Oceanic Fisheries Investigations*, pp. 344–347. Atlas No 33. La Jolla, CA: South West Fisheries Centre.

Weaver P.P.E., Thomson J., & Hunter P.M. (1987). Introduction. In: Weaver P.P.E. & Thomson J. (eds.) *Geology and Geochemistry of Abyssal Plains*, pp. vii–xii. Geological Society Special Publication No 31. Oxford: Blackwell.

Webb P. W. (1971). The swimming energetics of trout. II. Oxygen consumption and swimming efficiency. *Journal of Experimental Biology* 55: 521–540.

Webb T.J., Vanden Berghe E., & O'Dor R. (2010). Biodiversity's big wet secret: the global distribution of marine biological records reveals chronic under-exploration of the deep pelagic ocean. *PLoS ONE* 5(8): e10223. doi:10.1371/journal.pone.0010223

Weber R.E., Hourdez S., Knowles F., & Lallier F. (2003). Hemoglobin function in deep-sea and hydrothermal-vent endemic fish: *Symenchelis parasitica* (Anguillidae) and *Thermarces cerberus* (Zoarcidae). *Journal of Experimental Biology* 206: 2693–2702.

Wegner N.C., Sepulveda C.A., Bull K.B., & Graham J.B. (2010). Gill morphometrics in relation to gas transfer and ram ventilation in high-energy demand teleosts: scombrids and billfishes. *Journal of Morphology* 71: 36–49.

Wegner N.C., Snodgrass O.E., Dewar H., & Hyde J.R. (2015). Whole-body endothermy in a mesopelagic fish, the opah, *Lampris guttatus*. *Science* 348(6236): 786–789. doi: 10.1126/science.aaa8902.

Wei C.-L., Rowe G.T., Escobar-Briones E., Boetius A., Soltwedel T., et al. (2010). Global patterns and predictions of seafloor biomass using random forests. *PLoS ONE* 5(12): e15323. doi:10.1371/journal.pone.0015323

Wells H.G., Huxley J., & Wells G.P. (1931). *The Science of Life*. London: Cassel.

Wessel J.H. & Johnson R.K. (1995, 1998). The Sternoptychinae of the Somali Current region of the western Indian ocean: An introduction to Somali current mesopelagic fish studies. In: Pierrot-Bults A.C. & van der Spoel S. (eds.) *Pelagic Biogeography*, pp. 372–379. IOC Workshop Report No. 142 Paris: UNESCO.

Wessel P. (2001). Global distribution of seamounts inferred from gridded Geosat/ERS-1 altimetry. *Journal of Geophysical Research* B106: 19431–19442.

Wessel P., Sandwell D.T., & Kim S.-S. (2010). The global seamount census. *Oceanography* 23: 24–33.

Westbrook G.K. & Reston T.J (2002). The accretionary complex of the Mediterranean ridge: tectonics, fluid flow and the formation of brine lakes. *Marine Geology* 186: 1–8.

Wetherbee B.M. (2000). Assemblage of deep-sea sharks on Chatham Rise, New Zealand. *Fishery Bulletin* 98: 189–198.

White W.T., Fahmi, Dharmadi, Potter I.C. (2006). Preliminary investigation of artisanal deep-sea chondrichthyan fisheries in Eastern Indonesia. In: Shotton R. (ed.). *Deep Sea 2003: Conference on the Governance and Management of Deep-sea Fisheries. Part 2: Conference poster papers and workshop papers*, pp. 381–387. FAO Fisheries Proceedings 3/2, Rome: FAO.

Widder E.A. (1998). A predatory use of counterillumination by the squaloid shark, *Isistius brasiliensis*. *Environmental Biology of Fishes* 53(3): 267–273. doi:10.1023/A:1007498915860

Widder E.A., Robison B.H., Reisenbichler K.R., & Haddock S.H.D. (2005). Using red light for in situ observations of deep-sea fishes. *Deep-Sea Research I* 52: 2077–2085.

Wiebe P.H., Morton A.W., Bradley A.M. Backus R.H., Craddock J.E., Barber V., Cowles T.J., & Flier G.R. (1985). New developments in the MOCNESS, an apparatus for sampling zooplankton and micro-nekton. *Marine Biology* 87: 313–323.

Williams A., Koslow J.A., & Last P.R. (2001). Diversity, density and community structure of the demersal fish fauna of the continental slope off Western Australia (20 to 35° S). *Marine Ecology Progress Series* 212: 247–263.

Williams A., Last P.R., Gomon M.F., & Paxton J.R. (1996). Species composition and checklist of the demersal ichthyofauna of the continental slope off Western Australia (20–35°5) *Records of the Western Australian Museum* 18: 135–155.

Willughby F. (1686). *De Historia Piscium Libri Quatuor*. Oxford.

Wilson G. D. F. (1999). Some of the deep-sea fauna is ancient. *Crustaceana* 72: 1020–1030.

Wilson R.R. & Waples R.S. (1984), Electrophoretic and biometric variability in the abyssal grenadier *Coryphaenoides (N.) armatus* of the western North Atlantic, eastern Pacific and Eastern North Pacific Oceans. *Marine Biology* 80: 227–237.

Wippelhauser G.S., Miller M.J., & McCleave J.D. (1996). Evidence of spawning and the larval distributions of snipe eels (family nemichthyidae) in the Sargasso Sea. *Bulletin of Marine Science* 59(2): 298–309.

Wishner K.F. (1980). The biomass of the deep-sea benthopelagic plankton. *Deep-Sea Research Part A. Oceanographic Research Papers* 27(3–4): 205–216.

Wittmann A.C. & Pörtner H.-O. (2013). Sensitivities of extant animal taxa to ocean acidification. *Nature Climate Change*. 3: 995–1001 doi:10.1038/nclimate1982

Wolff T. (1961). The deepest recorded fishes. *Nature* 190: 283–284.

Wolff T. (1971). Archimede dive 7 to 4160 at Madeira: observations and collecting results. *Videnskabelige Meddelelser dansk naturhistorisk Forening Kobenhavn* 134: 127–147.

Wolff T. (2002). The Danish *Dana* Expedition, 1928–30: Purpose and accomplishments, mainly in the Indo-Pacific. In: Benson K.R. & Rehbock P.F. (eds.) *Oceanographic History, the Pacific & Beyond*, pp. 196–203. Proceedings of the 5th International Congress on the History of Oceanography. Seattle: University of Washington Press.

Woodward A.S. (1898). The antiquity of the deep-sea fish-fauna. *Journal of Natural Science* 12: 257–260.

Wright E.P. (1870). Notes on Sponges. 1, On *Hyalonema mirabilis*, Gray. 2, *Aphrocallistes Bocagei* sp. Nov. 3, On a new Genus and Species of Deep Sea Sponge. *Quarterly Journal of Microscopical Science* 10: 1–9, pls I–III.

Wright J.J., Konwar K.M., & Hallam S.J. (2012). Microbial ecology of expanding oxygen minimum zones. *Nature Reviews Microbiology* 10: 381–394.

Wynne-Edwards V.C. (1962). *Animal Dispersion in Relation to Social Behaviour*. Edinburgh: Oliver & Boyd.

Wyville T.C. (1873). *The Depths of the Sea*. London: Macmillan.

Yabe M. (1991). Bolinia euryptera, a new genus and species of sculpin (Scorpaeniformes; Cottidae) from the Bering Sea. *Copeia* 1991(2): 329–339. (Ref. 39612)

Yamaguchi M. (2000). Phylogenetic analyses of myctophid fishes using morphological characters: progress, problems, and future perspectives. *Japanese Journal of Ichthyology* 2000(47): 87–107.

Yamamoto J., Hirose M., Ohtani T., Sugimoto K., Hirase K., Shimamoto N., Shimura T., Honda N., Fujimora Y., & Mukai, T. (2008). Transportation of organic matter to the seafloor by carrion falls of the giant jellyfish *Nemopilema nomurai* in the Sea of Japan. *Marine Biology* 153: 311–317.

Yamanoue Y. & Yoseda K. (2001). A new species of the genus *Malakichthys* (Perciformes: Acropomatidae) from Japan. *Ichthyological Research* 48: 257–261.

Yanagimoto T. & Humphreys R.L., Jr. (2005). Maturation and reproductive cycle of female armorhead *Pseudopentaceros wheeleri* from the southern Emperor-northern Hawaiian Ridge Seamounts. *Fisheries Science* 71: 1059–1068.

Yancey P.H., Gerringer M.E., Drazen J.C., Rowden A.A., & Jamieson A. (2014). Marine fish may be biochemically constrained from inhabiting the deepest ocean depths. *Proceedings of National Academy of Science USA* 111(12): 4461–4465. doi: 10.1073/pnas.1322003111

Yancey P.H., Lawrence-Berrey R., & Douglas M.D. (1989). Adaptations in mesopelagic fishes. I. Buoyant glycosaminoglycan layers in species without diel vertical migrations. *Marine Biology* 103: 453–459.

Yancey P.H. & Siebenaller J.F. (2015). Co-evolution of proteins and solutions: protein adaptation versus cytoprotective micromolecules and their roles in marine organisms. *Journal of Experimental Biology* 218: 1880–1896. doi:10.1242/jeb.114355

Yang, J. & Huang, Z. (1986). The fauna and geographical distribution of deep-pelagic fishes in the South China Sea. In: Uyeno T., Arai R., Taniuchi T., & Matsuura K., (eds.) *Indo-Pacific Biology: Proceedings of the 2nd International Conference on Indo-Pacific Fishes*, pp. 461–164. Sado, Japan: Ichthyologic Society of Japan.

Yau C., Collins M.A., & Everson I. (2000), Commensalism between a liparid fish (*Careproctus* sp.) and stone crabs (Lithodidae) photgraphed *in situ* using a baited camera. *Journal of the Marine Biological Society of the UK* 80: 379–380.

Ye Y. & Cochrane K. (2011), Global overview of marine fishery resources. In: *Review of the state of world marine fishery resources*, pp. 3–18. FAO Fisheries and Aquaculture Technical Paper No. 569. Rome: FAO.

Yeh J. & Drazen J.C. (2009). Depth zonation and bathymetric trends of deep-sea megafaunal scavengers of the Hawaiian Islands. *Deep-Sea Research I* 56: 251–266.

Yesson C., Clark M.R., Taylor M., & Rogers A.D. (2011). The global distribution of seamounts based on 30-second bathymetry data. *Deep-Sea Research I* 58(4): 442–453.doi.org/10.1016/j.dsr.2011.02.004.

Yoshinaga T., Miller M.J., Yokouchi K., Otake T., Kimura S., Aoyama J., Watanabe S., Shinoda A., Oya M., Miyazaki S., Zenimoto K., Sudo R., Takahashi T., Ahn H., Manabe R., Hagihara S., Morioka H., Itakura H., Machida M., Ban K., Shiozaki M., Ai B., & Tsukamoto K. (2011). Genetic identification and morphology of naturally spawned eggs of the Japanese eel *Anguilla japonica* collected in the western North Pacific. *Fisheries Science* 77: 983–992. doi: 10.1007/s12562-011-0418-8

Zahl P.A. (1953). Fishing in the whirlpool of Charybdis. *National Geographic Magazine* 104(5): 579–618.

Zahuranec B.J. (2000). Zoogeography and systematics of the lanternfishes of the genus *Nannobrachium* (Myctophidae: Lampanyctini). *Smithsonian Contributions to Zoology* 607: 1–69.

Zenkevitch L.A. & Birstein J.A. (1960). On the problem of the antiquity of the deep-sea fauna. *Deep-Sea Research* 7: 10–23.

Zezina O.N. (1997). Biogeography of the bathyal zone. *Advances in Marine Biology* 32: 389–420.

Zintzen V., Roberts C.D., Anderson M.J., Stewart A.L., Struthers C.D., & Harvey E.S. (2011). Hagfish predatory behaviour and slime defence mechanism. *Scientific Reports*. 1, 131. doi:10.1038/srep00131 (2011).

Zintzen V., Rogers K.M., Roberts C.D., Stewart A.L., & Anderson M.J. (2013). Hagfish feeding habits along a depth gradient inferred from stable isotopes. *Marine Ecology Progress Series* 485: 223–234. doi: 10.3354/meps10341

Zylinski S. & Johnsen S. (2011). Mesopelagic cephalopods switch between transparency and pigmentation to optimize camouflage in the deep. *Current Biology* 21: 1937–1941. doi :10.1016/j.cub.2011.10.014

INDEX

Printed in the United States
By Bookmasters